T0342804

JACARANDA

Geoactive 2

STAGE 5 | FIFTH EDITION

JACARANDA Geoactive 2

STAGE 5 | FIFTH EDITION

LOUISE SWANSON

NICOLE GRAY

KAREN BOWDEN

ADRIAN HARRISON

KYMBERLY GOVERS

STEVEN NEWMAN

CONTRIBUTING AUTHORS AND CONSULTANTS

Hayley Saunders | Rob Bell | Laura Sproule | Adeba Qasim | Danielle Creeley
Natasha Craig | Cath McIntyre
Sandra Duncanson | Alex Kopp | Ben de Vries
Leah Truscott | Tracy Fynmore | Judy Mraz | Jill Price
Cathy Bedson | Jeana Kriewaldt | Denise Miles

jacaranda
A Wiley Brand

Fifth edition published 2021 by
John Wiley & Sons Australia, Ltd
42 McDougall Street, Milton, Qld 4064

First edition published 1998
Second edition published 2005
Third edition published 2010
Fourth edition published 2017

Typeset in 11/14 pt Times Ltd Std

ISBN: 978-0-7303-9428-0

Cover image: © Olivier Mesnage/Unsplash

Illustrated by various artists, diacriTech and Wiley Composition Services

Typeset in India by diacriTech

NATIONAL LIBRARY OF AUSTRALIA

A catalogue record for this book is available from the National Library of Australia

Printed in Singapore
M WEP295527 110724

The Publishers of this series acknowledge and pay their respects to Aboriginal Peoples and Torres Strait Islander Peoples as the traditional custodians of the land on which this resource was produced.

This suite of resources may include references to (including names, images, footage or voices of) people of Aboriginal and/or Torres Strait Islander heritage who are deceased. These images and references have been included to help Australian students from all cultural backgrounds develop a better understanding of Aboriginal and Torres Strait Islander Peoples' history, culture and lived experience.

It is strongly recommended that teachers examine resources on topics related to Aboriginal and/or Torres Strait Islander Cultures and Peoples to assess their suitability for their own specific class and school context. Teachers should also know and follow the guidelines laid down by the relevant educational authorities and local Elders or community advisors regarding content about all First Nations Peoples.

All activities in this resource have been written with the safety of both teacher and student in mind. Some, however, involve physical activity or the use of equipment or tools. All due care should be taken when performing such activities. Neither the publisher nor the authors can accept responsibility for any injury or loss that may be sustained when completing activities described in this resource.

The Publisher acknowledges ongoing discussions related to gender-based population data. At the time of publishing, there was insufficient data available to allow for the meaningful analysis of trends and patterns to broaden our discussion of demographics beyond male and female gender identification.

Contents

Chapter and lesson structure

Consistent, inclusive learning structure

Our strong instructional design has been developed and refined by experienced teachers. It provides best-practice pedagogy and easy navigation through each unit – for you and all of your students.

Each chapter begins with ...

Each chapter includes ...

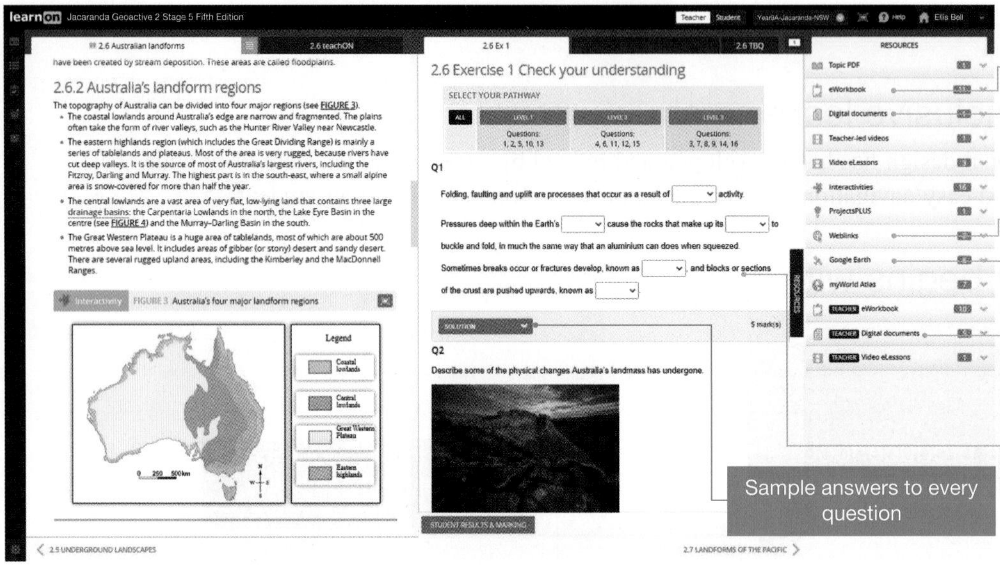

Worksheets for every lesson that you can adapt and customise

Printable maps for hard-copy mapping practice

Links to key sources of data

Easy links to Google Earth

School-based activities to build fieldwork skills (student worksheets and teacher notes)

Sample answers to every question

A range of carefully sequenced and worded self-marking and extended response questions

Each chapter ends with ...

Summative inquiry-based project, with marking rubric

Review of key concepts

Key terms glossary and revision activities

6.13 Thinking Big research project — Community garden design

Scenario

FIGURE 1 Rainforests may soon be gone.

Australia's largest urban areas — Sydney, Melbourne and Brisbane are continuing to show signs of urban sprawl. **Urban sprawl** inevitably affects food production, as areas that were once farmland are transformed into housing and commercial developments. One strategy that can reduce the impacts of urban sprawl is the creation of urban farms. Urban farms can range in size from small community gardens in our inner-city suburbs to large warehouses dedicated to horticulture in our outer suburbs. The best community gardens are small and providejust a moderate amount of fresh produce to the people who live in that area.

6.14 Review

6.14.1 Key knowledge summary

6.2 Feeding our future world population

- One in nine people on Earth do not have enough to eat; around a quarter of the population is overweight.
- The distribution of the world's arable land is uneven, and the fastest-growing parts of the world do not have enough land to grow sufficient food. As urban areas grow, the amount of available arable land decreases.
- Seventy per cent of the world's poorest people live in rural areas where trade is limited. Improving roads and other infrastructure would improve opportunities for trade.
- Farming yields are affected by a variety of factors: access to water, growing season, climate, soil types, finance and markets, insects and diseases, funds for agricultural research, government regulations, and weather events such as floods, storms and drought.
- Better use of farming areas and technology, and more efficient farming will improve food production.

6.3 Improving food production

- Strategies to improve food production include reducing the yield gap, developing genetically modified crops, expanding aquaculture, improving infrastructure and developing sustainable intense cropping.
- There are some concerns over the use of GM crops, including health risks and loss of seed variety.
- New technologies and systems are being developed to improve agricultural efficiency, such as the Sundrop Farm in Port Augusta, South Australia.

6.14.2 Key terms

active consumerism a movement that is opposed to the endless purchase of material possessions and the pursuit of economic goals at the expense of society or the environment
anthropogenic resulting from human activity
arable describes land that is suitable for growing crops
feedlot farming intensive farming practice in which high concentrations of animals are kept in confined areas to facilitate rapid growth and weight gain
fair trade trading exchanges that are sustainable and fair for both the seller and the buyer
food aid food, money, goods and/or services given for the specific purpose of helping those in need
food loss takes place at the production, post-harvest, processing, and distribution stages of food production
food waste takes place at the retail and consumption stages food production and consumption
genetically modified describes seeds, crops or foods whose DNA has been altered by genetic engineering techniques
hydroponic describes a method of growing plants using mineral nutrients, in water, without soil
malnourishment a condition that results from not getting the right amount of vitamins, minerals and other nutrients needed to maintain healthy tissues and organ function
mitigate to reduce negative impacts or severity
seasonal crops crops that are harvested in a certain season of the year rather than all year round
urban farming growing plants and raising animals within and around cities
urban sprawl the spreading of urban developments into areas on the city boundary
yield gap the gap between a certain crop's average yield and its maximum potential yield

6.14.3 Reflection

Complete the following to reflect on your learning.

Revisit the inquiry question posed in the Overview:

Will there come a time when we don't have enough food to feed everyone?

1. Now that you have completed this chapter, what is your view on the question? Discuss with a partner. Has your learning in this chapter changed your view? If so, how?
2. Write a paragraph in response to the inquiry question, outlining your views.

Structured student self-evaluation and reflection

Subtopic	Success criteria			
6.2	I can explain the prevalence and impacts of hunger.			
	I can outline the factors that influence food insecurity.			
6.3	I can compare and evaluate strategies for improving food production as a way to reduce food insecurity.			
6.4	I can compare and evaluate strategies for reducing food waste as a way to help reduce food insecurity.			
6.5	I can plot the data from two maps into a box scattergram.			
	I can describe the relationships that between the two data sets shown in a box scattergram.			
6.6	I can identify different types of food aid and outline how it is distributed.			
6.7	I can identify common reasons why Australians need food relief.			
	I can give examples of strategies for making food more widely available to people experiencing food insecurity in Australia.			
6.8	I can outline fair trade practices and evaluate their sustainability in terms of achieving food security.			
6.9	I can plot scaled circles on a map to show the pattern of data.			
	I can describe the patterns and any anomalies demonstrated by proportional circles on a map.			

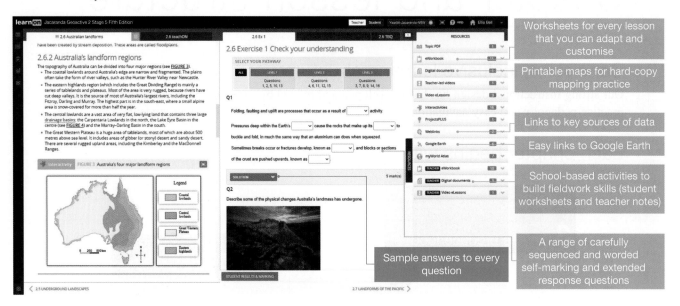

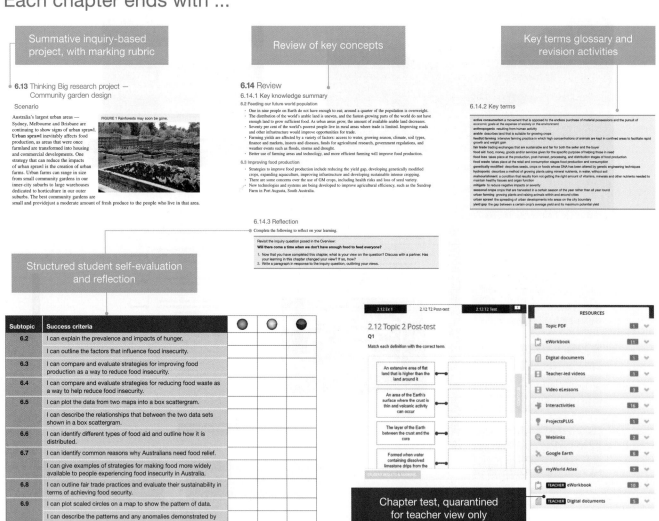

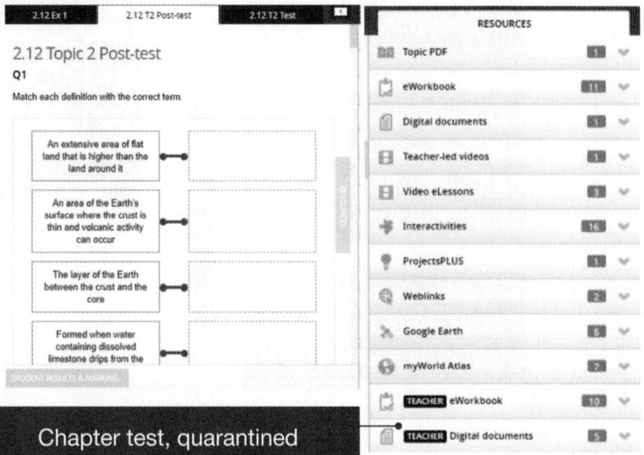

Chapter test, quarantined for teacher view only

Comprehensive teacher support

Teachers will find the resources and support they need, no matter what their level of experience of background in Geography. One simple click on the teachON tab or the Teacher Resources menu provides access a full suite of differentiated planning, assessment and teaching materials.

Each chapter is supported by ...

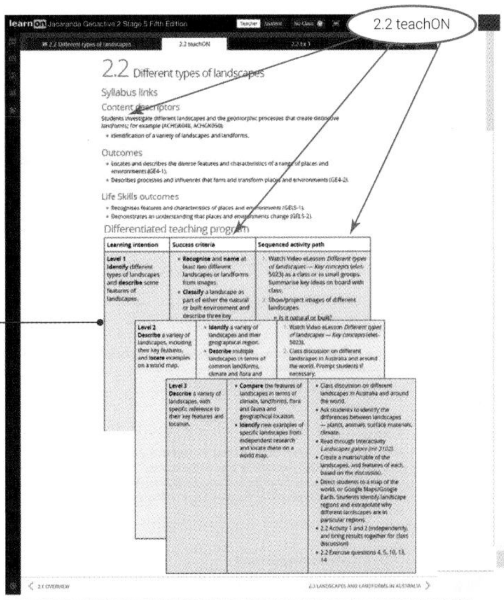

Differentiated planners and teaching notes that include learning intentions and success criteria for three achievement levels, all linked to syllabus outcomes

Teacher-only content including tests, worksheet sample answers and customisable course planners

Every task students complete provides ...

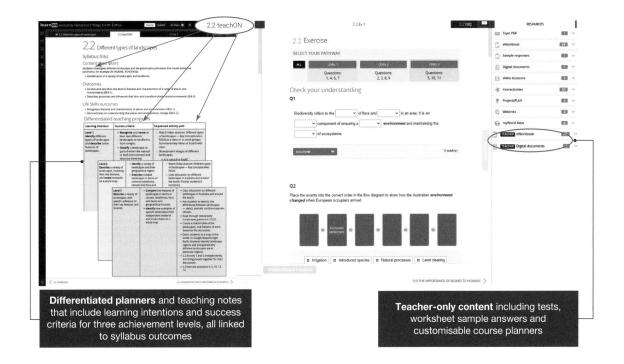

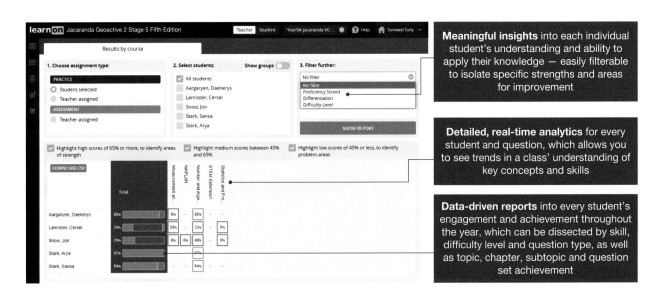

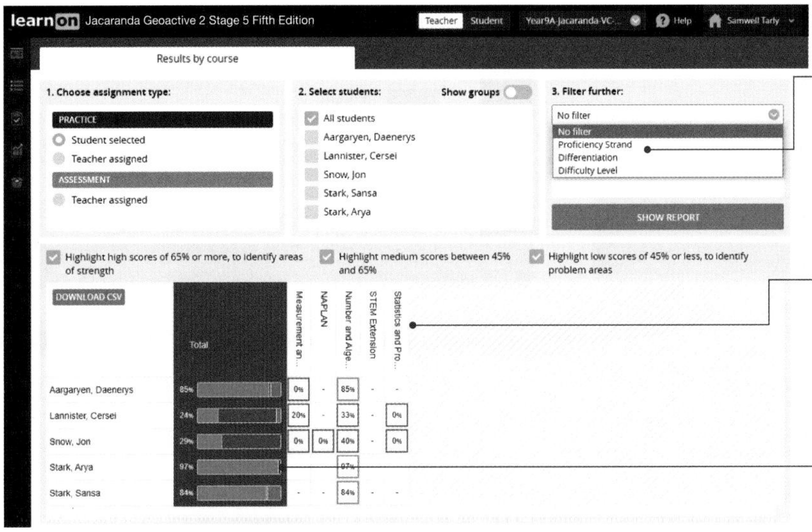

Meaningful insights into each individual student's understanding and ability to apply their knowledge — easily filterable to isolate specific strengths and areas for improvement

Detailed, real-time analytics for every student and question, which allows you to see trends in a class' understanding of key concepts and skills

Data-driven reports into every student's engagement and achievement throughout the year, which can be dissected by skill, difficulty level and question type, as well as topic, chapter, subtopic and question set achievement

The Geoactive package

Jacaranda Geoactive 2 Stage 5 NSW Ac Fifth Edition has been revised and reimagined to provide students and teachers with the most comprehensive and engaging Geography resource package on the market. This engaging and purposeful suite of resources is fully aligned to the NSW Australian curriculum Geography Stage 5. Jacaranda Geoactive is available in digital and print formats.

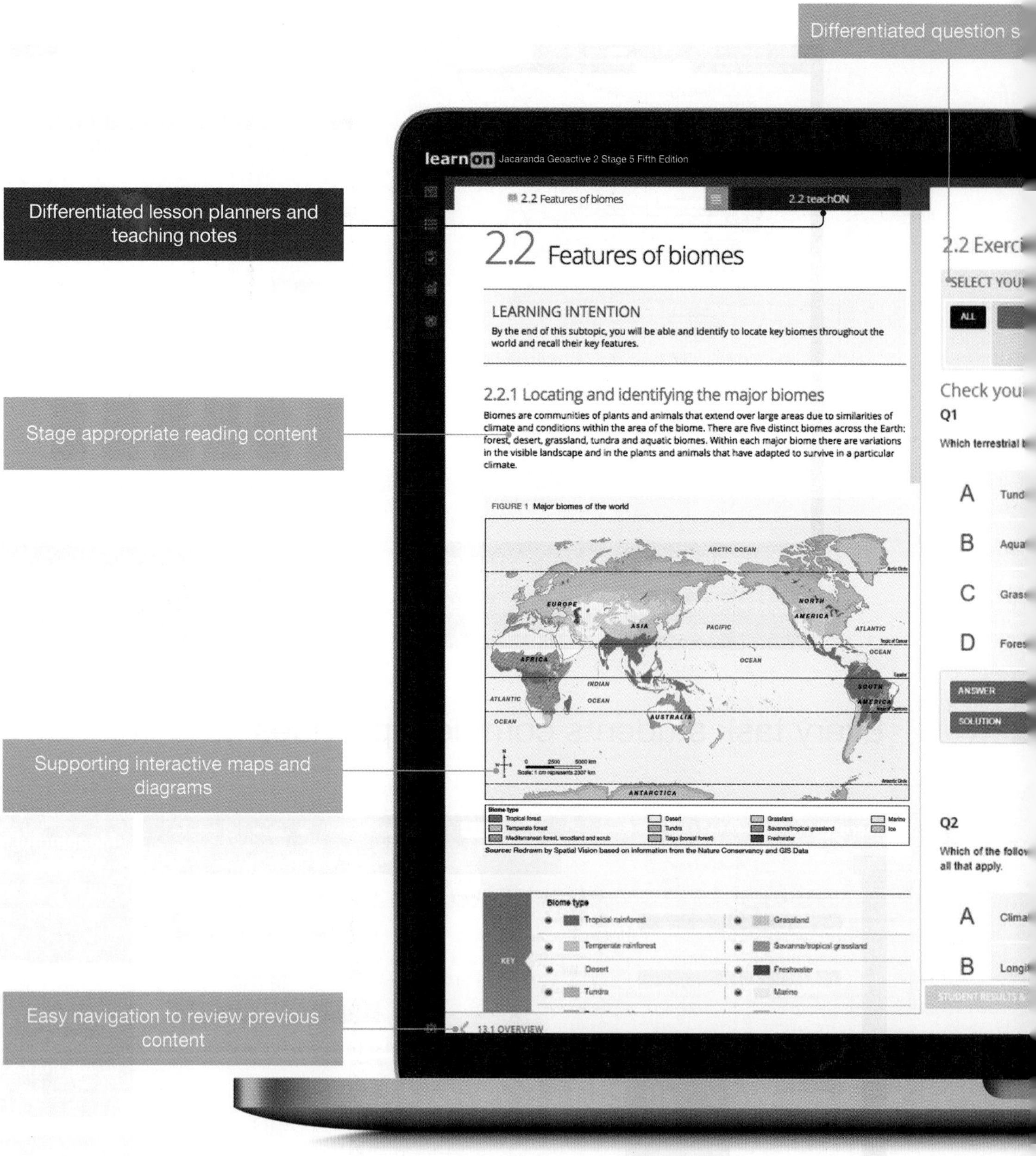

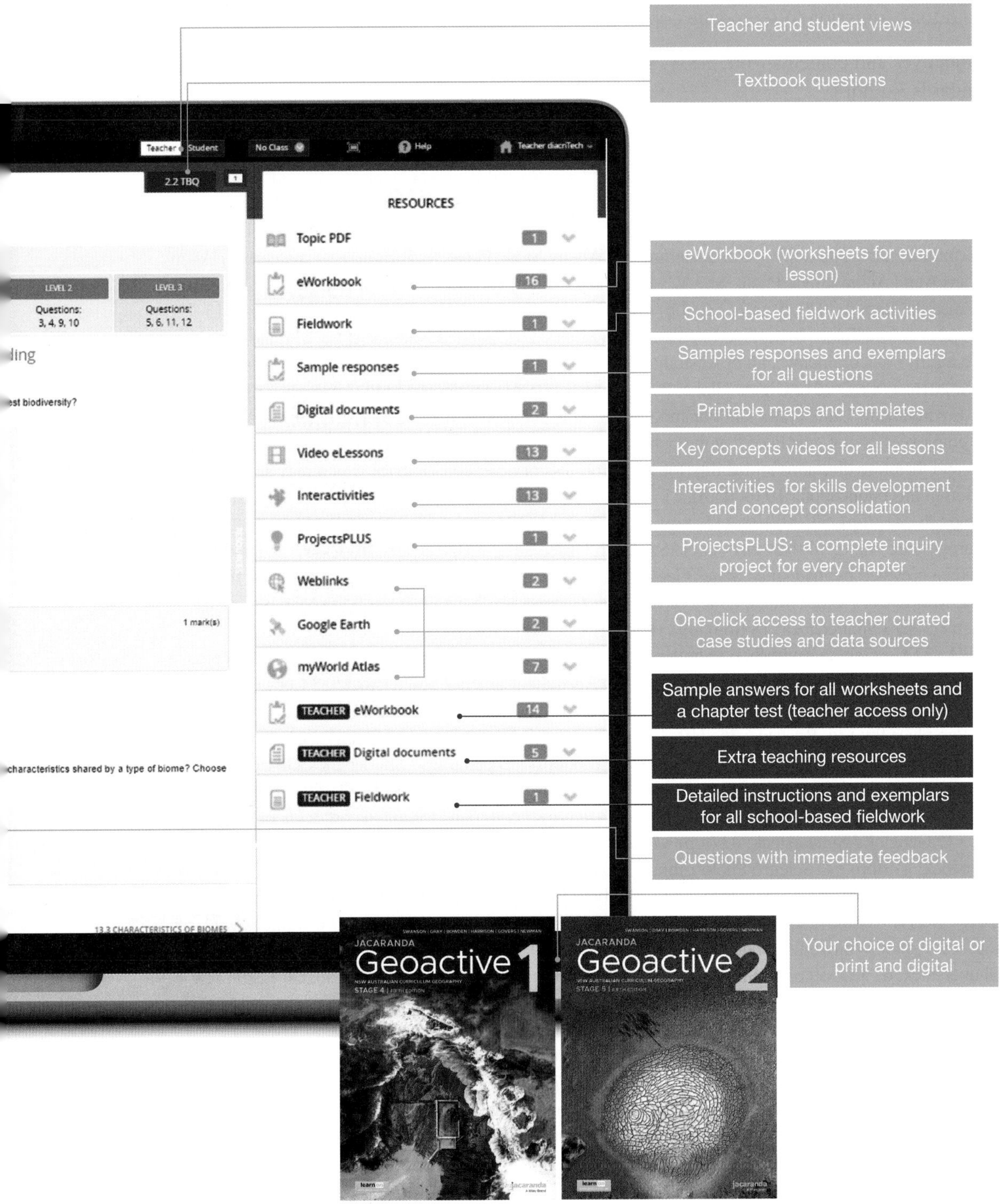
Teacher and student views
Textbook questions
Teacher Student
No Class
Help
2.2 TBQ
RESOURCES
Topic PDF 1
eWorkbook 16
Fieldwork 1
Sample responses 1
Digital documents 2
Video eLessons 13
Interactivities 13
ProjectsPLUS 1
Weblinks 2
Google Earth 2
myWorld Atlas 7
TEACHER eWorkbook 14
TEACHER Digital documents 5
TEACHER Fieldwork 1
LEVEL 2
Questions: 3, 4, 9, 10
LEVEL 3
Questions: 5, 6, 11, 12
1 mark(s)
eWorkbook (worksheets for every lesson)
School-based fieldwork activities
Samples responses and exemplars for all questions
Printable maps and templates
Key concepts videos for all lessons
Interactivities for skills development and concept consolidation
ProjectsPLUS: a complete inquiry project for every chapter
One-click access to teacher curated case studies and data sources
Sample answers for all worksheets and a chapter test (teacher access only)
Extra teaching resources
Detailed instructions and exemplars for all school-based fieldwork
Questions with immediate feedback
JACARANDA
Geoactive 1
JACARANDA
Geoactive 2
Your choice of digital or print and digital

Acknowledgements

The Geoactive team would like to thank those people who have played a key role in the production of this text. Their families and friends were always patient and supportive, especially when deadlines were imminent. This project was greatly enhanced by the generous cooperation of many academic colleagues and friends. The professionalism and expertise of Wiley staff is also appreciated.

The authors and publisher would like to thank the following copyright holders, organisations and individuals for their assistance and for permission to reproduce copyright material in this book.

Subject outcomes, objectives and contents from NSW Stage 4 & 5 Geography Syllabus. © Copyright 2021 NSW Education Standards Authority

Images

• © Oleksiy Mark / Shutterstock: 1 (top) • © Photodiem / Shutterstock: 3 • © Spatial Vision / Geophysical Fluid Dynamics Labartory, National Oceanic and Atmospheric Administration; Spatial Vision / USAID, FEWS NET 2011; Spatial Vision / NASA Earth Observatory; Golubovy / Shutterstock: 4 • © Data from World Trade Organization. Map by Spatial Vision.: 5 (top) • © c Karen Bowden: 5 (middle), 227, 382 (top) • © frans lemmens / Alamy Stock Photo: 115 (right) • © guatebrian / Alamy Stock Photo: 196 • © Kairi Aun / Alamy Stock Photo: 82 • © geogphotos / Alamy Stock Photo: 5 (bottom) • © c Clean Energy Council. Clean Energy Australia Report 2020, page 11.: 6 (top) • © Adapted from World Population Prospects The 2015 Revision, Key Findings and Advance Tables by DESA, Population Division, c 2015 United Nations. Used with the permission of the United Nations.: 6 (bottom) • © Commonwealth of Australia Geoscience Australia 2016.: 7, 147 • © (c) Institute for Sustainable Futures UTS, 2016: 178 • © 2016 Digital Globe: 8 • © TO BE PRINTED WITH MAP: Copyright © The State of Victoria, Department of Environment, Land, Water and Planning, 2016. This publication may be of assistance to you, but the State of Victoria and its employees do not guarantee that the publication is without flaw of any kind or is wholly appropriate for your particular purposes and therefore disclaims all liability for any error, loss or other consequence which may arise from you relying on any information in this publication.: 10 (bottom), 11 • © Den Edryshov / Shutterstock: 12 • © Spatial Vision: 14 (top), 31 (bottom), 32 (top), 40, 41 (top), 52, 90, 101, 115 (top), 133, 170 (top), 174, 177, 220, 302 (top), 453 (top), 494, 507, 568 • © Chronicle / Alamy Stock Photo: 14 (bottom) • © NASA Earth Observatory image created by Jesse Allen and Robert Simmon, using Landsat data provided by the United States Geological Survey. Caption by Michon Scott.: 16 (left), 16 (right) • © Government of India, Ministry of Home Affairs, Office of Registrar General Made with Natural Earth.Map by Spatial Vision: 17, 691 • © Caleb Holder / Shutterstock: 18 (top), 493 (bottom) • © Scott Prokop / Shutterstock: 19 (middle) • © Mmaxer / Shutterstock: 22 • © Galyna Andrushko / Shutterstock; Dmitry Pichugin / Shutterstock: 30 (top) • © Stephanie Jackson - Australian landscapes / Alamy Stock Photo: 32 (bottom) • © AustralianCamera / Shutterstock: 33 (top) • © Marc Anderson / Alamy Stock Photo: 33 (middle) • © Nicram Sabod / Shutterstock: 33 (bottom) • © aeropix / Alamy Stock Photo: 34 (top) • © MAPgraphics Pty Ltd, Brisbane: 36, 319, 335, 486 (top) • © Marco Saracco / Shutterstock: 37 (top) • © totajla / Shutterstock: 37 (middle) • © matteo_it / Shutterstock: 37 (bottom) • © paintings / Shutterstock: 38 (top), 476 (bottom) • © Ingo Oeland / Alamy Stock Photo: 38 (bottom) • © By Ville Koistinen user Vzb83 - the blank world map in Commons and WSOY Iso karttakirja for the information, CC BY-SA 3.0, https://commons.wikimedia.org/w/index.php?curid=1700408: 42 • © Gbuglok / Shutterstock: 43 (top) • © VectorMine / Shutterstock: 43 (bottom), 158, 442 • © Snaprender / Shutterstock: 47 (top) • © Science Photo Library / Alamy Stock Photo: 47 (bottom) • © gillmar / Shutterstock: 53 • © THPStock / Shutterstock: 54, 419 • © Brent Deuel / NOAA Photo Library; Derek Keats / Coralline algae un undersides of coral; Derek Keats / Coralline algae un undersides of coral: 56 • © The Australian Army © Commonwealth of Australia 1999: 60 (top) • © Ashley Whitworth / Shutterstock: 62, 509 (bottom) • © Heather Dine / NOAA Photo Library: 65 • © Bengker.Sakti / Shutterstock: 67 • © saiko3p / Shutterstock: 68 • © Data from FAO. Map drawn by Spatial Vision.: 69 • © Hurst Photo / Shutterstock: 70 • © Ivan Popovych / Shutterstock: 71 (top) • © CHEN WS / Shutterstock: 71 (middle) • © Data courtesy of the Institute on the Environment IonE, University of Minnesota. Map drawn by Spatial Vision: 71 (bottom) • © Source: Food and Agriculture Organization of the United Nations, 2015, World agriculture:towards 2015/2030 - Summary report, Table: Crop yields in developing countries, 1961 to 2030: 72 (top) • © Panos Pictures: 72 (bottom) • © pixelfusion3d / Getty Images: 74 • © c Laura Sproule: 75 (top) • © Elena Elisseeva / iStockphoto / Getty Images: 75 (bottom) • © Israel Hervas Bengochea / Shutterstock: 76 • © reproduced from www.S-cool.co.uk: 79 • © Nature Picture Library / Alamy Stock Photo: 80 (top), 81 (bottom), 172 • © Stefano Ravera / Alamy Stock Photo: 80 (bottom) • © Map by MAPgraphics Pty Ltd, Brisbane: 81 (top) • Alamy Stock Photo: 82, 105, 115 (bottom), 253, 261 (top), 270 (top), 270 (bottom), 278, 282 (top), 344 (bottom), 346, 367, 360, 533 • © David Beaumont: 84 • © Lycett, Joseph, approximately 1775-1828. Drawings of Aborigines and scenery, New South Wales, ca. 1820.: 85 (top) • © By CSIRO, CC BY 3.0, https://commons.

wikimedia.org/w/index.php?curid=35461031: 85 (bottom) • © Collection: Museum of Applied Arts and Sciences. Gift of Australian Consolidated Press under the Taxation Incentives for the Arts Scheme, 1985. Unattributed studio.: 86 (top) • © State of New South Wales Department of Premier and Cabinet 2019. A.W. Mullensplan of the Brewarrina fish traps drawn in 1906 for the NSW Western Lands Board from Hope and Vines CMP, 1994: 86 (bottom) • © denisbin: 87 • © Source: Australian Bureau of Statistics: 88 • © Patrina Malone / Newspix: 89 (top) • © Justin Sanson / Newspix: 89 (bottom) • © Commonwealth of Australia Geoscience Australia 2013. Map drawn by Spatial Vision.: 92 • © Simon Grosset / Alamy Stock Photo: 94 (bottom) • © Orientaly / Shutterstock: 94 (top) • © Rosamund Parkinson / Shutterstock: 94 (bottom) • © Commonwealth of Australia Geoscience Australia 2013. With the exception of the Commonwealth Coat of Arms and where otherwise noted, this product is provided under a Creative Commons Attribution 3.0 Australia Licence.: 97 • © Zzvet / Shutterstock: 98 • © Aubrey Wade / Panos Pictures: 99 • © Adwo / Shutterstock: 100 • © Tracing Tea / Shutterstock.com: 103 • © Lincoln Fowler / Alamy Stock Photo: 109 • © Russotwins / Alamy Stock Photo: 110 • © Data courtesy of the Institute on the Environment IonE, University of Minnesota. Map drawn by Spatial Vision.: 113 (top) • © Source: American Geophysical Union and Google Maps. Image created by Spatial Vision.: 113 (bottom) • © Sebastian Studio / Shutterstock: 114 • © Mongabay: 117 • © NASA Earth Observatory: 118, 134, 169 (top), 169 (bottom), 334, 535 • © Oscar_Romero_Photo / Shutterstock: 120 (top) • © Credit: U.S. Geological Survey Department of the Interior/USGS: 120 (middle) • © National Ocean and Atmospheric Administration: 120 (bottom) • © Paulo Vilela / Shutterstock: 121 • © Commonwealth of Australia Geoscience Australia.: 122 • © OECD-FAO Agricultural Outlook: 124 • © Andreas Altenburger / Shutterstock: 125 • © Source: Food and Agriculture Organization of the United Nations. The State of World Fisheries and Aquaculture 2018. Meeting the Sustainable Development Goals. Accessed: 3 March 2020. https://reliefweb.int/sites/reliefweb.int/files/resources/ca0191en.pdf: 126 (top) • © Sukpaiboonwat / Shutterstock: 126 (bottom) • © Jake Fuller / artizans.com: 128 • © Dirk Ercken / Shutterstock: 129, 480 • © European Union, 1995-2019. Cherlet, M., Hutchinson, C., Reynolds, J., Hill, J., Sommer, S.,von Maltitz, G. Eds., World Atlas of Desertification, PublicationOffice of the European Union, Luxembourg, 2018.: 130 • © Taras Vyshnya / Shutterstock: 131, 384 • © Data from the USGS. Map drawn by Spatial Vision.: 135 • © Kelly Cheng / Getty Images: 136 • © Cartoon by Nicholson from The Australian www.nicholsoncartoons.com.au: 138 • © Marcos Brindicci / Reuters / PICTURE MEDIA / BAS107: 139 (top) • © Lockenes / Shutterstock: 139 (bottom) • © Dario Sabljak / Shutterstock: 139 (top) • © Millennium Ecosystem Assessment, 2005. Ecosystems and Human Well-being: Synthesis. Island Press, Washington, DC. Copyright © 2005 World Resources Institute: 141, 142 • © Commonwealth of Australia Department of Sustainability, Environment, Water, Population and Communities 2013. Map drawn by Spatial Vision.: 143 • © Denton Rumsey / Shutterstock: 146 • © cybercrisi / Shutterstock: 149 (top) • © Serhii Ostapenko / Shutterstock: 149 (bottom) • © Clare Seibel-Barnes / Shutterstock: 152 • © MaxyM / Shutterstock: 155 • © Shuang Li / Shutterstock: 156 • © thirawatana phaisalratana / Shutterstock: 157 • © Data from the Centre for Environmental Systems Research, University of Kassel. Map drawn by Spatial Vision.: 159 (top) • © Paul Jeffrey / Alamy Stock Photo: 159 (bottom) • © Data from Australian Bureau of Statistics May 2020 Table: Sources of water used for agricultural production - original [data set], Water Use on Australian Farms, Australia, accessed 18 March 2021.: 163 (top) • © inga spence / Alamy Stock Photo: 163 (bottom) • © Commonwealth of Australia Geoscience Australia; © Commonwealth of Australia Geoscience Australia: 165 • © Food and Agriculture Organization of the United Nations, Reproduced with permission: 167 • © Frans Lanting Studio / Alamy Stock Photo: 168 (top) • © joloei / Shutterstock: 168 (bottom) • © Food and Agriculture Organisation, International Food Policy Research Institute: 170 (bottom) • © UNHCR: 173 • © Olga Kashubin / Shutterstock: 176 • © The origins and destinations of produce flows through Sydney Market Limited Flemington Site. May 2013 Urban Research Centre, University of Western Sydney: 178 • © c Institute for Sustainable Futures UTS, 2016: 179 (top) • © Source: Australian Bureau of Statistics, cat. no. 7503.0, Value of agricultural commodities produced, Australia 2020: 179 (bottom) • © Data sourced from UNHCR/OCHA; Image: FEWS NET.: 182 • © crbellette / Shutterstock; Taras Vyshnya / Shutterstock: 183 • © africa924 / Shutterstock: 185 • © Figure SPM.7 and Figure SPM.11 upper panel from IPCC, 2014: Climate Change 2014: Synthesis Report. Contribution of Working Groups I, II and III to the Fifth Assessment Report of the Intergovernmental Panel on Climate Change. [Core Writing Team, Pachauri, R.K. and Meyer, L. eds.]. IPCC, Geneva, Switzerland.: 186 • © Data from the European Commission. Map drawn by Spatial Vision.: 187 (top) • © Millward Shoults / Shutterstock: 187 (bottom) • © Data from Reducing climate change impacts on agriculture: Global and regional effects of mitigation, 2000–2080 by Tubiello F N, Fisher G in Technological Forecasting and Social Change 2007, 747: 1030-56. Map drawn by Spatial Vision.: 188 • © Fabio Alcini / Shutterstock: 190 • © Spatial Vision: 191 • © Vlad Karavaev / Shutterstock: 193 (top) • © ChameleonsEye / Shutterstock: 193 (bottom) • © Sodel Vladyslav / Shutterstock: 199 • © Elizabeth Wake / Alamy Stock Photo; Richard Levine / Alamy Stock Photo: 200 • © 2019 The Economist Intelligence Unit Limited, data from Global Food Security Index. Map drawn by Spatial Vision: 201 • © Re-drawn from an image by Global Harvest Initiative 2011 GAP Report®: Measuring Global Agricultural Productivity, data from Population Facts 2014/3. Department of Economics and Social Affairs, Population Division. © 2014 United Nations. Reprinted with the permission of the United Nations.: 203 • © Sam Hood / Alamy Stock Photo: 204 • © oticki / Shutterstock: 206 • © Data source from Hannah Ritchie and Max Roser, 2019. UN Food and Argriculture Organization FAO: 207 (top) • © BRONWYN GUDGEON / Shutterstock: 207 (bottom) • © Mr Privacy / Shutterstock: 207 (top) • © Sleeping cat / Shutterstock: 208 (top) • © Source: Food and Agriculture Organization of the United Nations. Reproduced with permission.: 208 (bottom), 215 (bottom), 219 • © J. Dennis Thomas / Alamy Stock Photo: 209 • © ISAAA. 2018. Global Status of

Commercialized Biotech/GM Crops: 2018. ISAAA Brief No. 54. ISAAA: Ithaca, NY.: 210 (top) • © MONOPOLY919 / Shutterstock: 210 (bottom) • © Michael Evans / Alamy Stock Photo: 211 • © RooM the Agency / Alamy Stock Photo: 212 • © attraction art / Shutterstock: 214 (top) • © Sally A. Morgan; Ecoscene / CORBIS: 214 (bottom) • © Source: Food and Agriculture Organization of the United Nations. The role of producer organizations in reducing food loss and waste. International Year of Cooperatives. Issues Brief Series. Accessed 23 April 2021. http://www.fao.org/3/ap409e/ap409e.pdf: 215 (top) • © 2019 Sustainability Victoria: 216 • © Suzanne Itzko / Alamy Stock Photo: 217 • © Source: Food and Agriculture Organization of the United Nations. 2020, State of food security and nutrition in the world, Accessed: 23 April 2021. http://www.fao.org/3/ca9692en/online/ca9692en.html#chapter-1_1: 222 • © Dana.S / Shutterstock: 223 • © AB Forces News Collection / Alamy Stock Photo: 224 (bottom) • © US Air Force Photo / Alamy Stock Photo: 224 (top) • © By United Nations - UN.org/SustainableDevelopment, Public Domain, https://commons.wikimedia.org/w/index.php?curid=65335693: 224 (bottom) • © Kim Haughton / Alamy Stock Photo: 225 • © goran_safarek / Shutterstock: 226 • © Fairtrade Australia: 228 • © MAPgraphics: 230 • © Copyright © 2016 International Center for Tropical Agriculture – CIAT: 231 (top) • © Source: FAO, 2008; FAOSTAT, 2009: 231 (bottom) • © MLA graph / OECD data: 232 • © c Meat Free Monday Ltd: 233 • © Cristobal Demarta / Getty Images: 234 (top) • © SIMPILI / Shutterstock: 234 (bottom) • © martinbertrand.fr / Shutterstock: 236 (top) • © Jeff Gilbert / Alamy Stock Photo: 236 (bottom) • © suprabhat / Shutterstock: 237 (top) • © epa european pressphoto agency b.v./ Alamy Stock Photo: 237 (bottom), 544 (bottom) • © Martchan / Shutterstock: 239 (top) • © Dietmar Temps / Shutterstock: 239 (bottom) • © Data sourced from U.S. Geological Survey, OpenStreetMap, and OpenStreetMap Foundation. Map drawn by Spatial Vision.: 240 • © Hannamariah / Shutterstock: 242 (top) • © Jakob Fischer / Shutterstock: 242 (bottom) • © Universal Images Group North America LLC / Alamy Stock Photo: 241 • © Kletr / Shutterstock: 247 • © nito / Shutterstock: 248 • © wavebreakmedia / Shutterstock: 249, 747 • Shutterstock / VIGO-S: 255 • Shutterstock / Tupungato: 258 • Shutterstock / John-james Gerber: 261 (bottom) • Shutterstock / Steve Mc Carthy: 268 • AFP via Getty Images: 271 • Alamy: 274 • Shutterstock: 276, 282 (bottom), 341, 625, 629, 648 (bottom), 650 • Shutterstock / amadeustx: 289 • Shutterstock / Yury Birukov: 293 • © MAGNIFIER / Shutterstock: 300 • © Greg Brave / Shutterstock: 302 (bottom) • © Data source from ABS. Map by Spatial Vision: 303 • © ABS data used with permission from the Australian Bureau of Statistics. www.abs.gov.au/ Map by MAPgraphics Pty Ltd, Brisbane: 304 • © Australian Bureau of Statistics, Experimental Estimates of Aboriginal and Torres Strait Islander Australians, June 2006 ABS cat. no. 3238.0.55.001: 306 • © Based on data from the ABS and US Population Bureau: 309 • © Glenn Young / Shutterstock: 310 • © Regional Plan Association: 311 • © Australian Bureau of Statistics / Spatial Vision: 312, 703 • © TierneyMJ / Shutterstock: 313 • © Photo Spirit / Shutterstock: 314 (top) • © Chart used with permission of the New York City Department of City Planning. All rights reserved: 314 (bottom) • © Created from data from City of New York, New Jersey Department of Environmental Protection, New Jersey Geographic Information Network 2012. Map drawn by Spatial Vision.: 314 (top) • © Willem Tims / Shutterstock: 315 • © slava17 / Shutterstock: 318 • © Aleksandar Todorovic / Shutterstock: 321 • © Doremi / Shutterstock: 322 • © Philip Lange / Shutterstock: 323 (top) • Shutterstock / Peter Gudella: 323 (left), 323 (right) • © Konstantin Solodkov / Shutterstock: 323 (bottom) • © Agus.d.wahyudi / Shutterstock: 325 • © Justin Adam Lee / Shutterstock: 326 (top) • © AnnZeruk / Shutterstock: 326 (bottom) • © Spatial Vision/ Dept. ELWP Victoria: 328 • © Photo by Sumana Wijeratne, VanLanka Planning: 330 • © Jeffrey Blackler / Alamy Stock Photo: 331 (top) • © Creative Commons: 331 (middle) • © Newscom / Alamy Stock Photo: 331 (bottom) • © lucarista / Shutterstock: 332 (top) • © Adam Calaitzis / Shutterstock: 332 (bottom) • © AsiaTravel / Shutterstock: 337 (left) • © Andrew Zarivny / Shutterstock: 337 (right) • Shutterstock / Grossinger: 342 • Shutterstock / Willjhunt: 344 (top) • Shutterstock / King Ropes Access: 345 • Shutterstock / Alecia Scott: 357 (top) • Shutterstock / Steven Tritton: 357 (bottom) • ©~c~Karen~Bowden: 382 (bottom) • © martin berry / Alamy Stock Photo: 383 • © Copyright Commonwealth of Australia Geoscience Australia 2006. Map drawn by Spatial Vision.: 386 (top) • © Johnny Lye / Shutterstock: 386 (bottom) • © Michael Willis / Alamy Stock Photo: 387 • © Bureau of Meteorology: 390 • © sljones / Shutterstock: 392 • © Map drawn by Spatial Vision: 387 • © John Wiley & Sons Australia: 393 • © Photo © A Kanck, Quality Freelancing: 394 (top) • © Perspective sketch & design by Paul F. Downton: 394 (middle) • © Photo © A Kanck, Print number 1942: 394 (bottom) • © Paul Lovelace / Alamy Stock Photo: 395 (top) • © c Louise Swanson; c Louise Swanson: 395 (middle), 407 (top), 440 • © Nils Versemann / Shutterstock: 395 (bottom), 431 • © Provided by Metropolitan Strategy, NSW Department of Planning & Infrastructure. Map re-drawn by Spatial Vision.: 398 • © John Martin - Fotografo / Alamy Stock Photo: 399 • © aiyoshi597 / Shutterstock: 392 • Shutterstock / Willowtreehouse: 405 (top) • © domonabike / Alamy Stock Photo: 405 (bottom) • © aiyoshi597 / Shutterstock; Historic Collection / Alamy Stock Photo: 393 • © c Development WA; c Development WA: 406 • © c Louise Swanson: 407 (bottom), 423, 446 (top), 446 (bottom), 452 (top) • © Material/information courtesy of Department of Climate Change and Energy Efficiency: 410 • © Richard Milnes / Alamy Stock Photo: 395 • © EEI_Tony / Getty Images: 418 (top) • © Don Mason / Getty Images, Inc.: 418 (bottom) • © Karen McFarland / Shutterstock: 398 • © Cromo Digital / Shutterstock; cvotography / Shutterstock: 419 • © Michael Brooks / Alamy Stock Photo; Tim Brown / Alamy Stock Photo: 422 • © RichardMilnes / Shutterstock: 401 • © c Dulwich Hill Action Group: 424 • © TGB / Alamy Stock Photo: 425 • © TO BE PRINTED WITH MAP: Copyright © The State of Victoria, Department of Environment, Land, Water and Planning, 2016.: 428 • © TO BE PRINTED WITH MAP: Copyright© The State of Victoria, Department of Environment, land, Water and Planning, 2016. This publication may be of assistance to you, but the State of Victoria and its employees do not guarantee that the publication is without flaw of any kind or is wholly appropriate for your particular purposes and therefore disclaims all liability for any error, loss or other consequence which may arise from you relying on any information in this publication. Copyright: The State of NSW, Land and Property Management Authority.: 429 • © oneinchpunch / Shutterstock: 403 • © Olga Kashubin / Alamy Stock Photo: 407 • © StreetVJ / Shutterstock: 427 • © Jackal Pan / Getty Images: 429 • © James LePage / Shutterstock: 436 • © Ozerov Alexander / Shutterstock: 438 (top) • © c Louise Swanson; c Louise Swanson; c Louise Swanson; c Louise Swanson: 438 (bottom) • © Mor65_Mauro Piccardi / Shutterstock: 451 • © Source: Licensed from the Commonwealth of Australia under a Creative Commons Attribution3.0 Australia Licence. The Commonwealth of Australia does not necessarily endorse the content of this publication.: 452 (bottom) • © Gingerss / Shutterstock: 453 (bottom) • © Wollnorth Wind Farm Holding Pty Ltd: 454 (top) • © brianafrica / Alamy Stock Photo: 454 (bottom) • © Reproduced by permission of Bureau of Meteorology, © 2021 Commonwealth of Australia.: 456 • © Data derived from NASA: 457 • © The Garnaut Climate Change Review 2008, published by the Commonwealth of Australia: 459 • © studiovin / Shutterstock: 461 • © vita khorzhevska / Shutterstock: 463 • © c Sustainable Society Foundation: 465 • © Global Footprint Network: 469 • © Senia Effe / Shutterstock: 467 • © Chris Putnam / Alamy Stock Photo: 473 • © Data sourced from Geoscience Australia 2006 GEODATA TOPO 250K Series 3. Bioregional Assessment Source Dataset. Viewed 13 March 2019 and Spatial Services, Department of Customer Service, State of New South Wales. Map drawn by Spatial Vision.: 474 • © idiz / Shutterstock: 476 (top) • © Pauline English, 2005: 484 • © Commonwealth of Australia Geoscience Australia 2013. © Commonwealth of Australia Department of Sustainability, Environment, Water, Population and Communities 2013. Map by Spatial Vision: 486 (bottom) • © Valentin Valkov / Shutterstock: 487 • © Hhelene / Shutterstock: 491 • © Mark Winfrey / Shutterstock: 492 (top) • © Neil Bradfield / Shutterstock: 492 (middle) • © Clem Sturmfel / Dept of Primary Industries Vic: 492 (bottom) • © John Ivo Rasic: 493 (top) • © '© State of Victoria Department of Environment and Primary Industries 2013. Reproduced with permission. Photograph by Stuart Boucher.': 495 (top) • © Stuart Boucher / Dept of Primary Industries Vic: 495 (bottom), 496 • © Mike Patterson / Alamy Stock Photo: 498, 504 • © UNEP World Conservation Monitoring Centre Made with Natural Earth. Map by Spatial Vision.: 499 • © Jane Kelly / Shutterstock; Yen Teoh / iStockphoto; bubaone / Getty Images; serazetdinov / Shutterstock: 500 • © SJ Allen / Shutterstock: 501 • © United States Department of Agriculture, Natural Resources Conservation Service, Soil Survey Division, World Soil Resources; Paul Reich, Geographer. 1998. Global Desertification Vulnerability Map. Washington, D.C.: 502 • © Image courtesy Jacques Descloitres, MODIS Rapid Response Team at NASA GSFC: 503 • © velvetweb / Shutterstock: 505 • © Pete Titmuss / Alamy Stock Photo: 509 (top) • © Commonwealth of Australia Geoscience Australia 2013. Map by Spatial Vision.: 510 • © Genevieve Vallee / Alamy Stock Photo: 511, 545 • © Made with Natural Earth. Vector Map Level 0 Digital Chart of the World Map by Spatial Vision: 512 • © doraclub / Shutterstock: 513 • © YuliiaKas/ Shutterstock: 515 • © Mike Keating / Newspix: 516 • © c Cynthia Wardle Newcastle: 517 • © Sam DCruz / Shutterstock: 519 • © J C Clamp / Alamy Stock Photo: 520 • © Paul Dymond / Alamy Stock Photo: 521 • © doug steley / Alamy Stock Photo: 525 • © Data sourced from Geoscience Australia 2006 GEODATA TOPO 250K Series 3. Bioregional Assessment Source Dataset. Viewed 13 March 2019 and Spatial Datamart, State Government of Victoria. Map drawn by Spatial Vision.: 526 • © edelmar / Getty Images: 528 • © Hypervision Creative / Shutterstock: 534 • © Warren Price Photography / Shutterstock: 536 • © Stephen Dwyer / Alamy Stock Photo: 537 • © DOUGLAS MAGNO / AFP/ Getty Images: 541 (top) • © NASA / Joshua Stevens/USGS; NASA / Joshua Stevens/USGS: 541 (bottom) • © Mario Tama / Getty Images: 544 (top) • © World Climate - http://www.worldclim.org/Made with Natural Earth.Map by Spatial Vision: 549 • © SCPhotos / Alamy Stock Photo: 550 (top) • © dpa picture alliance / Alamy Stock Photo: 550 (bottom) • © BGR & UNESCO 2008: Groundwater Resources of the World 1 : 25 000 000. Hannover, Paris. Made with Natural Earth.Map by Spatial Vision: 555 • © UNEP Global Environmental Alert Service GEAS Made with Natural Earth. Vector Map Level 0 Digital Chart of the WorldMap by Spatial Vision.: 556 • © BBC News, http://news.bbc.co.uk/2/hi/8545321.stm Made with Natural Earth. Map by Spatial Vision.: 557 • © Made with Natural Earth.Map by Spatial Vision.: 558 • © China Photos / Getty Images: 559 • © Sherrianne Talon / Shutterstock: 561 • © Source: United Nations Environment Programme UNEP: 563 (top) • © United Nations Environment Programme Made with Natural Earth. Vector Map Level 0 Digital Chart of the WorldMap by Spatial Vision: 563 (bottom) • © Jason Edwards / Alamy Stock Photo: 567 • © xpixel / Shutterstock: 570 • © Parilov / Shutterstock: 615 • © Mikadun / Shutterstock: 616 • © William Perugini / Shutterstock: 619 • © Tim Roberts Photography / Shutterstock: 620 • © Thorsten Rust / Shutterstock: 621 (bottom) • © Denis Crawford / Alamy Stock Photo: 621 (bottom) • © Auscape International Pty Ltd / Alamy Stock Photo: 621 (top) • © DavidWebb / Shutterstock: 621 (bottom) • Shutterstock / Anton_Ivanov: 626, 658 • Shutterstock / Monkey Business I: 637, 655 (top) • Shutterstock / paul prescott: 648 (top) • Shutterstock / Riccardo Mayer: 655 (bottom) • © Varavin88 / Shutterstock: 661 • © IndianFaces / Shutterstock: 662 • © United Nations Publications: 663, 664, 665 • © Public Domain: 666 • © Andrei Shumskiy / Shutterstock: 667 • © Matyas Rehak / Shutterstock: 669 • © December 2019 by PopulationPyramid.net, made available under a Creative Commons license CC BY 3.0 IGO: http://creativecommons.org/licenses/by/3.0/igo/; US Census Bureau: 670 (top) • © Government of India, Ministry of Home Affairs, Office of Registrar General Made with Natural Earth. Map by Spatial Vision: 670 (bottom), 671 (bottom) • © Percentage of Population Below Poverty Line URP Consumption, ChartsBin.com, viewed 27th August, 2013, <http://chartsbin.com/view/2797> Made with Natural Earth. Map by Spatial Vision: 671 (top) • © Used with permission from Microsoft.: 673, 674 • © Data based on United Nations, Department of Economic and Social Affairs, Population Division 2019. World Population Prospects 2019: 675 • © AFP / Stringer / Getty Images: 676 • © Sanjay JS / Shutterstock: 677 (top) • © ton koene / Alamy Stock Photo: 677 (left) • © Janelle Lugge /

Shutterstock: 677 (top), 677 (bottom) • © Sueddeutsche Zeitung Photo / Alamy Stock Photo: 679 • © Chris Warham / Shutterstock: 680 (top) • © Australian Government Department of Foreign Affairs and Trade/Jim Holmes: 680 (bottom) • © Kritsana Laroque / Shutterstock: 683 • © dbimages / Alamy Stock Photo: 685 • © December 2019 by PopulationPyramid. net, made available under a Creative Commons license CC BY 3.0 IGO: http://creativecommons.org/licenses/by/3.0/igo/; United States Census Bureau: 686 (top), 687 • © mykeyruna / Shutterstock: 686 (bottom) • © paul prescott / Shutterstock: 689 • © Government of India, Ministry of Home Affairs, Office of Registrar General Made with Natural Earth. Map by Spatial vision: 690 (top) • © Travel Stock / Shutterstock: 690 (bottom), 745 (top) • © zeber / Shutterstock: 692 • © Sk Hasan Ali / Shutterstock: 697 • © John Carnemolla / Shutterstock: 702 (bottom), 704 (bottom) • © superjoseph / Shutterstock: 702 (top) • © Elias Bitar / Shutterstock: 704 (top) • © Commonwealth of Australia. Australian Bureau of Statistics 2020 Table: Components of annual population changea, National, state and territory population, accessed 20 April 2020.: 705 (top) • © Unpublished ABS data,Treasury projections and Australian Bureau of Statistics 2020, Migration Australia, 2018-19 financial year, accessed 20 April 2021: 705 (bottom) • © Australian Bureau of Statistics: 707, 717, 718, 719 • © Copyright Commonwealth of Australia, Australian Bureau of Statistics http://www.abs.gov.au/AUSSTATS/abs@.nsf/DetailsPage/1270.0.55.005July%202011?OpenDocument © Commonwealth of Australia Geoscience Australia 2013.Map by Spatial Vision: 708 • © Based on Australian Institute of Health and Welfare material.: 709 (top) • © Data sourced from Department of Health, NHWDS Nurses and Midwife 2019, https://hwd.health.gov.au/resources/publications/factsheet-nrmw-2019.pdf: 709 (bottom) • © Country Women's Association: 710 (top) • © Marcel van den Bos / Shutterstock: 710 (bottom), 710 (top) • © Data sourced from © Crown in right of NSW through the Valuer General 2021. Map drawn by Spatial Vision.: 712 • © Australian Bureau of Statistics 2020 Table -Annual WPI growth - 1997 to 2020, Annual wage growth 1.8% in June quarter 2020, accessed 20 April 2021: 713 • © Drawn by Spatial Vision based on ATO and ABS data: 714 • © Australian Bureau of Statistics 2019 National Aboriginal and Torres Strait Islander Health Survey, accessed 20 April 2021: 720 • © Commonwealth of Australia, Department of the Prime Minister and Cabinet.: 721 • © Leonard Zhukovsky / Shutterstock: 724 • © Australian Human Rights Commission 2015.: 725 • © ArliftAtoz2205/Shutterstock: 726 • © ingehogenbijl / Shutterstock: 728 • © Data sourced from Geoscience Australia 2006 GEODATA TOPO 250K Series 3. Bioregional Assessment Source Dataset. Viewed 13 March 2019 and OpenStreetMap and OpenStreetMap Foundation. Map drawn by Spatial Vision.: 729 • © MintArt / Shutterstock: 731 • © Skreidzeleu / Shutterstock: 737 • © GNEs / Shutterstock: 739 • © Pacific Press Media Production Corp. / Alamy Stock Photo: 740 (top) • © Department of Foreign Affairs and Trade website – www.dfat.gov.au: 740 (bottom), 741 (top), 741 (bottom) • © OzHarvest: 744 • © nije salam / Shutterstock: 745 (left) • © Karve / Shutterstock: 745 (left) • © Goran Bogicevic / Shutterstock: 745 (right) • © Monkey Business Images / Shutterstock; ZUMA Press, Inc. / Alamy Stock Photo; Alex Hinds / Alamy Stock Photo; cbphoto / Alamy Stock Photo; martin berry / Alamy Stock Photo: 748 (top) • © phoelixDE / Shutterstock: 748 (bottom) • © Outback Stores: 754 • © Commonwealth of Australia Geoscience Australia 2013. © Commonwealth of Australia Australian Bureau of Statistics 2013.Map by Spatial Vision: 755 • © Instituto Brasileiro de Geografia e Estatística Made with Natural Earth. Map by Spatial Vision.: 758 • © Publio Furbino / Shutterstock: 759 • © Jan Sochor / Alamy Stock Photo: 760 • © UNEP-WCMC 2012. Data Standards for the World Database on Protected Areas. UNEP-WCMC: Cambridge, UK. Made with Natural Earth.: 761 • © Data sourced from Geoscience Australia 2006 GEODATA TOPO 250K Series 3. Bioregional Assessment Source Dataset. Viewed 13 March 2019 and Spatial Services, Department of Customer Service, State of New South Wales.: 765 • © John Wollwerth / Alamy Stock Photo: 767 • Deepak Tolange / Alamy Stock Photo: 770 • © 2xSamara.com / Shutterstock: 775

Text

• © Jacqueline Lynch.Anthony Pancia. ABC South West WA . Ni-Vanuatu workers arrive in Western Australia to combat labour shortages on farms. 9 January 2021, ABC News: 369 • © Hardie Grant. Langton, M 2018, Welcome to Country: A travel guide to Indigenous Australia, pp 9–10, Hardie Grant Books: 23 • © Commonwealth of Australia, 2013. Geographies of Human Wellbeing. Geography Teachers' Association of Victoria Inc. Global Education Project Victoria, Geography Teachers' Association of Victoria Inc: 748 • © Source: Food and Agriculture Organization of the United Nations, 2020, FAO, IFAD, UNICEF, WFP and WHO, The State of Food Security and Nutrition in the World 2020. Transforming food systems for affordable healthy diets, http://www.fao.org/3/ca9692en/ca9692en.pdf. Reproduced with permission, FAO: 233 • © Ida Kubiszewski. Beyond GDP: are there better ways to measure well-being? 2 December 2014. The Conversation. https://theconversation.com/beyond-gdp-are-there-better-ways-to-measure-well-being-33414, The Conversation: 747 • © Department of Foreign Affairs and Trade website – www.dfat.gov.au, Department of Foreign Affairs and Trade: 741 • © World Food Programme 2018, Public Domain: 222 • © Dana Cordell, Brent Jacobs, Laura Wynne. Urban sprawl is threatening Sydney's foodbowl. 25 February 2021, The Conversation: 177

1 The world of Geography

1.1 Overview

Numerous **videos** and **interactivities** are embedded just where you need them, at the point of learning, in your learnON title at www.jacplus.com.au. They will help you to learn the content and concepts covered in this topic.

How and why do we study Geography?

1.1.1 What is a geographical inquiry?

The world around us is made up of interesting places, people, cultures and environments. Geography is the subject that you study at school to learn about different places and how relationships between environments and people shape these places. Geographers question how environments function and why the world is the way it is. They explore geographic issues and challenges facing us today, predict outcomes and come up with possible solutions for the future. Geographers are active and responsible citizens who are informed about our world and are capable of shaping the future.

FIGURE 1 Our planet is made up of a large variety of fascinating places, peoples, cultures and environments.

on Resources

eWorkbook Chapter 1 eWorkbook (ewbk-8093)

1.2 Geographical inquiry

LEARNING INTENTION

By the end of this subtopic, you will be able to describe the process of geographical inquiry and identify geographical tools.

1.2.1 The process of geographical inquiry

Have you ever visited or gone on holidays to a place other than where you live? If so, you have probably noticed that some of the features and characteristics of the people and places are similar and some are different. Studying Geography at school provides you with the skills, the knowledge and the tools to learn about and understand the relationships between the world's people, places and environments.

As a geographer you get to ask questions and then seek to answer them. Geographers use what is called an inquiry approach to help them learn about and understand the world around them. This could involve you working individually, or as part of a group, to discover the answer to a geographical question, using a variety of geographical skills, tools and concepts.

Geographers also look at many interesting issues which face the world today; for example, different people have different viewpoints, or perspectives, about what we should do about climate change. The answer to this question might vary for an individual, a local area, a country or even on a world scale.

1.2.2 What are inquiry skills?

Have you ever noticed that young children ask many questions as they begin to learn because they are curious about the world around them? Below are some examples of questions which we can call *geographical* questions:

- Why are there many different types of landscapes around the world and how are they formed?
- Where is the best place to live?
- How can we look after our water resources so we have enough for the future?
- What are the effects of tourism in different places?

Geographical inquiry skills develop your ability to collect, process and communicate information.

Acquiring geographical information

Acquiring or collecting geographical information needs to be focused and well planned. Begin a geographical inquiry by developing a problem or issue to investigate. This will be the general theme of your inquiry. Develop a few geographical questions that will help you study your issue or problem. Ensure that your questions are not so broad that they will be difficult to investigate, for example water management in Australia, or so specific that you won't be able to find enough information to support your inquiry.

Think about how you will collect information about your inquiry. You should include both primary geographical data and information from secondary sources. Primary data is information that you have collected yourself using fieldwork. Secondary sources are data that has been collected and processed by someone else, or written by someone else. Secondary sources include websites, books and brochures. Once you have decided on the information you need, you can plan your investigation, carry out your fieldwork and collate information from secondary sources.

Processing geographical information

Before you begin processing the information you have collected, you should evaluate the sources and techniques you have used to determine whether they are reliable and free from bias. Can you trust the sources of information? Did you carry out your fieldwork techniques thoroughly and with care? Present your information in a range of different forms. This might include graphs, tables, diagrams, sketch maps and annotated photographs. You might also write paragraphs explaining your results. Look at the information you have collected and reflect on your research questions. At this stage you can start to interpret the information. Did you answer your research questions? What are the answers to your research questions? Analyse the findings of your research and draw conclusions.

FIGURE 1 Collecting geographical information involves a range of data collection techniques.

Communicating geographical information

You can choose to communicate your research findings in a range of ways. Consider who you will be presenting your findings to. Choose methods to communicate your information that are appropriate to your audience. Explain how you undertook your investigation and your findings. Propose actions that you think should be taken to address your problem or issue, and explain why you think this is the right course of action. If possible, take action yourself to address the geographical issue you have chosen.

1.2.3 What are geographical tools?

Geographers use a range of tools to help them collect information during a geographical inquiry, including

- maps
- fieldwork
- graphs and statistics
- spatial technologies
- visual representations.

Maps

Maps are the most basic tool of the geographer as they are possibly the most effective way to locate, represent, display and record spatial information. These days, geographers are able to use, and create, both digital and non-digital maps.

Political maps are common; they show the boundaries of countries, states and regions, and usually show major cities and bodies of water. Topographic maps and relief maps show the shape of the land on a map. Sketch maps are hand-drawn maps that show only the most basic details. Maps can be used to show information about particular themes, such as choropleth maps or flowline maps. Précis maps show a basic summary of information found on a topographic map.

It is important for geographers to develop skills in map reading to be able to use all the information found. Mapping skills include being able to determine direction and use the scale of the map to determine distance between different places. Geographers use lines drawn on maps to determine and communicate the location of different places. On topographic maps, grids are used to determine the area and grid reference of different places. On some maps, lines of latitude and longitude are shown to help us locate places.

FIGURE 2 Maps: a key tool for the geographer

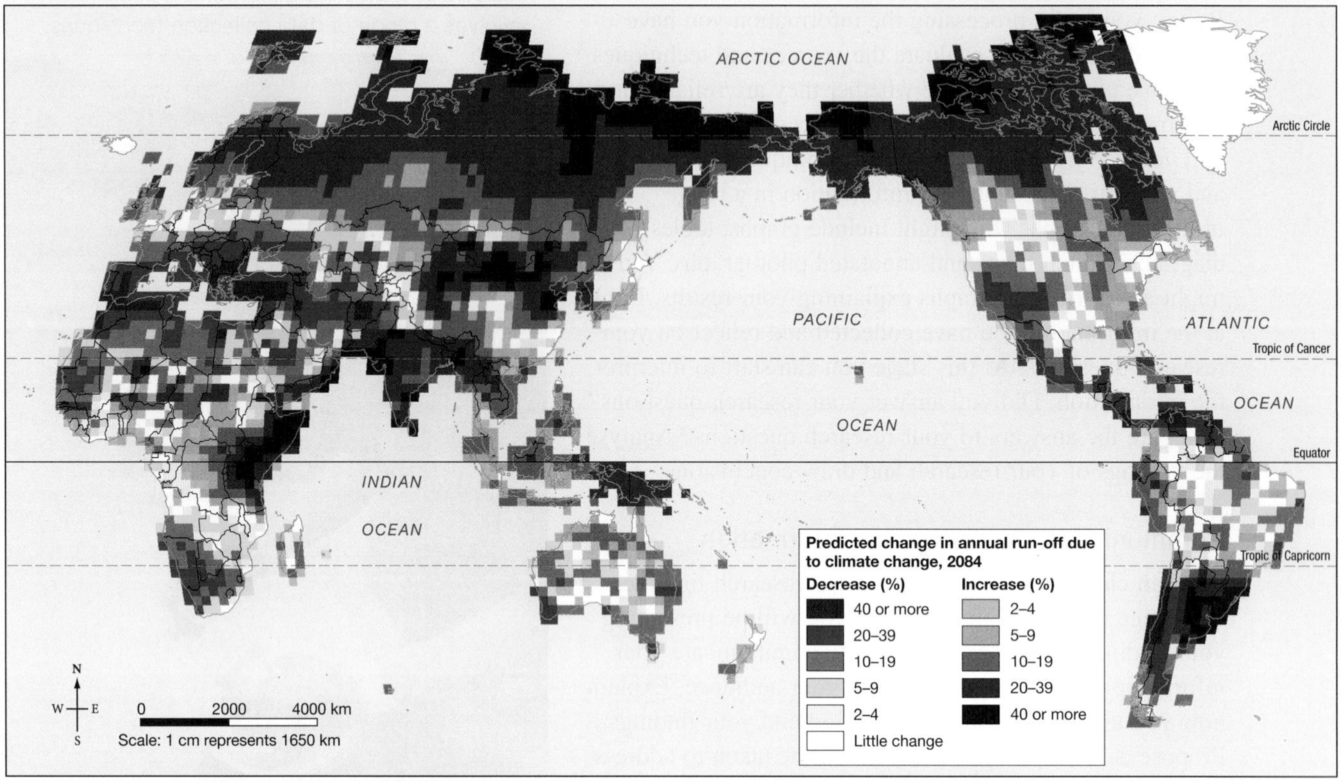

Source: Geophysical Fluid Dynamics Laboratory, National Oceanic and Atmospheric Administration, 2021

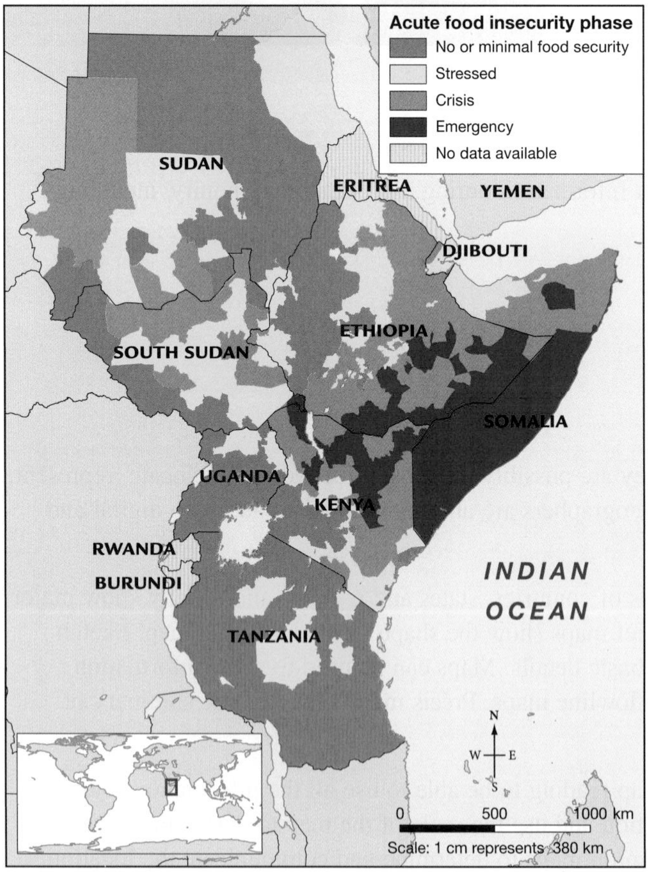

Source: USAID, FEWS NET 2011

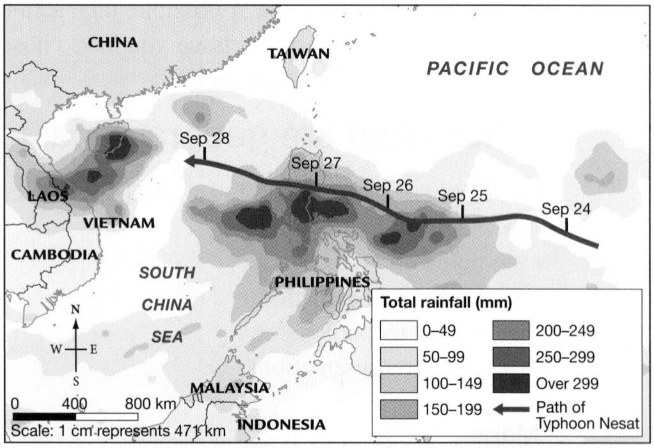

Source: NASA Earth Observatory, 2021

FIGURE 3 A flow line map shows agricultural trade around the world.

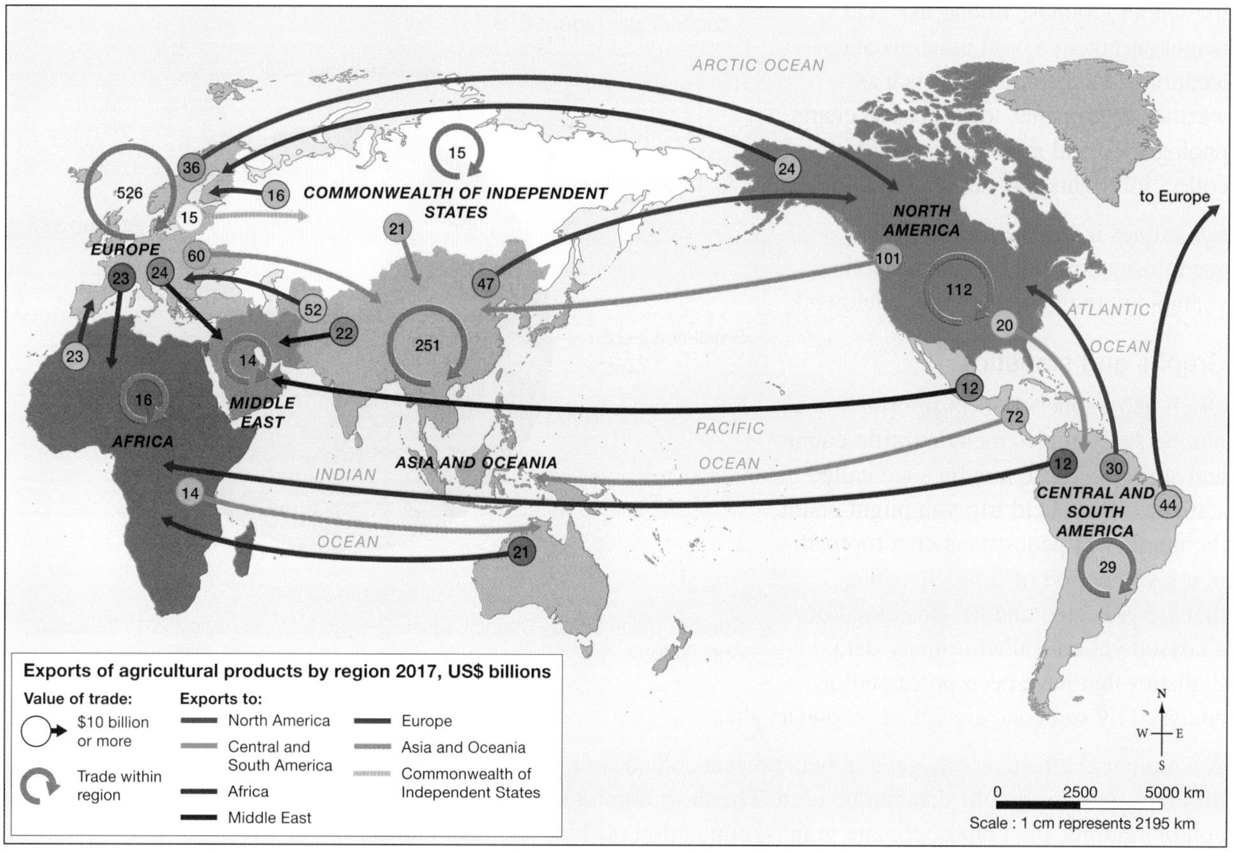

Source: Data from World Trade Organization.

Fieldwork

There is nothing better than going into an environment or to visit a place that you are studying. Seeing something first-hand provides a better understanding than reading about it or looking at it in photographs. That is why fieldwork is such an important and compulsory part of your studies.

Fieldwork involves observing, measuring, collecting and recording information and data outside the classroom.

FIGURE 4 Conducting a survey in the field

FIGURE 5 Collecting your own data and information

Fieldwork can be undertaken within the school grounds, around the local neighbouring area or at more distant locations. We can use tools such as weather instruments, identification charts, photographs and measuring devices to collect information about our environment.

Sometimes it may be necessary to use information and communication technology to undertake virtual fieldwork.

FIGURE 6 Percentage of renewable electricity generation by technology type, 2019

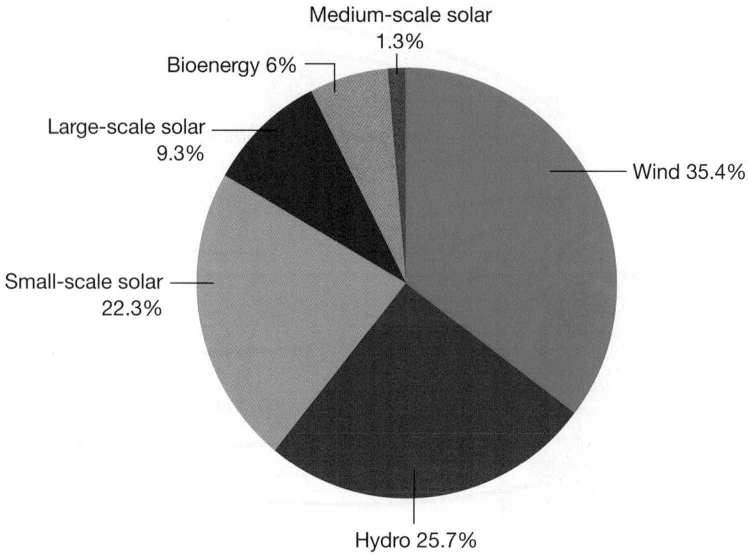

Source: Clean Energy Council, *Clean Energy Australia Report 2019* (released April 2020).

Graphs and statistics

Often geographers collect information as numbers. Examples include traffic counts and surveys. These numbers are called statistics. On a field trip you might count the number of pedestrians on a footpath in a given period of time. Statistics that are collected and not processed or analysed yet are called primary data. Statistics that have been processed or analysed by someone are called secondary data.

A simple and effective way geographers present statistics or data is through the use of graphs. There are many different types of graphs that can be used. The most common types of graphs you will use in this resource are column graphs, pie graphs, climate graphs, population profiles and data tables.

Graphs and statistics allow us to easily identify trends and patterns and to make comparisons.

Using statistics helps us to find patterns in the information we have collected. This will help us to draw conclusions about the themes we have investigated.

FIGURE 7 The growth of urban populations over time

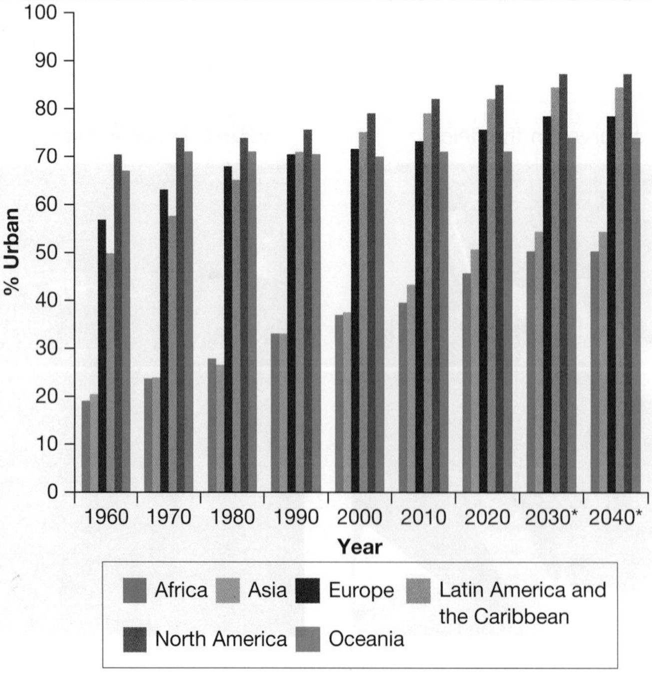

Source: World Population Prospects: 2015 Revision, United Nations (* projected)

Spatial technologies

Spatial technologies involve using satellite information and virtual maps to explore and record information. When you use the Global Positioning System (GPS) or Google Earth you are using a form of spatial technology. Spatial technologies are any software or hardware that interact with real-world locations. Geographic information systems (GIS) are another commonly used spatial technology. They help us analyse, display and record spatial data.

FIGURE 8 A false-colour satellite image of the Mt Lofty Ranges

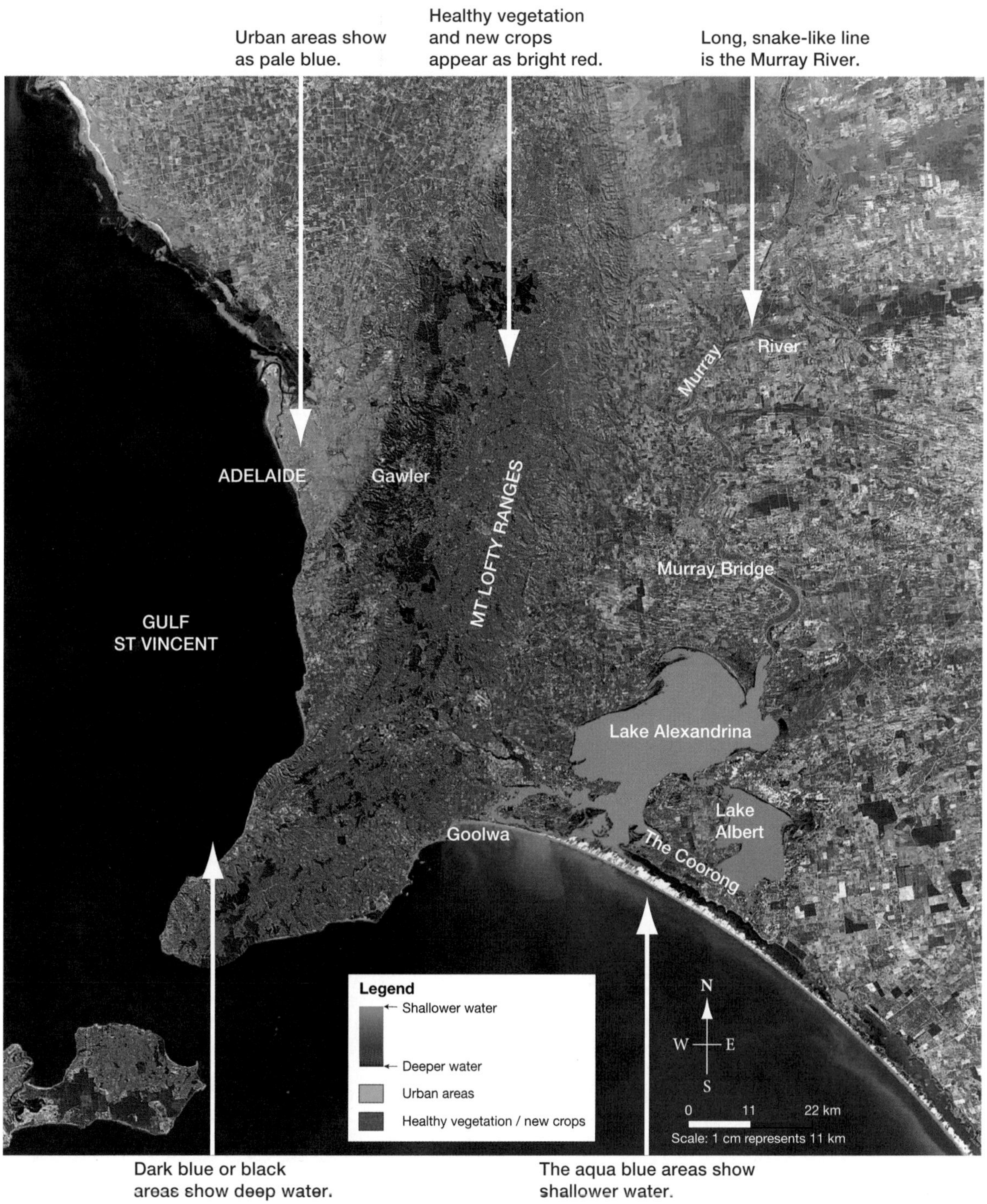

Source: © Commonwealth of Australia. Geoscience Australia 1982.

FIGURE 9 Satellite image of Canberra, by GeoEye, 26 September 2011. Satellite images show a realistic view like a photograph, providing a bird's-eye view of a place.

Source: © GeoEye

Visual representations

A visual representation is an effective way of showing complex information using pictures, symbols and diagrams. Examples of visual representations include photographs, field sketches, cartoons and infographics. They are used to display, analyse and communicate geographical data and information.

FIGURE 10 This visual representation of the water cycle and factors that affect flooding includes information and images to help you understand geographical processes.

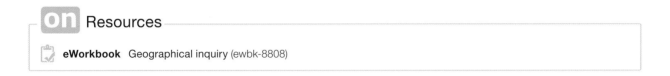

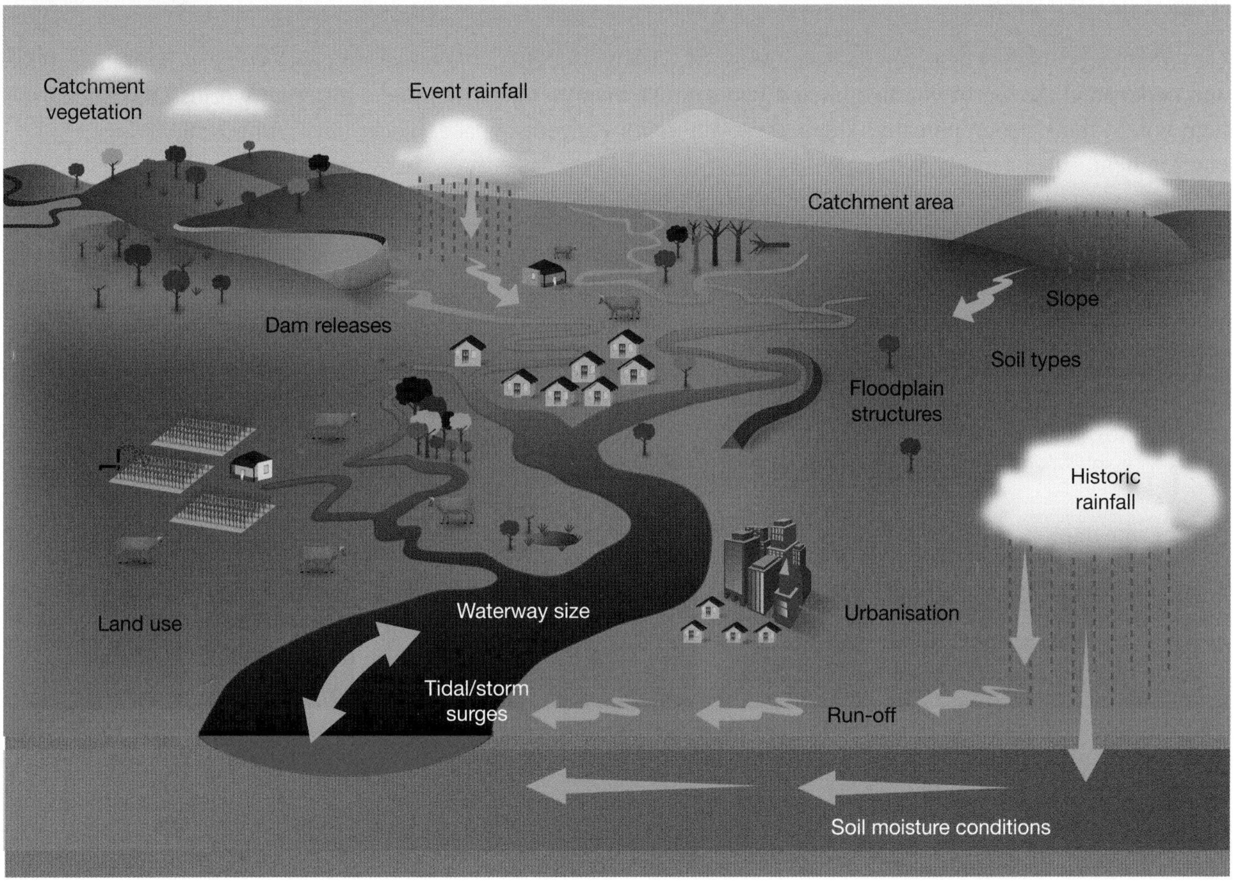

Resources

eWorkbook Geographical inquiry (ewbk-8808)

1.3 Geographical concepts

LEARNING INTENTION

By the end of this subtopic, you should be able to explain the seven key geographical concepts.

1.3.1 Overview

Geographical concepts help you to make sense of your world. By using these concepts you can both investigate and understand the world you live in, and you can use them to try to imagine a different world. The concepts help you to think geographically. There are seven major concepts: space, place, interconnection, change, environment, sustainability and scale.

A way to remember these seven concepts is to think of the term SPICESS (as shown in the diagram).

FIGURE 1 The seven geographical concepts

int-7765

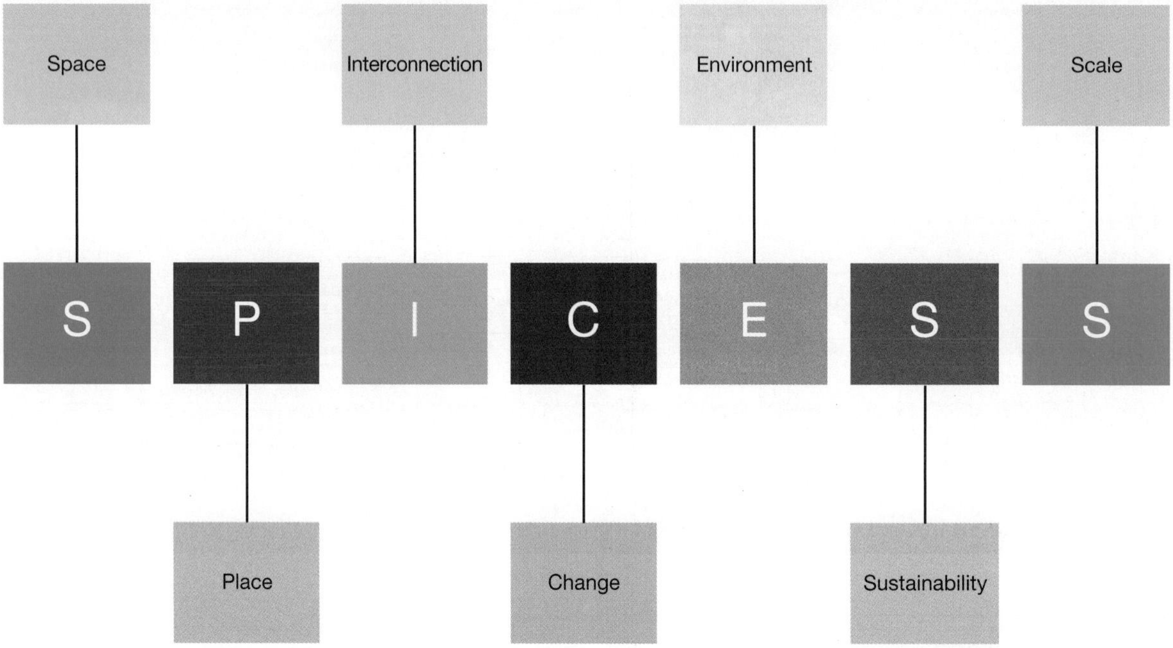

1.3.2 What is space?

Everything has a location on the space that is the surface of the Earth. Studying the effects of location, the distribution of things across this space, and how it is organised and managed by people helps us to understand why the world is like it is.

A place can be described by its **absolute location** (latitude and longitude), a grid reference, a street directory reference or an address. A place can also be described using a **relative location** — where it is in relation to another place in terms of distance and direction.

Geographers also study how features are distributed across **space**, the patterns they form and how they interconnect with other characteristics. For example, tropical rainforests are distributed in a broad line across tropical regions of the world, in a similar pattern to the distribution of high rainfall and high temperatures.

absolute location the latitude and longitude of a place

relative location the direction and distance from one place to another

space geographical concept concerned with location, distribution (spread) and how we change or design places

FIGURE 1 A topographic map extract of Narre Warren in 2016, a suburb on the rural–urban fringe of Melbourne

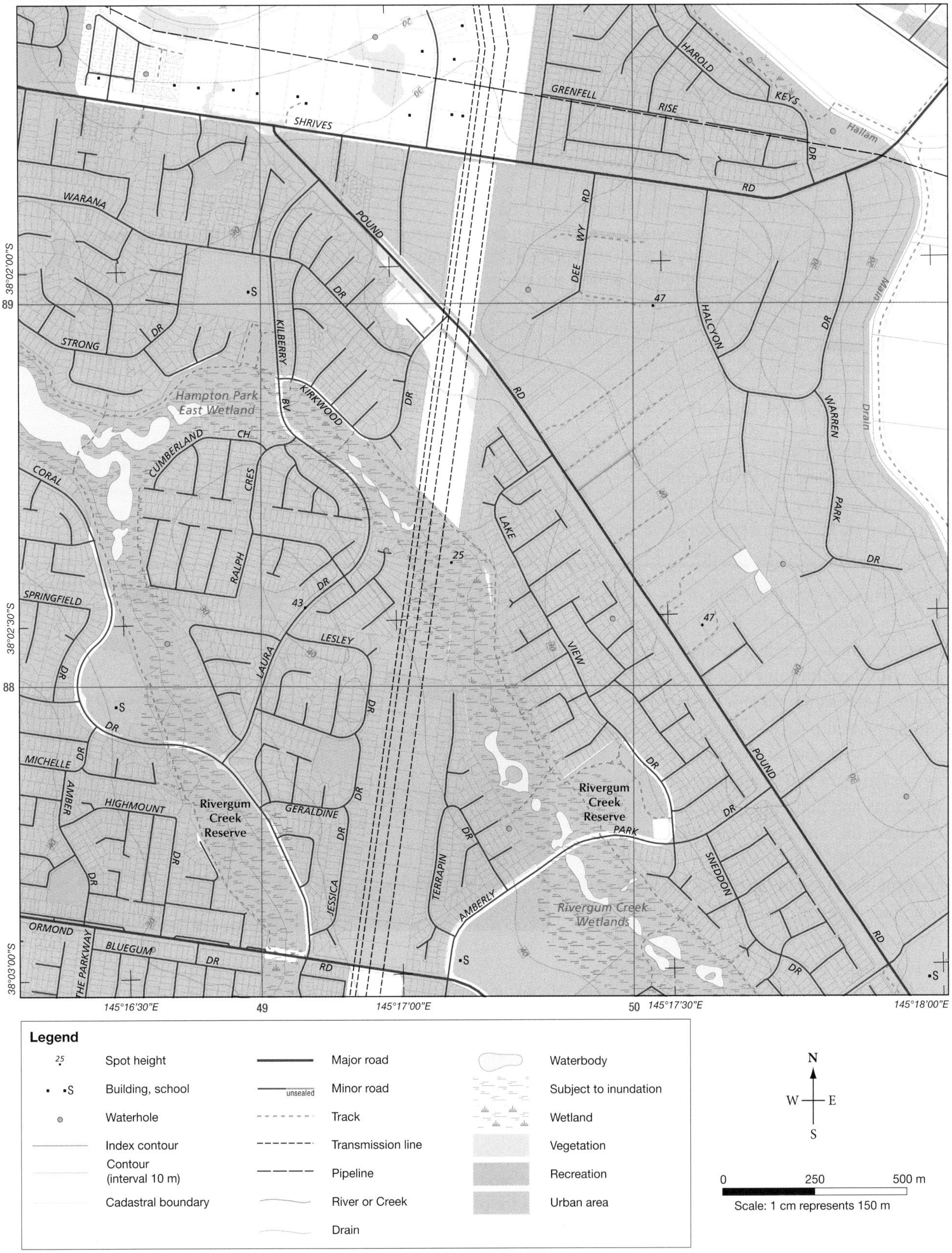

Source: © Vicmap Topographic Mapping Program / Department of Environment and Primary Industries. © The State of Victoria, Department of Environment, Land, Water and Planning, 2016. This publication may be of assistance to you, but the State of Victoria and its employees do not guarantee that the publication is without flaw of any kind or is wholly appropriate for your particular purposes and therefore disclaims all liability for any error, loss or other consequence which may arise from you relying on any information in this publication.

1.3.2 ACTIVITIES

1. Using an atlas, give the absolute location for Melbourne, Australia.

Refer to **FIGURE 1** to complete the following activities.

2. Identify the features at the following locations.
   a. GR490890
   b. GR499875
3. Using the grid references on the topographic map, give the absolute location for:
   a. where Coral Drive crosses the west side of Rivergum Creek Reserve
   b. the intersection of Shrives Road and Pound Road.
4. Describe the location of the recreation area on Amber Drive relative to the school near the intersection of Ormond Road and Amberley Road. Use distance and direction in your answer.
5. Describe the distribution pattern of creeks and drains in the map area.
6. Describe the use of space shown on this map.
7. Explain the influence of the creeks and drains on the distribution of streets and houses.

1.3.3 What is place?

The world is made up of places, so to understand our world we need to understand its places by studying their variety, how they influence our lives and how we create and change them.

Everywhere is a place. Each of the world's biomes — for example a desert environment — can be considered a place, and within each biome there are different places, such as the Sahara Desert. There can be natural places — an oasis is a good example — or man-made places such as Las Vegas. Places can have different functions and activities — for example, Canberra has a focus as an administration centre, the MCG is a place for major sporting events, and the Great Barrier Reef is a place of great natural beauty and a coral reef biome. People are interconnected to places and people in a wide variety of ways — for example when we move between places or connect electronically via computers. We are connected to the places that we live in or know well, such as our neighbourhood or favourite holiday destination.

FIGURE 2 Inside a greenhouse in Almeria, south-east Spain. Almeria has the largest concentrations of greenhouses in the world and is an important producer of vegetables. Located in Europe's biggest desert biome, the greenhouses cover more than 32 hectares.

1.3.3 ACTIVITIES

Refer to **FIGURE 2**.

1. Identify why people have changed this place by building greenhouses there.
2. Identify the characteristics of a desert biome that are being altered in this place.
3. Identify the features that this location might have for production of food.
4. Identify the advantages and disadvantages of greenhouse farming.
5. Suggest the types of crops that would be suitable for greenhouse farming.
6. List ways in which people living in other places in Europe may be interconnected to the greenhouses in Almeria.

1.3.4 What is interconnection?

People and things are connected to other people and things in their own and other places. Understanding these connections helps us to understand how and why places are changing.

The interconnection between people and environments in one place can lead to changes in another location. The damming of a river upstream can significantly alter the river environment downstream and affect the people who depend on it. Similarly, the economic development of a place can influence its population characteristics; for example, an isolated mining town will tend to attract a large percentage of young males, while a coastal town with a mild climate will attract retirees who will require different services. The economies and populations of places are interconnected.

FIGURE 3 Bangladesh is one of the most flood-prone countries in the world. This is due to a number of factors. Firstly, it is largely the floodplain for three major rivers (the Ganges, Brahmaputra and Meghna), which all carry large volumes of water and silt. Secondly, being a floodplain, the topography therefore is very flat, which allows for large-scale flooding. In addition, the country is located at the head of the Bay of Bengal, which is susceptible to typhoons and storm surges. It is expected that sea level rises associated with global warming will increase the flooding threat even further in the future.

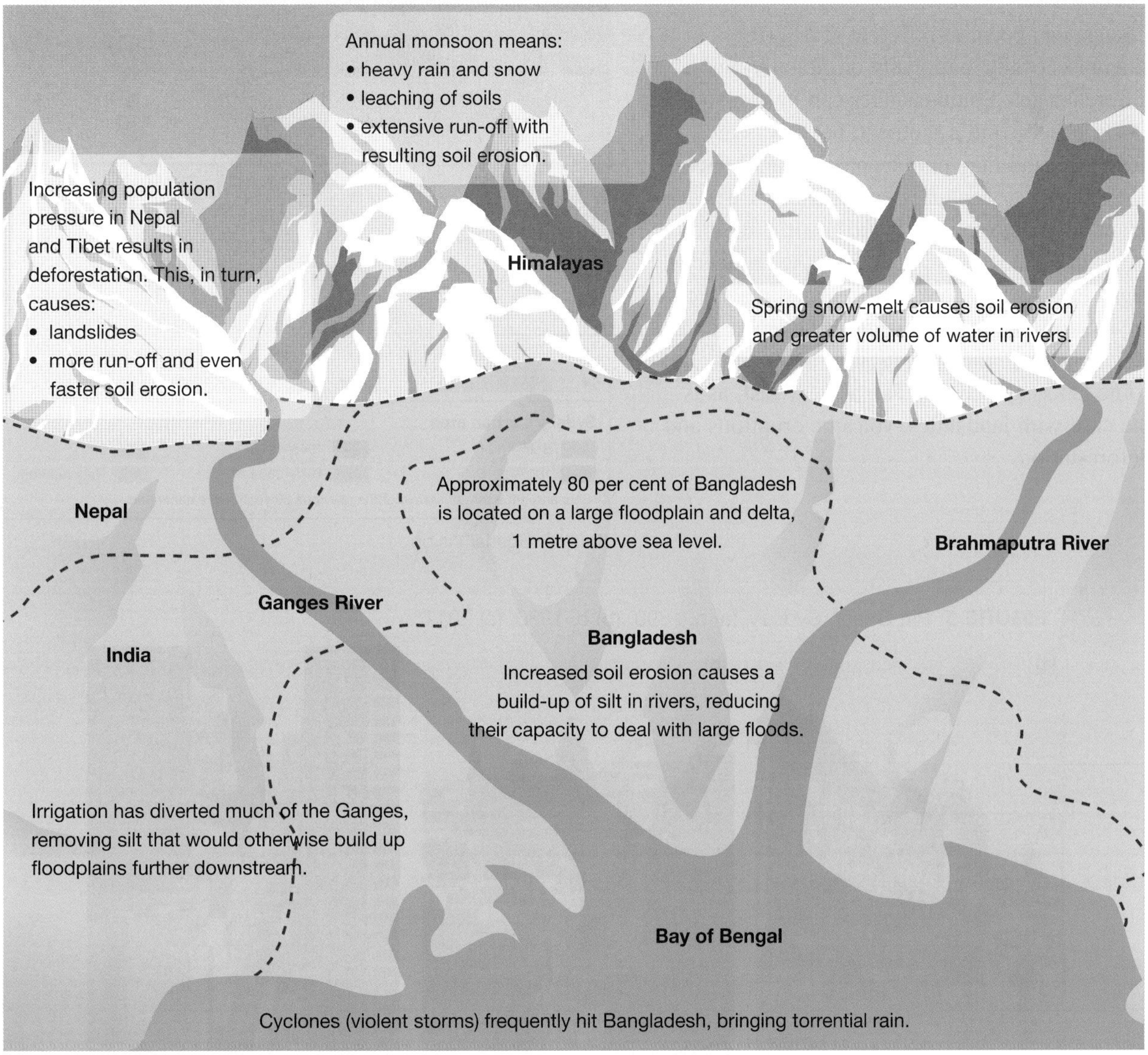
Annual monsoon means:
• heavy rain and snow
• leaching of soils
• extensive run-off with resulting soil erosion.
Increasing population pressure in Nepal and Tibet results in deforestation. This, in turn, causes:
• landslides
• more run-off and even faster soil erosion.
Himalayas
Spring snow-melt causes soil erosion and greater volume of water in rivers.
Nepal
Approximately 80 per cent of Bangladesh is located on a large floodplain and delta, 1 metre above sea level.
Brahmaputra River
Ganges River
India
Bangladesh
Increased soil erosion causes a build-up of silt in rivers, reducing their capacity to deal with large floods.
Irrigation has diverted much of the Ganges, removing silt that would otherwise build up floodplains further downstream.
Bay of Bengal
Cyclones (violent storms) frequently hit Bangladesh, bringing torrential rain.

1.3.4 ACTIVITIES

Refer to **FIGURE 3**.

1. Describe the interconnection between the physical characteristics of Bangladesh and its flood risk.
2. Identify the interconnection between human activities and Bangladesh's flood risk.
3. Use information from the figure to construct a flow diagram to show the interconnection between human activities and natural processes (increased risk of flooding) in Bangladesh.
4. Predict how an increase in the number and severity of floods might affect:
   a. people's wellbeing
   b. economic development in the country.
5. Considering the interconnections that you have identified, suggest some possible steps that could be taken to reduce the impact of flooding.

1.3.5 What is change?

The concept of change is about using time to better understand a place, an environment, a spatial pattern or a geographical problem.

Change can occur at different scales (for example, locally, nationally or globally) and at different rates. Some changes can be fast and easily observed in a moment, but others are very slow and may not be apparent until data is compared over centuries or millennia. Cities, for example, can expand outwards over a number of years or see a rapid growth in the number of high-rise buildings or developments in a specific area. Similarly, landforms generally change very slowly, as with the formation of mountains. But some landscape change can be very fast, as is the case with landslides, volcanic eruptions and deforestation.

FIGURE 4 The change in size of Sydney over time

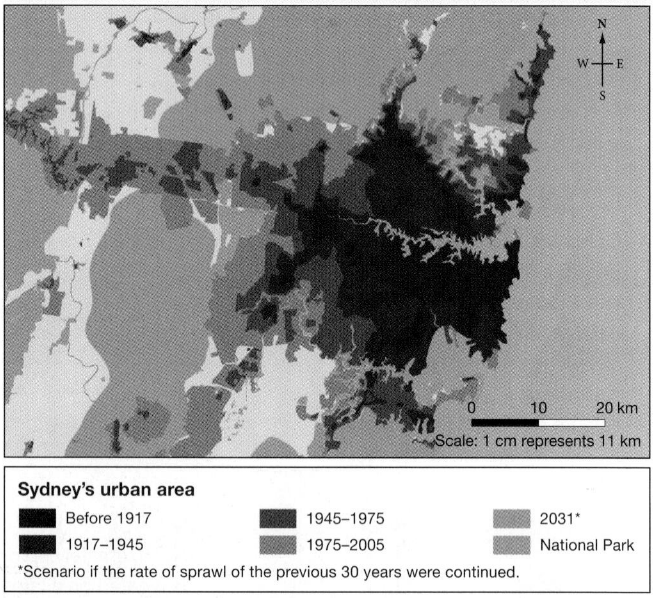

Source: Spatial Vision

FIGURE 5 Pitt Street, Sydney, (a) c. 1890, (b) c. 1930, (c) 2017

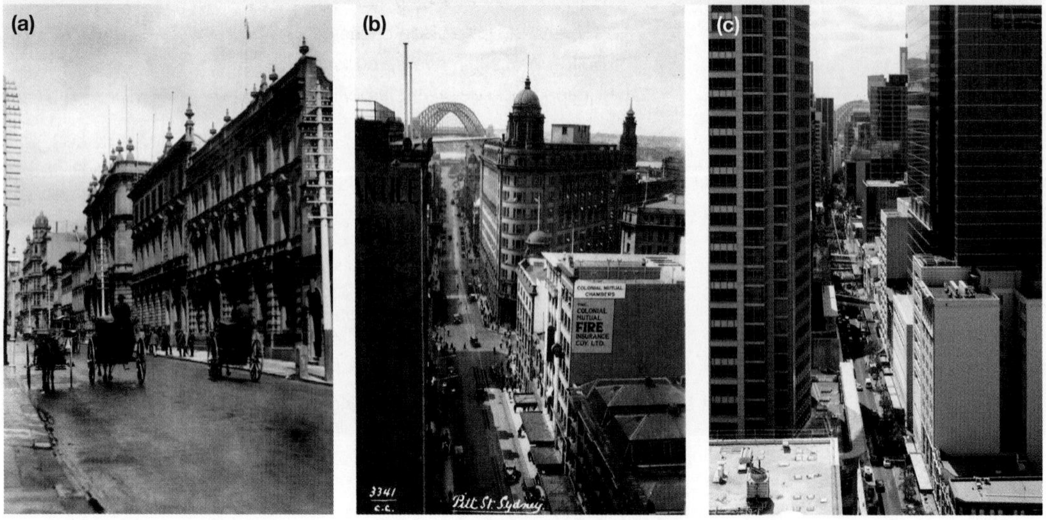

FIGURE 6(a) Landscape before deforestation

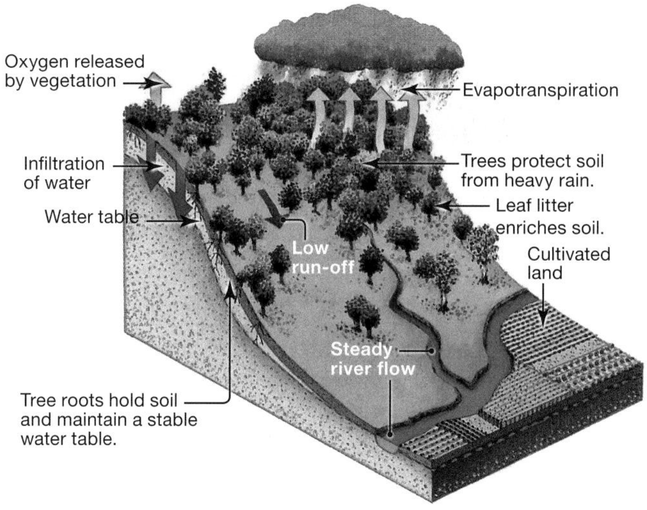

FIGURE 6(b) Landscape after deforestation

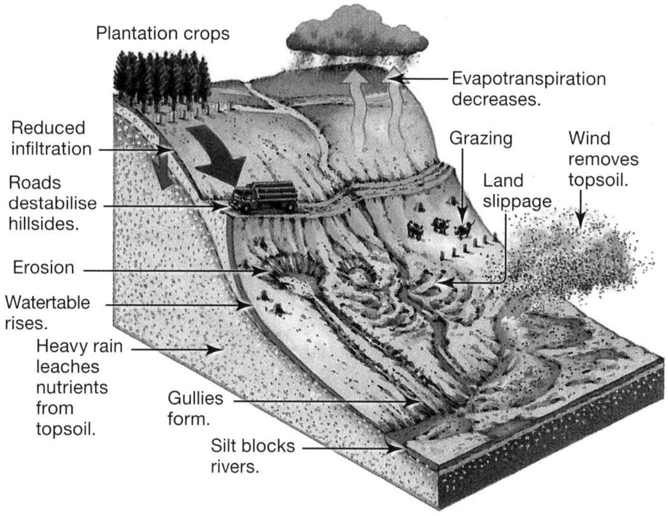

1.3.5 ACTIVITIES

Refer to **FIGURE 4** to answer questions 1–3.

1. Describe how Sydney has changed over time. How long has it taken for the city to spread to the furthest areas shown on the map?
2. Identify the main natural feature that attracted earliest settlement.
3. Describe the impact that this growth has had on the natural environment.
4. Describe the technological changes in transport that have allowed Sydney to spread and grow over time.

Refer to **FIGURES 5(a)**, **(b)** and **(c)** to answer questions 5–8.

5. What features of Pitt Street are consistent in all three images?
6. Suggest what technological changes or developments might have contributed to the changes that occurred. Provide examples from the images to support your answer.
7. In what ways might the natural features of central Sydney have led to the changes that occurred in the built environment?
8. Describe the pace and extent of change shown in these images in a short paragraph.

Refer to **FIGURES 6(a)** and **(b)** to answer questions 9–12.

9. List the changes that would have caused the slippage to occur.
10. Identify the interconnections between:
    a. vegetation cover and soil stability
    b. vegetation cover and infiltration
    c. high run-off and erosion.
11. List all the effects of the landslide on people and the environment.
12. Write a summary statement about the pace of change and the impact on people and the environment in these two examples.

1.3.6 What is environment?

People live and depend on the environment, so it has an important influence on our lives. There is a strong interrelationship between humans and natural and urban environments. People depend on the environment for the source, sink, spiritual and service functions it provides.

Humans significantly alter environments, causing both positive and negative effects. The building of dams to reduce the risk of flooding, the regular supply of fresh water and the development of large-scale urban environments to improve human wellbeing are examples. On the other hand, mismanagement has created many environmental threats such as soil erosion and global warming, which have the potential to have a negative impact on the quality of life for many people.

FIGURE 7 Lake Urmia — (a) in 1998 and (b) in 2011 — is the largest lake in the Middle East and one of the largest landlocked saltwater lakes in the world. Since 2005, the lake has lost over 65 per cent of its surface area due to over-extraction of water for domestic and agricultural needs. The lake and its surrounding wetlands are internationally important as a feeding and breeding ground for migratory birds.

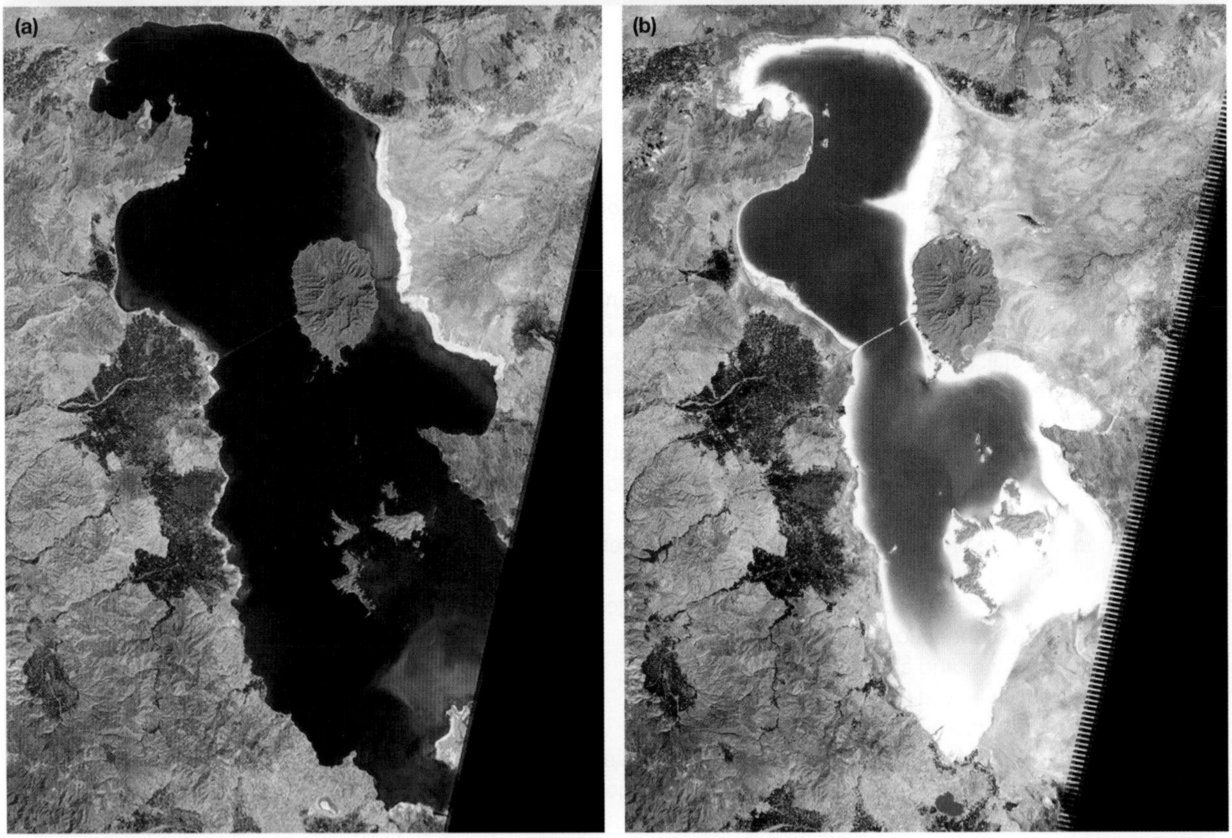

1.3.6 ACTIVITIES

Refer to **FIGURE 7**.

1. Identify the physical features that make up this environment.
2. Identify the features of the natural environment that are consistent across the two images.
3. Describe the changes to this environment over the time period of 1998 to 2011.
4. Describe the distribution of salt flats around the lake in 1998 compared with their distribution in 2011.
5. How might the loss of water and increase in salt flats affect the people and the environment in the surrounding region?
6. Suggest a possible future scenario for Lake Urmia:
   a. if water continues to be extracted and withdrawn
   b. if water withdrawals for irrigation are reduced and water conservation methods are introduced.

1.3.7 What is scale?

When we examine geographical questions at different spatial levels, we are using the concept of scale to find more complete answers.

Scale is a useful tool for examining issues from different perspectives; from the personal to the local, regional, national and global. It is also used to look for explanations or compare outcomes. For example, explaining the changing structure of the population in your local area may require an understanding of migration patterns at a national or even global scale.

FIGURE 8 A map of India showing the distribution of literacy levels (percentage) for 2011

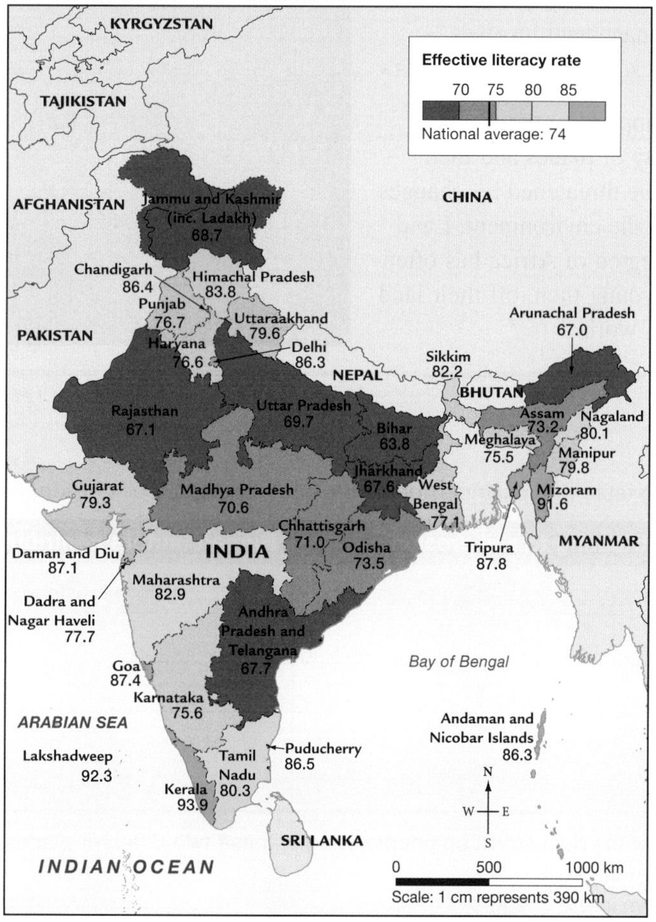

Source: Spatial Vision

1.3.7 ACTIVITIES

Refer to **FIGURE 8**.

1. At the national scale, identify the average literacy rate for India.
2. At the regional scale, identify the three states that have the lowest literacy levels.
3. Identify the factors that might contribute to a state's low literacy level.
4. Explain how literacy levels might affect the wellbeing of people.
5. Data, such as for literacy levels, is collected by governments during a census. Explain how knowing this sort of information would assist a government in planning for future populations.

1.3.8 What is sustainability?

Sustainability is about maintaining the capacity of the environment to support our lives and the lives of other living creatures.

Sustainability ensures that the source, sink, service and spiritual functions of the environment are maintained and managed carefully to ensure they are available for future generations. There can be variations in how people perceive sustainable use of environments and resources. Some people think that technology will provide solutions, while others believe that sustainable management involves environmental benefits and social justice.

This concept can also be applied to the social and economic sustainability of places and their communities, which may be threatened by changes such as the degradation of the environment. Land degradation in the Sahel region of Africa has often forced people, especially young men, off their land and into cities in search of work.

FIGURE 9 Dust storms are an extreme form of land degradation. Dry, unprotected topsoil is easily picked up and carried large distances by wind before being deposited in other places. Drought, deforestation and poor farming techniques are usually the cause of soil being exposed to the erosional forces of wind and water. It may take thousands of years for a new topsoil layer to form. Therefore, any land practices that lead to a loss of topsoil may be considered unsustainable.

1.3.8 ACTIVITIES

Refer to **FIGURE 9**.

1. Complete the following table with examples of factors contributing to soil erosion.

Natural factors contributing to soil erosion	Human factors contributing to soil erosion

2. Describe the impacts of the dust storm on people living in these two different places:
   a. rural areas (source of the soil)
   b. the urban area shown in the image.
3. Explain how the interconnection of human activities and natural processes can contribute to land degradation.
4. Explain the long-term implications of the unsustainable use of soil.
5. Explain how farming can be made more sustainable in terms of soil conservation.

on Resources

eWorkbook Geographical concepts (ewbk-8812)

myWorldAtlas Deepen your understanding of this topic with related case studies and questions.
- Space
- Place
- Interconnection
- Change
- Environment
- Scale
- Sustainability

1.4 Work and careers in Geography

online only

LEARNING INTENTION

By the end of this subtopic, you should be able to list some of the careers that Geography can lead to and outline some of the skills these jobs require.

The content in this subtopic is a summary of what you will find in the online resource.

Geographic skills will be useful for your future employment. In Geography, you develop an understanding of the world. The skills you develop in Geography are transferrable to the workplace and can be used as a basis for evaluating strategies for the sustainable use and management of the world's resources.

To learn more about careers related to Geography, go to your learnON resources at www.jacPLUS.com.au.

FIGURE 1 GIS (Geographic Information Systems) being used to manage spaces and plan escape routes during a fire

FIGURE 3 Geography pathways (a) Surveyor (b) Landscape architect

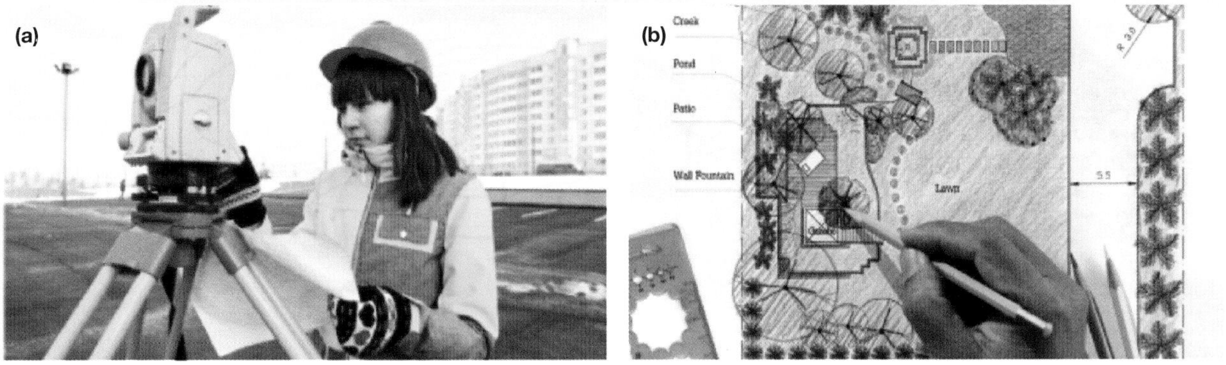

Contents

learnON

- 1.4.1 Geographic skills and future work
- 1.4.2 Where can Geography lead?
- 1.4.3 Geography in a changing world
- 1.4.4 The importance of work experience
- 1.4.5 Future careers and Geography

on Resources

eWorkbook Work and careers in Geography (ewbk-8816)

1.5 SkillBuilder: Reading contour lines on a map

online only

LEARNING INTENTION

By the end of this subtopic, you will be able to read contour lines on a map confidently.

The content in this subtopic is a summary of what you will find in the online resource.

1.5.1 Tell me

What are contour lines?

Contour lines drawn on a map join all places of the same elevation (height). These lines are usually brown and have a number written on them to indicate height above sea level.

Contour maps are used to show the relief (shape) of the land and the heights of the landscape. Land heights are identified from aerial photography. Natural features, such as rivers, lakes and beaches, and human features, such as towns, roads and power lines, are added to the map to complete the landscape picture. Symbols provided in a legend (or key) or labels on the map add information to complete the image of the environment.

FIGURE 1 A topographic map represents a three-dimensional landscape on a flat surface.

1.5.2 Show me

How to read contour lines

Step 1

To find the height of a particular area of land, identify a contour line in **FIGURE 2** and follow the line to find the number that states the height above sea level (in metres).

Step 2

Spot heights are dots that indicate the exact height at the highest point of a hill or the lowest point of a depression. For example, the hill in **FIGURE 2** is exactly 104 metres above sea level at its peak. This spot is higher than the last contour line (in this case 100 m), but lower than the height at which the next contour line would be drawn (110 m).

FIGURE 2 How contour lines show the shape of the land

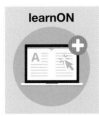

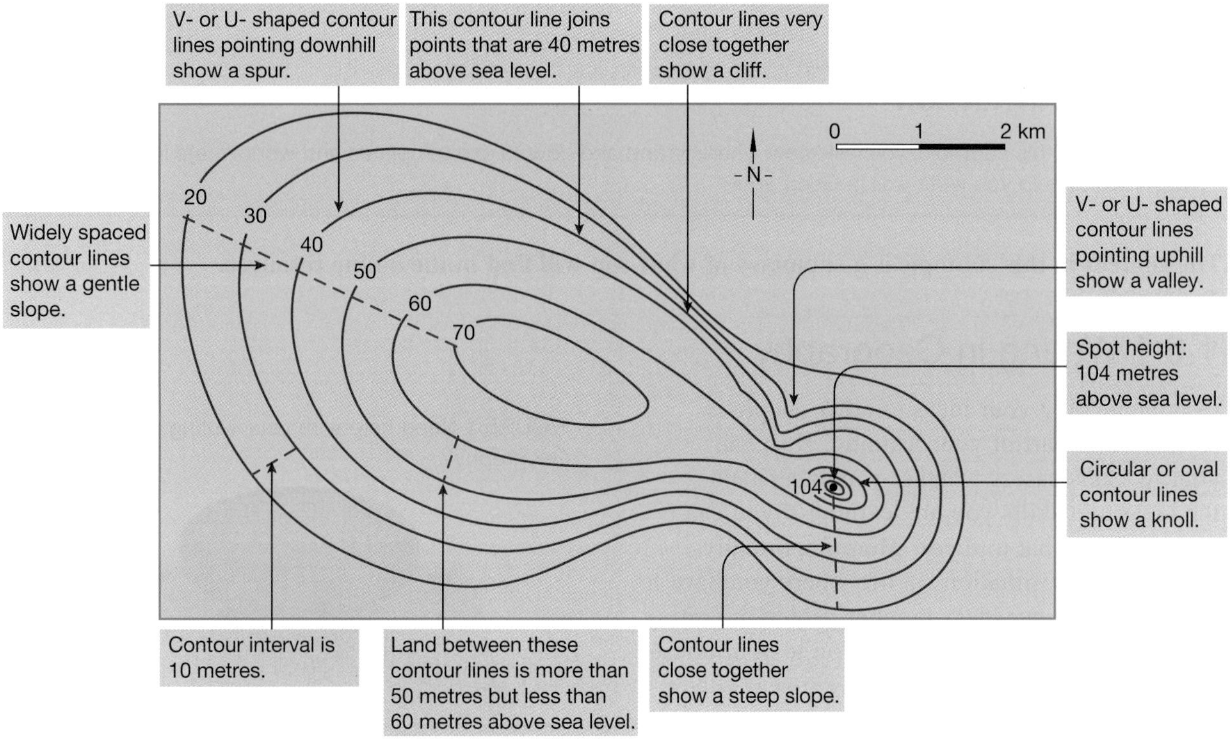

V- or U- shaped contour lines pointing downhill show a spur.
This contour line joins points that are 40 metres above sea level.
Contour lines very close together show a cliff.
0
1
2 km
-N-
20
30
40
50
60
70
104
Widely spaced contour lines show a gentle slope.
V- or U- shaped contour lines pointing uphill show a valley.
Spot height: 104 metres above sea level.
Circular or oval contour lines show a knoll.
Contour interval is 10 metres.
Land between these contour lines is more than 50 metres but less than 60 metres above sea level.
Contour lines close together show a steep slope.

Step 3

The contour interval of a map is the difference in metres between each of the contour lines. This interval is consistent across a map.

If the contour lines are too close and the numbers can't easily be written, then it is left to the reader to use the contour interval to calculate heights. The contour interval is often written in the legend as a guide.

1.5.3 Let me do it

learnON

Go to learnON to access the following additional resources to help you build this skill:

- a longer explanation of this skill and its application in Geography (Tell me)
- a video demonstrating the step-by-step process of this skills (Show me)
- an activity and interactivity for you to practise the skills (Let me do it)
- self-marking questions to consolidate your understanding of the skill.

on Resources

eWorkbook	SkillBuilder — Reading contour lines on a map (ewbk-10249)
Video eLesson	SkillBuilder — Reading contour lines on a map (eles-1651)
Interactivity	SkillBuilder — Reading contour lines on a map (int-3147)

1.6 Writing skills in Geography

online only

LEARNING INTENTION

By the end of this subtopic, you will know where to find and how to use templates and worksheets in your online resources to help you write well in Geography.

The content in this subtopic is a summary of what you will find in the online resource.

1.6.1 Writing in Geography

Communicating your ideas to other people is an important part of your learning. How can your teachers assess whether you understand the ideas and skills you are learning if you can't communicate that to them? More importantly, there are many situations in life where you have to communicate your ideas to other people in writing. Want to apply to build an extension to your house? Need to access help from the government after a bushfire? Want to protest against a new mine in your area? Most likely, you will need to do these things in writing.

FIGURE 1 Need help with your writing in Geography?

1.6.2 Writing templates and help sheets

Whether you are reading this in print or online, you have access to a range of SkillBuilder worksheets and templates online to help you improve your writing skills.

learnON

Go to your online Resources to download Word documents to help you:
- write an essay
- structure a paragraph
- interpret a short answer question
- interpret an extended answer question
- paraphrase well
- reference quotes and sources.

on Resources

eWorkbook
SkillBuilder — Writing an essay (structure) (ewbk-8591)
SkillBuilder — Writing a paragraph (ewbk-8655)
SkillBuilder — Interpreting short answer questions (ewbk-8658)
SkillBuilder — Interpreting extended answer questions (ewbk-8565)
SkillBuilder — Using the deconstruct–reconstruct method (ewbk-8567)
SkillBuilder — Referencing quotes (ewbk-8589)

Video eLessons
SkillBuilder — Writing an essay (structure) (eles-5338)
SkillBuilder — Writing a paragraph (eles-5339)
SkillBuilder — Interpreting short answer questions (eles-5340)
SkillBuilder — Interpreting extended answer questions (eles-5341)
SkillBuilder — Using the deconstruct–reconstruct method (eles-5342)
SkillBuilder — Referencing quotes (eles-5343)

1.7 Review

1.7.1 Key knowledge summary

1.2 Geographical inquiry

- Key inquiry skills in Geography are:
  - acquiring or collecting information
  - evaluating sources and processing information
  - communicating in a way that suits your audience.
- The geographical tools you will use to do this are:
  - maps
  - fieldwork
  - graphs and statistics
  - spatial technologies.

1.3 Geographical concepts

- The acronym SPICESS helps you remember the seven geographical concepts:
  - space
  - place
  - interconnection
  - change
  - environment
  - sustainability
  - scale.

1.4 Work and careers in Geography

- Many jobs in a wide variety of occupations have links with Geography.
- Geographers are typically curious, adaptable learners who can communicate well with others and think carefully through complex problems.
- Volunteering can help you to find out about different aspects of related jobs.

1.7.2 Key terms

absolute location the latitude and longitude of a place
relative location the direction and distance from one place to another
space geographical concept concerned with location, distribution (spread) and how we change or design places

1.7.3 Reflection

Complete the following to reflect on your learning.

Revisit the inquiry question posed in the Overview:

How and why do we study Geography?

1. Think about some of the challenges your community faces. Identify which of the SPICESS concepts that challenge relates to.
2. What research could you do, involving the geographical tools, to determine the extent of one of the challenges you identified in question 1? How might this research help you come up with possible solutions?

Subtopic	Success criteria	○	○	●
1.1	I can explain how studying Geography might help to develop important career skills.			
1.2	I can explain the three main stages in the geographical inquiry process: acquiring, processing and communicating information.			
	I can describe different types of maps and what they show.			
	I can explain the importance of fieldwork to studying Geography.			
	I can name several types of graphs used in Geography.			
	I can explain the purpose(s) of spatial technology.			
1.3	I can list the seven geographical concepts.			
	I can explain the geographical concept of space.			
	I can explain the geographical concept of place.			
	I can explain the geographical concept of interconnection.			
	I can explain the geographical concept of change.			
	I can explain the geographical concept of environment.			
	I can explain the geographical concept of sustainability.			
	I can explain the geographical concept of scale.			
1.4	I can outline some of the skills that jobs in geographical fields require.			
	I can recognise my strengths and evaluate how I might use them in my future career path.			
1.5	I can read contour lines on a map confidently.			

 Resources

📋 **eWorkbook**   Chapter 1 Reflection (ewbk-8521)
Chapter 1 Student learning matrix (ewbk-8520)

ONLINE RESOURCES

on Resources

Below is a full list of **rich resources** available online for this topic. These resources are designed to bring ideas to life, to promote deep and lasting learning and to support the different learning needs of each individual.

eWorkbook

- **1.1** Chapter 1 eWorkbook (ewbk-8093) ☐
- **1.2** Geographical inquiry (ewbk-8808) ☐
- **1.3** Geographical concepts (ewbk-8812) ☐
- **1.4** Work and careers in Geography (ewbk-8816) ☐
- **1.5** SkillBuilder — Reading contour lines on a map (ewbk-10249) ☐
- **1.6** SkillBuilder — Writing an essay (structure) (ewbk-8591) ☐
  - SkillBuilder — Writing a paragraph (ewbk-8655) ☐
  - SkillBuilder — Interpreting short answer questions (ewbk-8658) ☐
  - SkillBuilder — Interpreting extended answer questions (ewbk-8565) ☐
  - SkillBuilder — Using the deconstruct–reconstruct method (ewbk-8567) ☐
  - SkillBuilder — Referencing quotes (ewbk-8589) ☐
- **1.7** Chapter 1 Reflection (ewbk-8521) ☐
  - Chapter 1 Student learning matrix (ewbk-8520) ☐

Video eLessons

- **1.5** SkillBuilder — Reading contour lines on a map (eles-1651) ☐
- **1.6** SkillBuilder — Writing an essay (structure) (eles-5338) ☐
  - SkillBuilder — Writing a paragraph (eles-5339) ☐
  - SkillBuilder — Interpreting short answer questions (eles-5340) ☐
  - SkillBuilder — Interpreting extended answer questions (eles-5341) ☐
  - SkillBuilder — Using the deconstruct-reconstruct method (eles-5342) ☐
  - SkillBuilder — Referencing quotes (eles-5343) ☐

Interactivities

- **1.3** SPICESS (int-7765) ☐
- **1.5** SkillBuilder — Reading contour lines on a map (int-3147) ☐

myWorldAtlas

- **1.3** Space (mwa-1599) ☐
  - Place (mwa-1601) ☐
  - Interconnection (mwa-1600) ☐
  - Change (mwa-1602) ☐
  - Environment (mwa-1603) ☐
  - Scale (mwa-1604) ☐
  - Sustainability (mwa-1605) ☐

Weblinks

- **1.4** Job Outlook (web-0115) ☐
  - Worksite (web-1148) ☐
  - Careers 2030 (web-0129) ☐

Teacher resources

There are many resources available exclusively for teachers online.

UNIT

1 Sustainable biomes

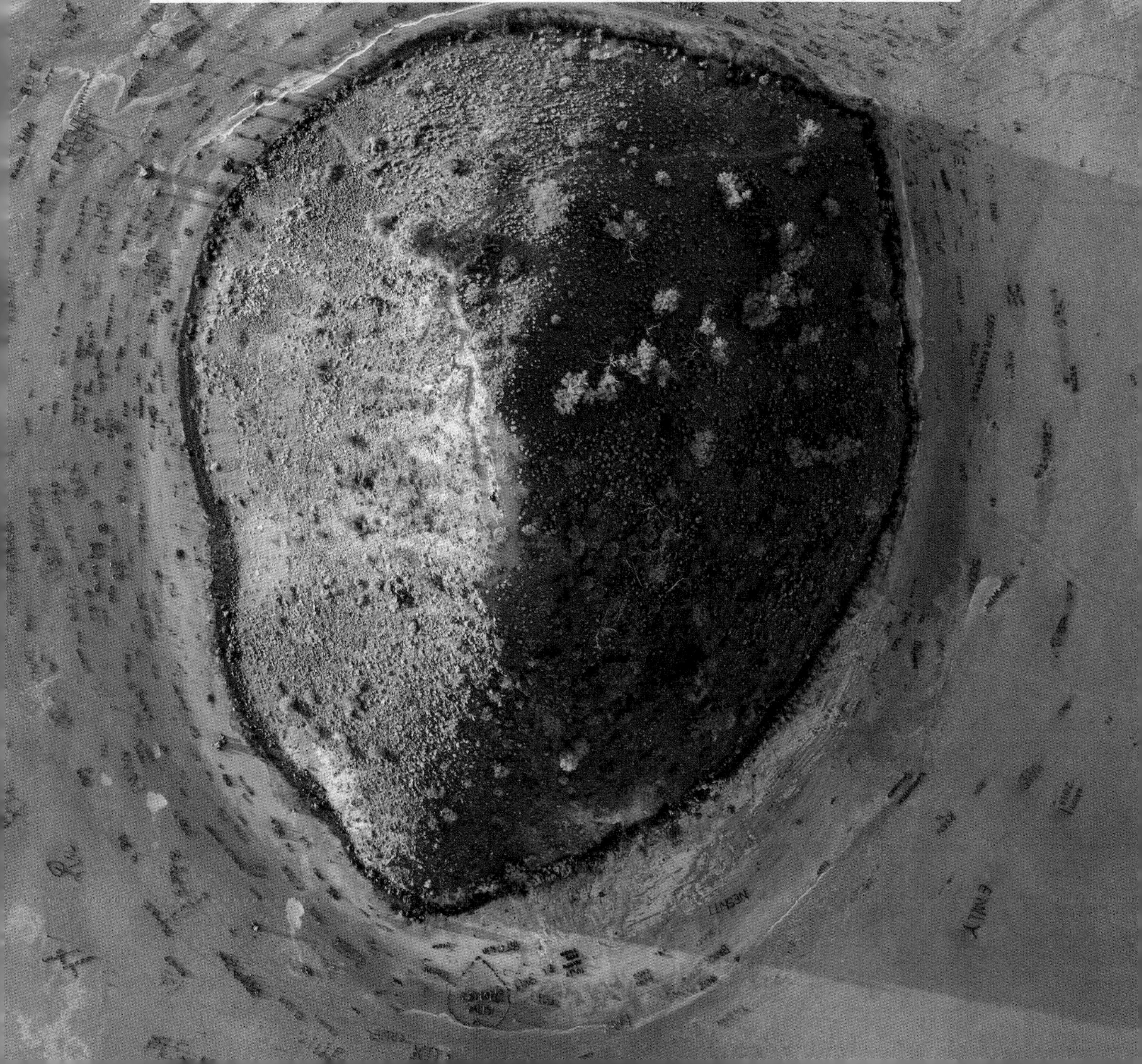

2 Biomes

INQUIRY SEQUENCE

To access a pre-test with **immediate feedback** and **sample responses** to every question in this chapter, select your learnON format at www.jacplus.com.au.

2.1 Overview

Numerous **videos** and **interactivities** are embedded just where you need them, at the point of learning, in your learnON title at www.jacplus.com.au. They will help you to learn the content and concepts covered in this topic.

What are biomes, where are they, why are they different, and what do humans use them for?

2.1.1 Introduction

Biomes are communities of plants and animals that extend over large areas. Some are dense forests; some are deserts; some are grasslands, like much of Australia; and so the variations continue. Within each biome, plants and animals have similar adaptations that allow them to survive. Biomes can be terrestrial (land based) or aquatic (water based). Understanding the diversity and functioning of biomes is essential to our survival and wellbeing, as they are responsible for the food we eat and the natural products we use on a daily basis.

STARTER QUESTIONS

1. As a class, list words that you think explain what the word biome means.
2. What information do the pictures shown in the **Biomes — Photo essay** video in your online Resources convey about biomes?
3. Create a list of the things you use and consume that come from biomes. Compare your list with other students in your class.

FIGURE 1 Understanding the way biomes function helps us to protect them.

on Resources

eWorkbook Chapter 2 eWorkbook (ewbk-8094)

Video eLessons Bountiful biomes (eles-1717)

Biomes — Photo essay (eles-5172)

2.2 Features of biomes

LEARNING INTENTION

By the end of this subtopic, you will be able to identify and locate key biomes throughout the world and recall their key features.

2.2.1 Locating and identifying the major biomes

Biomes are communities of plants and animals that extend over large areas due to similarities of climate and conditions within the area of the biome. There are five distinct biomes across the Earth: forest, desert, grassland, tundra and aquatic biomes. Within each major biome there are variations in the visible landscape and in the plants and animals that have adapted to survive in a particular climate.

int-8764

FIGURE 1 Major biomes of the world

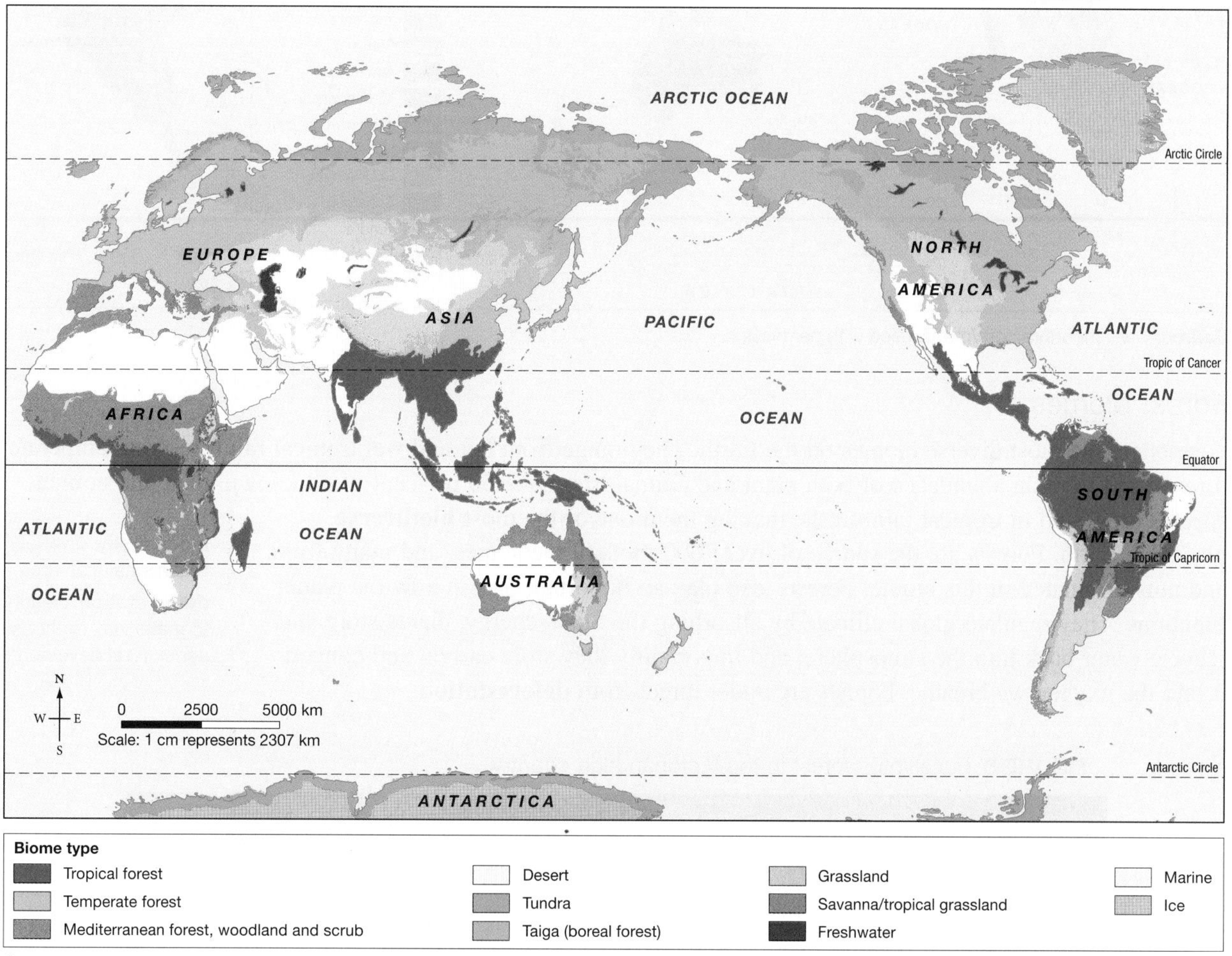

Source: Redrawn by Spatial Vision based on information from the Nature Conservancy and GIS Data

int-8765

FIGURE 2 Climatic zones of the world

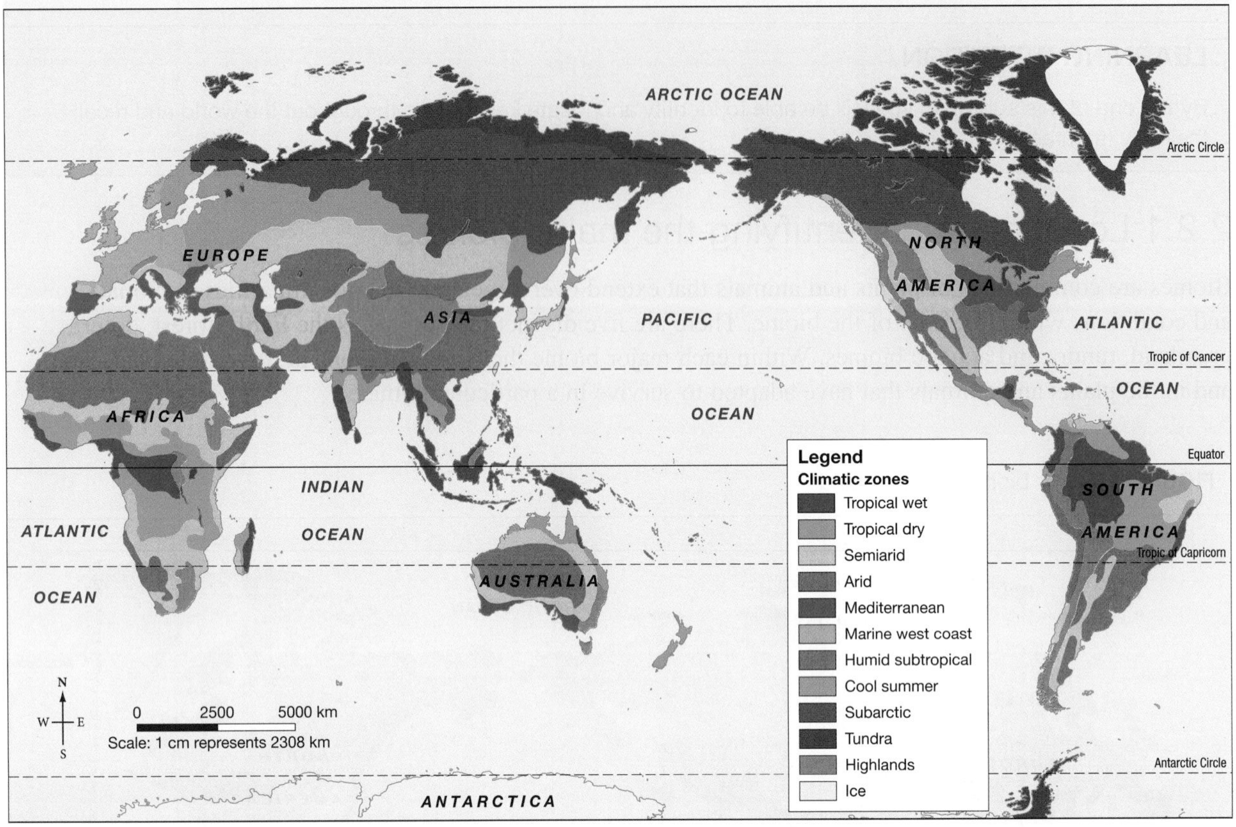

Source: www.theodora.com/maps, used with permission.

Forest biomes

Forests are the most diverse biomes on the Earth. They range from hot and wet tropical rainforests to temperate forests, and have an abundance of both plant and animal life. Over 50 per cent of all known plant and animal species are found in tropical rainforests, making them one of the most **biodiverse** places on Earth. Forests are the source of over 7000 modern medicines, and many fruits and nuts originated in this biome. Forests also play an important part in how the planet functions. They regulate global climate by absorbing the Sun's energy, plants store and recycle water back into the atmosphere, and importantly they store carbon and convert it into the oxygen we breathe. Forests are under threat from **deforestation**.

biodiversity a variety of living organisms in an area

deforestation clearing forests to make way for housing or agricultural development

FIGURE 3 Eucalyptus forest in the Victorian high country

Desert biomes

Deserts are places that experience low rainfall and can either be hot or cold, such as the hot deserts in Australia or the cold deserts of central Asia (the Gobi Desert). They are mainly located where the tropic and temperate zones meet. Generally they are places of temperature extremes, experiencing a large **diurnal temperature** range, meaning they are hot by day and cold by night. Most animals that inhabit deserts are **nocturnal**, and desert vegetation is sparse. Desert rain often evaporates before it hits the ground, or else it falls in short, heavy bursts. Following periods of heavy rain, deserts teem with life. Around 300 million people around the world live in desert regions.

FIGURE 4 The Simpson Desert, central Australia

Grassland biomes

Grasslands are biomes dominated by grass and may have small, widely spaced trees or no trees. There are many types of grasslands, including savanna, tropical grasslands and temperate grasslands. Examples include the Serengeti in Tanzania and the Great Plains in North America.

The coarseness and height of the grass varies with location. They are mainly inhabited by grazing animals, reptiles and ground-nesting birds, though many other animals can be found in areas with more tree cover. Grasslands have long been prized for livestock grazing, but overgrazing of grasslands is unsustainable and places them at risk of becoming deserts. Over one billion people inhabit the grassland areas of the world.

FIGURE 5 Grassland, 'Brigalow Belt', Queensland

Tundra biomes

Tundras are biomes characterised by cold climates and the absence of trees, but they have grasses, dwarf shrubs, mosses and lichens. They are located in the higher latitudes around the polar zones, such as the Arctic Tundra, and at high altitudes, such as the Alti near Siberia. Tundra falls into three distinct categories—Arctic, Antarctic and alpine—but they share the common characteristic of low temperatures. In Arctic regions there is a layer beneath the surface known as **permafrost**. The tundra biome is the most vulnerable to changes in climate, because plants and animals have little tolerance for environmental changes that reduce snow cover.

diurnal temperature the variation in high and low temperature on a given day

nocturnal active during the night

permafrost permanently frozen ground

FIGURE 6 Tundra, Dovrefjell National Park, Norway

Aquatic biomes

Water covers about three-quarters of the Earth, and aquatic biomes can be classified as freshwater or marine. Freshwater biomes contain very little salt and are found on land; these include lakes, rivers and wetlands. Marine biomes are the saltwater regions of the Earth and include oceans, coral reefs and estuaries. Both freshwater and marine environments are teeming with plant and animal life and are a major food source. Elements taken from the roots of mangroves have been used in the development of cancer remedies. Compounds from other marine life have also been used in cosmetics and toothpaste.

FIGURE 7 Aquatic environment — the Great Sandy Strait, Queensland

on Resources

eWorkbook	Features of biomes (ewbk-8820)
Video eLesson	Features of biomes — Key concepts (eles-5173)
Interactivities	Beautiful biomes (int-3317) Major biomes of the world (int-8764) Climatic zones of the world (int-8765)

2.2 ACTIVITIES

1. Using information in the text and **FIGURE 1**, create a table that lists the five major biomes, identifies the different types within each of them, and identifies their key characteristics and spatial distribution.

Major biome	Types	Characteristics	Spatial distribution (climate zones)

2. Tropical rainforests are biodiverse. Use the internet to research the names of some plants and organisms used to make medicines and the diseases they help cure/prevent.
3. a. For each of the major biomes outlined in this section, outline how the biome can be used by humans.
   b. Outline two ways that humans can change each biome.
   c. For one biome, describe the possible impacts of the changes that you outlined in part **b**.

2.2 EXERCISE

Learning pathways

■ LEVEL 1	■ LEVEL 2	■ LEVEL 3
1, 2, 9, 11, 13	3, 5, 8, 10, 14	4, 6, 7, 12, 15

Check your understanding

1. What is a biome?
2. What are the major types of biomes?
3. Define the term *biodiversity*.
4. Why do desert biomes have lower populations than grassland biomes?
5. Refer to **FIGURE 1**.
   a. Which biome covers the largest amount of the Earth's landmass?
   b. Which biome covers the smallest amount of the Earth's landmass?
6. Describe what a diurnal temperature range is like.

Apply your understanding

7. Why is there a tundra biome located in central Asia?
8. Consider **FIGURE 1**. What patterns can you identify as biomes move away from the equator?
9. Explain why desert-dwelling animals are more likely to be nocturnal.
10. Why might tundra be considered the most fragile of biomes?
11. Complete the following sentence to explain how forests help regulate global climate.
    The _____ in the forest store _____ and are often referred to as carbon sinks. They also absorb energy from the _____. When the forest is removed, both these factors are lost, causing _____ in _____.
12. Are all biomes equally important? Explain your answer.

Challenge your understanding

13. Consider the biomes shown in **FIGURE 1**.
    a. Which biome do you live in?
    b. Does it look like the pictures in **FIGURES 3–7**?
    c. Explain why this might be the case.
14. Predict what might happen if the permafrost beneath the Arctic surface thawed.
15. Select one of the major biomes outlined in this section. Suggest how this biome might become more or less useful to humans in the future because of climate change.

To answer questions online and to receive **immediate feedback** and **sample responses** for every question, go to your learnON title at www.jacplus.com.au.

2.3 Major Australian biomes

LEARNING INTENTION

By the end of this subtopic, you will be able to name and outline the features of Australia's major biomes.

2.3.1 Factors that shape Australian biomes

Australia is a large island continent and is the world's sixth largest country by physical size. It is a large landmass and extends over a number of latitudes from 10°S to 43°S of the equator. As such, it has a variety of climates and biomes (**FIGURE 1**). The northern part of the continent is within the tropics and the southern areas are located in the temperate zones. The vastness of the continent also impacts **precipitation** patterns, and the mountain ranges on the east coast create **rain shadows** inland and alpine areas in the higher altitudes.

However, since European colonisation, large-scale land clearing, irrigation of the land through water diversion from rivers, and drainage of wetlands have taken place. Despite the extensive changes made by humans in the past few hundred years, Australia's major biomes are still evident.

precipitation water droplets or ice crystals become too heavy to be suspended in the air and fall to Earth as rain, snow, sleet or hail

rain shadow the dry area on the leeward side of a mountain range

int-8588

FIGURE 1 Australia's vastness and latitudinal extent affect the diversity of biomes.

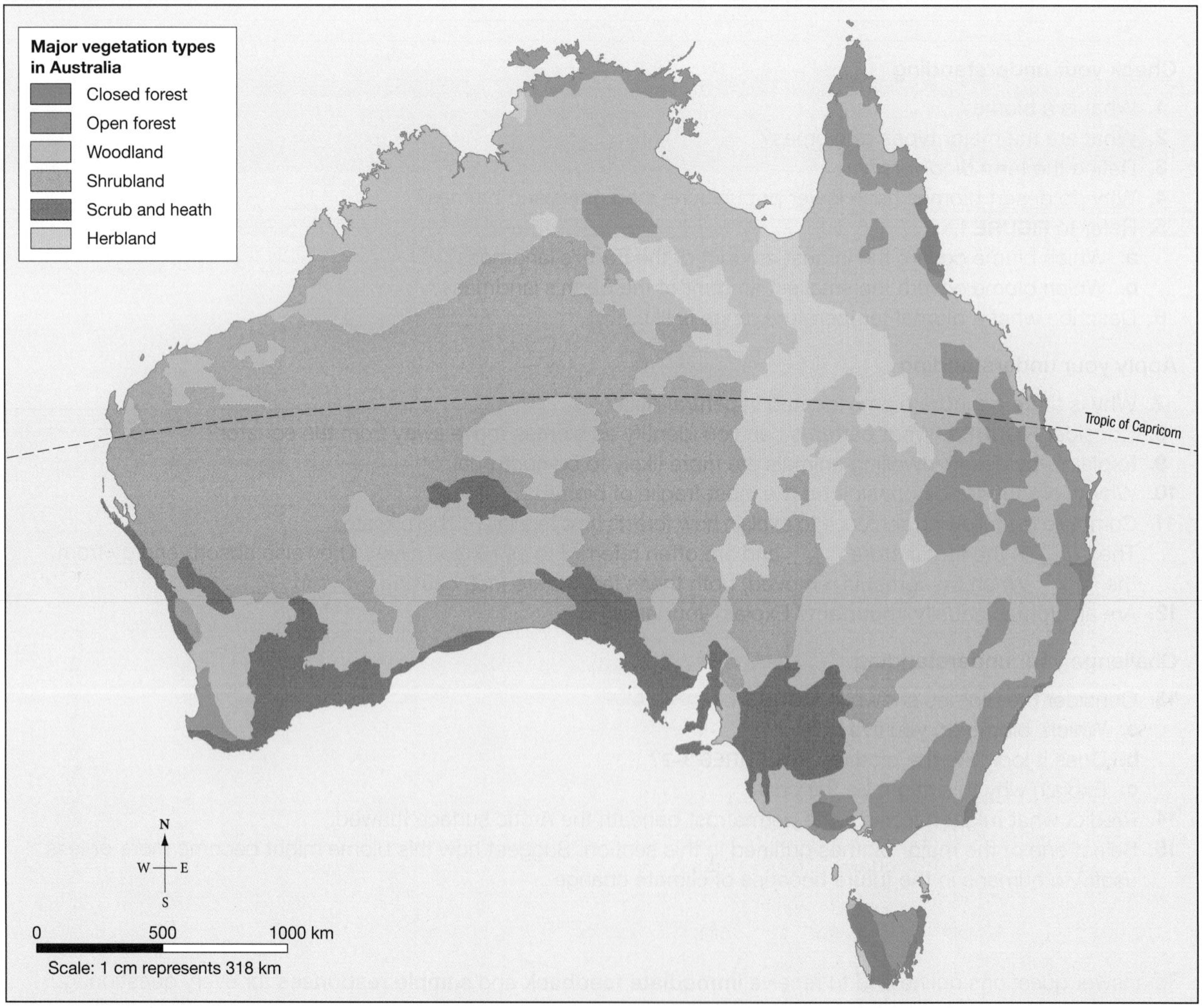

Source: Spatial Vision

Wetlands and rivers

The wetlands of mainland Australia were carefully managed as sustainable sources of food and water by Aboriginal Peoples for more than 60 000 years. Wetlands also provide important habitats for native and migratory birds and animals. Today, wetlands continue to provide Australians with food and water, but they are under threat in many parts of the country because water is diverted to produce commercial food crops and cotton. Farming practices can also damage wetlands and rivers when fertiliser, soil and insecticide run off the land and into water sources. Introduced species such as carp or weeds from overseas also have negative impacts.

FIGURE 2 Wetland near Kakadu National Park, Northern Territory

Grasslands

Grasslands are generally flat, having either few trees and shrubs or very open woodland. For many native species, grasslands provide vital habitat and protection from predators. Many grasslands depend on a regular cycle of burning to germinate their seeds and to revive the land. Periodic burning also prevents trees from gaining dominance in the landscape. Since before European settlement, and continuing today, Aboriginal Peoples have managed grasslands with a sophisticated practice of cultural burning and cultivation of plants and grasses to ensure the biome was sustained.

Grasslands often mark the transition between desert and forest, and are a very fragile biome. Without careful management they can quickly change to desert. Since colonisation, expanding towns and cities and extensive sheep and cattle grazing have meant that less than one per cent of Australia's original native grasslands survive today.

FIGURE 3 Grassland in the Kimberley, Western Australia

Seagrass meadows

Seagrasses are submerged flowering plants that form colonies off long, sandy ocean beaches, creating dense areas that resemble meadows. Of the 60 known species of seagrass, at least half are found in Australia's tropical and temperate waters. Western Australia alone is home to the largest seagrass meadow in the world. Seagrasses provide important habitats for a wide variety of marine creatures, including rock lobsters, dugongs and sea turtles. They also absorb nutrients from coastal run-off, slow water flow, help stabilise sediment (material that sinks to the bottom of water) and keep water clear. Activities such as boating impact these calm waters and can churn the water too much for some plants or animals to thrive.

FIGURE 4 Seagrass meadows at Green Island on the Great Barrier Reef

Old-growth forest

An old-growth forest is one in its oldest growth stage. It is multi-layered, and the trees are of mixed ages. Generally, there are few signs of human disturbance. These forests are biologically diverse, often home to rare or endangered species, and show signs of natural regeneration and decomposition. The trees within some old-growth forests have been felled for their timber and to create paper products; this is an example of using biomes to extract **industrial materials**. Logging can reduce biodiversity, affecting not only the forest itself but also the indigenous plant and animal species that rely on the old-growth habitat.

FIGURE 5 Old-growth forest, Cradle Mountain, Tasmania

It is estimated that **clearfelling** of Tasmania's old-growth forests would release as much as 650 tonnes of carbon per hectare into the atmosphere. In some places, for example in Victoria, near Melbourne, old growth forests lie within protected water supply catchments and help maintain the integrity of the city's water supply. However, Australia is one of the world's worst deforestation offenders. Much of the land on its east coast that was once forest has been destroyed for housing and industry.

Desert

Australian deserts are places of temperature extremes. During the day, temperatures sometimes exceed 50 °C, but at night this can drop to freezing. Australia's desert regions are often referred to as the outback, but they are not all endless plains of sand. Some, such as the Simpson and Great Sandy Deserts, are dominated by sand. The Nullarbor Plain and Barkly Tablelands are mainly smooth and flat, whereas the Gibson Desert and Sturt Stony Desert contain low rocky hills. In some areas, the landscape is dominated by spinifex and acacia shrubs (as shown in **FIGURE 6**).

industrial materials primary industry sources such as forestry (wood) or mining (iron ore) that can be used in the manufacturing of other goods

clearfelling the removal of all trees in an area

FIGURE 6 The red dunes of the Simpson Desert, Queensland

Resources

eWorkbook	Major Australian biomes (ewbk-8824)
Video eLesson	Major Australian biomes — Key concepts (eles-5174)
Interactivity	Major Australian biomes (int-8588)
myWorldAtlas	Deepen your understanding of this topic with related case studies and questions. • Australia's alpine biomes

2.3 ACTIVITIES

Select one of the biomes covered in this section.

1. Predict what might happen if the biome experienced change such as draining the wetlands or cutting down old-growth forests.
2. Describe the changes you consider may occur in this environment and outline what impacts they would have.
3. In pairs, select ONE biome to be proposed as an area for increased government protection.
   a. Locate and describe the biome.
   b. Identify the state and federal ministers responsible for the area and draft a letter outlining why the area should be protected.
   c. List the advantages and disadvantages of increased government protection of the chosen biome.

2.3 EXERCISE

Learning pathways

■ LEVEL 1	■ LEVEL 2	■ LEVEL 3
1, 2, 5, 11, 14	3, 4, 8, 9, 13	6, 7, 10, 12, 15

Check your understanding

1. Identify eight types of biomes that are found in Australia.
2. What latitudes does Australia sit between?
3. Which biome accounts for the majority of Australia's landmass?
4. Describe the climates in the north of Australia and the south of Australia.
5. Complete the following sentence: Biodiversity refers to the ______ of flora and ______ in an area. It is an ______ component of ensuring a ______ environment and maintaining the ______ of ecosystems.
6. Identify three environmental impacts of clearfelling Australia's old-growth forests.

Apply your understanding

7. Explain why Australia has a diversity of climates.
8. Explain why deserts have such large variations in temperature.
9. a. What is the best use of old-growth forests?
   b. If we were to continue to log these areas, who would benefit and who would not?
10. a. Explain why seagrass meadows would be cleared.
    b. Explain the consequences of clearing seagrass meadows.
11. Why might most of Australia's native grasslands have disappeared?
12. Explain how biomes and climate are interconnected.

Challenge your understanding

13. Predict what might happen if Australia's old-growth forests were logged.
14. Suggest why seagrass meadows might be referred to as the 'forests of the sea'.
15. Burning, including the cultural burning practices used by Aboriginal Peoples, is an essential element in maintaining the grassland biome. Explain why this is the case and suggest how controlled burning programs might help to prevent large-scale bushfires in Australia.

To answer questions online and to receive **immediate feedback** and **sample responses** for every question, go to your learnON title at www.jacplus.com.au.

2.4 SkillBuilder — Describing spatial relationships in thematic maps

online only

LEARNING INTENTION

By the end of this subtopic, you will be able to describe the spatial relationship between two or more thematic maps and the degree to which they influence each other's distribution in space.

The content in this subtopic is a summary of what you will find in the online resource.

2.4.1 Tell me

What are spatial relationships in thematic maps?

A spatial relationship is the interconnection between two or more pieces of information in a thematic map, and the degree to which they influence each other's distribution in space. Spatial relationships between features or information in thematic maps are the links between the distribution of those features. Finding these links can help us to see the world in an organised manner. They are useful in helping you, as a student, to understand how one thing affects another.

2.4.2 Show me

How to find and describe a spatial relationship in thematic maps

Step 1

Use an atlas to identify key places in the mapped area. For example, in **FIGURE 1(a)** and **(b)** locate India and China. (To access the full maps and zoom in to areas of interest, go to online resource.)

Step 2

Look for similar patterns in similar parts of the maps. In **FIGURE 1(a)**, rainforest biomes cover similar areas as the tropical wet climates shown in **FIGURE 1(b)**. The biomes and climate are linked; the climate provides the warm temperatures and high rainfall needed for a rainforest biome. Describe these strong interconnections.

FIGURE 1(a) Thematic map of Asia showing biomes

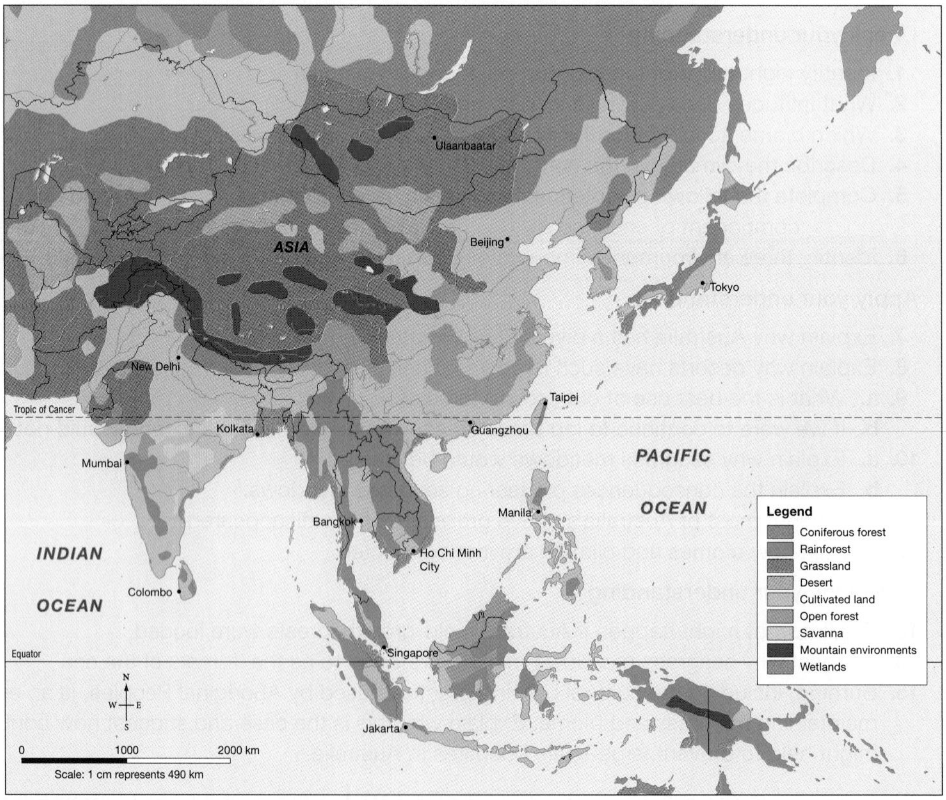

Source: Spatial Vision

FIGURE 1(b) Thematic map of Asia showing climatic zones

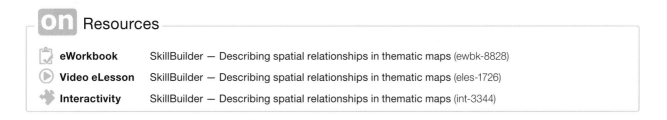

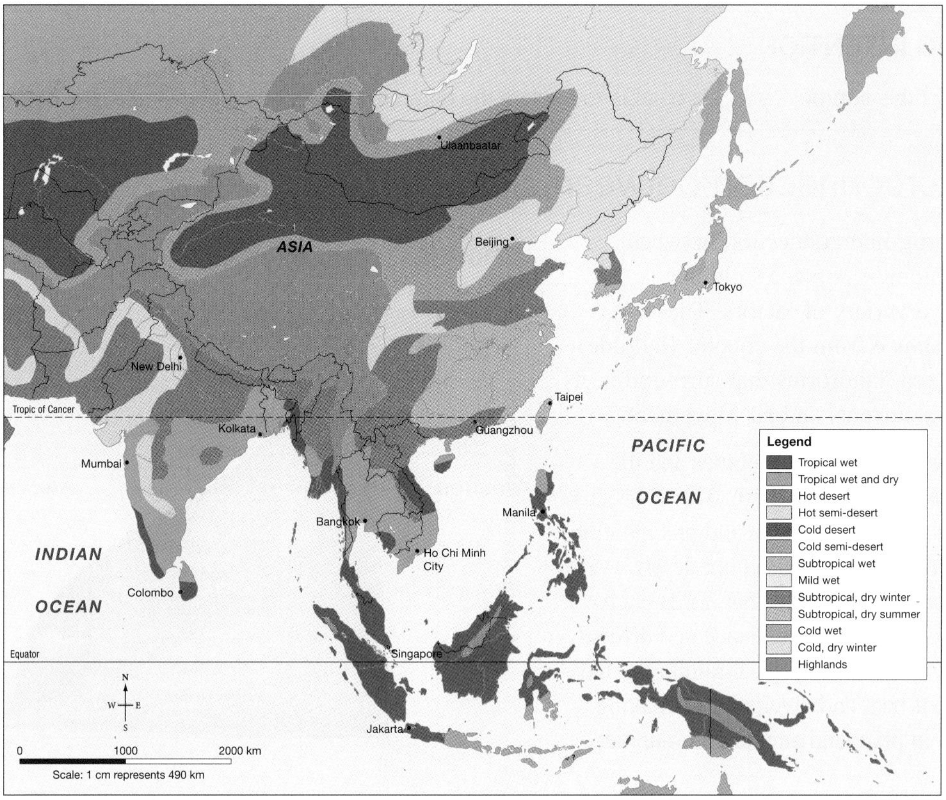

Source: Spatial Vision using Natural Earth

Step 3

Look for areas where there seem to be no connection between biomes and climate (anomalies). If necessary, write a few sentences outlining where there are no interconnections.

Step 4

Conclude your description with a statement about spatial relationships. Your description should clearly identify which features are interconnected, point out obvious anomalies and describe the extent of interconnections (for example, as strong or weak).

2.4.3 Let me do it

learnON

Go to learnON to access the following additional resources to help you build this skill:

- a longer explanation of this skill and its application in Geography (Tell me)
- a video demonstrating the step-by-step process of this skill (Show me)
- an activity and interactivity for you to practise the skill (Let me do it)
- self-marking questions to help you understand and use the skill.

on Resources

eWorkbook	SkillBuilder — Describing spatial relationships in thematic maps (ewbk-8828)
Video eLesson	SkillBuilder — Describing spatial relationships in thematic maps (eles-1726)
Interactivity	SkillBuilder — Describing spatial relationships in thematic maps (int-3344)

2.5 The role of climate in biomes

LEARNING INTENTION

By the end of this subtopic, you will be able to explain the different factors affecting climate and biomes.

2.5.1 Interconnection between climate and biomes

There is a strong interconnection between **climate** and biomes. A place's climate is influenced by a variety of factors. These include its distance from the equator (latitude) and from the sea, landforms that surround it, its altitude, ocean currents and air movement.

Climate affects the location of biomes and the flora and fauna found within them. This is because of the differences in temperature and precipitation patterns (**FIGURE 1**). For instance, hot desert biomes have high temperatures but not enough precipitation to sustain an abundance and diversity of life. On the other hand, tropical rainforests have an abundance of heat and precipitation, making them plentiful in plant and animal communities.

FIGURE 1 Interconnection between climate and biomes

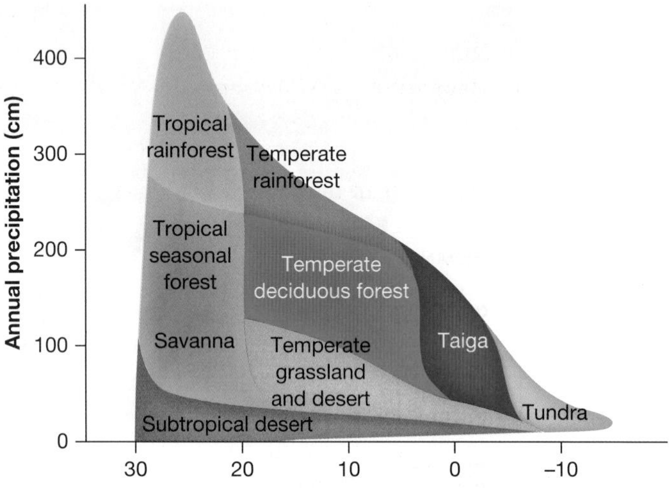

Latitude

The Sun's rays are more direct at the equator, so the land there receives higher levels of **insolation**. With more energy focused on that region, it heats up more quickly. As you move away from the equator to the mid and higher latitudes, the curvature of the Earth makes the Sun's rays travel through more atmosphere, creating lower levels of insolation in any specific place, as the energy spreads over a larger area. As a result the energy from the Sun does not heat up the Earth as effectively, and the closer to the poles you travel the cooler it becomes (**FIGURE 2**).

The tilt of the Earth on its axis also has a part to play. When a hemisphere tilts towards the Sun, the Sun's rays hit it more directly. This means that a larger space is in more intense sunlight for longer. The days are longer and warmer, and the hemisphere experiences summer. The reverse is true when a hemisphere tilts away from the Sun in winter. This has created distinct climate zones based on latitude. There is a relationship between climate zones and the type of biomes located within them (see **FIGURE 1**).

climate the long-term precipitation and temperature patterns of an area

insolation the level of solar energy that reaches the Earth's surface

FIGURE 2 The influence of latitude on climate

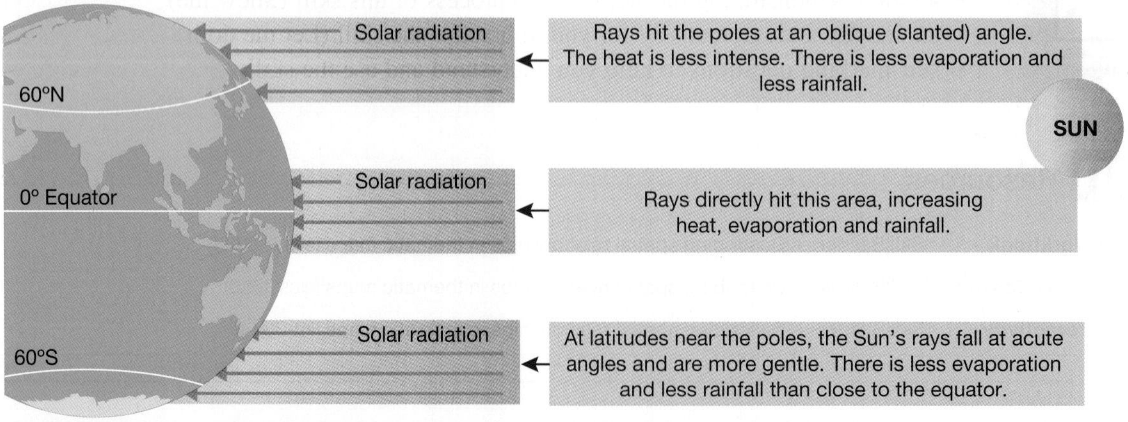

Altitude and landforms

Altitude also plays a significant role in determining climate. Temperatures fall by 0.65 °C for every 100 metres increase in elevation. This can be illustrated by Mt Kilimanjaro (**FIGURE 3**), which is located on the border of Tanzania and Kenya, in Africa, approximately 3 degrees south of the equator. Towering 5895 metres above sea level, Mt Kilimanjaro is the highest mountain in Africa. Depending on the time of the day, the temperature at the base of the mountain ranges from 21 °C to 27 °C. At the summit, temperatures can plummet to –26 °C. As you move from base to summit, variations occur in the landscape as it transitions from rainforest to alpine desert to desert tundra.

FIGURE 3 Mt Kilimanjaro is only three degrees south of the equator, but it is 5895 metres high; its altitude is the reason it has snow on its summit.

Another major geographical influence on climate is the location of mountain ranges (**FIGURE 4**). Although the altitude of ranges can affect temperature, the location of mountain ranges also affects the amount of precipitation that reaches inland areas. They prevent the moisture-laden **prevailing winds** from reaching inland areas by creating a barrier. Rain falls on the **windward** side of a mountain and rain shadows form on the **leeward** side of mountains. Desert biomes often form in rain shadows.

prevailing winds winds that blow from the direction that is typical at that time of year and place

windward describes the side of a mountain that faces the prevailing winds

leeward describes the area behind a mountain range, away from the moist prevailing winds

Ocean currents and air movement

There are other factors that influence climate and play a role in the development of biomes. Two of these are ocean currents and air movement.

When cold ocean currents flow close to a warm land mass, a desert is more likely to form. This is because cold ocean currents cool the air above, causing less evaporation and making the air drier. As this dry air moves over the warm land it heats up, making it less likely to release any moisture it holds and so making areas it reaches arid. For example, cold ocean currents flow off the coast of Western Australia, whereas on the east coast of Australia the Pacific Ocean currents are warmer (**FIGURE 4**). As a result, Perth on average receives less rainfall than Sydney.

FIGURE 4 The influence of mountains on climate, typical on the east coast of Australia

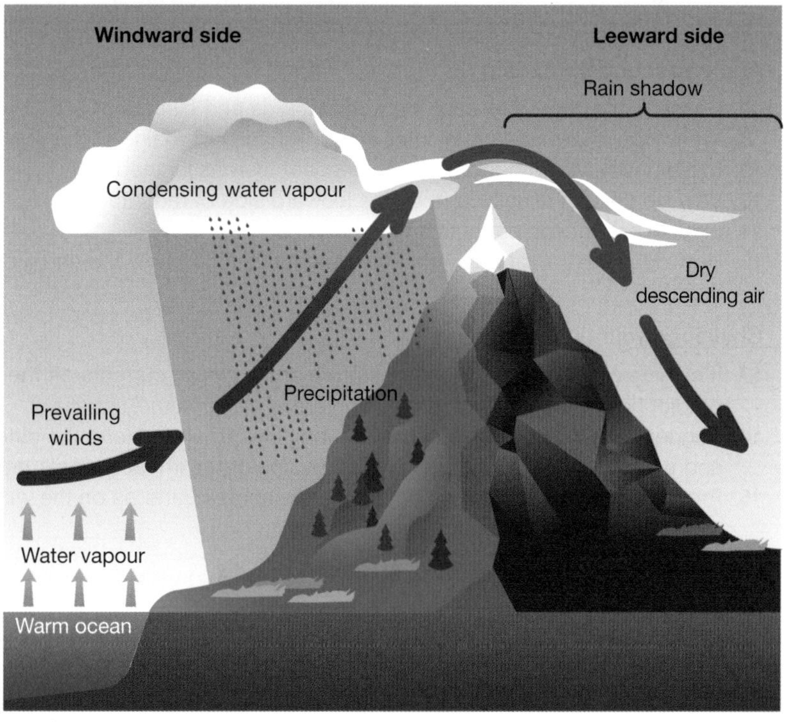

on Resources

eWorkbook	The role of climate in biomes (ewbk-8832)
Video eLesson	The role of climate in biomes — Key concepts (eles-5175)
Interactivity	The role of climate in biomes (int-8589)

2.5 ACTIVITIES

1. Research the average temperature and average annual precipitation of the area your school is in using the **Bureau of Meteorology (BOM)** weblink in your online Resources. Refer to **FIGURE 1** and determine what type of biome your school is in based on the climate data you researched.
2. Use an atlas to locate Rwanda in central Africa.
   a. Describe what you think the climate would be like in this country. Why?
   b. What type of biomes would you expect in Rwanda?
   c. Research Rwanda's climate and biomes, and test your predictions.

2.5 EXERCISE

Learning pathways

■ LEVEL 1	■ LEVEL 2	■ LEVEL 3
1, 2, 4, 8, 13	3, 5, 9, 10, 14	6, 7, 11, 12, 15

Check your understanding

1. Define the term *climate*.
2. Which biome has the greatest temperature and precipitation?
3. What is a rain shadow?
4. Describe the difference between the 'windward' and 'leeward' sides of mountain ranges.
5. At what rate does temperature fall as you ascend a mountain?
6. Why do areas closest to the equator get the most energy from the Sun?

Apply your understanding

7. Describe the interconnection between latitude and climate.
8. What type of climate and biomes would you expect to find at the equator? Why?
9. Explain why Mt Kilimanjaro has a variety of biomes located on it.
10. Why do deserts usually form on the leeward side of mountains?
11. Explain how ocean currents affect terrestrial biomes.
12. How does the curvature of the Earth and its tilt affect how the energy from the Sun impacts on the Earth's surface?

Challenge your understanding

13. If climate change raises the temperature of the ocean currents off the west coast of Australia significantly, predict how this will affect the farming areas of Western Australia.
14. Suggest why some people might find it difficult to understand the difference between climate and weather, and why this might affect their understanding of the impacts of climate change on biomes.
15. Propose one way to harness or change the rainfall patterns on the windward side of the Great Dividing Range to benefit farmers on the leeward side.

To answer questions online and to receive **immediate feedback** and **sample responses** for every question, go to your learnON title at www.jacplus.com.au.

2.6 The role of soil and vegetation in biomes

LEARNING INTENTION

By the end of this subtopic, you will be able to describe the importance of soil to vegetation and biomes.

2.6.1 The role of soil in biomes

Soil is important in determining which plants and animals inhabit a particular biome. Soils vary not only around the world but also within regions. The characteristics of soil are determined by:

- temperature
- rainfall
- the rocks and minerals that make up the bedrock, which is the basis of soil development.

These all play an important role in determining the quality of the soil. **FIGURE 1** shows a typical soil profile. The different soil layers are referred to as horizons.

Differences in soil composition

Biomes located in the high latitudes (those farthest from the equator) have lower temperatures and less exposure to sunlight than biomes located in the low latitudes (those close to the equator). There are also variations in the amount of precipitation that biomes receive. This is determined partly by their location in relation to the equator.

Temperature and precipitation patterns are important factors in determining the rate of soil development. However, soil moisture, its nutrient content and the length of the growing season also play key roles in soil development and, ultimately, the biodiversity of a biome.

Soil is more abundant in biomes that have both high temperatures and high moisture than in cold, dry regions. This is because erosion of bedrock is more rapid when moisture content is high, and organic material decomposes at a faster rate in high temperatures. The decomposition of organic matter provides the nutrients needed by plants, which in turn die and decompose in a continuous cycle. This is further demonstrated in **FIGURE 2**.

FIGURE 1 A typical soil profile has distinct layers.

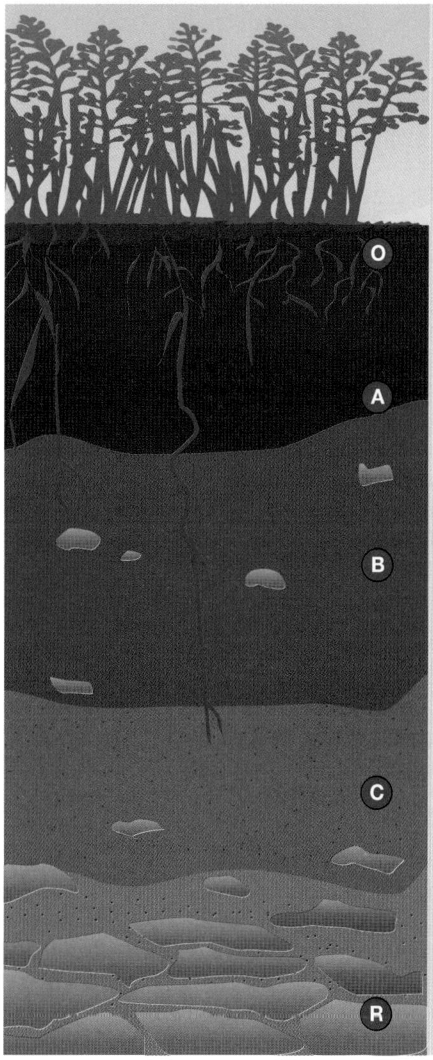

Horizon O (organic matter): A thin layer of decomposing matter, humus and material that has not started to decompose, such as leaf litter.

Horizon A (topsoil): The upper layer of soil, nearest the surface. It is rich in nutrients to support plant growth and usually dark in colour. Most plant roots and soil organisms are found in this horizon, which will also contain some minerals. In areas of high rainfall, such as tropical rainforests, minerals will be leached out of this layer. A constant supply of decomposing organic matter is needed to maintain soil fertility.

Horizon B (subsoil): Plant litter is not present in horizon B; as a result, little humus is present. Nutrients leached from horizon A accumulate in this layer, which will be lighter in colour and contain more minerals than the horizon above.

Horizon C (parent material): Weathered rock that has not broken down far enough to be soil. Nutrients leached from horizon A are also found in this layer. It will have a high mineral content; the type is determined by the underlying bedrock.

Horizon R (bedrock): Underlying layer of partly weathered rock.

FIGURE 2 Different biomes have different soil and vegetation characteristics.

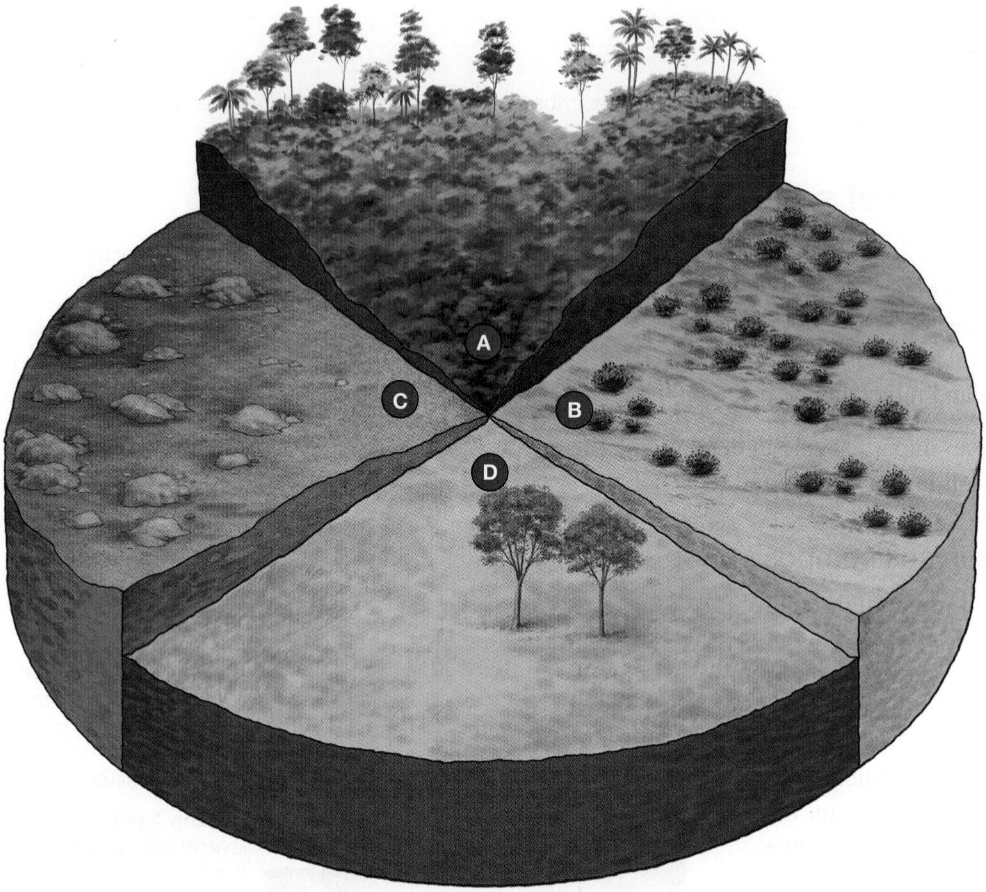

A Tropical rainforest
- High temperatures cause weathering, or breakdown, of rocks and organic matter.
- High rainfall **leaches** nutrients from the soil.
- Soil is often reddish because of high iron levels.
- Organic matter is often a shallow layer on the surface. Nutrients are constantly recycled, allowing the rainforest to flourish.
- Soil fertility is rapidly lost if trees are removed, as the supply of organic material is no longer present.

B Desert
- Limited vegetation means a limited supply of organic material for soil development.
- High temperatures rapidly break down any organic material.
- Soils are pale in colour rather than dark.
- Lack of rainfall limits plant growth.
- Lack of vegetation makes surface soil unstable and easily blown away.
- Soil does not have time to develop and mature.

C Tundra
- Soil is shallow and poorly developed.
- Includes layers that are frozen for long periods.
- Subsoil may be permanently frozen.
- It is covered by ice and snow for most of the year.
- Growing season may be limited to a few weeks.
- Soil may contain large amounts of organic material but extreme cold means it breaks down very slowly.
- Trees are absent; mosses and stunted grasses dominate.

D Temperate
- Generally brown in colour, soils have distinctive horizons and are generally around one metre deep.
- Ideal soils for agriculture; they are not subjected to the extremes of climate found in high and low latitudes.
- Moderate climate; temperature and rainfall are sufficient for plant growth.
- Dominated by temperate grasslands and deciduous forests.

leaching the process in which water runs through soil, dissolving minerals and carrying them into the subsoil

What else is in the soil?

Soil not only supports the plants and animals that we see on the surface of the land; the soil itself is also home to a variety of life forms such as bacteria, fungi, earthworms and algae.

Although most soil organisms are too small to be seen, there are some that are visible. For instance, more than 400 000 earthworms can be found on a hectare of land. Regardless of size, all soil organisms play a vital role in maintaining soil quality and fertility. For example, earthworms:

- compost waste and fertilise the soil
- improve drainage and aeration
- bring subsoil to the surface and mix it with topsoil
- secrete nitrogen and chemicals that help bind the soil.

FIGURE 3 There are more microbes (microorganisms) in a teaspoon of soil than there are people on Earth.

2.6.2 Net Primary Production

Net primary production (NPP) is a calculation of how much carbon dioxide plants take up during photosynthesis minus how much carbon dioxide plants release during respiration. Productivity will vary between seasons but tends to be higher in summer and highest in tropical forests throughout the world.

FIGURE 4 Earth's net primary productivity, measured in kilograms of carbon fixed via photosynthesis per square kilometre per year (based on satellite data). The colour coding shows this rate varying from close to zero (black, purple) through 1 and 2 (blue, green, yellow) to 3 (red).

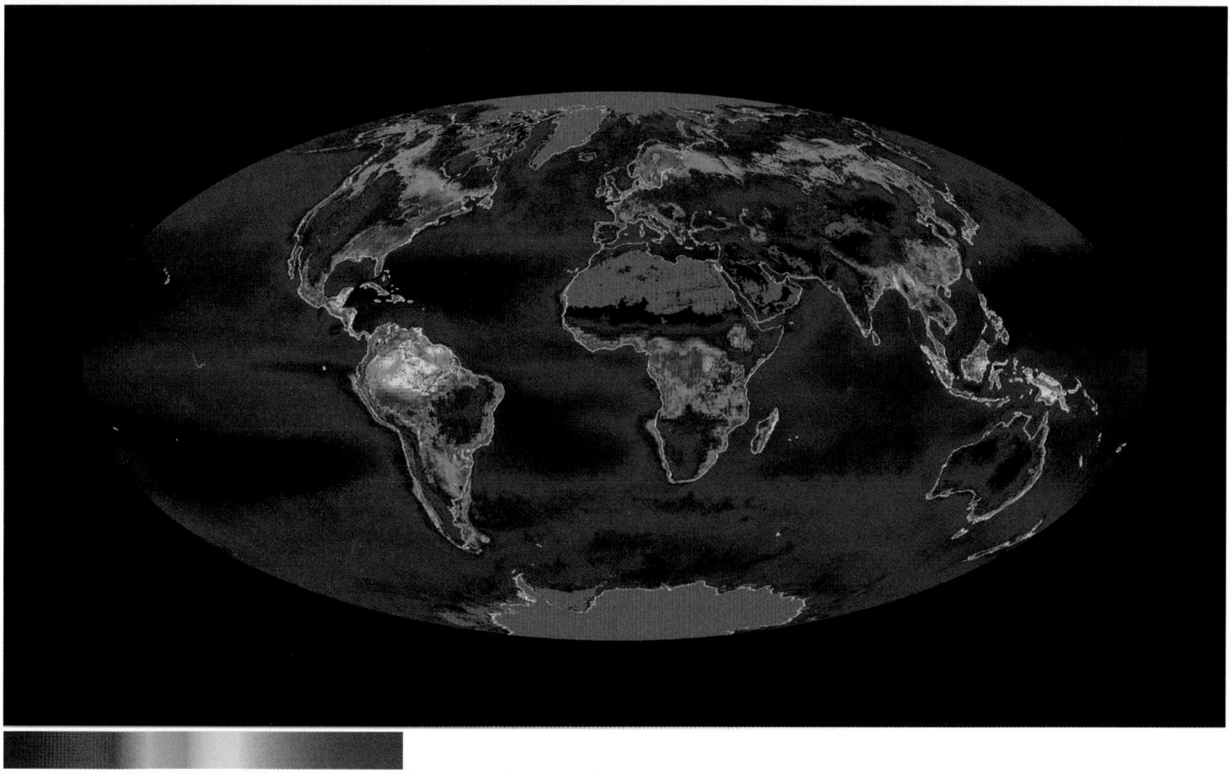

Source: Based on data from the MODIS sensors on the Terra and Aqua satellites

fdw-0027

FOCUS ON FIELDWORK

Soil testing

Soil is an important component within biomes. Soil is critical for the growing of plants; as such, determining its texture and its ability to hold nutrients helps humans grow crops more effectively.

Measuring soil pH

Determining a soil's pH identifies how acidic or alkaline it is using a scale from 1 to 14: a pH of 7 is neutral, less than 7 is acid and greater than 7 is alkaline. The pH level can interfere with how plants can absorb nutrients, thus affecting growth. The most effective way to test soil pH in the field is by using a soil testing kit (**FIGURE 5**).

FIGURE 5 A pH soil testing kit

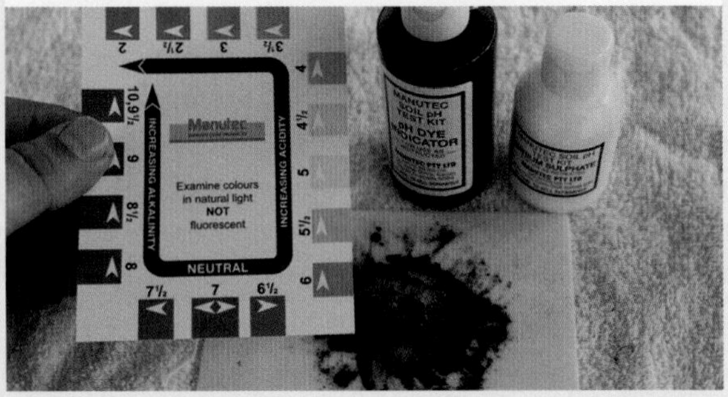

Learn more about what is in the soil in your local area using the **Soil testing** fieldwork activity in your learnON Fieldwork Resources.

FIGURE 6 Soil composition plays a crucial role in how well crops will grow, so many farmers use chemical fertilisers to ensure the balance of the soil's composition is right for their crops.

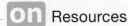

Resources

eWorkbook	The role of soil and vegetation in biomes (ewbk-8836)
Video eLesson	The role of soil and vegetation in biomes — Key concepts (eles-5176)
Interactivity	Why are biomes different? (int-3319)

2.6 ACTIVITIES

1. Construct a table comparing the differences between tropical rainforests and tundra biomes based on the following.
   a. Soil type
   b. Organic material
   c. Vegetation
   d. Location
   e. Human uses
2. In groups assign each member a different horizon based on the typical soil profile. Each member should conduct detailed research on their horizon. On an A3 page include a hand-drawn image and annotated notes. When your group has completed all 5 sections, join them together for a complete profile.

2.6 EXERCISE

Learning pathways

■ LEVEL 1	■ LEVEL 2	■ LEVEL 3
3, 4, 7, 12, 15	1, 5, 8, 10, 13	2, 6, 9, 11, 14

Check your understanding

1. Identify the characteristics that affect soils.
2. List the differences between horizon O, horizon A and horizon B in the typical soil profile shown in **FIGURE 1**.
3. How many earthworms can be found in a hectare of land?
4. Define the term *net primary production*.
5. What role do worms play in soil?
6. Refer to **FIGURE 2**. Describe how animals' use of the soil would vary between each of the four biomes shown.
7. Complete the following sentence to describe the role played by soil organisms in maintaining soil quality and fertility: ______ are important and help maintain soil quality by breaking down ______ material into ______, which are essential to plant growth. They secrete ______ and other chemicals that help bind the soil, improve soil drainage and aeration, and bring ______ to the surface and mix it with the ______.

Apply your understanding

8. Explain why soils vary in different biomes.
9. Explain the relationship between the location of biomes and their NPP.
10. What impact would deforestation have on soil and forest biomes? Analyse the impact for humans and their food sources.
11. Explain how temperature and rainfall can influence the development of soil.
12. Are all soils within a biome the same? Explain why or why not.

Challenge your understanding

13. Hot, dry conditions and strong winds can dry out and blow away top layers of soil when there is little or no vegetation. What impacts might this loss of topsoil have on future vegetation growth in the area?
14. Predict how the soil in Australia's temperate biomes would change with greater extremes of temperature — frequent days over 40 °C in summer and frequent overnight temperatures below 0 °C.
15. If you wanted to encourage healthy, fertile soil, would it be wise to use a pesticide that killed all of the earthworms? Give reasons to support your answer.

To answer questions online and to receive **immediate feedback** and **sample responses** for every question, go to your learnON title at www.jacplus.com.au.

2.7 SkillBuilder — Constructing and describing a transect on a topographic map

online only

LEARNING INTENTION

By the end of this subtopic, you will have learned how to construct and describe a transect on a topographic map.

The content in this subtopic is a summary of what you will find in the online resource.

2.7.1 Tell me

What is a transect?

A transect is a cross-section with additional detail which summarises information about the environment. In addition to the shape of the land, a transect shows what is on the ground, including landforms, vegetation, soil types, settlements and infrastructure.

2.7.2 Show me

Step 1

Choose the area on your topographical map that you would like to look at, and the two points that will give you the best line through that area.

Step 2

Place the straight edge of a piece of paper between the two points. In pencil, mark the two extremities of the transect on the edge. Label these with the place names or grid references.

Step 3

Create a mark where each contour line touches the edge of the paper. Beside each mark, write the height of the contour line. (Check the contour interval for how many metres the lines increase or decrease.)

Step 4

On another sheet of paper, use a ruler to draw:

- a horizontal axis as long as your cross-section from start to finish.
- a vertical axis with a scale to give a realistic impression of the slopes and landforms.

FIGURE 1 A completed transect of the Walls of Jerusalem National Park, Tasmania

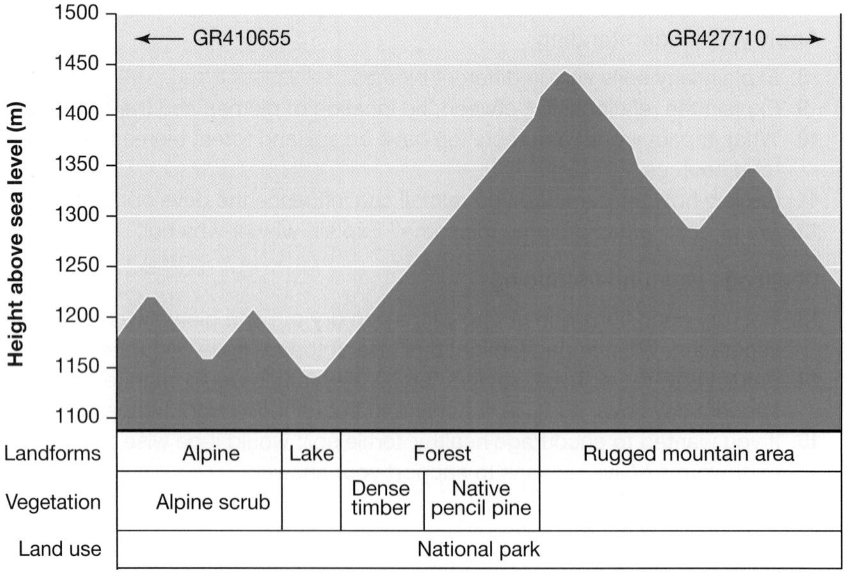

Step 5

Place the edge of the paper with the contour markings along the horizontal axis. At each contour marking, find the matching height on the vertical scale. Put a small dot directly across from that height and above the contour marked on the edge of the paper.

Step 6

Join the dots with a smooth line to show the slope of the land, as shown in **FIGURE 3**. A labelled notch has been used to show a river on the cross-section. Other features can be marked in a similar way.

FIGURE 3 Drawing up the shape of the transect

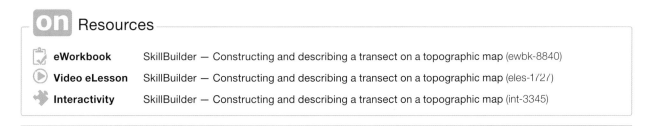

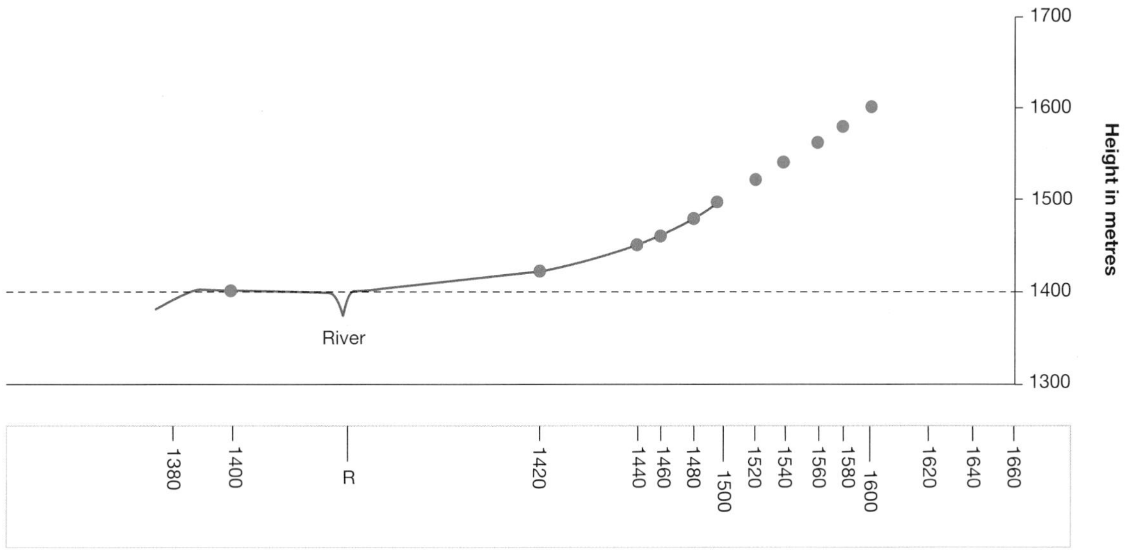

Step 7

Complete the cross-section with the geographical conventions of a title and labelled axes. Shade the area below the line of your cross-section. Include labels for the axes: height above sea level and distance.

Step 8

Beneath your cross-section, draw a table to indicate when a feature changes on the transect. Label each category to the left of the vertical axis. Common categories include landforms, vegetation, land use, transport, settlement and sometimes soils. Refer to the map to determine the exact place on the transect where the feature occurs.

Step 9

Land formations can become distorted or misshapen by the choice of scale, and this is referred to as vertical exaggeration (VE). To calculate the VE of your transect, you need to know the scale of your map and the scale of your transect. To find the vertical exaggeration divide the horizontal scale by the vertical scale: VE = 250/75 = 3.3.

2.7.3 Let me do it

learnON

Go to learnON to access the following additional resources to help you build this skill:

- a longer explanation of this skill and its application in Geography (Tell me)
- a video demonstrating the step-by-step process of this skill (Show me)
- an activity and interactivity for you to practise the skill (Let me do it)
- self-marking questions to help you understand and use the skill.

on Resources

eWorkbook SkillBuilder — Constructing and describing a transect on a topographic map (ewbk-8840)

Video eLesson SkillBuilder — Constructing and describing a transect on a topographic map (eles-1727)

Interactivity SkillBuilder — Constructing and describing a transect on a topographic map (int-3345)

2.8 Use of grassland biomes

LEARNING INTENTION

By the end of this subtopic, you will be able to describe different uses of grassland biomes.

2.8.1 Characteristics of grasslands

Grassland, pampas, savanna, chaparral, cerrado, prairie, rangeland and steppe all refer to a landscape that is dominated by grass. Once, grasslands occupied about 42 per cent of the Earth's land surface, but today they make up about 25 per cent of its land area. Grasslands are found on every continent except Antarctica (see **FIGURE 1**).

Grasslands are also one of the most endangered environments and are susceptible to **desertification**. The entire ecosystem depends on its grasses and their annual regeneration. It is almost impossible to re-establish a grassland ecosystem once desert has taken over.

Grasslands often depend on fire to germinate their seeds and generate new plant growth. Many grasslands around the world were managed with the use of fire for thousands of years. For example, Aboriginal Peoples lit carefully located low-burning fires to manage regeneration and to move wildlife out of specific areas for easier hunting, among other purposes.

Aboriginal Peoples also managed and cultivated native grasses, but since European colonisation, most grassland has been removed or changed by farming and other development. Vast areas of grassland were cleared for crops, and introduced grasses were planted for grazing animals. In some parts of Australia, cultural burning practices are still helping to maintain the grassland biome. However, less than one per cent of Australia's native grasslands survive, and they are now considered one of the most threatened of our habitats.

desertification the conversion of an area to have desert-like qualities, usually caused by overgrazing, prolonged drought or climate change

FIGURE 1 Grasslands occupy about a quarter of the Earth's land surface.

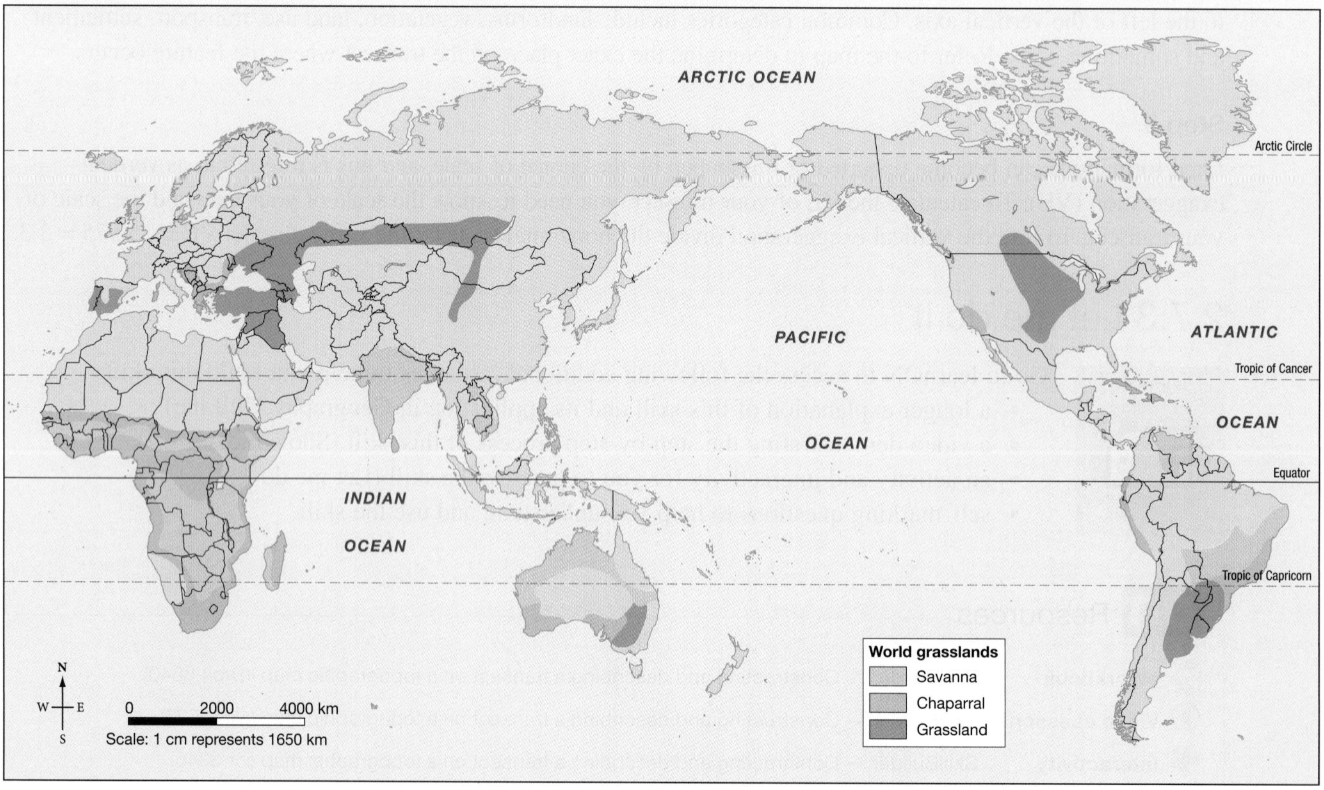

Source: Spatial Vision

CASE STUDY

Aboriginal Peoples' management of grasslands over time

Professor Marcia Langton is the Foundation Chair of Australian Indigenous Studies and Associate Provost at the University of Melbourne. The extract in **FIGURE 2** comes from her travel guide, *Welcome to Country*, which is aimed at teaching non-Aboriginal people about Aboriginal cultures and languages, and the significance and history of places many tourists visit in Australia.

FIGURE 2 Text extract from Langton, M (2018), *Welcome to Country: A travel guide to Indigenous Australia*, pp. 9–10 (Hardie Grant: Melbourne).

> The ancient peoples left traces of their lives in the rocks and on the rock faces everywhere they went. Perhaps their greatest legacy is to be found in the vegetation patterns and other evidence of their 'continent-wide Aboriginal management regime', as Bill Gammage put it … Australian vegetation communities in the wild have been shaped by fire and Aboriginal fire regime over millennia.
>
> There is no one better to explain this practice than the Aboriginal rangers who work across thousands of square kilometres and continue to burn the small, cool fires in the vast grass and spinifex plains and the savanna forests. They are continuing the ancient tradition of using fire against fire, working with fire-dependent species, such as eucalypts, to protect the vulnerable communities.

In contrast to Aboriginal grassland management, grazing large numbers of introduced species and failing to prevent frequent uncontrolled hot fires restrict the growth of tree seedlings and promote the uncontrolled spread of grasses. The growth points of grasses are low and close to the soil, so they can continue to grow even when they are continually grazed by animals. Because grasses are fast-growing plants, they can support a high density of grazing animals and can regenerate quickly after fire, whereas trees and many other plant species cannot.

2.8.2 The role of grasslands

Grasslands are the most useful biome for agriculture because the soils are generally deep and fertile. They are ideally suited for growing crops or creating pasture for grazing animals. The prairies of North America, for example, are one of the richest agricultural regions on Earth.

FIGURE 3 Wheat is a type of grass.

Almost one billion people depend on grasslands for their livelihood or as a food source. Grasslands have been used for livestock grazing and are increasingly under pressure from **urbanisation**. Grasslands have also become popular tourist destinations, because people flock to them to see majestic herds of animals such as wildebeest, caribou and zebra, as well as the migratory birds that periodically inhabit these environments.

urbanisation the process of economic and social change in which an increasing proportion of the population of a country or region lives in urban areas

2.8.3 The production of food

All the major food grains — corn, wheat, oats, barley, millet, rye and sorghum — have their origins in the grassland biome. Wild varieties of these grains are used to help keep cultivated strains disease free. Many native grass species have been used to treat diseases including HIV and cancer. Others have proven to have properties for treating headaches and toothache.

2.8.4 The production of fibre

Grasslands are also the source of a variety of plants whose fibres can be woven into clothing. The best known and most widely used fibre is cotton. Harvested from the cottonseed, it is used to produce yarn that is then knitted or sewn to make clothing. Lesser known fibres include flax and hemp. Harvested from the stalk of the plant, both fibres are much sturdier and more rigid than cotton but can be woven to produce fabric. Hemp in particular is highly absorbent and has UV blocking qualities.

FIGURE 4 Grasslands in Australia support a high density of grazing animals, for example sheep for fine wool production.

In Australia, it is estimated that less than one per cent of the grasslands that existed in 1788 still survive. Grasslands are now considered one of the most threatened of Australia's biomes. Since European colonisation, vast areas of grassland were cleared for crops, and introduced grasses were planted for grazing animals, such as sheep, which are able to provide both meat and fibre (wool).

on Resources

eWorkbook	Use of grassland biomes (ewbk-8844)
Video eLesson	Use of grassland biomes — Key concepts (eles-5177)
Interactivities	Grass, grains and grazing (int-3318)
myWorldAtlas	Deepen your understanding of this topic with related case studies and questions. • Wheat

2.8 ACTIVITIES

1. Working in teams, investigate one of the grassland biomes. Create a presentation on your chosen biome that covers the following:
   a. the characteristics of the environment, including climate and types of grasses that dominate this place
   b. the animals that are commonly found there
   c. how the environment is used and changed for the production of food, fibre and wood products
   d. threats to this particular grassland, including the scale of these threats
   e. what is being done to manage this grassland environment in a sustainable manner.
2. Research and construct a table that compares the value to the Australian economy of food production, fibre and industrial materials. Which contributes the highest value?
3. Research the different types of grasslands. Why do they have different names? Create an infographic to explain the different types and uses of grasslands for food and fibre around the world.

4. Use the **Good Food** weblink in your online Resources to explore how Aboriginal Peoples used native grasses in a similar way to how Europeans use wheat. What are the advantages and disadvantages of commercially growing and processing Australian grasses for food?
5. Few people realise that less than 1 per cent of Australia's native grasslands survive. Why does such a significant loss of grassland biome not attract the same attention as the loss of other biomes, such as our tropical rainforest and coral reefs?
   a. How would the following people perceive the value of grasslands?
      - people for whom grassland is part of their Country
      - graziers (sheep and cattle farmers)
      - city dwellers
      - environmentalists
   b. What factors might influence each person's views?
   c. Create a poster or meme to demonstrate why it is important to protect Australia's grasslands.

2.8 EXERCISE

Learning pathways

■ LEVEL 1	■ LEVEL 2	■ LEVEL 3
1, 4, 9, 10, 15	2, 6, 7, 8, 13	3, 5, 11, 12, 14

Check your understanding

1. Define the term *grassland*.
2. Describe the global distribution of grasslands.
3. Outline why grasslands are an important environment.
4. Describe three major threats to grasslands.
5. Outline the interconnection between urbanisation and grasslands.
6. Describe one way that Aboriginal Peoples managed grasslands.

Apply your understanding

7. Explain why grasslands are referred to as transitional landscapes.
8. Explain why introduced species had such a significant impact on Australia's grasslands.
9. Explain why so little of Australia's grasslands remain.
10. Identify and discuss one positive and one negative impact of clearing grasslands in Australia for sheep grazing.
11. Account for the relationship between grassland biomes and the different types of agriculture.
12. How might a high rate of population growth impact grassland biomes?

Challenge your understanding

13. Australia is becoming an increasingly urbanised country. Suggest how this might affect the distribution and extent of our grasslands.
14. Aboriginal Peoples in Australia manage and cultivate native grasses for food and other products. Suggest why British colonists planted new grass varieties instead of using the grasses that were already here.
15. Grasslands can be popular tourist destinations, especially when there are herds of wild animals and flocks of birds that people find interesting to watch. Do you think grasslands in Australia are popular tourist destinations for this reason? Explain why or why not.

To answer questions online and to receive **immediate feedback** and **sample responses** for every question, go to your learnON title at www.jacplus.com.au.

2.9 Productivity of coral reef biomes

LEARNING INTENTION

By the end of this subtopic, you will be able to explain how people use and change coral reef biomes.

2.9.1 Formation of coral reefs

Coral reefs are found in spaces around tropical and subtropical shores. They require specific temperatures to develop and sea conditions that are clean, clear and free from sediment. The upper layer is alive, growing on many layers of dead coral. They are one of the oldest ecosystems on Earth and also very vulnerable to human activity.

Coral reefs are one of the most biodiverse environments on Earth and are built by polyps that live in groups. A reef is a layer of living tiny animals called **coral polyps** that build and grow on the remains of previous generations coral. There are many different types of reefs, such as inner and outer reefs as well as coral cays (small islands of coral) and coral atolls (see **FIGURES 1** and **2**).

coral polyp a tube-shaped marine animal that lives in a colony and produces a stony skeleton. Polyps are the living part of a coral reef.

FIGURE 1 Anatomy of a continental island and fringing reef

- Corals form in warm shallow salt water where the temperature is between 18 °C and 26 °C.
- Water must be clear, with abundant sunlight and gentle wave action to provide oxygen and distribute nutrients.

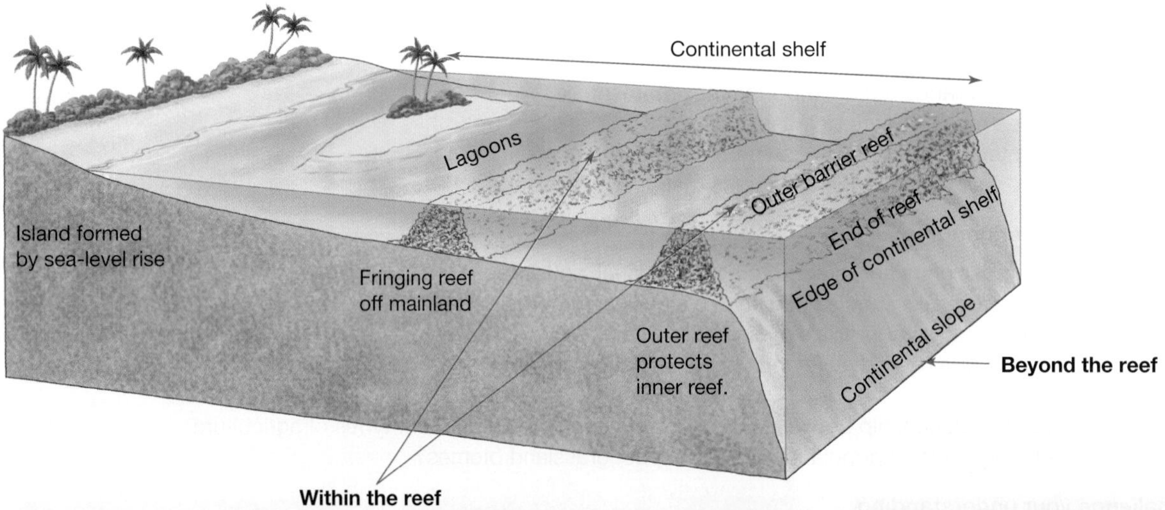

Coral polyps have soft, hollow bodies shaped like a sac with tentacles around the opening. They cover themselves in a limestone skeleton and divide and form new polyps.

Producers, such as algae, give coral its colour and provide a food source for marine life, such as fish. Coral reefs support at least one-third of all marine species. They are the marine equivalent of the tropical rainforest.

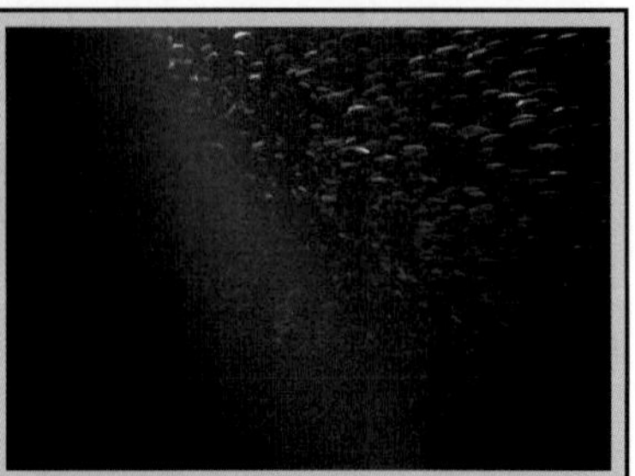

Beyond the continental shelf, the water is too deep and cold for coral. Sunlight cannot penetrate to allow coral growth.

FIGURE 2 The formation of fringing reefs and barrier reefs

(a) Fringing coral reefs develop along the shores of continents and islands.

(b) When sea levels rise, fringing reefs become barrier reefs.

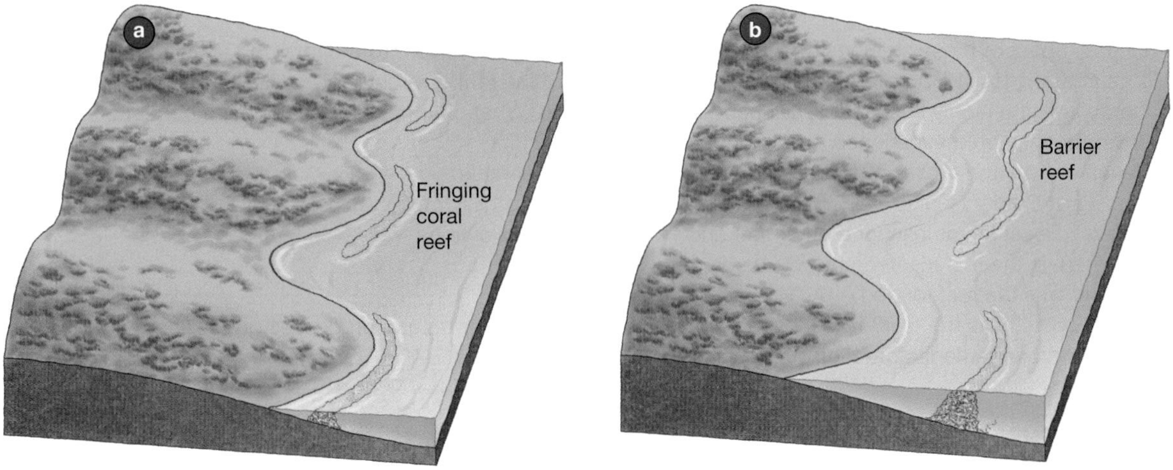

2.9.2 Benefits of coral reefs

Today, around 500 million people rely on reef systems, either for their livelihood, as a source of food, or as a means of protecting their homes along the coastline. Coral reefs help break up wave action, so waves have less energy when they reach the shoreline, thus reducing coastal erosion.

It is estimated that coral reefs contribute $375 billion to the global economy each year. Reefs are important to both the fishing and tourism industries. In 2014, approximately 1.88 million tourists visited Australia's Great Barrier Reef Marine Park alone. Nearly a third of all international tourists who visit Australia also visit the Great Barrier Reef.

Coral reefs have been found to contain compounds vital to the development of new medicines. For example:

- painkillers have been developed from the venom of cone shells
- some cancer treatments come from algae
- treatments for cardiovascular disease include compounds that were originally found in coral reefs.
- Corals form in warm shallow salt water where the temperature is between 18 °C and 26 °C.
- Water must be clear, with abundant sunlight and gentle wave action to provide oxygen and distribute nutrients.

2.9.3 Threats to coral reefs

Reefs also face a variety of threats.

- Urban development requires land clearing and wetland drainage, which increases erosion. Sediment washed into water prevents sunlight penetrating the water.
- Contamination by fossil fuels, chemical waste and agricultural fertilisers pollutes the sea.
- Tourism damages coral through boats dropping anchor or tourists directly removing coral or walking on it.
- Global warming increases water temperature, which bleaches the coral, turning it white and destroying the reef system.
- Predators, such as the crown of thorns starfish, prey on coral polyps, which affects the whole ecosystem.

on Resources

eWorkbook	Productivity of coral reef biomes (ewbk-8848)
Video eLesson	Productivity of coral reef biomes — Key concepts (eles-5178)
Interactivity	Productivity of coral reef biomes (int-8590)

2.9 ACTIVITY

1. Investigate two of the threats to coral reefs and prepare an annotated visual display that outlines:
   a. the nature of the threat
   b. the changes that will occur or have occurred as a result of this threat
   c. the impact of these changes on the environment, including references to the rate and scale of this change
   d. a strategy for the long-term sustainable management of the reef environment.
2. Coral reefs are highly susceptible to changes in the climate. What changes do you think the coral reef environment would experience if sea temperatures rise by 2 °C?
3. Research why the crown-of-thorns starfish is a threat to coral reefs.
4. The Great Barrier Reef is well known around the world. What other coral reefs can be found in Australian waters? Research at least one other reef and investigate the following.
   a. Where is the reef located and what area does it cover?
   b. How was the reef formed?
   c. How healthy is the reef? Has it been damaged by any human or natural processes or activities?
   d. What management strategies are in place to help protect the reef?

2.9 EXERCISE

Learning pathways

■ LEVEL 1	■ LEVEL 2	■ LEVEL 3
1, 2, 3, 9, 13	4, 5, 8, 10, 14	6, 7, 11, 12, 15

Check your understanding

1. What is a coral reef?
2. Outline how coral reefs form.
3. List the economic and medical benefits of coral reefs.
4. Describe the difference between a fringing reef and a barrier reef.
5. What are the major threats to coral reefs?
6. What is a coral polyp?
7. Describe two ways that humans have a negative impact on coral reefs.

Apply your understanding

8. Explain how coral reefs have a cultural significance to coastal communities.
9. Why are coral reefs considered to be of such economic importance?
10. Why are coral reefs particularly vulnerable to damage from humans?
11. Explain why coral is colourful.
12. Why don't coral reefs form in deep water?

Challenge your understanding

13. Suggest one strategy that might be used to assist in the protection of coral reefs at a:
    a. local level
    b. state level
    c. national level.
14. A volcano erupted and formed a new island off the coast of Iceland between 1963 and 1967. Could this island become a coral reef in the future? Provide reasons for your view.
15. Suggest how rising sea temperatures will affect coral reefs, and what might be done to protect them.

To answer questions online and to receive **immediate feedback** and **sample responses** for every question, go to your learnON title at www.jacplus.com.au.

2.10 Investigating topographic maps — Coastal wetland biome in Dalywoi Bay

LEARNING INTENTION

By the end of this subtopic, you will be able to explain how aspects of the wetland biome in Binydjarrŋa (Daliwuy Bay) function using examples from a topographic map.

2.10.1 Coastal wetlands: Binydjarrŋa (Dalywoi/Daliwuy Bay)

Wetlands are biomes where the ground is saturated, either permanently or seasonally. They are found on every continent except Antarctica. Wetlands include areas that are commonly referred to as marshes, swamps and bogs. In coastal areas they are often tidal and are flooded for part of the day. In the past they were often considered a 'waste of space', and in developed nations they were sometimes drained for agriculture or the spread of urban settlements.

Wetlands are a highly productive biome. They provide important habitats and breeding grounds for a variety of marine and freshwater species. In fact, a wide variety of aquatic species that we eat, such as fish, begin their life cycle in the sheltered waters of wetlands. They are also important nesting places for a large number of migratory birds.

Wetlands are also a natural filtering system and help purify water and filter out pollutants before they reach the coast. In addition, they help regulate river flow and stabilise the shoreline. **FIGURE 1** shows a cross-section through a mangrove wetland.

FIGURE 1 Cross-section of a mangrove wetland

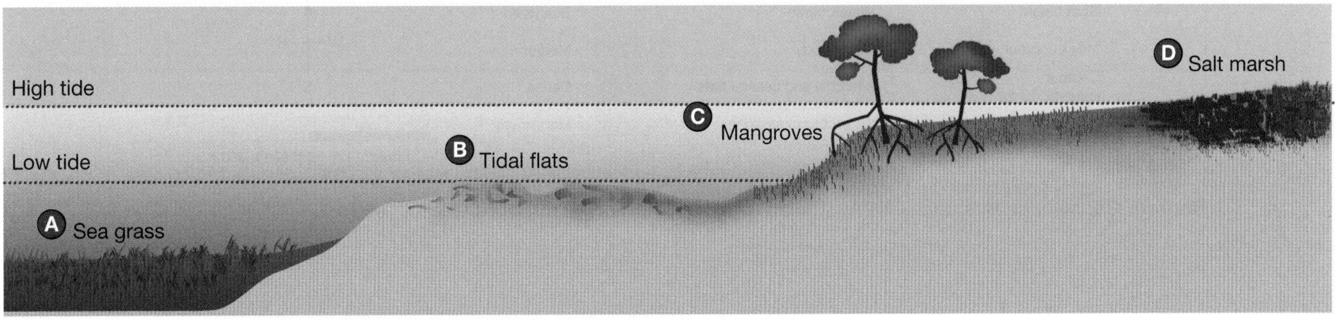

A Seagrass meadows:
- are covered by water all the time
- bind the mud and provide shelter for young fish
- produce **organic matter**, which is consumed by marine creatures (see **FIGURE 2**).

C Mangroves:
- have **pneumatophores** that trap sediment and pollutants from the land and sea
- change shallow water into swampland
- store water and release it slowly into the ecosystem
- have leaves that decompose and provide a food source for marine life
- provide shelter, breeding grounds and a nursery for marine creatures and birds.

B Tidal flats:
- are covered by tides most of the time
- are exposed for short periods of the day (low tide)
- are formed by silt and sand that has been deposited by tides and rivers
- provide a feeding area for birds and fish.

D Salt marshes:
- are covered by water several times per year
- provide decomposing plant matter — an additional food source for marine life
- have high concentrations of salt.

organic matter decomposing remains of plant or animal matter

pneumatophores exposed root systems of mangroves, which enable them to take in air when the tide is in

FIGURE 2 Topographic map extract, Binydjarrŋa (Daliwuy Bay), Northern Territory

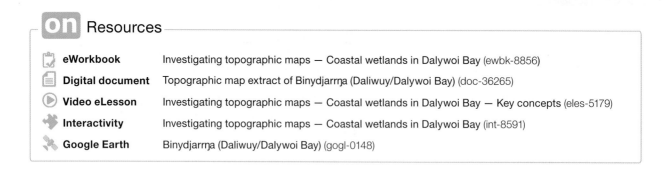

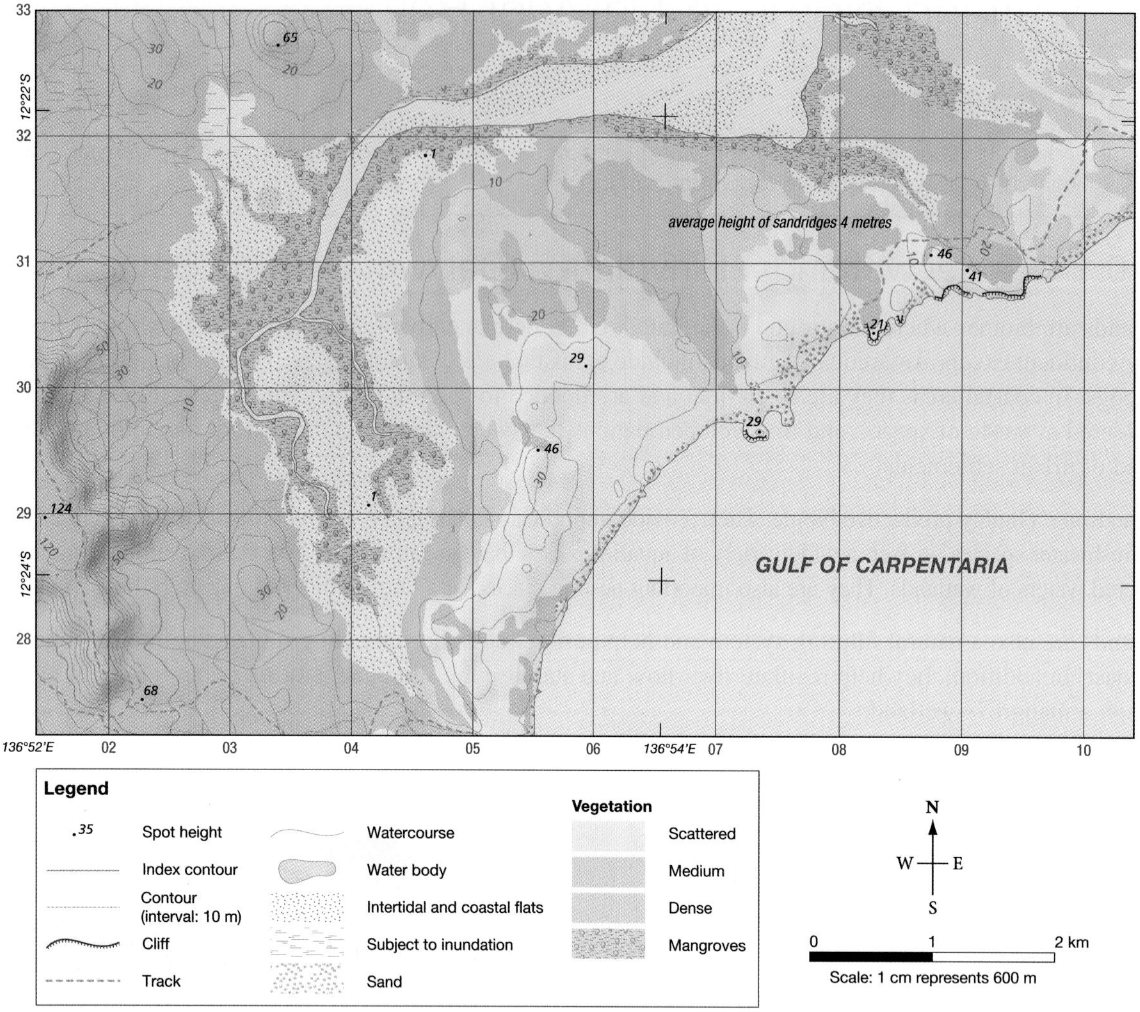

Source: Geoscience Australia, The Australian Army © Commonwealth of Australia (1999)

on Resources

eWorkbook	Investigating topographic maps — Coastal wetlands in Dalywoi Bay (ewbk-8856)
Digital document	Topographic map extract of Binydjarrŋa (Daliwuy/Dalywoi Bay) (doc-36265)
Video eLesson	Investigating topographic maps — Coastal wetlands in Dalywoi Bay — Key concepts (eles-5179)
Interactivity	Investigating topographic maps — Coastal wetlands in Dalywoi Bay (int-8591)
Google Earth	Binydjarrŋa (Daliwuy/Dalywoi Bay) (gogl-0148)

2.10 EXERCISE

Learning pathways

■ LEVEL 1	■ LEVEL 2	■ LEVEL 3
1, 2, 3, 9, 14	4, 5, 8, 10, 15	6, 7, 11, 12, 13

Check your understanding

1. What other names are wetlands known by?
2. Describe the natural functions of wetlands in the environment.
3. Identify the following features on the map.
   a. The highest point
   b. An area that is subject to inundation
   c. A track to the beach
4. Refer to **FIGURE 2** and describe the environment at the following locations.
   a. GR042309
   b. GR071329
   c. GR030320
   d. GR042285
5. What is the purpose of a pneumatophore?
6. Identify the location of the highest point shown on this map and calculate its distance from the nearest vehicle track or road.
7. How does the plant material rotting in salt marshes help marine life?

Apply your understanding

8. Refer to **FIGURE 2.**
   a. Locate the grid square bounded by the following grid references: GR030300, GR030310, GR040030, GR040310.
   b. Describe the natural environment in this area.
   c. Describe how this environment would change over the course of the day.
   d. Explain how this environment would be impacted if there was a cyclone in the area.
9. What are seagrass meadows and why are they important?
10. Why do salt marshes have high levels of salt?
11. Explain how wetlands help to purify water.
12. How do mangrove wetlands help to stop erosion of coastlines?

Challenge your understanding

13. A proposal has been put forward to construct a canal housing estate in the square bounded by the following grid references in **FIGURE 2**: GR030300, GR030310, GR040030, GR040310. Based on the features and topography in this area, is this proposal a good idea? (*Hint:* In addition to looking at the map in **FIGURE 2**, consider the photos of Binydjarrŋa (Daliwuy Bay) included in subtopic 2.12.3 **FIGURES 1** and **2**.)
14. Propose an organisation designed to protect wetland areas. Create a name, logo and description for your organisation that tell the public why wetlands are important and how you can help protect them.
15. If the mangroves were all cleared from a wetland, predict what impact this would have on the other parts of the biome.

To answer questions online and to receive **immediate feedback** and **sample responses** for every question, go to your learnON title at www.jacplus.com.au.

2.11 Thinking Big research project — Our world of biomes AVD

online only

The content in this subtopic is a summary of what you will find in the online resource.

Scenario

Biomes are not all the same. Across the Earth we recognise four distinct terrestrial biomes: forests, deserts, grasslands and tundra. In addition, there are also different aquatic biomes: freshwater and marine (saltwater). Within each of these biomes there are also distinct variations — the list seems to be endless! The Department of the Environment is keen to produce a display that explains these differences and the various influences on biome development.

Task

learnON

You have been commissioned by the Department of the Environment to carry out an in-depth study of biomes, their characteristics, the factors that influence their development and the variations that exist within them. You will create an engaging annotated visual display (AVD) to showcase your findings.

Go to your Resources tab to access the resources you need to complete this research project.

on Resources

ProjectsPLUS Thinking Big research project — Our world of biomes AVD (pro-1088)

2.12 Review

2.12.1 Key knowledge summary

2.2 Features of biomes

- Biomes are communities of plants and animals that extend over large areas due to similarities of climate within the area of the biome.
- There are five major terrestrial (land-based) biomes.
- Key biomes include forest, grassland, desert, tundra and aquatic, each with differing levels of biodiversity.

2.3 Major Australian biomes

- Australia has a variety of different biomes. In the north are tropical rainforests and savanna grasslands, and in the centre is a wide expanse of desert that is second in area only to the Sahara Desert in Africa.
- Differences in climate, latitude, altitude and proximity to the coast play a major role in differences between biomes.
- Australian biomes include wetlands and rivers, grasslands, seagrass meadows, forests and deserts.

2.5 The role of climate in biomes

- Climate is the strongest determining factor in the type of biome. It affects the location of the biome and the flora and fauna in the biome.
- Latitude, altitude and proximity to the coast influence climate, which in turn influences the different biomes.
- Latitude influences how much insolation is received by the land, with places closer to the equator receiving far greater amounts.
- Altitude influences temperatures: as elevation increases, temperatures fall.
- Cold ocean currents meeting warm land are more likely to lead to a desert.

2.6 The role of soil and vegetation in biomes

- Soils vary not only around the world but also within biomes.
- The characteristics of soil are determined by temperature, rainfall, and the rocks and minerals that make up the bedrock and the vegetation.
- Tropical rainforests have higher temperatures, which creates weathering. High rainfall leads to leaching, whereas tundra soils are shallow and poorly developed.
- Soil contains a variety of lifeforms including bacteria, earthworms and algae.
- Net Primary Production (NPP) refers to the energy or biomass produced by particular biomes and tends to be greatest in tropical rainforests.

2.8 Use of grassland biomes

- Grassland are landscapes dominated by grass, for example savanna, chaparral, cerrado, prairie, rangeland and steppe.
- Grasslands once occupied about 42 per cent of the Earth's land surface, but today they make up about 25 per cent of its land area. Grasslands are found on every continent except Antarctica.
- Grasslands are extremely susceptible to desertification.
- They are particularly important for the production of food grains, including corn, wheat, oats, barley, millet, rye and sorghum. They can also be used in the production of fibres, namely cotton, yarn, flax and hemp.

2.9 Productivity of coral reef biomes

- Coral reefs are one of the most biodiverse environments on Earth and are built by polyps that live in groups.
- Coral reefs have been found to contain compounds vital to the development of new medicines.
- The productivity of coral reefs can be hindered by a number of factors including predators, such as the crown-of-thorns starfish, that prey on coral polyps, which affects the whole ecosystem.

2.12.2 Key terms

biodiversity a variety of living organisms in an area
clearfelling the removal of all trees in an area
climate the long-term precipitation and temperature patterns of an area
coral polyp a tube-shaped marine animal that lives in a colony and produces a stony skeleton. Polyps are the living part of a coral reef.
deforestation clearing forests to make way for housing or agricultural development
desertification the conversion of an area to have desert-like qualities, usually caused by overgrazing, prolonged drought or climate change
diurnal temperature the variation in high and low temperature on a given day
industrial materials primary industry sources such as forestry (wood) or mining (iron ore) that can be used in the manufacturing of other goods
insolation the level of solar energy that reaches the Earth's surface
leaching the process in which water runs through soil, dissolving minerals and carrying them into the subsoil
leeward describes the area behind a mountain range, away from the moist prevailing winds
nocturnal active during the night
organic matter decomposing remains of plant or animal matter
permafrost permanently frozen ground
pneumatophores exposed root system of mangroves, which enable them to take in air when the tide is in
prevailing winds winds that blow from the direction that is typical at that time of year and place
urbanisation the process of economic and social change in which an increasing proportion of the population of a country or region lives in urban areas
windward describes the side of a mountain that faces the prevailing winds

2.12.3 Reflection

Complete the following to reflect on your learning.

Revisit the inquiry question posed in the Overview:

What are biomes, where are they, why are they different, and what do humans use them for?

1. Now that you have completed the chapter, create a mind map that addresses the inquiry question.
2. Write a paragraph in response to each part of the inquiry question.

FIGURE 1 Coastal mangroves at Binydjarrŋa (Daliwuy Bay), East Arnhem Land, Northern Territory

Subtopic	Success criteria			
2.2	I can identify key features of biomes.			
	I can locate examples of major biomes on a world map.			
2.3	I can outline Australia's major biomes.			
2.4	I can describe the spatial relationships between two or more thematic maps.			
	I can explain how the features shown on thematic maps influence each other's spatial distribution.			
2.5	I can explain the different factors affecting climate and biomes.			
2.6	I can describe the importance of soil in vegetation and biomes.			
2.7	I can construct and describe a transect on a topographic map.			
2.8	I can describe different uses of grassland biomes.			
2.9	I can explain how people use and change coral reef biomes.			
2.10	I can explain how the wetland biome in Binydjarrŋa (Dalywoi/ Daliwuy Bay) functions.			

on Resources

eWorkbook Chapter 2 Extended writing task (ewbk-8526)
Chapter 2 Reflection (ewbk-8525)
Chapter 2 Student learning matrix (ewbk-8524)

Interactivity Chapter 2 Crossword (int-8592)

FIGURE 2 The coastline at Binydjarrŋa (Daliwuy Bay), East Arnhem Land, Northern Territory

ONLINE RESOURCES

on Resources

Below is a full list of **rich resources** available online for this chapter. These resources are designed to bring ideas to life, to promote deep and lasting learning and to support the different learning needs of each individual.

eWorkbook

- **2.1** Chapter 2 eWorkbook (ewbk-8094) ☐
- **2.2** Features of biomes (ewbk-8820) ☐
- **2.3** Major Australian biomes (ewbk-8824) ☐
- **2.4** SkillBuilder — Describing spatial relationships in thematic maps (ewbk-8828) ☐
- **2.5** The role of climate in biomes (ewbk-8832) ☐
- **2.6** The role of soil and vegetation in biomes (ewbk-8836) ☐
- **2.7** SkillBuilder — Constructing and describing a transect on a topographic map (ewbk-8840) ☐
- **2.8** Use of grassland biomes (ewbk-8844) ☐
- **2.9** Productivity of coral reef biomes (ewbk-8848) ☐
- **2.10** Investigating topographic maps — Coastal wetlands in Dalywoi Bay (ewbk-8856) ☐
- **2.12** Chapter 2 Extended writing task (ewbk-8526) ☐
  Chapter 2 Reflection (ewbk-8525) ☐
  Chapter 2 Student learning matrix (ewbk-8524) ☐

Sample responses

- **2.1** Chapter 2 Sample responses (sar-0153) ☐

Digital document

- **2.10** Topographic map extract of Binydjarrŋa (Daliwuy/Dalywoi Bay) (doc-36265) ☐

Video eLessons

- **2.1** Bountiful biomes (eles-1717) ☐
  Biomes — Photo essay (eles-5172) ☐
- **2.2** Features of biomes — Key concepts (eles-5173) ☐
- **2.3** Major Australian biomes — Key concepts (eles-5174) ☐
- **2.4** SkillBuilder — Describing spatial relationships in thematic maps (eles-1726) ☐
- **2.5** The role of climate in biomes — Key concepts (eles-5175) ☐
- **2.6** The role of soil and vegetation in biomes — Key concepts (eles-5176) ☐
- **2.7** SkillBuilder — Constructing and describing a transect on a topographic map (eles-1727) ☐
- **2.8** Use of grassland biomes — Key concepts (eles-5177) ☐
- **2.9** Productivity of coral reef biomes — Key concepts (eles-5178) ☐
- **2.10** Investigating topographic maps — Coastal wetlands in Dalywoi Bay — Key concepts (eles-5179) ☐

Interactivities

- **2.2** Beautiful biomes (int-3317) ☐
  Major biomes of the world (int-8764) ☐
  Climatic zones of the world (int-8765) ☐
- **2.3** Major Australian biomes (int-8588) ☐
- **2.4** SkillBuilder — Describing spatial relationships in thematic maps (int-3344) ☐
- **2.5** The role of climate in biomes (int-8589) ☐
- **2.6** Why are biomes different? (int-3319) ☐
- **2.7** SkillBuilder — Constructing and describing a transect on a topographic map (int-3345) ☐
- **2.8** Grass, grains and grazing (int-3318) ☐
- **2.9** Productivity of coral reef biomes (int-8590) ☐
- **2.10** Investigating topographic maps — Coastal wetlands in Dalywoi Bay (int-8591) ☐
- **2.12** Chapter 2 Crossword (int-8592) ☐

Fieldwork

- **2.6** Soil testing (fdw-0027) ☐

Weblink

- **2.5** Bureau of Meteorology (BOM) (web-6418) ☐
- **2.8** Good Food (web-6417) ☐

myWorld Atlas

- **2.3** Australia's alpine biomes (mwa-7340) ☐
- **2.8** Wheat (mwa-7343) ☐

ProjectsPLUS

- **2.11** Thinking Big research project — Our world of biomes AVD (pro-1088) ☐

Google Earth

- **2.10** Google Earth Binydjarrŋa (Daliwuy/Dalywoi Bay) (gogl-0148) ☐

Teacher resources

There are many resources available for exclusively teachers online.

3 Biomes produce food

INQUIRY SEQUENCE

To access a pre-test with **immediate feedback** and **sample responses** to every question in this chapter, select your learnON format at www.jacplus.com.au.

3.1 Overview

Numerous **videos** and **interactivities** are embedded just where you need them, at the point of learning, in your learnON title at www.jacplus.com.au. They will help you to learn the content and concepts covered in this topic.

Everyone needs to eat. How does the world produce all the food it needs, and is there a better way?

3.1.1 Introduction

Food is a fundamental part of every person's life. For many people, what to eat each day can be a constant thought; for some, a constant worry. Food is an essential need for every human as it provides the essential nutrients for the body to sustain and maintain a healthy life. Biomes are key to producing the world's food.

STARTER QUESTIONS

1. Why do you think people around the world eat different kinds of food?
2. Do you think there is enough food in the world for everybody? Why? How do you know?
   a. Is there enough space for your family to grow fruit and vegetables in your home garden, if you had to?
   b. Does your family have a garden and, if so, does your family grow its own food?
   c. Conduct a class discussion to establish to what extent people in your class grow food at home.
   d. List the advantages and disadvantages of growing food at home.
3. Watch the **Biomes produce food — Photo essay** video in your online Resources. How much do you know about where your food comes from?

FIGURE 1 Fresh food market in Goa, India

on Resources

eWorkbook Chapter 3 eWorkbook (ewbk-8095)

Video eLessons A plate full of biomes (eles-1718)
Biomes produce food — Photo essay (eles-5180)

3.2 Feeding the world

LEARNING INTENTION

By the end of this subtopic, you will be able to describe the major food staples around the world and where they are located. You will also be able to explain why food demand is increasing.

3.2.1 The major food staples

Staple foods are those that are eaten regularly and in such quantities that they constitute a dominant portion of a diet. They form part of the normal, everyday meals of the people living in a particular place or country. The world has over 50 000 **edible** plants. Staple foods vary from place to place but are typically inexpensive or readily available. The staple food of an area is normally interconnected to the climate of that area and the type of land.

Most staple foods are cereals, such as wheat, barley, rye, oats, maize and rice, or root vegetables, such as potatoes, yams, taro and cassava. Maize, rice and wheat provide 60 per cent of the world's food energy intake; 4 billion people rely on them as their staple food.

Other staple foods include legumes, such as soya beans and sago; fruits, such as breadfruit and plantains (a type of banana); and fish. **FIGURE 1** shows the major food staples grown throughout the world.

staple an important food product or item that people eat or use regularly

edible fit to be eaten as food; eatable

int-7917

FIGURE 1 Staple foods around the world

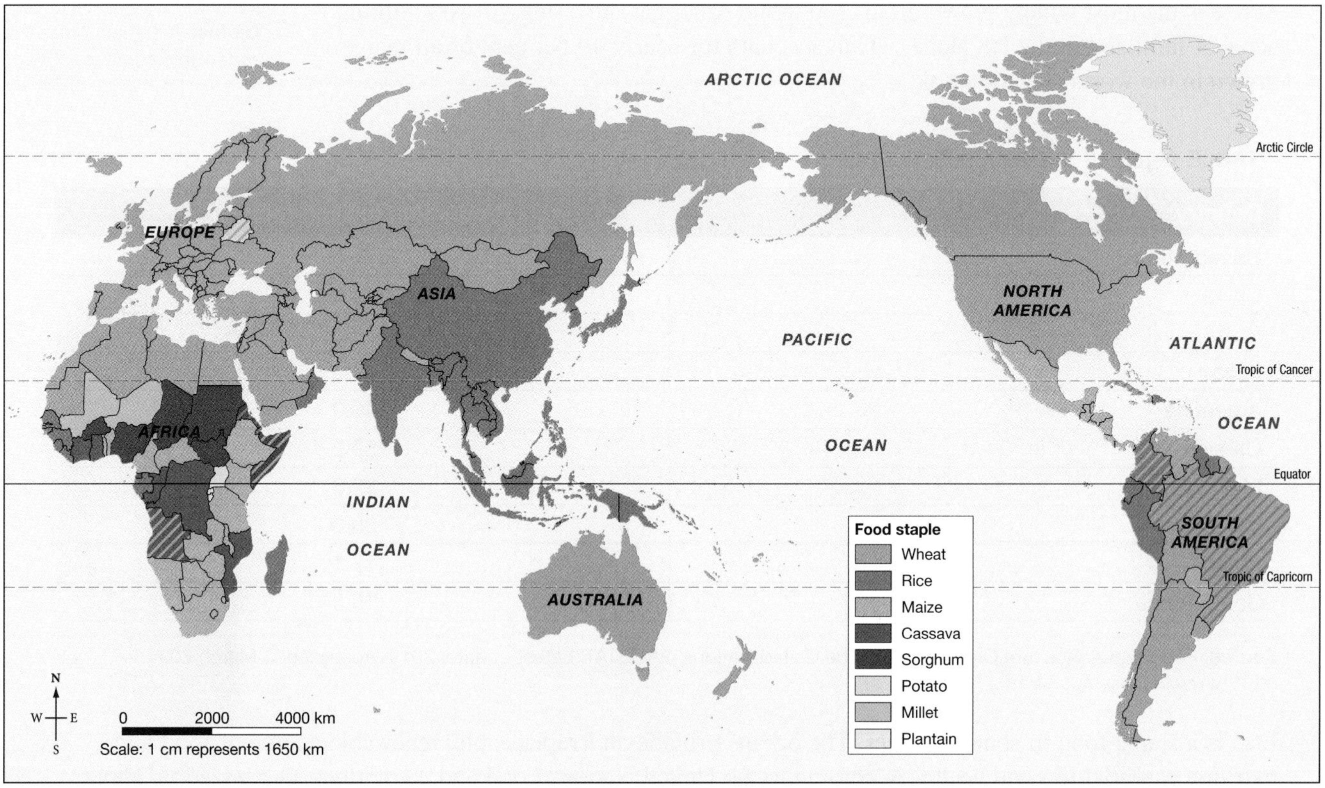

Source: Data from FAO, 2021

Wheat, maize and fish

Wheat is a cereal grain that is cultivated across the world. In 2019, world production of wheat was nearly 735 million tonnes, making it the second most produced cereal after maize (1.1 billion tonnes). (The third most produced cereal is rice, at 496 million tonnes.) World trade in wheat is greater than for all other crops combined.

Wheat was one of the first crops to be easily cultivated on a large scale, with the added advantage of yielding a harvest that could be stored for a long time. Wheat covers more land area than any other **commercial** crop (see **FIGURE 1**) and is the most important staple food for humans (**FIGURE 2**).

FIGURE 2 Wheat is used in a wide variety of foods such as breads, biscuits, cakes, breakfast cereals and pasta.

Maize, or corn (**FIGURE 3**), was commonly grown throughout the Americas. In the late fifteenth and early sixteenth centuries, explorers and traders carried maize back to Europe and introduced it to other countries. It then spread to the rest of the world, as it was a robust crop with the ability to grow in different environments. Sugar-rich varieties called sweet corn are usually grown for human consumption, while field corn varieties are used for animal feed and **biofuel**. Maize is the most widely grown grain crop in the Americas, covering 70–100 million acres of farmland in the US alone, which accounts for nearly 40 per cent of all maize grown in the world (see **TABLE 1**).

commercial an activity that is concerned with buying and/or selling of goods or services
biofuel fuel that comes from renewable sources

TABLE 1 Top 10 maize producers, 2019

Country	Production (million tonnes)
United States	377.5
China	224.9
Brazil	83.0
India	42.3
Argentina	40.0
Ukraine	39.2
Mexico	32.6
Indonesia	20.8
France	17.1
South Africa	15.5

Source: Food and Agriculture Organization of the United Nations. FAOSTAT. Latest update: 2019. Accessed: 3 March 2021. http://www.fao.org/faostat/en

Fish is a staple food in some societies. The oceans provide an irreplaceable, renewable source of food and nutrition essential to good health. According to the United Nations Food and Agriculture Organization, about 75 per cent of fish caught is used for human consumption. The remainder is converted into fishmeal and oil, used mainly for animal feed and farmed fish (**FIGURE 4**).

FIGURE 3 Maize, or corn

FIGURE 4 A fish haul in Bali, Indonesia

In general, people in developing countries, especially those in coastal areas, are much more dependent on fish as a staple food than those in the developed world. About one billion people rely on fish as their primary source of animal protein.

3.2.2 Food production

In the late eighteenth century, British economist and philosopher Thomas Malthus proposed a theory that the rate of food production would be unable to keep up with human population growth, with catastrophic consequences for humanity. However, technological advances and the **Green Revolution** allowed humans to feed the growing population, producing more food on a larger scale through **monoculture**. With the Earth's population projected to rise to nine billion people by 2050, what needs to be done to sustainably ensure there is enough food for everyone?

Green Revolution a significant increase in agricultural productivity resulting from the introduction of high-yield varieties of grains, the use of pesticides and improved management practices

monoculture the cultivation of a single crop on a farm or in a region or country

int-8599

FIGURE 5 World distribution of cropland, pasture and maize. More maize, for example, could be grown if improvements were made to seeds, irrigation, fertiliser and markets.

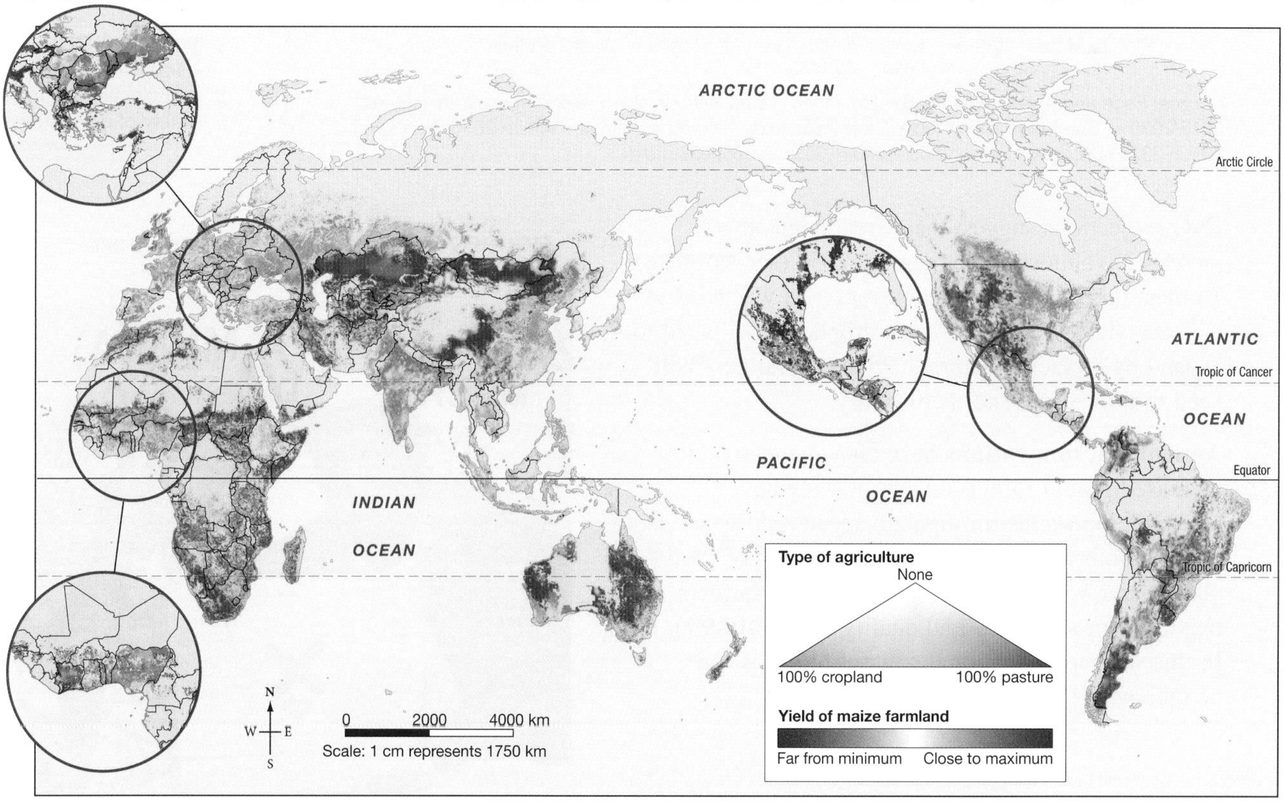

Source: Spatial Vision, 2021

3.2.3 Spatial issues with our food

FIGURE 5 shows the space that is used for agriculture around the world. With continued population growth, the availability of **arable** land is a concern. For example, there is currently about one-sixth of a hectare of arable land **per capita** in East and South Asia. The population of these regions is expected to experience rapid growth, but very little additional land is available for agricultural expansion. Consequently, arable land per capita will continue to decline in these areas, leading to a potential food crisis.

3.2.4 Food production increases

Agricultural **yields** vary widely around the world depending on climate, management practices and the types of crops grown. Globally, 15 million square kilometres of land are used for growing crops — altogether, that's about the size of South America. Approximately 32 million square kilometres of land around the world are used for pasture — an area about the size of Africa. Across the Earth, most land that is suitable for agriculture is already used for that purpose, and in the past 60 years we have increased our food production.

FIGURE 6 Crop yields (average) in developing countries, 1961 to 2030

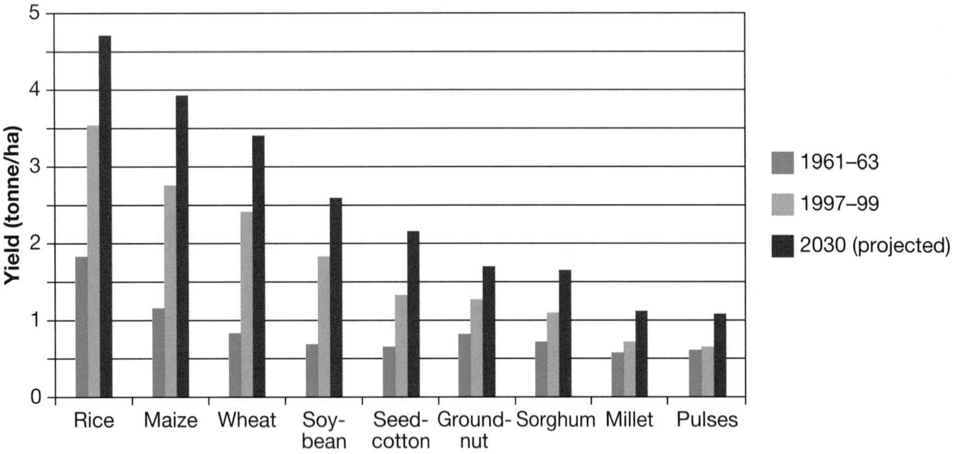

Source: Food and Agriculture Organization of the United Nations, 2015, *World agriculture: towards 2015/2030 — Summary report,* Table: Crop yields in developing countries, 1961 to 2030. Accessed: 3 March 2021.http://www.fao.org/docrep/004/y3557e/y3557e08.htm#l

arable describes land that can be used for growing crops
per capita per person
yield amount of agriculture produced or provided
FAO Food and Agricultural Organization of the United Nations
innovation new and original improvement to something, such as a piece of technology or a variety of plant or seed

FAO projections suggest that cereal demand will increase by almost 50 per cent by 2050 (see **FIGURE 6**). To meet this demand, either current land will need to increase yields per unit of area, croplands will need to expand by replacing natural habitats, or farmers will need to grow crops more efficiently.

Agricultural **innovations** have also changed and increased global food production. They have boosted crop yields through advanced seed genetics, agronomic practices (scientific production of food plants) and product innovations that help farmers maximise productivity and quality (see **FIGURE 7**). In this way, the nutritional content of crops can be increased.

FIGURE 7 Farmers in Kenya research plant diseases at a plant health clinic.

3.2.5 Increasing our food production

World food production has grown substantially over the past century. Increased fertiliser application and more water usage through irrigation have been responsible for over 70 per cent of crop yield increases. The Second Agricultural Revolution in developed countries after World War II and the Green Revolution in developing countries in the mid 1960s transformed agricultural practices and raised crop yields dramatically.

Since the 1960s agriculture has been more productive, with world per capita agricultural production increasing by 25 per cent in response to a doubling of the world population.

It is possible to get even more food out of the land we are already using. For example, **FIGURE 5** shows the places where maize yields could increase and become more **sustainable** by improving nutrient and water management, seed types and markets.

Environmental factors

In the past, growth in food production resulted mainly from increased crop yields per unit of land and to a lesser extent from expansion of cropland. From the early 1960s until 2014, total world cropland increased by only around 10 per cent, but total agricultural production grew by 60 per cent. Increases in yields of crops such as sweet potatoes and cereals were brought about by a combination of:

- increased agricultural inputs
- more intensive use of land
- the spread of improved crop varieties.

In some places, such as parts of Africa and South-East Asia, increases in fisheries (areas where boats are used to catch fish) and expansion of cropland areas were the main reasons for the increase in food supply. In addition, cattle herds became larger. In many regions — such as in the savanna grasslands of Africa, the Andes, and the mountains of Central Asia — livestock is a primary factor in food security today. Fertilisers have increased agricultural outputs and enabled more intensive use of the land. The global fertiliser use of 208 million tonnes in 2020 represents a 30 per cent increase since 2008.

sustainable describes the use by people of the Earth's environmental resources at a rate such that the capacity for renewal is ensured

TABLE 1 Fertiliser use, 1959–60, 1989–90 and 2020

Region/nutrient	Fertiliser use			Annual growth	
	1959–60	1989–90	2020	1960–90	1990–2020
	(million nutrient tonnes)			(per cent)	
Developed countries	24.7	81.3	86.4	4.0	0.2
Developing countries	2.7	62.3	121.6	10.5	2.2
East Asia	1.2	31.4	55.7	10.9	1.9
South Asia	0.4	14.8	33.8	12.0	2.8
West Asia/North Africa	0.3	6.7	11.7	10.4	1.9
Latin America	0.7	8.2	16.2	8.2	2.3
Sub-Saharan Africa	0.1	1.2	4.2	8.3	3.3
World total	27.4	143.6	208.0	5.5	1.2
Nitrogen	9.5	79.2	115.3	7.1	1.3
Phosphate	9.7	37.5	56.0	4.5	1.3
Potash	8.1	26.9	36.7	4.0	1.0

Source: Bumb, B. and C. Baanante. 1996. *World Trends in Fertilizer Use and Projections to 2020.* 2020 Brief 38, Table 1. Washington, DC: International Food Policy Research Institute. Reproduced with permission from the International Food Policy Research Institute http://www.ifpri.org/. The original brief in which this table appears is available online at http://www.ifpri.org/publication/world-trends-fertilizer-use-and-projections-2020.

Trade factors and economic factors

From the 1960s onwards, there has been significant growth of world trade in food and agriculture. Food and fertiliser imports by developing countries have grown, reducing the threat of famine in those countries.

TABLE 2 Percentage share of crop production increases, 1961–2030*

	Arable land expansion (1)		Increases in cropping intensity (2)		Harvested land expansion (1 + 2)		Yield increases	
	1961–99	1997/99–2030*	1961–99	1997/99–2030*	1961–99	1997/99–2030*	1961–99	1997/99–2030*
All developing countries	23	21	6	12	29	33	71	67
South Asia	6	6	14	13	20	19	80	81
East Asia	26	5	–5	14	21	19	79	81
East and North Africa	14	13	14	19	28	32	72	68
Latin America and the Caribbean	46	33	–1	21	45	54	55	46
Sub-Saharan Africa	35	27	31	12	66	39	34	61
World	15		7		22		78	

* projected

Source: Food and Agriculture Organization of the United Nations. *World agriculture: towards 2015/2030 — Summary report.* Rome, 2002. Table A7 and A8. Accessed: 3 March 2020. http://www.fao.org/3/y3557e/y3557e.pdf

3.2.6 The impact of the Green Revolution

The Green Revolution was a result of the development and planting of new **hybrids** of rice and wheat, which led to greatly increased yields. There have been a number of green revolutions since the 1950s, including those in:

- the United States, Europe and Australia in the 1950s and 1960s
- New Zealand, Mexico and many Asian countries in the late 1960s, 1970s and 1980s.

FIGURE 8 Applying fertiliser to crops in the Punjab, India

The Green Revolution saw a rapid increase in the output of cereal crops — the main source of calories in developing countries. Farmers in Asia and Latin America widely adopted high-yielding varieties. Governments, especially those in Asia, introduced policies that supported agricultural development. In the 2000s, cereal harvests in developing countries were triple those of 40 years earlier, while the population was only a little over twice as large. Planting of high-yield crop varieties coincided with expanded irrigation areas and fertiliser use, leading to significant increases in cereal output and calorie availability.

hybrid a plant or animal bred from two or more different species, breeds or varieties, usually to attain the best features from each type

fdw-0028

FOCUS ON FIELDWORK

Annotating a field sketch of your biome

How well do you know the characteristics and features of your local biome? Careful observation of the landscapes around us can be difficult when we see them everyday, but stopping to create an annotated field sketch can help to train our focus, so that we see things that we don't normally notice. Seeing new aspects of our everyday environment can also help us to consider how the features and characteristics formed, change and interconnect.

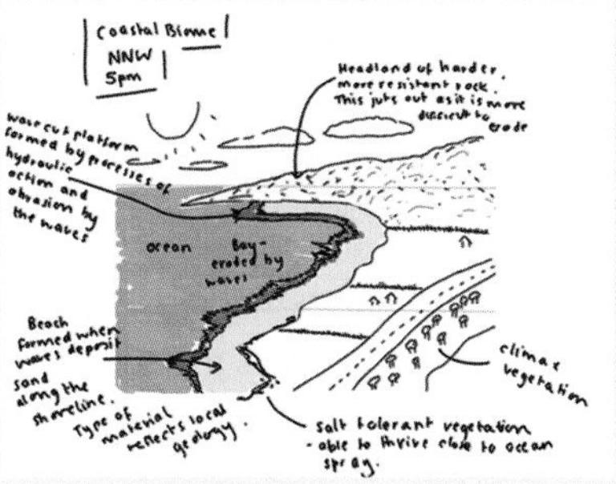

Learn more about field sketches using the **Annotating a field sketch of your biome** fieldwork activity in your online Resources.

on Resources

eWorkbook	Feeding the world (ewbk-8860)
Video eLesson	Feeding the world — Key concepts (eles-5181)
Interactivities	Staple foods around the world (int-7917) Distribution of cropland, pasture and maize (int-8599)
Weblink	United Nations Food and Agriculture Organization (web-6327) Feed the World (web-3303)
myWorldAtlas	Deepen your understanding of this topic with related case studies and questions. • Wheat

3.2 ACTIVITIES

1. Use the **United Nations Food and Agriculture Organization** (UN FAO) weblink in your online Resources to find out what is being done to promote sustainable aquatic biomes.
2. **FIGURE 5** shows where more crops could be grown. Investigate how Mexico or a country in West Africa or Eastern Europe could improve the sustainability of its agriculture. Create a mind map or flow chart diagram to represent your findings.
3. Research the background of the Green Revolution — why it occurred, the key places involved and the changes that resulted. Create a dot-point summary of your findings.
4. Some scientists are suggesting that there will be a new Green Revolution. Investigate current thinking and predict the potential scale of this possible agricultural change.
5. Use the **Feed the World** weblink in your online Resources to watch the interactive maps. Describe how the challenge of meeting the needs of a growing and increasingly affluent population can be met.
6. Use the **Focus on Fieldwork** link in your online Resources to undertake fieldwork on the soil around your school. Use SkillBuilder 3.3 to help with the ternary graph. (If you have completed this fieldwork before, compare your results. What might account for any similarities or differences in your data?)

FIGURE 9 How will agriculture need to change to meet global human needs?

3.2 EXERCISE

Learning pathways

■ LEVEL 1	■ LEVEL 2	■ LEVEL 3
2, 3, 6, 8, 9, 12	4, 5, 10, 11, 14, 16	1, 7, 13, 15, 17, 18

Check your understanding

1. Define what makes a food a staple.
2. List the main staple foods of the world and the places (continents) where they are grown.
3. What is biofuel?
4. Outline the interconnection between population and food production.
5. Outline where there are concerns for food production in East and South Asia.
6. Approximately how many people around the world rely on fish as their main source of animal protein?
7. Refer to **TABLE 1**. Describe the trends in the use of fertilisers from 1960 to 2020.
8. Match the following terms to their definitions.

Monoculture	A plant or animal bred from two or more different species, breeds or varieties, usually to attain the best features from each type
Yield	Land that is good for growing crops
Hybrid	An amount of agricultural produce provided or grown
Arable	Growing only one type of crop in an area

Apply your understanding

9. Explain why plants, rather than animals, dominate as the major staple foods of the world.
10. Australia is an exporter of wheat. Explain why Australia is able to produce such a surplus.
11. Explain how the increasing demand for cereals can be met.
12. Explain the impact on the environment if agricultural lands were to increase.
13. With reference to a specific place, suggest how increasing population densities might influence future crop production.
14. Although fish may be a staple food for many people, why is it not possible for fish to be a staple food for everyone?
15. Choose three strategies for improving the sustainability of crops over time. Evaluate and discuss the sustainability of each strategy.

Challenge your understanding

16. With the increase in world population and greater pressure on fish stocks, what could be done to sustain fish stocks in oceans and lakes?
17. Maize is currently used as feed for animals, as biofuel and as food for humans. Why might this be an unsustainable environmental practice in future?
18. Should countries that are more economically developed be supporting those who struggle to produce their own food?

FIGURE 10 Irrigation system for cornfields, León Province, Spain

To answer questions online and to receive **immediate feedback** and **sample responses** for every question, go to your learnON title at www.jacplus.com.au.

3.3 SkillBuilder — Constructing ternary graphs

online only

LEARNING INTENTION

By the end of this subtopic, you will be able to construct a ternary graph.

The content in this subtopic is a summary of what you will find in the online resource.

3.3.1 Tell me

What are ternary graphs?

Ternary graphs are triangular graphs that show the relationship or interconnection between three features (**FIGURE 1**). Most graphs you have seen show the relationship between two features, not three.

3.3.2 Show me

How to construct ternary graphs

Step 1

Create an equilateral triangle by drawing a 10-cm horizontal line, and draw 10 marks that are 1 cm apart. Label these 0 to 100, starting with 0 on the left-hand side. Angle these marks to the left at 60° (**FIGURE 2**).

Step 2

At the 50 per cent point, draw a faint vertical pencil line of about 9 cm, which will help you to draw the other two axes.

Step 3

Now from the 0 per cent point, draw a diagonal line that is 10 cm long and intersects with the vertical line. Draw 10 marks that are 1 cm apart along this diagonal axis. Mark 100 per cent at the bottom of the line and 0 at the top.

FIGURE 1 Economic activity in selected countries

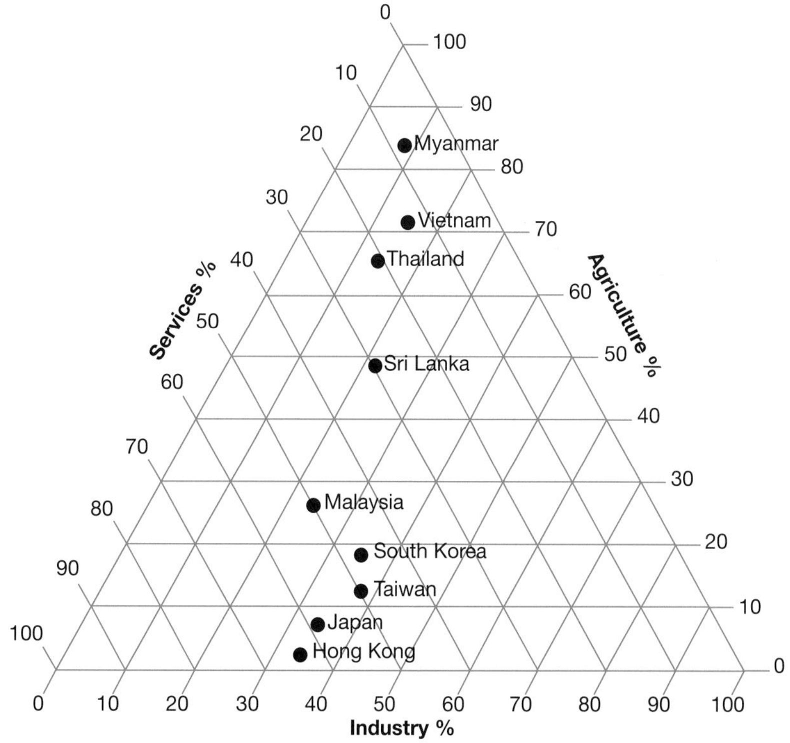

FIGURE 2 Horizontal line with 10 markings 1 cm apart

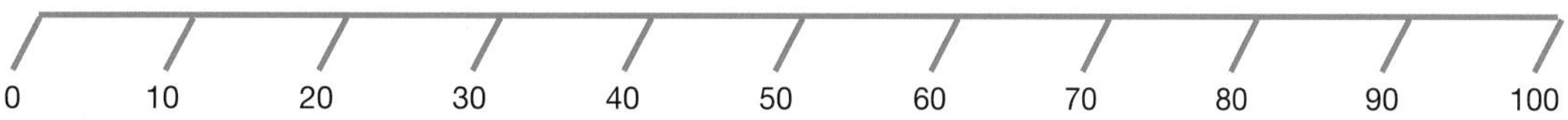

Step 4

Repeat this step on the other side of the vertical line to complete the triangle, but reverse the markings, so 0 is at the bottom of the line and 100 is at the top (**FIGURE 6**).

Step 5

Erase the vertical line that you drew to centre your graph.

Step 6

Use a ruler and carefully join points across the triangle that add to 100 per cent.

Step 7

Label the axes with the three features that you are going to plot.

Step 8

To plot data, you need to find the point where the percentages for the three features intersect. Follow the diagonal lines sloping down from left to right (\) from the left-hand axis, the diagonal lines sloping up from left to right (/) from the bottom axis, and the horizontal lines from the right-hand axis. Find the spot represented by the three sets of data and draw a small dot (**FIGURE 7**). Label each point and complete the graph with a title and source.

FIGURE 6 A ternary graph with its grid completed and axes labelled

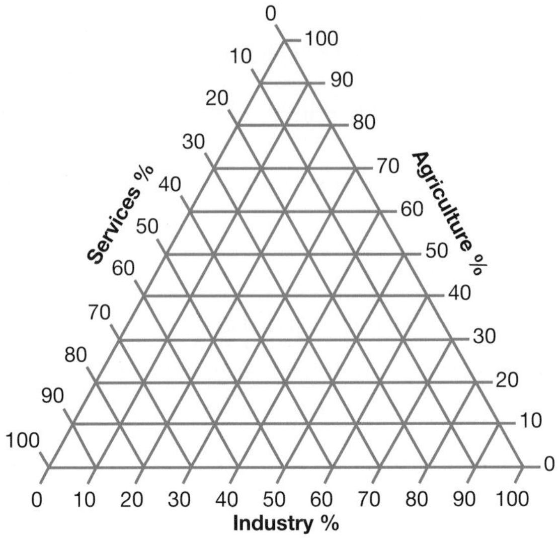

FIGURE 7 Reading the grid

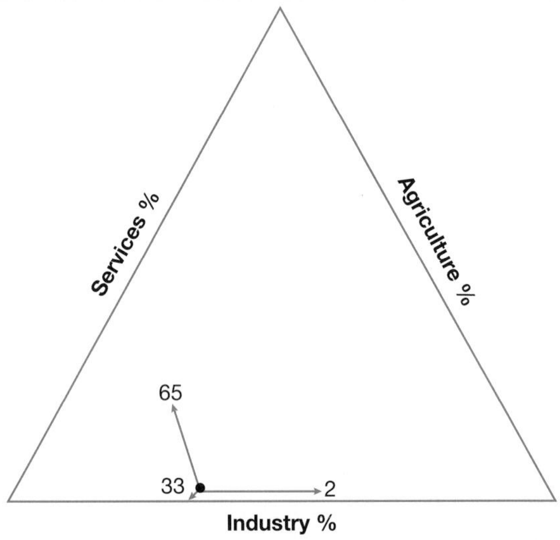

3.3.3 Let me do it

learnON

Go to learnON to access the following additional resources to help you build this skill:

- a longer explanation of this skill and its application in Geography (Tell me)
- a video demonstrating the step-by-step process of this skill (Show me)
- an activity and interactivity for you to practise the skill (Let me do it)
- self-marking questions to help you understand and use the skill.

on Resources

eWorkbook	SkillBuilder — Constructing ternary graphs (ewbk-8864)
Video eLesson	SkillBuilder — Constructing ternary graphs (eles-1728)
Interactivity	SkillBuilder — Constructing ternary graphs (int-3346)

3.4 Traditional agriculture

LEARNING INTENTION

By the end of this subtopic, you will be able to explain the traditional ways that different groups of people produce food.

3.4.1 Types of agriculture

In the more developed countries of the world, large-scale agriculture provides the food for the supermarkets to feed their populations. However, for many people in developing nations, the food produced on their large farms is exported to wealthy nations. To feed themselves they rely on **subsistence** agriculture and local markets to buy and/or exchange food. As seen in **FIGURE 1**, subsistence-farming practices predominantly occur in and around the tropics.

subsistence describes farming that provides food only for the needs of the farmer's family, leaving little or none to sell

FIGURE 1 World agricultural practices and food production

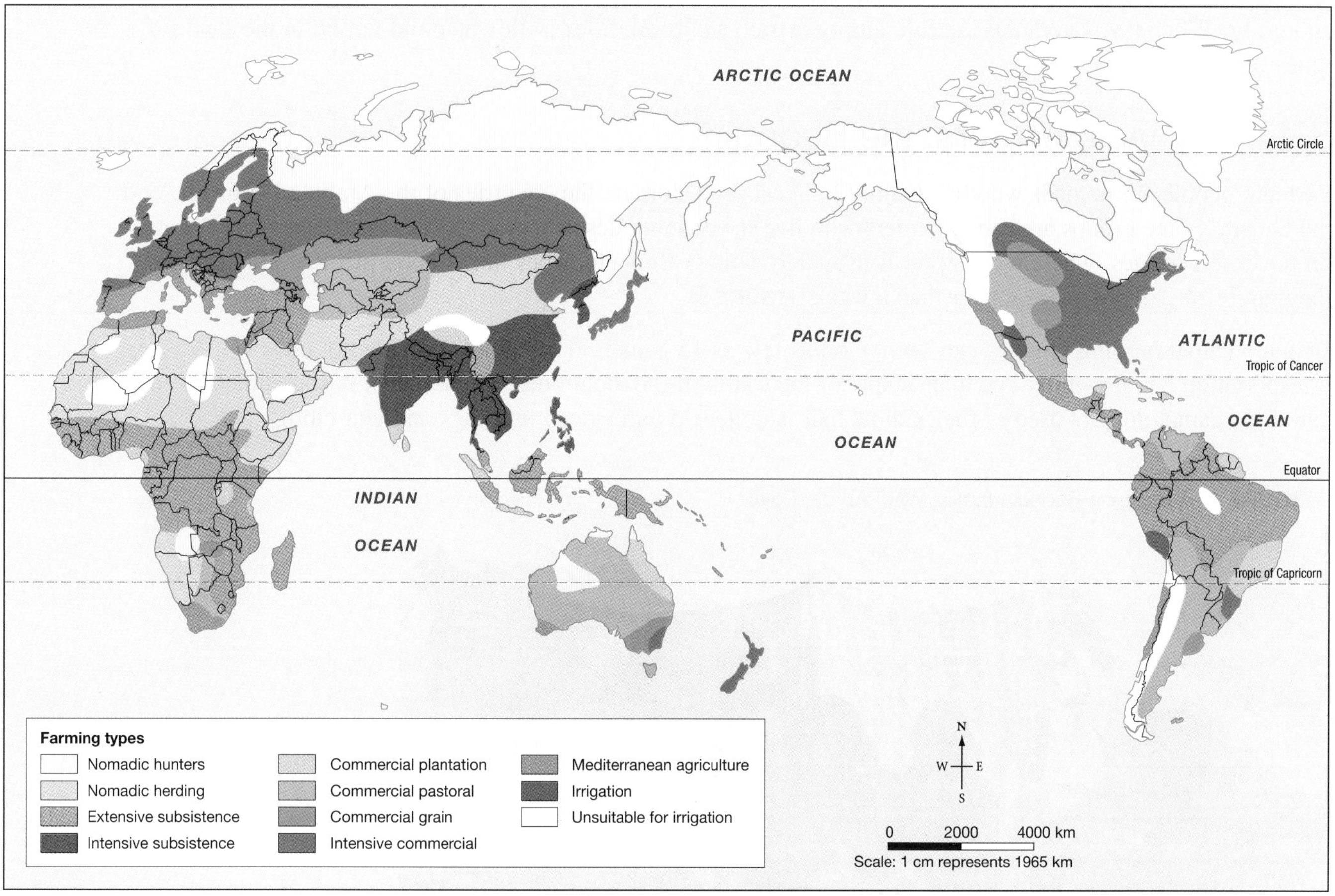

Source: FAO, 2021

3.4.2 Hunters and gatherers: the San

Today, around 55 500 San (or Kalahari Bushmen) live in the Kalahari Desert in southern Africa. Less than five per cent still live in the traditional way (**FIGURE 2**).

FIGURE 2 Naro San settlement, Ghanzi region, Botswana

Traditionally **nomadic** San people travel in small family groups, roaming over regions of up to 1000 square kilometres. They have no pack animals and carry few possessions — only spears, bows and arrows, bowls and water bags. The San's clothes are made from animal skins. When needed, they construct stick shelters thatched with grass.

The San are experts at finding water and tracking animals. The men hunt antelope and wildebeest, while the women hunt small game such as lizards, frogs and tortoises, and gather roots, berries and grubs. When the waterholes are full, empty ostrich shells are filled with water and buried in the sand for times of drought.

3.4.3 Nomadic herders: the Bedouin

Bedouin people are nomads who live mainly in Syria, Iraq, Jordan, the countries of the Arabian Peninsula, and the Sahara. Some groups are camel herders who live in the inner desert regions. Others herd sheep and goats on the desert fringes, where more water is available. Unless Bedouin tribes find a good piece of grazing land, they rarely stay in one place longer than a week (**FIGURE 3**).

Bedouin camel-herding families can survive on as few as 15 camels. The camels provide not only transportation but also milk — a main staple of the traditional Bedouin diet. Camel meat is sometimes eaten, and dried camel dung is used as fuel. Camel hair is collected and woven into rugs and tent cloth.

FIGURE 3 A Muzeina Bedouin family, Wadi Arada, Egypt

nomadic describes a group of people who have no fixed home and move from place to place according to the seasons, in search of food, water and grazing land

3.4.4 Shifting agriculture: the Huli

The Huli people live in the rainforests of the Papua New Guinea highlands (**FIGURE 4**). Many still lead a traditional way of life. The land on which they live has steep hillsides and dense rainforest.

The Huli people use a farming system known as **shifting agriculture**. The Huli clear a patch of rainforest and plant crops of sweet potato, sugar cane, corn, taro and green vegetables. When the soil of the garden no longer produces good crops, a new patch of rainforest is cleared, leaving the old one to recover naturally. The Huli's individual huts are built next to the gardens, and it is the women's responsibility to tend them. The garden crops are supplemented by food that the men have hunted (**FIGURE 5**). Wild and domesticated pigs are a common source of meat.

FIGURE 4 Map showing Huli land, Papua New Guinea

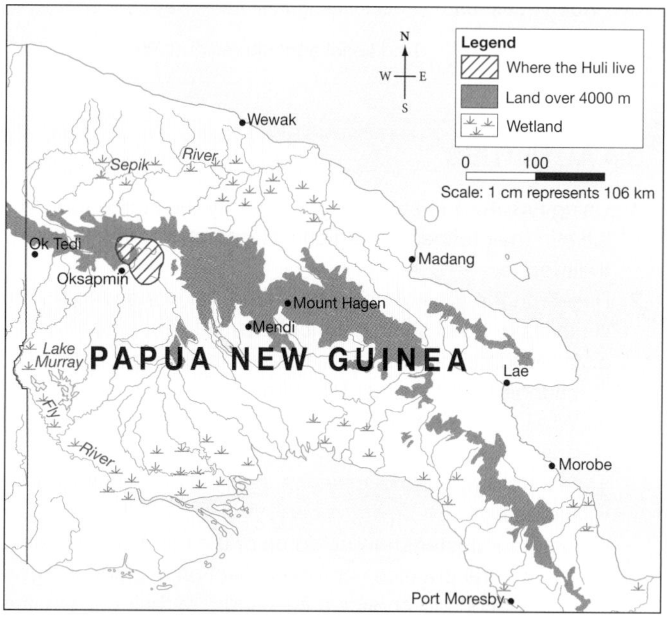

Source: MAPgraphics Pty Ltd Brisbane

FIGURE 5 Huli tribesmen from the Tari Valley, Southern Highlands of Papua New Guinea

shifting agriculture system in which small parcels of land are used to produce food for a period and abandoned when they become less productive so they can recover naturally, while the farmers move to another plot of land

Resources

eWorkbook	Traditional agriculture (ewbk-8868)
Video eLesson	Traditional agriculture — Key concepts (eles-5182)
Interactivity	Traditional agriculture (int-8593)

3.4 ACTIVITIES

1. Using **FIGURE 1** in subtopic 3.2, identify the staple foods that the San and Huli have in their respective regions. Research if farms in these regions are exporting their foods.
2. Research the lives of peoples around the world who traditionally lived nomadic lifestyles or used shifting agriculture.
   a. Investigate the lives of one group, including if and how their ways of life have changed over the last 50 years.
   b. What factors have affected how they grow and/or collect food? Consider possible positive and negative changes such as easier access to food stores, technology, urban expansion, political issues, war, famine, or changes in biomes.
   c. Create a poster showing some of the most significant changes to their agricultural practices or food collection. Include images and short explanations to provide a full picture of their current lifestyle.

FIGURE 6 Yurt phone call, Bayan-Ölgii, Mongolia

3.4 EXERCISE

Learning pathways

■ LEVEL 1	■ LEVEL 2	■ LEVEL 3
2, 3, 6, 11, 14	1, 5, 8, 9, 12	4, 7, 10, 13

Check your understanding

1. What is subsistence agriculture?
2. Refer to **FIGURE 1**. Identify three farming types found in Australia.
3. Name and describe the three types of traditional agriculture.
4. Describe the shifting agriculture farming system used by the Huli.
5. Describe what the term 'nomadic' means.
6. List at least five ways that camels are useful to the Bedouin people.
7. How might people who practise shifting agricultural techniques obtain meat?

Apply your understanding

8. Explain why many people in countries that are less economically developed use subsistence agriculture.
9. In what ways are traditional agricultural practices more sustainable than modern practices?
10. Refer to **FIGURE 1**. Explain why commercial food production is concentrated in the places and spaces bordering the tropical zones.
11. How might the lives of nomadic herders change as an area becomes more populated?

Challenge your understanding

12. How might modern technology affect the Huli people in the next 25 years with respect to their traditions and food production practices?
13. Predict what changes may occur to the way of life of nomadic herders in the future. Consider social/cultural, economic and environmental changes.
14. Why do you think that fewer than five per cent of the San people still live a traditional lifestyle?

To answer questions online and to receive **immediate feedback** and **sample responses** for every question, go to your learnON title at www.jacplus.com.au.

3.5 Food security over time in Australia

LEARNING INTENTION

By the end of this subtopic, you will be able to describe some of the land management strategies and food sources used by Aboriginal Peoples and Torres Strait Islander Peoples, and identify and explain the reasons for some of the challenges to food security that exist in remote Australian communities today.

3.5.1 Land stewardship and land ownership

The sustainable land and resource management practices of Aboriginal Peoples and Torres Strait Islander Peoples, carried out over many thousands of years, ensured food security for the people and respect for the lands, waterways, lakes and marine environments that sustained them. At the time of European occupation in 1788, Aboriginal Peoples' food supplies were sourced and maintained through their deep knowledge of and close association with the land. These knowledges allowed for sustainable management of the ecosystems and biomes in which they lived. The 'world view' that describes this sustainable lifestyle is called an 'earth-centred' approach. This means people's interaction with the environment is one of caring stewardship rather than ownership.

3.5.2 Sourcing food before 1788

Aboriginal Peoples sourced their foods from a wide range of plants and wild animals. Food sources were greatly influenced by both the season and geographic location. Cereals, fruits and vegetables were collected and cultivated; game and fish were hunted and systematically trapped; and the land was managed with a variety of techniques including cultural burning practices that controlled and encouraged germination and healthy growth. The new growth after a fire also attracted animals such as kangaroos to the area to feed. Fire was also used to control the movement of animals in dense bushland to enable more efficient hunting.

To ensure food security, communities developed a range of sustainable techniques for cultivating and harvesting food. For example, some plants or their seeds were left behind after harvesting to allow for new growth, and a few eggs were always left in nests to hatch. This ensured that species would survive and communities would have ongoing access to food sources in the future. **FIGURE 1** provides examples of typical food types from both tropical and temperate regions of Australia, including arid and desert regions.

FIGURE 1 A selection of different foods and water resources

Cereal foods: Grass seeds from the clover fern were ground to form flour for damper. Many other seed types were similarly treated.

Fruit and vegetables: Fruits, berries, orchids and pods were available, depending on the region and seasonal availability (for example sow thistle, lilly pilly, pigface fruit, kangaroo apple, wild raspberry, quandong and native cherry) as well as wild figs, plums, grapes and gooseberries. Also eaten were plant roots such as bull rushes, yams and bulbs; the heart of the tree fern and the pith of the grass tree; and the blister gum from wattles, native truffles and mushrooms.

Eggs: Emu, duck, pelican and many other birds' eggs were eaten.

Meat: Meats included insects such as the larval stage of the cossid moth or witchetty grub and the Bogong moth, honey ants, native bees and their honey, and scale insects; animals such as kangaroos, emus, eels, crocodiles, sea turtles, snakes, goannas and other lizards; and birds such as ducks, gulls and pelicans.

Fish and shellfish: Freshwater fish, such as perch, yabbies and mussels in creeks and rock holes, and all varieties of saltwater fish were caught.

Medicines: Over 120 native plants were used as sedatives, ointments, diarrhoea remedies, and cough and cold palliatives, as well as for many other known treatments.

Water: Water was obtained from rivers, lakes, rock holes, soaks, beds of intermittent creeks and dew deposited on surfaces. Moisture obtained from foods such as tree roots and leaves also provided water.

3.5.3 Torres Strait Islander Peoples' food management

Torres Strait Islander Peoples' food sources, both historically and today, are largely based on fishing, **horticulture** and trading. Torres Strait Islander Peoples have a profound understanding of the sea, including its tides and sea life. Although their food sources vary from island to island, their lifestyle can best be described as subsistence agriculture with seafood, garden foods and other produce stored and preserved for both local use and trade. For example, recent research has found that the Goemulgal people of Mabuyag Island in the Torres Strait were building terraced plantations for banana crops over 2000 years ago.

FIGURE 2 Cooking in a Kup Murrie (ground oven)

3.5.4 Aboriginal land management techniques

Agriculture

Cultivating crops

Aboriginal Peoples' cultivation of crops and land management technologies were documented in the written observations of many early European explorers. Major Thomas Mitchell wrote of seeing grasses being harvested and piled into haystacks along eastern Australia. From these harvests, seeds were collected and made into a type of bread. Charles Sturt recorded similar observations in South Australia and Queensland. George Grey reported similar farming in the Gascoyne region of Western Australia, noting large wells that were 3–4 metres deep and large areas of cultivated yams. In Victoria, John Batman and his men observed and documented yams being harvested by the Wathaurong people. Tools, dams and irrigation systems were also used to aid in the cultivation of the plants and harvesting.

Cultural burning

The use of fire was a significant aspect of Aboriginal Peoples' agricultural systems. What has been described as the 'park-like' landscape of the Australian bush, as represented in early colonial art such as **FIGURE 4**, was purposely created by clearing undergrowth using cultural burning practices. The landscape encountered by the early settlers in the Paramatta and Liverpool areas, for example, was described as 'lightly timbered and so clear of undergrowth that you could ride a gig in any direction without hinderance' (quoted in Bruce Pascoe, *Dark Emu*, p. 164).

Cultural burning practices served many purposes: reducing the likelihood of dangerous, hot uncontrolled fires; cultivating germination and growth of plant species, such as yams and grains; facilitating easier hunting; and protecting important places. New growth also attracted animals, so timing cultural burning to attract kangaroos and other animals to grasslands away from food crops was also an important food security measure.

Burning was conducted on a rotational basis, with different areas being burned in different years. This enabled the rejuvenation of soil and new plant growth, and provided sanctuary for animals in the areas not being burned in any given year. This attracted all types of birds and animals to the area, and the landscape became ideal for hunting. Burning also flushed animals out into the open where they could be more easily hunted.

horticulture the practice of growing fruit and vegetables

Fires were also lit at very specific times of the year, when the conditions provided the best opportunity to keep the fires low and cool. For example, fires would be lit in cool weather with low winds, when there was a lot of moisture in the soil. This also allowed for the protection of species that had seasonal growing periods by not burning during the key growth periods.

When cultural burning practices were stopped, fuel accumulated in the undergrowth, for example dense plant growth, leaves, sticks and fallen branches. This meant that when fires occurred, for example due to lightning strikes, they were much hotter and far more intense than they had been before. This also meant that the fires reached the canopy or tops of the trees and spread in a way that people could not control.

FIGURE 4 An early European representation of using fire for hunting and to manage the land

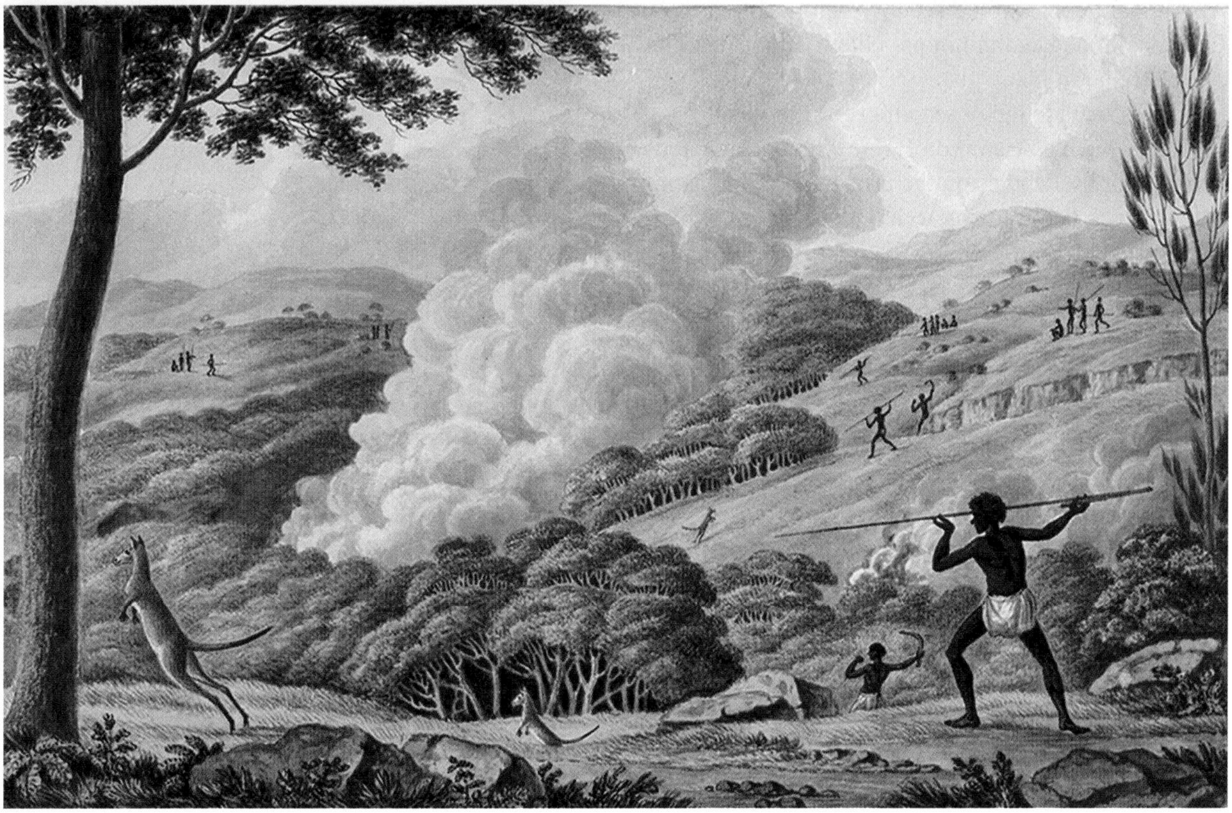

Source: Lycett, Joseph, approximately 1775–1828. Drawings of Aborigines and scenery, New South Wales, ca. 1820.

Managing seasonal food sources

One of the key aspects of food management was understanding and managing food sources that were plentiful at specific times of the year. These food sources were protected and managed for practical reasons — to ensure ongoing food supply — but also for important cultural reasons.

FIGURE 5 Massed Bogong moths on a rock face

Although there were many other sources of food for Aboriginal communities that lived near the south-eastern Australian highlands, the Bogong moth was a particularly important seasonal specialty. The Bogong moths, which lived in the ground as larvae in Queensland, migrated in millions to the south-eastern highlands to seek out cool, rocky overhangs and crevices where they could sleep through the long, hot summer months, surviving off the fat in their bodies (see **FIGURE 5**).

The Bogong moths were a rich source of fat and protein for Aboriginal Peoples who lived adjacent to the highlands of Victoria and New South Wales. Many nations would move from the valleys and foothills into the highlands to feast on the moths when they migrated. Moths were smoked out and collected by the thousands and cooked over hot rocks. In addition to savouring this important seasonal food source, making the annual pilgrimage to the high country presented people with an important opportunity to interact socially, participate in ceremonies and to arrange inter-community marriages.

3.5.5 Aquaculture

Fish and eel farming also played a crucial role in Aboriginal Peoples' food security before colonisation. Evidence of aquaculture in pre-colonisation Australia can be found in many places across the country. Examples include the observations of European explorers, such as Hamilton Hume who documented net-making by Peoples on the Darling River. Whaling and complex fish-trapping techniques were used by the Yuin people near present-day Pambula and Eden.

Large-scale aquaculture was practiced by the Ngemba people, whose fish trapping system at Brewarrina, east of Bourke in NSW, is described as one of Australia's largest and oldest fish trapping systems, and was a hub for communities to gather.

FIGURE 6 A glass-plate negative of the fish traps dating from 1880–1923.

Source: Collection: Museum of Applied Arts and Sciences. Gift of Australian Consolidated Press under the Taxation Incentives for the Arts Scheme, 1985. Unattributed studio.

FIGURE 7 A.W. Mullens' plan of the Brewarrina fish traps drawn in 1906 for the NSW Western Lands Board

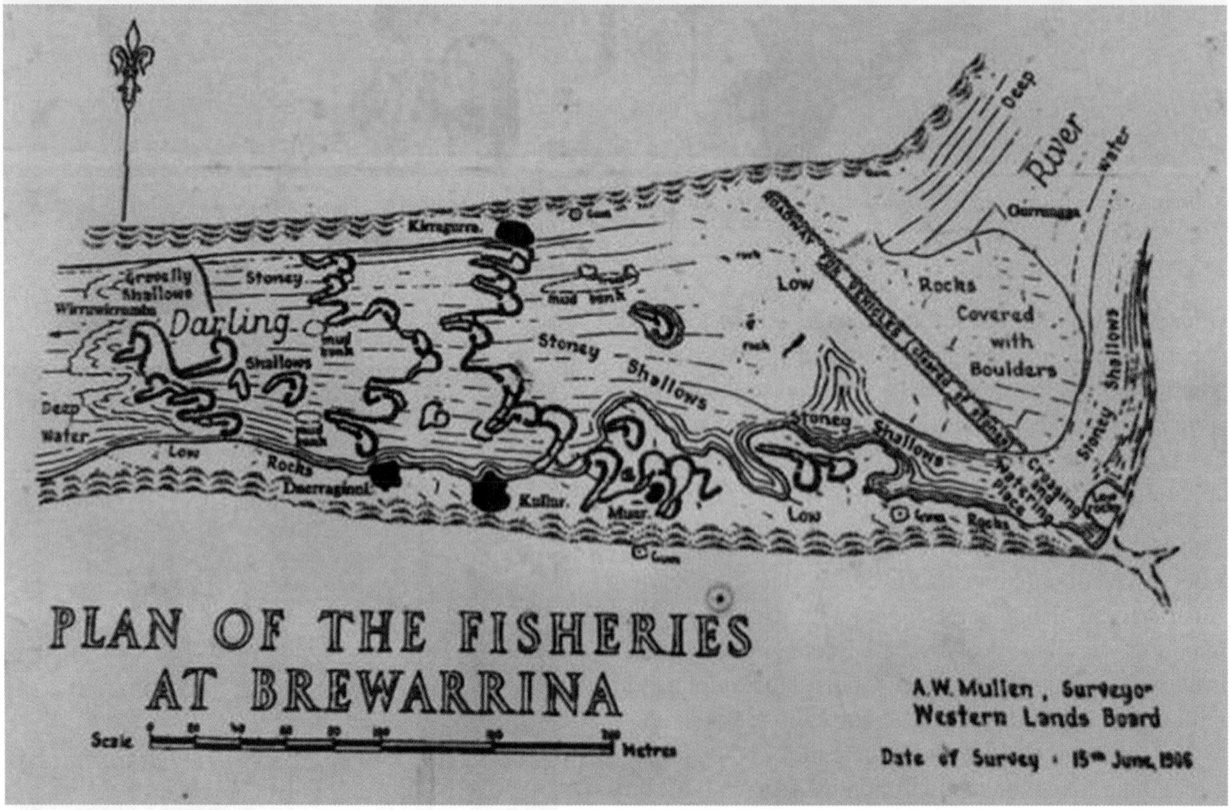

Source: Heritage Council of NSW

FIGURE 8 Budj Bim (Lake Condah), Victoria

Source: Tyson Lovett-Murray, © Gunditj Mirring Traditional Owners Aboriginal Corporation

The home of the Gunditjmara people, the Budj Bim Cultural Landscape, is the site of one of Australia's largest ancient aquaculture systems. This area, which is part of the Mount Eccles National Park near Portland in Victoria, shows evidence of a large, permanent settlement of stone huts and channels used for farming and the local trade of eels (see **FIGURE 8**). The Gunditjmara people managed this landscape by digging channels and constructing weirs to bring water and young eels from Darlot Creek to local ponds and wetlands. Woven baskets placed at the weirs were used to harvest the mature eels. The area provided an abundance of food, ensuring food security for all.

Following European occupation of the area in the 1830s, the Gunditjmara people fought for their lands in the Eumerella Wars, which lasted for more than 20 years. By the 1860s the remaining Gunditjmara people were forcibly moved to a government mission at Lake Condah. The mission lands were returned to the Gunditjmara people in 1987. The Deen Maar Indigenous Protected Area (IPA) was declared in 1999 and the area was listed on the Australian National Heritage register in 2004. The Budj Bim Cultural Landscape was added to the World Heritage List in July 2019.

Today the Gunditjmara people, as part of the Winda-Mara Aboriginal Corporation, manage the 248-hectare Darlots Creek (Killara), which flows from Lake Condah in the Budj Bim Cultural Landscape.

The wetlands and manna gum woodlands have been largely re-established through works to control weeds and feral animals. There are also prospects of restarting the eel aquaculture industry as a sustainable business. To further ecotourism, boardwalks have been built, signage put in place, and a range of tours of the wetlands, lakes and woodlands are offered by Gunditjmara guides. These tours examine rebuilt channels, weirs and eel traps, and sourcing of food stocks.

3.5.5 Human impacts of preventing Aboriginal land management practices

With European occupation and directly resulting from government policies, Aboriginal Peoples were forcibly dispossessed and displaced from Country, which prevented them from managing and accessing vital food sources. This restriction also limited people's access to their familiar, varied diet of nutritious and fresh food, resulting in significantly detrimental consequences for Aboriginal Peoples' health and access to food, especially in remote areas without strong transport links to major centres. This is particularly significant because nearly 20 per cent of Aboriginal Peoples and Torres Strait Islander Peoples live in remote and very remote areas of Australia (see **FIGURE 9**).

A prolonged and continuing lack of sufficient healthy food contributes to significant health issues, including a higher risk of respiratory diseases, cardiovascular disease, diabetes, chronic kidney disease and mental health issues. The importance of ongoing access to fresh and healthy food supplies cannot be understated.

FIGURE 9 Distribution of remote Aboriginal and Torres Strait Islander communities

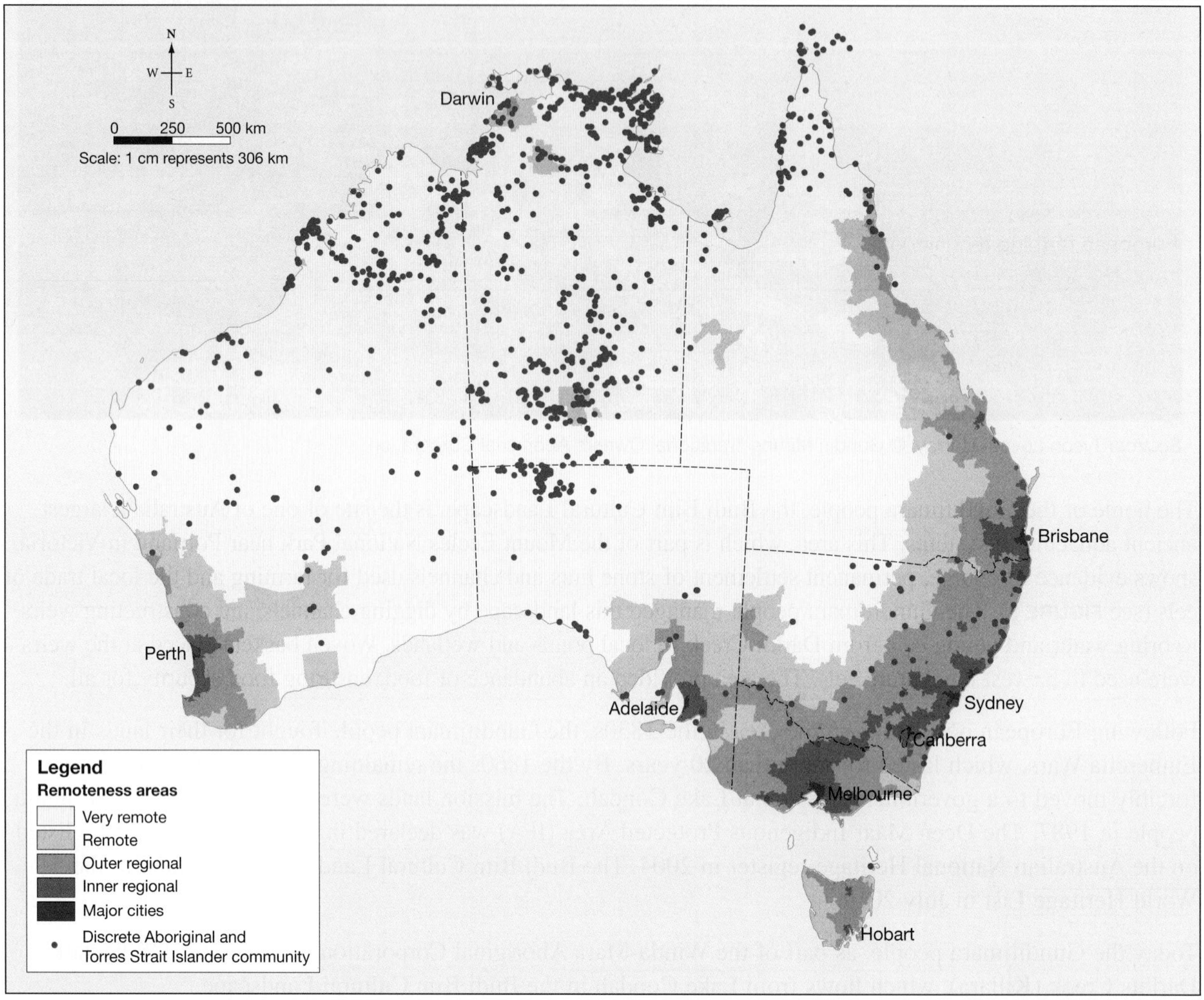

Source: Australian Bureau of Statistics

With the passing of the *Native Title Act 1993* and involvement of federal and state governments, some Aboriginal communities are managing Country through collaborative land and water management projects. Funding has been made available for initiatives to return management of Country back to the peoples of the area, including park and ranch management, ecotourism, and aquaculture ventures. This also helps to bring revenue into remote areas for further community development, employment and training opportunities. These initiatives also further strengthen Aboriginal Peoples spiritual ties to Country and place and improves mental health that results from caring for Country. One such initiative is Fish River Station in the Northern Territory.

CASE STUDY

Fish river station in the Daly River region

The Daly River area is important to the Wagiman, Labarganyan, Malak Malak and Kamu clans. This Country is remote, so contact with non-Aboriginal people came relatively late to this region when attempts were made in the 1880s to establish agriculture and mining operations.

What the farmers did not realise was that the biomes of the region were not suitable for European farming techniques or crops. By the 1920s, the planned tobacco and peanut farms failed because of the dominance of local animals and insects, native grasses, acidic soil and flooding in the wet season (**FIGURE 10**).

FIGURE 10 Daly River in the wet season

FIGURE 11 Fish River Station rangers Desmond Daly and Jeff Long patrol 178 000 hectares of land.

These problems were also compounded by poor communication links and transport, and the fact that farm labourers were difficult to find and keep. In 1967, the Tipperary Land Corporation cleared large tracts of land around the settlement and started growing sorghum, but this operation closed down in 1973. At that time, Aboriginal people worked in the area as labourers on farms in or in fishing and crocodile-shooting enterprises. Many had also moved to nearby towns and communities.

The Fish River cattle station covers about 180 000 hectares in the region (see **FIGURE 12**). The station has been run collaboratively since 2010 by the Indigenous Land and Sea Corporation (ILSC), The Nature Conservancy, the Pew Environment Group and Greening Australia to foster cultural and sustainable economic development. It is also protected under the National Reserve System.

FIGURE 12 Fish River Cattle Station is located in the remote Daly River region of the Northern Territory.

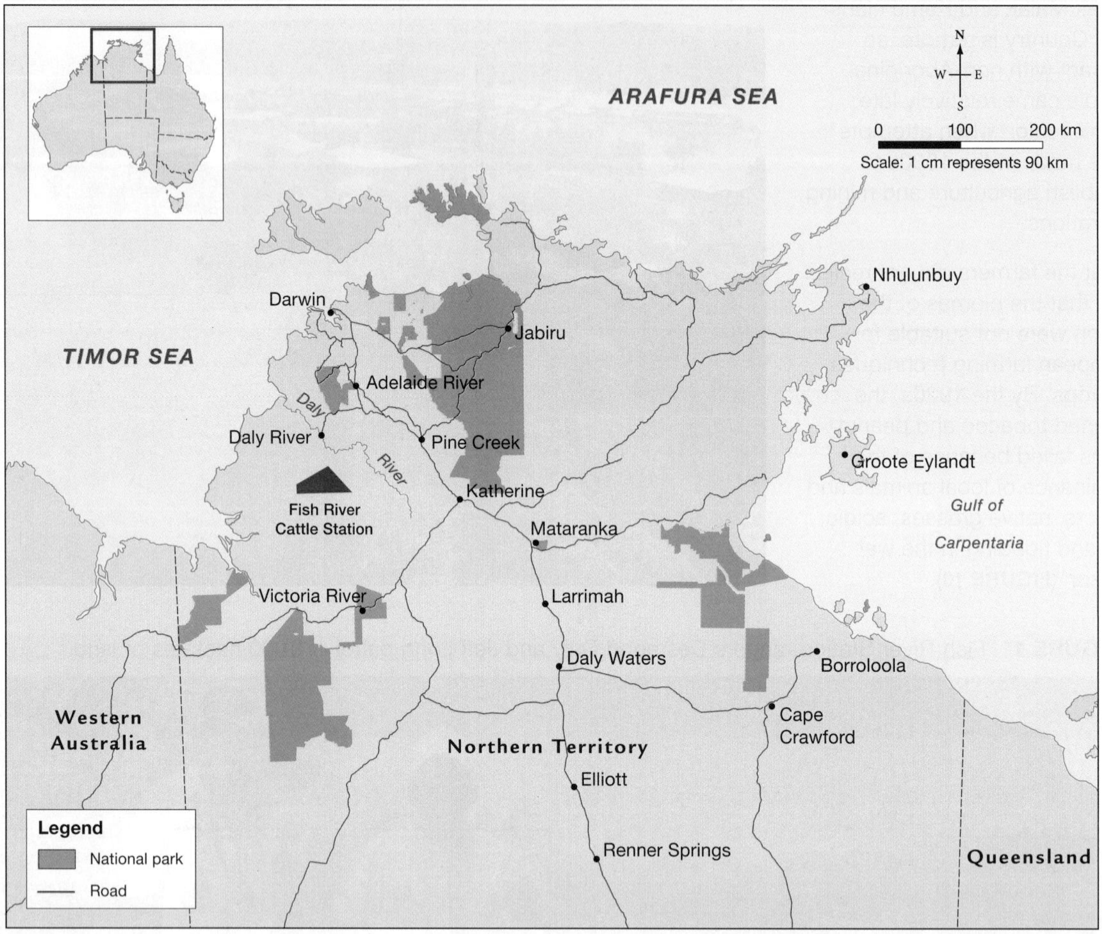

Source: Spatial Vision

The management of the station combines modern science with thousands of years of collective local knowledges to encourage biodiversity and pass on knowledge to future generations. The station now employs around 25 Aboriginal rangers who run projects such as controlling feral animals, monitoring levels of biodiversity, and managing a cultural burning program that is not only reducing the numbers of destructive bushfires but is also reducing overall greenhouse emissions. The Fish River Rangers also played a key role in combatting dangerous wildfires in the area in 2020.

on Resources

eWorkbook	Food security over time in Australia (ewbk-8872)
Video eLesson	Food security over time in Australia — Key concepts (eles-5183)
Interactivities	Growing more! (int-3320) Food security over time in Australia (int-8594)

3.5 ACTIVITIES

1. As a class, or in groups, discuss what we can learn from the land and resource management practices of Aboriginal Australians in relation to food security.
2. With a partner, use the **Indigenous food sources**, **Outback Stores** and **Fish River Station** weblinks in your online Resources to research and create an infographic poster of one of these topics.
3. Use the **Daly River food calendar** weblink in your online Resources to investigate the similarities and differences between modern food consumption and that shown in the Daly River seasonal calendar.
4. Use the **United Nations Food and Agriculture Organization (UN FAO)** weblink in your online Resources to investigate the idea of biodiversity through sustainable food production.

3.5 EXERCISE

Learning pathways

■ LEVEL 1	■ LEVEL 2	■ LEVEL 3
1, 2, 7, 8, 13	3, 5, 10, 12, 14	4, 6, 9, 11, 15

Check your understanding

1. Describe two ways of using fire to source food.
2. Who are the traditional owners of the Budj Bim region of Victoria?
3. Why did Aboriginal Peoples and Torres Strait Islander Peoples lose access to many traditional food sources after European occupation?
4. Why was the Bogong moth a good food source?
5. Identify two issues faced by communities living in remote locations in Australia today.
6. Examine **FIGURE 6**. Describe the distribution of remote and regional Aboriginal communities across Australia.

Apply your understanding

7. Explain what it means to be caretakers and traditional custodians of the land.
8. Explain one way in which the Gunditjmara people were able to ensure food security in their community.
9. Use the **Indigenous food sources, Outback Stores** and **Fish River Station** weblinks in the Resources panel in your learnON format to outline the various measures in place to address food insecurity issues in remote Indigenous communities.
   Analyse how two of these measures may improve food security.
10. How might a higher level of food security improve the health of people living in remote communities?
11. When they arrived, British colonists believed that Australia's landscape had never been cultivated or managed. Explain why their belief was incorrect and why they might not have been able to see the evidence that their belief was wrong.
12. Explain and give one example of how seasonal food sources can have a significant cultural or social importance for Aboriginal Peoples.

Challenge your understanding

13. How could tourism on the Fish River Station help with its economic development in the future?
14. Suggest one way that food security could be improved for people living in remote communities in Australia.
15. The Goemulgal people were building terraced banana farms at least 1800 years before the arrival of British colonists in Australia. Suggest why this information might be referred to as being 'discovered' by researchers in 2020.

To answer questions online and to receive **immediate feedback** and **sample responses** for every question, go to your learnON title at www.jacplus.com.au.

3.6 Food production in Australia

LEARNING INTENTION

By the end of this subtopic, you will be able to outline the different types of agriculture practised in Australia and explain the interconnections between climate, soils and land use.

3.6.1 Farming in Australia

Modern farming in Australia is mainly commercial and produces food for local consumption and exports for global markets. Australian farms may produce single crops (monoculture), such as sugar cane, or they may be mixed farms that produce cereal and graze sheep, for example. Many Australian farms have an **agribusiness** approach, are often run by large corporations and use sophisticated technology to help produce higher yields.

Why are farms found in certain locations?

There is a wide range of agriculture types in Australia, as shown in **FIGURE 1**. They occupy space across all biomes found in Australia, from the tropics to the temperate zones.

agribusiness business set up to support, process and distribute agricultural products

int-5582

FIGURE 1 Types of agriculture in Australia

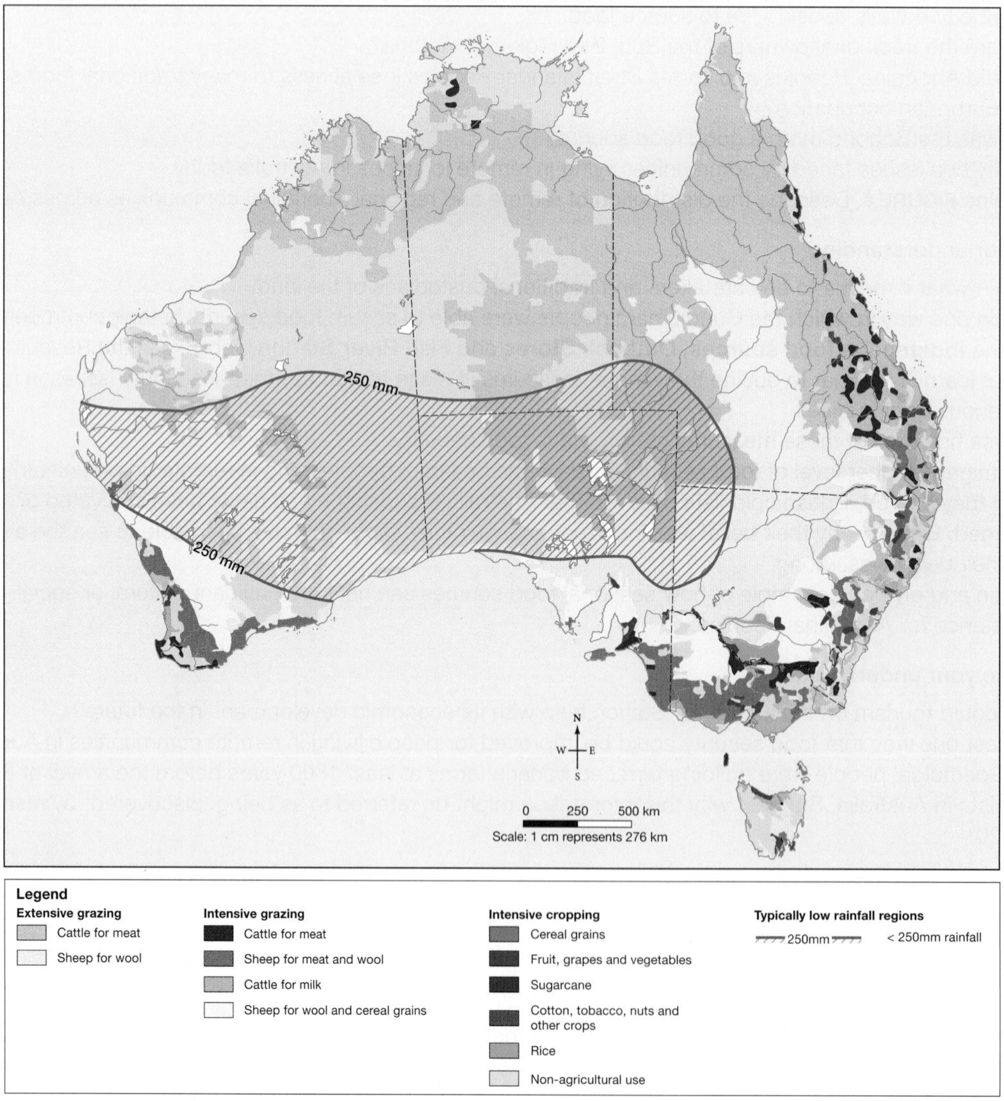

Source: Spatial Vision

The location of farms in Australia shows that there is a change in the pattern of farming types from the well-watered urban coastal regions towards the arid interior. Because much of Australia's inland rainfall is less than 250 millimetres, farm types in these places are limited to open-range cattle and sheep farming.

The pattern of land use and transition of farm types is shown in **FIGURE 2**. It illustrates that **intensive farms**, which produce perishables such as fruit and vegetables, are located on high-cost land close to urban markets. At the other extreme, **extensive farms**, which manage cattle for meat and sheep for wool, are found on the less expensive lands distant from the market.

FIGURE 2 Changes in agricultural land use

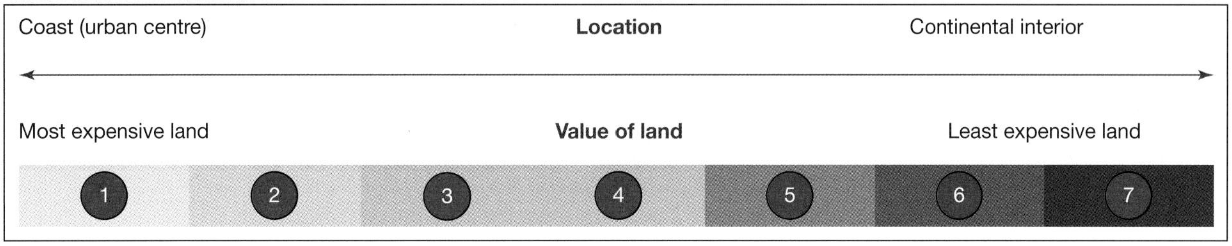

Key

1. Vegetables and fruit (including grapes)
2. Dairy cattle
3. Sheep for meat and wool
4. Cattle for meat (intensive farming)
5. Cereal grains and sheep for wool
6. Sheep for wool
7. Cattle for meat (extensive ranching)

3.6.2 Some farm types in Australia

Extensive farming of sheep or cattle

Sometimes known as livestock farming or grazing, these sheep and cattle stations are found in semi-arid and desert grassland biomes, with rainfall of less than 250 millimetres (see **FIGURE 1**). In 2017, Australia's 26 million cattle were predominantly farmed in Queensland, New South Wales and Victoria, while our 72 million sheep were found mainly in New South Wales, Victoria and Western Australia. Farms are generally large in scale, sometimes covering hundreds of square kilometres. These days, they have very few employees and often use helicopters and motor vehicles for mustering (**FIGURE 3**). Meat and wool products go to both local and overseas markets for cash returns.

Wheat farms

About 30 000 farms in Australia grow wheat as a major crop. The average farm size is 910 hectares, or just over nine square kilometres. As in other areas of the world, extensive wheat farming is found in mid-latitude temperate climates that have warm summers and cool winters, and annual rainfall of approximately 500 millimetres. In Australia, these conditions occur away from the coast in the semi-arid zone. The biome associated with this form of food production is generally open grassland, **mallee** or savanna that has been cleared for the planting of crops.

Soils can be improved by the application of fertilisers, and crop yields increased by the use of disease-resistant, fast-growing seed varieties. Wheat farms are highly mechanised, using large machinery for ploughing, planting and harvesting (**FIGURE 4**). The farm produce, which can amount to two tonnes per hectare, is sold to large corporations on local and international markets.

intensive farm farm that requires a lot of inputs, such as labour, capital, fertiliser and pesticides

extensive farm farm that extends over a large area and requires only small inputs of labour, capital, fertiliser and pesticides

mallee vegetation areas characterised by small, multi-trunked eucalypts found in the semi-arid areas of southern Australia

FIGURE 3 Cattle mustering

FIGURE 4 Wheat harvesting

Mixed farms

Mixed farms combine both grazing and cropping practices. They are located closer to markets in the wetter areas, and are generally smaller in scale but operate in much the same way as cattle and sheep farms.

Intensive farming

Intensive farms are close to urban centres, producing dairy, horticulture and market gardening crops (**FIGURE 5**). They produce milk, fruit, vegetables and flowers, all of which are perishable, sometimes bulky, and expensive to transport. The market gardens are capital- and labour-intensive, because the cost of land near the city is high, and many workers are required for harvesting.

FIGURE 5 Strawberries are typically grown in market gardens near cities and towns.

Plantation farming

This form of agriculture is often found in warm, well-watered tropical places. Plantations produce a wide range of produce such as coffee, sugar cane, cocoa, bananas, rubber, tobacco and palm oil. Farm sizes can be 50 hectares or more in size. Although many such farms in Australia are family owned, in other parts of the world they are often operated by large multinational companies. Biomes that contain plantations are mainly tropical forests or savanna, and require large-scale clearing to allow for farming. Cash returns are high, and markets are both local and global.

on Resources

eWorkbook	Food production in Australia (ewbk-8876)
Video eLesson	Food production in Australia — Key concepts (eles-5184)
Interactivity	Types of agriculture in Australia (int-5582)

3.6 ACTIVITIES

1. Investigate what foods are grown closest to you. Create a map infographic showing locations and types of food grown.
2. Collect information on the percentage of land used for the different forms of farming in Australia and show this data in a graph. Comment on the details shown in your graph.
3. Various plantations in Queensland (such as pineapple, sugar cane and banana plantations) are associated with fertiliser run-off, which is affecting the Great Barrier Reef. Investigate this issue and find out what effects fertiliser has on this marine environment.

3.6 EXERCISE

Learning pathways

■ LEVEL 1	■ LEVEL 2	■ LEVEL 3
1, 2, 3, 7, 15	4, 5, 9, 11, 13	6, 8, 10, 12, 14

Check your understanding

1. What is an agribusiness?
2. Describe the difference between intensive and extensive farming.
3. Based on **FIGURE 2**, order the following land uses from least to most expensive:
   - dairy farming
   - growing what
   - growing pineapples
   - raising merino sheep for their wool
   - raising beef cattle
   - growing cotton.
4. Which type of agricultural land use is closest to urban centres, and which is the furthest away?
5. How does the environment in the centre of Australia affect farming types?
6. What is the interconnection between climate and farm type in Australia?

Apply your understanding

7. Explain why extensive, large-scale cattle and sheep farms are typically located in remote and arid regions of Australia.
8. Using **FIGURE 2**, explain how the economic value of land interconnects with land use.
9. Why is so much of Australia's food production available for export?
10. It used to be said that Australia's economy 'rode on the sheep's back'. What do you think this means, and do you think it is still true today?
11. Intensive farming requires a lot of people to harvest crops. Discuss how this need for workers might have posed problems for farmers during the COVID-19 pandemic.
12. Why are most of Australia's extensive grazing cattle farms found in South Australia, Queensland, the Northern Territory and Western Australia?

Challenge your understanding

13. What would be the impact of flood or drought on any of the commercial methods of food production?
14. Predict the impact of the growth of Australian capital cities on the sustainability of surrounding market gardens.
15. Choose one of the areas shown in **FIGURE 1** as non-agricultural use. Suggest why the area is not used for agriculture.

To answer questions online and to receive **immediate feedback** and **sample responses** for every question go to your learnON title at www.jacplus.com.au.

3.7 SkillBuilder — Describing patterns and correlations on a topographic map

online only

LEARNING INTENTION

By the end of this subtopic, you will be able to describe patterns and correlations on a topographic map.

The content in this subtopic is a summary of what you will find in the online resource.

3.7.1 Tell me

Why consider patterns on a topographic map?

A pattern is the way in which features are distributed or spread. A correlation shows how two or more features are interconnected — that is, the relationship between the features. Patterns and correlations in a topographic map can show us cause-and-effect connections.

3.7.2 Show me

How to describe patterns and correlations in a topographic map

Step 1

Take the time to carefully analyse the topographic map, particularly its legend. Visualise the landforms and land use of the mapped place.

Step 2

Systematically look for connections between features, beginning with places that have strong connections; for example, between landforms and water drainage, vegetation types or land use. In **FIGURE 1**, you can see that the eastern ridge slopes are used for growing grapes, suggesting a connection between the landform and ideal grape-growing conditions. After you have identified these connections, write a few sentences describing any connections that are obvious.

Step 3

Systematically look for any anomalies that are evident. You are looking for things that seem unusual or show no connections.

Step 4

Complete your description with a concluding statement about the place.

3.7.3 Let me do it

learnON

Go to learnON to access the following additional resources to help you build this skill:

- a longer explanation of this skill and its application in Geography (Tell me)
- a video demonstrating the step-by-step process of this skill (Show me)
- an activity and interactivity for you to practise the skill (Let me do it)
- self-marking questions to help you understand and use the skill.

FIGURE 1 Topographic map extract showing the Clare Valley, South Australia

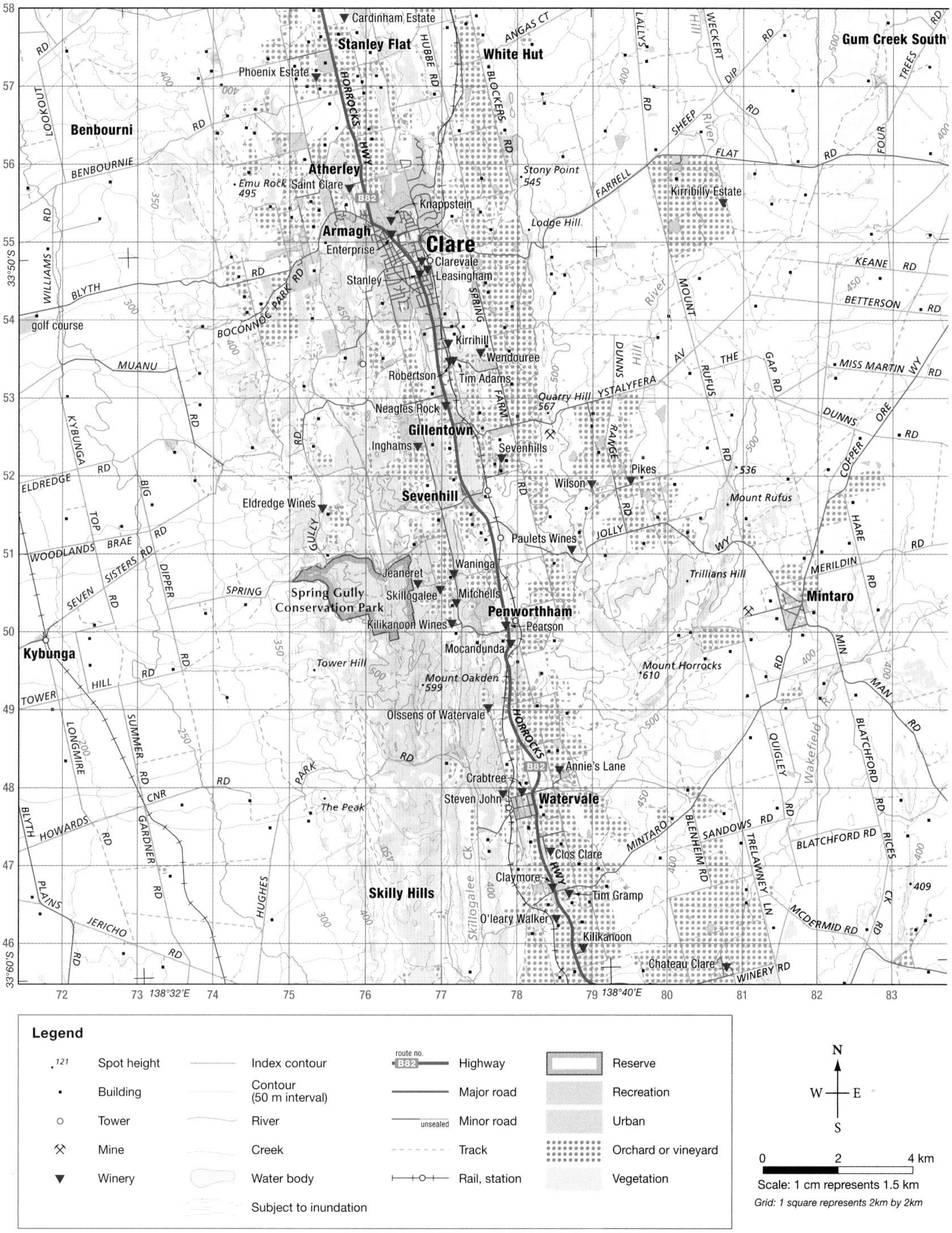

Resources

eWorkbook	SkillBuilder — Describing patterns and correlations on a topographic map (ewbk-8880)
Video eLesson	SkillBuilder — Describing patterns and correlations on a topographic map (eles-1729)
Interactivity	SkillBuilder — Describing patterns and correlations on a topographic map (int-3347)
Digital document	Topographic map of Clare Valley (doc 27426)
Google Earth	Clare Valley, South Australia (gogl-0068)

3.8 Rice — An important food crop

online only

LEARNING INTENTION

By the end of this subtopic, you will be able to explain the importance of rice as a staple food crop, and the various factors that affect its production in Asia and in Australia.

The content in this subtopic is a summary of what you will find in the online resource.

Rice is one of the most important staple foods of more than half the global population, and it influences the livelihoods and economies of several billion people. The majority of the world's rice production occurs in Asia, but Australia also has a rice-growing industry, with sophisticated farming processes producing varieties more suited to the Australian climate and growing conditions.

To learn more about the production of rice and its importance as a staple food crop, go to your learnON resources at www.jacPLUS.com.au.

FIGURE 2 Spectacular rice terraces in Yunnan Province, China. These terraces are at an elevation of 1570 metres.

Contents

learnON

- 3.8.1 Factors affecting rice production in Asia
- 3.8.2 Factors affecting rice production in Australia`

on Resources

eWorkbook	Rice — An important food crop (ewbk-8884)
Video eLesson	Rice — An important food crop — Key concepts (eles-5185)
Interactivity	How is rice grown? (int-3322)
myWorldAtlas	Deepen your understanding of this topic with related case studies and questions. • Rice

3.9 Cacao — A special cash crop

online only

LEARNING INTENTION

By the end of this subtopic, you will be able to explain how cacao is used for chocolate production, and its economic importance for the people who grow this crop throughout the world.

The content in this subtopic is a summary of what you will find in the online resource.

Chocolate is made from the beans of the cacao plant, which is grown in tropical regions across the globe. It is an important cash crop, providing essential income for farmers and their communities. Although demand for cacao is increasing, threats from pests and infections, climate change, and other yield-limiting factors need to be addressed in order to sustain this industry.

To learn more about cacao production and the importance of this crop to the communities that farm it, go to your learnON resources at www.jacPLUS.com.au.

FIGURE 3 A cacao farmer from Ghana carrying cacao pods

Contents

learnON

- 3.9.1 Cacao farming regions
- 3.9.2 Factors affecting the growth and production

on Resources

eWorkbook	Cacao — A special cash crop (ewbk-8888)	
Video eLesson	Cacao — A special cash crop — Key concepts (eles-5186)	
Interactivity	Cacao — A special cash crop (int-8596)	

3.10 Investigating topographic maps — Horticulture in Carnarvon

LEARNING INTENTION

By the end of this subtopic, you will be able to describe why Carnarvon is an important food growing area.

Modern-day food production relies heavily on technology to create ideal farming conditions. This may involve reshaping the land to allow for large agricultural machinery and for the even distribution and drainage of water. Uneven or unreliable rainfall can be supplemented by irrigation. As a result of such changes, large areas can become important farmland.

Carnarvon, located in the Gascoyne region of Western Australia, is an important horticulture and food-growing centre for the state. The farmland in the area around the Gascoyne River delta is very fertile, but because the river does not regularly flow, fruit and vegetable production relies on irrigation from aquifers. Some pastoral leases close to the coast also have access to the Carnarvon Artesian Basin for irrigating food crops for stock.

The wider Gascoyne region has a diverse agricultural sector. The most important commodities in the region include fruit and vegetables (about $97 million annually) and livestock, predominantly cattle (about $27 million annually). The 2016 census showed that of the 999 businesses in the area, 28.5 per cent were in the agriculture, forestry or fishing industries, with construction the next highest at just under 16 per cent. The 2000 hectares of zoned horticultural land close to Carnarvon produces a range of fresh produce including avocados, bananas, capsicums, tomatoes and mangoes.

FIGURE 1 Banana plantation, Carnarvon, Western Australia

FIGURE 2 Topographic map extract of Carnarvon, Western Australia

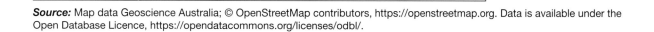

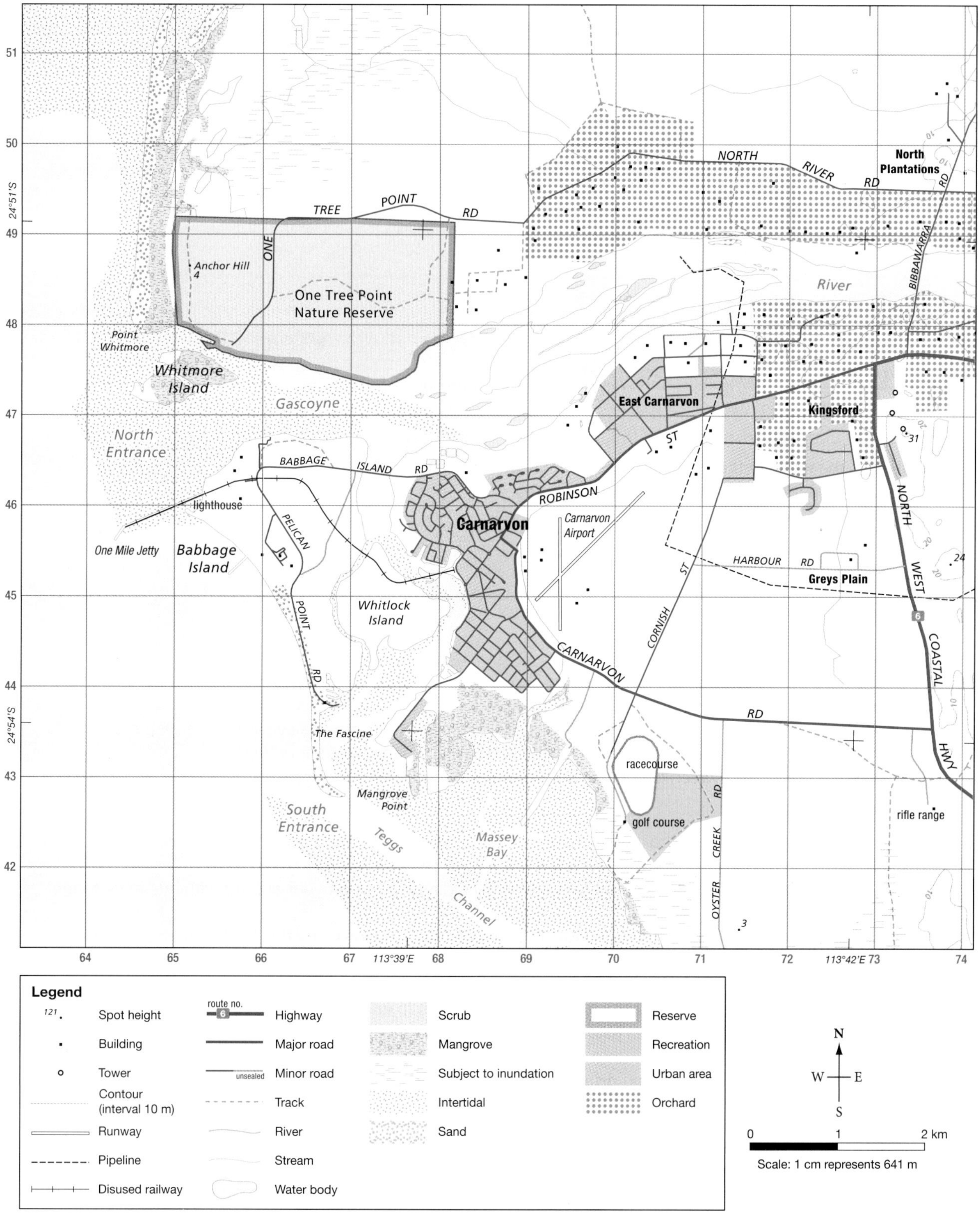

Source: Map data Geoscience Australia; © OpenStreetMap contributors, https://openstreetmap.org. Data is available under the Open Database Licence, https://opendatacommons.org/licenses/odbl/.

Resources

eWorkbook	Investigating topographic maps — Horticulture in Carnarvon (ewbk-8892)
Digital document	Topographic map of Carnarvon (doc-36266)
Video eLesson	Investigating topographic maps — Horticulture in Carnarvon — Key concepts (eles-5187)
Interactivity	Investigating topographic maps — Horticulture in Carnarvon (int-8597)
Google Earth	Carnarvon, Western Australia (gogl-0147)

3.10 EXERCISE

Learning pathways

■ LEVEL 1	■ LEVEL 2	■ LEVEL 3
1, 2, 7, 9, 14	3, 4, 8, 11, 15	5, 6, 10, 12, 13

Check your understanding

1. What is the name of the river that flows through Carnarvon?
2. Provide area references for the following features.
   a. Carnarvon airport
   b. Carnarvon golf course
   c. Kingsford's urban area
   d. Whitlock Island
3. Provide grid references for the following features.
   a. Babbage Island lighthouse
   b. Anchor Hill
   c. The road bridge to Babbage Island
   d. Mangrove Point
4. Using the contour lines and spot heights as a guide, estimate the average elevation of the map area.
5. Approximately what percentage of the map area is labelled as land used for orchards?
6. Is there a correlation between land slope and agricultural land use?

Apply your understanding

7. Why are the water channels straight? Is there an interconnection between slope and water resources? Explain your answer.
8. How do we know that the irrigated orchards are smallholdings? Support your answer with reference to the features shown in **FIGURE 2**.
9. Examine the location of the orchards in **FIGURE 2**. Suggest two reasons why this land might have been chosen for growing fruits and vegetables. Consider the environment and the need for transporting produce to consumers.
10. What is the importance of topography (the shape of the land) to irrigation?
11. Why might orchards not have been established on the land immediately north of the One Tree Point Nature Reserve?
12. What would be the advantages and disadvantages of locating processing factories close to growing areas?

Challenge your understanding

13. What types of environment might have existed in the Carnarvon area when pastoralists first arrived?
14. In 2015, Tropical Cyclone Olwyn caused significant damage to plantations in Carnarvon. Based on the location of farm allotments shown in **FIGURE 2**, what other hazards, apart from strong winds, might fruit and vegetable growers near Carnarvon have faced as a result of TC Olwyn?
15. If you were going to establish a banana farm in Carnarvon, which piece of land that is not currently being used for an orchard would you want to buy? In your answer, provide the area size and area reference for the land, and provide reasons why you think this land would be ideal.

To answer questions online and to receive **immediate feedback** and **sample responses** for every question, go to your learnON title at www.jacplus.com.au.

3.11 Thinking Big research project — Subsistence living gap-year diary

online only

The content in this subtopic is a summary of what you will find in the online resource.

Scenario

Across the globe there are communities that rely on traditional subsistence approaches to food security, such as nomadic herding and shifting agriculture. As an end-of-school gap-year experience, you have been given the opportunity to travel overseas to live with a community that practises one or more of these subsistence approaches.

Task

learnON

You will create a gap-year diary to help you remember forever the people you've met and their methods of managing their environment and food/water supply. Go to your online Resources to access the resources you need to complete this research project.

on Resources

ProjectsPLUS Thinking Big research project — Subsistence living gap-year diary (pro-0189)

3.12 Review

3.12.1 Key knowledge summary

3.2 Feeding the world

- Staple foods are those that are eaten regularly and in such quantities that they constitute a dominant portion of a diet. Most staple foods are cereals, such as wheat, barley, rye, oats, maize (corn) and rice; or root vegetables, such as potatoes, yams, taro and cassava. Other staple foods include legumes, such as soya beans and sago; fruits, such as breadfruit and plantain; and fish.
- Staple food production is interconnected with climate, environment, culture and traditions.
- The three main factors that have affected recent increases in world crop food production are increased cropland and rangeland area, increased yield per unit area and greater cropping intensity.
- The Green Revolution was a result of the development and planting of new hybrids of rice and wheat, combined with expanded irrigation and use of fertilisers, which have led to greatly increased yields.

3.4 Traditional agriculture

- Subsistence agriculture predominantly occurs in countries that are less economically developed. Examples include the Huli people, who use shifting agriculture; the San people, who survive on their hunting and gathering skills; and the Bedouin people, who herd camels.

3.5 Food security over time in Australia

- Food security for Aboriginal Peoples before colonisation was based on a profound knowledge of the land and water resources and a sustainable approach to sourcing and cultivating the essentials for living.
- Restrictions enforced by Europeans in the occupation of Australia resulted in many sources of food no longer being available to Aboriginal Peoples.
- Recent changes in Native Title and government food security schemes have offered more opportunities for Aboriginal Peoples to reconnect to the land and manage Country in a culturally appropriate way.

3.6 Food production in Australia

- Climate and distance to markets are major factors for all forms of agriculture in Australia.
- Types of farming in Australia include: extensive farming of sheep or cattle, extensive cereal crop farming, intensive farming such as dairy, horticulture and market garden cropping.

3.8 Rice — An important food crop

- Rice is a staple food crop for more than half the world's population.
- Rice production has increased significantly due to technological advances. The world's top rice-producing countries are in Asia, where growing conditions are favourable.
- Rice production in the Australian Riverina relies on irrigation and thus has environmental impacts on the water resources of the Murray–Darling Basin.

3.9 Cacao — A special cash crop

- Although cacao does not 'feed the world', it is an essential cash crop, connected to the livelihoods of 40–50 million people worldwide, providing them with income to purchase or cultivate crops for food.
- Most cacao is grown in a narrow belt between 10 degrees north and 10 degrees south of the equator.
- Consumer demand for chocolate is on the rise, but the cacao tree is under threat from pests, fungal infections, climate change, and farmers' lack of access to fertilisers and other products that enhance yields.

3.12.2 Key terms

agribusiness business set up to support, process and distribute agricultural products
agroforestry the use of trees and shrubs on farms for profit or conservation; the management of trees for forest products
arable describes land that can be used for growing crops
biofuel fuel that comes from renewable sources
cash crop a crop grown to be sold so that a profit can be made, as opposed to a subsistence crop, which is for the farmer's own consumption
commercial an activity that is concerned with buying and/or selling of goods or services
crop rotation a procedure that involves the rotation of crops, so that no bed or plot sees the same crop in successive seasons
edible fit to be eaten as food; eatable
extensive farm farm that extends over a large area and requires only small inputs of labour, capital, fertiliser and pesticides
FAO Food and Agricultural Organization of the United Nations
Green Revolution a significant increase in agricultural productivity resulting from the introduction of high-yield varieties of grains, the use of pesticides and improved management practices
horticulture the practice of growing fruit and vegetables
hybrid a plant or animal bred from two or more different species, breeds or varieties, usually to attain the best features from each type
income diversity income that comes from many sources
innovation new and original improvement to something, such as a piece of technology or a variety of plant or seed
intensive farm farm that requires a lot of inputs, such as labour, capital, fertiliser and pesticides
mallee vegetation areas characterised by small, multi-trunked eucalypts found in the semi-arid areas of southern Australia
monoculture the cultivation of a single crop on a farm or in a region or country
nomadic describes a group of people who have no fixed home and move from place to place according to the seasons, in search of food, water and grazing land
per capita per person
shifting agriculture system in which small parcels of land are used to produce food for a period and abandoned when they become less productive so they can recover naturally, while the farmers move to another plot of land
sustainable describes the use by people of the Earth's environmental resources at a rate such that the capacity for renewal is ensured
subsistence describes farming that provides food only for the needs of the farmer's family, leaving little or none to sell
staple an important food product or item that people eat or use regularly
yield amount of agriculture produced or provided

FIGURE 1 Road bridge crossing the dry Gascoyne River, Carnarvon, Western Australia

3.12.3 Reflection

Complete the following to reflect on your learning.

Revisit the inquiry question posed in the Overview:

Everyone needs to eat. How does the world produce all the food it needs, and is there a better way?

1. Now that you have completed this chapter, what is your view on the question? Discuss with a partner. Has your learning in this chapter changed your view? If so, how?
2. Write a paragraph in response to the inquiry question, outlining your views.

Subtopic	Success criteria			
3.2	I can describe the different major food staples around the world.			
	I can describe the global spatial distribution of the major food staples (where they are located).			
	I can explain why food demand is increasing.			
3.3	I can construct a ternary graph.			
3.4	I can explain the traditional ways that different groups of people produce food.			
3.5	I can describe traditional land management strategies used by Aboriginal Peoples and Torres Strait Islander Peoples.			
	I can describe traditional food sources of Aboriginal and Torres Strait Islander Peoples.			
	I can describe some of the challenges to food security that exist in remote communities today.			
3.6	I can outline the different types of agriculture practised in Australia.			
	I can explain the interconnections between climate, soils and land use.			
3.7	I can describe patterns and correlations on a topographic map.			
3.8	I can explain the importance of rice as a staple food crop.			
	I can describe the various factors that affect rice production in Asia and Australia.			
3.9	I can explain how cacao is used for chocolate production.			
	I can describe the economic importance of cacao for people who grow this crop and throughout the world.			
3.10	I can identify and describe features of a food-growing area from a topographic map.			

on Resources

eWorkbook
Chapter 3 Extended writing task (ewbk-8530)
Chapter 3 Reflection (ewbk-8529)
Chapter 3 Student learning matrix (ewbk-8528)

Interactivity
Chapter 3 Crossword (int-7645)

ONLINE RESOURCES

Resources

Below is a full list of **rich resources** available online for this chapter. These resources are designed to bring ideas to life, to promote deep and lasting learning and to support the different learning needs of each individual.

eWorkbook

- 3.1 Chapter 3 eWorkbook (ewbk-8095) ☐
- 3.2 Feeding the world (ewbk-8860) ☐
- 3.3 SkillBuilder — Constructing ternary graphs (ewbk-8864) ☐
- 3.4 Traditional agriculture (ewbk-8868) ☐
- 3.5 Food security over time in Australia (ewbk-8872) ☐
- 3.6 Food production in Australia (ewbk-8876) ☐
- 3.7 SkillBuilder — Describing patterns and correlations on a topographic map (ewbk-8880) ☐
- 3.8 Rice — an important food crop (ewbk-8884) ☐
- 3.9 Cacao — A special cash crop (ewbk-8888) ☐
- 3.10 Investigating topographic maps — Horticulture in Carnarvon (ewbk-8892) ☐
- 3.12 Chapter 3 Extended writing task (ewbk-8530) ☐
- Chapter 3 Reflection (ewbk-8529) ☐
- Chapter 3 Student learning matrix (ewbk-8528) ☐

Sample responses

- 3.1 Chapter 3 Sample responses (sar-0154) ☐

Digital documents

- 3.7 Topographic map of Clare Valley (doc-27426) ☐
- 3.10 Topographic map of Carnarvon (doc-36266) ☐

Video eLessons

- 3.1 A plate full of biomes (eles-1718) ☐
- Biomes produce food — Photo essay (eles-5180) ☐
- 3.2 Feeding the world — Key concepts (eles-5181) ☐
- 3.3 SkillBuilder — Constructing ternary graphs (eles-1728) ☐
- 3.4 Traditional agriculture — Key concepts (eles-5182) ☐
- 3.5 Food security over time in Australia — Key concepts (eles-5183) ☐
- 3.6 Food production in Australia — Key concepts (eles-5184) ☐
- 3.7 SkillBuilder — Describing patterns and correlations on a topographic map (eles-1729) ☐
- 3.8 Rice — An important food crop — Key concepts (eles-5185) ☐
- 3.9 Cacao — A special cash crop — Key concepts (eles-5186) ☐
- 3.10 Investigating topographic maps — Horticulture in Carnarvon — Key concepts (eles-5187) ☐

Interactivities

- 3.2 Staple foods around the world (int-7917) ☐
- Distribution of cropland, pasture and maize (int-8599) ☐
- 3.3 SkillBuilder — Constructing ternary graphs (int-3346) ☐
- 3.4 Traditional agriculture (int-8593) ☐
- 3.5 Growing more! (int-3320) ☐
- Food security over time in Australia (int-8594) ☐
- 3.6 Types of agriculture in Australia (int-5582) ☐
- 3.7 SkillBuilder — Describing patterns and correlations on a topographic map (int-3347) ☐
- 3.8 How is rice grown? (int-3322) ☐
- 3.9 Cacao — A special cash crop (int-8596) ☐
- 3.10 Investigating topographic maps — Horticulture in Carnarvon (int-8597) ☐
- 3.12 Chapter 3 Crossword (int-7645) ☐

ProjectsPLUS

- 3.11 Thinking Big research project — Subsistence living gap-year diary (pro-0189) ☐

Weblinks

- 3.2 United Nations Food and Agriculture Organization (web-6327) ☐
- Feed the World (web-3303) ☐
- 3.5 Daly River food calendar (web-1154) ☐
- Fish River Station (web-4338) ☐
- Indigenous food sources (web-4337) ☐
- Outback stores (web-4339) ☐
- United Nations Food and Agriculture Organization (UN FAO) (web-0117) ☐

Fieldwork

- 3.2 Annotating a field sketch of your biome (fdw-0028) ☐

myWorld Atlas

- 3.2 Wheat (mwa-7343) ☐
- 3.8 Rice (mwa-7342) ☐

Google Earth

- 3.7 Clare Valley, South Australia (gogl-0068) ☐
- 3.10 Carnarvon (gogl-0147) ☐

Teacher resources

There are many resources available exclusively for teachers online.

[illegible]
Food security in Australia [illegible]
Types of agriculture in Australia [illegible]
SkillBuilder — Describing patterns and correlations
on a topographic map [illegible]
[illegible]

[illegible]
Food security in Australia [illegible]
Food production in Australia [illegible]
SkillBuilder — Describing patterns and correlations
[illegible]

4 Changing biomes

INQUIRY SEQUENCE

To access a pre-test with **immediate feedback** and **sample responses** to every question in this chapter, select your learnON format at www.jacplus.com.au.

4.1 Overview

Numerous **videos** and **interactivities** are embedded just where you need them, at the point of learning, in your learnON title at www.jacplus.com.au. They will help you to learn the content and concepts covered in this topic.

Our planet works hard to feed the ever-growing human population ... but at what cost?

4.1.1 Introduction

Food is essential to human life, and over the past centuries we have been able to produce more food to feed our growing population. Although technology has enabled us to increase production, it has come at a price. Large-scale clearing of our forests, the overfishing of our oceans, and constant overuse of soils has resulted in a significant decline in our biophysical world.

STARTER QUESTIONS

1. How can feeding the world be destroying the world?
2. Do you know where your food comes from?
3. What food items do you eat most of within a typical day? Are they animal-based, plant-based or fish-based?
4. Do you or your family grow any of your own food?
5. Use an online mapping program such as Google maps and choose the satellite view over Mackay, Queensland. Discuss with your class how the environment around Mackay may have changed over time. What is the main crop being grown in this region? What might have grown here before European colonisation?
6. Watch the **Changing biomes — Photo essay** video in your online Resources. Consider each of the photos and how the landscape has changed.

FIGURE 1 Fruit orchard in rural New South Wales

on Resources

eWorkbook	Chapter 4 eWorkbook (ewbk-8096)
Video eLessons	Trashing our biomes (eles-1719) Changing biomes — Photo essay (eles-5188)

4.2 Food production's effect on biomes

LEARNING INTENTION

By the end of this subtopic, you will be able to describe why biomes are altered to grow food and explain the effect this has on biomes.

4.2.1 Our biophysical world

Biomes are created by the interactions of the four spheres of the **biophysical environment**. A change in any of the spheres will impact the others at varying levels. The large-scale production of food requires modifications to the environment and as a consequence, biomes have been impacted.

Planet Earth is made up of four spheres: the atmosphere, lithosphere, hydrosphere and biosphere (see **FIGURE 1**).

All these spheres are interconnected and make up our biophysical or natural environment. For example, rain falling from a cloud (atmosphere) may soak into the soil (lithosphere) or flow into a river (hydrosphere) before being taken up by a plant or animal (biosphere) where it may be evaporated and returned to the atmosphere.

biophysical environment all elements or features of the natural or physical and the human or urban environment, including the interaction of these elements; made up of the Earth's four spheres — the atmosphere, biosphere, lithosphere and hydrosphere

int-7919

FIGURE 1 The Earth's four spheres

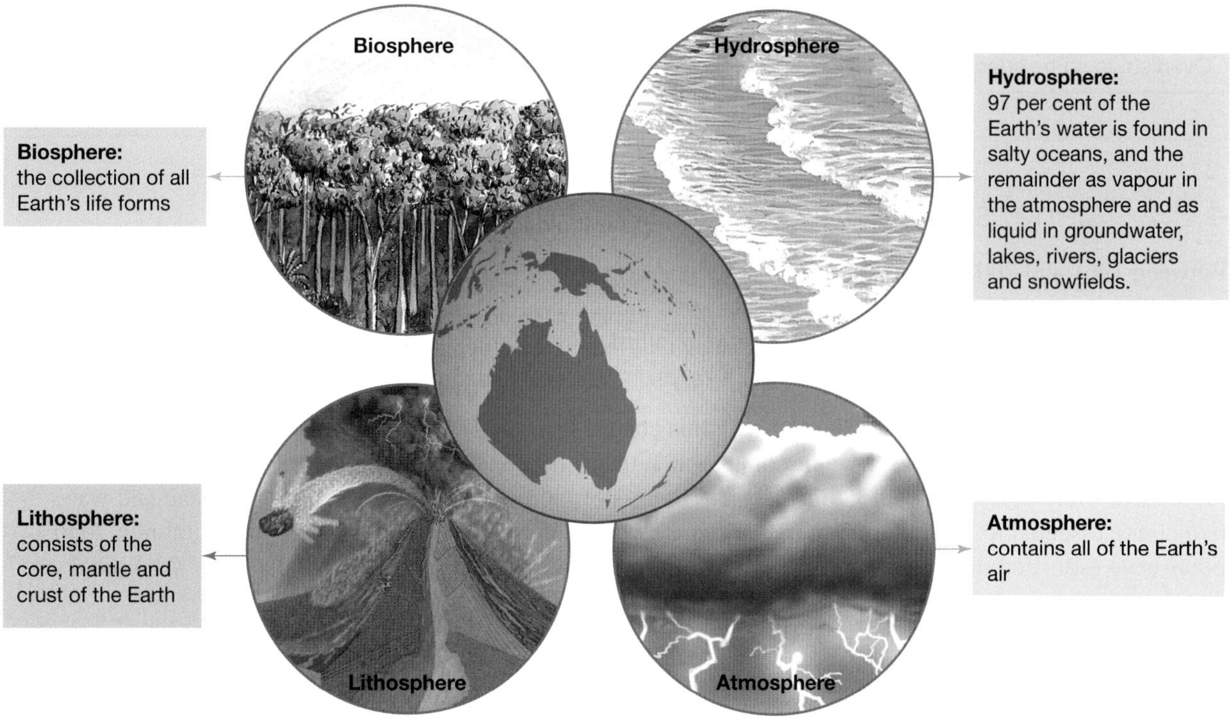
Biosphere
Hydrosphere
Lithosphere
Atmosphere
Biosphere: the collection of all Earth's life forms
Hydrosphere: 97 per cent of the Earth's water is found in salty oceans, and the remainder as vapour in the atmosphere and as liquid in groundwater, lakes, rivers, glaciers and snowfields.
Lithosphere: consists of the core, mantle and crust of the Earth
Atmosphere: contains all of the Earth's air

Natural events, such as storms or earthquakes, or human activities can create changes to one or all of these spheres. The production of food, whether from the land or sea, has the potential to change the natural environment and, in doing so, increases the likelihood of food insecurity. **TABLE 1** shows how food production can affect the biophysical world.

TABLE 1 Food production's effects on the biophysical world

Activities	Atmosphere	Lithosphere	Biosphere	Hydrosphere
Clearing native vegetation for agriculture	x	x	x	x
Overgrazing animals		x	x	x
Overusing irrigation water, causing saline soils		x	x	x
Burning forests to clear land for cultivation	x	x	x	x
Releasing pesticides and fertilisers into streams in run-off		x	x	x
Producing greenhouse gases by grazing animals and farming rice	x			
Changing from native vegetation to cropping		x	x	x
Withdrawing water from rivers and lakes for irrigation	x	x	x	x
Overcropping soils		x	x	x
Overfishing some species			x	x

4.2.2 Changes in our biophysical world

Between 1961 and 2008, the world's population increased by 117 per cent, or by 3.5 billion, while food production increased by 179 per cent. This has been the result of improved farming methods; the increased use of fertilisers and pesticides; large-scale irrigation; and the development of new technologies, ranging from farm machinery to better quality seeds.

There have been many benefits associated with this change, especially in terms of human wellbeing and economic development. However, at the same time, humans have changed the Earth's biomes more rapidly and more extensively than in any other time period. The loss of biodiversity and **degradation** of land and water (which are essential to agriculture) are not sustainable. With an expected population of nine billion in 2050, it has been estimated that food production will need to increase by approximately 70 per cent. The global distribution of environmental risks associated with food production can be seen in **FIGURE 2**.

4.2.3 Modifying biomes for agriculture

Throughout the twentieth century rapid population growth and the development of new technologies allowed for larger scale agriculture to occur. To accommodate the increase in size and amount of farms to grow crops and graze animals, biomes were altered to meet the higher production demands. In general, the focus of agriculture is to modify water, climate, soils, land and crops.

degradation reduced quality of land and water resources caused by over-use

'920

FIGURE 2 Global land and water resources for food and agriculture

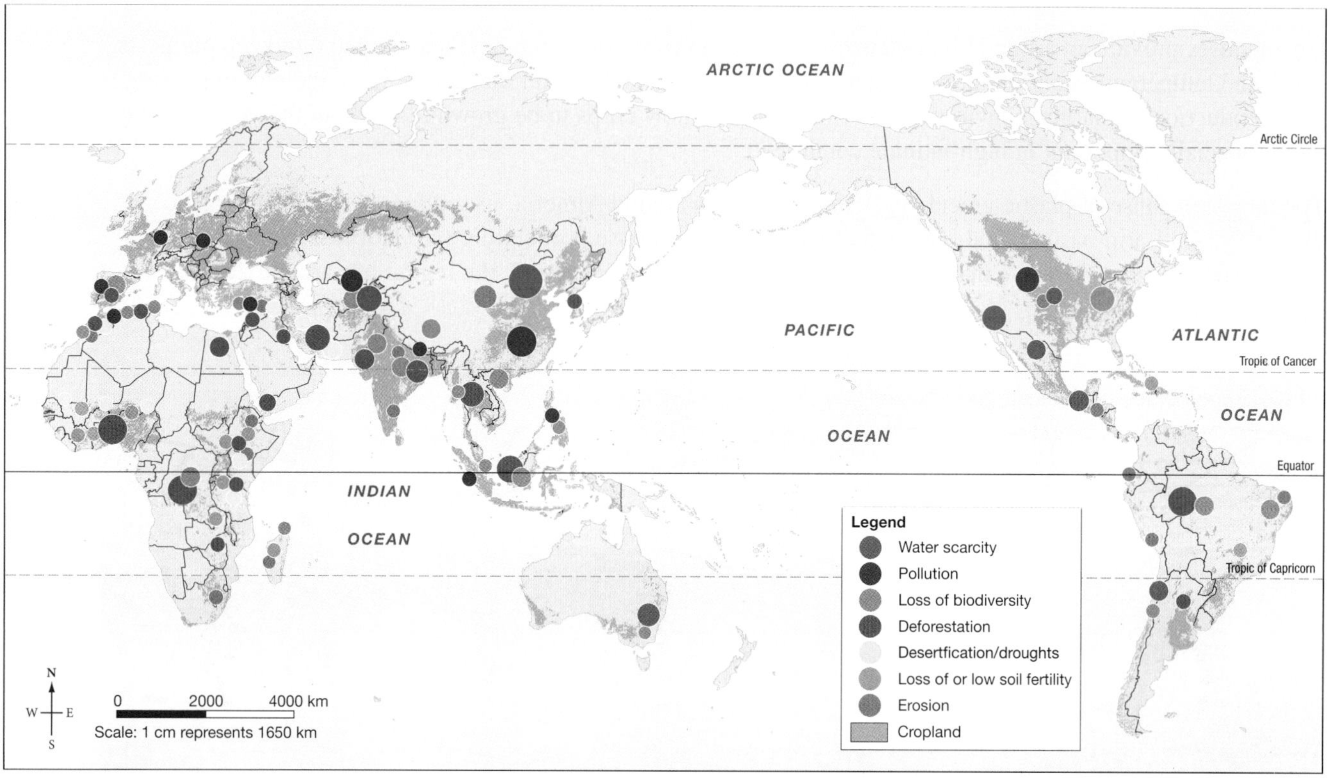

Source: Spatial Vision

Climate

Irrigation is the artificial application of water to the land or soil to supplement natural rainfall. It is used to assist in the growing of agricultural crops to increase food production in dry areas and during periods of inadequate rainfall.

In flood irrigation, water is applied and distributed over the soil surface by gravity. It is by far the most common form of irrigation throughout the world and has been practised in many areas, virtually unchanged, for thousands of years. Modern irrigation methods include computer-controlled drip systems that deliver precise amounts of water to a plant's root zone.

FIGURE 3 False-colour satellite image of greenhouses in the Almeria region, Spain

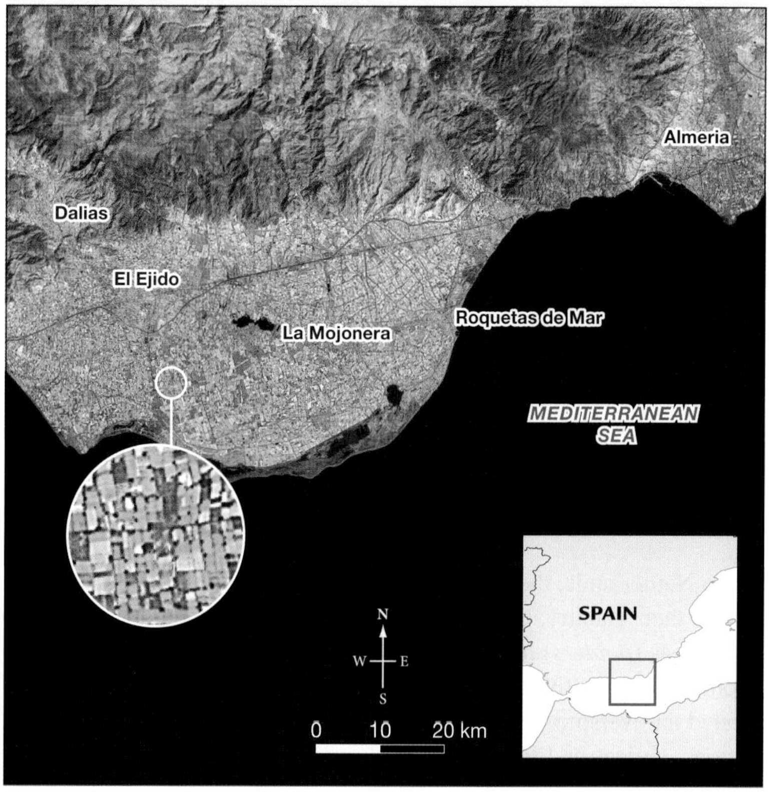

Source: American Geophysical Union and Google Maps.

Another way of modifying climate is with the use of greenhouses (or glasshouses) used for growing flowers, vegetables, fruits and tobacco (see **FIGURE 4**).

Greenhouses provide an artificial biotic environment to protect crops from heat and cold and to keep out pests. Light and temperature control allows greenhouses to turn non-arable land into arable land, thereby improving food production in marginal environments. Greenhouses allow crops to be grown throughout the year, making them especially important in high-latitude countries.

The largest expanse of plastic greenhouses in the world is around Almeria, in south-east Spain. Here, since the 1970s, semi-arid pasture land has been replaced by greenhouse horticulture (see **FIGURES 3** and **4**). Today, Almeria has become Europe's market garden. To grow food all year round, the region has around 26 000 hectares of greenhouses.

FIGURE 4 Inside an Almerian greenhouse

Soils and land

Fertilisers are organic or inorganic materials that are added to soils to supply one or more essential plant nutrients. Fertilisers are essential for high-yield harvests, and it is estimated that about 40 to 60 per cent of crop yields are due to fertiliser use. It is estimated that almost half the people on Earth are currently fed as a result of adding fertiliser to food crops.

People change landscapes in order to produce food. **Undulating** land can be flattened, steep slopes terraced or stepped, and wetlands drained. Land reclamation is the process of creating new land from seas, rivers or lakes. In addition, it can involve turning previously unfarmed land, or degraded land, into arable land by fixing major deficiencies in the soil's structure, drainage or fertility.

In the Netherlands, the Dutch have tackled huge reclamation schemes to add land area to their country. One such scheme is the IJsselmeer (see **FIGURE 5**), where four large areas (*polders*) have been reclaimed from the sea, adding an extra 1650 square kilometres for cultivation. This has increased the food supply in the Netherlands and created an overspill town for Amsterdam. The schemes include adding **dykes**, pumping stations, **sluices** and **locks** to control water.

undulating describes an area with gentle hills

dyke a barrier or levee to hold back water

sluice a constructed channel for water

lock a gateway across a channel of water that can be used to raise or lower water on either side; allows boats to travel between waterways at different levels

FIGURE 5 Land reclamation in the Netherlands

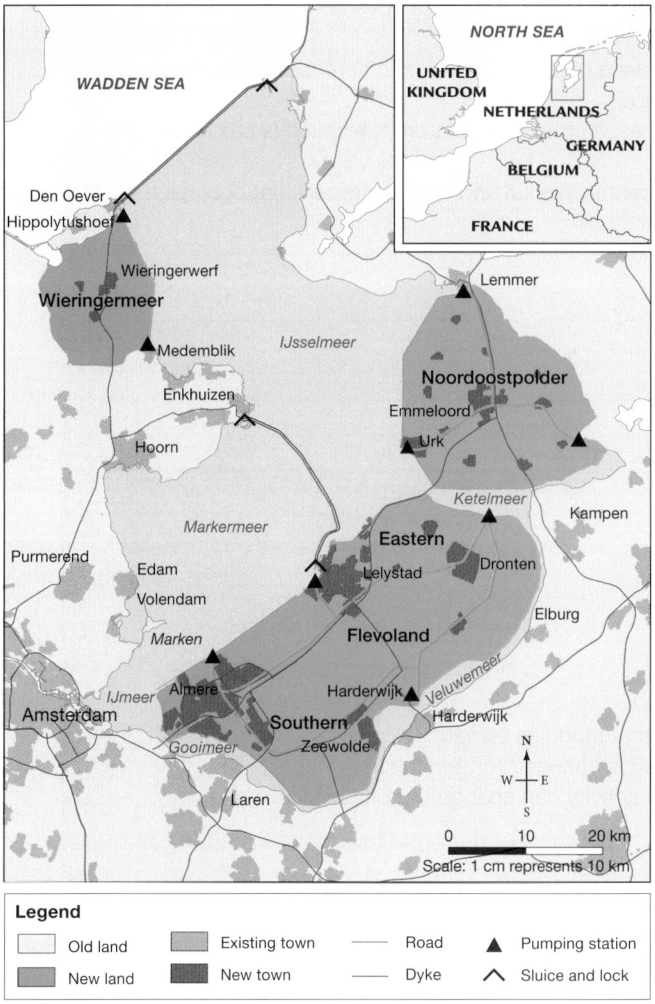

Source: Spatial Vision

FIGURE 6 Farms and farmland, Noordoostpolder, The Netherlands

Resources

eWorkbook Food production's effect on biomes (ewbk-8900)

Video eLesson Food production's effect on biomes – Key concepts (eles-5189)

Interactivities Degrading our farmland (int-3323)
State of the world's land and water resources for food and agriculture (int-7920)
The Earth's four spheres (int-7919)

myWorldAtlas Deepen your understanding of this topic with related case studies and questions.
- World: Climate
- World: Vegetation

4.2 ACTIVITIES

1. Use the following labels to create a flow diagram showing how the clearing of native vegetation can affect all four of the Earth's spheres.
   - Soil is left bare and exposed to wind and water erosion.
   - There is less evaporation of water from vegetation.
   - There is a loss of habitat for birds, animals and insects.
   - Increased water runs off from exposed land.
   - Increased sediment builds up in streams.

2. a. Rank what you consider to be the three most serious environmental issues shown on the **FIGURE 2** map.
   b. What criteria did you use to make your decisions (for example extent of area covered, number of people affected, economic impacts)?
   c. Share your list with at least two other students. What were the similarities and differences in the rankings?
   d. Would you now alter your own ranking? Why or why not?
3. Select one agricultural product in Australia and conduct research to find data on how much is produced and how this has changed over time.
4. Research the land reclamation project in IJsselmeer (Zuiderzee Works) and create a report that outlines the scope of the project.

4.2 EXERCISE

Learning pathways

■ LEVEL 1	■ LEVEL 2	■ LEVEL 3
1, 2, 7, 8, 14	3, 4, 10, 11, 13	5, 6, 9, 12, 15

Check your understanding

1. Describe the spheres of the biophysical environment.
2. Why has food production increased so rapidly over time?
3. Describe irrigation and how it is an agricultural method that modifies climate.
4. Describe how soils are modified and how undulating land is changed for agriculture.
5. Select one example from **TABLE 1**. Describe how human activity can change the biophysical world.
6. What is the purpose of a sluice?

Apply your understanding

7. Referring to **FIGURE 1**, explain how change in one sphere of the biophysical environment could impact the other spheres.
8. Select one example from **TABLE 1**. Describe how human activity can change the biophysical world.
9. Refer to **FIGURE 2**.
   a. What are the main environmental issues facing Australia's food production?
   b. In which places in the world is deforestation a major concern?
   c. Which continents suffer from water scarcity?
   d. What do you notice about the location and distribution of regions that do not have environmental problems relating to food production?
10. Refer to **FIGURES 3** and **4**. How do greenhouses modify spaces and places on the Earth's surface?
11. Refer to **FIGURE 5**. What might be the purpose of the pumping station?
12. What are the strengths and weaknesses of the strategies being used in the Netherlands to provide enough land for food production? In your answer, refer to the methods shown in **FIGURE 5**.

Challenge your understanding

13. What major challenges do you think the land reclamation program in the Netherlands will face in the next 50 years? Predict one challenge, outline its likely causes and suggest one way any negative impacts might be prevented.
14. Can greenhouses be used anywhere to produce food efficiently? Consider the equipment and supplies required to produce commercial amounts of food in greenhouses in your answer.
15. Predict whether a reduction in biodiversity might have a negative impact on our ability to produce enough food to feed the world. Give reasons to support your view.

To answer questions online and to receive **immediate feedback** and **sample responses** for every question, go to your learnON title at www.jacplus.com.au.

4.3 Modifying our forests

LEARNING INTENTION

By the end of this subtopic, you will be able to describe how humans have modified forest biomes and explain the impacts that this has had.

4.3.1 The importance of forests

In pre-industrial times, nearly 45 per cent of the world's land surface was covered in forest. Today, this figure is only 31 per cent. With industrialisation, technological development and population growth, large-scale **deforestation** has occurred as a result of the increasing need over time for timber products and land for food. It is estimated that of the forest cover lost, 85 per cent can be readily attributed to human activity — with 30 per cent due to clearing, 20 per cent through degradation and 35 per cent through fragmentation. Agricultural use now accounts for 37 per cent of the Earth's land surface.

Human society, the global economy and forests are interconnected, with more than one billion people depending on forests and forest products. Forest biomes offer us many goods and services, from providing wood and food products to supporting biological diversity. They provide habitat for a wide range of animals, plants and insects. Forests contribute to soil and water conservation, and they absorb **greenhouse gases**. Despite the growing awareness of the value of preserving forests, large-scale clearing continues. **FIGURE 1** shows the annual rate of deforestation in Brazil, and **FIGURE 2** shows the cumulative amount of forest lost over time.

deforestation clearing forests to make way for housing or agricultural development

greenhouse gases any of the gases that absorb solar radiation and are responsible for the greenhouse effect. These include water vapour, carbon dioxide, methane, nitrous oxide and various fluorinated gases.

FIGURE 1 Annual loss of Amazon forest, Brazil, 1988–2020

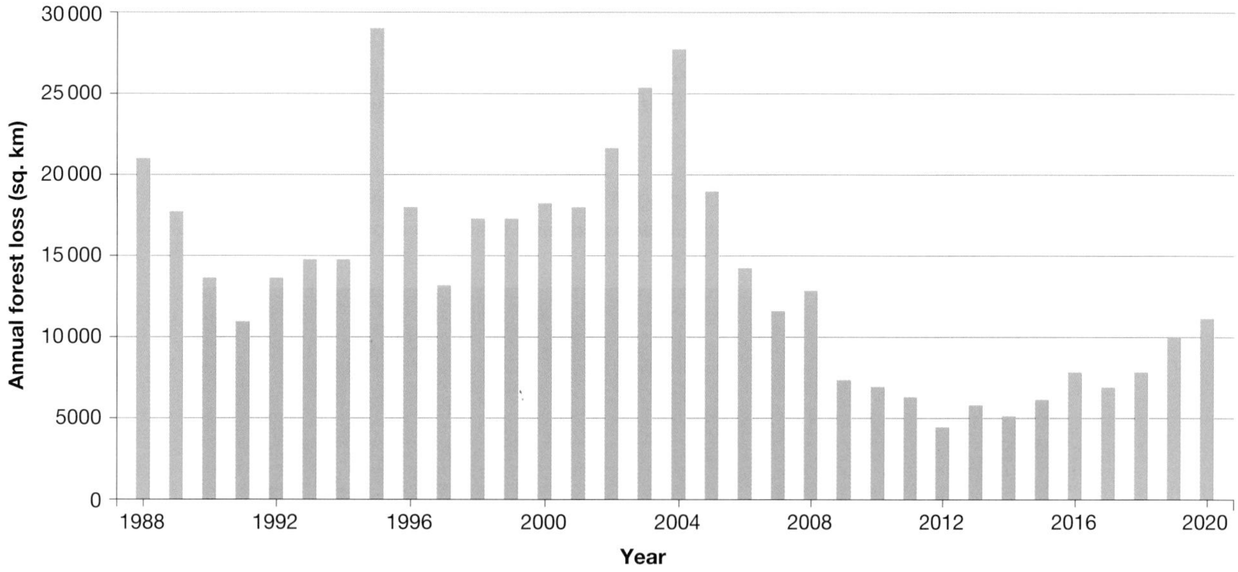

Source: © Mongabay

FIGURE 2 Total loss of Amazon forest, Brazil, since 1970

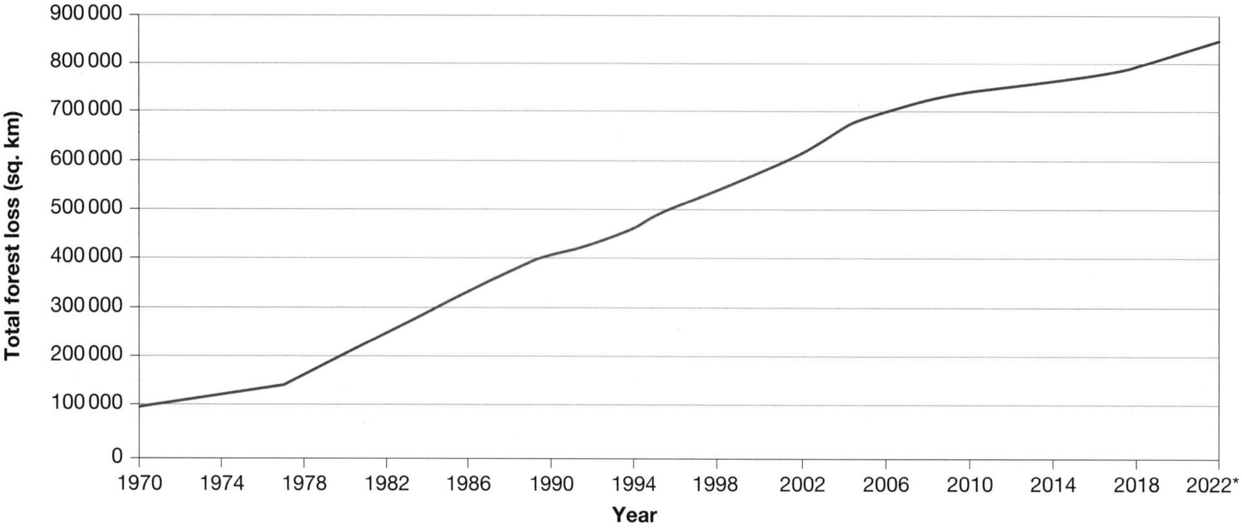

Source: © Mongabay
* projected

4.3.2 Clearing forests

By clearing forests, trees can be harvested for timber and paper production, and valuable ores and minerals can be mined from below the Earth's surface. Sometimes, forests are flooded rather than cleared in order to construct dams for hydroelectricity. Forests may also be cleared for food production, such as small-scale subsistence farming, large-scale cattle grazing, and for **plantations** and crop cultivation.

Road construction, usually funded by governments, also plays a part in changing rainforest environments (see **FIGURE 3**). Roads help to improve access and make more land available, especially to the landless poor. They also reduce population pressures elsewhere by encouraging people to move to new places. At the same time, businesses benefit from improved access to mining resources and forest timbers, and are better able to establish large cattle ranches and farms.

plantation an area in which trees or other large crops have been planted for commercial purposes

FIGURE 3 The effects of road building in the Amazon. Settlements tend to follow a linear pattern along the roads and then gradually move inland, opening up the forests.

4.3.3 Effects of forests clearing

FIGURE 4 illustrates some changes that forest clearing in the Amazon can have on the environment.

FIGURE 4 Impacts of clearing the Amazon forest

- A New farmland with mixed crops established
- B Smoke from clearing and burning
- C Newly cleared land, trees cut down and burned. This is called slash-and-burn agriculture.
- D Weeds and exotic species invade edges of remaining forest.
- E New road gives access to more settlers and to animal poachers.
- F Large cattle ranch
- G Introduced cattle erode the fragile topsoil with their hard hooves.
- H Erosion of topsoil increases, caused by rain on exposed soils.
- I Flooding increases as the stream channel is clogged with sediment.
- J The river carries more sediment as soil is washed into streams.
- K Fences stop movement of rainforest animals in search of food.
- L Pesticides and fertilisers wash into the river.
- M Farm is abandoned as soil fertility is lost.
- N Weeds and other species dominate bare land.
- O Harvesting of timber reduces forest biodiversity.

Resources

eWorkbook	Modifying our forests (ewbk-8904)
Video eLesson	Modifying our forests — Key concepts (eles-5190)
Interactivity	Impacts of clearing the Amazon Rainforest (int-5578)
myWorldAtlas	Deepen your understanding of this topic with related case studies and questions. • Forest environments

fdw-0030

FOCUS ON FIELDWORK

Remote sensing

Remote sensing is a high-tech and sophisticated method of collecting data by geographers, in which data is collected without their being physically present in the field. Common forms of remote sensing equipment that geographers utilise are weather instruments on satellites or aircraft that scan the Earth to collect information about it.

FIGURE 5 Drones are a more cost-effective way than using satellites or aircraft for geographers to collect data in the field. Drones are also used to inspect wind turbines for maintenance issues.

FIGURE 6 Bolivia Deforestation II Sensor: L5 TM acquisition date: July 2, 1986 and June 22, 2000. These false colour satellite images show the progression of deforestation in Bolivia caused by settlement expansion and agricultural development.

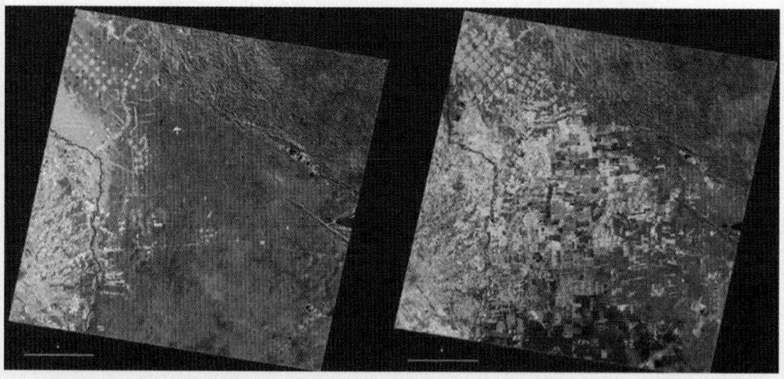

FIGURE 7 Lidar (Light Detection and Ranging) is a remote sensing device used to detect the topography of an area.

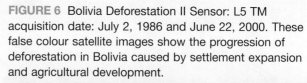

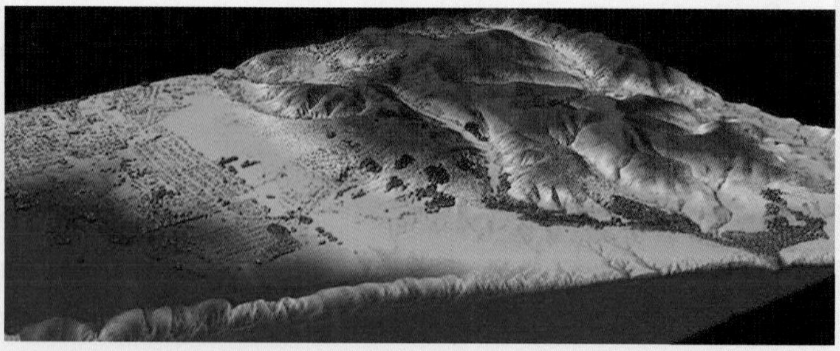

Learn how to use remote sensing images to create timelapse sequences using the **Remote sensing** fieldwork activity in your online Resources.

4.3 ACTIVITIES

1. Research soya-bean farming in the Amazon. How does it compare with cattle ranching in terms of environmental sustainability?
2. Examine the illustration of rainforest destruction shown in **FIGURE 4**. Draw a sketch of what you predict the area will look like in ten years' time. Use labels and arrows to show important features.
3. What alternatives exist to clearing forests for agriculture? Research the latest innovations in farming. How might they help to reduce the amount of forest that is cut down each year for farmland?

FIGURE 8 Soya-bean farm, São Paulo region, Brazil

4.3 EXERCISE

Learning pathways

■ LEVEL 1	■ LEVEL 2	■ LEVEL 3
3, 4, 7, 8, 14	1, 2, 9, 10, 15	5, 6, 11, 12, 13

Check your understanding

1. Refer to **FIGURE 2**. Describe the total loss in Brazilian forests since 1970. Use data in your answer.
2. What are the advantages and disadvantages of road building in the Amazon?
3. Why would subsistence farming in the Amazon be referred to as slash-and-burn farming?
4. Refer to **FIGURE 4**. Identify three changes to the river that have resulted from forest clearing.
5. Describe how a plantation is different to native forest.
6. How do new roads lead to greater deforestation, even after they are completed?

Apply your understanding

7. In what ways would the environmental changes of small-scale subsistence farming differ from those of large-scale soya-bean cropping?
8. Why might the large-scale clearing of tropical rainforests be considered an unsustainable practice?
9. Opening up the rainforest with roads can lead to fragmentation of the forest. What might the effect of this be on:
   a. native animals
   b. local indigenous populations?
10. Compare how a small-scale farmer from the Amazon and an environmentalist from another country might view the resources of a rainforest.
11. How might changes as a result of forest clearing affect farming downstream?
12. Discuss the impacts of forest fragmentation. In your response, you should outline at least one impact on native flora or fauna, one impact on farming practices, and one impact for Indigenous communities.

Challenge your understanding

13. Predict whether the trends in annual forest loss and total forest loss will continue. In your answer, outline three of the most important factors that will determine forest loss trends in Brazil in the future.
14. Propose one strategy to minimise the amount of run-off from the farm shown in **FIGURE 4** from entering the river. Draw a diagram or annotate **FIGURE 4** to show how you would implement this strategy.
15. Suggest two strategies that could be used to allow native animals to move across cleared land and roads safely when they move between sections of their habitat.

To answer questions online and to receive **immediate feedback** and **sample responses** for every question, go to your learnON title at www.jacplus.com.au.

4.4 SkillBuilder — GIS — Deconstructing a map

online only

LEARNING INTENTION

By the end of this subtopic, you will be able to deconstruct a topographic map to create a simple GIS overlay map.

The content in this subtopic is a summary of what you will find in the online resource.

4.4.1 Tell me

What is GIS?

A geographical information system (GIS) is a storage system for information or data, which is stored as numbers, words or pictures. The data has the location attached so that it may be viewed as a map or an image.

4.4.2 Show me

How to deconstruct maps to build a simple GIS

Step 1

Point features on the map have a location that may be defined using either a grid reference or latitude and longitude. Overlay one piece of tracing paper on the map and, using an appropriate colour, mark the homesteads (point data).

Step 2

Line features on the map may be straight or winding. Their location is determined by joining multiple points. Overlay a second piece of tracing paper on the map and, with an appropriate colour, trace the rivers and creeks (line features).

Step 3

A polygon is a shape that has many sides. Its location on the map is determined by joining multiple points. Overlay a third piece of tracing paper on the map and, with an appropriate colour, trace the forests (polygon data).

Step 4

Place the three tracing paper layers in the following order: point features on top, line features underneath and polygon features at the bottom. Provide BOLTSS for your map. In GIS, the finished map would be called a layout (see **FIGURE 6**).

FIGURE 6 The three layers of tracing paper are now combined, and BOLTSS is added.

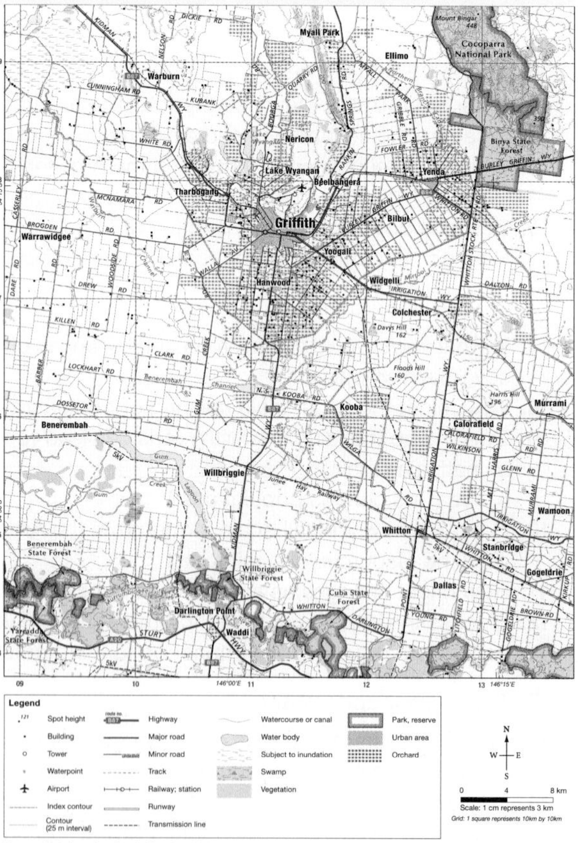

Source: Data based on Spatial Services 2019

4.4.3 Let me do it

learnON

Go to learnON to access the following additional resources to help you build this skill:
- longer explanation of this skill and its application in Geography (Tell me)
- a video demonstrating the step-by-step process of this skill (Show me)
- an activity and interactivity for you to practise the skill (Let me do it)
- self-marking questions to help you understand and use the skill.

on Resources

eWorkbook	SkillBuilder — GIS — Deconstructing a map (ewbk-8908)
Digital document	Topographic map of Griffith, New South Wales (doc-36316)
Video eLesson	SkillBuilder — GIS — deconstructing a map (eles-1730)
Interactivity	SkillBuilder — GIS — Deconstructing a map (int-3348)
Google Earth	Griffith, New South Wales (gogl-0149)

4.5 Paper production and its effect on biomes

online only

LEARNING INTENTION

By the end of this subtopic, you will be able to explain how biomes are altered to produce paper and describe effects of this.

The content in this subtopic is a summary of what you will find in the online resource.

Paper is a renewable, recyclable material, and yet something so common and so useful also poses significant environmental consequences to biomes at all stages of its usage cycle.

To learn more about the impact paper
use and production has on the environment, go to your learnON resources at www.jacPLUS.com.au.

Contents

learnON

- 4.5.1 Everyday paper use
- 4.5.2 The 'paperless' society
- 4.5.3 The impact of making and using paper
- 4.5.4 The future sustainability of paper

on Resources

eWorkbook	Paper production and its effect on biomes (ewbk-8912)
Video eLesson	Paper production and its effect on biomes — Key concepts (eles-5191)
Interactivity	Paper production and its effect on biomes (int-7921)

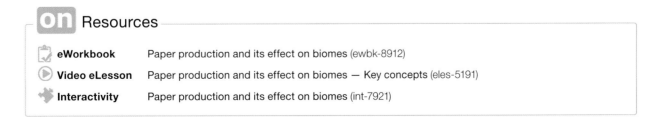

4.6 Overfishing our ocean biome

LEARNING INTENTION

By the end of this subtopic, you will be able to describe how humans have changed the ocean biome and the effects that this has. You will also be able to explain the benefits and problems of aquaculture.

4.6.1 Overfishing — causes and consequences

The ocean biome has often been seen as an unlimited resource of food for humans. However, overfishing is causing the collapse of many important marine ecosystems, and threatens the main source of protein for over one billion people. **Aquaculture** is a possible solution but, at the same time, it contributes to the decline in fish stocks.

Overfishing is simply catching fish at a rate higher than the rate at which fish species can repopulate. It is an unsustainable use of our oceans and freshwater biomes. Massive improvements in technology have enabled fish to be located and caught in larger numbers and from deeper, more inaccessible waters. The use of spotter planes, radar and factory ships ensures that fish can be caught, processed and frozen while still at sea.

Fish is the most common animal protein consumed around the world (see **FIGURE 1**). Historically, a lack of conservation and management of fisheries, combined with rising demand for fish products, has seen a 'boom and bust' mentality (see **FIGURE 2**). Large marine animals and fish species were fished until the species were extinct, endangered or so rare that they could no longer be caught in the numbers needed to sustain an income; then the next species was fished to the same extent. Examples of species that were over-fished include blue whales, Atlantic cod and bluefin tuna.

aquaculture the farming of aquatic plants and aquatic animals such as fish, crustaceans and molluscs

FIGURE 1 Global protein demand, 1980–2030* (million tonnes)

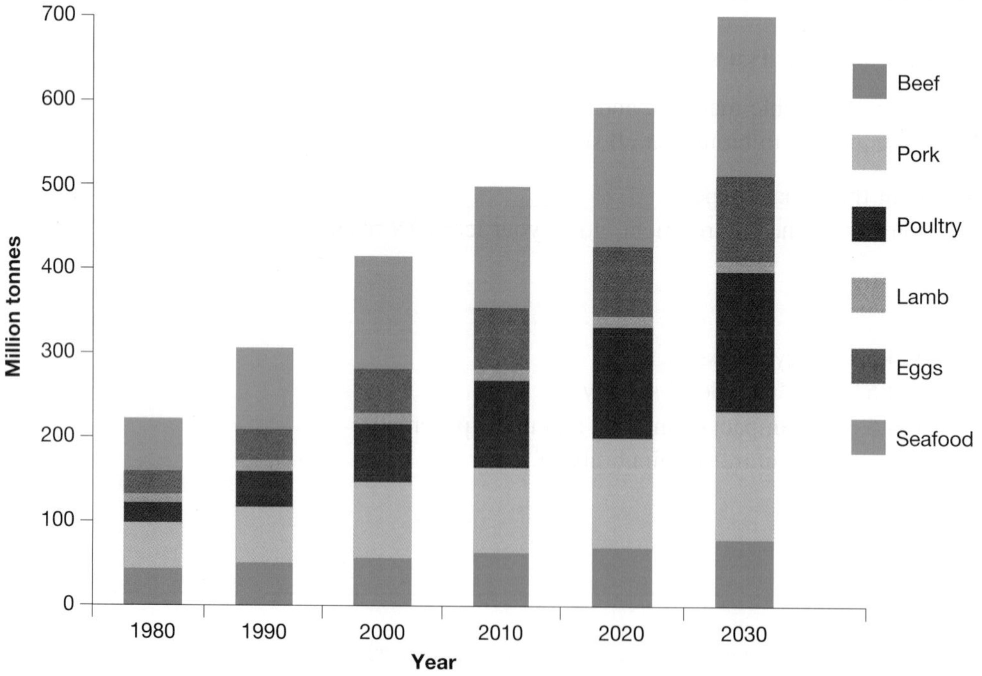

Source: OECD-FAO Agricultural Outlook

* projected

FIGURE 2 Unsustainable fishing

What happens when we overfish?

- With overfishing there are often large quantities of bycatch. This means that juvenile fish and other animals, such as dolphins and sea birds, are swept up in nets or baited on hooks before being killed and discarded. For every kilogram of shrimp caught in the wild, 5 kilograms of bycatch are wasted (see **FIGURE 3**).
- Destructive fishing practices such as cyanide poisoning, dynamiting of coral reefs and bottom trawling (which literally scrapes the ocean floor) cause continual destruction to local ecosystems.
- A large quantity of fish that could have been consumed by people is converted to fishmeal to feed the aquaculture industry, to fatten up pigs and chickens, and to feed pet cats. In Australia, the average cat eats 13.7 kilograms of fish a year. The average Australian eats 15 kilograms per year.
- Coastal habitats are under pressure. Coral reefs, mangrove wetlands and seagrass meadows, all critical habitats for fish breeding, are being reduced through coastal development, overfishing and pollution.

FIGURE 3 Up to 80 per cent of some fish catches is bycatch.

4.6.2 Aquaculture

Aquaculture (fish farming) is one of the fastest-growing food industries, providing fish for domestic and export markets. It brings economic benefits and increased food security (see **FIGURE 4**).

Since 2014, fish farming has produced more fish than fish caught in the wild; a harvest of 114.5 million metric tons was recorded in 2018. China is the largest farmed-fish producer, followed by India, Indonesia, Vietnam, Bangladesh, Egypt and Norway.

FIGURE 4 Global fish and aquaculture 1990–2030

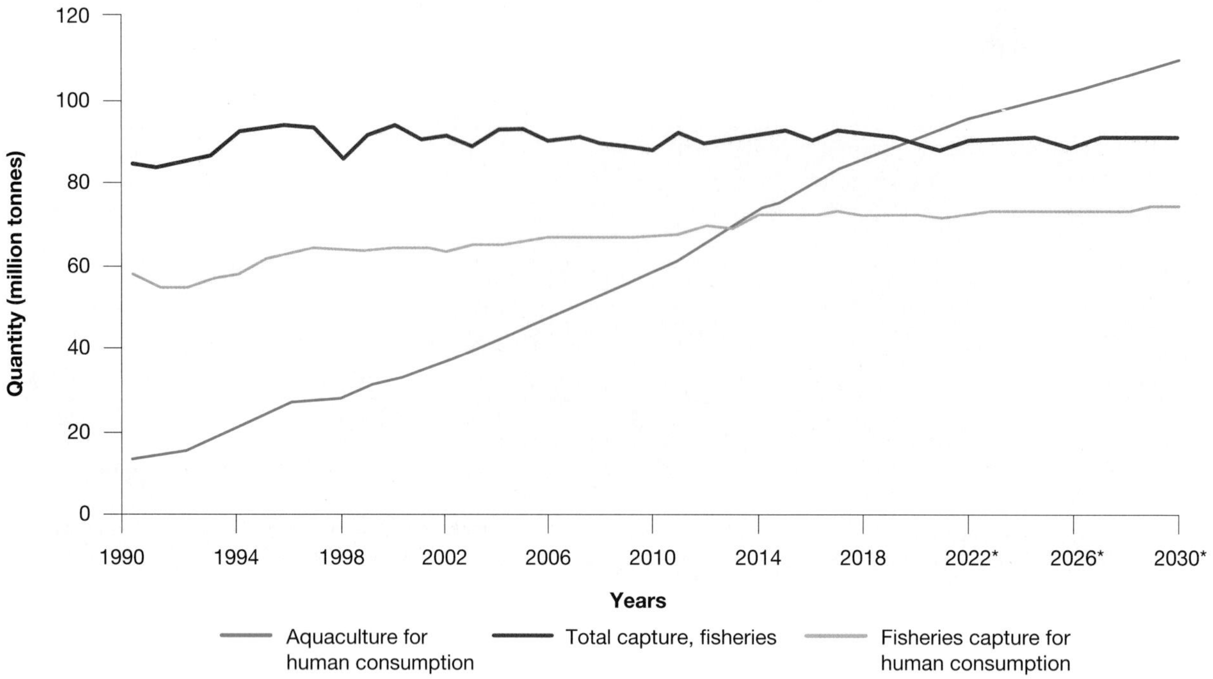

Source: Food and Agriculture Organization of the United Nations. *The State of World Fisheries and Aquaculture 2018. Meeting the Sustainable Development Goals*. Accessed: 3 March 2020. https://reliefweb.int/sites/reliefweb.int/files/resources/ca0191en.pdf

* projected

Australia's history of fish farming started more than 6600 years ago, when the Gunditjmara people created a series of fish traps in Lake Condah, in south-west Victoria, to capture a reliable source of eels. This eel trapping system is now protected as part of the Budj Bim National Heritage Landscape. Today, aquaculture is Australia's fastest growing primary industry, producing more than 40 per cent of Australia's seafood.

Negative impacts of aquaculture

Although aquaculture is often seen as a sustainable and eco-friendly solution to overfishing, its rapid growth and poor management in many places has created large-scale environmental change. Some of these changes are described below.

FIGURE 5 Feeding fish in pens, Thailand

- *Pollution.* Many fish species are fed a diet of artificial food in dry pellets (see **FIGURE 5**). Chemicals in the feed, and the massive waste generated by fish farms, can pollute the surrounding waters.
- *Loss of fish stock.* Food pellets are usually made of fish meal and oils. Much of this comes from bycatch; we are catching fish to feed fish. It can take 2 to 5 kilograms of wild fish to produce one kilogram of farmed salmon. Other ingredients in the food pellets include soybeans and peanut meal — products that are suitable for human consumption and grown on valuable farmland.
- *Loss of biodiversity.* Many of the fish species farmed are selectively bred to improve growth rates. If accidentally released into the wild, they can breed with native species and change their genetic makeup. This can lead to a loss of biodiversity. Capture of small ocean fish, such as anchovies, depletes food for wild fish and creates an imbalance in the food chain.

- *Loss of wetlands*. Possibly the greatest impact of aquaculture is in the loss of valuable coastal wetlands. In Asia, over 400 000 hectares of mangroves have been converted into shrimp farms. Coastal wetlands provide important ecological functions, such as protecting the shoreline from erosion and providing breeding grounds for native fish.

on Resources

eWorkbook	Overfishing our ocean biome (ewbk-8916)
Video eLesson	Overfishing our ocean biome — Key concepts (eles-5192)
Interactivity	Hook, line and sinker (int-3324)

4.6 ACTIVITIES

1. Investigate and write a newspaper article on the collapse of the Atlantic cod fishery in Newfoundland. What lessons in the sustainability of fishing can be learned from the case of the Atlantic cod?
2. Collect photographs and other information to create an annotated poster showing one of the destructive fishing practices mentioned in this subtopic.

4.6 EXERCISE

Learning pathways

■ LEVEL 1	■ LEVEL 2	■ LEVEL 3
2, 3, 8, 10, 13	1, 4, 7, 9, 14	5, 6, 11, 12, 15

Check your understanding

1. Explain how overfishing can lead to a loss of biodiversity.
2. What is aquaculture?
3. What is bycatch?
4. List three benefits and three drawbacks of fish farming.
5. Why is it difficult to manage wild fish capture and prevent overfishing?
6. How does the aquaculture industry pose a threat to wetlands?

Apply your understanding

7. Refer to **FIGURE 1**. How important is fish as a source of protein compared with other sources? Use figures in your answer.
8. Examine the photograph in **FIGURE 3** and describe the bycatch that you see.
9. Refer to **FIGURE 4**. Compare the predicted growth of fisheries capture (fish caught in the ocean) with aquaculture production to 2030.
10. Explain how fish food pellets used in aquaculture can cause pollution.
11. Explain how aquaculture can reduce the biodiversity of fish stocks in the wild.
12. Do the benefits of aquaculture outweigh the negative impacts on the environment? Provide reasons for your answer.

Challenge your understanding

13. Suggest one reason why wild fish capture will not increase greatly in the future.
14. What do you think the future of aquaculture might be? Explain your view.
15. Develop a fact-based advertising slogan to encourage people to feed their pets less fish.

To answer questions online and to receive **immediate feedback** and **sample responses** for every question, go to your learnON title at www.jacplus.com.au.

4.7 SkillBuilder — Interpreting a geographical cartoon

online only

LEARNING INTENTION

By the end of this subtopic, you will be able to outline some of the ways that a cartoon can illustrate a geographical theme or issue, and explain the point of view being expressed in a specific geographical cartoon.

The content in this subtopic is a summary of what you will find in the online resource.

4.7.1 Tell me

What are geographical cartoons?

Geographical cartoons are humorous or satirical drawings on topical geographical issues, social trends and events. A cartoon conveys the artist's perspective on a topic, generally simplifying the issue.

4.7.2 Show me

How to interpret a cartoon

Step 1

Study the cartoon to determine the overall idea.

Step 2

Consider your general knowledge of the topic. How does the cartoon relate to the topic? Things to look for in a cartoon include the following: symbolism, stereotyping, caricatures, visual metaphors, exaggeration or distortion, human, perspective, captions.

FIGURE 1 Cartoon on overfishing

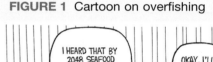

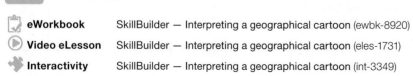

Step 3

To systematically analyse the cartoon, answer the following:

- What issue does the cartoon explore?
- What geographical concepts are related to the issue in the cartoon?
- What are the geographical implications of the cartoon?
- What point of view or opinion is the cartoonist expressing?

Step 4

Write a statement on how you feel about the topic of the cartoon and the cartoonist.

4.7.3 Let me do it

learnON

Go to learnON to access the following additional resources to help you build this skill:

- a longer explanation of this skill and its application in Geography (Tell me)
- a video demonstrating the step-by-step process of this skill (Show me)
- an activity and interactivity for you to practise the skill (Let me do it)
- self-marking questions to help you understand and use the skill.

on Resources

eWorkbook	SkillBuilder — Interpreting a geographical cartoon (ewbk-8920)
Video eLesson	SkillBuilder — Interpreting a geographical cartoon (eles-1731)
Interactivity	SkillBuilder — Interpreting a geographical cartoon (int-3349)

4.8 The impact of farming on land and water

LEARNING INTENTION

By the end of this subtopic, you will be able to outline the causes and effects of land degradation on our food producing lands, and discuss how poor food-production practices are creating environmental degradation.

4.8.1 Land degradation

The land, or lithosphere, is one of our most basic resources and is often taken for granted. In our quest to produce as much as possible from any one area of land, we have often failed to manage it sustainably. Land degradation is the result of such poor management.

FIGURE 1 Soil erosion as a result of overgrazing in Australia

Land degradation is a decline in the quality of the land to the point where it is no longer productive. Land degradation covers such things as soil **erosion**, invasive plants and animals, salinity and desertification. Degraded land is less able to produce crops, feed animals or renew native vegetation. There is also a loss in soil fertility because the top layers, rich in **humus**, can be easily eroded by wind or water. In Australia, it can take up to 1000 years to produce just three centimetres of soil, which can be lost in only a few minutes during a dust storm.

Globally, degradation has caused the loss of more than 350 million square kilometres of the Earth's surface. In Australia, of the five million square kilometres of land used for agriculture, more than half has been affected by or is in danger of degradation.

Land degradation is common to both the developed world and the developing world, and results from both human and biophysical causes.

Human causes

Human causes of land degradation involve unsustainable land management practices, such as:

- *land clearance* — deforestation or excessive clearing of protective vegetation cover
- *overgrazing of animals* — plants are eaten down or totally removed, exposing bare soil, and hard-hoofed animals such as cows and sheep compact the soil (see **FIGURE 1**)
- *excessive irrigation* — can cause watertables to rise, bringing naturally occurring salts to the surface, which pollute the soil
- *introduction of exotic species* — animals such as rabbits and plants such as blackberries become the dominant species
- *decline in soil fertility* — caused by continual planting of a single crop over a large area, a practice known as **monoculture**
- *farming on marginal land* — takes place on areas such as steep slopes, which are unsuited to ordinary farming methods.

erosion the wearing down of rocks and soils on the Earth's surface by the action of water, ice, wind, waves, glaciers and other processes

humus an organic substance in the soil that is formed by the decomposition of leaves and other plant and animal material

monoculture the cultivation of a single crop on a farm or in a region or country

Biophysical causes

Natural processes such as prolonged drought can also lead to land degradation. However, land can sometimes recover after a drought period. Topography and the degree of slope can also influence soil erosion. A steep slope will be more prone to erosion than flat land.

4.8.2 Impacts of land degradation

As land becomes degraded, productivity, or the amount of food it can produce, is lost. Some countries in sub-Saharan Africa have lost up to 40 per cent productivity in croplands over two decades, while population has doubled in the same time period. Farmers may choose to abandon the land or try to restore it; if the pressure to produce food is too great, they may have no choice but to continue using the land. Unproductive land will be exposed to continual erosion or weed invasion.

If extra fertilisers are applied to try to improve fertility, the excessive nutrients can create pollution and algae build-up in nearby streams. Airborne dust creates further hazards for both people and air travel. Land degradation is a classic example of human impact on all spheres of the environment — atmosphere, biosphere, lithosphere and hydrosphere.

FIGURE 2 shows the total amount of land across the Earth's surface (not just farmland) and the extent of land degradation. Twenty per cent of the world's cropland shows declining or stressed land productivity, despite the efforts and resources being used to maintain food production.

FIGURE 2 Global land productivity for selected land uses

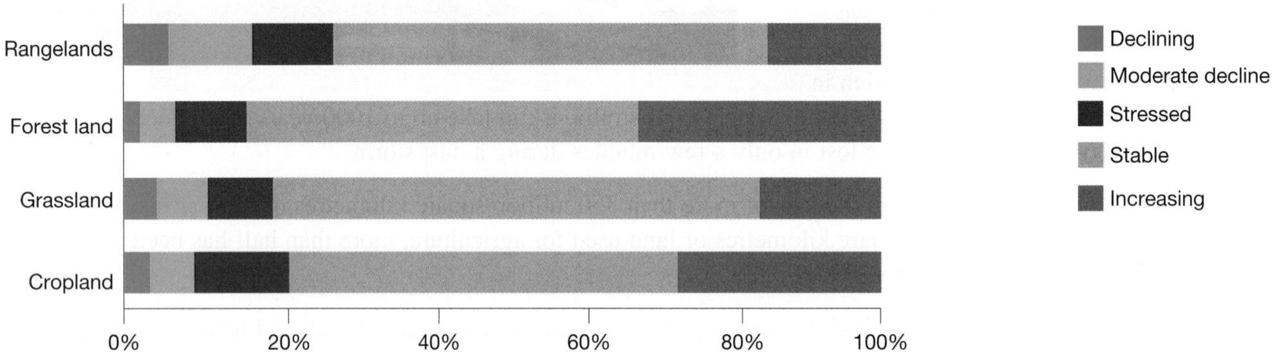

Notes:

Rangelands refers to shrublands mostly used for grazing.

Forest land applies to land with more than 40 per cent tree cover.

Grassland includes natural grasslands and pasture for grazing.

Cropland includes all arable land and where 50 per cent of land is used for crops.

Source: © European Union, 1995–2019. Cherlet, M., Hutchinson, C., Reynolds, J., Hill, J., Sommer, S.,von Maltitz, G. Eds., *World Atlas of Desertification*, Publication Office of the European Union, Luxembourg, 2018

About 40 per cent of degraded lands are found in places that experience widespread poverty, which is a contributing factor to food insecurity. Poor farmers with degraded land and few resources often have little choice but to continue to work the land. There is a strong interconnection between land degradation, migration and political instability. If declining soil quality and an increase in droughts due to climate change continue, between 50 and 700 million people could be forced to move by 2050.

Desertification is an extreme form of land degradation. It usually occurs in semi-arid regions of the world, and the result gives the appearance of spreading deserts. Desert biomes, or arid regions, are harsh, dry environments where few people live. In contrast, semi-arid regions, or drylands, occupy 41 per cent of the Earth's surface and support over two billion people, 90 per cent of whom live in developing nations. Although traditional grazing and cropping has taken place in dryland regions for centuries, population growth and the demand for food has put enormous pressure on land resources. Overclearing of vegetation, overgrazing and overcultivation are a recipe for desertification.

FIGURE 3 The Hunter River and Valley with the contrast of dry landscape and green irrigated fields, 2019

4.8.3 Positive impacts of irrigation

Food production and security is directly related to water availability. As the population increases, so too does demand for water. Moreover, there are always competing demands for water from the domestic, industrial and environmental sectors. In many places in the world, water is becoming increasingly scarce. Consequently, the development of water resources is becoming more expensive and, in some cases, environmentally destructive.

Most of the world's food production is rain-fed; that is, dependent on naturally occurring rainfall. Only a small proportion of agricultural land is irrigated, yet irrigation is now the biggest user of water in the world, consuming 70 per cent of the world's freshwater resources. Irrigation brings many benefits, such as:

- supplementing or replacing rain, especially in places where rainfall is low or unreliable. In many parts of the world, it is not possible to produce food without irrigation.
- increasing crop yields, up to three times higher than rain-fed crops. Only 20 per cent of the world's farmland is irrigated, but this land produces over 40 per cent of our food.
- enabling a wide variety of crops to be grown, especially those with high water needs, such as rice, or with high value, such as fruit and wine grapes
- flexibility, being used at different times according to crop needs, for example during planting and growing or close to harvest time.

4.8.4 Negative impacts of irrigation

Although irrigation has resulted in increased food production and greater food security, it has also created major changes to the biomes where it is used. Irrigation changes the natural environment by extracting water from rivers and lakes and through the building of structures to store, transfer and dispose of water. The topography, or shape of the land, is often changed too, such as when terraces are built for **paddy fields**. In addition, irrigation water is often applied to the land in much larger quantities than naturally occurs, which can lead to changes in soil composition, and **waterlogging** and salinity problems.

How does irrigation create salinity problems?

Overwatering of shallow-rooted crops adds excess water to the **watertable**, causing it to rise (see **FIGURE 4**).

If the subsoils are naturally salty, much of this salt can be drawn to the surface. Most crops and pasture will not grow in salty soils, so the land becomes useless for farming. Land that is affected by salinity is also more prone to wind and water erosion.

Globally, some 62 million hectares of land (an area the size of France) has been lost due to such issues. Salinity is also a major cause of land degradation in Australia (see **FIGURE 5**).

paddy field a growing area that is flooded to farm semi-aquatic plants such as rice

waterlogging saturation of the soil with groundwater such that it hinders plant growth

watertable the upper level of the groundwater, below which all pores in the soils and rock layers are saturated with water

FIGURE 4 The development of irrigation salinity

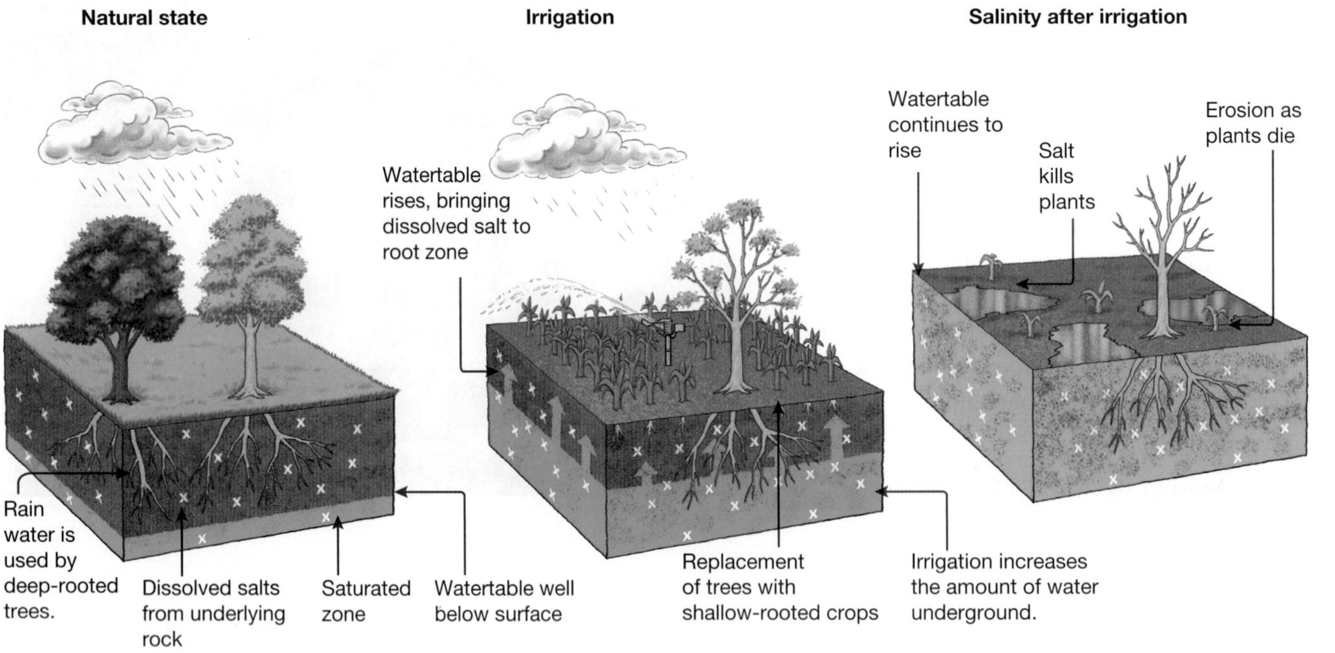

Global population and demand for water is constantly increasing. Moreover, there are always competing demands for water from the domestic, industrial and environmental sectors. For countries that have growing populations and limited water resources, water deficits and food insecurity are a growing concern. In many places in the world, water is becoming increasingly scarce. Consequently, the development of water resources is becoming more expensive and, in some cases, environmentally destructive.

FIGURE 5 The distribution of salinity in Australia

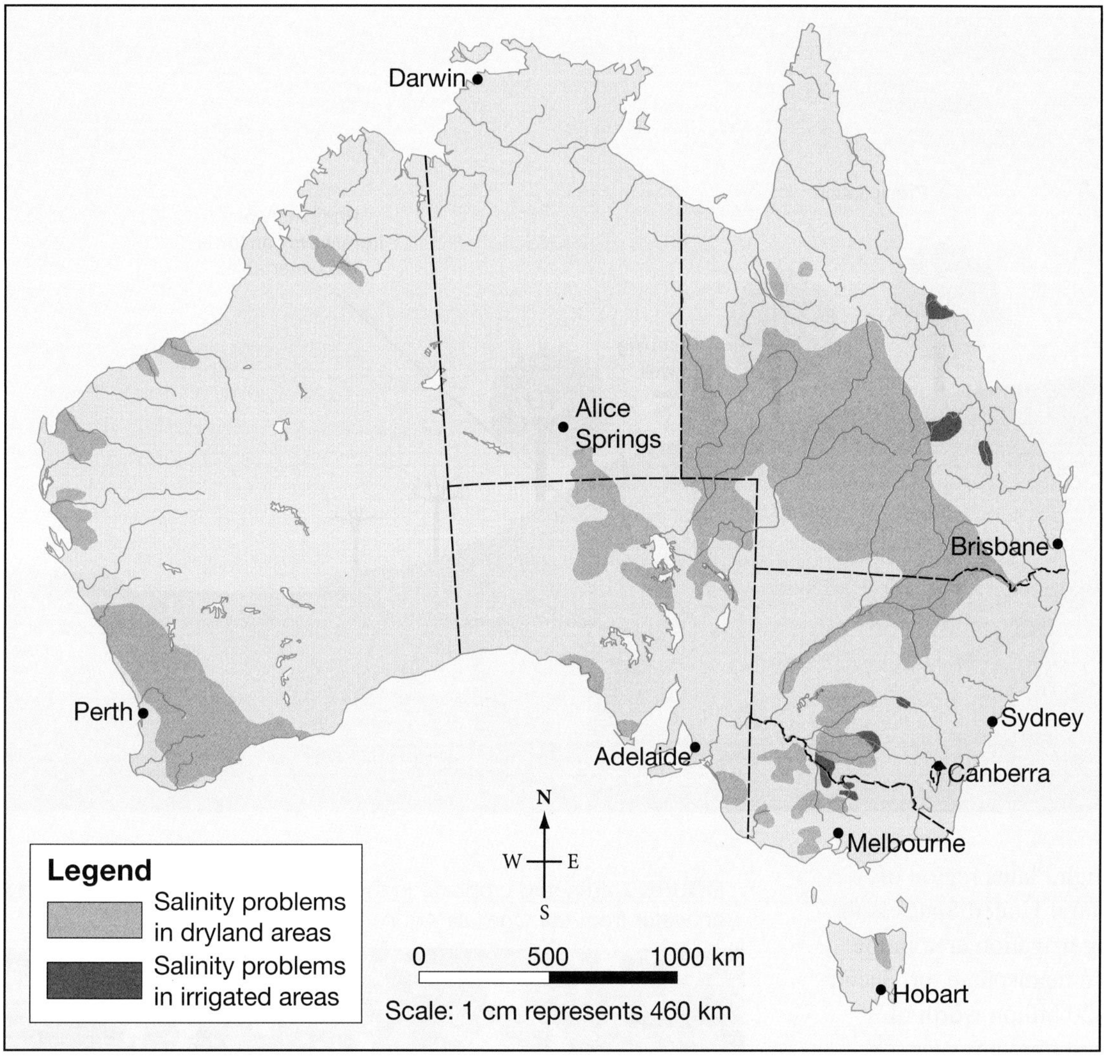

Source: Spatial Vision

For thousands of years, farmers have diverted water from rivers, lakes and wetlands for watering crops and pastures in dry areas. Large-scale irrigation schemes can effectively 'water' our deserts but, if too much water is used, wetlands can dry out, rivers cease to flow, and lakes and underground **aquifers** shrink. It is estimated that between three and six times more water is held in reservoirs around the world than exists in natural rivers. It is possible that the level of water extraction will nearly double by 2050.

As surface water resources become fully exploited, people turn to underground water sources. Improvements in technology have also enabled farmers to pump water from aquifers deep underground (see **FIGURE 6**).

Groundwater levels do not respond to changes in the weather as rapidly as rivers and lakes do. If the water is removed unsustainably (at a rate that is faster than the rate of replenishment by rainfall, run-off or underground flow), then watertables fall. Water then becomes harder and more expensive to extract. Water stored in aquifers can take thousands of years to replenish. Over-extraction of groundwater can result in wells running dry, reduced stream flow, and even land subsidence (sinking).

aquifer a body of permeable rock below the Earth's surface that contains water, known as groundwater

FIGURE 6 The use of groundwater as a water source in farming

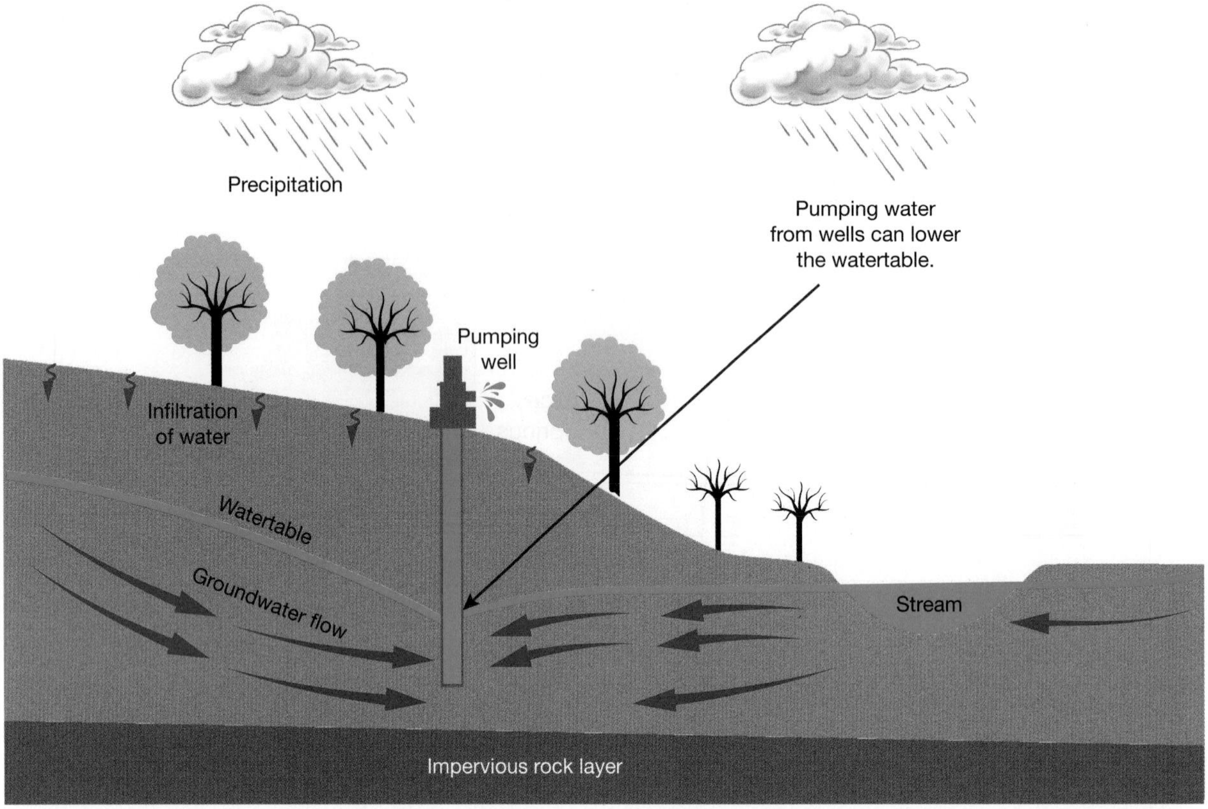

The High Plains region of the central United States is the leading irrigation area in the western hemisphere, producing over $20 billion worth of food and fibre per year (see **FIGURE 7**). In all, 5.5 million hectares of semi-arid land is irrigated using water pumped from the huge Ogallala Aquifer (see **FIGURE 8**). Since large-scale irrigation was developed in the 1940s, groundwater levels have dropped by more than 30 metres. Pesticides and other pollutants from farming have also infiltrated underground aquifers. Scientists estimate that if the aquifer was pumped dry, it would take over 6000 years to refill it naturally.

FIGURE 7 Irrigated cropland in the central United States relies heavily on water from the Ogallala Aquifer.

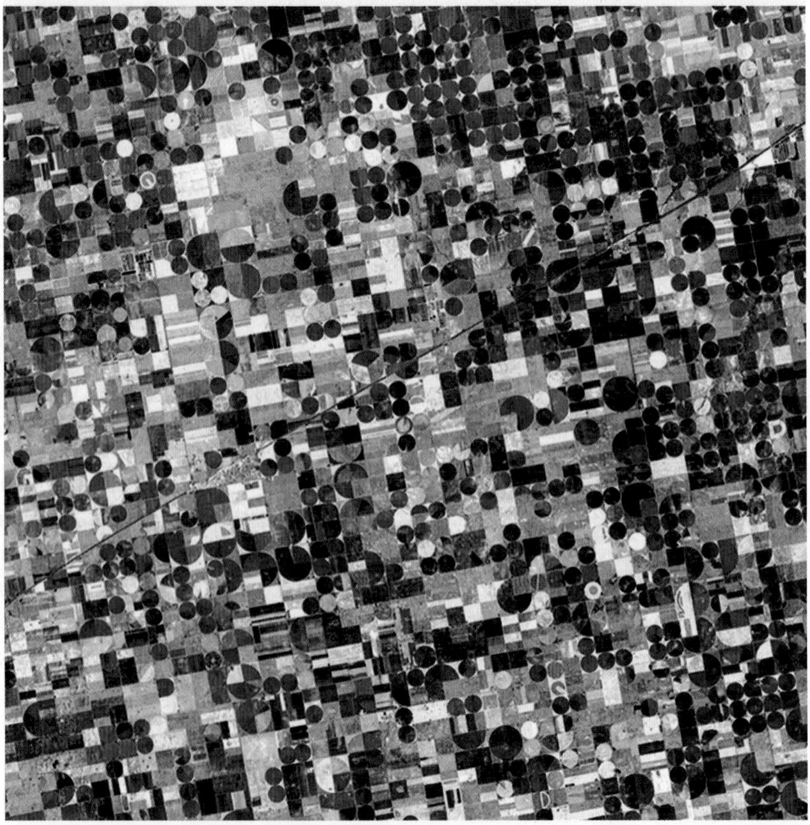

FIGURE 8 The size of the Ogallala Aquifer in the central United States

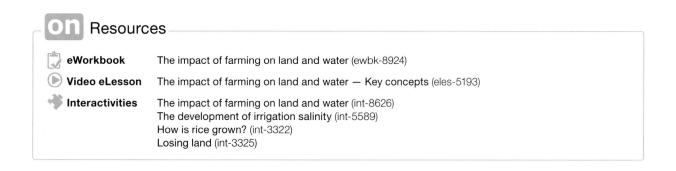

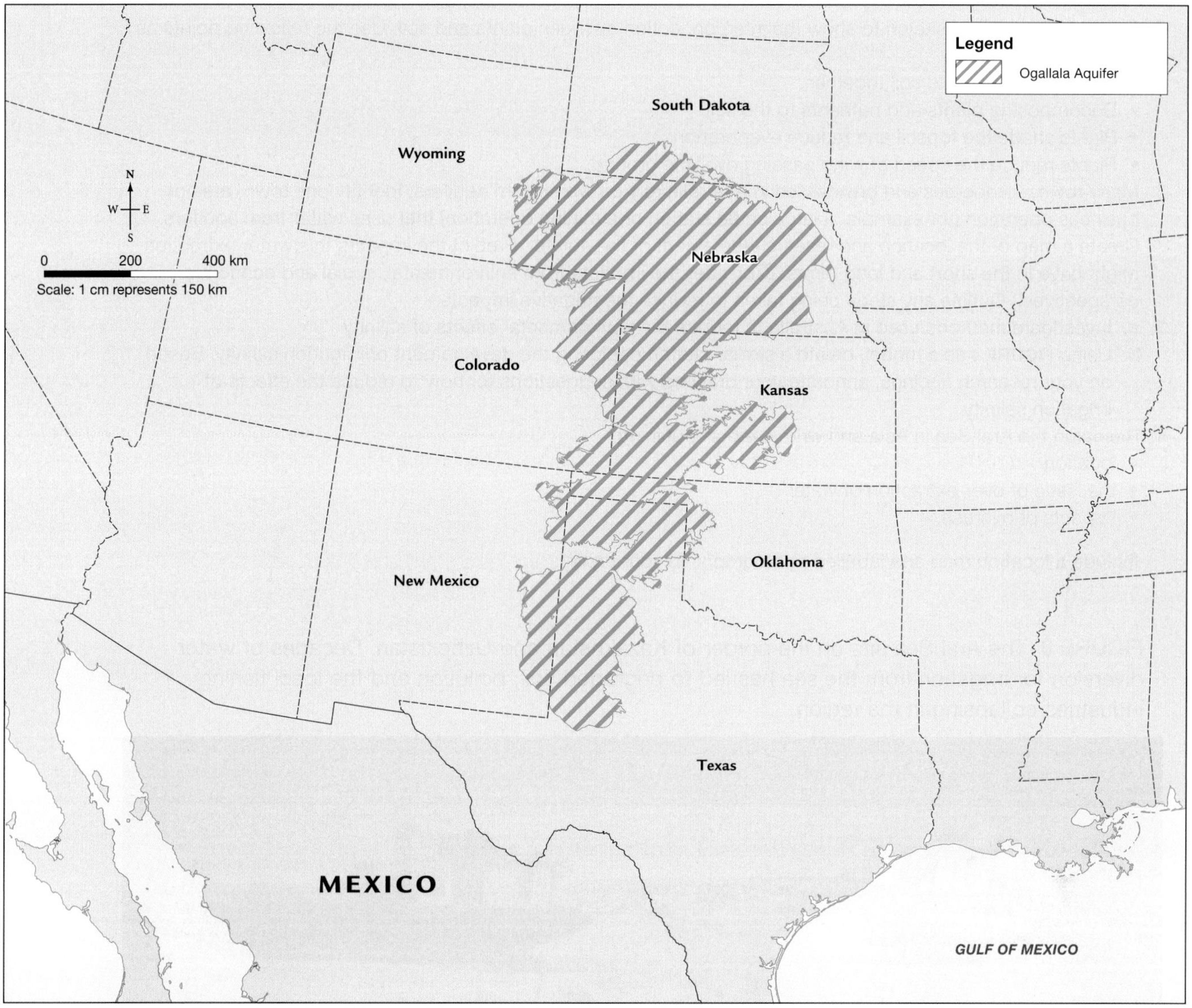

Source: USGS

on Resources

eWorkbook	The impact of farming on land and water (ewbk-8924)
Video eLesson	The impact of farming on land and water — Key concepts (eles-5193)
Interactivities	The impact of farming on land and water (int-8626) The development of irrigation salinity (int-5589) How is rice grown? (int-3322) Losing land (int-3325)

4.8 ACTIVITIES

1. Create an annotated sketch to show the interconnection between plants and soil. Use the following points as labels on your sketch.
   - Plant roots help hold soil together.
   - Decomposing plants add nutrients to the soil.
   - Plants shade the topsoil and reduce evaporation.
   - Plants reduce the speed of wind passing over the ground.
2. Many towns, properties and businesses in Australia rely on water from aquifers. Identify one town, area or business operation (for example, a large cattle station or a mining operation) that uses water from aquifers. Create a map of the location and research the extent of the aquifer. Predict the impacts this water extraction might have in the short and long terms. (Consider the impacts from environmental, social and economic perspectives.) Outline any steps being taken to reduce the negative impacts.
3. a. Investigate methods used in Australia to reduce the environmental effects of salinity.
   b. Using **FIGURE 4** as a model, create a similar sketch depicting the development of irrigation salinity. Based on your research findings, annotate your drawing with suggestions for how to reduce the effects of irrigation salinity.
4. Research the Aral Sea in Asia and write a report outlining:
   - location
   - the issue of over-extraction of water
   - impacts of overuse.

   Include a location map and labelled photographs in your report.

FIGURE 9 The Aral Sea sits on the border of Kazakhstan and Uzbekistan. Decades of water diversion for irrigation from the sea has led to degraded soil, pollution and the local fishing industries collapsing in the region.

4.8 EXERCISE

Learning pathways

■ LEVEL 1	■ LEVEL 2	■ LEVEL 3
1, 2, 4, 10, 15	3, 5, 8, 9, 11	6, 7, 12, 13, 14, 16

Check your understanding

1. Identify four human causes of land degradation.
   a. Over-cultivation
   b. Steep slopes
   c. Over-irrigation
   d. Farming on marginal lands (for example, steep slopes or poor soils)
   e. The introduction of feral plant species
   f. Floods
2. Identify three natural causes of land degradation.
   a. The introduction of feral animal species
   b. Steep slopes
   c. Drought
   d. Clearing of vegetation
   e. Flood
   f. Overgrazing
3. Outline three changes to the environment that are needed in order to irrigate land for agriculture.
4. Consider the photograph in **FIGURE 1**. Why would it be difficult to either graze animals or grow crops on this land?
5. Identify the different types of water resources that can be used to supply water for food production.
6. What is meant by the term waterlogging?
7. What percentage of the world's fresh water is consumed by irrigation?

Apply your understanding

8. Explain why land degraded by drought may recover, whereas land degraded by cultivation may not.
9. Study the **FIGURE 5** map, which shows the distribution of salinity in Australia. Why do you think dryland salinity covers a larger area than irrigation salinity?
10. Study **FIGURE 8**. Explain how pumping groundwater can lower watertables.
11. Compare the advantages and disadvantages of using groundwater and surface water for farming.
12. Examine the photograph in **FIGURE 1**. If this was your property and your livelihood, what steps would you take to reduce the erosion problem?
13. Refer to **FIGURE 2**.
    a. Which land cover has the greatest percentage of stressed and declining productivity?
    b. What type of farming activities could explain the increased productivity in croplands?

Challenge your understanding

14. Soil salinity from over-cultivation was not an issue before European colonisation. What does this suggest about land management practices in this country before 1788?
15. Has irrigation been a success or failure in Australia? Write a paragraph expressing your viewpoint.
16. Select one of the examples shown in this subtopic and consider what steps water managers could take to reduce the impact of unsustainable water use in the region.

To answer questions online and to receive **immediate feedback** and **sample responses** for every question, go to your learnON title at www.jacplus.com.au.

4.9 The impact of farming on the atmosphere

LEARNING INTENTION

By the end of this subtopic, you will be able to explain how farming effects the atmosphere and describe the negative impacts of this.

4.9.1 Farming's impact on the climate

Agriculture's dependence on the atmosphere, in particular climate, is evident by the interconnection of rain and temperature patterns, and where food can be grown. However, as the climate changes, so will the distribution of cropland. As agriculture has grown globally and made significant modifications to many global biomes, there is evidence to suggest that agriculture has contributed to climate change.

It has been observed that the Earth's climate is experiencing changes that have been influenced by human activities. In particular, the Earth's temperature is warming, partly due to increases in greenhouse gases such as water vapour, carbon dioxide, methane and nitrous oxide that are entering the atmosphere. These increases in greenhouse gases have been attributed to human activities, and agriculture has had a significant part to play. Food production has contributed to changes in the climate in two ways:

FIGURE 1 A cartoonist's view of livestock and global warming

- Grazing animals and flooded rice paddies produce the greenhouse gas methane. Methane is 20 times more effective at warming the planet than CO_2. Livestock are thought to be responsible for 29 percent of the world's methane output. The next largest sources, in order, are oil and gas, landfill, rice paddies and wastewater treatment systems.
- Food production changes the surface of the Earth, which then alters the planet's ability to absorb or reflect heat and light. Large-scale deforestation and desertification can significantly alter the **microclimate** of a region. Around 80 per cent of global deforestation is caused by clearing the land for agriculture such as grazing, slash and burn farming and cropping.

microclimate the climate of a small area

4.9.2 Agricultural sources of pollution

Cows emit large quantities of methane through belching and flatulence (caused by digestive gases). The gas is produced by bacteria digesting grass in one of the four stomachs that cows have. It has been estimated that one cow could produce somewhere between 100 and 500 litres of methane per day (see **FIGURE 2**). This amount is similar to the pollution produced by one car in one day. When you consider there are over 1.5 billion cows in the world, this equates to a lot of gas. Scientists today are working on 'fuel-efficient cows' — cows that convert feed more efficiently into milk rather than methane.

FIGURE 2 Argentine scientists are strapping plastic tanks to cows to assess how much methane they produce.

Rice farming is one of the biggest sources of human-produced methane, averaging between 50 and 100 million tonnes per year. The gas is produced in the warm, waterlogged soils of the rice paddies (see **FIGURE 3**).

FIGURE 3 Methane is produced and released by rice paddies.

FIGURE 4 Factory farming produces large quantities of waste products.

The practice of **factory farming**, in which a very high number of animals are concentrated in one place, produces an unmanageable amount of waste (see **FIGURE 4**). On a sustainable farm, animal manure can be used as a natural fertiliser, but on a factory farm the large quantity becomes a source of methane, because the waste is often mixed with water and stored in large ponds or lagoons. An additional problem can occur if these ponds leak, as they create soil and water pollution. The use of nitrogen-based fertilisers on farms also releases nitrous oxide, another greenhouse gas.

4.9.3 Deforestation's impact on the climate

Trees are 50 per cent carbon, so when they are burned or felled, the CO_2 they store is released back into the atmosphere. On average, 13 million hectares of the world's forests are lost each year, mostly in tropical regions of South-East Asia, Latin America and Africa. Deforestation accounts for 30 per cent of greenhouse gases released into the atmosphere each year. Forests also act as carbon sinks, the most effective way of storing carbon. Large areas of cleared land absorb more heat than native vegetation, which can lead to changes in local weather conditions.

factory farming the raising of livestock in confinement, in large numbers, for profit

Resources

eWorkbook The impact of farming on the atmosphere (ewbk-8928)

Video eLesson The impact of farming on the atmosphere — Key concepts (eles-5194)

Interactivity The impact of farming on the atmosphere (int-8627)

4.9 ACTIVITIES

1. Research the sources of methane gas and find out what percentage each contributes to world methane output. Present your information in a pie or bar graph. Is the biggest source natural or human?
2. If one cow produces the equivalent of 16 kWh of energy per day, how many cows would be needed to power your own home per day? You will need to check your household electricity bill.
3. How might climate change affect agriculture? Do some online research to investigate some of the possible effects of climate change on food production.

4.9 EXERCISE

Learning pathways

LEVEL 1	LEVEL 2	LEVEL 3
1, 2, 3, 10, 13	4, 5, 8, 11, 14	6, 7, 9, 12, 15

Check your understanding

1. List three greenhouse gases.
2. Outline two ways in which deforestation can contribute to changes in greenhouse gases.
3. What is the biggest global emitter of methane: livestock, cars or rice paddies?
4. Outline two ways that farming creates pollution.
5. Why might a researcher want to measure the greenhouse gas production of a cow?
6. What is a carbon sink?
7. How are agriculture and climate interconnected?

Apply your understanding

8. Is factory farming a sustainable form of food production? Give reasons for your answer.
9. The building of large-scale dams and subsequent flooding of forests in the Amazon is also contributing to increases in greenhouse gas emissions. What is the reason for this?
10. Why are trees cut down for farming? Explain what this achieves for farm productivity and management.
11. Study **FIGURE 4**.
    a. How does factory farming differ from traditional farming methods? Create a table with two columns, one headed 'Key features of a traditional dairy farm' and the other 'Key features of a factory farm'. List the features of both styles of farming and then compare your lists.
    b. Write a paragraph summarising the similarities and differences between the two the two methods.
12. Explain how deforestation can change the microclimate of a region, especially temperatures and moisture levels. (A microclimate is a local set of atmospheric conditions that differ from those in the surrounding areas.)

Challenge your understanding

13. Suggest some ideas for reducing agriculture's contribution to global warming.
14. Agriculture and climate change are interconnected processes. Suggest possible ways that increased temperatures and increased frequency of storms could impact on food production in Australia.
15. Propose one strategy that would reduce the negative environmental impact of a dairy farm. Explain what impact your strategy would reduce and why.

To answer questions online and to receive **immediate feedback** and **sample responses** for every question, go to your learnON title at www.jacplus.com.au.

4.10 Diminishing global biodiversity

LEARNING INTENTION

By the end of this subtopic, you will be able to describe and give examples of the negative impacts of agriculture on biodiversity.

4.10.1 Diminishing biodiversity

The last few centuries have seen the greatest rate of species extinction in the history of the planet (see **FIGURE 1**). The population of most species is decreasing, and genetic diversity is declining, especially among species that are cultivated for human use. Six of the world's most important land biomes have now had more than 50 per cent of their area converted to agriculture (see **FIGURE 2**).

FIGURE 1 Extinctions per thousand species per millennium

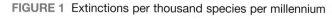

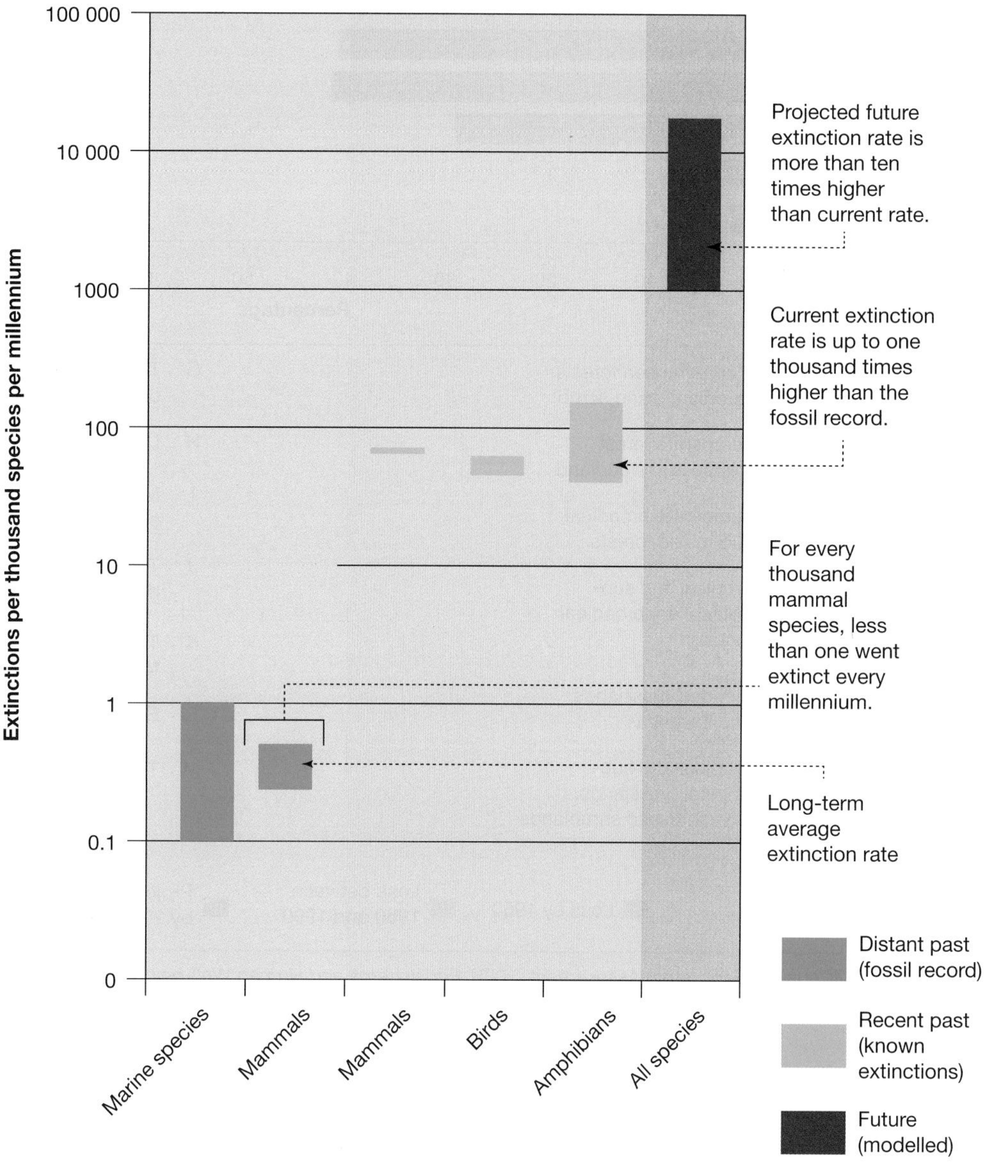

Source: Millennium Ecosystem Assessment, 2005. *Ecosystems and Human Well-being: Synthesis.* Island Press, Washington, DC. Copyright © 2005 World Resources Institute

In places where there has been very little industrial-scale farming, a huge variety of crops are still grown. In Peru, for example, over 3000 different potato varieties are still cultivated. Elsewhere, biodiversity as well as agricultural biodiversity (biodiversity that is specifically related to food items) is in decline. In Europe, 50 per cent of all breeds of domestic animals have become extinct, and in the USA, 6000 of the original 7000 varieties of apple no longer exist. The International Union for Conservation of Nature's Red List estimates that about 640 species (combined) of plant, animal and fungus have become extinct globally since the year 2000.

FIGURE 2 Percentage of biomes converted to agriculture over time

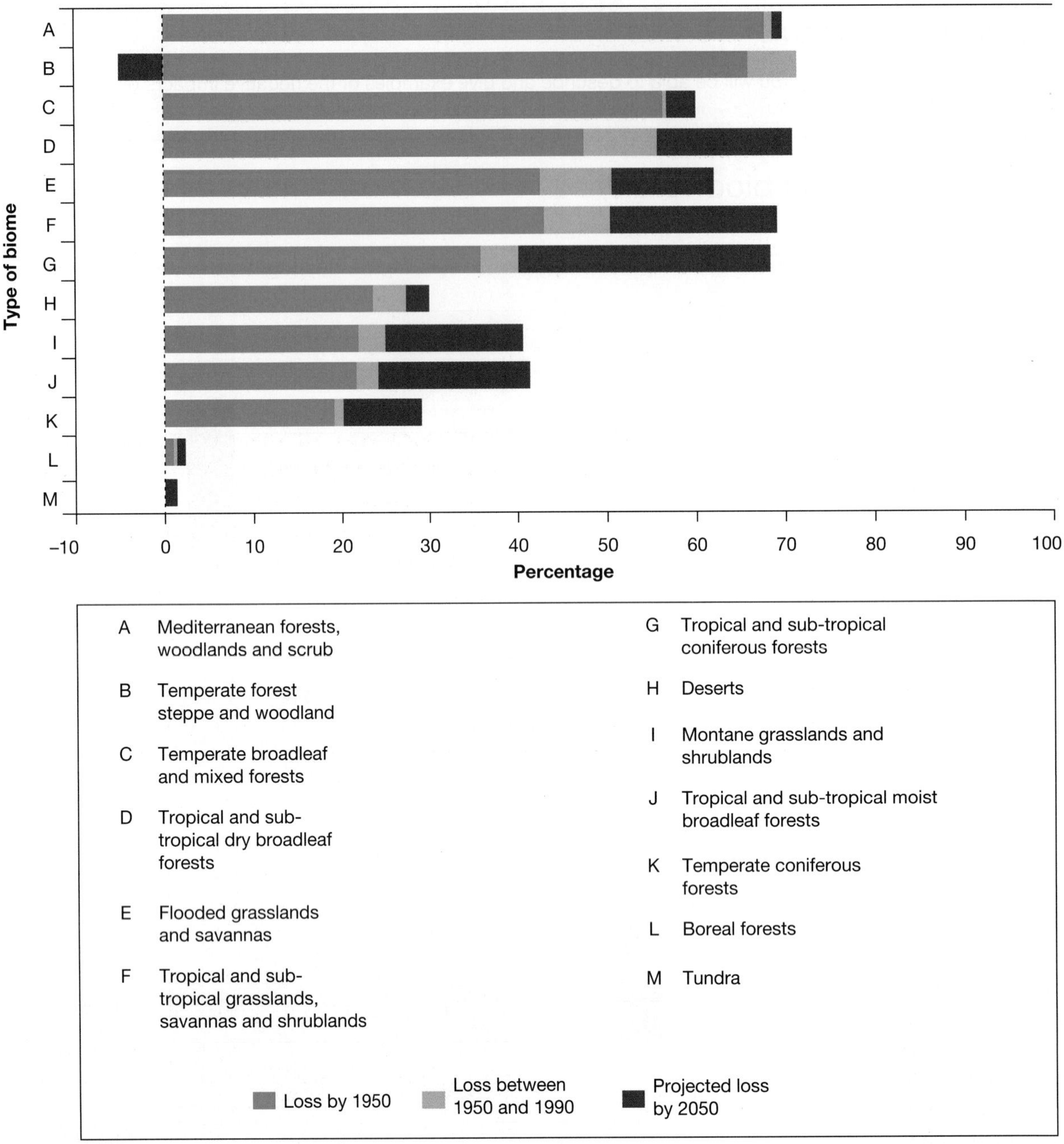

Source: Millennium Ecosystem Assessment, 2005. *Ecosystems and Human Well-being: Synthesis.* Island Press, Washington, DC. Copyright © 2005 World Resources Institute

Reasons for diminishing biodiversity

- Converting natural habitats to cropland and other uses replaces systems that are rich in biodiversity with monoculture systems that are poor in diversity (see **FIGURE 4**).
- Industrial-scale farming and new high-yielding, genetically uniform crops replaces thousands of different traditional species. Two new rice varieties in the Philippines account for 98 per cent of cropland.
- Uniform crops are vulnerable to pests and diseases, which then require large inputs of chemicals that ultimately pollute the soil and water. Traditional ecosystems have many natural enemies to combat pest species.
- The introduction of modern breeds of animals has displaced indigenous breeds. In the space of 30 years, India has lost 50 per cent of its native goat breeds, 30 per cent of sheep breeds and 20 per cent of indigenous cattle breeds.

4.10.2 Australia's biodiversity

Australia has a high number of **endemic** species and has seven per cent of the world's total species of plants, animals and micro-organisms. That makes Australia one of only 17 countries in the world that are classified as megadiverse — having high levels of biodiversity. These 17 nations combined contain 75 per cent of the Earth's total biodiversity (see **FIGURE 3**). Australia's unique biodiversity is due to its 140 million years of geographic isolation. However, Australia has experienced the largest documented decline in biodiversity of any continent over the past 200 years. It is thought that 50 species of animals (27 mammal species and 23 bird species) and 48 plant species are now extinct.

The sustainable land and resource management practices of Aboriginal and Torres Strait Islander Peoples carried out over many thousands of years ensured food security for people and respect for the lands, waterways, lakes and marine environments that sustained them. Rotational land occupation, sustainable fishing practices and cultural burning ensured that both biodiversity and food security were maintained. However, since 1788, farming practices such as land clearing by European settlers have led to significant habitat loss and reduction in biodiversity in Australia.

In 2019–2020, agriculture was worth 65 billion dollars to the Australian economy. While farming is important economically and for providing food to fccd thc population, 30 million hectares of farming land is also being protected by farmers to conserve native vegetation, and in turn the habitats of many animals. Farmers are using strategies such as excluding or reducing access to livestock, managing pests and feral animals, managing weeds, planting seeds of native vegetation and retaining existing native vegetation. Farmers have been able to achieve this by working closely with the government and other organisations such as Landcare.

endemic describes species that occur naturally in only one region

FIGURE 3 Distribution of megadiverse countries

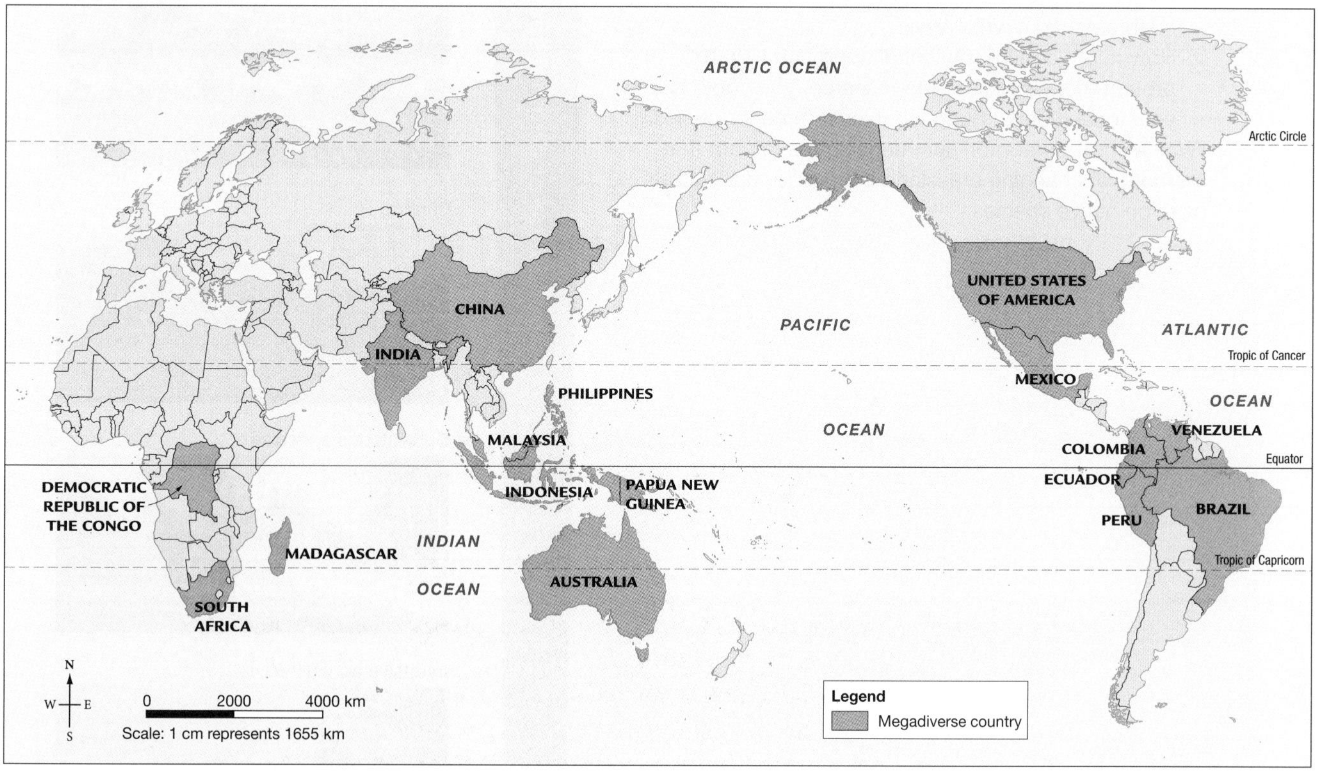

Source: Spatial Vision

Resources

eWorkbook Diminishing global biodiversity (ewbk-8932)

Video eLesson Diminishing global biodiversity — Key concepts (eles-5195)

Interactivity Diminishing global biodiversity (int-8628)

myWorldAtlas Deepen your understanding of this topic with related case studies and questions.
- World species
- Endangered species in Australia
- Introduced species in Australia
- Tasmanian devil

4.10 ACTIVITIES

1. Investigate the issue of whaling and conflicting viewpoints held by Australia and Japan.
   a. What were the factors (reasons) involved in:
      i. Australia's decision to ban whaling?
      ii. Japan's decision to continue whaling?
   b. How would you suggest the two counties could come to a resolution?
2. Australia has experienced the largest documented decline in biodiversity of any continent over the past 200 years.
   a. Investigate:
      i. land management practices that were used in your area before European colonisation
      ii. plant and animal species that are endangered or threatened in your area.
   b. How might a more traditional approach help to protect these species? (As research, you could ask your local Elders.) Suggest at least three strategies.
3. Write a short paragraph summarising the impact that methods of changing grassland biomes for agriculture have on native species.

FIGURE 4 Changes to percentage of original species according to changes in biomes for food production

4. Research genetically modified foods and their impact on the environment. Complete a table like the one shown to help you form your opinion of whether they are, overall, a positive or negative development.

Impacts of genetically modified food							
Environmental		Social		Economic		Ethical	
Positive	Negative	Positive	Negative	Positive	Negative	Positive	Negative

4.10 EXERCISE

Learning pathways

■ LEVEL 1	■ LEVEL 2	■ LEVEL 3
1, 2, 9, 10, 14	3, 4, 7, 8, 13	5, 6, 11, 12, 15

Check your understanding

1. Describe the ways in which human activities can lead to a loss of biodiversity.
2. What is a megadiverse country?
3. Why is Australia considered a megadiverse country?
4. Outline three ways that farmers can help to protect Australia's biodiversity.
5. How does farming based on monoculture systems contribute to reduced biodiversity?
6. Describe the spatial distribution of megadiverse countries shown in **FIGURE 3**.

Apply your understanding

7. Study **FIGURE 2**.
   a. Which three biomes have seen the greatest percentage change in areas converted to agriculture? Use figures in your answer.
   b. Why might these three have had the most change?
8. Study the information in **FIGURE 4**. Describe the changes to the grassland biome as seen over time.
9. In what ways would Aboriginal Peoples' practice of rotational land occupation have helped maintain biodiversity before the European occupation of Australia?
10. Does it matter that we have fewer species of apple or goats? Explain your view with reference to what you have learned in this subtopic.
11. What factors might account for less land in desert and tundra areas being converted to agriculture than in some other biomes?
12. Compare the rates of extinction in the distant and recent past, based on **FIGURE 1**. What factors might account for the differences and/or similarities you found?
13. What are the potential consequences of an animal's extinction? Using one specific example, expand on what some of the short- and long-term consequences might be if a specific animal were to die out.

Challenge your understanding

14. What impacts might genetically modified crops have on species diversity?
15. Suggest how the environmental impacts of a traditional small-scale farm might compare with the environmental impacts of a large-scale producer.
16. Do you think it will be possible, in the future, for Australia to maintain its megadiverse status? What actions might contribute to this?

To answer questions online and to receive **immediate feedback** and **sample responses** for every question, go to your learnON title at www.jacplus.com.au.

4.11 Investigating topographic maps — Food production in the Riverina

LEARNING INTENTION

By the end of this subtopic, you will be able to describe how the Murrumbidgee Irrigation Area has been changed to produce food, giving examples from a topographic map.

4.11.1 The Murrumbidgee Irrigation Area (MIA)

The Murrumbidgee Irrigation Area (MIA) is located within the Riverina region of New South Wales. The climate in this area is semi-arid (warm, with unreliable rainfall). The land only became productive after irrigation was available. The MIA was established in 1912 to control and divert water from local rivers and creeks to produce food. An elaborate series of weirs, canals and holding ponds, fed by upstream rivers and dams, has been established throughout the region. Today approximately a quarter of the area of MIA's 660 000 hectares is irrigated.

Irrigation water is used to produce much of the rice, citrus, walnuts, livestock, vegetables, wine, cereal crops, pulses and oilseeds in the MIA, which contributes over $5 billion annually to the Australian economy.

Before 1912, Barren Box Swamp (also known as Barren Box Storage and Wetlands) was only filled with water for short periods following rain or flooding from Mirrool Creek. Once irrigation commenced, the swamp received irrigation drainage water. Excess water arriving at Willow Dam was diverted into the swamp for storage. This irrigation water was later used downstream. Today, it is the main irrigation and urban drainage water recycling point for the Murrumbidgee Irrigation Area (MIA).

FIGURE 1 Irrigation water used for grape production

FIGURE 2 Topographic map of Griffith, New South Wales

Source: Data based on Spatial Services 2019

Resources

eWorkbook	Investigating topographic maps — Food production in the Riverina (ewbk-8936)
Digital document	Topographic map of Griffith, New South Wales (doc-36316)
Video eLesson	Investigating topographic maps — Food production in the Riverina — Key concepts (eles-5196)
Interactivity	Investigating topographic maps — Food production in the Riverina (int-8629)
Google Earth	Griffith, New South Wales (gogl-0149)

4.11 EXERCISE

Learning pathways

■ LEVEL 1	■ LEVEL 2	■ LEVEL 3
1, 2, 3, 12	4, 5, 6, 10	7, 8, 9, 11

Check your understanding

1. Name the river that is located in the southern part of the map, in AR0994.
2. Name the creek at AR1297.
3. Name the hill located in AR1396.
4. Identify the types of land use at the following locations.
   a. GR110970
   b. GR130980
5. Identify the features that have been constructed at the following locations.
   a. the western end of N.S. Kooba Road
   b. in AR1294, running parallel to Point Road
6. What is the height of the land and grid reference at the following locations?
   a. Mount Bingar
   b. Griffith Airport
   c. The intersection of Irrigation Way and the railway line heading south east from Griffith
7. How is the elevation of the land in the north-east corner of the map different from the elevation for the rest of the area in general?
8. Approximately how much of the area shown is orchard? Give your answer as a proportion of the total area shown.

Apply your understanding

9. Compare the pattern of irrigation channels and buildings in AR1196 with the pattern of irrigation channels and buildings in AR0995
10. Explain two ways that people have changed the land to overcome the low and unreliable water availability in the MIA region.
11. How has the environmental factor of topography of the region been an advantage for food production?
12. How have the creeks and rivers of this area been changed to provide water for food production?

To answer questions online and to receive **immediate feedback** and **sample responses** for every question, go to your learnON title at www.jacplus.com.au.

4.12 Thinking Big research project — Fished out!

online only

The content in this subtopic is a summary of what you will find in the online resource.

Scenario

Overfishing is one of the largest human-caused ecological threats to the world's food supply. An average person now eats 20 kilograms of fish each year, which is more than twice the amount consumed 50 years ago. Combined with a global population that has quadrupled in the same time period, the result is that 60 per cent of fishing waters are fully fished out and 30 per cent of commercially fished waters are overfished. Globally, fish accounts for 17 per cent of all animal protein consumed, and the fishing industry provides employment for more than 60 million people, especially in developing countries where 97 per cent of the world's fisherfolk live. Not only are we threatening an important food source, we are also causing damage to marine ecosystems and people's livelihoods. Progress is being made in some fishing grounds, with commitment from fishers, governments, scientists and the Marine Stewardship Council, to establish science-based standards for environmentally responsible and sustainable fishing.

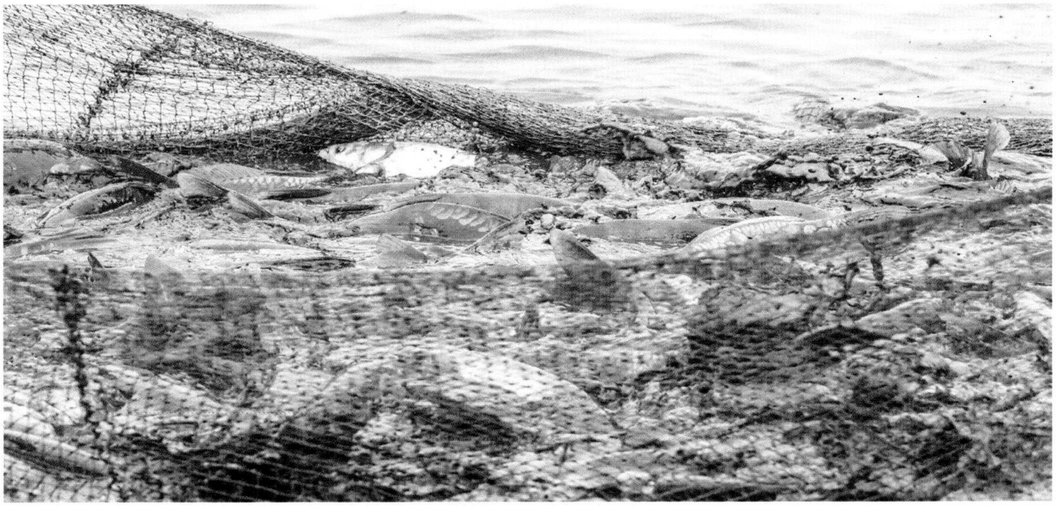

Task

learnON

A conference has been organised for secondary Geography students studying biomes and food security. You have been invited to give a presentation on a current issue relating to food security, in this case overfishing, and to outline some of the responses that take into account economic, social and environmental factors.

Go to learnON to access all of the resources you need to complete this research project. When you are satisfied with your work, present your conference presentation to your teacher and class.

on Resources

ProjectsPLUS Thinking Big research project — Fished out! (pro-0190)

4.13 Review

4.13.1 Key knowledge summary

4.2 Food production's effect on biomes

- Earth is made up of four interconnected spheres: the atmosphere, hydrosphere, lithosphere and biosphere.
- Natural events and human activities can create changes to these spheres.
- New technologies and improvements in farming methods have increased our rate of food production but have also caused loss of biodiversity and unsustainable degradation of land and water.
- Rapid population growth has an impact on food production and the consequent modification of biomes.

4.3 Modifying our forests

- Forest biomes support wide biodiversity but also provide resources for a range of goods and services.
- The need for farmland and forest products has seen large-scale clearing of the world's forests.
- Deforestation creates a range of environmental impacts, as can be seen in the Amazon rainforest.

4.5 Paper production and its effect on biomes

- Biomes provide products for manufacturing, such as fibre for paper products.
- Global consumption of paper products continues to rise; societies in more developed countries are very dependent on paper.
- The clearing of forest biomes, manufacturing of paper products and the disposal of these products all have environmental impacts.
- Changes are taking place to reduce these impacts and alternatives to wood pulp have been developed.

4.6 Overfishing our ocean biome

- Fish are an important source of food for over one billion people around the world.
- Improvements in technology have enabled larger quantities of fish to be captured, processed and stored at greater distances from the coast.
- A 'boom and bust' mentality has seen large-scale overfishing and the decline in fish species.
- Aquaculture is now supplying more fish and fish products than wild fish capture.
- Aquaculture, if poorly managed, can create environmental change.

4.8 The impact of farming on land and water

- Land that is poorly managed or overworked is susceptible to degradation. Erosion, salinity and pest invasions are all causes of land degradation.
- Land degradation can result from both natural and human causes, and can lead to a loss of productivity.
- There is a strong interconnection between land degradation and food insecurity.
- Food production and security is linked to water availability.
- Irrigation is the biggest user of water in the world, consuming 70 per cent of freshwater resources. It contributes to an increase in type, yield and seasonality of food production; however, poorly managed irrigation has environmental costs, such as soil salinity and waterlogging.
- Diversion of surface water and extraction of underground water need to be carried out in a sustainable manner, or watertables will fall and groundwater sources will run dry.

4.9 The impact of farming on the atmosphere

- Food production can contribute to global warming.
- Methane, an important greenhouse gas, is a by-product of farming, especially in rice and cattle farming.
- Deforestation also contributes to global warming through the release of carbon.

4.10 Diminishing global biodiversity

- Globally, there is a decline in the number and population of most species.
- Changes in agriculture, large-scale changes to habitats and modern breeding of plants and animals all contribute to a loss of biodiversity
- Australia is considered a megadiverse country, with one of the highest levels of biodiversity in the world.

4.13.2 Key terms

aquaculture the farming of aquatic plants and aquatic animals such as fish, crustaceans and molluscs

aquifer a body of permeable rock below the Earth's surface that contains water, known as groundwater

biophysical environment all elements or features of the natural or physical and the human or urban environment, including the interaction of these elements; made up of the Earth's four spheres — the atmosphere, biosphere, lithosphere and hydrosphere

deforestation clearing forests to make way for housing or agricultural development

degradation reduced land quality and water resources caused by over-use

endemic describes species that occur naturally in only one region

erosion the wearing down of rocks and soils on the Earth's surface by the action of water, ice, wind, waves, glaciers and other processes

factory farming the raising of livestock in confinement, in large numbers, for profit

greenhouse gases any of the gases that absorb solar radiation and are responsible for the greenhouse effect. These include water vapour, carbon dioxide, methane, nitrous oxide and various fluorinated gases.

humus an organic substance in the soil that is formed by the decomposition of leaves and other plant and animal material

kenaf a plant in the hibiscus family that has long fibres; useful for making paper, rope and coarse cloth

monoculture the cultivation of a single crop on a farm or in a region or country

old-growth forests natural forests that have developed over a long period of time, generally at least 120 years, and have had minimal unnatural disturbance such as logging or clearing

ores raw material that minerals can be extracted from

plantation an area in which trees or other large crops have been planted for commercial purposes

pulp the fibrous material extracted from wood or other plant material to be used for making paper

undulating describes an area with gentle hills

waterlogging saturation of the soil with groundwater such that it hinders plant growth

watertable the upper level of the groundwater, below which all pores in the soils and rock layers are saturated with water

4.13.3 Reflection

Complete the following to reflect on your learning.

Revisit the inquiry question posed in the Overview:

Our planet works hard to feed the ever-growing human population ... but at what cost?

1. Now that you have completed this topic, what is your view on the question? Discuss with a partner. Has your learning in this topic changed your view? If so, how?
2. Write a paragraph in response to the inquiry question, outlining your views.

Subtopic	Success criteria	●	●	●
4.2	I can describe why biomes are altered.			
	I can explain the effect this has on biomes.			
4.3	I can describe why forest biomes have been altered.			
	I can describe ways in which forest biomes have been altered.			
	I can explain the impact of modifying forest biomes.			
4.5	I can explain how biomes are used to create paper.			
	I can describe the effects of paper production on biomes.			

(continued)

Subtopic	Success criteria	●	●	●
4.6	I can describe how humans have modified the ocean biome.			
	I can explain the effects that this has had.			
	I can explain the benefits and problems with aquaculture.			
4.8	I can outline the causes and effect of land degradation.			
	I can describe how poor agricultural practices are creating environmental degradation.			
4.9	I can explain how farming has negative effects on the atmosphere.			
4.10	I can explain how agriculture reduces biodiversity.			
4.11	I can describe how the Murrumbidgee Irrigation Area has been modified to produce food.			

on Resources

eWorkbook Chapter 4 Extended writing task (ewbk-8534)
Chapter 4 Reflection (ewbk-8533)
Chapter 4 Student learning matrix (ewbk-8532)

Interactivity Chapter 4 Crossword (int-8630)

FIGURE 1 Grape picking, Fordwich, New South Wales, 2021

ONLINE RESOURCES

on Resources

Below is a full list of rich resources available online for this chapter. These resources are designed to bring ideas to life, to promote deep and lasting learning and to support the different learning needs of each individual.

eWorkbook

- 4.1 Chapter 4 eWorkbook (ewbk-8096) ☐
- 4.2 Food production's effect on biomes (ewbk-8900) ☐
- The impact of human activity (ewbk-1049) ☐
- 4.3 Modifying our forests (ewbk-8904) ☐
- 4.4 SkillBuilder — GIS — Deconstructing a map (ewbk-8908) ☐
- 4.5 Paper production and its effect on biomes (ewbk-8912) ☐
- Paper profits, global losses (ewbk-1051) ☐
- 4.6 Overfishing our ocean biome (ewbk-8916) ☐
- 4.7 SkillBuilder — Interpreting a geographical cartoon (ewbk-8920) ☐
- 4.8 The impact of farming on land and water (ewbk-8924) ☐
- 4.9 The impact of farming on the atmosphere (ewbk-8928) ☐
- 4.10 Diminishing global biodiversity (ewbk-8932) ☐
- 4.11 Investigating topographic maps — Food production in the Riverina (ewbk-8936) ☐
- 4.13 Chapter 4 Extended writing task (ewbk-8534) ☐
- Chapter 4 Reflection (ewbk-8533) ☐
- Chapter 4 Student learning matrix (ewbk-8532) ☐

Sample responses

- 4.1 Chapter 4 Sample responses (sar-0155) ☐

Digital documents

- 4.4 Topographic map of Griffith, New South Wales (doc-36316) ☐
- 4.11 Topographic map of Griffith, New South Wales (doc-36316) ☐

Video eLessons

- 4.1 Trashing our biomes (eles-1719) ☐
- Changing biomes — Photo essay (eles-5188) ☐
- 4.2 Food production's effect on biomes — Key concepts (eles-5189) ☐
- 4.3 Modifying our forests — Key concepts (eles-5190) ☐
- 4.4 SkillBuilder — GIS — Deconstructing a map (eles-1730) ☐ ☐
- 4.5 Paper production and its effect on biomes — Key concepts (eles-5191) ☐
- 4.6 Overfishing our ocean biome — Key concepts (eles-5192) ☐
- 4.7 SkillBuilder — Interpreting a geographical cartoon (eles-1731) ☐
- 4.8 The impact of farming on land and water — Key concepts (eles-5193) ☐
- 4.9 The impact of farming on the atmosphere — Key concepts (eles-5194) ☐
- 4.10 Diminishing global biodiversity — Key concepts (eles-5195) ☐
- 4.11 Investigating topographic maps — Food production in the Riverina — Key concepts (eles-5196) ☐

Interactivities

- 4.2 Degrading our farmland (int-3323) ☐
- State of the world's land and water resources for food and agriculture (int-7920) ☐
- The Earth's four spheres (int-7919) ☐
- 4.3 Impacts of clearing the Amazon rainforest (int-5578) ☐
- 4.4 SkillBuilder — GIS — Deconstructing a map (int-3348) ☐
- 4.5 Paper production and its effect on biomes (int-7921) ☐
- 4.6 Hook, line and sinker (int-3324) ☐
- 4.7 SkillBuilder — Interpreting a geographical cartoon (int-3349) ☐
- 4.8 The impact of farming on land and water (int-8626) ☐
- The development of irrigation salinity (int-5589) ☐
- How is rice grown? (int-3322) ☐
- Losing land (int-3325) ☐
- 4.9 The impact of farming on the atmosphere (int-8627) ☐
- 4.10 Diminishing global biodiversity (int-8628) ☐
- 4.11 Investigating topographic maps — Food production in the Riverina (int-8629) ☐
- 4.13 Chapter 4 Crossword (int-8630) ☐

ProjectsPLUS

- 4.12 Thinking Big research project — Fished out! (pro-0190) ☐

Weblink

- 4.5 Paper production (web-4341) ☐

Fieldwork

- 4.3 Remote sensing (fdw-0030) ☐

myWorld Atlas

- 4.2 World: Climate (mwa-4428) ☐
- World: Vegetation (mwa-4429) ☐
- 4.3 Forest environments (mwa-4491) ☐
- 4.10 World species (mwa-4485) ☐
- Endangered species in Australia (mwa-4486) ☐
- Introduced species in Australia (mwa-4487) ☐
- Tasmanian devil (mwa-4489) ☐

Google Earth

- 4.4 Griffith, New South Wales (gogl-0149) ☐
- 4.11 Griffith, New South Wales (gogl-0149) ☐

Teacher resources

There are many resources available exclusively for teachers online.

5 Challenges to food production

INQUIRY SEQUENCE

To access a pre-test with **immediate feedback** and **sample responses** to every question in this chapter, select your learnON format at www.jacplus.com.au.

5.1 Overview

Numerous **videos** and **interactivities** are embedded just where you need them, at the point of learning, in your learnON title at www.jacplus.com.au. They will help you to learn the content and concepts covered in this topic.

The world produces enough food to feed everyone. So why do nearly 800 million people around the world each year experience hunger?

5.1.1 Introduction

Food needs sunlight, water, land and good soil to be able to grow. Easy, right? The increasing scarcity of water, and demands on land for housing and other fuel, as well as climate change, have all made the production of food both today and in the future increasingly complex.

The term 'hunger' means not having enough of the right food to cover your basic needs for physical survival.

Global levels of hunger across the world rose in 2020 with the COVID-19 pandemic, but this number has been rising steadily since 2015, so there are other factors to consider. What are the challenges to food production?

FIGURE 1 Community garden in the suburbs, Kensington, Victoria

STARTER QUESTIONS

1. How long has it been since you had anything to eat? How many meals do you have on a typical day?
2. How many different food items have you eaten today?
3. How many of these did your family grow? Do you have a garden at home to grow food?
4. Do you know when and where your next meal is coming from? How might it feel if you didn't?
5. Why do you think we have so many people hungry when there is enough food produced in the world?
6. Watch the **Challenges to food production — Photo essay** video in your online Resources. Discuss the factors that affect whether people in Australia have enough food to eat. What should we do to help people who don't have enough to eat in our community?

on Resources

eWorkbook Chapter 5 eWorkbook (ewbk-8097)

Video eLessons Food for thought (eles-1720)
Challenges to food production — Photo essay (eles-5197)

5.2 Water security and food production

LEARNING INTENTION

By the end of this subtopic, you will be able to define the term *water security* and explain the impact of water scarcity on food production.

5.2.1 Water security

There is no substitute for water. Without water there is no food, and agriculture already consumes 70 per cent of the world's fresh water. Every type of food production — cropping, grazing and processing — requires water. Thus, a lack of water is possibly the most limiting factor for increasing food production into the future to meet our growing needs.

To feed an additional two billion people by 2050, the world will need to generate more food and find more efficient ways to source water for food production. The two main concerns that threaten future water security are water quantity and water quality.

FIGURE 1 Water scarcity is a serious threat to food security — without water, plants will not grow.

5.2.2 Factors leading to water scarcity

In theory, the world has enough water; it is just not available where we want it or when we want it, and it is not easy to move from place to place. We already use the most accessible surface water, and now we are searching for underground water reserves. Underground aquifers (shown in **FIGURE 2** as groundwater) hold 100 times more water than surface rivers and lakes. However, groundwater is not always used at a sustainable rate, with extraction exceeding natural recharge or filling. This occurs in many of the world's major food-producing places, in countries such as the United States, China and India.

FIGURE 2 Global water budget and water accessibility

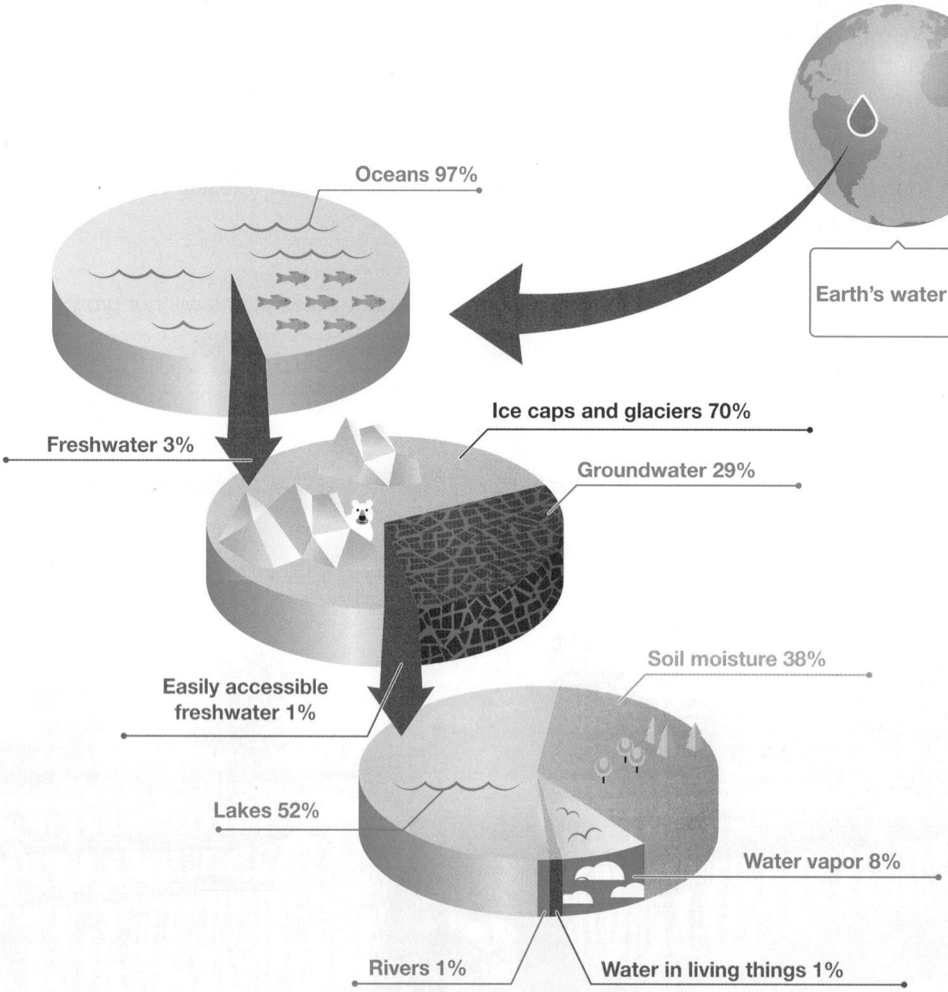

Water insecurity and food insecurity are closely connected to each other. According to the United Nations in 2020, approximately 748 million people currently suffer from water scarcity.

- **Water stress**: water supplies drop below 1700 m^3 per person per year.
- **Water scarcity**: water supplies drop below 1000 m^3 per person per year.
- Absolute water scarcity: water supplies drop below 500 m^3 per person per year.

water stress situation that occurs when water demand exceeds the amount available or when poor quality restricts its use

water scarcity situation that occurs when water supplies drop below 1000 m^3 per person per year

FIGURE 3 Projected water availability based on temperature, population and industrialisation increases, 2050s

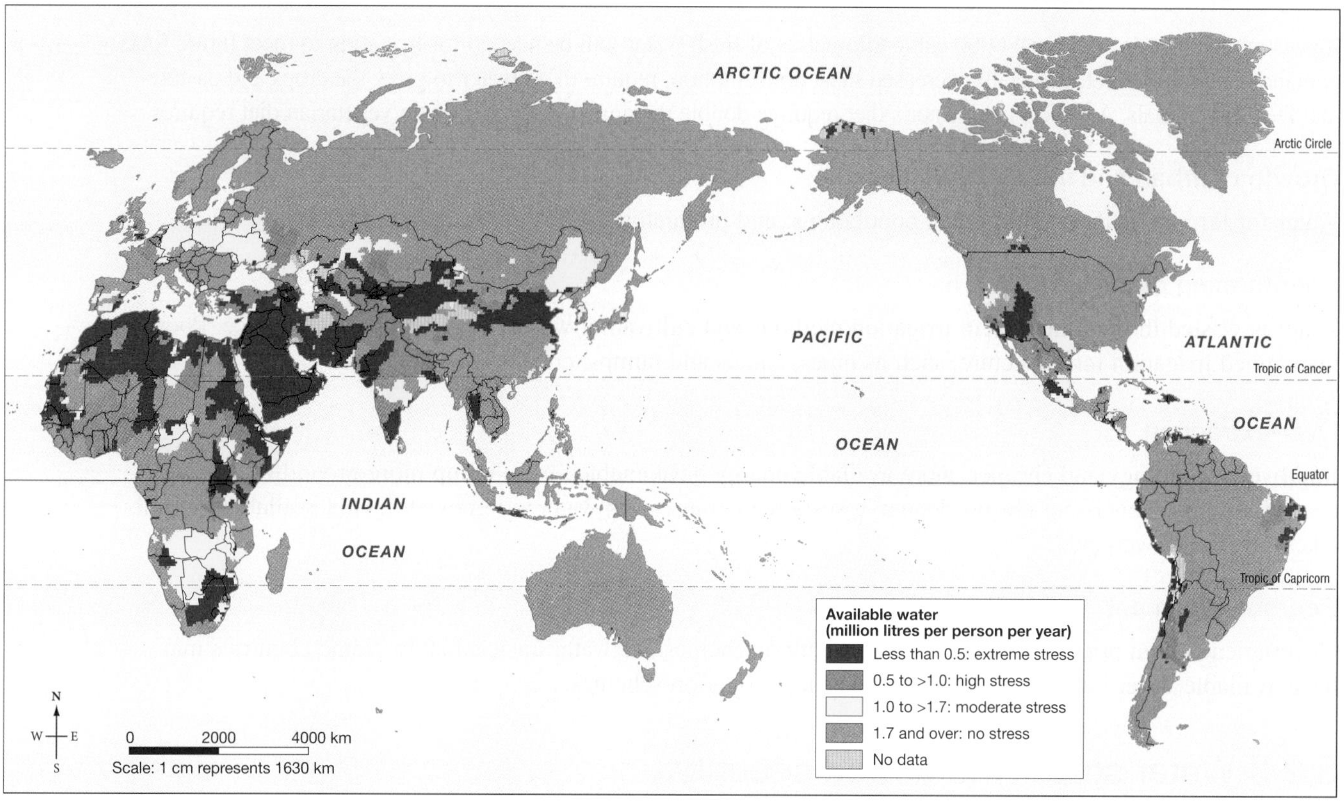

Source: Spatial Vision

It is expected that by 2025, 1.8 billion people will be living in conditions of absolute water scarcity and that 67 per cent of people will be experiencing water stress. By 2030, it is projected that half of the world's population will experience high water stress conditions. **FIGURE 3** shows an interconnection between increased demand for water and predicted climate change, population increase and greater industrialisation in the 2050s.

When water availability drops below 1.5 million litres per person per year, a country needs to start importing food, which makes the country susceptible to changes in global prices. Developing countries that experience water stress cannot afford to import food. They are also more vulnerable to environmental disasters. Seventy per cent of food emergencies in developing countries are brought on by drought.

FIGURE 4 Drought makes water scarce, but floods can also influence water security by contaminating communal wells (such as this one in South Sudan) and destroying crops.

The main causes of the growing water shortage

Food production

It is estimated that an additional 6000 cubic kilometres of fresh water will be needed for irrigation to meet future food demand. Changes in diet, especially increased meat consumption, require more water to grow the crops and pasture that feed the animals. A typical meat eater's diet requires double the amount of water that a vegetarian diet requires.

Growth of urban and industrial demand

Water for farming is diverted to urban populations, and productive land is converted to urban use.

Poor farming practices

Water is wasted through inefficient irrigation methods and cultivating water-hungry crops such as rice. Poorly maintained irrigation infrastructure, such as pipes, canals and pumps, creates leakage.

Over-extraction

Improved technology and cheaper, more available energy have enabled us to pump more groundwater from deeper aquifers. This is not always done at a sustainable rate, so as water is removed, less is available to refill lakes, rivers and wetlands.

Poor management

Governments often price water cheaply, so irrigation schemes use water unsustainably. Some countries may have available water but lack the money to develop irrigation schemes.

5.2.3 Water scarcity and food security

Water security has a direct impact on food security. For the 800 million people who do not have enough food to eat, the issue of finding sufficient and nutritious food must be faced daily.

At least 75 per cent of the world's people are **undernourished**, with diets that are minimal or below the level of sustenance. People who do not have a regular and healthy diet often have shortened life expectancy and an increased risk of disease. Children are especially vulnerable to poor diet, and their growth, weight, and physical and mental development suffer. Almost 50 per cent of India's children are **malnourished**, and it is estimated that there are 146 million children in the world suffering from chronic hunger.

This is often referred to as the 'double burden' of disease. This means that individuals who are most at risk of having poor diets are more likely to suffer from the effects of disease. Once a disease has been contracted, it has an impact on a person's health and functioning.

undernourished describes a person who is not getting enough calories in their diet; that is, not enough to eat

malnourished describes a person who is not getting the right amount of vitamins, minerals and other nutrients to maintain healthy tissues and organ function

TABLE 1 Global food security rankings, December 2019

Rank	Least food secure countries	Rank	Most food secure countries
1	Venezuela	1	Singapore
2	Burundi	2	Ireland
3	Yemen	3	United States
4	Congo (Dem. Rep.)	4	Switzerland
5	Chad	=5	Finland
6	Madagascar	=5	Norway

Source: https://foodsecurityindex.eiu.com/Index

Resources

eWorkbook	Water security and food production (ewbk-9718)
Video eLesson	Water security and food production — Key concepts (eles-5198)
Interactivity	The last drop (int-3328)

5.2 ACTIVITIES

1. Use the **Water use** weblink in your online Resources to find out more about countries' water usage.
   a. Choose four countries — one from each continent of Europe, Africa, Asia and South America.
   b. Using the data from the **Water use** weblink, construct a table to compare water usage for your four countries. (Try to select different countries from those chosen by other students.)
   c. Write a paragraph to summarise your findings.
2. **FIGURE 2** tells us about freshwater accessibility. Use this data to explain why there is so little of the world's water available for our use. Investigate where your water comes from.

5.2 EXERCISE

Learning pathway

■ LEVEL 1	■ LEVEL 2	■ LEVEL 3
1, 3, 9, 10, 14	2, 5, 7, 8, 13	4, 6, 11, 12, 15

Check your understanding

1. Describe the difference between water scarcity and water stress.
2. With reference to data, describe the projected changes in the numbers of people affected by water shortages (both scarcity and stress) over the period from now until 2030.
3. If a country has an average of 0.5 to <1.0 million litres of water per person, per year, would they be considered to be water stressed? Why?
4. Refer to **FIGURE 3**. Describe those places in the world that are predicted to be in high to extreme water stress in the 2050s.
5. At what stage of water scarcity does a country need to import food?
6. Approximately what percentage of food emergencies in developing countries are the result of drought conditions affecting food production?

Apply your understanding

7. Are areas that are predicted to be suffering high to extreme stress by 2050 also areas of low rainfall? Refer to data to support your hypothesis.
8. How are meat consumption and water scarcity interconnected?
9. What challenges does a country face when it is forced to import food due to lack of water?
10. Explain at least two ways in which population increases and demand for water are interconnected.
11. How are urbanisation and water stress interconnected?
12. Looking at **TABLE 1** and **FIGURE 3**, what trends can you identify in the location of countries at risk of food insecurity?

Challenge your understanding

13. What do you think water managers could do to help prevent water scarcity affecting future food security?
14. Restructure the information in **FIGURE 2** into a flowchart or tree diagram showing the breakdown of water in the world.
15. In **FIGURE 3**, it is shown that Australia, as a nation, is not experiencing water stress. Suggest whether this might be true if Australia were divided up into smaller regions. Do you think there are parts of Australia that are likely to be suffering water stress? Give reasons for your view.

To answer questions online and to receive **immediate feedback** and **sample responses** for every question, go to your learnON title at www.jacplus.com.au.

5.3 Pollution and food production

LEARNING INTENTION

By the end of this subtopic, you will be able to explain the impact of pollution on food production.

5.3.1 Impacts of pollution

Pollution affects agricultural activity, or food production, almost as much as does water scarcity. In fact, pollution of air and water poses one of the greatest threats to the production of food. The irony is that much of the pollution in the water comes from agricultural processes themselves.

Water supplies can become contaminated by excess nutrients, pesticides, sediment and other pollutants that run off farmland or leach into soils and groundwater. Industrial waste, untreated sewage and urban run-off may also pollute water that is used to irrigate farmland. Food that is irrigated with polluted water may pass on diseases to people. Pollution is an important contributor to the scarcity of clean, **potable** water.

FIGURE 1 demonstrates the variety of ways in which the sources of pollution can be transported or transformed into harmful effects. These can have a considerable impact on our food production, not only in a direct manner in the form of agricultural products (bottom right) but also indirectly through some of the other effects (shown with black dots).

In Australia, irrigation was the most sourced supply of water for agricultural production in 2018–19, followed by groundwater and then rivers, creeks and lakes. Excessive irrigation can cause **waterlogging** or soil **salinity**. This salty water not only poisons the soil, causing reduced levels of agricultural production, but also drains into river systems, leading to salinity problems downstream.

potable drinkable; safe to drink

waterlogging saturation of the soil with groundwater such that it hinders plant growth

salinity the presence of salt on the surface of the land, in soil or rocks, or dissolved in rivers and groundwater

int-8631

FIGURE 1 Global pollutants circuit

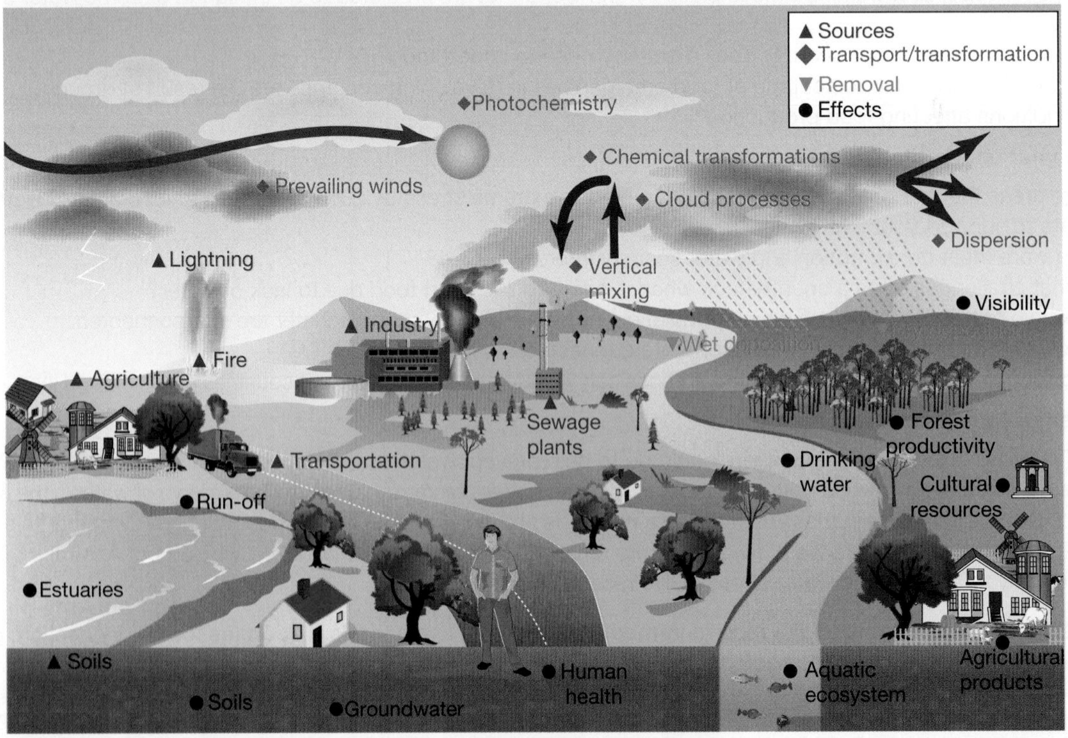

FIGURE 2 Australian sources of water for agriculture 2018–19

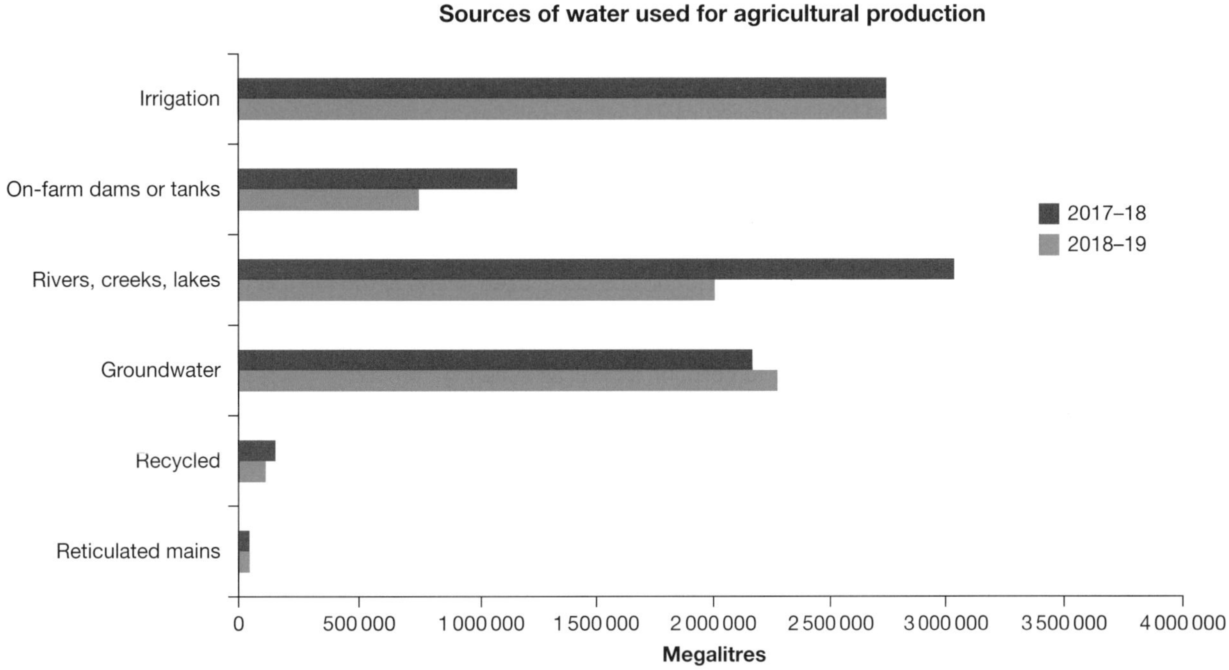

Source: ABS, https://www.abs.gov.au/statistics/industry/agriculture/water-use-australian-farms/latest-release

5.3.2 Reducing pollution levels in food production

Methods of food production that aim to reduce levels of environmental pollution are becoming increasingly popular. In Australia, food that has been grown or produced without any contact with artificial fertilisers and chemicals is known as **organic**.

Australian food produced for export must be produced using a set of food production standards and requirements including:

- the use of renewable resources
- the conservation of energy, soil and water
- recognition of livestock welfare needs
- environmental maintenance and enhancement.

organic products grown or created with ingredients that were grown without any contact with artificial fertilisers or chemicals

FIGURE 3 A citrus farm using organic food production requirements

Resources

eWorkbook Pollution and food production (ewbk-9722)

Video eLesson Pollution and food production — Key concepts (eles-5199)

Interactivity Pollution and food production (int-8631)

5.3 ACTIVITIES

1. Research to investigate the food production techniques being used in one of the top five most food-secure countries listed in **TABLE 1** in subtopic 5.2.
2. Select one the countries identified in **TABLE 1** in subtopic 5.2 and find out more about its agricultural production. Does agriculture contribute to pollution as well as suffering from it?
3. Use the **UN — Actions to clean the air** weblink in your online Resources to create an infographic of ways that individuals can act to reduce their contribution to air pollution.

5.3 EXERCISE

Learning pathways

■ LEVEL 1	■ LEVEL 2	■ LEVEL 3
1, 7, 8, 9, 13	2, 5, 6, 10, 14	3, 4, 11, 12, 15

Check your understanding

1. List five sources of pollution shown in **FIGURE 1**.
2. Describe how two of the sources shown in **FIGURE 1** indirectly affect food production.
3. Choose one source of pollution shown in **FIGURE 1** and describe how it is transported/transformed and what its effects are.
4. In what way is agriculture both a contributor to and victim of water pollution?
5. What makes a product 'organic'?
6. What is 'potable' water?
7. Order the following sources of water supply for Australian agriculture from most to least used: groundwater; creeks, rivers and lakes; irrigation

Apply your understanding

8. Explain two ways in which salinity creates problems for food producers.
9. Explain the four basic standards that apply to food produced in Australia that is intended for export.
10. How might agriculture contribute to contaminating water supplies?
11. What are the dangers of people consuming food that has been irrigated with polluted water?
12. Explain the following statement: 'The irony is that much of the pollution in the water comes from agricultural processes themselves.'

Challenge your understanding

13. Suggest why organic foods might be increasing in popularity.
14. In Australia, do you think that air or water pollution presents a great threat to our food production? Give reasons to support your answer.
15. Predict whether the quality of water available to food producers in Australia will improve, stay the same as it is now, or get worse over time? Give reasons for your answer.

To answer questions online and to receive **immediate feedback** and **sample responses** for every question, go to your learnON title at www.jacplus.com.au.

5.4 SkillBuilder — Interpreting satellite images to show change over time

online only

LEARNING INTENTION

By the end of this subtopic, you will be able to identify features on a satellite image with false colours and use your observations to describe how the area has changed over time.

The content in this subtopic is a summary of what you will find in the online resource.

5.4.1 Tell me

What is a satellite image?

A satellite image is an image taken from a satellite orbiting the Earth. Satellite images allow us to see very large areas—much larger than those that can be visualised using vertical aerial photography.

A satellite image often does not use the natural colours that we expect. This is referred to as using false colours. These are applied in the computer processing of the images in order to highlight spatial patterns more clearly.

FIGURE 1 Change over time around the Yangtze River, China: (a) in 1987 and (b) in 2004, after the building of the Three Gorges Dam

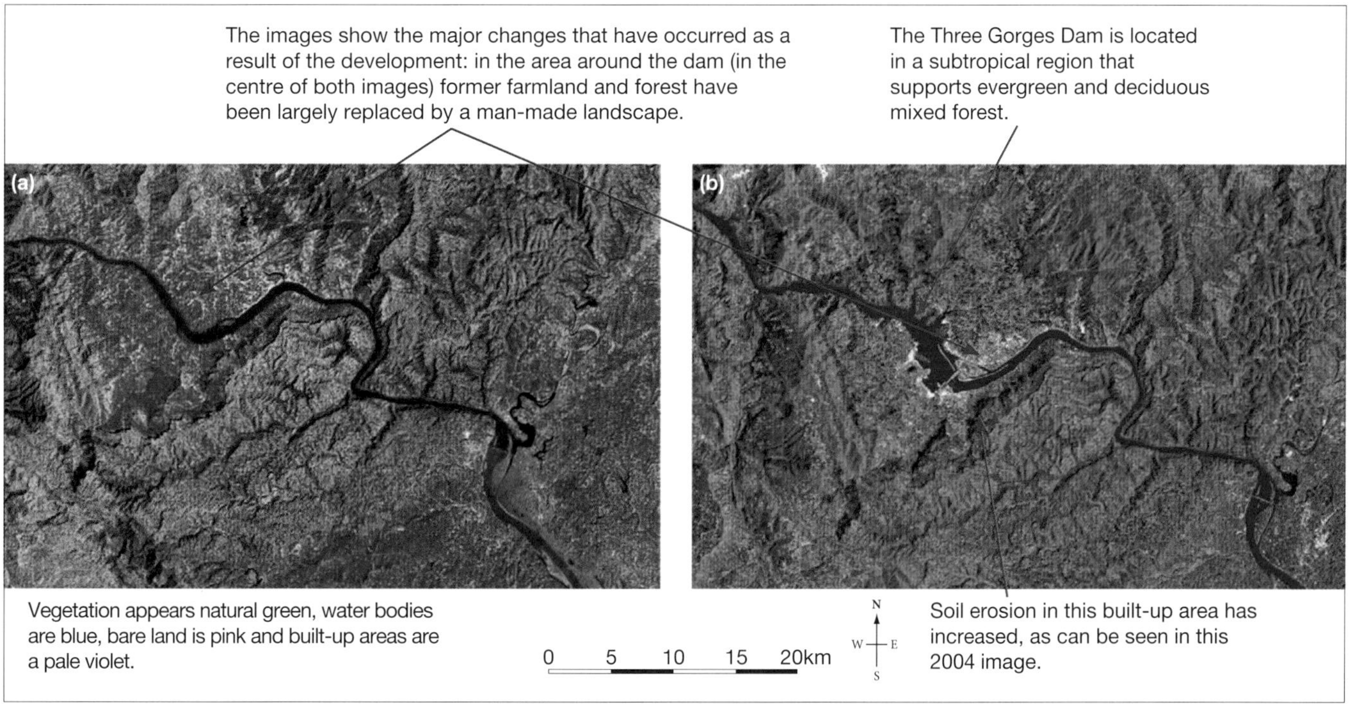
The images show the major changes that have occurred as a result of the development: in the area around the dam (in the centre of both images) former farmland and forest have been largely replaced by a man-made landscape.
The Three Gorges Dam is located in a subtropical region that supports evergreen and deciduous mixed forest.
(a)
(b)
Vegetation appears natural green, water bodies are blue, bare land is pink and built-up areas are a pale violet.
0 5 10 15 20km
N
W E
S
Soil erosion in this built-up area has increased, as can be seen in this 2004 image.

Source: Geoscience Australia

5.4.2 Show me

Step 1

Consider the time span between the images. In **FIGURE 1**, it is 17 years. Check that the satellite images are at the same scale. In **FIGURE 1**, the two images show slightly different but overlapping areas at the same scale.

Step 2

Identify the key features of the place by noting where the colours appear. Commonly used colours in false-colour imagery are shown in the table.

TABLE 1 Colours commonly used in false-colour satellite imagery

Colour	Ground feature
Green	Vegetation
Dark blue	Water — the deeper the water, the darker the colour
Bright blue to mauve/grey	Housing and industrial areas
White to cream	Beaches and sands
Yellow	Barren areas, heavily grazed or fallow land
Pink to red	Recent plant growth, suburban parklands
Red	Flourishing vegetation, including forests (mangroves appear brown)

Step 3

Use the same location in both images to identify change. Use compass directions, scale or features to help reference the place that you are discussing. For example, in **FIGURE 1** the bends in the Yangtze River help to identify common points.

Step 4

Can you draw inferences from the satellite image? Changes in topography may relate to changes in land use. For example, in 1987 the area of the present-day dam was surrounded by high, barren hills (yellow) and areas of dense subtropical forest (green); by 2004 a human-constructed landscape (mauve) surrounded the dam.

5.4.3 Let me do it

learnON

Go to learnON to access the following additional resources to help you build this skill:

- a longer explanation of this skill and its application in Geography (Tell me)
- a video demonstrating the step-by-step process of this skill (Show me)
- an activity and interactivity for you to practise the skill (Let me do it)
- self-marking questions to help you understand and use the skill.

on Resources

eWorkbook	SkillBuilder — Interpreting satellite images to show change over time (ewbk-9726)
Video eLesson	SkillBuilder — Interpreting satellite images to show change over time (eles-1733)
Interactivity	SkillBuilder — Interpreting satellite images to show change over time (int-3351)

5.5 Land loss and food production

LEARNING INTENTION

By the end of this subtopic, you will be able to identify the key features of land degradation and to explain its impact on food production.

5.5.1 Loss of productive land

Land is essential for food production, and the world has more than enough **arable** land to meet future demands for food. Nevertheless, we need to find a balance between competing demands for this finite resource.

The loss of productive land has two main causes. First, there is the degradation of land quality through such things as erosion, **desertification** and salinity. Second, there is the competition for land from non-food crops, such as **biofuels**, and from expanding urban areas. As **FIGURE 1** shows, the growth in world population is inversely proportional to the amount of arable land available. This does not even take into consideration the land that is degraded and no longer suitable for growing food.

FIGURE 1 Comparison of world population growth and arable land per capita

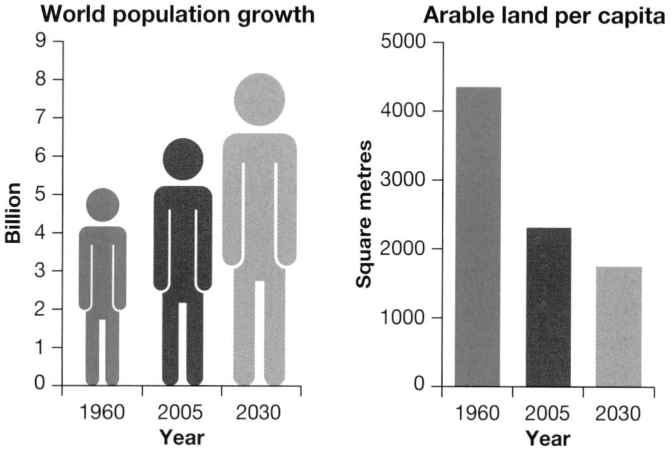

Source: Food and Agriculture Organization of the United Nations, Reproduced with permission.

Land degradation

Although there have been significant improvements in crop yields, seeds, fertilisers and irrigation, they have come at a cost. Environmental degradation of water and land resources places future food production at risk.

The main forms of land degradation are:

- erosion by wind and water
- salinity
- pest invasion
- loss of biodiversity
- desertification.

Land degradation occurs in all food-producing biomes across the globe. Some degradation occurs naturally; for example, a heavy rainstorm can easily wash away topsoil. However, the most extensive degradation is caused by overcultivation, overgrazing, overwatering, overloading with chemicals and over-clearing (see **FIGURE 2**). Currently, 25 per cent of the world's land is highly degraded, while only 10 per cent is improving in quality. In South-East Asia, 50 per cent of cultivated land has severe soil quality problems, which prevent increases in food production. The Ministry for Agriculture in China estimates that 3.3 million hectares of arable land is polluted with chemicals and heavy metals, mostly in regions that grow grains.

arable describes land that is suitable for growing crops

desertification the transformation of land once suitable for agriculture into desert by processes such as climate change or human practices such as deforestation and overgrazing

biofuel fuel that has been produced from renewable resources, such as plants and vegetable oils

5.5.2 Competition for land

There has been a growing global trend to convert valuable cropland to other uses. **Urban expansion**, industrialisation and energy production is taking over land once used to produce food. This is a major global issue that affects the future production of our food.

For example, in less than 16 years, China lost more than 14.5 billion hectares of arable land to other uses. This land no longer produces food, which then puts pressure on existing land resources to make up the loss.

Biofuel

Biofuel refers to fuel that has been produced from renewable resources, such as plants and vegetable oils, and treated municipal and industrial wastes.

Traditionally, the main forms of biofuel have been wood and charcoal. Almost 90 per cent of wood harvested in Africa and 40 per cent harvested in Asia is used for heating and cooking. Today, people are seeking more renewable energy sources and wanting to reduce CO^2 emissions associated with deforestation, so there is greater demand for alternative energy sources. Consequently, the use of agricultural crops to produce biofuels is increasing. Ethanol (mostly used as a substitute for petrol) is extracted from crops such as corn, sugar cane and cassava. Biodiesel is derived from plantation crops such as palm oil, soya beans and **jatropha**.

FIGURE 2 Land degradation caused by deforestation in Madagascar

FIGURE 3 Jatropha

urban expansion the increasing size of urban areas

jatropha any plant of the genus *Jatropha*, but particularly *Jatropha curcas*, which is used as a biofuel

The growth of the biofuel industry has the potential to threaten future food security by:

- changing food crops to fuel crops, so less food is produced and crops have to be grown on marginal land rather than arable land
- increasing prices, which makes staple foods too expensive for people to purchase
- forcing disadvantaged groups, such as women and the landless poor, to compete against the might of the biofuel industry.

Urban expansion

Cities tend to develop in or near places that are agriculturally productive. However, as they expand due to population growth, the city encroaches on valuable farmland. This farmland is then converted into housing and infrastructure to support the population. Additional pressure is placed on farmers to produce more food for the increasing number of people on less arable land.

int-8770

FIGURE 4 Satellite image of the city of Tehran (a) in 1985; (b) in 2009 — the expansion of the city has taken over valuable arable land, reducing the availability of farmland

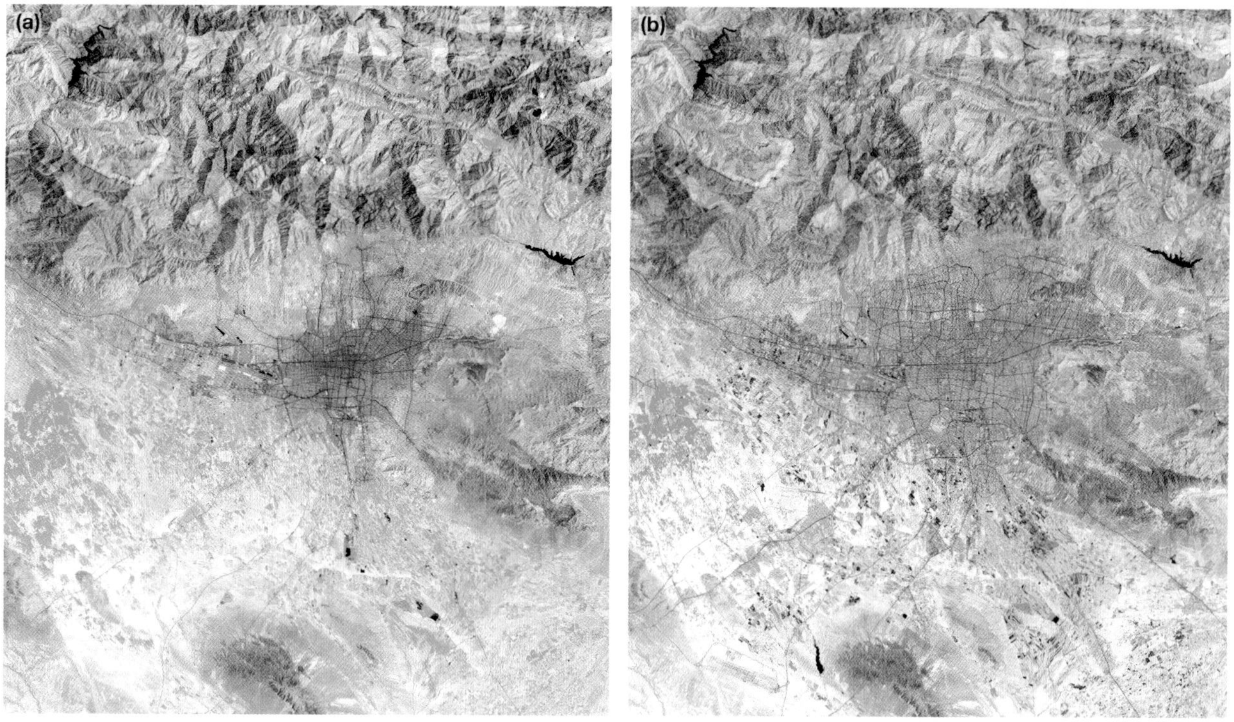

Land grabs

A growing challenge to world food security is the purchase or lease of land, largely in developing nations, by resource-poor but wealthier nations. Large-scale 'land grabs', as they are known, have the potential to improve production and yields, but at the same time there is growing concern over the loss of land rights and food security for local populations.

Since 2000, foreign investors have acquired over 26 million hectares around the world to produce food crops and biofuels. **FIGURE 5** shows the extent of China's expansion into other countries with investments in land and agricultural businesses.

Forty-two per cent of global acquisitions have occurred in Africa, examples of which can be seen in **FIGURE 6**. Africa's appeal is based on the fact the continent accounts for 60 per cent of the world's arable land and yet most countries within it currently achieve less than 25 per cent of their potential yield.

int-8766

FIGURE 5 Global map of China's land and food footprint

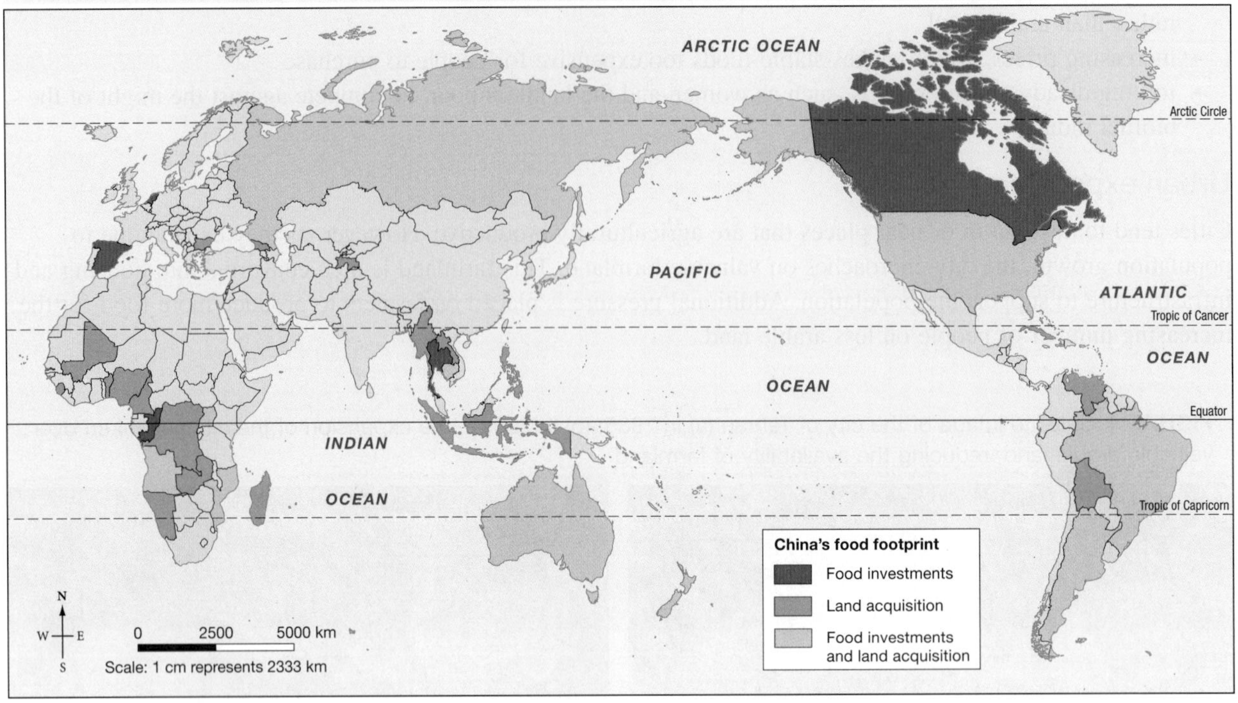

Source: The Heritage Foundation, GRAIN.org, Bloomberg. Map drawn by Spatial Vision.

FIGURE 6 Examples of land grabs in Africa

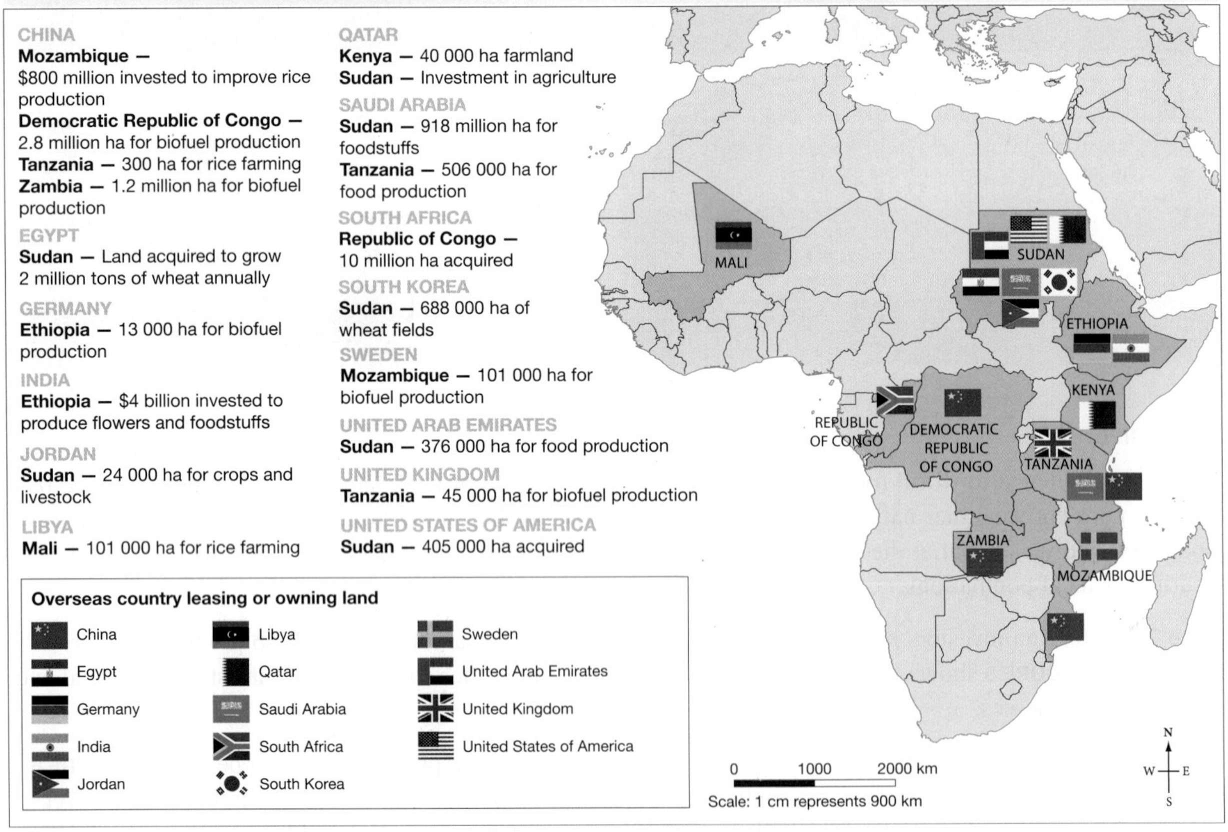

Source: Food and Agriculture Organization, International Food Policy Research Institute

'Triple-F' crisis

The rise of land grabs came about as a result of the 'triple-F' crisis — food, fuel and finance.

Food crisis

Massive increases in world food prices in 2007–08 emphasised the need for those countries heavily reliant on importing food, such as Saudi Arabia and China, to improve their food security by obtaining land in other countries to produce food to meet their own needs.

Fuel crisis

Rising and fluctuating oil prices in 2007–09 created an incentive for countries to acquire land to produce their own biofuels (see **FIGURE 5**).

Financial crisis

The global financial crisis in 2008 saw organisations switch from investing in stocks and shares to land in overseas countries, especially land that could be used to generate food and fuel crops.

5.5.3 The risk to food security

Investors in farmland are, understandably, seeking large expanses of land that has fertile soils and good rainfall or access to irrigation water. In many instances, land that is purchased is already occupied and used by small-scale farmers, often women who rarely benefit from any compensation. Prices for land can be much lower and there is frequently corruption, with much money going to local and government officials. People can also be forced off their land by governments keen to make deals with wealthy nations and corporations. Many land grabs have neglected the social, economic and environmental impacts of the deals.

With the purchase of land can come the right to withdraw the water linked to it, and this can deny local people access to water for fishing, farming and watering animals. Withdrawal of water can reduce flow downstream. The Niger River, West Africa's largest river, flows through three countries and sustains over 100 million people, so any large-scale water reductions create significant impact to downstream environments and people.

Not all farmland grab projects have been successful. At least 17.5 million hectares of foreign-controlled land have failed. There are several interconnected reasons, including a lack of understanding of local conditions, natural disasters, failed accounting and, increasingly, challenges from local communities that have been displaced. When projects collapse, communities rarely get their lands back or are compensated for their loss. Promises of new schools, health clinics, infrastructure and jobs simply disappear.

It has been estimated that the land taken up by foreign investors for biofuel projects could feed as many as 190 to 370 million people, or even more, if yields were raised to the level of industrialised western farming. In addition to these human costs, there are important concerns about environmental risks that are associated with monoculture farming and the loss of biodiversity in the region.

on Resources

eWorkbook	Land loss and food production (ewbk-9730)
Video eLesson	Land loss and food production — Key concepts (eles-5200)
Interactivity	Losing land (int-3325)

5.5 ACTIVITIES

1. Find Tehran (as featured in **FIGURES 4a** and **4b**) using the **Google time lapse** weblink in your online Resources.
   a. Choose the 0.25x speed on the date view bar.
   b. Create a screen shot of the city in modern times using an updated Google image of the place.
   c. Compare the city over time. What changes do you see?
   d. Suggest possible reasons for the changes that have taken place over time.
2. What is happening in Australia? Investigate which foreign companies own farmland here, what they are using it for and where it is located.
3. Research how deforestation can lead to land degradation, such as has occurred in Madagascar. Has land degradation occurred on this kind of scale in Australia? Share your findings with the class.

FIGURE 7 Deforestation and erosion near Maevatanana, Madagascar, 2013

5.5 EXERCISE

Learning pathways

■ LEVEL 1	■ LEVEL 2	■ LEVEL 3
1, 3, 7, 10, 15	2, 4, 8, 9, 13	5, 6, 11, 12, 14

Check your understanding

1. List the two main ways that productive farmland can be lost.
2. Why is the use of corn as a biofuel a threat to food security?
3. Define the term *land grab*.
4. Refer to **FIGURE 1**. Describe the changes in population growth and the arable land per person between 1960 and 2030, making use of figures.
5. What does **FIGURE 1** suggest about food security?
6. Refer to **FIGURE 5**. Describe the distribution of countries in which China has acquired land.

Apply your understanding

7. What is jatropha? What are the benefits of growing this rather than corn and other biofuels?
8. Compare the advantages and disadvantages of using biofuels, such as wood and charcoal, instead of oil and gas in developing and developed nations.
9. Explain the interconnection between food security and land grabs.
10. What are the three main causes of the most extensive land degradation that occurs in the world?
11. Refer to **FIGURE 5**. What patterns or connections do you notice about the use of land in Africa that is being acquired by foreign countries? Write two detailed paragraphs explaining your observations and suggesting reasons why these patterns or connections have come about.
12. Discuss the points for and against the following statement: 'Australia will not need to purchase farmland overseas as it has enough land to produce food for the whole population.'

Challenge your understanding

13. Suggest why China might invest in food production and land in Australia.
14. Are 'land grabs' an effective, long-term solution for establishing a country's food security? Outline your view.
15. What do you notice about the use of land in Africa that is being acquired by foreign countries? What might these places have in common? Predict where governments might look to acquire land next.

To answer questions online and to receive **immediate feedback** and **sample responses** for every question, go to your learnON title at www.jacplus.com.au.

5.6 Features of a famine — Somalia

online only

LEARNING INTENTION

By the end of this subtopic, you will be able to explain the reasons for the 2011 famine in Somalia and identify the challenges of addressing famines.

The content in this subtopic is a summary of what you will find in the online resource.

FIGURE 1 Famine in Somalia, May 2011

Although many countries across the globe face food insecurity, it is rare for a country or region to be officially declared in famine. At the same time that countries in the Horn of Africa were suffering drought, much of southern Australia was similarly affected. Yet we did not suffer from food insecurity, nor was a famine declared. Why was there such a difference?

To learn more about the famine experienced in Somalia, go to your learnON resources at www.jacPLUS.com.au.

Contents

learnON

- 5.6.1 Causes of the 2011 famine in Somalia
- 5.6.2 The current situation in Somalia

FIGURE 3 The extent of the drought experienced in Somalia, 2016-17

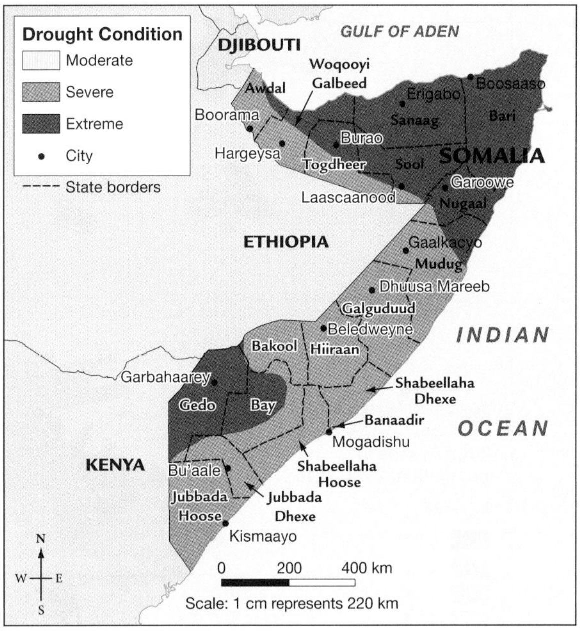

Source: Data from FAO, Relief Web.

Resources

eWorkbook	Features of a famine — Somalia (ewbk-9734)
Video eLesson	Features of a famine — Somalia — Key concepts (eles-5201)
Interactivity	Features of a famine — Somalia (int-8632)

5.7 SkillBuilder — Constructing and describing complex choropleth maps

online only

LEARNING INTENTION

By the end of this subtopic, you will be able to create a choropleth map from a data set, and identify and describe patterns and anomalies on a choropleth map.

The content in this subtopic is a summary of what you will find in the online resource.

5.7.1 Tell me

What is a complex choropleth map?

A complex choropleth map is a map that is shaded or coloured to show the average density or concentration of a particular feature or variable. The least dense or lowest concentration is usually the lightest shade. Average values are attached to the colour shadings in the key or legend. A complex choropleth map is used to show values in a pictorial way.

FIGURE 1 Population change in Sydney, 2001–06

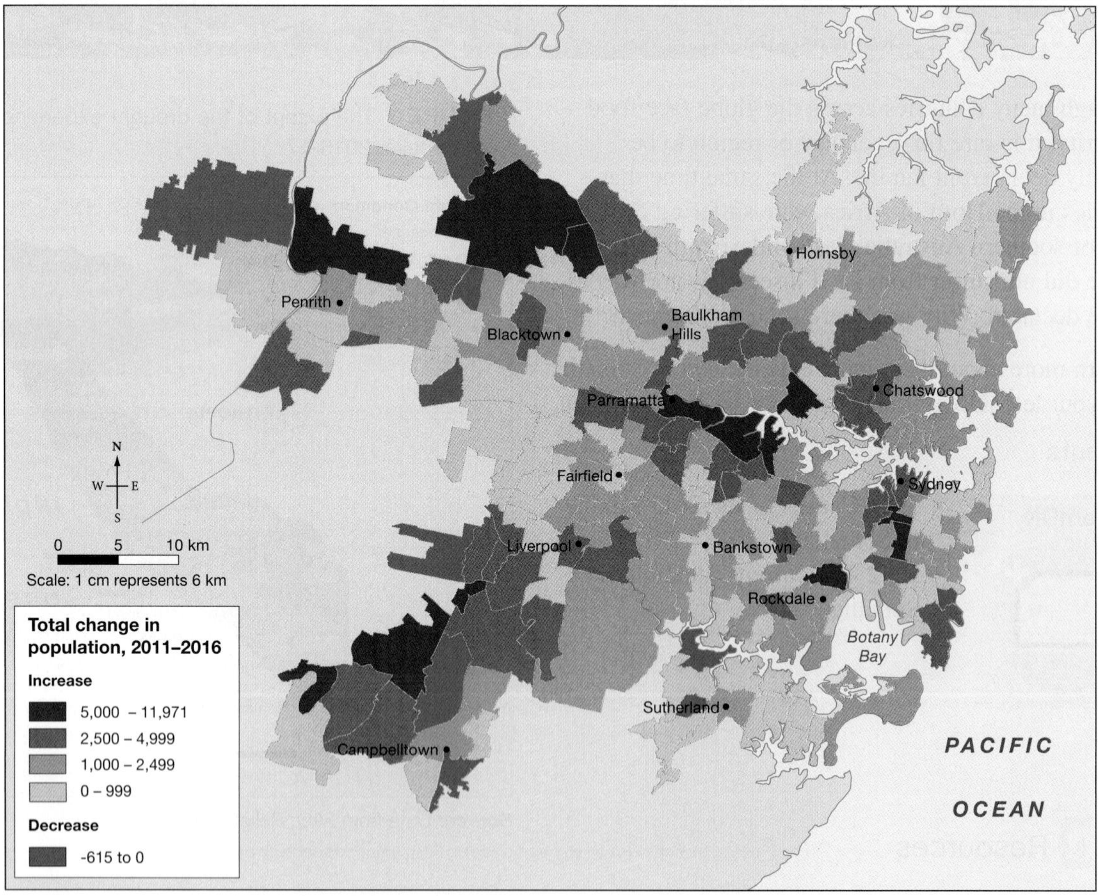

Source: Based on Australian Bureau of Statistics data

5.7.2 Show me

Creating a choropleth map

Step 1

Select a set of data to map, and ensure that you have a base map to match the area.

Step 2

Divide the data into approximately five categories and select a colour shade for each. The darkest colour represents the greatest value and the lightest colour represents the lowest value.
Create a legend.

Step 3

Colour all areas on the base map as represented by their colour value in your legend.

Step 4

Check that the geographical conventions are complete: border, orientation, legend, title, scale and source.

Describing a choropleth map

Step 5

Comment on where the areas of most intense colours occur. For example, in **FIGURE 1** the areas of greatest change (1440 to 4780 people) are often on the urban fringes, showing that the city's boundaries are spreading.

Step 6

Comment on where the areas of least intense colours occur. For example, in **FIGURE 1** most of the areas that have shown only moderate increase (0 to 290) are found 5 to 15 kilometres from the city.

Step 7

To identify a subtle change, look for small patches of different colour within an area that is predominantly another colour. In **FIGURE 1**, in the middle of the northern area of 1440 to 4780 increase there are two census areas that show a 290 to 1440 increase.

Step 8

Are there any anomalies? Identify any places that are different from the surrounding places. For example, there are no areas of 1440 to 4780 increase in the Sutherland Shire, south of the river that flows into Botany Bay.

5.7.3 Let me do it

learnON

Go to learnON to access the following additional resources to help you build this skill:

- a longer explanation of this skill and its application in Geography (Tell me)
- a video demonstrating the step-by-step process of this skill (Show me)
- an activity and interactivity for you to practise the skill (Let me do it)
- self-marking questions to help you understand and use the skill.

on Resources

eWorkbook	SkillBuilder — Constructing and describing complex choropleth maps (ewbk-9738)
Digital document	Blackline master — Map of Africa (doc-12012)
Video eLesson	SkillBuilder — Constructing and describing complex choropleth maps (eles-1732)
Interactivity	SkillBuilder — Constructing and describing complex choropleth maps (int-3350)

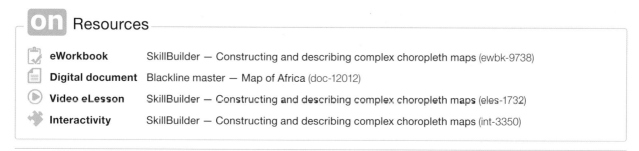

5.8 Urban expansion into food production areas

LEARNING INTENTION

By the end of this subtopic, you will be able to describe the impact of Sydney's urban expansion on food production.

5.8.1 Sydney's expansion

The population of Greater Sydney reached 5.312 million in 2019 and is projected to keep growing. Due to urban expansion, Sydney's surrounding farmland is increasingly being transformed into housing estates. Food once grown in the Sydney region is now grown interstate and transported to the markets. Into the future, there are plans to develop urban farming precincts and indoor farms within the Sydney Basin.

By 2055, it is estimated that more than 8 million people will call Sydney home. Challenges come with living in the city such as the cost of living, cost of housing, traffic and commute times, and employment opportunities. The State and Federal Governments have to plan sustainably to support the needs of the whole community while addressing economic growth and environmental concerns.

FIGURE 1 The Sydney Royal Botanic Gardens is the oldest botanic garden and scientific institution in Australia. Covering 30 hectares near the CBD, it is protected from urban expansion and sprawl.

FIGURE 2 Predicted urban growth within the Greater Sydney region, 2016–2041

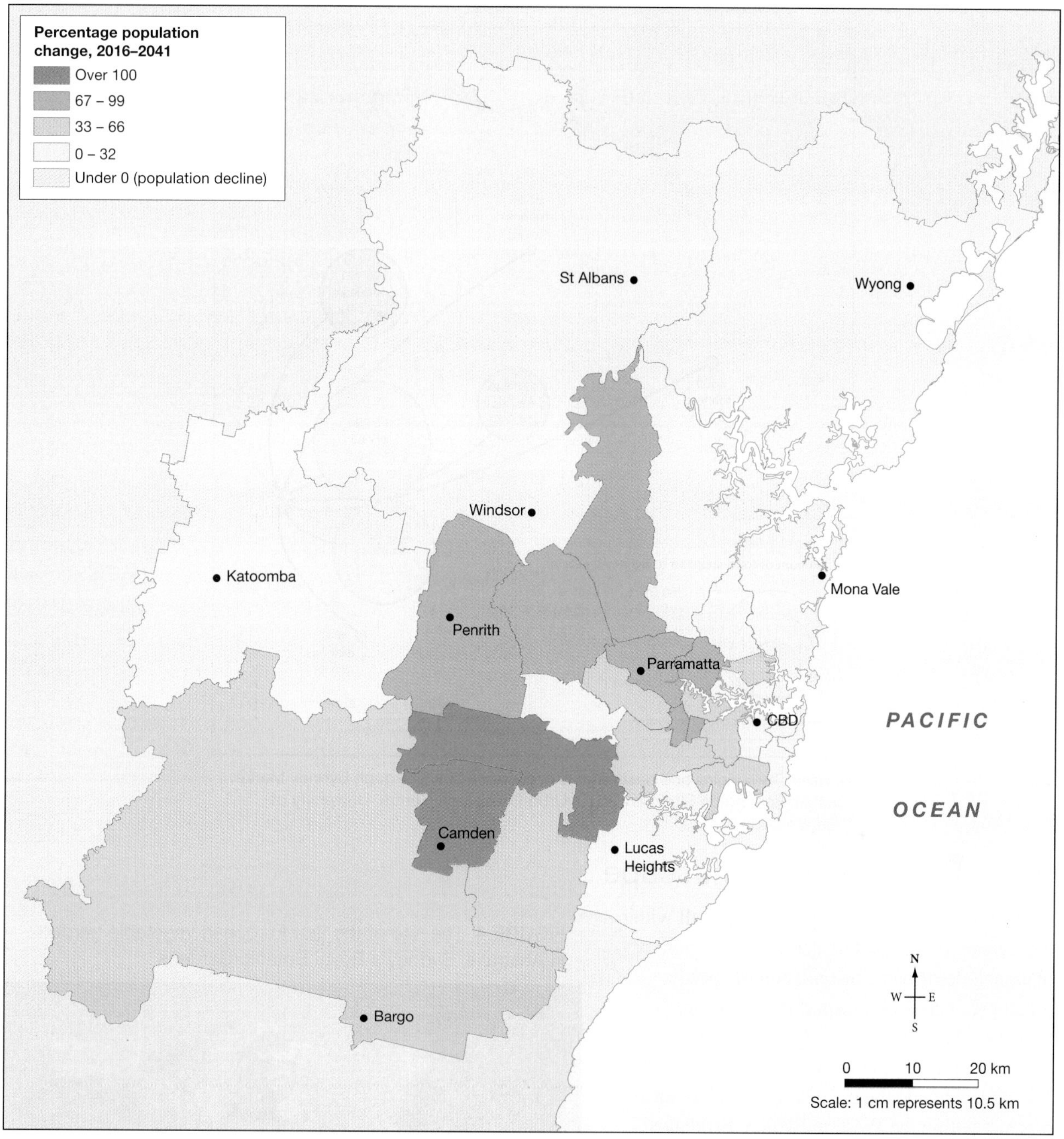

Source: NSW Department of Planning, Industry and Environment (2019). Map drawn by Spatial Vision.

The Institute for Sustainable Futures assesses the changes in the Greater Sydney Metropolitan area. Currently, the Sydney Basin produces approximately 500 000 tonnes of food, including eggs, fruit, vegetables, meat and dairy. This supports about 20 per cent of the Sydney population. If current development trends continue, by 2031, Sydney will lose 60 per cent of its total food production including up to 90 per cent of its fresh fruit and vegetable production. The Sydney basin will produce about 220 000 tonnes of food by 2031, which is enough to support only about 6 per cent of the projected population. Areas that will experience a decline in food production will see an increase in urbanisation.

FIGURE 3 Sydney's Food Footprint: where the city's fresh fruit and vegetables come from

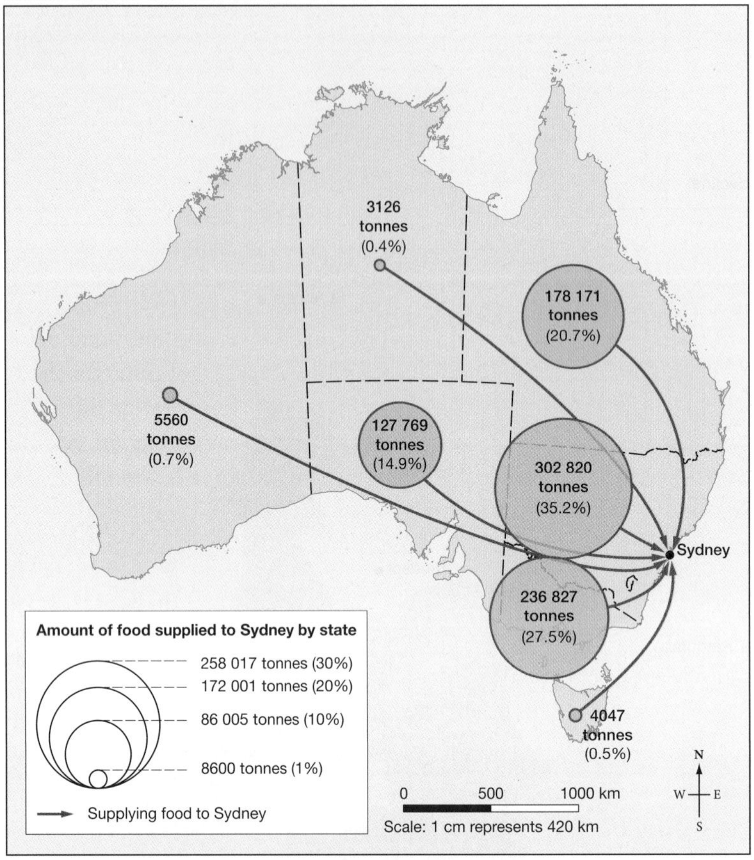

Source: The origins and destinations of produce flows through Sydney Market Limited Flemington Site. May 2013. Urban Research Centre, University of Western Sydney.

5.8.2 The changing landscape

To accommodate future population growth within the Sydney basin, the Greater Sydney region plan includes urban consolidation of specific urban precincts and transport infrastructure projects. The new Western Sydney airport is being constructed in the south-western part of the city. The plans include the creation of an urban farm precinct within the Western Sydney Parklands development, encouraging urban farming benefits such as those listed in **FIGURE 5**. Other urban farming options growing in popularity throughout the city include indoor farms, rooftop farms and the repurposing of former sporting facilities into city farms.

FIGURE 4 The site of the first European vegetable garden in Australia, Sydney's Royal Botanic Gardens

Greater Sydney currently supports a diverse range of agricultural activity. Outdoor vegetable-growing farms account for 31% of all farms in the Greater Sydney region. **FIGURE 6** shows the value of a range of farming activities taking place within the region.

FIGURE 5 Reasons to continue producing food within the Sydney basin

Why grow fresh food in the Sydney basin?
Close to market improves quality and reduces spoilage
High-value produce
Good rainfall and fertile soils
Reduce food waste in supply chain
Reduce food miles and buffer against fuel price shocks
Local employment for farmers and food processors
Nutrients and organic matter readily available in urban waste
Biodiversity, agro-tourism and urban heat mitigation benefits

Source: © Institute for Sustainable Futures UTS, 2015

FIGURE 6 Value of agricultural production in the Greater Sydney region, 2018–19

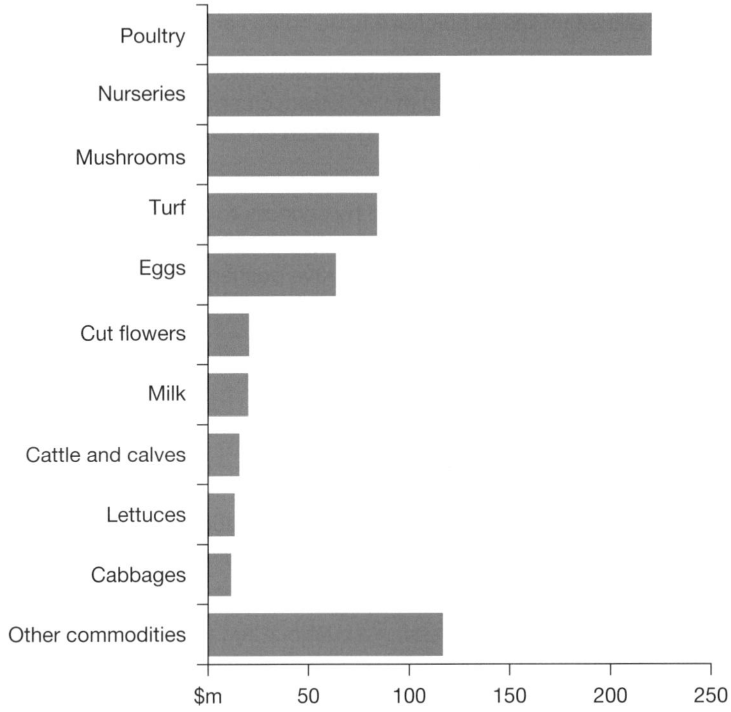
Poultry
Nurseries
Mushrooms
Turf
Eggs
Cut flowers
Milk
Cattle and calves
Lettuces
Cabbages
Other commodities
$m
50
100
150
200
250

Source: https://www.agriculture.gov.au/abares/research-topics/aboutmyregion/nsw-sydney#agricultural-sector

CASE STUDY

Urban sprawl is threatening Sydney's foodbowl

The Conversation, 25 February 2016
by Dana Cordell, Brent Jacobs and Laura Wynne

Sydney loves to talk about food, and the housing market. But rarely do we talk about the threat that housing poses to the resilience of Sydney's food system.

If we continue along the path we're on, Sydney stands to lose more than 90% of its current fresh vegetable production. Total food production could drop by 60% and the city's supply of food from within the basin could drop from 20% of total food demand to a mere 6%.

Like most Australian cities, Sydney is facing an influx of people — 1.6 million new residents are expected over the next 15 years.

Competing priorities for land are compounded by this growing population, as well as by planning laws that favour development over agriculture — not to mention a changing climate. Cities worldwide are facing the same issues as we try to feed a growing population with limited resources.

Protecting farmers

Sydney's fertile soils are being paved over at a rapid rate. Large portions of areas that currently grow Sydney's fresh produce are earmarked for release for housing development.

Currently, the planning system does not prioritise agriculture as a land use, meaning urban sprawl into potential farmland continues relatively unchecked. Instead, planning tends to focus on whichever use has the greatest economic value.

In an overheated housing market such as Sydney's, this tends to mean agricultural land is allowed to be rezoned for houses or other higher-value land uses.

As city land prices rise, more people are moving further out for a 'tree change'. Lower land prices on the city's fringe allow families to purchase large homes and lots at a lower price than in the city.

But many of these new rural residents don't like the early morning sound of tractors and the smell of manure on neighbouring farms, and make nuisance complaints to their local council. These complaints often result in tough operating restrictions being placed on farmers' activities, such as limits on hours of operation and types of fertiliser that can be used.

These restrictions are introduced by councils to appease local residents and are in accordance with noise pollution laws designed for urban residential areas. But they can have significant impacts on farm viability. In several instances, such restrictions have pushed marginal farms into the red, eventually forcing farmers off their land and out of the basin.

The New South Wales government is interested in taking steps to ameliorate this problem, as demonstrated through its recently tabled Right to Farm policy. This seeks to ensure that farmers' right to operate their business is protected against nuisance complaints.

Why growing food in Sydney is important

There are enormous benefits to growing fresh food in the Sydney basin — and, indeed, near any city. Perishable foods such as Asian greens and eggs can be grown close to market, reducing spoilage, waste and food miles, and buffering against spikes in fuel prices.

Agriculture and food processing are labour-intensive, providing significant local job opportunities. In fact, the benefit of Sydney's agriculture to the economy is estimated at upwards of A$4.5 billion. This includes jobs in storage, processing, transport and retail.

A changing climate will mean many of Australia's important foodbowls, such as the Murray-Darling Basin, are likely to be more vulnerable to droughts and floods. Sydney's higher rainfall and fertile soils will become even more suitable for growing food, meaning their importance to Sydney's food supply will grow.

FIGURE 7(a) Food production areas around Sydney, 2011

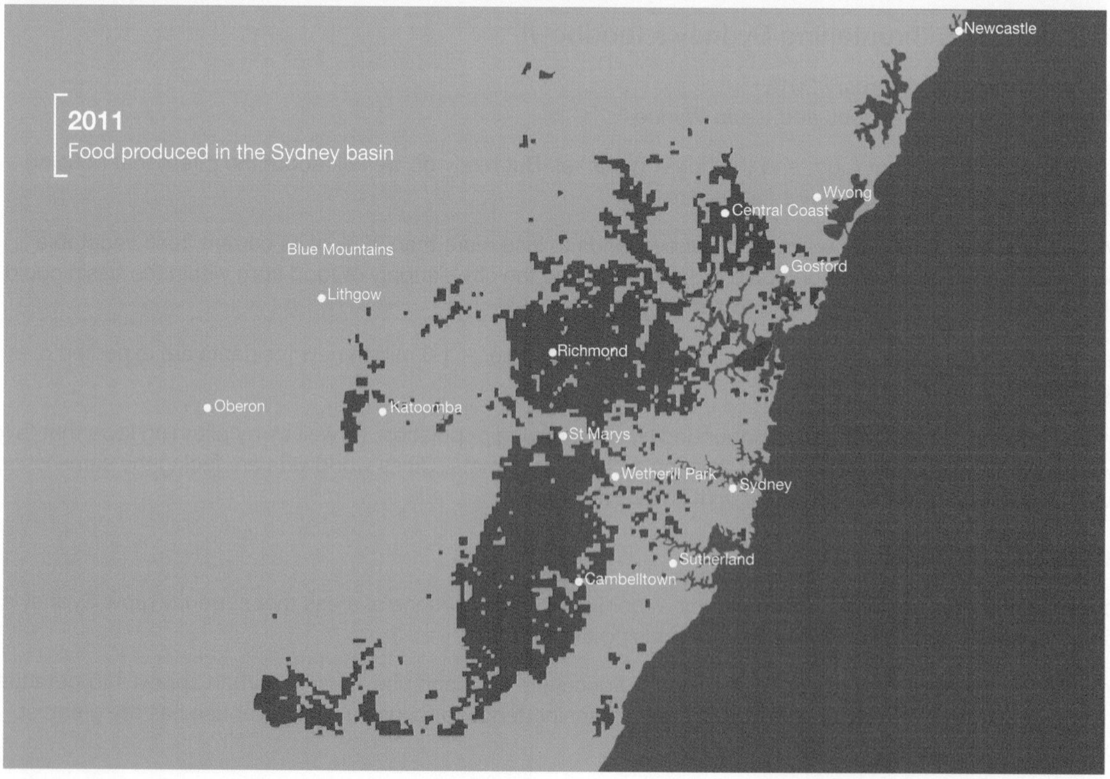

Source: © Institute for Sustainable Futures UTS, 2016

Farms on the fringes of our city will help buffer the city against the impacts of climate change, by cooling the city and helping wildlife move between habitat.

Food produced in close proximity to the city can also be fertilised by nutrients and organics in urban food waste, garden waste and wastewater. Accounting for these sources, Sydney actually has 15 times more phosphorus supply than agricultural demand. That means local food systems can better buffer against the growing global threat of phosphorus fertiliser scarcity, a threat that could lead to further fertiliser price spikes and supply disruptions.

Sydney's farms also help buffer the city against disruptions to food supply. For example, if a bushfire or fuel shortage cut transport routes into Sydney, the city would have only two days' stock of fresh produce. Our research shows that in the face of dramatically increasing population, Sydney stands to lose these benefits. A similar study in Melbourne found their city's foodbowl could plummet from meeting 41% of Melburnians' food demand to 20%.

Unlike Melbourne, Sydney is geographically constrained by mountains on one side and ocean on the other, meaning there is nowhere for agricultural production in the basin to go. Our agricultural production is literally being chased to the hills — and this at a time when we face the challenge of feeding over a million extra mouths. We've mapped Sydney's current and future food production. The pink areas of the images below are areas where food is produced. As the maps indicate, the areas producing food in 2031 will dramatically decrease if we continue along the path we're on.

FIGURE 7(b) Food production areas around Sydney, 2031 (projected)

Source: © Institute for Sustainable Futures UTS, 2016

A better food future

Our city plans need to value and better protect agriculture from urban sprawl. Planners need to make decisions based on evidence to balance competing land uses.

These decisions need to take account of the full suite of values and benefits we gain from Sydney farmers, not just the economic gains we stand to achieve by converting the land to houses.

Farmers in the basin need better commercial conditions, a fair price for commodities, land security and support from other residents.

Sydneysiders need access to affordable housing, jobs and infrastructure.

But, equally, they need access to nutritious and affordable food, reversing the high rate of obesity and diabetes, and 'food deserts' without access to groceries particularly prevalent in Western Sydney.

Through increased awareness and accessibility, food shoppers can also support local food producers, increasing the resilience of Sydney's food system and simultaneously reducing the environmental footprint of food.

However, strategic policies and plans are needed to ensure that agriculture is valued and prioritised as an important land use and economic activity within our city, to ensure that buying local food is a choice that consumers can make in future.

fdw-0031

FOCUS ON FIELDWORK

Extracting information from media interviews

When you are conducting research into important geographical issues it can be difficult to focus on the right or relevant information in a media interview. We are all used to seeing influential people interviewed, but because of their expertise and language, as well as the questioning style of the interviewer, it isn't always easy to separate the useful information. This can be especially difficult if the issue is one that provokes a lot of strong opinions. This might even make it hard to set aside your own views and listen to what is being said objectively.

Learn more about field sketches using the **Extracting information from media interviews** fieldwork activity in your online Resources.

on Resources

eWorkbook	Urban expansion into food production areas (ewbk-9742)
Video eLesson	Urban expansion into food production areas — Key concepts (eles-5202)
Interactivity	Urban expansion into food production areas (int-8633)

5.8 ACTIVITIES

1. Use the **Sprout stack** weblink in your online Resources to investigate indoor farming technology in a northern suburb of Sydney. Brainstorm the advantages and disadvantages of growing food for the Sydney markets in this way. Use this to create an infographic to encourage people to buy locally grown food.
2. Read the case study *Urban sprawl is threatening Sydney's foodbowl* in section 5.8.2. As you read, highlight and look up any terms that you don't understand.
   a. Use Google maps to calculate the shortest distance by road between the following places and Sydney's central business district.
      i. Wyong
      ii. Gosford

FIGURE 8 (a) Wyong station (b) Gosford CBD

   b. What are the advantages of food production within the Sydney region for consumers?
   c. Research the perspective of one of the following groups (individually, in pairs or small groups). Use the **Sydney food futures**, **The future of Sydney's food bowl** and **A Metropolis of Three Cities — the Greater Sydney Region Plan** weblinks in your online Resources to help begin your research.
      - Food farmers in Sydney's food bowl area
      - Aboriginal Peoples whose Country is in Sydney's food bowl area
      - Real estate developers
      - Food retailers and market stall owners in Sydney
      - 'Tree changers'
      - Consumers in Sydney
      - Residents in Sydney's food bowl area who are not farmers but have lived and worked in the area for a long time
      - City planners
      - Food transport workers
      - Restaurant and café owners in coastal New South Wales
   d. Conduct a class role play in which your teacher is the moderator of a meeting to discuss whether a large area west of the Pacific Motorway between Gosford and Wyong should be rezoned for medium density residential land. The land is currently being used for intensive food farming of vegetables for the Sydney market. How would commuter numbers or traffic at these stations be affected by the proposed rezoning?

5.8 EXERCISE

Learning pathways

■ LEVEL 1	■ LEVEL 2	■ LEVEL 3
1, 2, 3, 10, 15	4, 5, 6, 9, 14	7, 8, 11, 12, 13

Check your understanding

1. By 2055, what is the population of Sydney expected to be?
2. State reasons why food has to be transported to markets in Sydney.
3. Why is agricultural land being replaced in semi-rural areas of Sydney?
4. Describe the pattern of where the highest percentage of population growth is occurring in Sydney.
5. Outline three reasons why it is important to grow food close to urban centres.
6. What were the three highest-value agricultural products grown in the Greater Sydney region in 2018–19?
7. Approximately how much does Sydney food bowl agriculture contribute to the economy annually?
8. Without significant change in current land development trends, how much of each of the following would be lost in the next 15 years?
   a. Fresh vegetable production
   b. Total food production
   c. Percentage of food demand grown in the Sydney region

Apply your understanding

9. Why is there a declining number of farmers producing food in the areas around Sydney?
10. Explain the benefits and drawbacks of moving to live in a farming area.
11. Why do you think local council might place noise restrictions on farmers when new residents complain about their machinery noise?
12. How can individuals help to reduce the impact of a shrinking local farming industry?

Challenge your understanding

13. With limited agricultural land, there is a lot of pressure placed on farmers to feed people. As the Australian population continues to grow in urban areas, less agricultural land is available. Suggest some solutions to meet the food requirements. Consider what individuals, groups and the government could do to assist.
14. Predict how the COVID-19 pandemic might have changed the way Australians feel about food production, food security and food shopping.
15. How might a smaller Sydney food bowl affect people in New South Wales outside the Greater Sydney area? Consider the possible impact on north and south coastal areas, regional towns and people living remotely.

To answer questions online and to receive **immediate feedback** and **sample responses** for every question, go to your learnON title at www.jacplus.com.au.

5.9 Climate change and food production

LEARNING INTENTION

By the end of this subtopic, you will be able to describe and explain the possible impacts of climate change on food production

5.9.1 Impacts of climate change

The impacts of climate change on future world food security are a case of give and take. Some regions of the world will benefit from increases in temperature and rainfall, while others will face the threat of greater climatic uncertainty, lower rainfall and more frequent drought. In either case, food production will be affected.

FIGURE 1 The majority of the workforce in sub-Saharan Africa works in the agriculture industry.

Agriculture is important for food security because it provides people with food to survive. It is also the main source of employment and income for 26 per cent of the world's workforce. In heavily populated countries in Asia, between 40 and 50 per cent of the workforce is engaged in food production, and this figure increases to over 54 per cent in sub-Saharan Africa.

It is difficult to predict the likely impacts of climate change, because there are many environmental and human factors involved (see **FIGURE 2**), as well as different predictions from scientists (see **FIGURE 3**).

FIGURE 2 Some of the possible impacts of climate change on food production

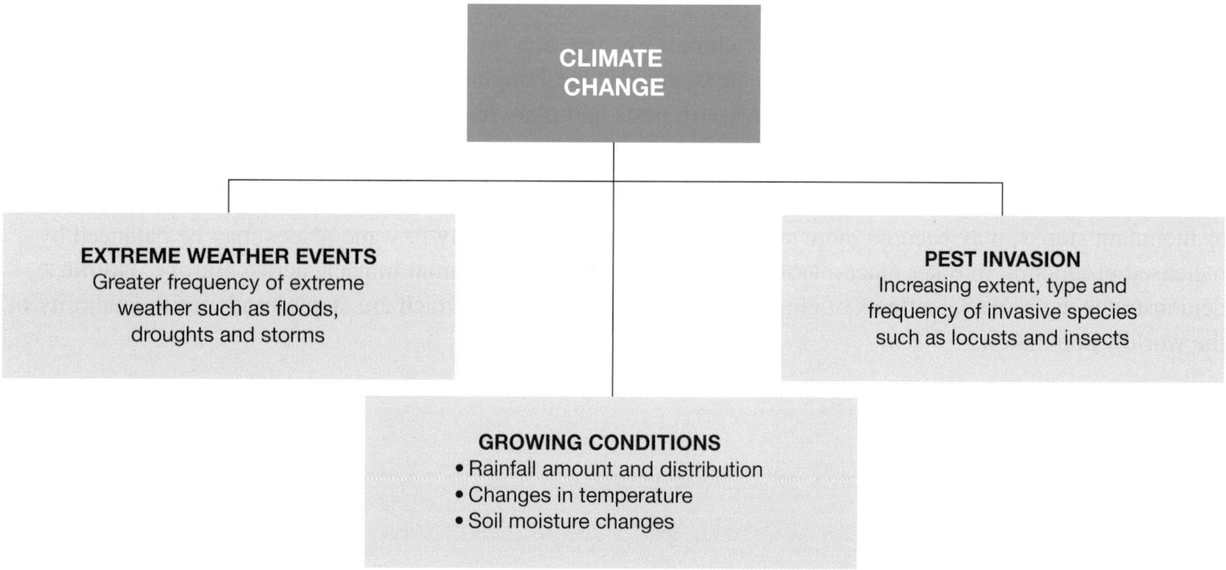
CLIMATE CHANGE
EXTREME WEATHER EVENTS
Greater frequency of extreme weather such as floods, droughts and storms
PEST INVASION
Increasing extent, type and frequency of invasive species such as locusts and insects
GROWING CONDITIONS
• Rainfall amount and distribution
• Changes in temperature
• Soil moisture changes

FIGURE 3 Projected consequences of climate change

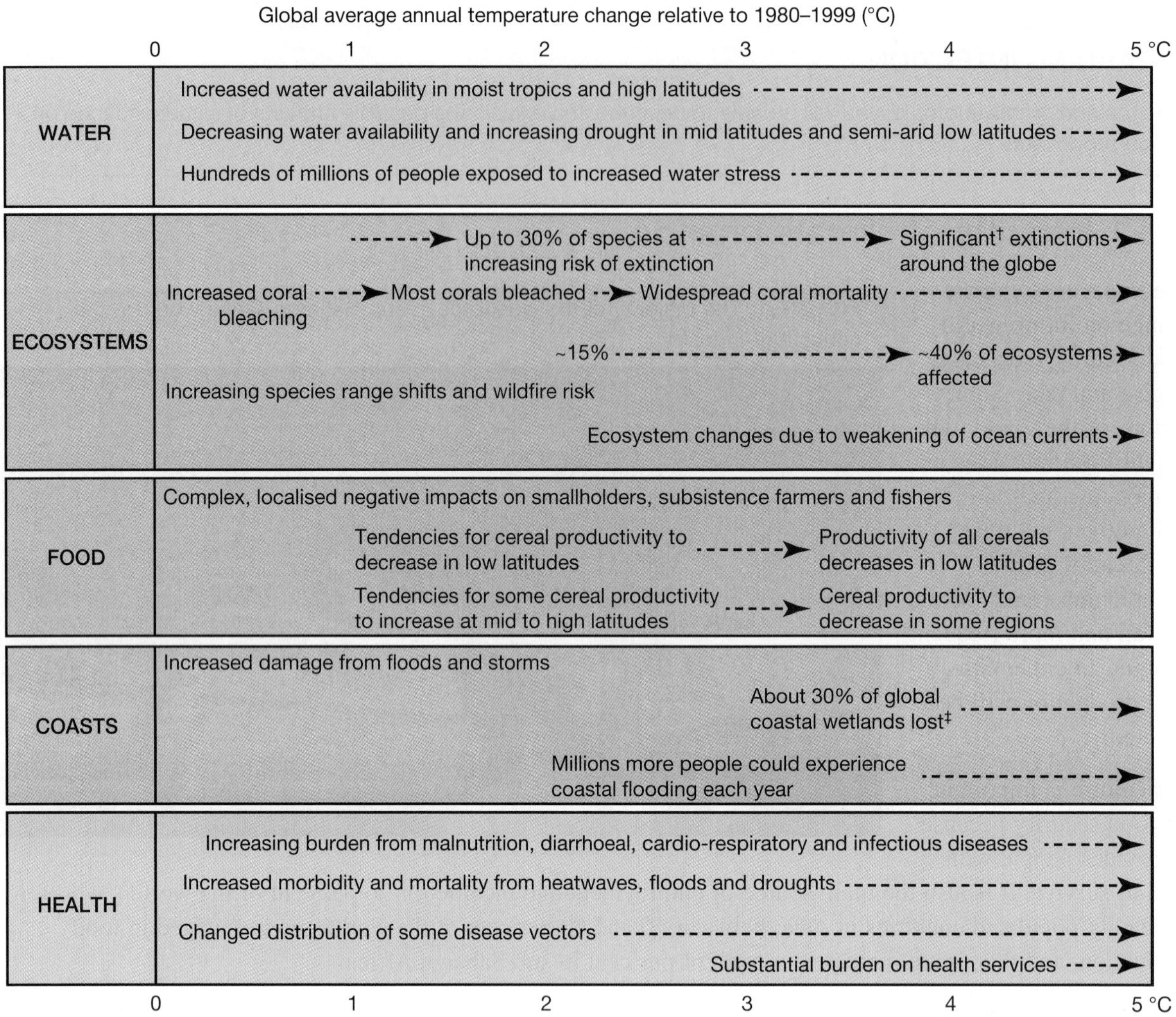
Global average annual temperature change relative to 1980–1999 (°C)
0
1
2
3
4
5 °C
WATER
Increased water availability in moist tropics and high latitudes
Decreasing water availability and increasing drought in mid latitudes and semi-arid low latitudes
Hundreds of millions of people exposed to increased water stress
ECOSYSTEMS
Up to 30% of species at increasing risk of extinction
Significant† extinctions around the globe
Increased coral bleaching
Most corals bleached
Widespread coral mortality
~15%
~40% of ecosystems affected
Increasing species range shifts and wildfire risk
Ecosystem changes due to weakening of ocean currents
FOOD
Complex, localised negative impacts on smallholders, subsistence farmers and fishers
Tendencies for cereal productivity to decrease in low latitudes
Productivity of all cereals decreases in low latitudes
Tendencies for some cereal productivity to increase at mid to high latitudes
Cereal productivity to decrease in some regions
COASTS
Increased damage from floods and storms
About 30% of global coastal wetlands lost‡
Millions more people could experience coastal flooding each year
HEALTH
Increasing burden from malnutrition, diarrhoeal, cardio-respiratory and infectious diseases
Increased morbidity and mortality from heatwaves, floods and droughts
Changed distribution of some disease vectors
Substantial burden on health services
0
1
2
3
4
5 °C

† Significant is defined here as more than 40%. ‡ Based on average rate of sea level rise of 4.2 mm/year from 2000 to 2080.

Source: Figure SPM.7 and Figure SPM.11 (upper panel) from IPCC, 2014: *Climate Change 2014: Synthesis Report. Contribution of Working Groups I, II and III to the Fifth Assessment Report of the Intergovernmental Panel on Climate Change.* [Core Writing Team, Pachauri, R.K. and Meyer, L. (eds.)]. IPCC, Geneva, Switzerland.

There is a wide range of possible impacts of climate change. Sea-level rises may cause flooding and the loss of productive land in low-lying coastal areas, such as the Bangladesh and Nile River deltas. Changes in temperatures and rainfall may cause an increase in pests and plant diseases.

However, agriculture is also adaptable. Crops can be planted and harvested at different times, and new types of seeds and plants or more tolerant species can be used. Low-lying land may be lost, but higher elevations, such as mountain slopes, may become more suitable. The loss in productivity in some places may be balanced by increased production in other places. **FIGURE 4** shows the range of potential impacts across Europe; **FIGURE 7** demonstrates the projected effects of climate change on cereal crops, which are staple foods for the majority of the world's population.

Essentially, hundreds of millions of people are at risk of increased food insecurity if they have to become more dependent on imported food. This will be evident in the poorer countries of Asia and Sub-Saharan Africa, where agriculture dominates their economy. There is also a risk of an increase in **environmental refugees** or people fleeing places of food insecurity.

Global climate strikes (**FIGURES 5** and **6**) have also helped to highlight the potential of climate change to reduce food security significantly around the world. People living in wealthy developed nations might not be affected by food shortages, but they will feel the impact of greater numbers of environmental refugees needing help.

FIGURE 5 Global Climate Strike, London 2019

FIGURE 4 Examples of potential consequences of climate change in Europe

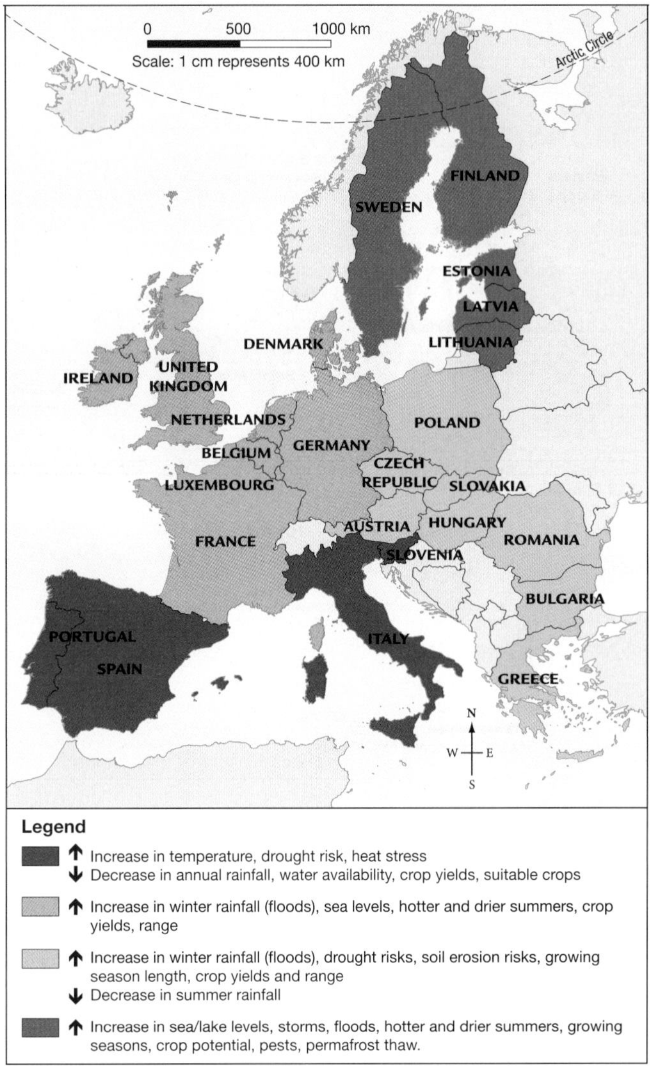

Source: Spatial Vision

FIGURE 6 Global Climate Strike, London 2019

environmental refugees people who are forced to flee their home region due to environmental changes (such as drought, desertification, sea-level rise or monsoons) that affect their wellbeing or livelihood

int-8634

FIGURE 7 Predicted regional effects of climate change on cereal production

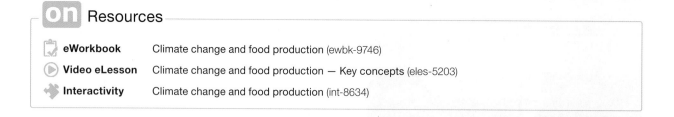

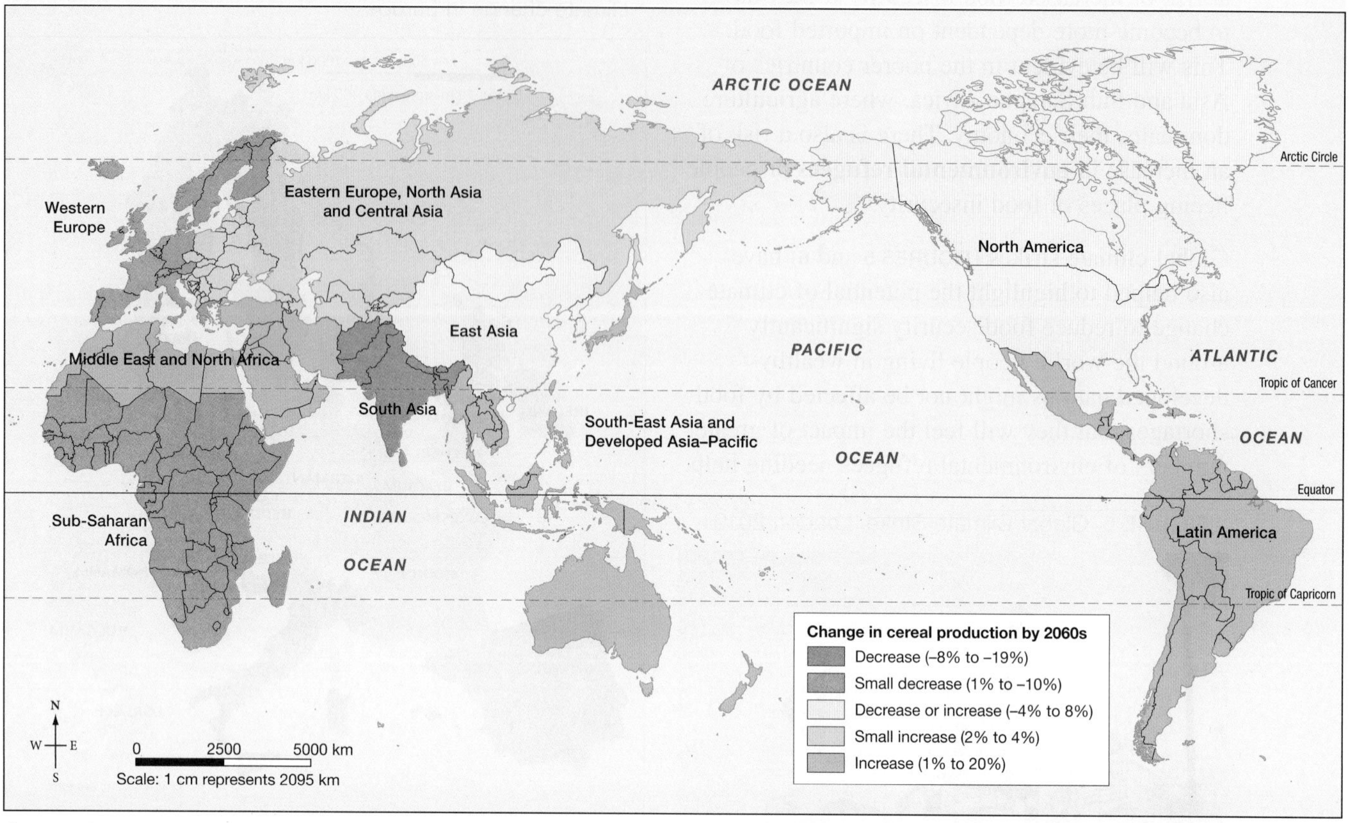

Source: Spatial Vision

on Resources

eWorkbook	Climate change and food production (ewbk-9746)
Video eLesson	Climate change and food production — Key concepts (eles-5203)
Interactivity	Climate change and food production (int-8634)

5.9 ACTIVITIES

1. Use the **How to feed the world in 2050** weblink in your online Resources to find out more about this topic. Research potential impacts of climate change on Australia. Create an annotated map to illustrate your findings.
2. As the climate changes, who should be responsible for making sure that people have food security? As a class, discuss the role that each of the following has to combat the impact of climate change on peoples' access to food:
   - governments
   - global organisations and NFPs
   - industry
   - individuals.

5.9 EXERCISE

Learning pathways

■ LEVEL 1	■ LEVEL 2	■ LEVEL 3
1, 2, 3, 10, 14	4, 7, 8, 9, 15	5, 6, 11, 12, 13

Check your understanding

1. Approximately what proportion of the world's workers rely on agriculture for their main income?
2. Define the term *environmental refugee*.
3. List three countries in Europe that are likely to:
   a. be at greater risk of drought and heat stress
   b. have more winter flooding and less summer rainfall
   c. suffer from the permafrost thawing
   d. have more land suitable for cropping but experience hotter drier summers
4. Outline two ways that rising global temperatures are expected to affect the places that grain can be grown. Refer to longitude and latitude in your answer.
5. How is climate change predicted to affect cereal production in Australia by the 2060s?
6. Refer to **FIGURE 7**. Which places have the potential to be cereal exporters, and which places are likely to become dependent on grain imports? Use data in your answer.
7. Describe the interconnection between environmental refugees and climate change.

Apply your understanding

8. Which countries of Europe will benefit from climate change in terms of food production, and which countries are likely to suffer negative outcomes?
9. Choose one of the regions that is likely to experience a 1–20% increase in cereal production. What are the implications of countries in that region producing more grain? Explain one economic and one social implication.
10. Refer to **FIGURE 3**. For each of the following statements, determine whether the statement is true or false, and explain why you came to this conclusion. Use data in your answer.
    a. If temperatures increase by 3 °C, you expect to see crop yields rising around the equator.
    b. Changes in extreme weather events are unlikely unless temperatures increase by at least 1 °C.
    c. Food insecurity will be felt greatly in developing regions if temperatures rise by more than 4 °C.
    d. Places that are likely to experience decreasing crop yields will be found in the higher latitudes.
11. Would increases in irrigation be a sustainable solution to growing food in Spain? Explain your answer.
12. Is it accurate to suggest that there are some places for which climate change will bring more benefits than problems? Evaluate the impacts of climate change and write a paragraph, including data as supporting evidence, explaining your response.

Challenge your understanding

13. Predict how one type of extreme weather event could change food production. Explain the interconnection between the event and levels of food production in a well-structured paragraph.
14. Suggest what countries around the world might do to share food more equitably.
15. How might a country such as Australia best prepare its food production systems to cope with potential changes in climate?

To answer questions online and to receive **immediate feedback** and **sample responses** for every question, go to your learnON title at www.jacplus.com.au.

5.10 Investigating topographic maps — Yarra Yarra Creek Basin food bowl

LEARNING INTENTION

By the end of this subtopic, you will be able to describe the areas of the Yarra Yarra Creek Basin that are suitable for food production, identify them on a topographic map, and explain how these places might change over time.

5.10.1 Yarra Yarra Creek Basin (near Holbrook), NSW

The Yarra Yarra Creek Basin is renowned for producing high-quality beef and sheep meat. The Yarra Yarra Basin is located east of Holbrook, a small agricultural town with a population of approximately 1300. Holbrook is located 492 kilometres south-west of Sydney along the Hume Highway between Melbourne and Sydney.

The Yarra Yarra Basin is in a high-rainfall region at the foothills of the Great Dividing Range. Many small creeks feed water into the Yarra Yarra Creek (AR 3951). The high levels of beef and lamb production in the Basin are due to good water resources and the use of **improved pasture**. Crops such as oats, triticale and canola are grown mostly for livestock grazing and sometimes harvested for hay and grain for **fodder**. Farms in the basin are relatively large, in the range of 800–1200 hectares.

improved pasture pasture that has been specially selected and sown, which is usually more productive than the local native pasture

fodder food such as hay or straw for cattle and other livestock

FIGURE 1 Silos: tall cylindrical structures used to store grain.

FIGURE 2 Topographic map of the Yarra Yarra Creek Basin

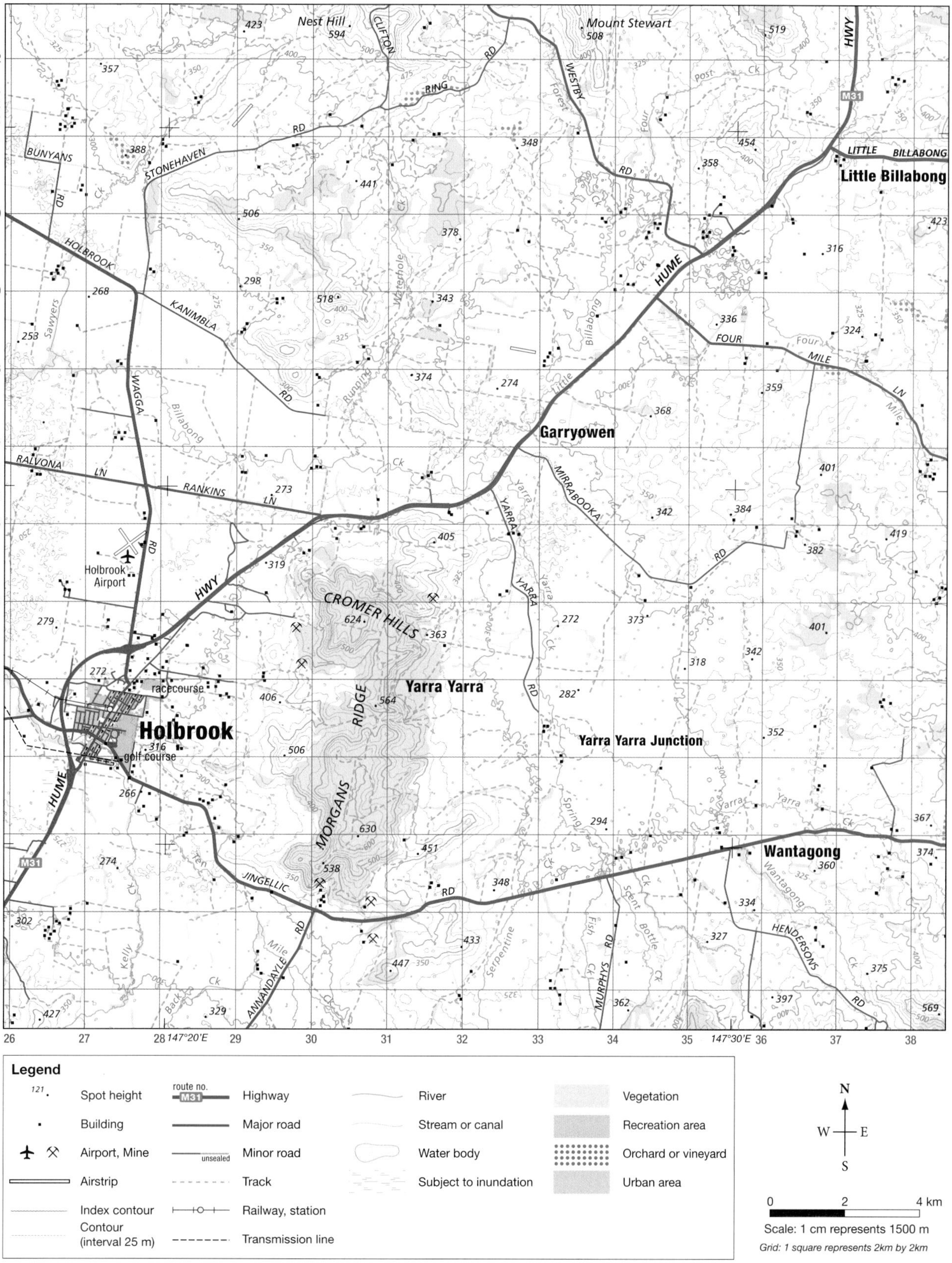

Source: Data based on Spatial Services 2019.

Resources

eWorkbook	Investigating topographic maps — Yarra Yarra Creek Basin food bowl (ewbk-9750)
Digital document	Topographic map of Yarra Yarra Creek Basin (doc-36317)
Video eLesson	Investigating topographic maps — Yarra Yarra Creek Basin food bowl — Key concepts (eles-5204)
Interactivity	Investigating topographic maps — Yarra Yarra Creek Basin food bowl (int-8635)
Google Earth	Yarra Yarra Creek Basin (gogl-0137)

5.10 EXERCISE

Learning pathway

■ LEVEL 1	■ LEVEL 2	■ LEVEL 3
1, 2, 7, 11	3, 4, 8, 10	5, 6, 9, 12

Check your understanding

Refer to **FIGURE 2** to answer the following questions.

1. What is the contour interval of the map?
2. What is the area reference for the spot height of 538 metres on Morgans Ridge?
3. What is the aspect of the slope of the spot height 538 m?
4. Write the scale of the map as a ratio.
5. In which general direction would you need to drive to get from Holbrook to Yarra Yarra Junction?
6. Describe the natural environment on top of Morgans Ridge.

Apply your understanding

7. Which area is more likely to produce food — AR2847 or AR3042? Justify your answer.
8. If you were looking to clear new land to start an orchard, would AR3650 be a good place? Explain your reasoning.
9. Consider the features of the area north of the Hume Highway. Is this land suitable land for farming? Refer to three features of that area to support your view.

Challenge your understanding

10. Predict how climate change is likely to affect the water security of the Yarra Yarra Creek Basin in the future.
11. Suggest actions that could be taken to ensure food production levels in the area are maintained in the future. Consider one thing that each of the following could do.
    a. Farmers
    b. Local government
    c. Water authorities.
12. Is it likely, considering the size of properties and the types of food produced in this area, that the same types of farms will exist in the Yarra Yarra Creek Basin in 50 years? Give reasons for your answer.

To answer questions online and to receive **immediate feedback** and **sample responses** for every question, go to your learnON title at www.jacplus.com.au.

5.11 Thinking Big research project — Famine crisis report

online only

The content in this subtopic is a summary of what you will find in the online resource.

Scenario

Although many countries across the globe face food insecurity, it is rare for a country or region to be officially declared in famine, the worst form of food insecurity. By definition, a famine is an extreme crisis of access to adequate food, resulting in widespread malnutrition and loss of life due to starvation and infectious disease. The effects of climate change could mean that these numbers begin to increase.

People who are experiencing famine have lost the means of earning an income and have few, if any, resources to sustain themselves. In general, there is no one cause of famine; rather, it is a series of overlapping factors including climate extremes, crop failures, poor governance and, most importantly, conflict. Conflicts, such as civil wars, can prevent people from producing food, create large-scale movement of people fleeing the fighting, and prevent aid from reaching people. Often governments do not have the resources, planning or will to deal with the issue, and international assistance is needed.

Task

The United Nations (UN) has asked you, a world-leading specialist in food security, to write a report that will assist them in organising a response to a famine. You will need to present your report to a famine taskforce panel at the UN headquarters in New York. Go to learnON to access the resources you need to complete this research project.

on Resources

ProjectsPLUS Thinking Big research project — Famine crisis report (pro-0191)

5.12 Review

5.12.1 Key knowledge summary

5.2 Water security and food production

- Water is an essential ingredient for food production.
- Water quantity and water quality threaten future water security.
- Water stress occurs when water supplies fall below 1700 m^3 per year.
- Food insecurity is a symptom of water stress. It will be a challenge in the future as we experience growing population numbers and the effects of climate change.

5.3 Pollution and food production

- Agriculture is a major contributor to water source contamination as nutrient, chemical and sediment pollutants run off and seep into local water supplies.
- Irrigation may lead to soil salinity and waterlogging of farmland and surrounding areas.
- Airborne pollutants may impact food quality delivered to consumers.
- Approximately 75% of the world's population are undernourished.

5.5 Land loss and food production

- Biofuels offer a renewable energy source. Increasingly, agricultural crops are being used for biofuel production.
- The biofuel industry has the potential to threaten future food security by increasing the purchase price of crops and encouraging farmers to choose fuel crop production over food crops.
- Expanding urban centres compete for agricultural land and compromise local food production.
- Some countries are investing in agricultural land to expand their crop production capabilities.

5.6 Features of a famine

- A famine is a drastic, widespread food shortage leading to extreme hardship for people.
- Emergency measures are required to sustain the population and provide a food relief plan.
- Many factors may contribute to a famine, including drought, political instability, corruption, lack of infrastructure, and/or insurgent groups redirecting resources.
- Humanitarian aid is one measure taken to address the effects of famine.

5.8 Urban expansion into food production areas

- Sydney's urban growth is shaped by the planned green space zones to the city's north, south and west.
- Urban policies encourage population growth in concentrated precincts (called urban consolidation) rather than outward expansion of suburbs to manage the outward expansion of the city.
- Market gardens and farming activities within the Greater Sydney area are being squeezed out of suburbs due to more expensive land prices and competing land uses.

5.9 Climate change and food production

- Agriculture provides a country with food security and employment.
- Key predicted changes to productive land include loss of low-lying coastal areas to rising sea levels, changes to average rainfall and temperature totals, greater incidence of severe storm activity in selected areas, and possible increases in pest infestations within areas not currently impacted.
- Climate refugee movements have started to take place in low-lying areas of the world.

5.12.2 Key terms

arable describes land that is suitable for growing crops

biofuel fuel that has been produced from renewable resources, such as plants and vegetable oils

desertification the transformation of land once suitable for agriculture into desert by processes such as climate change or human practices such as deforestation and overgrazing

environmental refugees people who are forced to flee their home region due to environmental changes (such as drought, desertification, sea-level rise or monsoons) that affect their wellbeing or livelihood

famine a drastic, widespread food shortage

fodder food such as hay or straw for cattle and other livestock

humanitarian aid assistance provided in response to a human crisis caused by natural or man-made disasters, to save lives and alleviate suffering

improved pasture pasture that has been specially selected and sown, which is usually more productive than the local native pasture

jatropha any plant of the genus *Jatropha*, but particularly *Jatropha curcas*, which is used as a biofuel

malnourished describes a person who is not getting the right amount of vitamins, minerals and other nutrients to maintain healthy tissues and organ function

organic products grown or created with ingredients that were grown without any contact with artificial fertilisers or chemicals

potable drinkable; safe to drink

salinity the presence of salt on the surface of the land, in soil or rocks, or dissolved in rivers and groundwater

undernourished describes a person who is not getting enough calories in their diet; that is, not enough to eat

urban expansion the increasing size of urban areas

waterlogging saturation of the soil with groundwater such that it hinders plant growth

water scarcity situation that occurs when water supplies drop below 1000 m^3 per person per year

water stress situation that occurs when water demand exceeds the amount available or when poor quality restricts its use

5.12.3 Reflection

Complete the following to reflect on your learning.

Revisit the inquiry question posed in the Overview:

The world produces enough food to feed everyone. So why do more than 800 million people around the world each year experience hunger?

1. Now that you have completed this chapter, what is your view on the question?
   Discuss with a partner. Has your learning in this chapter changed your view?
   If so, how?
2. Write a paragraph in response to the inquiry question, outlining your views.

Subtopic	Success criteria			
5.2	Define *water security* and identify its key features.			
	Explain the impact of water scarcity on food production.			
5.3	Explain the impact of pollution on food production.			
5.4	Identify features on a satellite image with false colours.			
	Describe how the area shown in multiple satellite images has changed over time.			
5.5	Identify the key features of land degradation.			
	Explain the impact of land degradation on food production.			
5.6	Explain the reasons for the 2011 famine in Somalia.			
	Identify the challenges of addressing famines.			
5.7	Create a choropleth map from a data set.			
	Identify and describe patterns and anomalies on a choropleth map.			
5.8	Describe the impact of Sydney's urban expansion on food production.			
5.9	Explain the possible impacts of climate change on food production.			
5.10	Identify the areas of the Yarra Yarra Creek Basin that are suitable for food production from a topographic map.			
	Explain how places where food is produced might change over time.			

on Resources

eWorkbook Chapter 5 Extended writing task (ewbk-8538)
Chapter 5 Reflection (ewbk-8537)
Chapter 5 Student learning matrix (ewbk-8536)

Interactivity Chapter 5 Crossword (int-8661)

ONLINE RESOURCES

on Resources

Below is a full list of **rich resources** available online for this chapter. These resources are designed to bring ideas to life, to promote deep and lasting learning and to support the different learning needs of each individual.

eWorkbook

- 5.1 Chapter 5 eWorkbook (ewbk-8097) ☐
- 5.2 Water security and food production (ewbk-9718) ☐
- 5.3 Pollution and food production (ewbk-9722) ☐
- 5.4 SkillBuilder — Interpreting satellite images to show change over time (ewbk-9726) ☐
- 5.5 Land loss and food production (ewbk-9730) ☐
- 5.6 Features of a famine — Somalia (ewbk-9734) ☐
- 5.7 SkillBuilder — Constructing and describing complex choropleth maps (ewbk-9738) ☐
- 5.8 Urban expansion into food production areas (ewbk-9742) ☐
- 5.9 Climate change and food production (ewbk-9746) ☐
- 5.10 Investigating topographic maps — Yarra Yarra Creek Basin food bowl (ewbk-9750) ☐
- 5.12 Chapter 5 Extended writing task (ewbk-8538) ☐
  - Chapter 5 Reflection (ewbk-8537) ☐
  - Chapter 5 Student learning matrix (ewbk-8536) ☐

Sample responses

- 5.1 Chapter 5 Sample responses (sar-0156) ☐

Digital document

- 5.7 Blackline master — Map of Africa (doc-312012) ☐
- 5.10 Topographic map of Yarra Yarra Creek Basin (doc-36317) ☐

Video eLessons

- 5.1 Food for thought (eles-1720) ☐
  - Challenges to food production — Photo essay (eles-5197) ☐
- 5.2 Water security and food production — Key concepts (eles-5198) ☐
- 5.3 Pollution and food production — Key concepts (eles-5199) ☐
- 5.4 SkillBuilder — Interpreting satellite images to show change over time (eles-1733) ☐
- 5.5 Land loss and food production — Key concepts (eles-5200) ☐
- 5.6 Features of a famine — Somalia — Key concepts (eles-5201) ☐
- 5.7 SkillBuilder — Constructing and describing complex choropleth maps (eles-1732) ☐
- 5.8 Urban expansion into food production areas — Key concepts (eles-5202) ☐
- 5.9 Climate change and food production — Key concepts (eles-5203) ☐
- 5.10 Investigating topographic maps — Yarra Yarra Creek Basin food bowl — Key concepts (eles-5204) ☐

Interactivities

- 5.2 The last drop (int-3328) ☐
- 5.3 Pollution and food production (int-8631) ☐
- 5.4 SkillBuilder — Interpreting satellite images to show change over time (int-3351) ☐
- 5.5 Global map of China's land and food footprint (int-8766) ☐
  - Satellite image of the city of Tehran (a) in 1985; (b) in 2009 (int-8770) ☐
  - Losing land (int-3325) ☐
- 5.6 Features of a famine — Somalia (int-8632) ☐
- 5.7 SkillBuilder — Constructing and describing complex choropleth maps (int-3350) ☐
- 5.8 Urban expansion into food production areas (int-8633) ☐
- 5.9 Climate change and food production (int-8634) ☐
- 5.10 Investigating topographic maps — Yarra Yarra Creek Basin food bowl (int-8635) ☐
- 5.12 Chapter 5 Crossword (int-8661) ☐

ProjectsPLUS

- 5.11 Thinking Big Research project — Famine crisis report (pro-0191) ☐

Weblinks

- 5.2 Water use (web-4342) ☐
- 5.3 UN — Actions to clean the air (web-6363) ☐
- 5.5 Google time lapse (web-6364) ☐
- 5.6 Fleeing Somalia's drought (web-1160) ☐
  - Current famine and recovery status of Somalia (web-6365) ☐
- 5.8 Sydney food futures (web-6366) ☐
  - The future of Sydney's food bowl (web-6367) ☐
  - A Metropolis of Three Cities — the Greater Sydney Region Plan (web-6368) ☐
  - Sprout stack (web-6376) ☐
- 5.9 How to feed the world in 2050 (web-4344) ☐

Fieldwork

- 5.8 Extracting information from media interviews (fdw-0031) ☐

Google Earth

- 5.10 Yarra Yarra Creek Basin (gogl-0137) ☐

Teacher resources

There are many resources available exclusively for teachers online.

6 Food security

INQUIRY SEQUENCE

To access a pre-test with **immediate feedback** and **sample responses** to every question in this chapter, select your learnON format at www.jacplus.com.au.

6.1 Overview

Numerous **videos** and **interactivities** are embedded just where you need them, at the point of learning, in your learnON title at www.jacplus.com.au. They will help you to learn the content and concepts covered in this topic.

Will there come a time when we don't have enough food to feed everyone?

6.1.1 Introduction

Currently we produce enough food to adequately feed everyone in the world. However, it is estimated that the world's population will grow by another two billion people in the next 30 years. A greater proportion of people will live in urban areas, and it is estimated that almost one in seven people will go hungry. If we want to stop the number of hungry people from increasing, we will need improvements in food production, new sources of food, better aid programs, and different attitudes to food consumption and waste.

STARTER QUESTIONS

1. Make some predictions about the type of food you might eat in the year 2050 and how it might be produced.
2. Ask your parents or grandparents what type of food they ate when they were young. Did they grow any of their own food? Are there places in your community where people can grow their own food to share with the community?
3. Do you think we have food shortages in Australia? Why or why not?
4. Watch the **Food security — Photo essay** video in your online Resources. Have you ever thought about the ethical or social justice aspects of food supply? If you couldn't access enough food to stay healthy and well, how would this affect other aspects of your life?

FIGURE 1 (a) Battery Urban Farm is an educational forest and vegetable farm in Battery Park, Lower Manhattan, New York. (b) Battery Urban Farm has 80 organic vegetable plots available to students and the public.

on Resources

eWorkbook	Chapter 6 eWorkbook (ewbk-8098)
Video eLessons	Future food (eles-1721) Food security — Photo essay (eles-5205)

6.2 Feeding our future world population

LEARNING INTENTION

By the end of this subtopic, you will be able to explain the prevalence and impacts of hunger in the world, and outline the factors that influence food insecurity.

6.2.1 Measuring food security

According to the United Nations Food and Agriculture Organization, 'Food security exists when all people, at all times, have physical and economic access to enough safe and nutritious food to meet their dietary needs and food preferences for an active and healthy lifestyle.'

FIGURE 1 shows the countries of the world according to the Global Food Security Index. This is based on a range of 12 different indicators (data that provides a pointer, especially to a trend), including:

- affordability of food
- accessibility of food
- the nutritional value of food
- safety of food
- the nutritional and health status of the population.

Countries that have a high rating on the index can produce more food than they require, so they export their surplus, or they are able to afford to import all of their food needs, as is the case for Singapore.

int-8636

FIGURE 1 Global Food Security Index, 2018

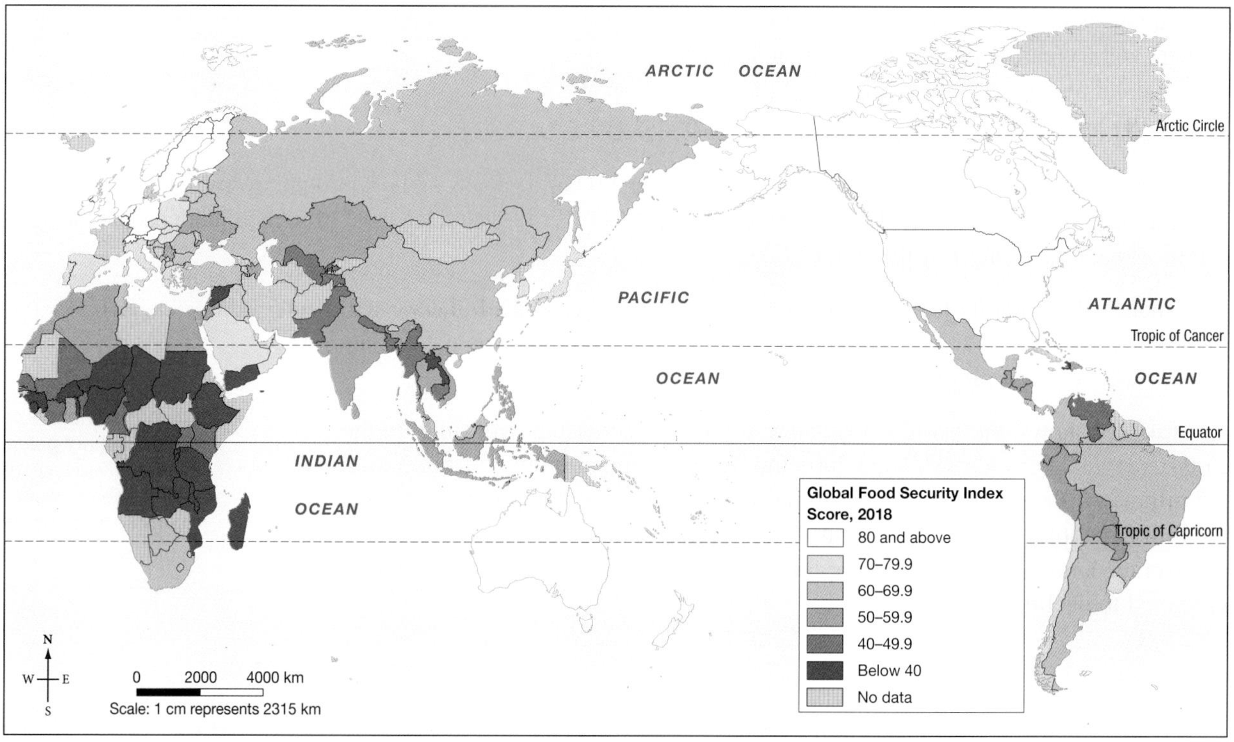

Source: © 2019 The Economist Intelligence Unit Limited, data from Global Food Security Index. Map drawn by Spatial Vision.

Reasons for food insecurity

Some of the reasons for food insecurity include:

- poverty
- population growth
- weak economy and/or political systems
- conflict or war
- natural disasters such as drought or a pandemic.

FIGURE 2 Australian dairy foods are exported to many countries

In Australia, we produce three times as much food as we consume. We are a major exporter of both fresh and processed food, and can trade competitively in cereals, oil seeds, beef, lamb, sugar and dairy products. About 90 per cent of our food is grown here in Australia. Of the remaining 10 per cent that we import, many foods are either processed or out of season in Australia; oranges are an example. Global trade is an important component of food security because it is almost impossible to exactly match food production to food demands.

As a country, Australia does not have a lack of food, but it has a humanitarian interest in the food security of developing nations. As a major food producer, Australia does face future challenges. There is declining growth in agricultural productivity, the threat of climate change, and increasing competition for land and water

6.2.2 The prevalence and impacts of hunger

According to the World Health Organization (WHO), over 1.9 billion adults in the world are overweight, while 821 million go hungry each day. The WHO predicted that COVID-19 would lead to 130 million more people facing chronic (long-term) hunger by the end of 2020.

The impact of hunger on people cannot be overstated. It is estimated that we will need to produce between 70 and 100 per cent more food to feed future populations and to prevent increase in **malnourishment**.

As **FIGURE 3** shows, the world's population growth is increasing over time. In the forty-five years between 1960 and 2005 the world's population grew from 3 billion to 6.5 billion. It is predicted by 2050 that the world will support approximately 9.6 billion people. Between 1950 and 2050, while the population has been growing, the average amount of land available for crops (**arable** land) has been declining and is predicted to decline further.

malnourishment a condition that results from not getting the right amount of vitamins, minerals and other nutrients needed to maintain healthy tissues and organ function

arable describes land that is suitable for growing crops

FIGURE 3 Population growth by world and region, 1950–2050*

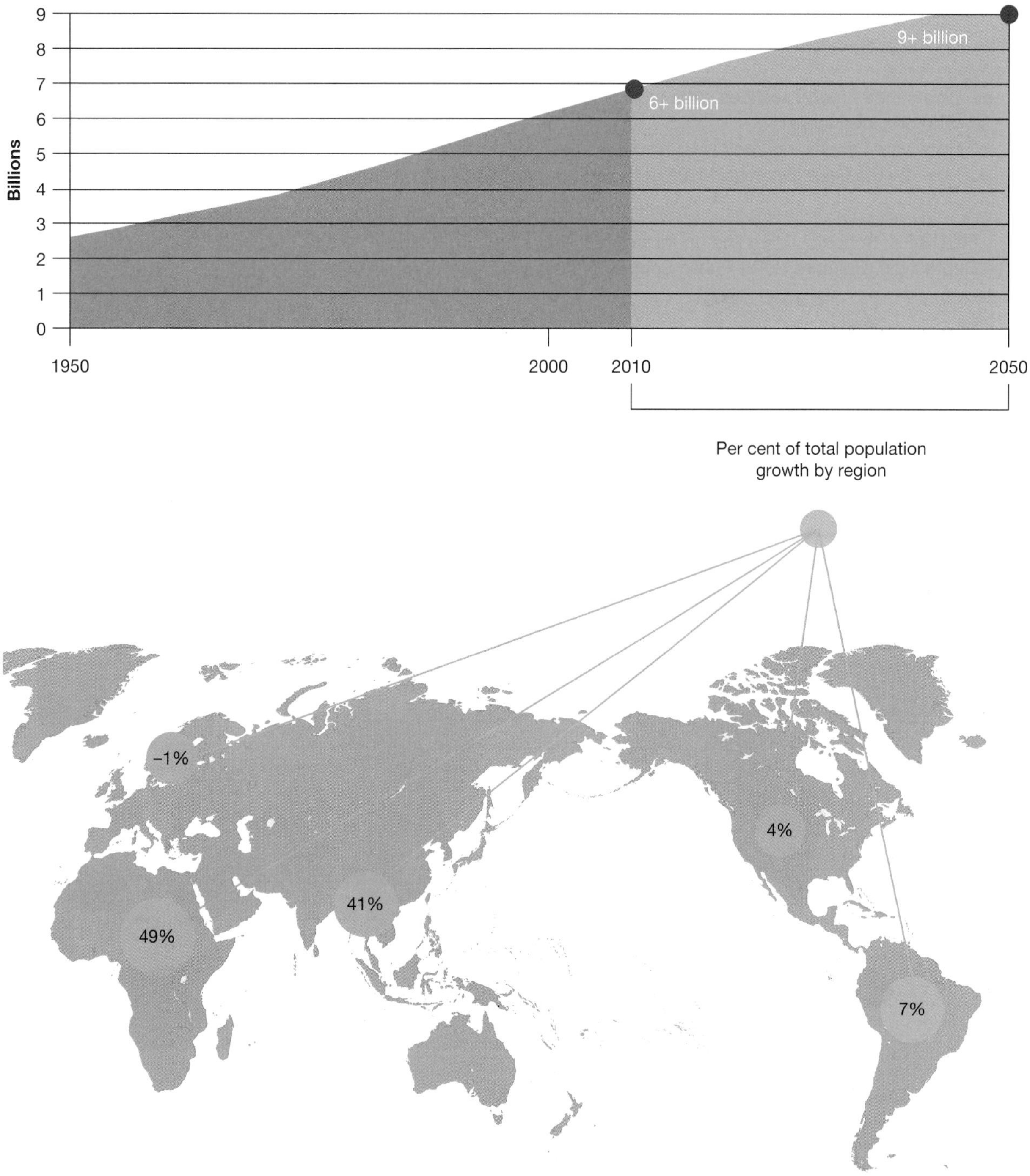

*projected

Source: Redrawn from an image by Global Harvest Initiative *2011 GAP Report®: Measuring Global Agricultural Productivity*, data from Population Facts 2014/3. Department of Economics and Social Affairs, Population Division. © 2014 United Nations. Reprinted with the permission of the United Nations

on Resources

eWorkbook	Feeding our future world population (ewbk-9758)
Video eLesson	Feeding our future world population — Key concepts (eles-5206)
Interactivity	Global Food Security Index (int-8636)

6.2 ACTIVITIES

1. Times of great social, political and economic upheaval such as the Great Depression have led to widespread food insecurity in Australia and other wealthy developed nations. Research the impact of the Great Depression in your area.
   a. What kinds of food services were provided to people?
   b. Who provided them?
   c. How were they distributed?
   d. How did the Great Depression affect the population trends and employment opportunities in your area?

FIGURE 4 Schoolchildren line up for free issue of soup and a slice of bread during the Great Depression, Belmore North Public School, Sydney, 1934

2. Refer to **TABLE 1**. Enter these figures into a spreadsheet and create a graph to demonstrate the predicted changes to rural and urban populations.

TABLE 1 Rural–urban population, 1960–2050

Population	1960	2005	2050
% urban	34%	49%	66%
% rural	66%	51%	34%

3. Lack of food has been a factor in pushing people to leave their homes and move to cities in search of employment and food. Consider the trends in availability of arable land, the amount of land needed to sustain specific global diets, and the impact of the COVID-19 pandemic using the **Arable land**, **Global diets** and **Food systems and COVID-19** weblinks in your online Resources.
   a. Predict the places of the world where people are most likely to leave their homes because of food insecurity.
   b. Predict which sections of a society this is likely to affect.
   c. Consider whether this is or might become an issue in Australia, and what the possible causes might be.
   d. Share your predictions and reasons as a class.

6.2 EXERCISE

Learning pathways

■ LEVEL 1	■ LEVEL 2	■ LEVEL 3
1, 4, 8, 11, 15	2, 5, 6, 10, 13	3, 7, 9, 12, 14

Check your understanding

1. Define the term *food insecurity*.
2. How much more food is it estimated that we will need to produce to feed future populations?
3. Describe the relationship between fast-growing populations and the amount of arable land per person.
4. Which region is predicted to have the largest population growth between 2010 and 2050?
5. Examine **FIGURE 3**.
   a. Which region is predicted to decrease in population by 2050?
   b. Which two continents are expected to have the greatest increase in population?
   c. What is the predicted world population in 2050?
6. What is malnourishment?
7. Outline five of the 12 indicators that determine rankings on the Global Food Security Index.

Apply your understanding

8. Explain why hunger is such a serious issue.
9. Analyse **FIGURE 1**.
   a. What patterns can you see in the locations of countries with low food security index rankings?
   b. What patterns can you see in the locations of countries with high food security index rankings?
10. How does a growing world population put pressure on food supplies?
11. Is it possible for someone to be overweight but also malnourished? Explain why or why not.
12. Why might the COVID-19 pandemic have increased the number of people in the world suffering from chronic hunger?

Challenge your understanding

13. What may need to happen to ensure there is enough food in the future for people who live in places with growing populations and limited arable land? Propose and explain one strategy to help manage the situation, and identify who should be responsible for its implementation.
14. Singapore is not able to produce any of its own food. How might a global pandemic or other extreme event affect food security in Singapore? Identify a specific extreme event, and give specific examples of what aspects of food security might change and why.
15. Would conflict or war affect all countries' food security to the same degree? Consider and suggest two factors that might influence the levels to which a conflict affects food security.

To answer questions online and to receive **immediate feedback** and **sample responses** for every question, go to your learnON title at www.jacplus.com.au.

6.3 Improving food production

LEARNING INTENTION

By the end of this subtopic, you will be able to compare and evaluate various strategies for improving food production as a way to reduce food insecurity.

6.3.1 Food yield

Improving the lives of over 70 per cent of the world's poor live in rural areas would create greater food security. If farmers living in poverty can produce more food, they can then feed themselves and provide for local markets. Improved infrastructure, such as roads in rural regions, would enable them to transport their produce to market and increase their incomes. Preventing hunger at a global scale is important, but action aimed at improving food yield at a local scale will deliver more sustainable results.

Farming is a complex activity, and farmers around the world face many challenges in producing enough food to feed themselves and to create surpluses they can sell to increase their incomes. Some of these are outlined in **FIGURE 1**.

FIGURE 1 Factors affecting farming yields

6.3.2 Land availability

As urban areas grow, the amount of available arable land decreases. According to the United Nations Food and Agriculture Organization (FAO), the world has an extra 2.8 billion hectares of unused potential farmland. This is almost twice as much as is currently farmed. However, only a fraction of this extra land is realistically available for agricultural expansion, owing to inaccessibility and the need to preserve forest cover and land for infrastructure. **FIGURE 2** shows the proportion of available land on the Earth's surface that is available for food production.

FIGURE 2 Global land use for food production

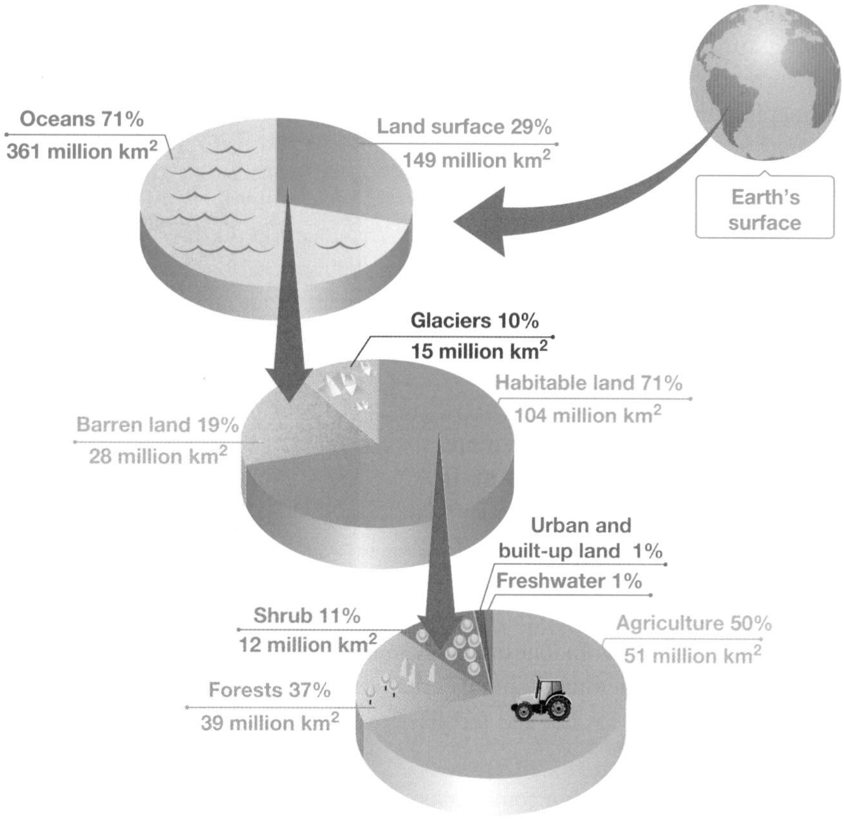

Source: http://ourworldindata.org/agricultural-land-by-global-diets

The growing populations of the future will be found in places where expansion of land for agriculture is already limited. Consequently, increased food production will need to come from better use of current agricultural areas, better use of technology, and new ways of thinking about food production and approaches to farming. One such example is the Ord River irrigation scheme in the East Kimberley region of Western Australia (**FIGURE 3**), which is transforming this semi-arid region and growing a large quantity of food that is primarily exported to Asia.

These schemes also have negative impacts on the environment and have been criticised for being expensive to build with little benefit to food production. In 2013, an estimated 60 per cent of farms drawing from the Ord River scheme were growing sandalwood rather than food crops. The damming of the Ord River has also had impacts downstream, such as changes to sediment levels and a reduction of fish habitat. The damming of the river also reduced the size of floodplains that were important traditional food sources for the Muriuwung and Gajerrong Peoples. Important cultural sites that had been used for over 40 000 years are now under water.

FIGURES 3 (a) and (b) The Ord River Irrigation Scheme has allowed great expansion of the available farming area in the East Kimberley region of Western Australia.

(a)

(b)

6.3.3 Improving sustainable production

The strategy that is likely to be the most important in increasing future crop production is called closing the **yield gap**. This means that farmers who are currently less productive will need to increase their yields so that their outputs are closer to those of the more productive farmers. Changing crops and sources of protein that are farmed can also help to ensure larger yield. For example, insects such as locusts and grasshoppers are consumed in some cultures as an important source of sustainable protein (**FIGURE 4**).

FIGURE 4 Food vendor selling deep-fried insects, Bangkok, Thailand, 2019

There is a serious yield gap in more than 157 countries (see **FIGURE 5**). If this gap was closed, greater amounts of food would be available without needing more land. There are wide geographic variations in crop and livestock productivity. Indonesia, China and India have all made great progress in increasing their agricultural output. Much of the increase has been achieved through more efficient use of water and fertilisers.

yield gap the gap between a certain crop's average yield and its maximum potential yield

int-8637

FIGURE 5 Yield gap for a combination of major crops

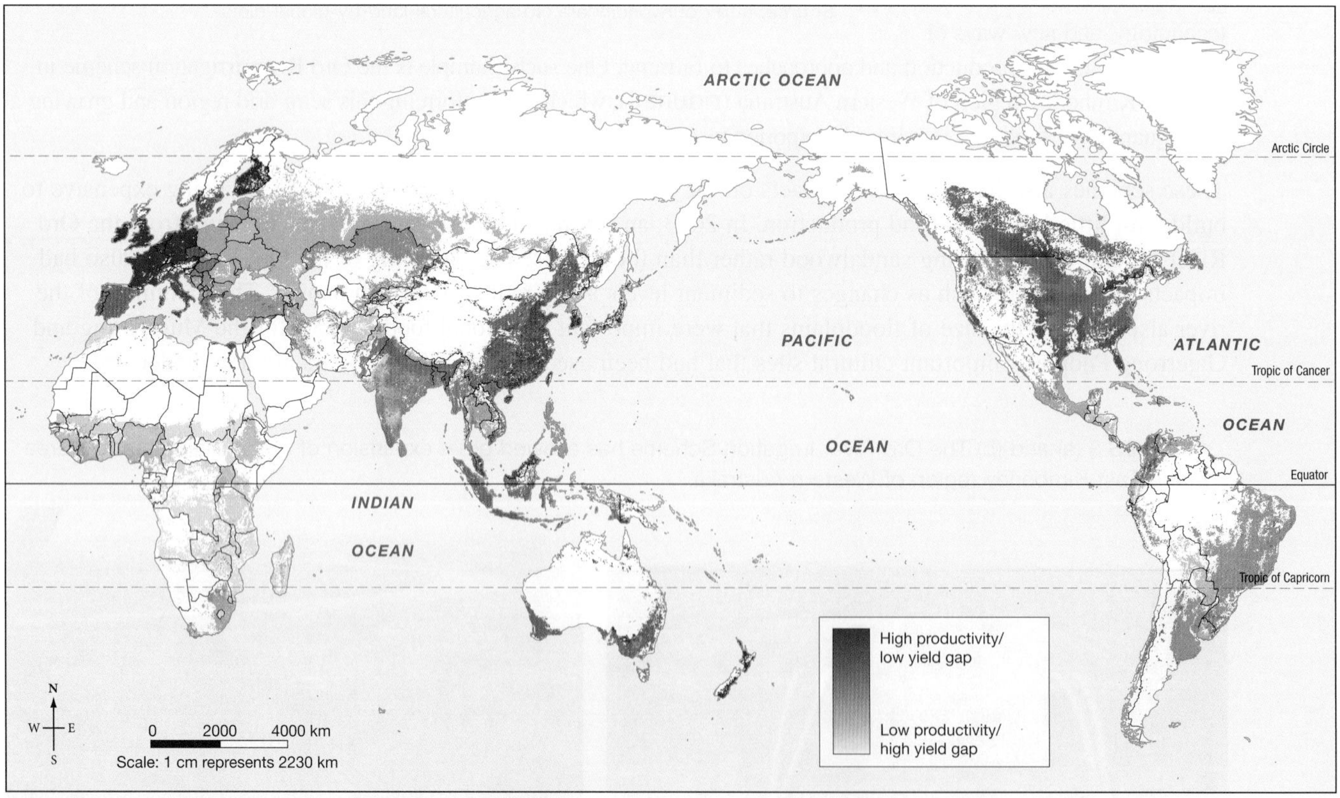

Source: Food and Agriculture Organization of the United Nations. Reproduced with permission.

FIGURE 6 Strategies for improving yield

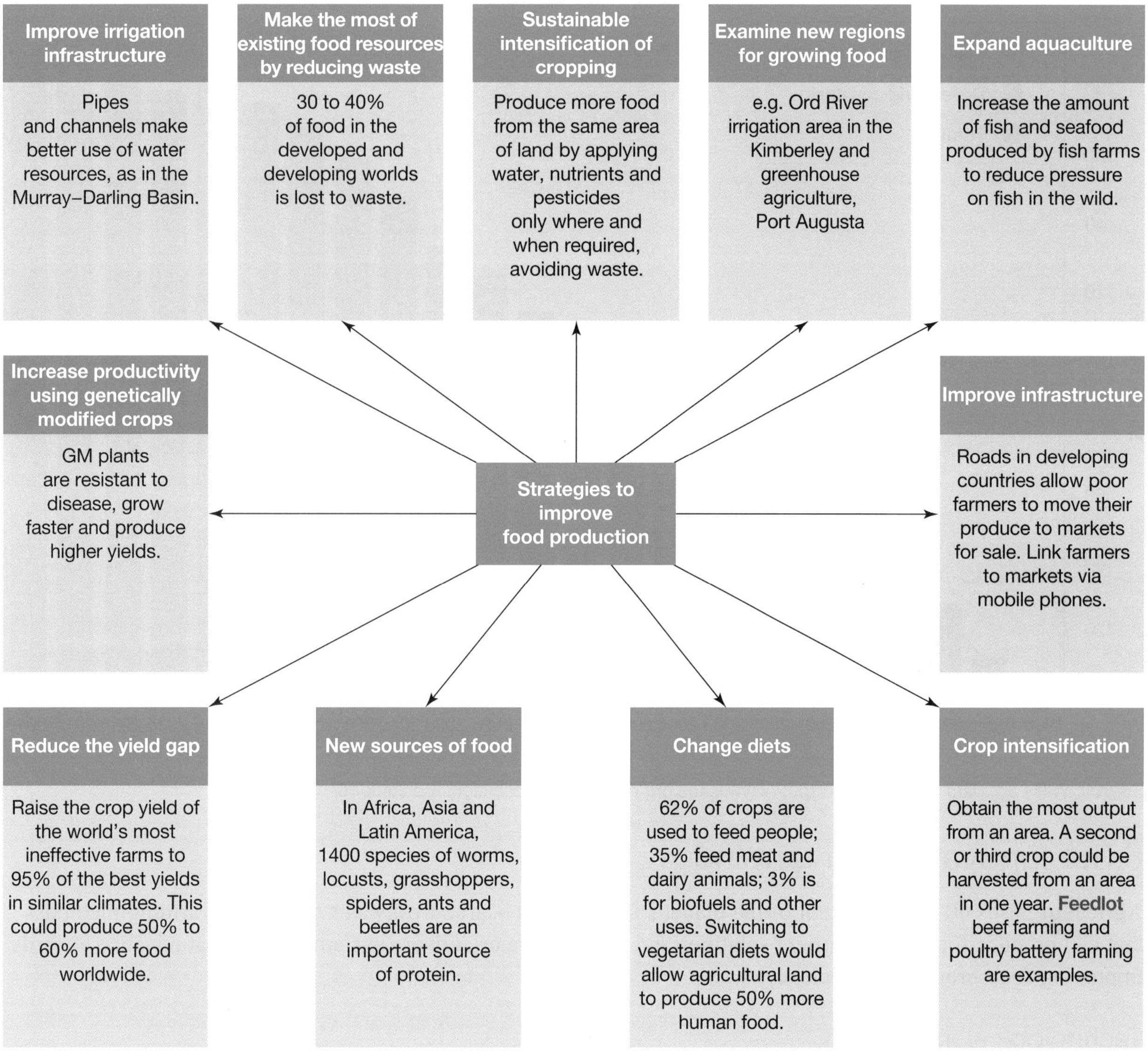
Strategies to improve food production
Improve irrigation infrastructure
Pipes and channels make better use of water resources, as in the Murray–Darling Basin.
Make the most of existing food resources by reducing waste
30 to 40% of food in the developed and developing worlds is lost to waste.
Sustainable intensification of cropping
Produce more food from the same area of land by applying water, nutrients and pesticides only where and when required, avoiding waste.
Examine new regions for growing food
e.g. Ord River irrigation area in the Kimberley and greenhouse agriculture, Port Augusta
Expand aquaculture
Increase the amount of fish and seafood produced by fish farms to reduce pressure on fish in the wild.
Increase productivity using genetically modified crops
GM plants are resistant to disease, grow faster and produce higher yields.
Improve infrastructure
Roads in developing countries allow poor farmers to move their produce to markets for sale. Link farmers to markets via mobile phones.
Reduce the yield gap
Raise the crop yield of the world's most ineffective farms to 95% of the best yields in similar climates. This could produce 50% to 60% more food worldwide.
New sources of food
In Africa, Asia and Latin America, 1400 species of worms, locusts, grasshoppers, spiders, ants and beetles are an important source of protein.
Change diets
62% of crops are used to feed people; 35% feed meat and dairy animals; 3% is for biofuels and other uses. Switching to vegetarian diets would allow agricultural land to produce 50% more human food.
Crop intensification
Obtain the most output from an area. A second or third crop could be harvested from an area in one year. Feedlot beef farming and poultry battery farming are examples.

Genetic modification

The use of **genetically modified** (GM) foods has increased, and this has increased crop yields. However, there is some opposition to GM crops because of concerns about:

- loss of seed varieties
- potential risks to the environment, species involved in the pollination of plants such as bees, and people's health
- the lack of regulations for labelling of food that has been genetically modified
- the fact that large companies hold the copyright to the seeds of GM plants that are food sources.

FIGURE 7 Protest against GM foods, Austin, Texas, 2015

GMO
IS KILLING
My BuZZ!
We Need
Bees
not
Killer

feedlot farming intensive farming practice in which high concentrations of animals are kept in confined areas to facilitate rapid growth and weight gain

genetically modified describes seeds, crops or foods whose DNA has been altered by genetic engineering techniques

FIGURE 8 Growth of genetically modified crops, 1996–2018

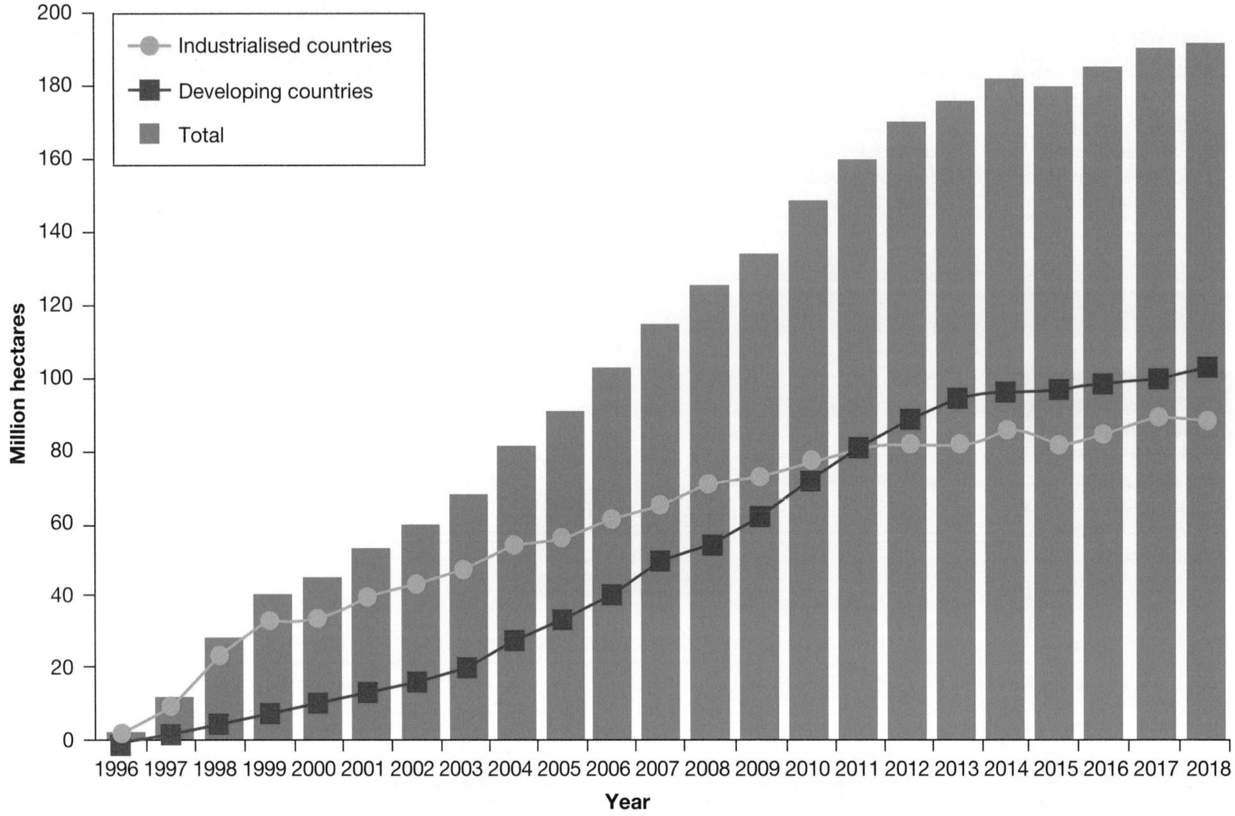

Source: Wiley art, data from ISAAA, https://www.isaaa.org/resources/publications/pocketk/16/

6.3.4 Innovations in farming

Agriculture uses 60 to 80 per cent of the planet's increasingly scarce freshwater resources. Farming methods that can produce food using nutrient solutions and efficient water delivery systems are a great advance towards improving food production efficiency.

Technological innovation

Australian farmers see technology as a means of decreasing production costs and increasing crop production. Additional technologies in Australian agriculture include the following.

- Robots are being tested to determine whether they can be used in complex jobs such as watering or harvesting. This would be of advantage in the horticultural sector, which is the third largest sector in agriculture, with an export trade worth $2.2 billion in 2017–2018.
- Automated and self-driving tractors and harvesters, assisted by satellite guidance, ensure precise application of fertilisers and pesticides.
- Robots and drones can be fitted with sensors, vision, laser, radar, and conductivity sensors including GPS and thermal sensors. These are used to help with harvesting. Increasingly, robots rather than people are harvesting food. Infrared cameras and thermal sensing are also being used to monitor water stress and determine when underground crops are ready for harvest.

FIGURE 9 Drones apply smart technology to farmland and assist with field management

CASE STUDY

Sundrop vertical farming

Port Augusta is located in a hot, arid region of South Australia, and is not normally associated with agriculture. However, one company, Sundrop Farms, is using this region's abundant renewable resources of sunlight and sea water to produce year-round quantities of high quality, pesticide-free **hydroponic** vegetables, including tomatoes, capsicums and cucumbers.

FIGURE 10 The world's first Sundrop Farm is situated in Port Augusta, South Australia.

The 20-hectare greenhouse is powered by a 115-metre-tall solar tower with 23 000 mirrors that concentrate the Sun's energy. The collected heat creates steam to drive electricity production, heat the greenhouse, and desalinate up to one million litres of sea water from the Spencer Gulf a day for crop irrigation.

Learn more about the Port Augusta Sundrop Farm greenhouse with the **Sundrop Farm** weblink in your online Resources.

hydroponic describes a method of growing plants using mineral nutrients, in water, without soil

on Resources

eWorkbook	Improving food production (ewbk-9762)
Video eLesson	Improving food production — Key concepts (eles-5207)
Interactivity	Yield gap for a combination of major crops (int-8637)

6.3 ACTIVITIES

1. Research other farming innovations online. (Begin with the **Vertical farming** weblink in your online Resources.) Consider the following in your research.
   a. What is the food-producing potential of innovative farming methods?
   b. How might the farming innovation help feed future populations? Could it be developed quickly enough to meet predicted future demands for food?
   c. Where can this type of farming be conducted?
   d. What resources are needed to build and maintain the practice? Is it sustainable? Is it cost effective?
   e. Does this practice solve problems such as water scarcity or land degradation?
2. If you were to design a vertical farm to provide fresh vegetables for your school canteen, what would it look like? Where in the school would you position it? Draw a diagram to show what your vertical farm might look like.
3. Classify each of the following factors affecting farming yields as either environmental, economic, or social/political.
4. Research one of the strategies for improving farming yield shown in **FIGURE 6**. Construct a similar diagram that shows at least one negative impact (social/ethical, environmental or economic) that might result from farmers using that strategy. For example, feedlot and battery farming is often criticised for being cruel to animals.

FIGURE 11 Vertical farming using hydroponics in Singapore

Factors	Environmental, economic, or social/political
Access to finance	
Availability of insects for pollination	
Soil fertility and type	
Money spent on agricultural research	
Availability of surface water and groundwater	
Access to markets	
Length of growing season	
Government regulations	
Rainfall amounts	
Impact of natural disasters	
Seasonality and temperature	
Impact of insects and diseases	

6.3 EXERCISE

Learning pathways

■ LEVEL 1	■ LEVEL 2	■ LEVEL 3
1, 2, 6, 9, 13	3, 5, 8, 10, 15	4, 7, 11, 12, 14

Check your understanding

1. How much of the Earth's:
   a. surface is land
   b. land is habitable
   c. habitable land is used for agriculture?
2. What is meant by the term *yield gap*?
3. Identify three factors affecting farming yields: one environmental, one economic and one social/political.
4. Why is it important that the yield gap be narrowed to increase future crop yields?
5. List three different strategies, other than closing the yield gap, for improving food production.
6. What is meant by the term *genetically modified* (GM)?

Apply your understanding

7. Examine **FIGURE 8**. What changes have there been in the production of genetically modified foods? Provide reasons for your answer and use the data from the graph to support your answer. Discuss changes in:
   a. developed countries
   b. developing countries.
8. Explain how two factors that affect production levels can be either harnessed or mitigated to help ensure food security.
9. Explain one way that local governments can help rural farmers produce more food.
10. Refer to the Sundrop vertical farming case study. Explain one advantage and one disadvantage of locating a large greenhouse near Port Augusta.
11. Select one of the strategies outlined in **FIGURE 6** that can be used to improve food production. Explain this strategy in your own words and outline some of the strengths and weaknesses of this strategy.
12. Examine **FIGURE 10**.
    a. Identify at least three geographic features of the Port Augusta landscape.
    b. Sundrop distributes its hydroponic produce across Australia. What locational advantages does the Port Augusta site have?

Challenge your understanding

13. **FIGURE 2** shows that 37% of the habitable land on Earth is forest. Should this land be used to help prevent food insecurity? Assess the benefits and problems of this idea.
14. Predict what the impact might be on people and places if the greenhouse method of farming shown in **FIGURE 12** were to become more readily available. What might be the effects on places where the yield gap is large compared to places that are currently more productive?
15. Suggest reasons why countries in the Middle East might be interested in Sundrop technology.

FIGURE 12 Greenhouse farming near Almeria, Spain, 2020

To answer questions online and to receive **immediate feedback** and **sample responses** for every question, go to your learnON title at www.jacplus.com.au.

6.4 Reducing food waste

LEARNING INTENTION

By the end of this subtopic, you will be able to compare and evaluate various strategies for reducing food waste as a way to help reduce food insecurity.

6.4.1 Food loss and food waste

What food have you thrown out today? Across the world, one-third of all food produced is wasted. Each year, around 1.6 million tonnes of food, worth up to $1.2 trillion, is discarded, while more than 850 million people remain undernourished. According to the United Nations' Food and Agriculture Organization, one-quarter of the food wasted each year could feed all the world's hungry people.

To meet the growing demand for food by the middle of this century, it has been calculated that the world will need to produce as much food as has been produced over the past 8000 years. Although the world does produce sufficient food for everyone, distribution and affordability prevent it from getting to everyone who needs it. However, dealing with **food loss** and **food waste** could certainly help to improve food security. **FIGURE 1** shows an example of food loss where the cheapest method of disposal available to farmers is to dump their produce.

FIGURE 1 Vegetables rot on a garbage dump in Turkey.

FIGURE 2 Surplus tomatoes, Tenerife, Canary Islands

6.4.2 Consequences of food loss and waste

The UN International Year of Fruits and Vegetables 2021 is an initiative aimed at improving nutrition and reducing food loss and waste.

Food wastage represents misuse of the resources used in production, such as land, freshwater, energy and fertilisers. Up to 50 per cent of fruits and vegetables produced in developing countries are lost in the supply chain between harvest and consumption. Waste can increase food prices and make food less affordable. The World Bank has calculated that in sub-Saharan Africa, a region prone to food insecurity, a reduction of only one per cent wastage could save $40 million per year, with most of this saving going to the farmers.

A consequence of wastage is the need to dispose of the food, usually by discarding or burning. Food waste now contributes 8 per cent of global greenhouse gas emissions.

food loss takes place at the production, post-harvest, processing, and distribution stages of food production

food waste takes place at the retail and consumption stages food production and consumption

Food waste exists in all countries, regardless of their levels of development, although the causes of wastage vary. **FIGURE 3** shows the breakdown of food wastage on a regional basis.

FIGURE 3 World food loss and food waste per region

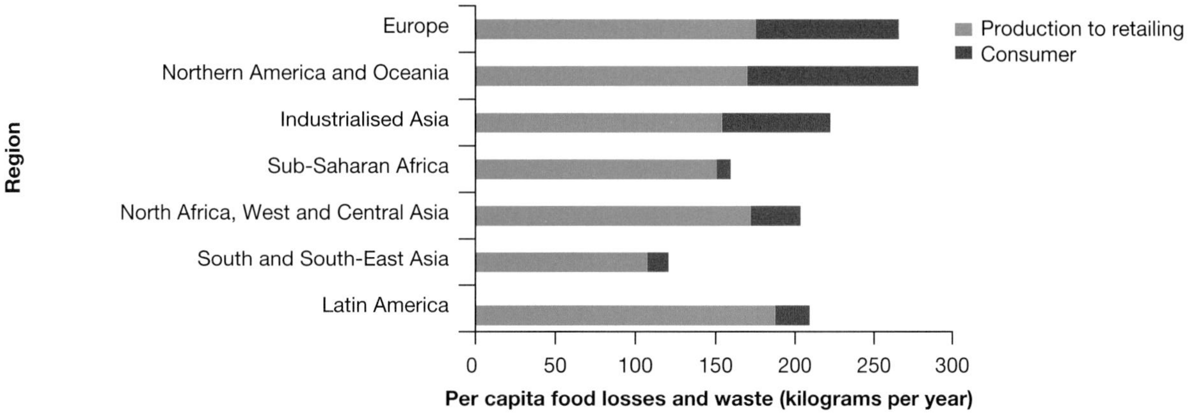

Source: Food and Agriculture Organization of the United Nations. *The role of producer organizations in reducing food loss and waste.* International Year of Cooperatives. Issues Brief Series. Accessed 23 April 2021. http://www.fao.org/3/ap409e/ap409e.pdf

In developing nations, food waste and loss are generally related to a lack of food-chain infrastructure and a poor knowledge of or investment in storage technologies on farms. Other causes of waste include:

- lack of refrigeration
- limited or non-existent road and rail networks to deliver food to markets
- a shortage of processing and packaging facilities.

In India, up to 40 per cent of fresh food is lost due to a lack of cold storage in wholesale and retail outlets. Over one-third of the rice harvest in South-East Asia can be destroyed by pests or spoilage.

In contrast, in the developed world, food waste is more evident at the retail and home stages of the food chain. In developed countries, food is relatively cheap, so there is little incentive to avoid waste. Consumers are used to purchasing food that is visually appealing and unblemished, so retailers end up throwing out perfectly edible but slightly damaged food. More and more people rely on 'use by' dates, so despite the food still being suitable to eat, it is discarded.

FIGURE 4 Specific food loss categories

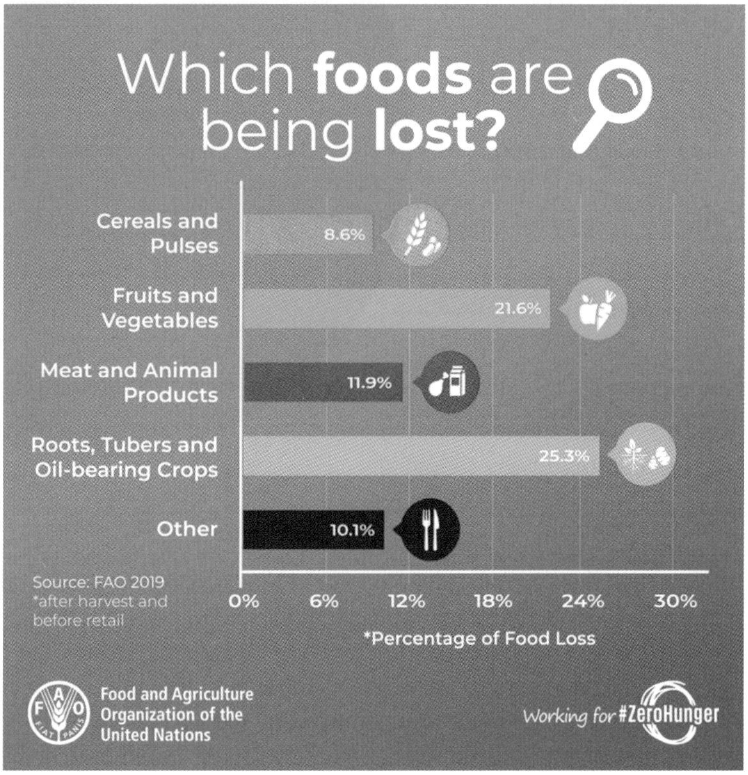

*after harvest and before retail

Source: Food and Agriculture Organization of the United Nations, 2019. Reproduced with permission.

Waste is also a part of the growing culture of 'supersize' or 'buy one get one free' advertising. Further waste can occur if the discarded food is sent to landfill when it could be used for animal feed or even compost.

Food loss and waste in Australia

Australia produces enough food for 60 million people, and this enables us to trade the surplus. Yet each person wastes an average of 361 kg of food each year. This costs the economy $20 billion annually. At the same time, four million Australians have experienced some form of food insecurity in the past year. This means that around 18 per cent of the population have not had enough food for themselves and their family or could not afford to purchase food at some stage over the twelve-month period.

As an example of the extent of the waste in Australia, consider the example of Victoria, where food wastage costs $5.4 billion annually. **FIGURE 5** shows the composition of the 255 000 tonnes of food thrown into rubbish bins in Victoria each year.

FIGURE 5 Household food waste in Victoria, tonnes per year

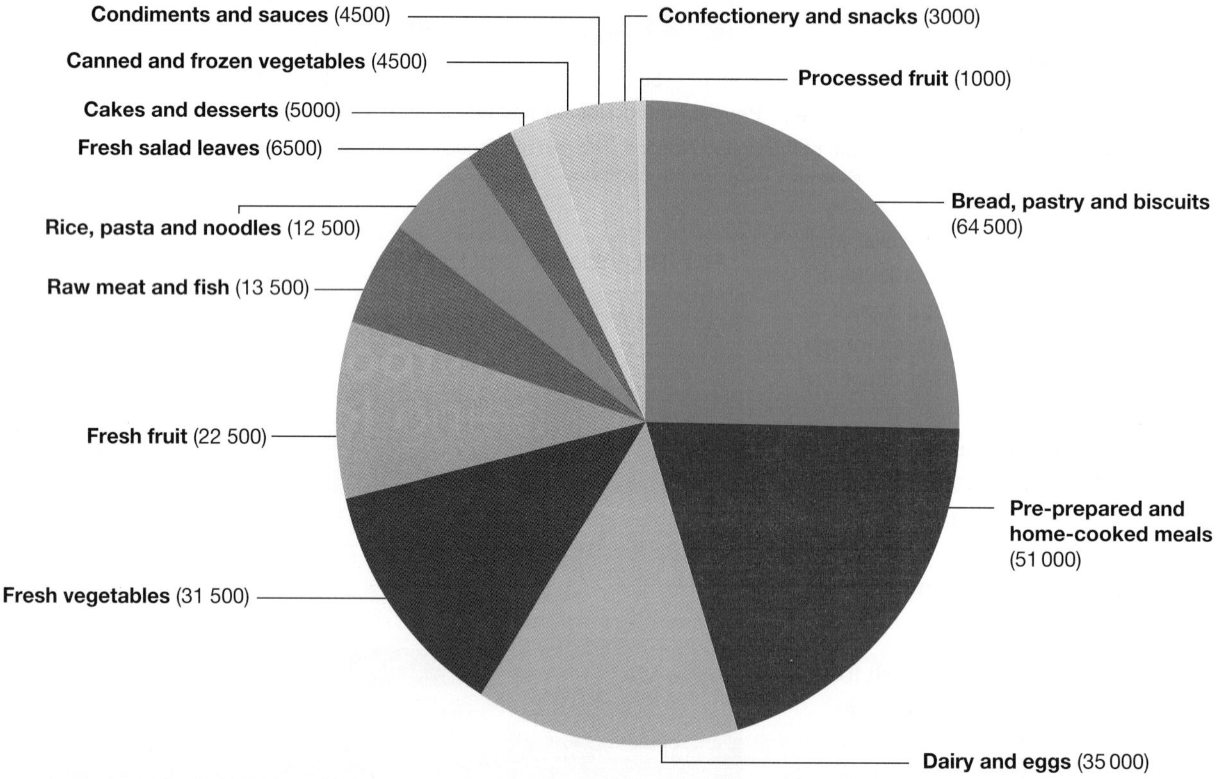

Source: © 2019 Sustainability Victoria

In New South Wales, the Environmental Protection Agency (EPA) conducts surveys of households across the state to measure people's attitude and understanding how much they throw away — including food. The 2017 survey (the last survey before the COVID-19 pandemic) found the following.

- Twenty per cent of people believed that food was not safe to eat after the 'best by date' and must be thrown away, even if it seemed to be in good condition. This means they were more likely to throw food away that was still safe to eat. This was most common with 18–34-year-olds and least common with those over 55.
- Forty-six per cent of people agreed that food waste is a contributing factor to climate change.
- On average, households estimated that they wasted about 5.46 litres of food a week, including 2.18 litres of fresh food, 1.39 litres of packaged or long-life food, and 1.9 litres of leftovers. This was estimated to be about $73.19 worth of food a week, or $3805.88 per year.
- Food was most commonly thrown away because it had been left in the fridge for too long (40%), a meal was not finished (29%) and food had reached its use-by/best-before date (29%).

6.4.3 Reducing food waste

Reducing global food waste supports the United Nations Sustainable Development Goals (SDGs), which were designed to create a more sustainable future for the world (see 8.9.2 'Sustainable urban development' for the complete list). SDG 12 aims to reduce food loss and per capita food waste by 50 per cent by 2030. SDG 2 aims to achieve food security and promote sustainable agriculture. If these specific targets can be achieved, food security will be improved, greenhouse gases can be reduced, and valuable land and water resources will not be wasted. **FIGURE 6** highlights the efforts of retailers to change customer preferences for perfectly formed and unblemished produce by offering fruit and vegetables of varying shape and form. These 'odd'-shaped products are marketed to customers in an attempt to reduce food waste at the retail end of the food supply chain.

FIGURE 6 Retailers promote naturally formed fruit and vegetables to customers.

Here is a snapshot of what is happening around the world.

- Farmers in Ghana are trialling a new phone app that shows farmers, food transporters and traders the fastest route to market, which reduces food spoilage. In addition, the app can identify illegal roadblocks set up to take bribes from drivers.
- In France, an estimated 10 million tonnes of food are wasted each year. A new law now compels restaurants to provide containers in which customers can take home uneaten food. Shops are also banned from destroying food products, and supermarkets must give away unsold food that has reached its use-by date to charities. By 2020, all Parisian households were expected to have a biowaste recycling bin for food scraps. Waste will be collected and converted into fertiliser or biofuels.
- Seoul in South Korea has taken a different approach to reduce its food waste by 20 per cent. It is trialling a program whereby people are charged according to the weight of the garbage they produce. The more kilograms they generate, the higher their bill. In South Korea 95 per cent of food waste is recycled into compost, animal feed or fuel. Landfilling of food waste is banned.
- Australia has now set a target to reduce the amount of food waste by 50 per cent by 2030. Much of this will come from supporting food rescue operations such as Second Bite and Foodbank Australia. These organisations collect and redistribute surplus food. Foodbank provides relief to 710 000 Australians every month, 26 per cent of whom are under 19 years old.

on Resources

eWorkbook	Reducing food waste (ewbk-9766)
Video eLesson	Reducing food waste — Key concepts (eles-5208)
Interactivity	Reducing food waste (int-8638)

6.4 ACTIVITIES

1. In groups, wearing disposable gloves and using tongs or grippers, conduct a survey of the school rubbish bins after lunch. You may need to lay out newspaper onto which you can tip the contents of the bins. Some groups could also deal with food litter. (You could also do a home bin audit and follow the same procedure.)
   a. Construct a table to record the different food types, such as fruit, cakes and biscuits. Collate your results with the other groups in your class, and then graph your data.
   b. Write a summary of your findings. What food types were most and least represented and why?
   c. Use the **Love food hate waste** weblink in your online Resources to investigate the amount and types of food wasted in New South Wales. Do your findings fit the patterns across the state or for your area?
   d. What strategies could be implemented to reduce levels of food being thrown away? Create an infographic to display around the school to encourage people to waste less food. Use the example from the **UN Food loss** weblink in your online Resources for ideas.
2. Use the **UN SDG2 Indicator** weblink in your online Resources to explore which places around the world are experiencing moderate or severe food insecurity. Discuss your findings as a class, including any patterns or anomalies you noticed.

6.4 EXERCISE

Learning pathways

■ LEVEL 1	■ LEVEL 2	■ LEVEL 3
1, 3, 10, 12, 13	2, 4, 7, 8, 14	5, 6, 9, 11, 15

Check your understanding

1. Outline the difference between food loss and food waste.
2. What proportion of greenhouse gasses are produced by food that has been thrown away?
3. Approximately how many tonnes of food does the average Australian throw away every year?
4. Identify three different strategies for improving food production.
5. Why is there more food wasted by retailers and in homes in developed countries than in developing countries?
6. Refer to **FIGURE 3**. Which regions of the world are shown to waste the greatest amount of food in the production to retailing and consumer sectors? Use data in your answer.

Apply your understanding

7. Explain why food waste is a global problem.
8. Explain the interconnection between food waste and global warming.
9. Explain the interconnection between Australia's food waste and food insecurity in other parts of the world.
10. Consider South Korea's and Australia's plans to reduce food waste.
    a. Create a dot point summary to compare the strengths and weaknesses of the plans.
    b. Which of the two plans do you think will be most effective, and why?
11. Explain the aim of SDG 12 and provide an example of one action being taken in Australia to achieve this aim.
12. What are the main causes of food waste and food loss in Australia?

Challenge your understanding

13. Goal 12 of the United Nations Sustainable Development Goals aims to 'by 2030, halve per capita global food waste at the retail and consumer levels and reduce food losses along production and supply chains...' Do you think this is possible? Why or why not?
14. Respond to the following statement: 'People are suffering from food insecurity because the world's biomes do not have the ability to feed such a large global population.'
15. Predict whether the world will achieve SDG 2 of having no food insecurity by 2030. Provide reasons for your answer.

To answer questions online and to receive **immediate feedback** and **sample responses** for every question, go to your learnON title at www.jacplus.com.au.

6.5 SkillBuilder — Constructing a box scattergram

online only

LEARNING INTENTION

By the end of this subtopic, you will be able to plot the data from two maps into a box scattergram and describe the relationships that between the two data sets.

The content in this subtopic is a summary of what you will find in the online resource.

6.5.1 Tell me

What is a box scattergram?

A box scattergram is a table with columns and rows that displays the relationship between two sets of data that have been mapped (see **TABLE 3**). The distribution becomes clear, although in a generalised way, as there are usually only four to five categories of data. Box scattergrams are a useful way of summarising data from maps.

6.5.2 Show me

How to construct a box scattergram

Step 1

Construct a table with enough cells for all the categories of the data shown in the legends of **FIGURE 1** (six categories; list these down column 1) and **FIGURE 2** (seven categories; list these across row 1).

FIGURE 1 World hunger map, 2015

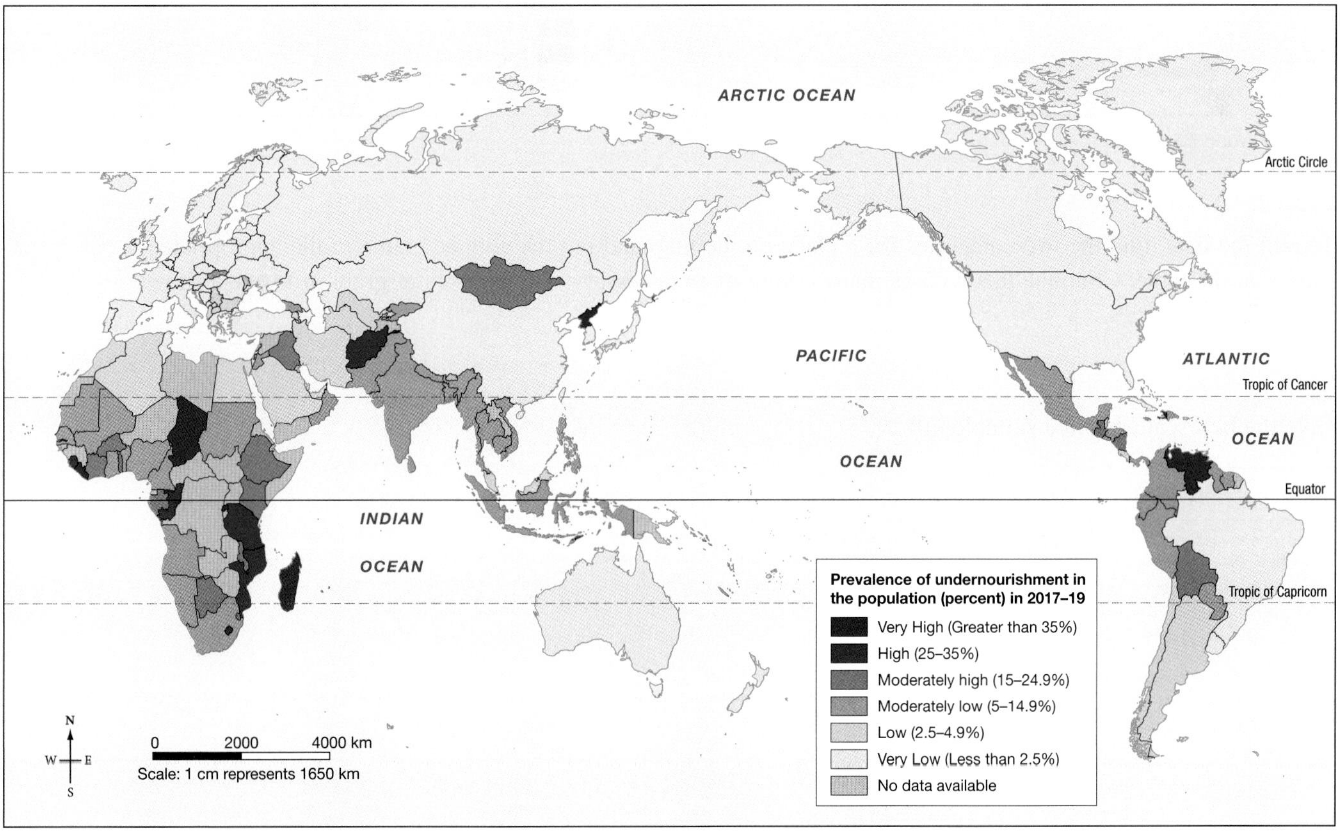

Source: Food and Agriculture Organization of the United Nations. Reproduced with permission.

Step 2

In the left-hand column, enter the title and units of measurement from **FIGURE 1**. Place the lowest numbered category at the base of the column (in this case 'No data'). See **TABLE 3** left column.

Step 3

In the header rows, enter the title and units of measurement from **FIGURE 2**. Place the lowest numbered category on the left-hand side (in this case 'No data'). See the top two rows of **TABLE 3**.

FIGURE 2 Net official development assistance and official aid received

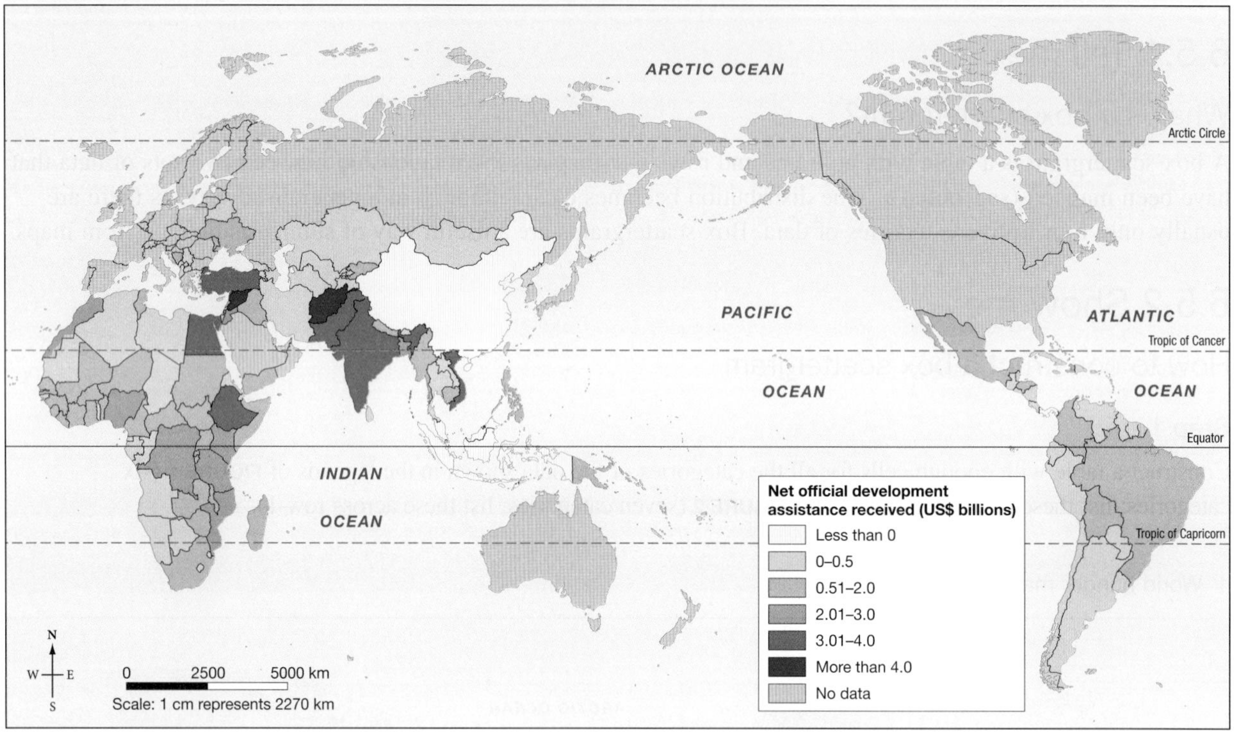

Source: World Bank

Step 4

To plot the data, find the two categories for a place on the map and put the country name in the appropriate square of the table. Continue this with as many countries as necessary. The box scattergram in **TABLE 3** shows ten African countries.

Step 5

Give the box scattergram a suitable title.

TABLE 3 Ten selected African countries, showing the relationship between undernourishment and international aid received

Hunger level (% undernourished)	Aid received (US$ billions)						
	No data	Less than 0	0–0.5	0.51–2.0	2.01–3.0	3.01–4.0	More than 4.0
More than 35			Namibia	Zambia			
25–35			Chad		Tanzania	Ethiopia	
15–24.9			Botswana				
5–14.9							
Less than 5			Algeria	South Africa		Egypt	
No data					Democratic Republic of the Congo		

6.5.3 Let me do it

learnON

Need more help? Go to your online Resources to access:

- a longer explanation of this skill and its application in Geography (Tell me)
- a video demonstrating the step-by-step process of this skill (Show me)
- an activity and interactivity for you to practise the skill (Let me do it)
- self-marking questions to help you understand and use the skill.

on Resources

eWorkbook SkillBuilder — Constructing a box scattergram (ewbk-9770)

Video eLesson SkillBuilder — Constructing a box scattergram (eles-1734)

Interactivity SkillBuilder — Constructing a box scattergram (int-3352)

6.6 Food aid

LEARNING INTENTION

By the end of this subtopic, you will be able to identify different types of food aid and how it is distributed, and explain why people need access to food relief.

6.6.1 Understanding food aid

According to the World Health Organization, the three main factors that may determine food security levels are:

1. **Food access:** access to nutritious food at an affordable price
2. **Food availability:** supply of food within a community
3. **Food use:** knowledge of nutritious food preparation

On a global scale, **food aid** is food, money, goods and services given by wealthier, more developed nations to less developed nations for the specific purpose of helping those in need. **FIGURE 1** shows that during 2019, over half of the people affected by food insecurity were living in Asia and more than one-third were living in Africa. Since 2020, rates of food insecurity escalated due to the COVID-19 pandemic. Job insecurity, disruption to food production and distribution channels, widespread illness and high infection rates led to widespread food insecurity. Record numbers of vulnerable people in both developed and developing nations needed food relief.

FIGURE 1 (a) Global food security, 2019 (millions of people), and (b) locations of people affected by food insecurity, 2019 (millions of people)

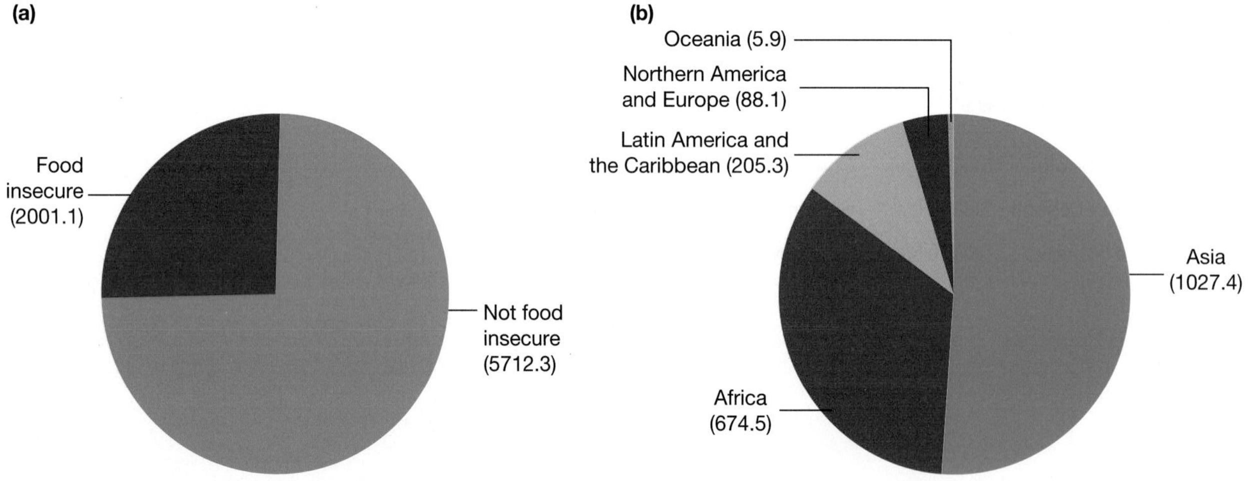

Note: Total world population (2019): 7713 million

Source: Food and Agriculture Organization of the United Nations. 2020, *State of food security and nutrition in the world.* Accessed: 23 April 2021. http://www.fao.org/3/ca9692en/online/ca9692en.html#chapter-1_1

To learn more about the prevalence of undernourishment in specific countries, use the **World Hunger Map 2015** weblink in your online Resources.

People needing food aid may include:

- people who are struggling financially and who cannot afford to buy food even if it is available
- people who have fled violence or civil conflict
- people devastated by natural disasters.

food aid food, money, goods and/or services given for the specific purpose of helping those in need

There are three general categories of food aid:

1. program food aid, which is organised between national governments and provides resources that offer budgetary support to countries in need
2. project food aid, which is targeted at specific areas or groups and provides support for disaster prevention activities and poverty alleviation measures
3. relief (crisis or emergency) food aid, which assists victims of man-made and natural disasters

6.6.2 The United Nations World Food Programme

The United Nations World Food Programme (WFP) is a voluntary arm of the United Nations. It provides more than 80 million people in more than 92 countries with food assistance after conflicts and natural disasters including cyclones, floods and droughts. Some relief aid is provided in the short term as emergency food. Project food relief is often required over lengthy periods, typically after civil war or prolonged drought.

FIGURE 2 How the WFP works

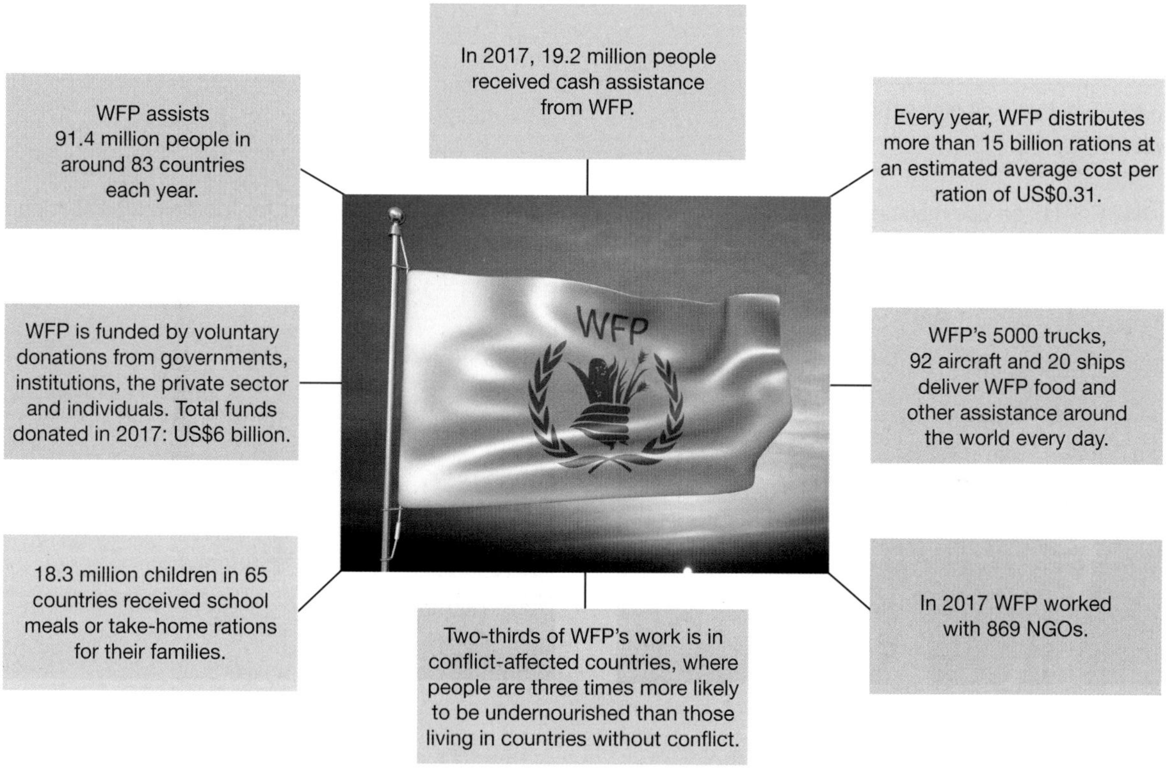

In 2017, 19.2 million people received cash assistance from WFP.
WFP assists 91.4 million people in around 83 countries each year.
Every year, WFP distributes more than 15 billion rations at an estimated average cost per ration of US$0.31.
WFP
WFP is funded by voluntary donations from governments, institutions, the private sector and individuals. Total funds donated in 2017: US$6 billion.
WFP's 5000 trucks, 92 aircraft and 20 ships deliver WFP food and other assistance around the world every day.
18.3 million children in 65 countries received school meals or take-home rations for their families.
In 2017 WFP worked with 869 NGOs.
Two-thirds of WFP's work is in conflict-affected countries, where people are three times more likely to be undernourished than those living in countries without conflict.

Natural and **anthropogenic** disasters are the drivers of hunger and malnutrition. The WFP works to prevent, **mitigate** and prepare for such disasters. In 2018, the WFP worked on five major humanitarian disasters:

- In the Kasai region of the Democratic Republic of the Congo, over 7.7 million people were at risk of not having access to enough nutritious food.
- In Borne, Yobe and Adamawa in north-east Nigeria almost 3 million people were facing hunger.
- Since South Sudan gained independence in 2011, approximately 60 per cent of the population has suffered from the effects of famine.
- In Syria, people facing hunger caused by ongoing civil unrest were provided with support.
- In Yemen, people facing hunger caused by ongoing civil war were provided with support.

anthropogenic resulting from human activity
mitigate to reduce negative impacts or severity

Sources of food aid

Overall, the WFP's disaster response has helped more than 80 million people worldwide. The major donor countries to the WFP in 2018 are shown in **TABLE 1**.

TABLE 1 Major funding contributors to the WFP, 2018*

	Contributor (All donors and sources)	Amount ($US)
1	USA	2 541 479 166
2	European Commission	1 113 106 906
3	Germany	849 141 329
4	United Kingdom	617 188 873
5	Saudi Arabia	247 907 959
6	United Arab Emirates	226 215 581
7	Canada	222 172 109
8	UN Other Funds and Agencies (excl. CERF)	151 703 536
9	Sweden	148 185 097
10	UN Central Emergency Response Fund (UN CERF)	138 632 047

*Figures current as at 28 April 2019

Source: WFP

FIGURE 3 US service members unload WFP aid in Haiti after Hurricane Matthew, 2016.

FIGURE 4 Children wait for food aid after Cyclone Idai in Mozambique, 2019.

6.6.3 Zero Hunger target

A plan to end hunger is being driven by the United Nations through its Zero Hunger campaign. It encourages the global community to work together on solving the issues of poverty, inequality and climate change. Zero Hunger by 2030 is Goal 2 of the 17 UN Sustainable Development Goals and is based on the following five elements.

- Zero children less than two years old showing signs of stunted growth or development
- 100 per cent access to adequate food all year round
- All food systems are sustainable
- 100 per cent increase in small landholder productivity and income
- Zero loss or waste of food.

FIGURE 5 Sustainable Development Goal 2

These goals aim to address the causes of poverty and hunger, and to build resilience levels of those nations most at risk. Unfortunately, COVID-19 has increased levels of food insecurity across the world. Availability and cost of nutritious and healthy foods have been restricted, forcing some of the most vulnerable people into severe hunger. Progress has been severely delayed for this SDG target. In 2020, FAO reported that globally 21.3 per cent of children under 5 years showed signs of stunted growth or development, with relatively higher numbers found in rural areas.

6.6.4 Solutions to malnutrition

CASE STUDY

Plumpy'Nut

In 2005 a revolutionary approach to treating malnutrition was released. This is a ready-to-use therapeutic food (RUTF) called Plumpy'Nut.

It is a sweet, edible paste made of peanut butter, vegetable oils, powdered milk, sugar, vitamins and minerals. Its advantages are that it:

- is easy to prepare
- is cheap (a sachet costs about $1.40)
- needs no cooking, refrigeration or added water
- has a shelf life of two years.

Children suffering from malnutrition can be fed at home without having to go to hospital. It is specially formulated to help malnourished children regain body weight quickly, because malnutrition leads to stunting of growth, brain impairment, frailty and attention deficit disorder in children under two years of age.

FIGURE 6 Plumpy'Nut is a therapeutic food that helps treat malnourished children.

Plumpy'Nut is not a miracle cure for hunger or for malnutrition; it only treats extreme food deprivation, mainly associated with famines and conflicts. It is not designed to reduce chronic hunger resulting from long-term poor diets or malnutrition. Since its introduction, Plumpy'Nut has lowered mortality rates during famines in Malawi, Niger and Somalia.

Most of the world's peanuts are grown in developing countries, where allergies to them are relatively uncommon. Manufacturing plants have been established in several developing countries, including Mali, Niger and Ethiopia. These factories provide employment and ensure ease of access when needed. The patent for Plumpy'Nut is owned by the French company Nutriset, which works with UNICEF to save the lives of millions of children with this simple solution to childhood hunger.

on Resources

eWorkbook	Food aid (ewbk-9774)
Video eLesson	Food aid — Key concepts (eles-5209)
Interactivity	Food aid — Interactivity (int-8639)

6.6 ACTIVITIES

1. Select a major food aid donor from **TABLE 1**. Research the main population characteristics of that country, such as life expectancy, literacy levels and death rates. Discuss your findings as a class.
2. Use the **World Food Programme** weblink in your online Resources to learn about the WFP's involvement in Syria and surrounding places since 2012. What action is the WFP taking there and why?

FIGURE 7 The World Food Programme, Damascus, Syria, 2016

6.6 EXERCISE

Learning pathways

■ LEVEL 1	■ LEVEL 2	■ LEVEL 3
2, 4, 7, 11, 14	1, 3, 8, 10, 15	5, 6, 9, 12, 13

Check your understanding

1. Use **FIGURE 1** to calculate the percentage of the world's population that experienced food insecurity in 2019.
2. How many countries receive WFP assistance each year?
3. Outline the three key factors that affect food security levels.
4. What are the three different categories of food aid?
5. Use **TABLE 1** to calculate the amount of money contributed to the WFP by the top five donors.
6. Describe the impacts of COVID-19 on food security in one country outside of Australia using specific data to demonstrate your learning.

Apply your understanding

7. Explain why the WFP is so active in school feeding and emergency aid programs.
8. Identify one organisation that delivers food aid. Assess the effectiveness of their program. Give reasons for your answer.
9. Explain why there might be difficulties with access to food in 2050 if 25 per cent of the population is over 65 years old.
10. How might a country's ability to donate to international food aid and relief programs change during a major economic downturn?
11. Explain two advantages and two disadvantages of using Plumpy'Nut or other RUTFs to treat childhood malnutrition.
12. Should Australia increase its levels of food aid to help combat the decrease in food security because of the COVID-19 pandemic? If so, how? If not, why?

Challenge your understanding

13. Do you think that all of the elements of SDG 2 are achievable? Predict which is likely to be the most difficult to achieve and provide reasons to support your view.
14. What can you do as an individual to contribute to addressing SDG 2 Zero Hunger?
15. Predict the likely consequences for children who suffer from malnutrition. Suggest two strategies that might be employed internationally to reduce the impact of childhood malnutrition.

To answer questions online and to receive **immediate feedback** and **sample responses** for every question, go to your learnON title at www.jacplus.com.au.

6.7 Food aid in Australia

online only

LEARNING INTENTION

By the end of this subtopic, you will be able to identify reasons why people in Australia need food relief, and give examples of strategies for making food more widely available to people experiencing food insecurity in Australia.

The content in this subtopic is a summary of what you will find in the online resource.

6.7.1 Access to food in Australia

In 2020, 3.24 million people, or 13.6 per cent of Australians, were living below the internationally accepted poverty line. The prices of essentials — food, health, education, housing, utilities and transport — increased so much that people who were already struggling became unable to cope. The effects of COVID-19 increased the rate of food insecurity in Australia as people lost jobs and faced uncertainty about the future.

To learn more about food aid in Australia, go to your learnON resources at www.jacPLUS.com.au.

FIGURE 1 Foodbank delivery truck

Contents

learnON

- 6.7.1 Access to food in Australia
- 6.7.2 Access to food in remote communities
- 6.7.3 Access to food for people who need support to live independently

on Resources

eWorkbook	Food aid in Australia (ewbk-8159)
Video eLesson	Food aid in Australia — Key concepts (eles-5210)
Interactivity	Food aid in Australia (int-8640)

6.8 Fair trade

online only

LEARNING INTENTION

By the end of this subtopic, you will be able to outline fair trade practices and evaluate their sustainability in terms of achieving food security.

The content in this subtopic is a summary of what you will find in the online resource.

6.8.1 Features of fair trade

As people become more concerned about the level of poverty and hunger in the world, they sometimes seek ways to improve the situation. Trade is the way countries sell what they have produced and buy what they need. On a global scale, this does not mean that trade is mutually beneficial. Fair trade is a consumer-driven movement to promote fair prices and reasonable conditions for producers in developing regions.

To learn more about fair trade, go to your learnON resources at www.jacPLUS.com.au.

FIGURE 1 Advertising for Fairtrade certification

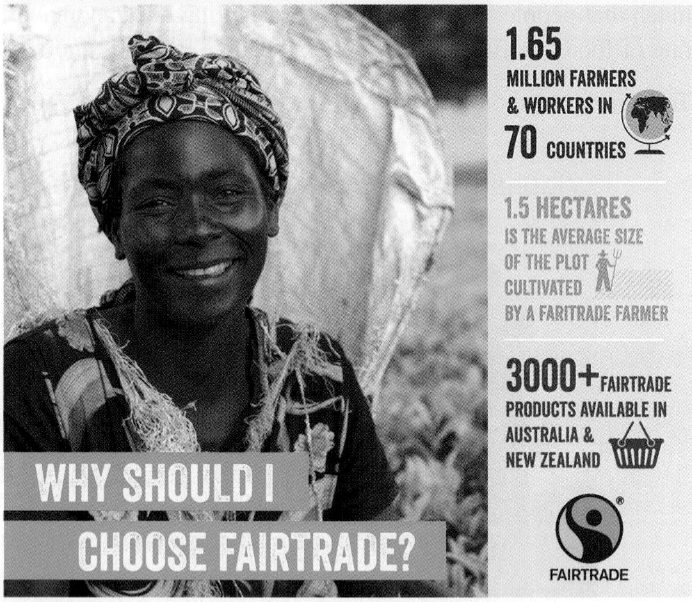

1.65
MILLION FARMERS & WORKERS IN
70 COUNTRIES
1.5 HECTARES
IS THE AVERAGE SIZE OF THE PLOT CULTIVATED BY A FAIRTRADE FARMER
3000+ FAIRTRADE PRODUCTS AVAILABLE IN AUSTRALIA & NEW ZEALAND
WHY SHOULD I
CHOOSE FAIRTRADE?
FAIRTRADE

Source: Fairtrade

Contents

learnON

- 6.8.1 Features of fair trade
- 6.8.2 Fairtrade certification

on Resources

eWorkbook	Fair trade (ewbk-9778)
Video eLesson	Fair trade — Key concepts (eles-5211)
Interactivity	Fair trade (int-8641)

6.9 SkillBuilder — Constructing and describing proportional circles on maps

online only

LEARNING INTENTION

By the end of this subtopic, you will be able to plot scaled circles on a map to show the visual pattern of data and describe the patterns and any anomalies that exist.

The content in this subtopic is a summary of what you will find in the online resource.

6.9.1 Tell me

What are proportional circle maps?

Proportional circle maps are maps that incorporate circles, drawn to scale, to represent data for particular places. They provide an immediate visual pattern, especially when the figures being handled are large. Different-sized circles on a map reflect different values or amounts of something. Proportional circles provide an easy way to interpret patterns, give an instant impression and allow us to compare data for different places.

6.9.2 Show me

How to draw and describe proportional circles on a map

Step 1

Study the data; decide how many categories you need to include the highest and the lowest values to be represented by the circles — use no more than five categories in a key. The key in the **FIGURE 1** has only four categories, but there are more circle sizes shown on the map, representing populations between the tens of millions. For example Tokyo's population in 2000 was 26.4 million, so the orange circle is bigger than the 20 million circle but not as big as the 30 million circle.

Step 2

Rank the values in your data table from highest to lowest. Work out the square root (√) of each value. **TABLE 1** shows the projected population data (in millions) for selected megacities in 2025. The largest figure is 6.09 for Tokyo, and the smallest is 3.24 for Lahore. These numbers give us the measurement of the radius of the proportional circles for our map. (*Note:* When working with population figures, you would leave off the millions and work simply with the base number, e.g. '36' for 36 million.)

TABLE 1 Projected megacity population for selected cities, 2025 (millions)

Megacity	2025	√ (radius, mm)	Megacity	2025	√ (radius, mm)
Tokyo	37.1	6.09	Kolkata	17.3	4.16
Delhi	32.7	5.72	Kinshasa	16.9	4.11
New York	19.3	4.39	Lahore	10.5	3.24

Step 3

Construct a scaled group of circles as seen in the legend for **FIGURE 1**. Allow one millimetre to represent one unit. Ensure the largest circle has a radius big enough to encompass the largest figure in your set of square root data. The data for the largest population (Tokyo) has a square root of 6.09, so draw the largest circle with a radius of seven millimetres to ensure that Tokyo can be plotted with a radius of 6.09 mm. All other data in the table will fit somewhere between these two sizes.

Set your mathematical compass to seven millimetres and draw a circle with a seven-millimetre radius on your base map near Tokyo. Your smallest circle would need a radius of three millimetres to include Lahore.

FIGURE 1 Growth of megacities over time, 1950–2025 (projected)

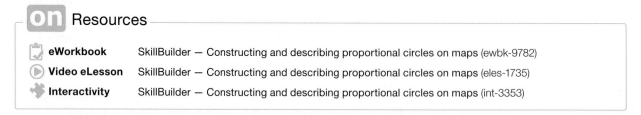

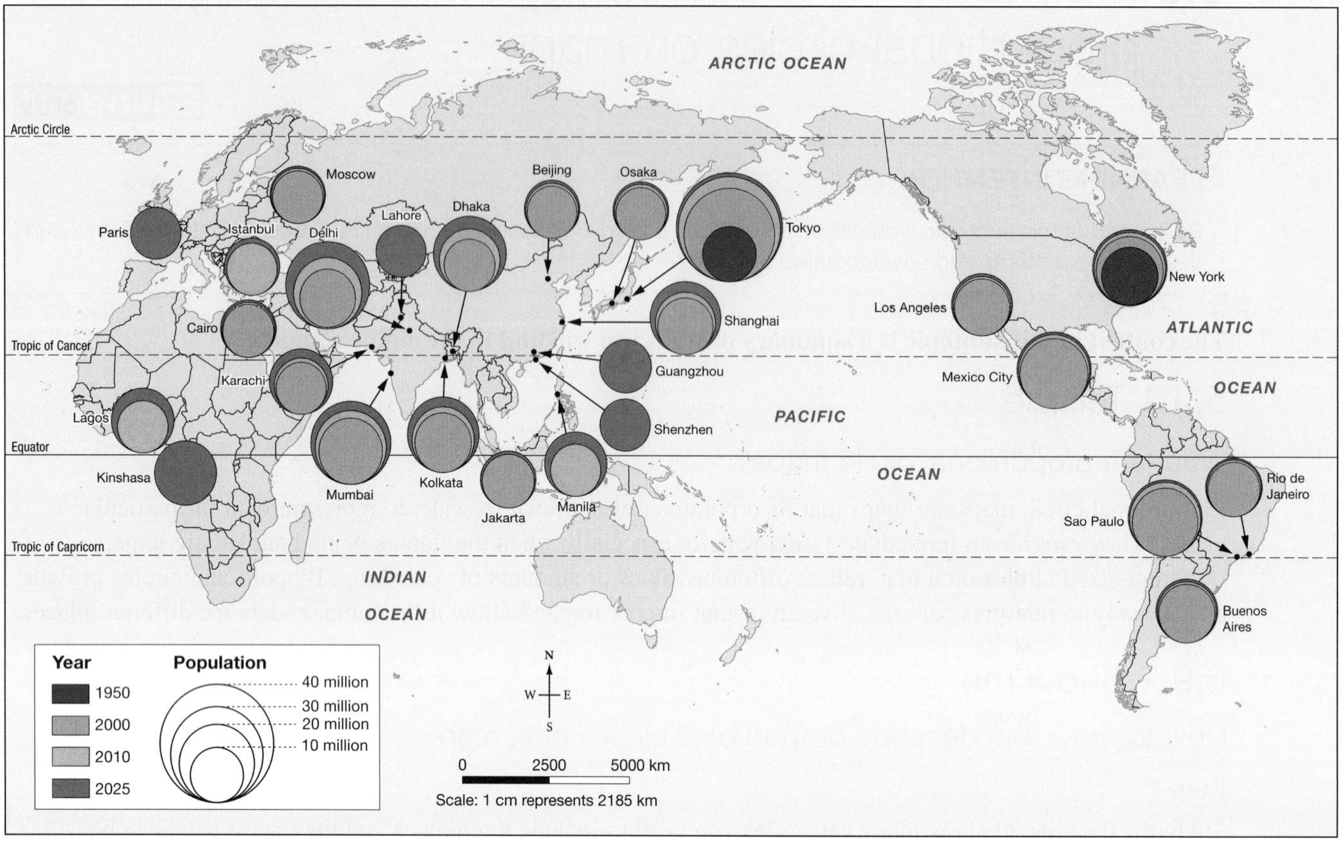

Source: Australian Bureau of Statistics

Step 4

Draw the circles representing all of the megacities on the base map, using an atlas as a reference to place circles close to the location they represent. (Use arrows if there are too many circles near each other.) Complete the map with the geographical conventions of BOLTSS.

Describing a proportional circle map

Step 5

To interpret your mapped data, you need to look for patterns. Where are the largest circles? Where are the smallest circles? Are there any groupings of circles? Are there any patterns that can be identified, such as radial, linear, clustered or sporadic?

6.9.3 Let me do it

learnON

Need more help? Go to your online Resources to access the following:

- a longer explanation of this skill and its application in Geography (Tell me)
- a video demonstrating the step-by-step process of this skill (Show me)
- an activity and interactivity for you to practise the skill (Let me do it)
- self-marking questions to help you understand and use the skill.

on Resources

eWorkbook	SkillBuilder — Constructing and describing proportional circles on maps (ewbk-9782)
Video eLesson	SkillBuilder — Constructing and describing proportional circles on maps (eles-1735)
Interactivity	SkillBuilder — Constructing and describing proportional circles on maps (int-3353)

6.10 The effects of dietary changes on food supply

LEARNING INTENTION

By the end of this subtopic, you will be able to outline the interconnection between dietary trends and food security, and discuss likely impacts of population projections on diet and food availability.

6.10.1 Changes in diet over time

The human diet has changed throughout history and will continue to do so. Since the 1960s, the total calories consumed per day globally and the proportion of the diet comprised of animal products, oils and sweeteners have increased. These food types are typically found in higher amounts in the Western-style diet eaten by much of the population of developed countries. **FIGURE 1** shows the changing global diet as recorded in a study of dietary trends from 1961 to 2009.

Since 1960, diets around the world have become more alike. The amount of food people eat and the levels of calories, protein and, fat have all increased. Although animal products, oils and sweeteners have long been a feature of the diet in developed countries, they are increasingly becoming part of the diet in developing countries too. These trends are predicted to continue (see **FIGURE 2**). This is especially the case in countries such as India and China, where the standard of living is rising and people can increasingly afford access to a wider variety of foods.

FIGURE 1 Changes to global diet, 1961–2009

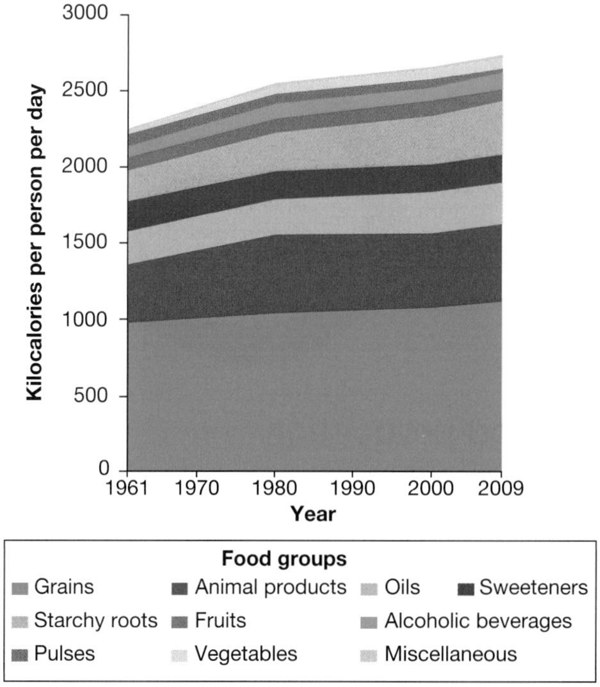

Source: © 2016 International Center for Tropical Agriculture — CIAT

FIGURE 2 Changing diets in developing countries, 1964–66, 1997–99, 2030*

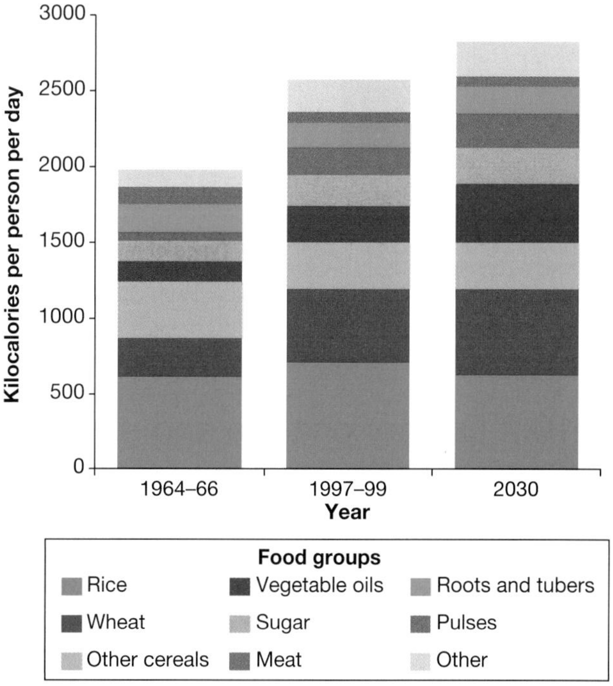

*projected

Source: FAO, 2008; FAOSTAT, 2009

For centuries, the typical Chinese diet was rice and vegetables, supplemented by fish and small amounts of other meat. Rice is a valuable source of protein, but as people's incomes grow, per capita rice consumption is expected to decline and consumption of protein from meat sources is expected to increase accordingly.

In 1962, the average Chinese person ate just 4 kilograms of meat per year. By 2015, this figure was closer to 80 kilograms (around 220 grams per day) and rising. In contrast, Australians and Americans are the world's highest consumers of meat, eating an average of around 110 kilograms per year (300 grams per person per day), which is significantly more than the global average of around 42 kilograms per year (115 grams per day). The predicted growth in global meat production is shown in **FIGURE 4**.

FIGURE 3 Global impacts of economic growth and dietary change

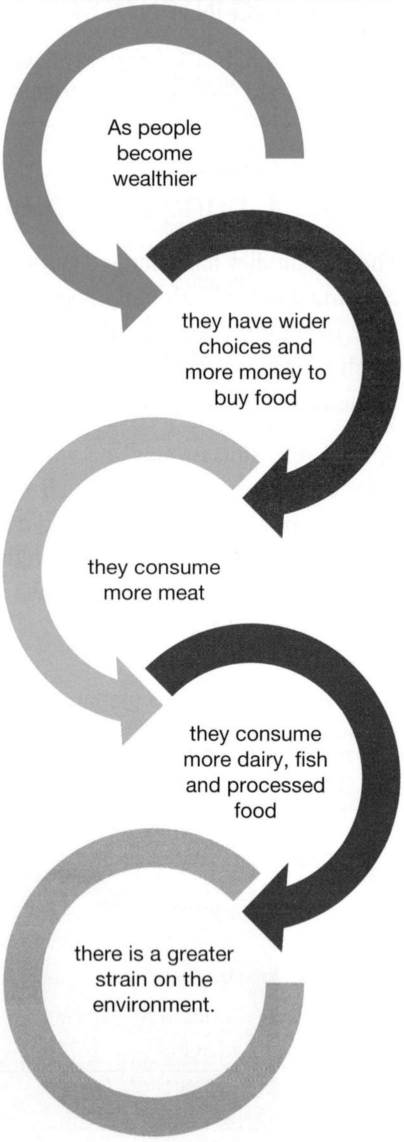

FIGURE 4 Expected growth in meat consumption, 2018–2028

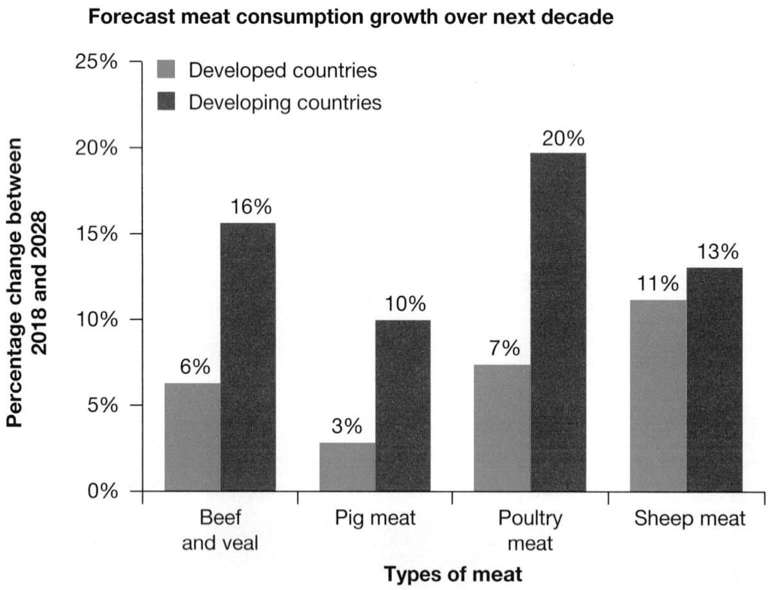

Source: Meat and Livestock Australia, MLA graph/OECD data

6.10.2 Dietary change and sustainable food production

The countries of the Asian region are home to more than half the world's population. With significant economic growth occurring throughout much of the area and over four billion people to be fed, ensuring food security in Asia presents a significant challenge. On the other hand, it also offers unparalleled opportunities for Australian farmers and the Australian economy.

But the opposite is also true. As societies become more affluent, people's rising standards of living mean they expect to have access to more and a wider range of goods — including food. Australia is well placed to provide a wide variety of desirable foods and quality fibres, such as wool and cotton. Particularly, considering Australia's reputation for 'clean and green' agricultural systems and our geographic proximity, Australian farmers are ideally placed to capitalise on the economic opportunities that the fast-developing Asian region presents to become the 'food bowl of Asia'.

FIGURE 5 The UN Food and Agriculture 2020 report into food security and nutrition in the world demonstrates that 'diet quality worsens with increasing severity of food insecurity.'

While … most of the poor around the world can afford an energy sufficient diet they cannot afford either a nutrient adequate or a healthy diet. A healthy diet is far more expensive than the full value of the international poverty line of USD 1.90 PPP per day, let alone the portion of the poverty line that can credibly be reserved for food (63 percent), to end up with a threshold of USD 1.20 PPP per day … It is estimated that based on estimated income distributions more than 3 billion people in the world could not afford a healthy diet in 2017. Most of these people are found in Asia (1.9 billion) and Africa (965 million), although there are also millions that live in Latin America and the Caribbean (104.2 million), and in Northern America and Europe (18 million).

Source: Food and Agriculture Organization of the United Nations, 2020, FAO, IFAD, UNICEF, WFP and WHO, *The State of Food Security and Nutrition in the World 2020. Transforming food systems for affordable healthy diets,* http://www.fao.org/3/ca9692en/ca9692en.pdf. Reproduced with permission.

Although the COVID-19 pandemic disrupted flows of agricultural produce to export markets, the ongoing challenge for Australian farmers will be in meeting this booming global need for food and fibre by increasing production at a time when we have decreasing arable land, less water and fewer people working in agriculture.

Sustainable practices for future food security

One-third of the world's grain crop is fed to animals to produce meat. From a sustainability perspective, this can be considered wasteful, as the amount of grain used to feed a cow for the purposes of meat production is 11 times what would be needed to adequately feed a person with grain alone. Similarly, while 1500 litres of water are needed to produce 1 kilogram of cereal, 15 000 litres are needed to produce 1 kilogram of meat.

Meeting the needs of future populations is not just the responsibility of farmers and producers. We as consumers can also contribute. Attitudes may need to change towards how and what we eat. If we are to feed nine billion people sustainably in 2050, it is unlikely we'll be eating a meat-rich, Western-style diet. The world produces enough food to feed 10 billion people. However, a significant portion of our crops is used to feed animals or is used as biofuel to produce energy.

A switch to a plant-based diet would allow land currently used to produce animal feed to instead grow crops to feed humans. Although such a huge change is unlikely, even a small shift can have an impact. The Meat Free Monday campaign encourages people to go without meat for one day per week. This small change would benefit human health and the health of the planet. Meat production requires a large amount of land, water and energy. Cattle are also the largest source of methane gas, which is one of the main contributors to greenhouse gases.

FIGURE 6 The Meat Free Monday campaign encourages people to eat a plant-based diet one day a week as a sustainable food practice.

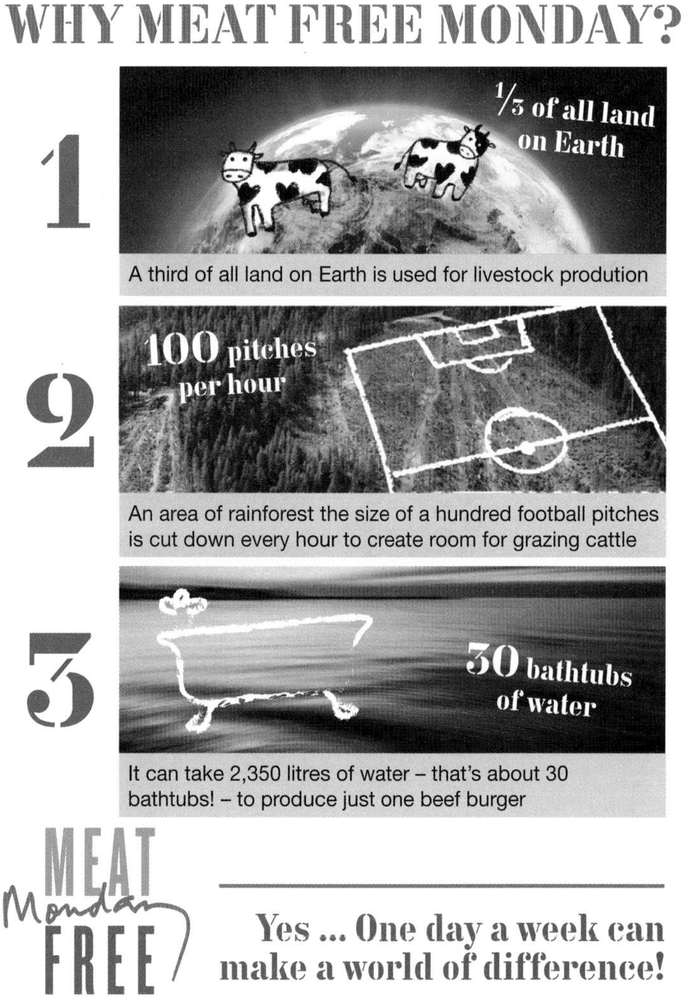
WHY MEAT FREE MONDAY?
1
1/3 of all land on Earth
A third of all land on Earth is used for livestock prodution
2
100 pitches per hour
An area of rainforest the size of a hundred football pitches is cut down every hour to create room for grazing cattle
3
30 bathtubs of water
It can take 2,350 litres of water – that's about 30 bathtubs! – to produce just one beef burger
MEAT Monday FREE
Yes … One day a week can make a world of difference!

Source: Meat Free Monday Ltd

It is estimated that there are more than 20 000 edible plants on the planet that we do not currently eat. Exploring ways of developing and introducing these into our diets may provide additional, sustainable food sources for future generations. One example of an 'old food' that has become increasingly popular in the modern diet is quinoa (pronounced *keen-wah*). A crop from South America, quinoa was being eaten over 4000 years ago by people living in the Andes. It has high nutritional value and grows in a wide variety of climatic conditions. Another advantage of the crop is that all parts of it can be eaten. Peru and Bolivia supply 99 per cent of the world's quinoa demand, and many other countries are now investigating its suitability for their locations. More plant-based foods from other cultures are also being introduced into the diets of many Australians, for example tofu (which originated in China but is used in many Asian cuisines) and tempeh (from Indonesia), both of which are made from soybeans.

FIGURE 6 Quinoa fields in Chile

Increasing consumption of fruits and vegetables, whole grains, legumes and nuts, and limiting intake from animal sources, fats and sugars will not only have health benefits for individuals but will also benefit the planet, as more land and water resources can be directed to sustainable food crop development.

FIGURE 7 Protein-rich plant-based food alternatives

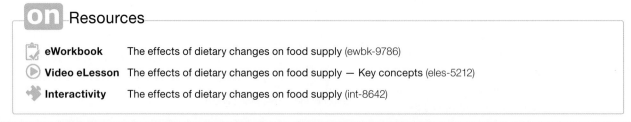

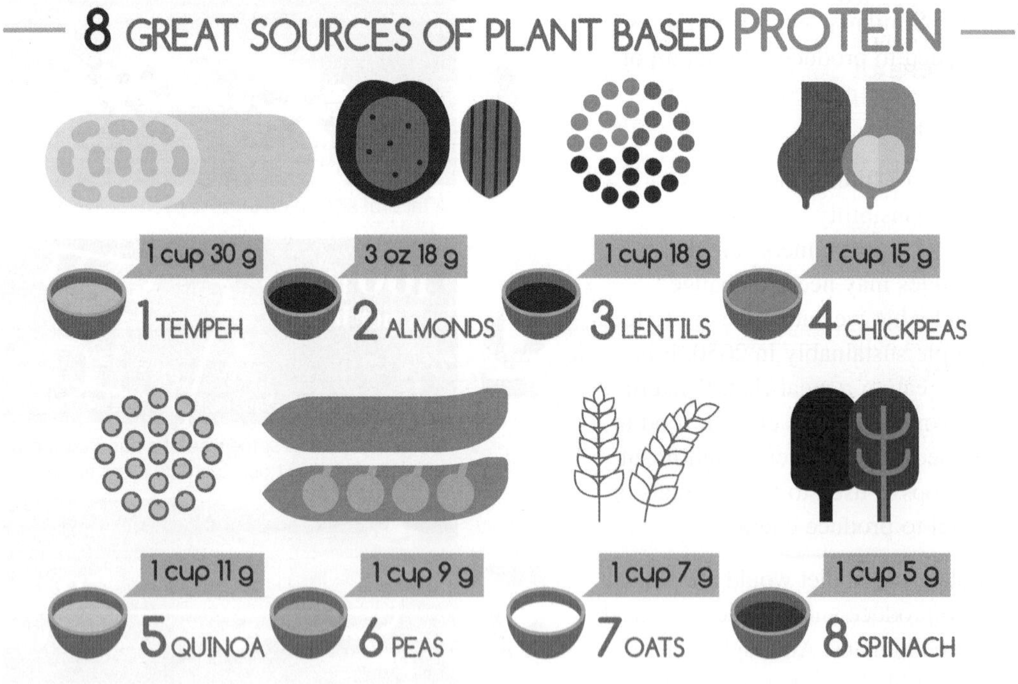

on Resources

eWorkbook	The effects of dietary changes on food supply (ewbk-9786)
Video eLesson	The effects of dietary changes on food supply — Key concepts (eles-5212)
Interactivity	The effects of dietary changes on food supply (int-8642)

6.10 ACTIVITIES

1. In teams of three, you have 20 minutes to research and to prepare a 45-second pitch to convince your classmates that they should adopt one of the following future food sources into their future diet. You are permitted one slide with no words as a backdrop to your presentation.
   Food choices:
   - Rodents
   - Bugs and insects
   - 3D printed food
   - Cultured meat
   - Seaweed
   - Vegan
2. Design a television commercial to promote a Meat Free Monday campaign.
3. Use the **Changing diet** and **Food supply** weblinks in your online Resources to create a summary of Australia's diet and food supply trends over time. With the data you have collected, write a paragraph to outline how the makeup of your diet is likely change over the next twenty years.

6.10 EXERCISE

Learning pathways

■ LEVEL 1	■ LEVEL 2	■ LEVEL 3
1, 6, 8, 9, 15	2, 5, 7, 10, 13	3, 4, 11, 12, 14

Check your understanding

1. Refer to **FIGURE 1**.
   a. Which crops have become more popular?
   b. Which crops have become less popular?
2. Describe the connection between diet and economic development.
3. Refer to **FIGURE 4**. Describe the changes in meat consumption in developed and developing countries over time.
4. On average, how much meat does an Australian eat every day?
5. Draw a five-stage flow chart to show the interconnection between people becoming wealthier and depletion of key food production in our biomes.
6. Approximately how many people in the world cannot afford a healthy diet?
7. What does the term *food bowl of Asia* mean?

Apply your understanding

8. What might be some of the issues confronting Australia as it attempts to become the 'food bowl of Asia'? What advantages does Australia have in this attempt? How might a farmer react to this suggestion?
9. Explain how eating a completely plant-based diet is more sustainable for our grassland and forest biomes than a diet including beef and sheep.
10. How is Australia uniquely positioned to help feed the growing populations of Asian nations?
11. Why have people's diets changed over time?
12. Consider the possible difficulties that Australian farmers will face in meeting growing demand for food production for export to Asia.
    a. Identify and analyse three potential difficulties to assess the extent of the challenge they pose.
    b. Which do you think will provide the most significant challenge?

Challenge your understanding

13. Predict what you will be eating and where your foods might come from in 2050.
14. Should developed, food-exporting countries like Australia play a greater role in encouraging developing nations to eat less meat as they build their economies? Give reasons for your answer.
15. What would the impact be on food security if people around the world started eating twice as much meat? In your answer, consider whether this is a possibility considering current consumption and production levels.

To answer questions online and to receive **immediate feedback** and **sample responses** for every question, go to your learnON title at www.jacplus.com.au.

6.11 Urban farms

LEARNING INTENTION

By the end of this subtopic, you will be able to explain the role urban farms can play at a local level to feed urban populations.

6.11.1 The advantages of urban farming

Farming is usually associated with rural areas, but a growing trend in food production is **urban farming**. This involves growing plants and raising animals within and around cities, often in unused spaces — even the rooftops of buildings.

In many industrialised countries, it takes over four times more energy to move food from the farm to the plate than is used in the farming practice itself. Properly managed, urban agriculture can turn urban waste (from humans and animals) and urban wastewater into resources, rather than sources of serious pollution.

Urban agriculture is becoming more productive. In 2000, about 15 to 20 per cent of the world's food supply came from urban gardens; in 2018, more than 800 million people practised urban agriculture, contributing to over 20 per cent of all global agricultural production.

Benefits of urban farming include:
- increasing the amount, variety and freshness of vegetables and meat available to people in cities through sustainable production methods
- improving community spirit through community participation, often including disadvantaged people
- incorporating exercise and a better diet into people's lives, leading to improved physical and mental health
- using urban wastewater as a resource for irrigation, rather than making it a source of serious pollution
- reducing the percentage of income spent on food.

Urban farming could become more important with rapid urbanisation. With the developing countries in Africa, Asia and Latin America home to 75 per cent of all urban dwellers, they will face the problems of providing enough food and disposing of urban waste.

urban farming growing plants and raising animals within and around cities

FIGURE 1 (a) Roof top gardening in the Causeway Bay district of Hong Kong. (b) Former air raid shelters, 100 metres below Clapham in London, have been repurposed as a vertical farm.

CASE STUDY

East Kolkata wetlands

The East Kolkata wetlands in India cover 12 500 hectares and contain sewage farms, vegetable fields, pig farms, rice paddies and over 300 fishponds. With a population of almost 15 million, the Indian city of Kolkata produces a high volume of sewage daily. The wetlands system treats this sewage, and the nutrients contained in the wastewater then sustain the fish ponds and agriculture.

FIGURE 2 East Kolkatta wetlands and sewage ponds

About one-third of the city's daily fish supplies come from the wetlands, which are the world's largest system for converting waste into consumable products. The wetlands are also a protected Ramsar site for migratory birds. A Ramsar site is a wetland of international importance as defined by the Ramsar Convention — an intergovernmental treaty on the protection and sustainable use of wetlands. However, the area is now under pressure from urban growth and from the subsequent increase in waste that it needs to treat.

CASE STUDY

Container fish farm

On a smaller scale, a German company has developed a sustainable form of aquaculture that can be used in small spaces in cities. It is called aquaponics. Fish swim in large tanks in a recycled shipping container (see **FIGURE 3**), and their waste fertilises tomatoes, salad leaves and herbs growing in a greenhouse mounted above the tank. The plants purify the water, which is returned to the tanks.

FIGURE 3 Urban farming — fish and agriculture

These structures can be set up anywhere, such as on rooftops and in car parks. The sustainably produced fresh vegetables and fish can be delivered to nearby city markets and shops, reducing the distance that the products must travel. Farmers only need to feed the fish and keep the fish-tank water topped up to sustain the efficient aquaponic system.

on Resources

eWorkbook	Urban farms (ewbk-9790)
Video eLesson	Urban farms — Key concepts (eles-5213)
Interactivity	Urban farms (int-8643)

6.11 ACTIVITIES

1. Use the **Sprout stack** weblink in your online Resources to investigate a sustainable food production initiative operating out of a Sydney suburb.
   a. What is produced at the 'farm'?
   b. What makes the food production unit sustainable?
   c. What are the environmental, economic and social advantages and disadvantages of growing food in this way?
2. Use the **Aquaponics** weblink in your online Resources to watch a video about aquaponics. Outline the advantages of aquaponics presented in the video.
3. Use the **Vertical farming** weblink in your online Resources to help you understand vertical farming. Draw an annotated diagram to illustrate vertical farming.

6.11 EXERCISE

Learning pathways

■ LEVEL 1	■ LEVEL 2	■ LEVEL 3
5, 7, 8, 10, 15	1, 2, 3, 9, 14	4, 6, 11, 12, 13

Check your understanding

1. List the main features of urban farming.
2. What proportion of food production across the world occurs in urban areas?
3. Outline the functions that the East Kolkata wetlands perform.
4. What proportion of the fish eaten every day in Kolkata comes from the wetland fish ponds?
5. How do communities benefit from urban farms? Outline three benefits.
6. Outline two advantages and two disadvantages of producing food on the rooftop spaces of city buildings.
7. List four types of food that are farmed in the East Kolkata wetlands.

Apply your understanding

8. What factors might influence the types of food that could be produced on rooftops?
9. When investigating urban farms and people's gardening activities in Denver, USA, researchers found that:
   - people's community pride improved
   - graffiti and vandalism decreased
   - gardeners felt a greater connection with their local place.

   Evaluate whether these are worthwhile benefits, even if there was no significant environmental benefit.
10. Draw a diagram to demonstrate how urban aquaculture and horticulture work together in aquaponics.
11. Write a draft email to the Minister for Planning, suggesting that food production spaces should be included in every new urban development in New South Wales.
12. How might urban farms encourage agricultural tourism?

Challenge your understanding

13. Where in the world would you expect to find vertical farms? Suggest at least three cities and explain why you would expect to find vertical farming in these places.
14. An old air raid shelter from World War II has been repurposed for vertical farming in London. Suggest one place in your area that is no longer used and could be used for a similar project. Provide reasons why that place would be suitable (such as access to water, transport or power) and what might be grown there.
15. Predict how people in your area would respond to a plan to release sewage into wetlands as food for fish farms. Do you think people would approve or disapprove of the idea? Do you think people in your community would be happy to eat fish farmed in this way?

To answer questions online and to receive **immediate feedback** and **sample responses** for every question, go to your learnON title at www.jacplus.com.au.

6.12 Investigating topographic maps — Lake Victoria as a food source

LEARNING INTENTION

By the end of this subtopic, you will be able to locate and describe the features of the Lake Victoria area from a topographic map, and discuss how changes to the environment might affect food production in the area.

6.12.1 Lake Victoria

Lake Victoria, with a surface area of 68 800 km^2, is Africa's largest freshwater lake and a source of water for the Nile River. The lake supports a population of over 30 million people in east Africa through fishing, agriculture, local industry, forestry, hydro-electric power, transport and tourism. Lake Victoria is shared by Uganda, Kenya and Tanzania, and its catchment area includes Rwanda and Burundi.

High levels of hunger exist within the densely populated communities (see **FIGURE 1**) that live around Lake Victoria. Crops grown within the river catchment include beans, coffee, cotton, maize, sisal (a fibrous plant used for making rope), sugarcane and tobacco. The lake also supports a productive fishing industry (**FIGURE 2**); however, fish stocks in recent years have declined due to overfishing and increased environmental pressures. Invasive weeds, such as the water hyacinth, have contributed to a decline in fish stock, increased the incidence of waterborne diseases, reduced water quality and increased turbidity.

FIGURE 1 Densely populated areas along the shores of Lake Victoria.

FIGURE 2 Fishermen carrying their catch from Lake Victoria onshore.

FIGURE 3 Topographic map extract, of Lake Victoria

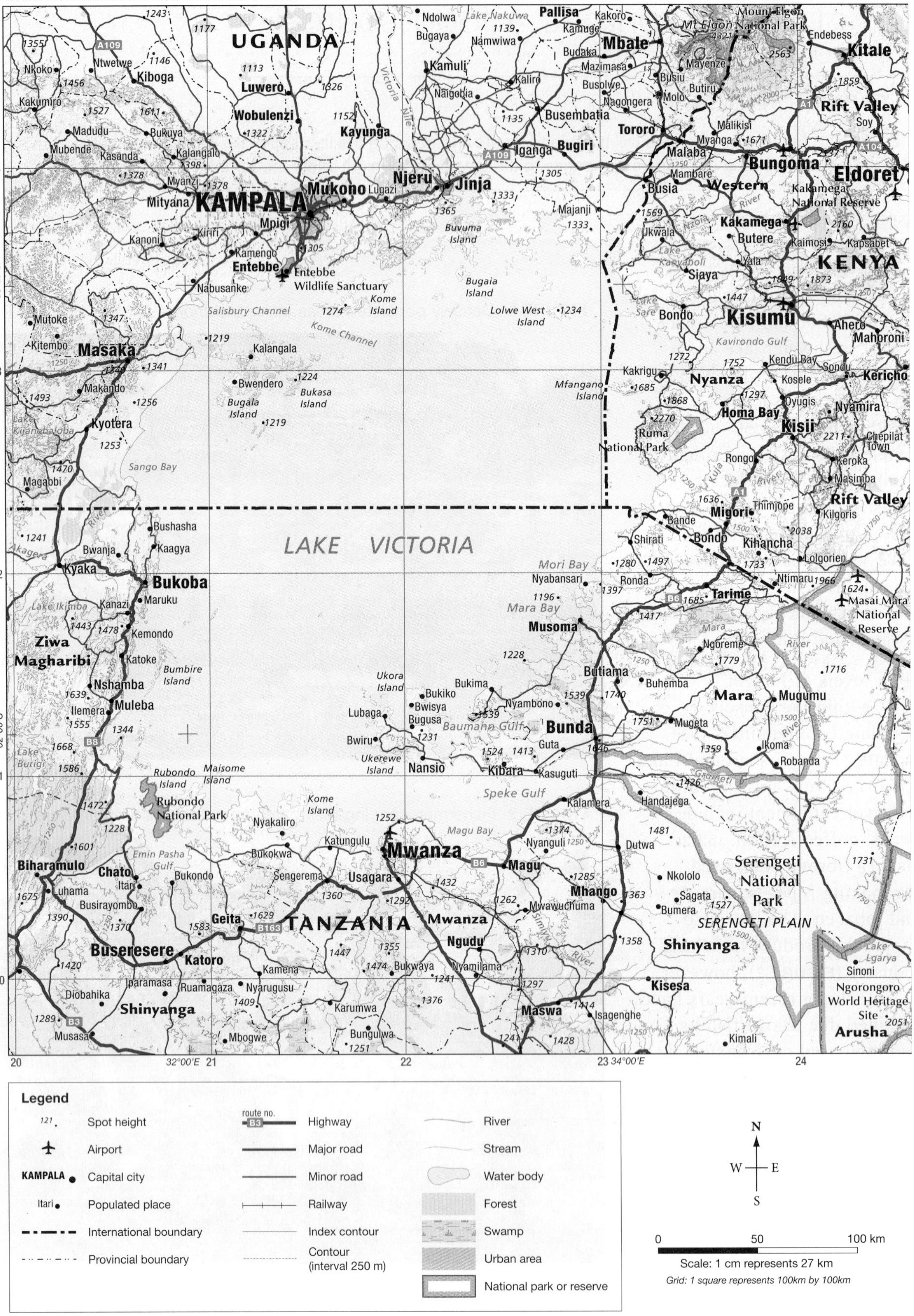

Source: Map data © OpenStreetMap contributors, https://openstreetmap.org. Data is available under the Open Database Licence, https://opendatacommons.org/licenses/odbl/; elevation data sourced from USGS.

Resources

eWorkbook	Investigating topographic maps — Lake Victoria as a food source (ewbk-9794)
Digital document	Topographic map of Lake Victoria (doc-36318)
Video eLesson	Investigating topographic maps — Lake Victoria as a food source — Key concepts (eles-5214)
Interactivity	Investigating topographic maps — Lake Victoria as a food source (int-8644)
Google Earth	Lake Victoria (gogl-0133)

6.12 EXERCISE

Learning pathways

LEVEL 1	LEVEL 2	LEVEL 3
1, 2, 7, 12	3, 5, 9, 11	4, 6, 8, 10

Check your understanding

Refer to **FIGURE 3** to answer the following questions.

1. List the names of each of the countries that share the edge of Lake Victoria.
2. Which country is to the east of Lake Victoria?
3. If a fishing boat was located at 2° 00' S 32° 00' E, what country would it be in?
4. What is the distance across the widest part of Lake Victoria?
5. How many airports and landing grounds can you see in the region shown in this map?
6. How many kilometres would a plane travel on a trip from Entebbe airport to Mwanza airport, then to Kisumu airport, before a return trip to Entebbe?

Apply your understanding

7. How might human impact affect the ability of the lake to support a fishing industry?
8. Explain the impact of invasive weeds on food production in the area. Focus your answer on two ways that introduced plant species lower the region's food production capacity.
9. Explain the factors that lead to the high levels of hunger in the communities around Lake Victoria. In your explanation, outline whether the factor could be considered political, economic, environmental or social/cultural.

Challenge your understanding

10. What impacts are humans likely to be having on the water quality of Lake Victoria? Predict whether and how the water quality will change in the next 50 years.
11. Refer to **FIGURE 3**. Propose one strategy that would help the three countries in which Lake Victoria is located to manage the resources in the lake equitably.
12. Based on the features shown in **FIGURES 3** and **4**, if you were going to build a tourist resort for international tourists to visit the Serengeti National Park and Lake Victoria, where would you locate it and why?

FIGURE 4 True colour satellite image, Lake Victoria, Africa

To answer questions online and to receive **immediate feedback** and **sample responses** for every question, go to your learnON title at www.jacplus.com.au.

6.13 Thinking Big research project — Community garden design

online only

The content in this subtopic is a summary of what you will find in the online resource.

Scenario

Australia's largest urban areas — Sydney, Melbourne and Brisbane — are continuing to show signs of urban sprawl. **Urban sprawl** inevitably affects food production, as areas that were once farmland are transformed into housing and commercial developments. One strategy that can reduce the impacts of urban sprawl is the creation of urban farms. Urban farms can range in size from small community gardens in our inner-city suburbs to large warehouses dedicated to horticulture in our outer suburbs. The best community gardens are small and provide just a moderate amount of fresh produce to the people who live in that area.

Task

Your school may already have a small community farm or garden. Your task is to research and design (and potentially build) a small new community garden for your school or neighbourhood, to produce fresh vegetables and fruit to sell locally or to be used in your school canteen.

Go to learnON to access the resources you need to complete this research project.

urban sprawl the spreading of urban developments into areas on the city boundary

on Resources

ProjectsPLUS Thinking Big research project — Community garden design (pro-0192)

6.14 Review

6.14.1 Key knowledge summary

6.2 Feeding our future world population

- One in nine people on Earth do not have enough to eat; around a quarter of the population is overweight.
- The distribution of the world's arable land is uneven, and the fastest-growing parts of the world do not have enough land to grow sufficient food. As urban areas grow, the amount of available arable land decreases.
- Seventy per cent of the world's poorest people live in rural areas where trade is limited. Improving roads and other infrastructure would improve opportunities for trade.
- Farming yields are affected by a variety of factors: access to water, growing season, climate, soil types, finance and markets, insects and diseases, funds for agricultural research, government regulations, and weather events such as floods, storms and drought.
- Better use of farming areas and technology, and more efficient farming will improve food production.

6.3 Improving food production

- Strategies to improve food production include reducing the yield gap, developing genetically modified crops, expanding aquaculture, improving infrastructure and developing sustainable intense cropping.
- There are some concerns over the use of GM crops, including health risks and loss of seed variety.
- New technologies and systems are being developed to improve agricultural efficiency, such as the Sundrop Farm in Port Augusta, South Australia.

6.4 Reducing food waste

- Food loss takes place at the production and distribution stages of food production.
- Food waste takes place at the retail and consumer stages of food production.
- Up to 50 per cent of fruits and vegetables produced in developing countries are lost in the supply chain between harvest and consumption. Waste can increase food prices and make food less affordable.
- Australia produces enough food for 60 million people, and this enables us to trade the surplus. Yet each person in Australia wastes an average of 361 kg of food each year

6.6 Food aid

- Food aid is food, money, goods and services given by more developed nations to less developed nations.
- The three broad types of food aid are program food aid, project food aid and relief food aid.
- The key organisation that provides food aid worldwide is the United Nations World Food Programme (WFP). The WFP reaches over 80 million people in more than 92 countries and supports those who are affected by natural disasters, conflict and the cycle of poverty.
- The WFP's school feeding program provides meals to children in schools, and cash vouchers for food and other essentials.

6.7 Food aid in Australia

- As the cost of living increases, food security is becoming a major issue in Australia. Housing costs, including rent, are the largest contributor to household poverty.
- For many people, programs provided by charity food agencies are a vital source of daily food needs.
- Communities in remote areas experience food insecurity due to the cost of transporting and storing fresh food.

6.8 Fair Trade

- Trade usually favours countries that are the strongest economically.
- Fair trade promotes fair prices and reasonable conditions for producers in developing regions.
- Fairtrade organisations operate in poorer nations, particularly in Asia, Africa and South America. Fairtrade farmers enjoy the certainty of having a fixed price and a guaranteed market for their product.

6.10 The effects of dietary changes on food supply

- The human diet has changed over time and continues to change; average calorie intake has increased significantly since the 1960s. Diets around the world have also become more similar and larger in terms of calories, protein, fat and food weight. Animal products, oils and sweeteners are increasingly becoming part of the diet in developing countries.
- As people's incomes grow in Asian nations such as China and India, per capita rice consumption is expected to decline and consumption of protein from meat sources is expected to increase accordingly.
- There is opportunity for Australia to become a significant food source for growing Asian nations.
- One-third of the world's grain crop is fed to animals to produce meat, and 15 000 litres of water are needed to produce 1 kilogram of meat. This is unsustainable. To help meet the future food needs of the world's population, it will be necessary to eat fewer animal products and more plant-based foods.

6.11 Urban farms

- Urban farms are becoming more common around the world. Over 800 million people practise urban agriculture, contributing over 20 per cent of global agricultural production.
- Benefits of urban farming include increased food freshness, reduced transportation costs, reuse of urban waste water and increased community spirit. With rapid urbanisation, urban farming will become increasingly important as a sustainable food source for growing populations.

6.14.2 Key terms

active consumerism a movement that is opposed to the endless purchase of material possessions and the pursuit of economic goals at the expense of society or the environment

anthropogenic resulting from human activity

arable describes land that is suitable for growing crops

feedlot farming intensive farming practice in which high concentrations of animals are kept in confined areas to facilitate rapid growth and weight gain

fair trade trading exchanges that are sustainable and fair for both the seller and the buyer

food aid food, money, goods and/or services given for the specific purpose of helping those in need

food loss takes place at the production, post-harvest, processing, and distribution stages of food production

food waste takes place at the retail and consumption stages food production and consumption

genetically modified describes seeds, crops or foods whose DNA has been altered by genetic engineering techniques

hydroponic describes a method of growing plants using mineral nutrients, in water, without soil

malnourishment a condition that results from not getting the right amount of vitamins, minerals and other nutrients needed to maintain healthy tissues and organ function

mitigate to reduce negative impacts or severity

seasonal crops crops that are harvested in a certain season of the year rather than all year round

urban farming growing plants and raising animals within and around cities

urban sprawl the spreading of urban developments into areas on the city boundary

yield gap the gap between a certain crop's average yield and its maximum potential yield

6.14.3 Reflection

Complete the following to reflect on your learning.

Revisit the inquiry question posed in the Overview:

Will there come a time when we don't have enough food to feed everyone?

1. Now that you have completed this chapter, what is your view on the question? Discuss with a partner. Has your learning in this chapter changed your view? If so, how?
2. Write a paragraph in response to the inquiry question, outlining your views.

Subtopic	Success criteria	●	●	●
6.2	I can explain the prevalence and impacts of hunger.			
	I can outline the factors that influence food insecurity.			
6.3	I can compare and evaluate strategies for improving food production as a way to reduce food insecurity.			
6.4	I can compare and evaluate strategies for reducing food waste as a way to help reduce food insecurity.			
6.5	I can plot the data from two maps into a box scattergram.			
	I can describe the relationships that between the two data sets shown in a box scattergram.			
6.6	I can identify different types of food aid and outline how it is distributed.			
6.7	I can identify common reasons why Australians need food relief.			
	I can give examples of strategies for making food more widely available to people experiencing food insecurity in Australia.			
6.8	I can outline fair trade practices and evaluate their sustainability in terms of achieving food security.			
6.9	I can plot scaled circles on a map to show the pattern of data.			
	I can describe the patterns and any anomalies demonstrated by proportional circles on a map.			
6.10	I can describe the interconnection between dietary trends and food security.			
	I can discuss likely impacts of population projections on diet and food availability.			
6.11	I can explain the role urban farms can play at a local level to feed urban populations.			
6.12	I can locate and describe the features of the Lake Victoria area from a topographic map.			
	I can discuss how changes to the environment might affect food production in the Lake Victoria area.			

on Resources

eWorkbook Chapter 6 Student learning matrix (ewbk-8540)
Chapter 6 Reflection (ewbk-8541)
Chapter 6 Extended writing task (ewbk-8542)

Interactivity Chapter 6 Crossword (int-8645)

ONLINE RESOURCES

on Resources

Below is a full list of **rich resources** available online for this chapter. These resources are designed to bring ideas to life, to promote deep and lasting learning and to support the different learning needs of each individual.

eWorkbook

- **6.1** Chapter 6 eWorkbook (ewbk-8098) ☐
- **6.2** Feeding our future world population (ewbk-9758) ☐
- **6.3** Improving food production (ewbk-9762) ☐
- **6.4** Reducing food waste (ewbk-9766) ☐
- **6.5** SkillBuilder — Constructing a box scattergram (ewbk-9770) ☐
- **6.6** Food aid (ewbk-9774) ☐
- **6.7** Food aid in Australia (ewbk-8159) ☐
- **6.8** Fair trade (ewbk-9778) ☐
- **6.9** SkillBuilder — Constructing and describing proportional circles on maps (ewbk-9782) ☐
- **6.10** The effects of dietary changes on food supply (ewbk-9786) ☐
- **6.11** Urban farms (ewbk-9790) ☐
- **6.12** Investigating topographic maps — Lake Victoria as a food source (ewbk-9794) ☐
- **6.13** Thinking Big Research project — Community garden design (ewbk-9798) ☐
- **6.14** Chapter 6 Student learning matrix (ewbk-8540) ☐
  - Chapter 6 Reflection (ewbk-8541) ☐
  - Chapter 6 Extended writing task (ewbk-8542) ☐

Sample responses

- **6.1** Chapter 6 Sample responses (sar-0157) ☐

Digital document

- **6.12** Topographic map of Lake Victoria (doc-36318) ☐

Video eLessons

- **6.1** Future food (eles-1721) ☐
  - Food security — Photo essay (eles-5205) ☐
- **6.2** Feeding our future world population — Key concepts (eles-5206) ☐
- **6.3** Improving food production — Key concepts (eles-5207) ☐
- **6.4** Reducing food waste — Key concepts (eles-5208) ☐
- **6.5** SkillBuilder — Constructing a box scattergram (eles-1734) ☐
- **6.6** Food aid — Key concepts (eles-5209) ☐
- **6.7** Food aid in Australia — Key concepts (eles-5210) ☐
- **6.8** Fair trade — Key concepts (eles-5211) ☐
- **6.9** SkillBuilder — Constructing and describing proportional circles on maps (eles-1735) ☐
- **6.10** The effects of dietary changes on food supply — Key concepts (eles-5212) ☐
- **6.11** Urban farms — Key concepts (eles-5213) ☐
- **6.12** Investigating topographic maps — Lake Victoria as a food source — Key concepts (eles-5214) ☐

Interactivities

- **6.2** Global Food Security Index (int-8636) ☐
- **6.3** Yield gap for a combination of major crops (int-8637) ☐
- **6.4** Reducing food waste (int-8638) ☐
- **6.5** SkillBuilder — Constructing a box scattergram (int-3352) ☐
- **6.6** Food aid — Interactivity (int-8639) ☐
- **6.7** Food aid in Australia (int-8640) ☐
- **6.8** Fair trade (int-8641) ☐
- **6.9** SkillBuilder — Constructing and describing proportional circles on maps (int-3353) ☐
- **6.10** The effects of dietary changes on food supply (int-8642) ☐
- **6.11** Urban farms (int-8643) ☐
- **6.12** Investigating topographic maps — Lake Victoria as a food source (int-8644) ☐
- **6.14** Chapter 6 Crossword (int-8645) ☐

ProjectsPLUS

- **6.13** Thinking Big research project — Community garden design (pro-0192) ☐

Weblinks

- **6.2** Arable land (web-1206) ☐
  - Global diets (web-6369) ☐
  - Food systems and COVID-19 (web-6370) ☐
  - Arable land by country (web-6371) ☐
- **6.3** Sundrop Farm (web-6372) ☐
  - Vertical farming (web-3317) ☐
- **6.4** Love food hate waste (web-6382) ☐
  - UN food loss (web-6373) ☐
  - UN SDG2 Indicator (web-6383) ☐
- **6.6** World Hunger Map 2015 (web-3182) ☐
  - World Food Programme (web-3318) ☐
- **6.7** Foodbank (web-1174) ☐
  - OzHarvest (web-1168) ☐
  - SecondBite (web-1164) ☐
  - Meals on Wheels (web-1163) ☐
  - ACOSS (web-2398) ☐
  - Australian poverty (web-3315) ☐
- **6.8** Fairtrade (web-4501) ☐
- **6.10** Changing diet (web-6374) ☐
  - Food supply (web-6375) ☐
- **6.11** Sprout stack (web-6376) ☐
  - Aquaponics (web-3314) ☐
  - Vertical farming (web-3317) ☐

Fieldwork

- **6.8** Fair trading food (fdw-0032) ☐

Google Earth

- **6.12** Lake Victoria (gogl-0133) ☐

Teacher resources

There are many resources available exclusively for teachers online.

7 Geographical inquiry — Sustainable biomes

7.1 Overview

Numerous **videos** and **interactivities** are embedded just where you need them, at the point of learning, in your learnON title at www.jacplus.com.au. They will help you to learn the content and concepts covered in this topic.

LEARNING INTENTION

By completing this subtopic, you will be able to plan a geographical inquiry, conduct a geographical inquiry to investigate sustainable biomes, and process and communicate the findings of the geographical inquiry.

7.1.1 Scenario and task

Task: Create a website that informs people of the importance of one particular biome as a producer of food, and the current threats to food production.

Everyone in the world depends completely on the Earth's biomes for the services they provide — from our food and water supply to the regulation of our climate. Over the past fifty years, people have had a more rapid and more extensive impact on these biomes than during any other time in human history. Our demand for food, water, fibres, timber and fuel has driven these changes. The results have contributed to improvements in human wellbeing and economic development, but there has also been detrimental change to many of our major food-producing biomes.

Your task

Your team has been selected to create a website that not only grabs people's attention but also informs them of the importance of one particular biome as a producer of food, and the current threats to food production. Looking into the future, you will also suggest more sustainable ways of managing this biome.

7.2 Inquiry process

7.2.1 Process

Open the ProjectsPLUS application for this chapter located in your learnON Resources. Watch the introductory video lesson and then click the 'Start Project' button and set up your project group. You can complete this project individually or invite members of your class to form a group. Save your settings and the project will be launched.

- **Planning:** Navigate to your Research Forum. You will need to research the characteristics of a biome and address the four key inquiry questions supplied in the Research Forum. Each group should decide how to divide the workload so that each of the four inquiry questions is studied.

7.2.2 Collecting and recording data

Once you have chosen your biome and divided the key questions among the team, it is time to start researching information. For your own key question, break it down into several minor questions that can become subheadings to form the structure of your research. As a group, check each person's research structure to ensure that it follows the inquiry sequence.

When researching, look for maps, graphs and images that supports your key question or that of another team member. You should also look for data or statistics that you can show visually in the form of maps, diagrams or graphics.

7.2.3 Processing and analysing your information and data

Once you have researched and collected relevant information, you need to review it, ensure that you understand the material and then use it to answer your key questions. From maps and graphs, describe any patterns or trends that you identify. If using photographs, write clear annotations for each one, highlighting particular features.

Visit your Media Centre and download the website model and website planning template to help you build your website. Your Media Centre also includes images and audio files to help bring your site to life.

Use the website planning template to create design specifications for your site. You should have a home page and at least three link pages per topic. You might want to insert features such as 'Amazing facts' and 'Did you know?' into your interactive website. Remember the three-click rule in web design—you should be able to get anywhere in a website (including back to the homepage) with a maximum of three clicks.

7.2.4 Communicating your findings

Use website-building software to build your website. Remember that less is more with website design. Your mission is to engage and inform people about a topic they may never have thought about. You want people to take the time to read your entire website.

7.3 Review

7.3.1 Reflecting on your work

Think back over how well you worked with your partner or group on the various tasks for this inquiry. Reflect on your contribution to the team by completing the Reflection template in your Media Centre. Determine strengths and weaknesses and recommend changes you would make if you were to repeat the exercise. Identify one area where you were pleased with your performance and one area where you would like to improve. Write two sentences outlining how you might be able to do this.

Print out your Research Report from ProjectsPLUS and hand it in with your website and reflection notes.

on Resources

ProjectsPLUS Geographical inquiry — Sustainable biomes (pro-0148)

UNIT

2 Changing places

8 Causes and consequences of urbanisation

INQUIRY SEQUENCE

To access a pre-test with **immediate feedback** and **sample responses** to every question in this chapter, select your learnON format at www.jacplus.com.au.

8.1 Overview

Numerous **videos** and **interactivities** are embedded just where you need them, at the point of learning, in your learnON title at www.jacplus.com.au. They will help you to learn the content and concepts covered in this topic.

Why do millions of people choose to live so close to other people in busy urban areas?

8.1.1 Introduction

There are many advantages to living in large cities — for example, the economic benefit brought about by sharing the costs of providing fresh water, electricity or other energy sources and public transport between many people. There can be social benefits, because cities provide a wider choice of sporting, recreational and cultural events. However, there are also disadvantages of living in a large city environment. In this topic we will explore and compare urbanisation around the world.

STARTER QUESTIONS

1. Why do people choose to live in large cities?
2. What cultural, recreational or social activities are only found in larger cities?
3. What are some of the advantages and disadvantages of living in the world's largest cities, for example New York, Tokyo or São Paulo? Can a city be 'too big'?
4. Would you rather live in a large city or a small town? Explain why.
5. Watch the **Causes and consequences of urbanisation — Photo essay** video in your online Resources. Is this experience of daily life very different from yours?

FIGURE 1 Tokyo, Japan — the most populated city in the world

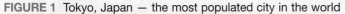

Resources

eWorkbook	Chapter 8 eWorkbook (ewbk-8100)
Video eLessons	Our urban world (eles-1628) Causes and consequences of urbanisation — Photo essay (eles-5215)

8.2 The development of urban environments

LEARNING INTENTION

By the end of this subtopic, you will be able to define urbanisation and specific types of urban environments, and outline global patterns of urbanisation over time.

8.2.1 Expansion of cities

The earliest cities emerged about 5000 years ago in Mesopotamia (part of present-day Iran, Iraq and Syria). Originally these cities depended on agriculture. As cities grew and trade developed, urban areas became centres for merchants, traders, government officials and craftspeople. However, in 1800, 98 per cent of the global population still lived in rural areas and most were dependent upon farming and livestock production — only 2 per cent of people lived in urban areas.

FIGURE 1 The region between the Euphrates and Tigris Rivers in Mesopotamia; it is here that the first urban centres developed.

Before 1800, over 90 per cent of the world's population lived in rural agriculture-based societies. With the **Industrial Revolution**, people began to move from rural areas to find employment in the factories of the rapidly expanding industrialised cities. In 1850, only two cities in the world — London and Paris — had a population above one million. By 1900 there were 12, by 1950 there were 83 and by 1990 there were 286.

By 2008, the proportion of people living in urban areas had increased to 50.1 per cent. In 2018, more than 500 cities had populations of a million or more people; over half the global population now lives in urban areas. In 2020 the figure had risen again to nearly 56.2 per cent, with projections of it reaching 68 per cent in 2050. The rate of growth has varied in different regions (see **FIGURE 2**).

Industrial Revolution the period from the mid-1700s into the 1800s that saw major technological changes in agriculture, manufacturing, mining and transportation, with far-reaching social and economic impacts

int-7857

FIGURE 2 The growth in urban population over time

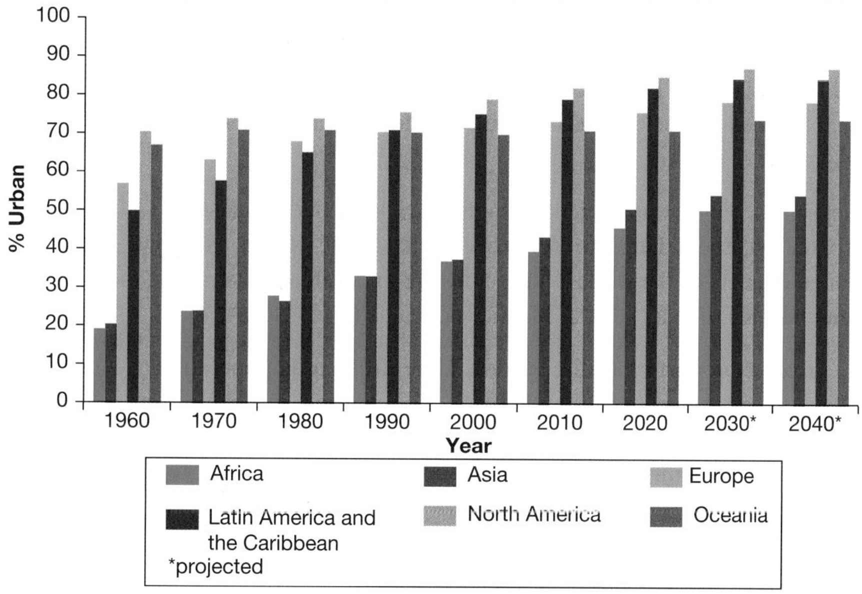

Source: © 2018 United Nations, DESA, Population Division. Licensed under Creative Commons license CC BY 3.0 IGO

As the world's population increases, urban areas continue to grow. In some regions, people are moving from rural to urban areas at very high rates. **Urbanisation** is the growth and expansion of urban areas, and involves the movement of people to towns and cities.

8.2.2 Megacities

int-7966

FIGURE 3 Distribution of the world's population, 2018

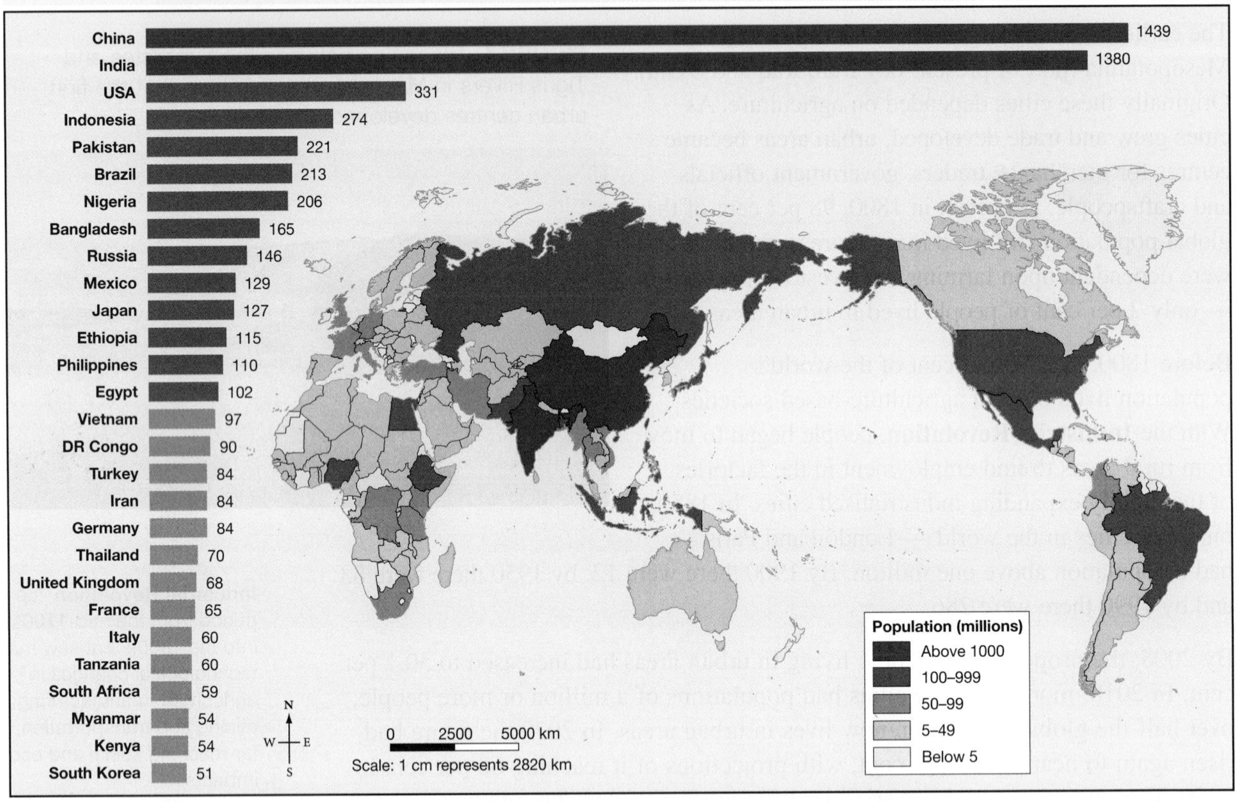

Source: United Nations, Department of Economic and Social Affairs, Population Division (2019). World Population Prospects: The 2019 Revision, Medium Variant.

The term **megacity** commonly refers to urban settlements of 10 million inhabitants or more. Currently, about 530 million people live in 33 megacities across the world. By 2030, the world is projected to have 43 megacities that will be home to more than 750 million people. Nineteen of these cities are projected to have a population greater than 15 million.

Of the 20 largest megacities, only six are located in highly developed industrialised countries: Tokyo–Yokohama, Seoul–Incheon, New York, Osaka–Kobe–Kyoto, Moscow and Los Angeles. Three-quarters of the world's megacities are in developing countries; they include gigantic **conurbations** such as Jakarta, Manila and Karachi (see **FIGURE 4**).

Asia has by far the greatest proportion of the world's large urban area population. Regions such as Oceania and Africa are less urbanised. For example, in Papua New Guinea (Oceania) and Burundi (East Africa) only 10 per cent of the population is urbanised, whereas in Singapore this figure is 100 per cent.

urbanisation the social and economic processes whereby an increasing proportion of the population of a country or region live in urban areas

megacity a settlement with 10 million or more inhabitants

conurbations an urban area formed when two or more towns or cities (e.g. Tokyo and Yokohama) spread into and merge with each other

FIGURE 4 Urban areas with more than 10 million population, 2019

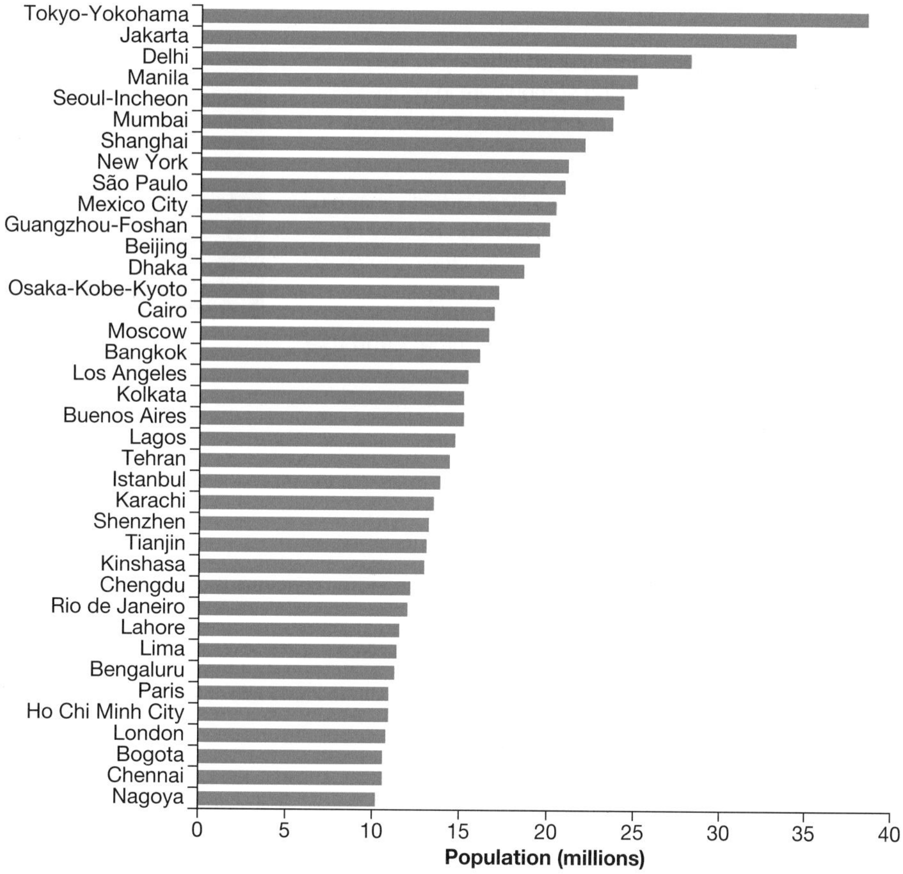

Source: Demographia World Urban Areas http://www.demographia.com/db-worldua.pdf

8.2.3 Impacts of urban growth on people

The United Nations predicts that by the year 2050, 70 per cent of the population in developed nations and 40 per cent in **developing nations** will live in large urban complexes. Rapid growth in city populations has led to problems such as **urban sprawl**, traffic congestion and air and water pollution, with significant impacts on the natural environment. Social problems related to urbanisation include unemployment; inadequate housing, **infrastructure**, water, sewerage and electricity supplies; pollution; and the spread of slums and crime.

In addition, with prospects of climate change through global warming, many of the world's coastal cities are under threat from rising sea levels. The application of **human–environment systems thinking** will be the key to evaluating and solving these economic and social issues.

Both small and large urban areas can provide people with positive and negative experiences. Cities attract people to them with the opportunity of work and the possibility of better housing, education and health services. There is a strong interconnection between the wealth of a country and how urbanised it is. Generally, countries with a high per capita income tend to be more urbanised, whereas low-income countries are the least urbanised.

This happens because people grouped together create many chances to move out of poverty, generally because of increased work opportunities. There are often better support networks from governments and local councils. It is also cheaper to provide facilities such as housing, roads, public transport, hospitals and schools to a population concentrated into a smaller area.

developing nation a country whose economy is not well developed or diversified, although it may be showing growth in key areas such as agriculture, industries, tourism or telecommunications

urban sprawl the spreading of urban developments into areas on the city boundary

infrastructure the facilities, services and installations needed for a society to function, such as transportation and communications systems, water and power lines

human–environment systems thinking using thinking skills such as analysis and evaluation to understand the interaction of the human and biophysical or natural parts of the Earth's environment

fdw-0042

FOCUS ON FIELDWORK

Land use in your area

How much do you know about the way land is used where you live? Have you ever had the experience of travelling somewhere in your neighbourhood and seeing a place that you never noticed before, or a building that has a mysterious purpose?

We can learn a lot about a place from either physically or virtually mapping out the kinds of land use. For example, if an area has a high density of apartment blocks and very few houses, but also has very little open green space, what might this suggest about the liveability of the area?

Learn how to measure and discuss urban land use using the **Land use in your area** activity in your online Resources.

on Resources

eWorkbook	The development of urban environments (ewbk-9851)
Video eLesson	The development of urban environments — Key concepts (eles-5216)
Interactivities	The growth in urban population over time (int-7857) Distribution of the world's population (int-7966)
myWorldAtlas	Deepen your understanding of this topic with related case studies and questions. • World urbanisation

8.2 ACTIVITY

Use **FIGURE 3** to create a graph of the top ten most populous countries. Research what their projected populations will be in 2050 and represent this information on the same graph. Give reasons for the changes that are shown in your graph.

8.2 EXERCISE

Learning pathways

■ LEVEL 1	■ LEVEL 2	■ LEVEL 3
1, 2, 7, 11, 14	4, 5, 8, 10, 13	3, 6, 9, 12, 15

Check your understanding

1. When and where did the first cities develop?
2. Why did cities experience rapid growth and development after the Industrial Revolution?
3. What factors are driving the process of urbanisation in the world?
4. In which regions of the world is urbanisation occurring most quickly? Why?
5. Define urbanisation in your own words.
6. Look at **FIGURE 2**. Which region's urbanisation rate has consistently been the highest over time?
7. Refer to **FIGURE 3**. What are the world's two most populous nations?

Apply your understanding

8. What are some of the major economic and social issues facing rapidly developing cities in the world?
9. How has urbanisation changed from 1960 to the present? How is this different around the world?
10. Explain how **FIGURE 2** shows that urbanisation has varied in different regions of the world. Which two regions have the greatest rural population?
11. Consider **FIGURE 2**. Which two continents have the lowest urbanisation rates?
12. What is expected to happen with urbanisation in the future?

Challenge your understanding

13. What impact do you think global warming and rising sea levels will have on coastal cities around the world? What are some other urban problems, besides those mentioned in this subtopic, that arise as cities develop?
14. What do you think are some of the advantages of living: (a) in a large city (b) in a small town (c) on a farm? Which would you prefer? Why?
15. Will Australian cities ever reach megacity size? Predict whether or when this might happen, and which Australian city is the most likely to reach 10 million in population first. Justify your point of view.

To answer questions online and to receive **immediate feedback** and **sample responses** for every question go to your learnON title at www.jacplus.com.au.

8.3 Distribution of urban areas

LEARNING INTENTION

By the end of this subtopic, you will be able to identify and describe patterns in where cities and urban areas are located around the world.

8.3.1 Distribution of the world's cities

How is a city different from other urban areas such as towns and villages? A city is a large and permanent settlement, and is usually quite complex in terms of transport, land use and **utilities** such as water, power and **sanitation**.

The image of the Earth at night (**FIGURE 1**) shows where lights are shining. The brightest areas on the map are the most urbanised but might not be the most populated. If you compare this image with **FIGURE 2**, you can make some comparisons. For example, there are very bright lights in western Europe (Belgium, The Netherlands, France, Spain and Portugal, Germany, Switzerland, Italy and Austria) and yet more people living in China and India. Refer to your atlas to locate these countries.

utilities services provided to a population, such as water, natural gas, electricity and communication facilities

sanitation services provided to remove waste such as sewage and rubbish

FIGURE 1 Satellite image of the Earth at night

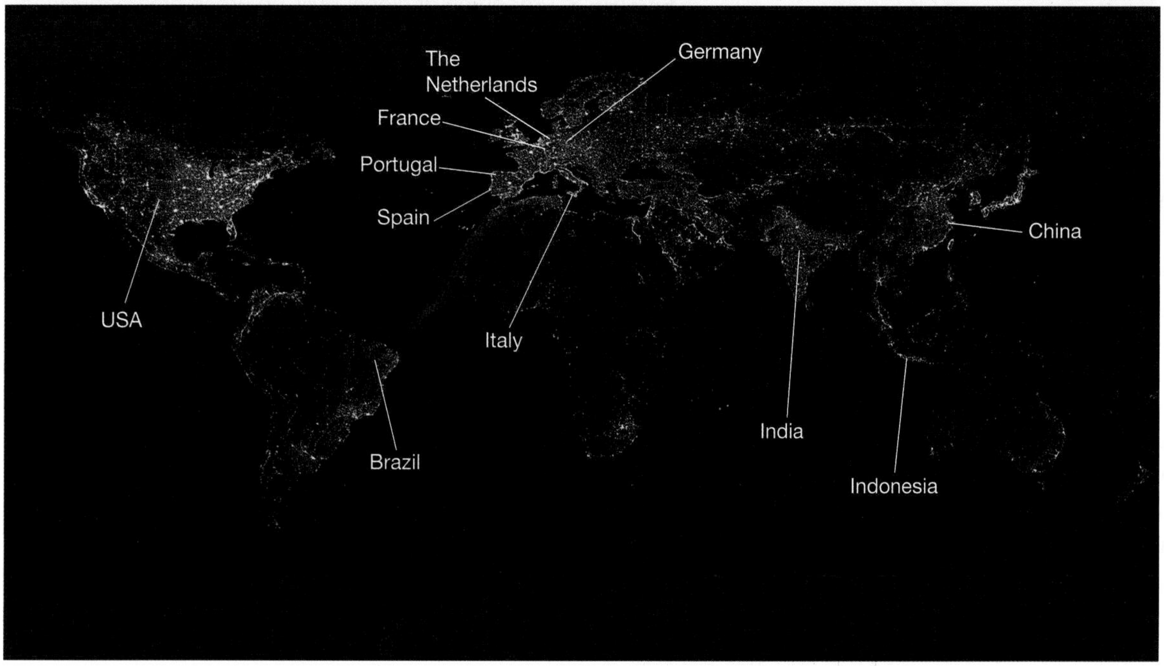

Some regions remain thinly populated and unlit. Antarctica is entirely dark. The interior forests of Africa and South America are mostly dark, but lights are beginning to appear there. Deserts in Africa, Arabia, Australia, Mongolia and the United States are poorly lit as well, although there are some lights along coastlines. Other dark areas include the forests of Canada and Russia, and the great mountains of the Himalayan region and Mongolia.

int-7967

FIGURE 2 Percentage of urban population and urban agglomerations, 2018

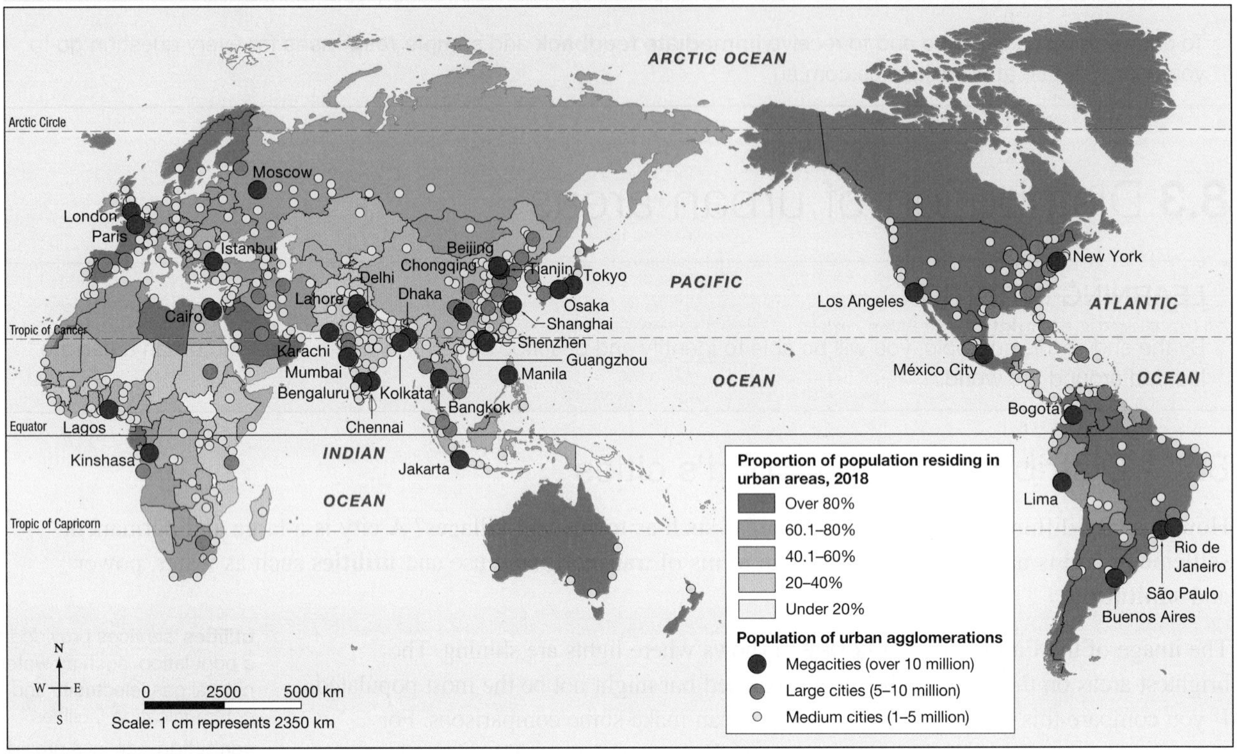

Source: United Nations Department of Economic and Social Affairs; Worldatlas.com

8.3.2 Uneven urbanisation

As **FIGURES 1** and **2** show, urbanisation has not occurred evenly across the world. Some countries are predominantly rural, such as Cambodia and Papua New Guinea (populations 76 per cent and 86 per cent rural respectively), whereas others are almost completely urban, such as Belgium and Kuwait (98 per cent urban for both). In fact, some countries have 100 per cent urbanisation, including Bermuda, Cayman Islands, Hong Kong, Macau, Monaco, Vatican City and Singapore. South America is becoming one of the most urbanised regions in the world and currently has a population of around 422 million people. It is estimated that by 2050, 91.4 per cent of its population will be residing in urban areas.

FIGURE 3 Proportion of global built-up urban area population, by region, 2019

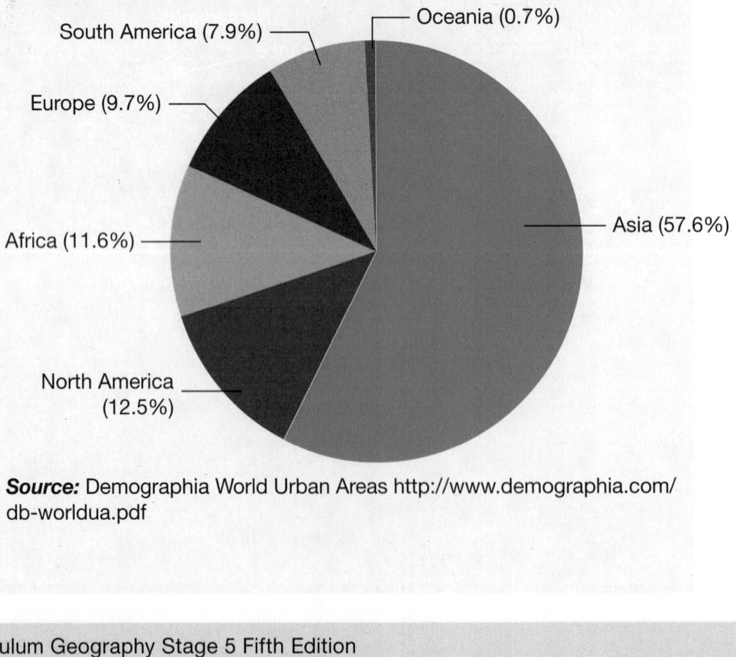

Source: Demographia World Urban Areas http://www.demographia.com/db-worldua.pdf

The United Nations estimates that a staggering 90 per cent of the worlds' population growth is taking place in the cities of developing nations. For many people in these countries, **push factors** such as extreme poverty, famine and civil unrest often 'push' them away from rural areas and 'pull' them towards cities. **Pull factors** include the promise of jobs, shelter and protection.

Coastal urbanisation

Coastlines have been important to many human settlements throughout history. Often at the mouth of rivers, coastal settlements became centres of trade and commerce and quickly grew into cities. Today, about half the world's population lives along or within 200 kilometres of a coastline (see **FIGURE 5**). According to the European Commission, 95 per cent of the world's population lives on only 10 per cent of the Earth's land area.

Countries that have over 80 per cent of their population living within 100 kilometres of a coastline include the United Kingdom, Senegal, Portugal, Belgium, the Netherlands, Sweden, Norway, Tunisia, Greece, Oman, the United Arab Emirates, Kuwait, Qatar, Sri Lanka, Japan, Singapore, Indonesia, Malaysia, the Philippines, Australia and New Zealand.

FIGURE 4 Urban housing in Kuwait

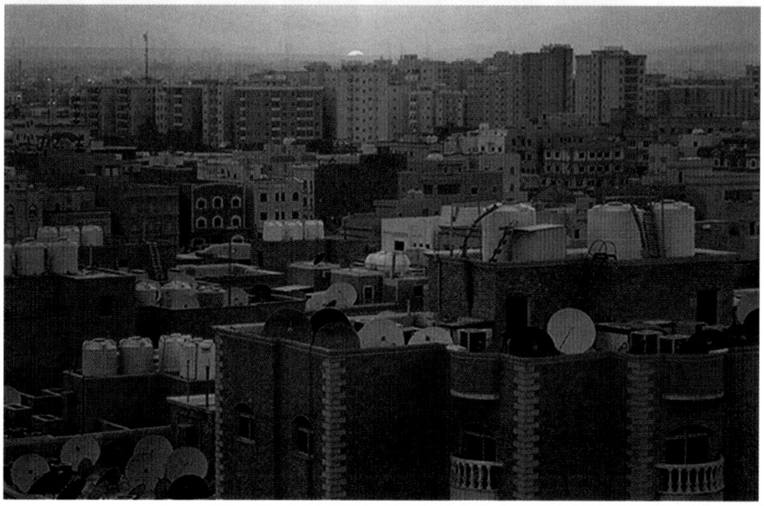

FIGURE 5 Cape Town in South Africa

8.3.3 Spatial distribution of megacities

Spatial distribution is the way that features or events are spread over the surface of the Earth — how they are distributed across a space. The spatial distribution of megacities can be seen in **FIGURE 2**. Each red dot represents a city that had a population of over 10 million in 2018.

The spatial distribution of megacities has also changed. In 1975, there were three megacities: New York, Tokyo and Mexico City. In 2014, more than half (15) of all megacities were located in Asia; it is predicted that, in 2030, 23 of the 41 megacities will be located in Asia. There is also a change in terms of the wealth of countries that contain megacities, with the majority now located in developing countries.

push factor unfavourable quality or attribute of a person's current location that drives them to move elsewhere

pull factor favourable quality or attribute that attracts people to a particular location

FIGURE 6 Regional population distribution in different city sizes in 2008 and projected to 2030

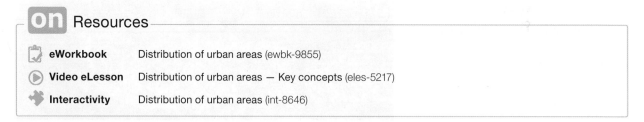

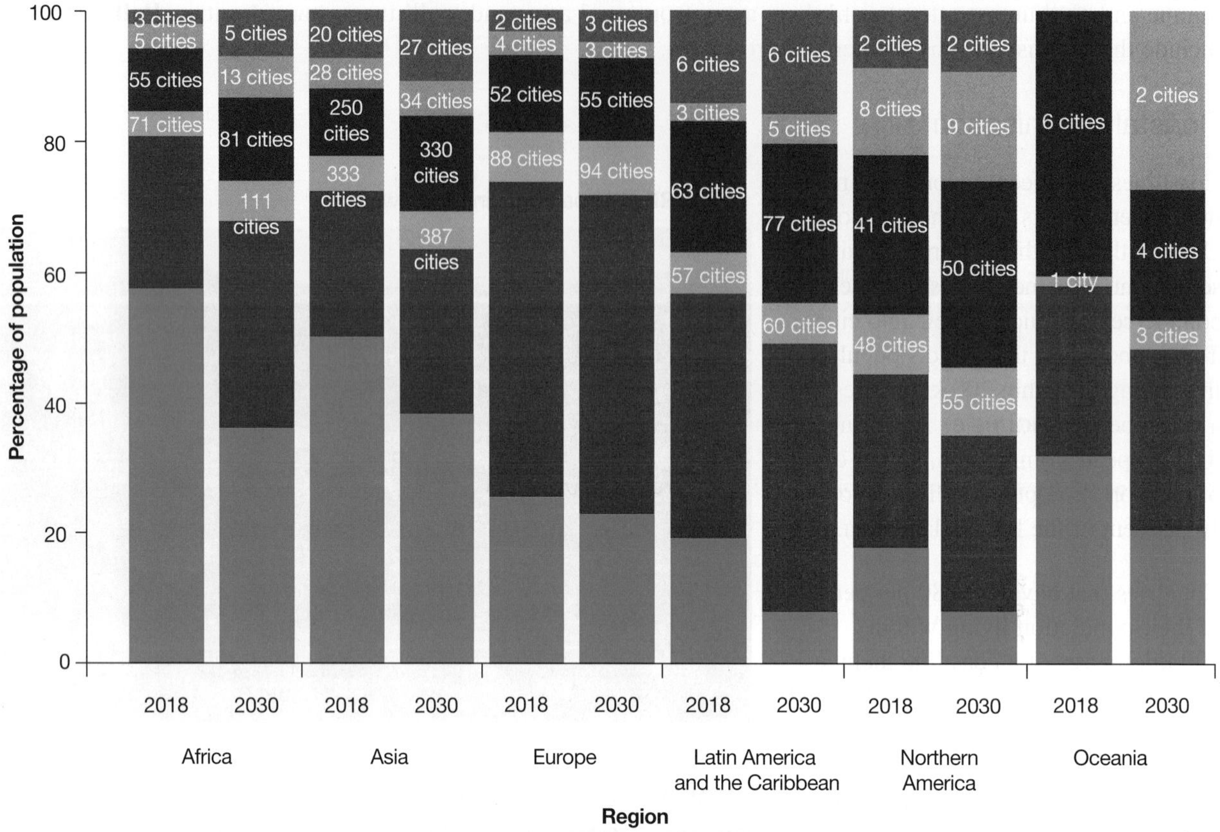

***Source*:** United Nations

This is in contrast to the development of urbanisation, when North America and Europe were the focus of historic urban growth. By 2030, it is predicted that 23 megacities will exist in less developed countries.

on Resources

eWorkbook	Distribution of urban areas (ewbk-9855)
Video eLesson	Distribution of urban areas — Key concepts (eles-5217)
Interactivity	Distribution of urban areas (int-8646)

8.3 ACTIVITIES

1. Refer to a world population density map in your atlas or online. Compare this map with the two regions that have the highest rural population. What pattern do you see?
2. Look at a physical map in an atlas to locate the countries with more than 80 per cent of their population located on the coast. Study the location of each country and create a table to record possible reasons for this pattern.

8.3 EXERCISE

Learning pathways

■ LEVEL 1	■ LEVEL 2	■ LEVEL 3
1, 3, 8, 9, 13	2, 5, 7, 11, 14	4, 6, 10, 12, 15

Check your understanding

1. Describe what makes a city different from a town.
2. Identify two countries that have 100 per cent urbanised population.
3. Consider why people move from rural to urban areas.
   a. Outline two push factors that cause people to move from rural areas
   b. Outline two pull factors that cause people to move to urban areas
4. Describe how the distribution of megacities has changed since 1975.
5. Consider the way that the distribution of population is expected to change between 2018 and 2030 in **FIGURE 6**.
   a. Rank the six regions from highest to lowest rural population in 2018.
   b. Rank the six regions from highest to lowest rural population in 2030.
   c. Rank the six regions from highest to lowest number of megacities in 2018.
   d. Rank the six regions from highest to lowest number of megacities in 2030.
   e. Identify the region with the largest increase in number of cities over 1 million from 2018 to 2030.
   f. Identify the region with the smallest increase in number of cities over 1 million from 2018 to 2030.
6. Describe the pattern of where megacities are growing most quickly around the world. In your answer, refer to data and use the words developed and developing.

Apply your understanding

7. Study **FIGURE 2**, which shows the population in urban areas. Identify and name the three countries with the highest and the three with the lowest percentage of people living in urban areas. Write a description of the general pattern shown in the map. Include patterns within different continents in your description.
8. Study **FIGURE 3**. What proportion of the Australian population lives in urban areas? Explain why this is so, despite the fact that Australia has no megacities.
9. Consider the Cape Town area shown in **FIGURE 5**.
   a. Draw a sketch of the photograph of Cape Town.
   b. Annotate the sketch, identifying possible advantages and disadvantages to the natural environment when cities and towns are located on the coast.
10. Explain how **FIGURE 6** shows that urbanisation has varied in different regions of the world.
11. Explain two social consequences of people moving from rural to urban areas: one consequence for the place people leave and one for the place they move to.
12. Explain two economic consequences of people moving from rural to urban areas: one consequence for the place people leave and one for the place they move to.

Challenge your understanding

13. Rural areas are where most food is produced. What are two possible outcomes for food production if urbanisation continues?
14. What impact do you think global warming and rising sea levels will have on coastal cities around the world?
15. Africa is the second most populated continent after Asia. Suggest why Africa has comparatively fewer megacities in contrast with continents with smaller populations (for example, South America has about a third of the population of Africa).

To answer questions online and to receive **immediate feedback** and **sample responses** for every question go to your learnON title at www.jacplus.com.au.

8.4 SkillBuilder — Comparing population profiles

online only

LEARNING INTENTION

By the end of this subtopic, you will be able to analyse and identify similarities and differences in the data shown in a population profile.

The content in this subtopic is a summary of what you will find in the online resource.

8.4.1 Tell me

What is a population profile?

A population profile is a bar graph that provides information about the age and gender of a population. The bars identify the proportion of a country's population within a particular age group. The graph is split to show information about males and females. The shape of the population profile tells us about a particular population. Traditionally, population profiles were called population pyramids because they were shaped like a pyramid or triangle — wide at the base and narrow at the top.

FIGURE 1 Population profile of Niger, 2009

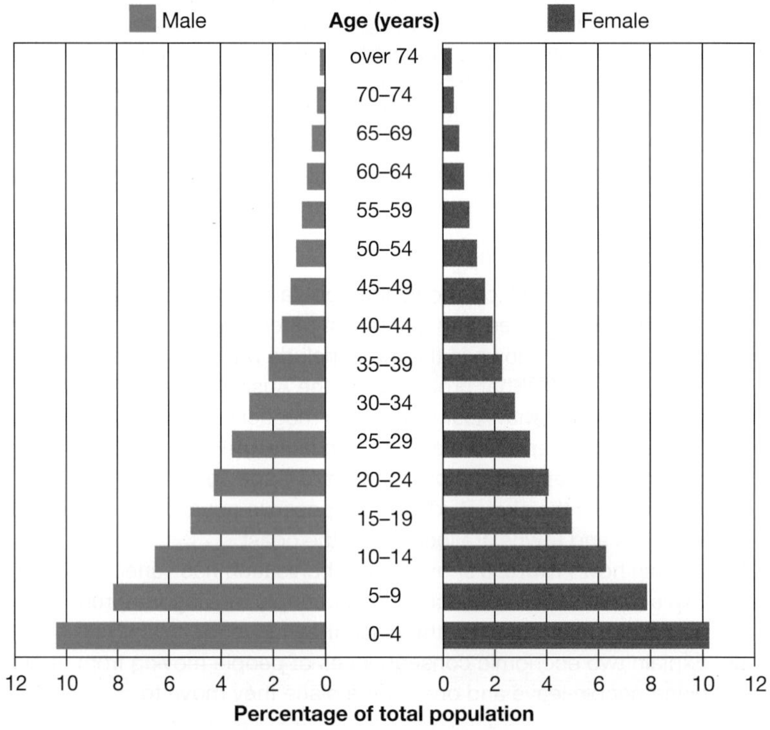

Source: © Central Intelligence Agency (CIA) World Factbook

8.4.2 Show me

How to compare population profiles

Step 1

To compare populations, you must have two or more profiles for the same place at different times, or for different places at the same time.

Step 2

Calculate the percentages of males and females in each of these three categories:
- children (0–14 years)
- adults (15–64 years)
- aged (65 years and over).

A population is considered to be old when less than 30 per cent of the population is younger than 15 years and more than 6 per cent is aged 65 years and over. A population is considered to be young when more than 30 per cent of the population is younger than 15 years and less than 6 per cent is aged 65 years and over.

Step 3

Look for patterns revealed by each population profile. Is it a pyramid shape? Are the profile shapes similar? If not, at what age groupings do the variations appear? Write a few statements to summarise your findings.

Step 4

Consider any unusual aspects. Are there any indents (places where the graph narrows unexpectedly) or extended age groupings? Can you suggest why these may occur?

FIGURE 2 Population profile of Germany, 2009

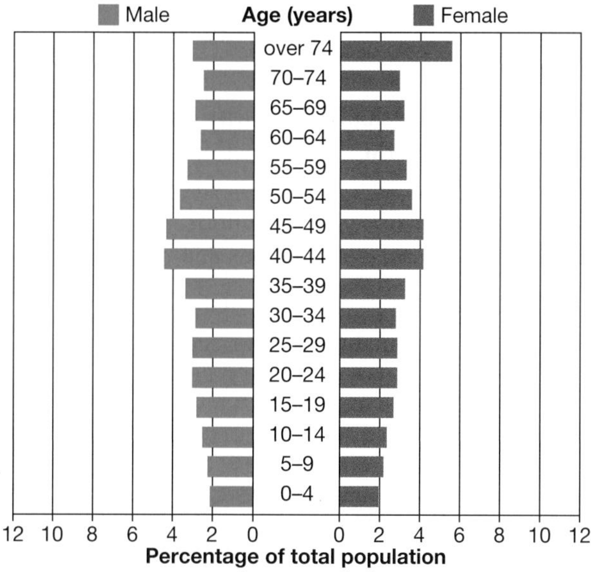

Source: © Central Intelligence Agency (CIA) World Factbook

FIGURE 3 Projected population profile of Germany, 2050

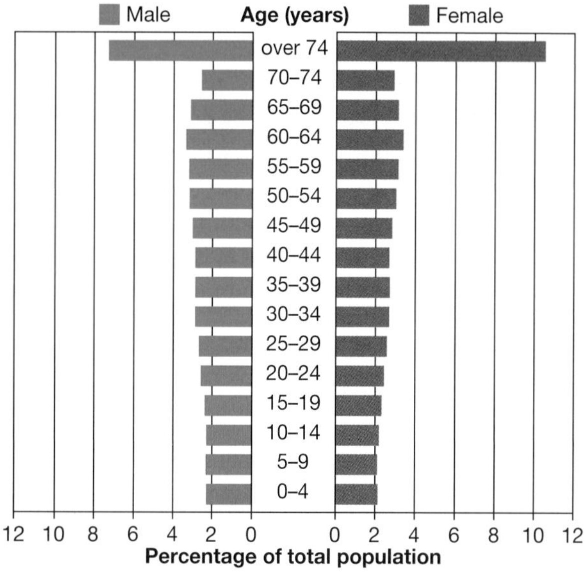

Source: © Central Intelligence Agency (CIA) World Factbook

8.4.3 Let me do it

Go to learnON to access the following additional resources to help you build this skill:
- a longer explanation of this skill and its application in Geography (Tell me)
- a video demonstrating the step-by-step process of this skill (Show me)
- an activity and interactivity for you to practise the skill (Let me do it)
- self-marking questions to help you understand and use the skill.

on Resources

eWorkbook	SkillBuilder — Comparing population profiles (ewbk-9859)
Video eLesson	SkillBuilder — Comparing population profiles (eles-1704)
Interactivity	SkillBuilder — Comparing population profiles (int-3284)

8.5 Urban expansion

LEARNING INTENTION

By the end of this subtopic, you will be able to identify and describe how urban places are growing, and explain the consequences of different growth patterns for people.

8.5.1 Growth rate of urban places

In 2008, for the first time in history, the world's urban population outnumbered its rural population. In 2019, the world's population exceeded 7.7 billion; it is expected to reach 9.2 billion by 2050. Where will all these people live? What challenges will cities and communities face in trying to ensure a decent standard of living for all of us?

Global population growth will be concentrated mainly in urban areas of developing countries. It is forecast that by 2030, 3.9 billion people will be living in cities of the developing world. The impact of expanding urban populations will vary from country to country and could prove a great challenge if a country is not able to produce or import sufficient food. Hunger and starvation may increase the risk of social unrest and conflict. On the other hand, farmers can help satisfy the food needs of expanding urban populations and provide an economic livelihood for people in the surrounding region.

One of the biggest challenges we face in Australia is ensuring that the sustainability of our economy, communities and environment is compatible with our growing urban population (see **TABLE 1**).

TABLE 1 Percentage of population residing in urban areas by country, 1950–2050 (projected)

	1950	1975	2000	2025*	2050*
Australia	77.0	85.9	87.2	90.9	92.9
Brazil	36.2	60.8	81.2	87.7	90.7
Cambodia	10.2	4.4	18.6	23.8	37.6
China	11.8	17.4	35.9	65.4	77.3
France	55.2	72.9	76.9	90.7	93.3
India	17.0	21.3	27.7	37.2	51.7
Indonesia	12.4	19.3	42.0	60.3	72.1
Japan	53.4	75.7	78.6	96.3	97.6
Papua New Guinea	1.7	11.9	13.2	15.1	26.3
United Kingdom	79.0	77.7	78.7	81.8	85.9

*projected
Source: UN Population Division, 2018

8.5.2 Urban sprawl

Across the world, cities are expanding at a rapid rate, putting pressure on the built environment and the infrastructure that supports it, as well as the natural environments that are increasingly diminished. With metropolitan populations growing, cities around the world need to accommodate more people while reducing their urban sprawl or expansion outwards into areas surrounding the city.

The Perth Metropolitan Area, or Greater Perth, was the fastest growing capital city in Australia between 2011 and 2015, with an 11.2 per cent growth rate; Melbourne and Sydney recorded much lower growth rates for the same period of 8.6 per cent and 6.8 per cent, respectively. However, while Melbourne and Sydney saw the majority of the growth in the urban centre, Perth experienced most growth in the outer suburbs, resulting in ever increasing urban sprawl (see **FIGURE 1**).

FIGURE 1 Growth rates of (a) Melbourne, (b) Sydney and (c) Perth, 2011–15

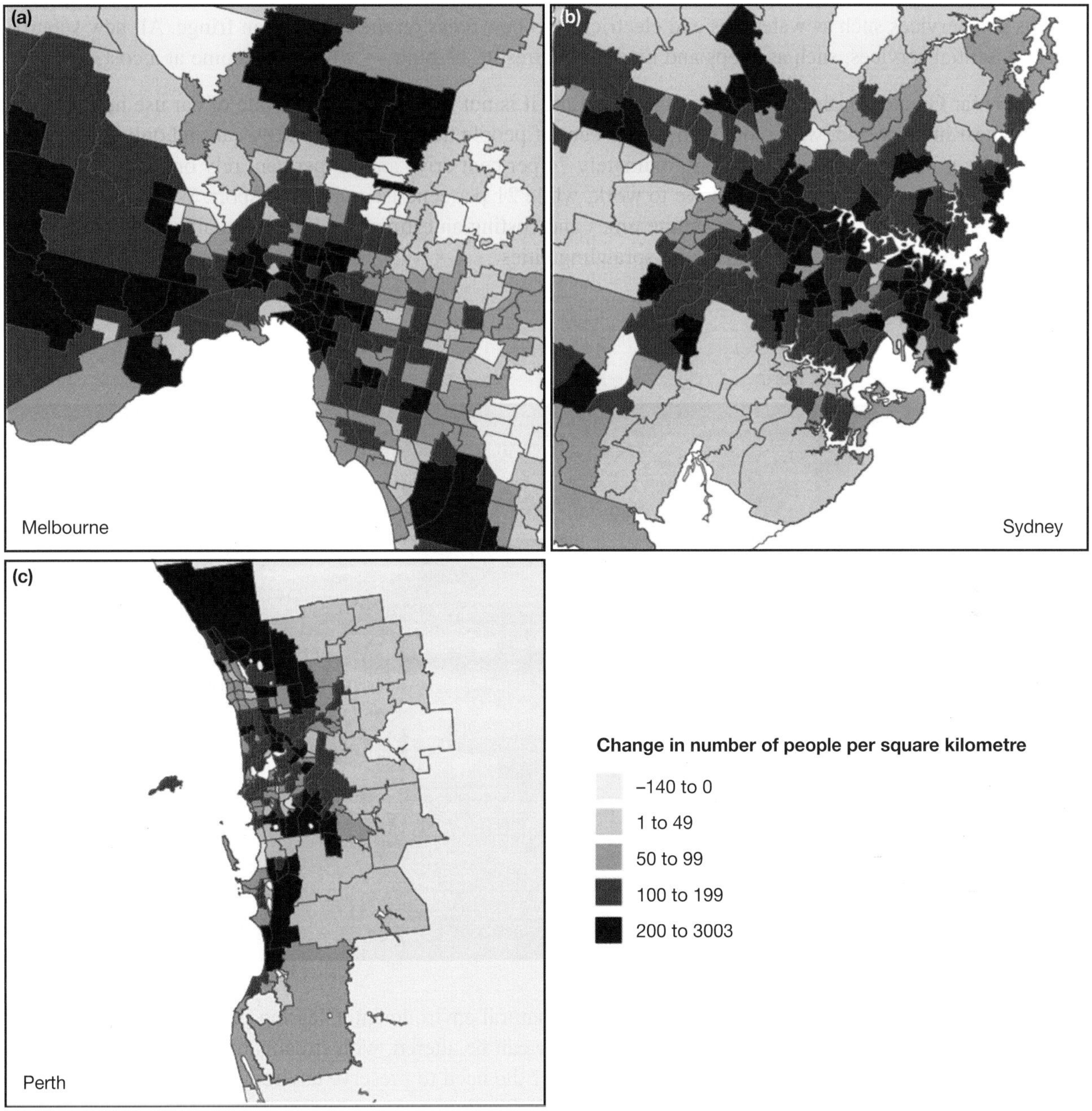

Source: © Commonwealth of Australia. Australia State of the Environment 2016. Retrieved from https://soe.environment.gov.au/theme/built-environment/topic/2016/increased-urban-footprint [online resource].

Housing choice can also be attributed to Perth's urban sprawl, with 77 per cent of dwellings being a separate house, compared to 56.9 per cent in Sydney and 67.8 per cent in Melbourne.

8.5.3 Impacts of urban sprawl

The **rural–urban fringe** is typically the urban zone that is undergoing the most rapid change in expanding cities. Former farmland, market gardens and orchards are often sold off, and new housing and industrial estates are built. These are usually low-density planned estates sometimes built around a theme or geographical feature such as a built lake or wetland. Urban expansion into the rural–urban fringe has environmental, economic and social impacts.

rural–urban fringe the transition zone where rural (country) and urban (city) areas meet

Pressure on infrastructure

A major problem of urban sprawl is the cost to provide infrastructure (for example, roads and other transport systems and services such as water, gas and electricity) to new areas on the rural–urban fringe. All new suburbs require essential services, such as shops and medical centres, to be built — which also come at a cost.

Given how far Greater Perth spreads along the coastline, it is not surprising that the rate of car use here is one of the highest in the world, with more than 80 per cent of people travelling to work by car and only 8 per cent taking public transport. In Melbourne, approximately 74 per cent drive and 13 per cent rely on public transport. In Sydney, approximately 65 per cent drive to work, while 21 per cent use public transport. With increasing traffic congestion creating increased commute times and costing an estimated $2.1 billion annually, road infrastructure is under significant pressure in sprawling cities.

Loss of biodiversity

FIGURE 2 Forrestdale Lake, Perth

The expansion of urban areas can significantly alter the natural environment. Clearing of vegetation can reduce habitat and biodiversity. Natural drainage and topography can be altered, with streams redirected or even converted to pipes. Today there is a growing awareness of the need to preserve environments as they provide important functions for wildlife and people. Perth is located in the south-west of Australia and the wetlands in this region support an extensive range of plants and animals (see **FIGURE 2**). However, significant urban development in Perth has seen the wetlands continue to disappear to make way for houses and industrial areas, and only 10 per cent of these wetlands survive. Many endemic species (species native to that area) will be at risk if the wetlands continue to be altered.

8.5.4 Managing urban sprawl

With Greater Perth already stretching 150 kilometres from the north to the south, more needs to be done to curb urban sprawl while also providing housing for the 3.5 million people who are predicted to live there by 2050. The 'Perth and Peel @ 3.5 million' plan has been developed by the Western Australia Government to sustainably manage Perth's urban growth using **urban infilling** and by increasing the population density. The project aims to increase density and infilling around public transport hubs, which will allow people to be better connected to the CBD while also taking pressure off the road infrastructure.

urban infilling the division of larger house sites into multiple sites for new homes

Increasing density in established suburbs

To increase the density of housing in older suburbs, one idea is to establish activity centres. These consist of higher-density housing in specific locations, where people shop, work, meet, relax and live in the local environment. They are centred around existing infrastructure, transport networks, population shopping centres, employment opportunities and community facilities. New housing tends to be medium-rise apartments (three to five storeys) built along main transport routes.

Urban infilling

Urban infilling is the process of using the existing space in the CBD, either green space or manufacturing spaces that are no longer used, to create high-density living. Infill development within urban areas can contribute significantly to housing diversity and choice, providing opportunities for affordable living and a connected community.

8.5.5 The growth of megacities

CASE STUDY

The expansion of Mumbai

Located in the western Indian state of Maharashtra, Mumbai is the most populous city in India and the ninth most populous city in the world, with a population of over 20.4 million in 2020. In 2009, Mumbai was named an **alpha world city**. It is the richest city in India, with the highest **gross domestic product** (GDP) of any city in south, west or central Asia; however, many of its residents live in **slum** conditions. The large numbers of people and rapid population growth have contributed to serious social, economic and environmental problems for Mumbai. Mumbai's business opportunities and its potential to offer a higher standard of living attract migrants from all over India seeking employment and a better way of life. In turn, this has made the city a melting pot of many communities and cultures. In 2019, the population density was estimated to be around 20 000 persons per square kilometre and the living space just 8 square metres per person.

alpha world city a city generally considered to be important in the global economic system

gross domestic product (GDP) a measurement of the annual value of all the goods and services bought and sold within a country's borders

slum an area of a city characterised by poor housing and poverty

Despite government attempts to discourage the influx of people, the city's population grew by more than 4 per cent between 2001 and 2011, and by a staggering 61.8 per cent between 2011 and 2019 (see **TABLE 2**). The number of migrants to Mumbai from outside Maharashtra during the ten-year period from 2001 to 2011 was over one million, which amounted to 54.8 per cent of the net addition to the population of Mumbai.

TABLE 2 Population growth in Mumbai

Census	Population	% change
1971	5970575	–
1981	8243405	38.1
1991	9925891	20.4
2001	11914398	20.0
2011	12478447	4.7
2019*	20185064	61.8

Source: *2019 data: Mumbai Metropolitan Region Development Authority. All other data is based on Government of India Census, conducted every ten years.

Many newcomers end up in poverty, often living in slums or sleeping in the streets when they cannot find work. By 2017, an estimated 62 per cent of the city's inhabitants lived in slum conditions. Some areas of Mumbai city have population densities of around 46 000 per square kilometre — among the highest in the world.

Challenges

Mumbai suffers from the same major urbanisation problems that are seen in many fast-growing cities in developing countries: widespread poverty and unemployment, urban sprawl, traffic congestion, inadequate sanitation, poor public health, poor civic and educational standards, and pollution. These pose serious threats to the quality of life in the city for a large section of the population. Vehicle exhausts and industrial emissions, for example, contribute to serious air pollution, which is reflected in a high incidence of chronic respiratory problems.

With available land at a premium, Mumbai residents often reside in cramped, relatively expensive housing, usually far from workplaces, and therefore commute long distances on crowded public transport or clogged roadways (see **FIGURE 3**). Although many people live in close proximity to bus or train stations, suburban residents spend a significant amount of time travelling southwards to the main commercial district.

FIGURE 3 Chhatrapati Shivaji Terminus, Mumbai

The Dharavi slum

Dharavi, Asia's second-largest slum, is located in central Mumbai. Stretching across 220 hectares of land, it is home to more than one million people (see **FIGURE 4**). In Dharavi, it is estimated that there is only one toilet for every 200 people.

This results in floods of human excrement during the monsoon season. Much of the water becomes contaminated because of this, and death rates tend to be significantly higher in Mumbai's slums than in upper- and middle-class areas.

FIGURE 4 Aerial view of Dharavi, and the suburbs of Mumbai

Dharavi's recycling entrepreneurs

Hidden amid Dharavi's labyrinth of ramshackle huts and squalid open sewers are an estimated 20 000 single-room factories, employing around a quarter of a million people and turning over a staggering US$1 billion each year through recycling and other trades such as the production of pottery, textiles and leather goods.

In developed countries, communities recycle because there is the understanding that it contributes to sustaining the planet's resources. However, for some of the poorest people in the developing world, recycling often isn't a choice, but rather a necessity of life.

In India, it is estimated that anywhere between 1.5 million and 4 million people make their living by recycling waste. At least 300 000 of these live and work in Mumbai. These people are known as 'ragpickers' and are made up of India's poorest and most marginalised groups (see **FIGURE 5**). The ragpickers wade through piles of unwanted goods to salvage easily recyclable materials such as glass, metal and plastic, which are then sold to scrap dealers who process the waste and sell it on either to be recycled or to be used directly by the industry.

FIGURE 5 Salvaging recyclables, Mumbai, India

Due to the lack of formal systems of waste collection, it falls to Mumbai's ragpickers to provide this basic service for fellow citizens. Without them, solid waste and domestic garbage would not even be collected, let alone sorted or recycled. Despite many of the social and ethical controversies surrounding the recycling industry in India, Dharavi is seen as the 'ecological heart of Mumbai', recycling up to 85 per cent of all waste material produced by the city, an excellent example of human–environment systems thinking in action.

The future of Dharavi

There are plans to demolish and redevelop Dharavi. This redevelopment would transform the slum into a series of high-rise housing facilities, and each of Dharavi's 57 000 registered families would get 21 square metres of living space.

However, some Dharavi residents do not support this plan or are nervous about having to move from a place that is very familiar to them. Many have lived their entire lives in Dharavi and do not want to trade their culture for a redeveloped lifestyle. Others worry that they will be required to pay large sums of money to move or will have to live in a new place away from their community.

on Resources

eWorkbook	Urban expansion (ewbk-9863)
Video eLesson	Urban expansion — Key concepts (eles-5218)
Interactivity	Urban expansion (int-8647)

8.5 ACTIVITIES

1. Research Dharavi. Create a one-page infographic detailing life in the slum. Consider questions such as the following:
   a. How many people live there and what are living conditions like?
   b. What work is done here?
   c. What are the risks to health and what could be done to improve the situation?
2. Research the natural and human influences on the development of cities. Examples for research include Canberra, Australia (a planned city); Cape Town, South Africa (a port city); Rothenburg, Germany (a walled city); Geneva, Switzerland (where a river meets a lake); Johannesburg, South Africa (near a mining site); Chicago, United States (where north–south and east–west railway routes cross); Jerusalem, Israel (an ancient religious city); and Bath, England (located at the site of a natural supply of mineral waters).

8.5 EXERCISE

Learning pathways

■ LEVEL 1	■ LEVEL 2	■ LEVEL 3
1, 2, 7, 8, 13	3, 4, 9, 10, 14	5, 6, 11, 12, 15

Check your understanding

1. Refer to **TABLE 1**.
   a. In 2050, which countries will be the most and the least urbanised?
   b. Which three countries are predicted to experience the greatest percentage change in their urban population?
   c. Are there any countries that have not seen a gradual increase in their percentage of urban population since 1950?
2. What is urban infilling?
3. Study **FIGURE 1**. Describe the areas of Sydney, Melbourne and Perth that experience the most growth.
4. Outline one argument for and one argument against demolishing Dharavi.
5. List the advantages and disadvantages of replacing slums with high-rise low-income housing.
6. Outline three challenges that Mumbai faces as the city continues to expand.

Apply your understanding

7. Explain the two main ways that additional housing can be established in an expanding city.
8. Using the data in **TABLE 1**, create a bar graph that shows the change over time for four countries of your choice.
9. Explain why urban expansion patterns occur differently across Sydney, Melbourne and Perth.
10. Explain how urban sprawl has affected the environment in Perth.
11. What would be the social, environmental and economic benefits of ragpickers?
12. How are urban sprawl and food production interconnected?

Challenge your understanding

13. Predict some challenges that can come from developing land next to a river.
14. Suggest how urban planners can reduce some of the environmental, social and economic impacts of expansion into the rural–urban fringe.
15. The populations of Perth, Melbourne and Sydney are expected to continue growing. By 2050, the population of Perth is projected to be approaching 4 million; Melbourne almost 8 million; and Sydney about 7.5 million. Predict how (or if) the use of public transport in these places will change if the cities reach their projected populations.

To answer questions online and to receive **immediate feedback** and **sample responses** for every question go to your learnON title at www.jacplus.com.au.

8.6 Urban decline

LEARNING INTENTION

By the end of this subtopic, you will be able to explain how and why urbanised areas decrease in size or are abandoned.

8.6.1 Environmental causes

Over time, all forms of urban environments deteriorate with age and require renovation or renewal. Extreme atmospheric events such as cyclones, hurricanes and tornadoes, which exhibit strong winds and flooding rains, can have devastating short-term impacts on urban environments. Longer-term events such as **desertification** and climate change can also have negative impacts.

Movements of the Earth such as those due to earthquakes and volcanic eruptions can also destroy urban environments. One well-documented example is the eruption of Mt Vesuvius in Italy in 79 CE, which completely buried the cities of Pompeii and Herculaneum under volcanic ash (see **FIGURE 1**).

FIGURE 1 Ruins of the city of Pompeii, near Naples, Italy, with Mt Vesuvius in the background

Tsunamis are also a significant hazard that can lead to the destruction of settlements. On 22 December 2018, for example, the coastline regions of Banten and Lampung in Indonesia were devastated by a 3-metre tsunami triggered by an underwater landslide following the volcanic eruption of Anak Krakatau. Hundreds of lives were lost and many villages were destroyed. Residents were forced to relocate until restoration works could be completed.

Similarly, as a result of a massive earthquake and resultant tsunami in Japan in 2011, towns such as Otsuchi underwent significant change. Thousands of people simply left the region — the lack of employment opportunities and the risk associated with living in a disaster-prone region combined to drive people to move elsewhere.

8.6.2 Human causes

Human factors, which include changes in the social, economic and political elements of a region, can also be a cause of the decline of cities and their **urban environment**. There are many modern examples of towns and cities with extensive urban environments that have declined. Some reasons for change include depletion of mineral supplies and mining operations, changes in demand for industrial production and manufactured goods, and **economic downturn**. Declining populations can also lead to de-urbanisation, for example when birth rates drop too low. The destructive effects of war and civil unrest also have significant impacts on people's willingness or ability to stay in urban environments.

desertification the transformation of land once suitable for agriculture into desert by processes such as climate change or human practices such as deforestation and overgrazing

urban environment the humanmade or built structures and spaces in which people live, work and recreate on a day-to-day basis

economic downturn a recession or downturn in economic activity that includes increased unemployment and decreased consumer spending

Political causes

Political mismanagement or conflict can also be causes of urban decline.

Angkor, the capital of the Khmer Empire in Cambodia, is believed to have been the largest pre-industrial city in the world by area. The city was abandoned in the fifteenth century due to a combination of wars and a series of droughts. The destruction of its economy, which was based on management of water and rice production, meant that Angkor was no longer viable. The elaborate Khmer temples constructed in the twelfth century (see **FIGURE 2**) have now become popular tourist attractions; more than two million people visit these sites each year.

FIGURE 2 Main temple complex, Angkor Wat, Cambodia

Economic and social causes

The economic failure of planned communities or towns can lead to their abandonment. An example of an urban project that failed due a downturn in the Turkish economy is that of Burj Al Babas. The project started in 2014 but went into **bankruptcy** in 2018; the chateau-style houses remained unoccupied in 2019 (see **FIGURE 3**).

Economic downturn or changes in employment opportunities can also lead to urban decline. Some cities in the United States that relied heavily on automotive manufacturing were severely impacted by **off-shoring** of operations and the global financial crisis. One example is Detroit, Michigan, which used to be known as the hub of American car manufacturing. In the last 60 years, the population has decreased from just under 2 million to about 700 000 people as large industrial employers closed. This resulted in high unemployment rates, housing **foreclosures** and rising crime. In 2013, the City of Detroit filed for bankruptcy, citing debts of US$18.5 billion. Some estimates suggest one third of the buildings in Detroit were abandoned at one point. The city has embarked on a renewal program but still has an unemployment rate of around 25 per cent – a figure that was also affected by the COVID-19 pandemic.

There are also social causes of urban decline. For example, when unemployment rises, people are more likely to move elsewhere to find work. Crime rates and perceptions of safety can also influence people to move from one area to another.

bankruptcy a legal process that declares that a person cannot pay their debts and allows them to make a fresh start

off-shoring the practice of relocating a business's processes from one country to another, to take advantage of lower costs

foreclosure a legal process that allows a lender to sell a borrower's asset to recover their money if the borrower has stopped making repayments

FIGURE 3 Cities abandoned due to changing human and physical factors

(a) Burj Al Babas, Turkey, became a ghost town due to a downturn in the economy.

(b) Wittenoom, Australia, was abandoned in 2006 after asbestos caused health issues.

(c) Oradour-sur-Glane, France, was deserted in 1944 but kept as a memorial to people who were massacred there during World War II.

(d) Pripyat, Ukraine, was abandoned in 1986 after nuclear accident and radiation contamination at nearby Chernobyl nuclear reactor.

(e) Bodie, United States, was abandoned after gold-mining boom concluded in 1915.

(f) Kowloon, Hong Kong, was demolished by the government in 1993 to 'clean up the city'.*

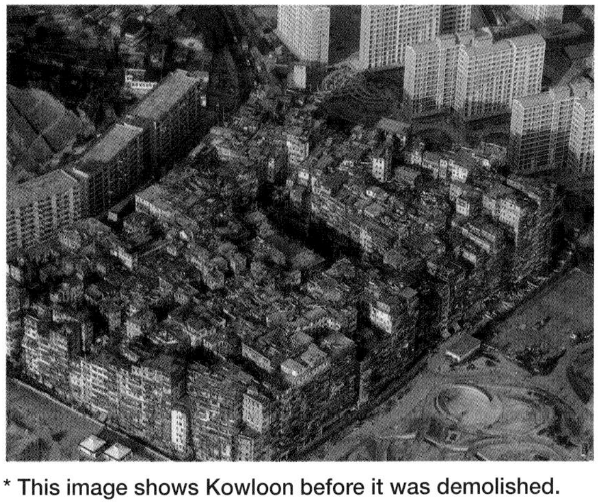

* This image shows Kowloon before it was demolished.

8.6.3 Impacts of shrinking cities

Some cities around the world are experiencing de-urbanisation due to declines in fertility rates (the number of children that people are having), reductions in manufacturing and mining hubs, and resource depletion. Although the concept of urban decline is not new, it was predominantly found in more regional and rural areas that are dependent on the agricultural or resource industry. More recently, however, populations have begun to decline in city areas, and these changes are presenting new problems for cities as they downsize existing infrastructure to accommodate smaller populations. Leipzig, Germany, for example, has water infrastructure designed to provide 200 litres of clean water per person per day, but de-urbanisation has meant that the water usage has dropped to 92 litres per person per day. Lower demand means that water stays in the pipes for longer, increasing the risk of bacterial growth and microbial contamination. Additionally, lower water use means lower waste water generation, leading to increased sedimentation in the pipes from low water flow. The result is higher maintenance costs, which place financial pressure on the metropolitan area.

CASE STUDY: DAEGU, SOUTH KOREA — A CITY IN DECLINE

The city of Daegu is located in the south-eastern area of South Korea, 80 kilometres from the coast (see **FIGURES 4** and **5**).

FIGURE 4 Industrial area, Daegu, Republic of Korea, April 2021

Daegu grew rapidly with a booming textile industry when South Korea went through industrialisation in the 1960s. However; over the past twenty years it has been affected by a slowing down of the economy and population growth. Despite having a large population of 2 566 540, Daegu is experiencing urban decline (see **FIGURE 6**). More recently, this decline has accelerated; during the period of December 2019 to May 2020, the population fell by 12 522, with an additional decline of 2678 in the month of October 2020.

Responses

As an initial step to combating the decline, local tourism was heavily promoted within the metropolitan area. Daegu City also invited 1000 international students from China to study in January 2020. However, the COVID-19 pandemic halted these plans.

Daegu is not the only example of population decline in South Korea. The country overall is experiencing very low fertility rates of 1.17 and a rapidly ageing population. Successful responses require a country-wide response to boosting population levels, including boosting immigration. However, South Korea does not typically have a lot of long-term migrants due to the difficulty of obtaining citizenship, and migration rates away from the city (for study and work) are also high.

The future

With ever declining fertility rates, South Korea and its major cities will continue to experience de-urbanisation into the future. February 2020 saw births (23 049) outstripped by deaths (26 353), and although this coincided with coronavirus, it has been a very real threat for South Korea for a long time. This is because the population over 65 (16.3 per cent) is greater than the population under 14 (12.2 per cent), and the fertility rates are among some of the lowest in the world. Future policies, not only for South Korea but other cities around the world experiencing de-urbanisation, must meet the challenges of declining populations and extra infrastructure while balancing the related financial costs.

FIGURE 5 Daegu, South Korea

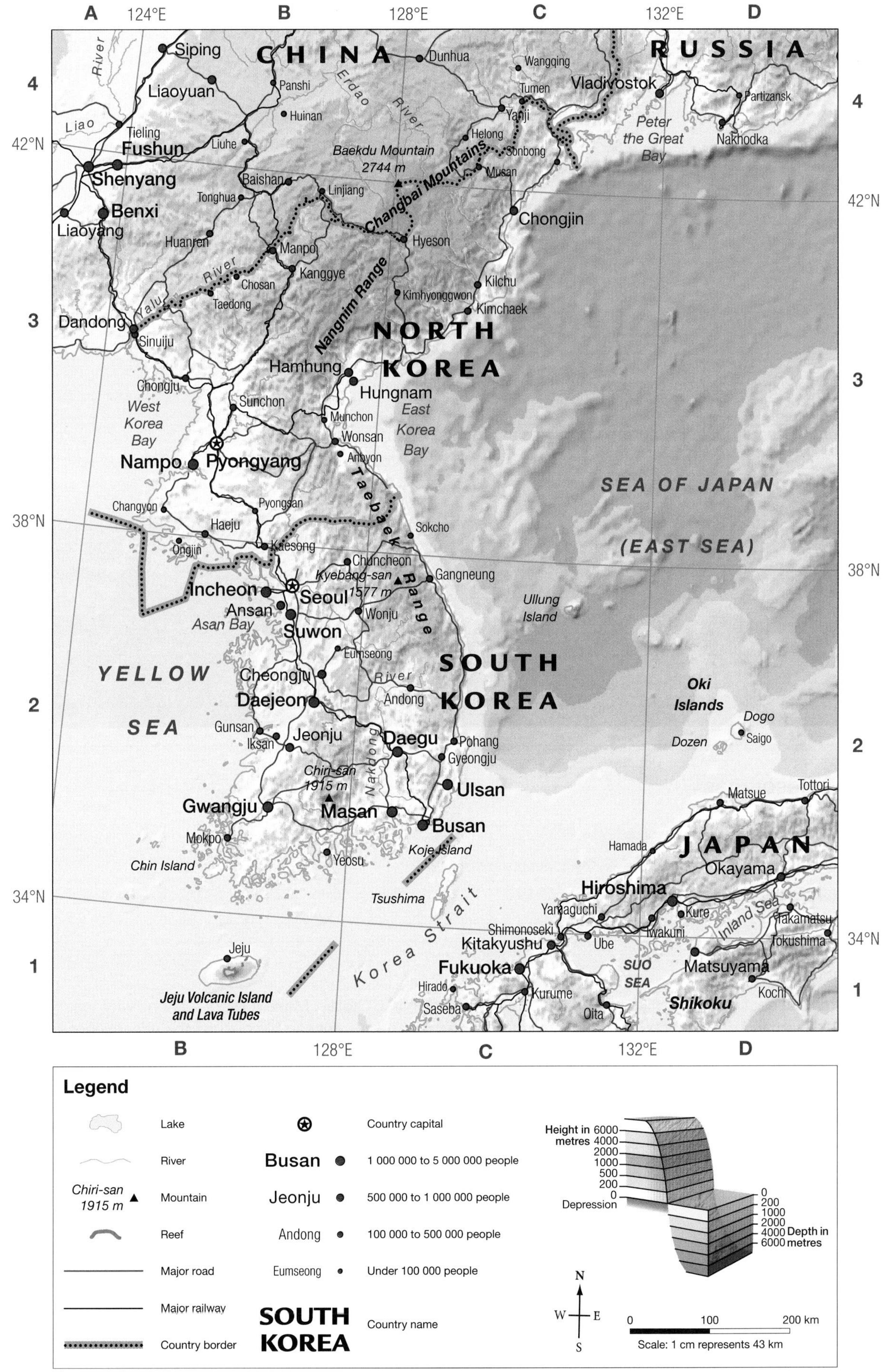

Source: Natural Earth Data, www.naturalearthdata.com; ETOPO1; WDPA.

FIGURE 6 Population of Daegu, 1990–2020

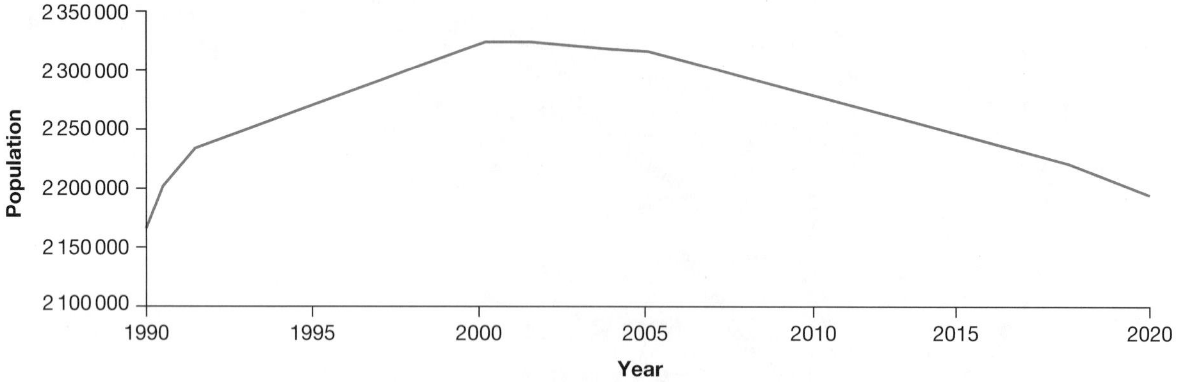

Source: Macrotrends, www.macrotrends.net/cities/21763/daegu/population

FIGURE 7 Daegu, Republic of Korea from Apsan Mountain, 2019

on Resources

eWorkbook	Urban decline (ewbk-9867)
Video eLesson	Urban decline — Key concepts (eles-5219)
Interactivity	Urban decline (int-8648)

8.6 ACTIVITIES

Research the decline of the cities of Gary and Detroit in the United States. Make a list of reasons why these cities have declined and outline plans for their reinvigoration. As part of your research, consider the effectiveness of the strategies being put in place.

8.6 EXERCISE

Learning pathways

■ LEVEL 1	■ LEVEL 2	■ LEVEL 3
1, 2, 7, 9, 14	3, 4, 8, 10, 13	5, 6, 11, 12, 15

Check your understanding

1. What environmental hazards can lead to destruction of or damage to urban environments?
2. Why was the town of Pripyat near Chernobyl in Ukraine abandoned?

3. Where is Daegu located?
4. Outline three reasons for population decline in urban areas.
5. Describe how an urban place might change when people move away.
6. What characteristics of the South Korean population mean that it is more vulnerable to the impacts of COVID-19?

Apply your understanding

7. Why was water essential to the survival and decline of the city of Angkor?
8. Explain what is happening to the population in Daegu. What impact would a continued decline of population have on Daegu and on South Korea as a whole?
9. What reasons can you put forward to explain why some cities decline over time?
10. What consequences might arise from urban decline? Identify and explain one economic, one social and one environmental consequence.
11. Explain how inviting international students to study in a city might help to combat urban decline.
12. Assess whether inviting international immigrants to move to Daegu is a good long-term solution to their population decline.

Challenge your understanding

13. If ore bodies are depleted in the mining town of Broken Hill, how might the town sustain its existence into the future?
14. Is there a future for ghost towns? Explain your view.
15. Predict what will happen to other urban areas around the world if global fertility rates continue to drop.

To answer questions online and to receive **immediate feedback** and **sample responses** for every question go to your learnON title at www.jacplus.com.au.

8.7 Impacts of urbanisation on the environment

LEARNING INTENTION

By the end of this subtopic, you will be able to describe the four spheres that comprise the Earth's biophysical environment and explain the ways in which urban environments interconnect with these elements.

8.7.1 Interconnections between urban and biophysical environments

The earliest forms of urban environments consisted of shelters to protect people from the elements and provide security from the attacks of predators. From these simplest forms, the highly complex modern urban environment has developed.

All forms of urban environments are interconnected with the **biophysical environment**. The 'bio' elements are all forms of plant and animal life, including people and all their activity and industry. The 'physical' elements are the atmosphere, hydrosphere and lithosphere.

The biophysical elements impose limits on the development and sustainability of all forms of urban environment. Conversely, the urban environment imposes significant human-induced change on the biophysical world. The understanding of this interconnection is particularly important in a world of increasing human numbers and pressure for resources on the biophysical environment (see **FIGURE 1**).

biophysical environment all elements or features of the natural or physical and the human or urban environment, including the interaction of these elements

FIGURE 1 Interaction between the urban environment and the biophysical environment

8.7.2 Effects on the atmosphere

Where sources of potentially dangerous gaseous emissions are high, such as from buildings, transport systems and industry, atmospheric pollution can be a problem. Examples such as hazy conditions, **photochemical smogs**, light and noise pollution and acid rain are significant problems that need to be addressed. Thus, the development of clean air policies controlling emissions of gases into the atmosphere is important. Examples of such measures include the introduction of lead-free petrol, banning the burning of household waste, and emission-control systems on factory furnaces.

Cities and industries have huge energy demands, and the byproduct of this is heat. The **heat island effect**, whereby urban environment structures such as buildings and roads absorb heat from the sun, raises the temperature of the city environment compared to rural surrounds (see **FIGURE 2**).

The production of greenhouse gases such as carbon dioxide and methane by urban environments is recognised as probably the greatest contemporary climate issue. Global warming leading to climate change is largely the result of emissions of these gases into the atmosphere, particularly in large urban centres.

photochemical smog air pollution that can be seen as a smoky haze

heat island effect structures in urban environments absorb heat from the sun and raise the temperature of the city environment compared to rural surrounds

FIGURE 2 The heat island effect of cities

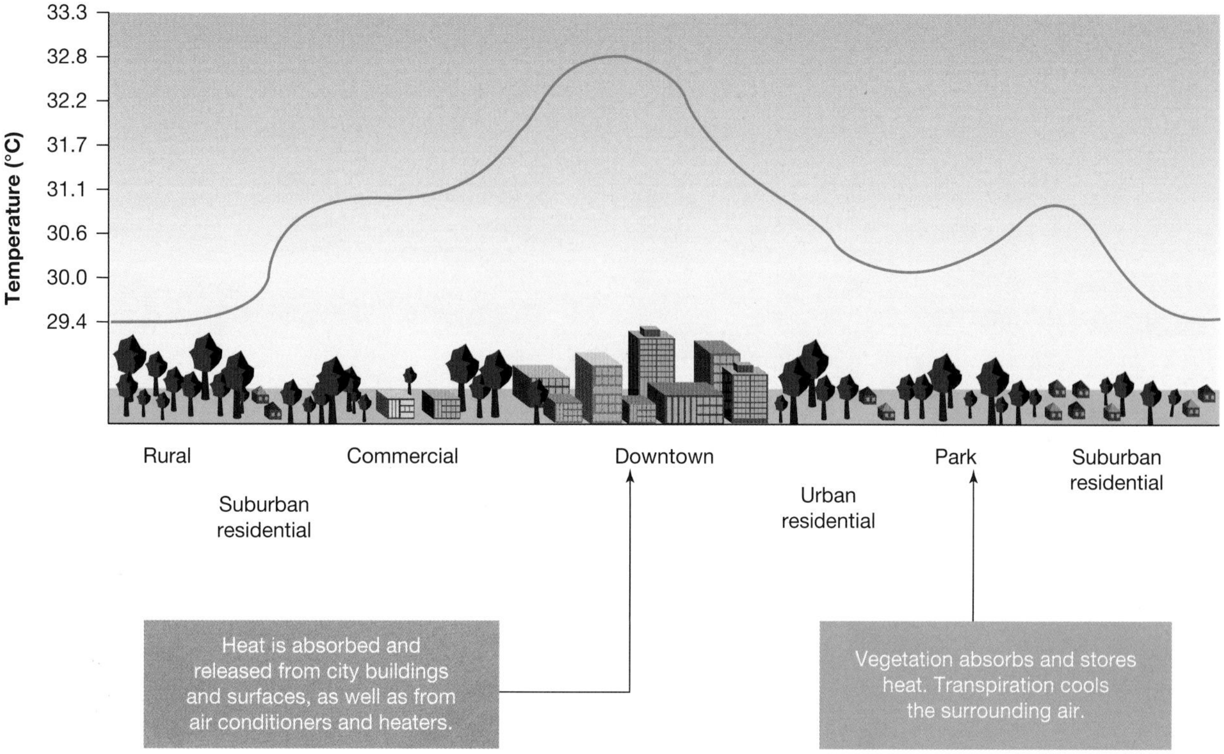

8.7.3 Effects on the hydrosphere

As the urban environment is closely dependent on the hydrosphere, it is not surprising that **water security** and **water rights** are important management objectives for a sustainable future. One of the most important aims for urban planners is to ensure the supply of clean water and to manage the waste water from cities.

In general, all urban centres are trying to find increasing supplies of water for domestic and industrial use from rivers, groundwater and, more recently, desalinisation — a process that removes the salt from sea water. Water is gathered by infrastructure in the form of dams, pipelines and artesian waters at the local level, and major water management schemes such as the Snowy Mountains Scheme in Australia.

Water pollution caused by urban environments is also important, as polluted waters are a risk to all life forms in any environment. In environmental management and the planning of settlements, the biomes and ecosystems of rivers, wetlands and swamps need to be considered in terms of protecting habitats and maintaining biodiversity.

8.7.4 Effects on the lithosphere

The 'tar and cement' structures that are our cities often cover vast areas of land. The built-up areas of greater Sydney, for instance, cover more than 4196 square kilometres. Associated problems include the disposal of the enormous amount of waste that cities produce, and the impacts on agriculture, plants and animal life in adjacent habitats and ecosystems.

Urban environment surfaces, such as footpaths, roads and carparks, generate two to six times more run-off than a natural surface. Rain that falls on roads, carparks, airports and open industrial areas can be contaminated with chemicals such as petroleum residues, which can then find their way into waterways.

water security the reliable availability of acceptable quality water to sustain a population

water rights refers to the right to use water from a water source such as a river, stream, pond or groundwater source

FIGURE 3 Daxing International Airport, in Beijing, China, covers 47 square kilometres of land, including (at the time of opening in 2019) the world's largest single-building terminal.

fdw-0044

FOCUS ON FIELDWORK

The impact of traffic

Traffic plays a significant role in our experience of urban spaces. We have to navigate road networks, cross roads safely, choose between modes of transport and live with negative impacts that some of our transport choices have on the biophysical environment: air pollution, noise and depletion of non-renewable resources.

Learn how to measure and discuss the impact of traffic in your area using the **Impact of traffic** fieldwork activity in your learnON Fieldwork Resources.

on Resources

eWorkbook Impacts of urbanisation on the environment (ewbk-9871)

Video eLesson Impacts of urbanisation on the environment — Key concepts (eles-5220)

Interactivity Impacts of urbanisation on the environment (int-8649)

myWorldAtlas Deepen your understanding of this topic with related case studies and questions.

- Polluted cities
- Mexico City

8.7 ACTIVITIES

1. As a class, create a survey to discover the issues existing in your local area that need to be improved in order to achieve a sustainable urban environment. Distribute the survey and collate all the data to analyse your findings. Use graphs and statistics to support your analysis. Complete the following questions.
   a. Describe the range of issues that exist in the biophysical environment.
   b. Discuss whether the issues are easy to manage.
   c. Describe the causes of issues in your local area.
   d. Discuss whether these issues are a direct result of urban growth.

FIGURE 4 Cape Town, South Africa.

2. Research a place where human activity has had significant negative effects on the environment, making the place unhealthy or uninhabitable for humans. Some suggestions include:
   a. the impact of air pollution in Beijing, China
   b. the impact of water scarcity in Chennai, India
   c. the impact of the heat island effect in Tokyo, Japan.
3. Research the water crisis of 2015-18 experienced in Cape Town, South Africa, during which the city almost reached 'day zero' (the day the city was predicted to run out of water). Investigate how the city reached this point, and what strategies were put in place to avoid 'day zero'.

8.7 EXERCISE

Learning pathways

■ LEVEL 1	■ LEVEL 2	■ LEVEL 3
3, 4, 8, 9, 15	1, 6, 10, 12, 13	2, 5, 7, 11, 14

Check your understanding

1. What does the 'bio' part of the 'biophysical environment' refer to?
2. Give reasons why urban environments can have such a major impact on the Earth's atmosphere.
3. Why do most of the large urban centres of the world have high-rise buildings?
4. What is meant by the 'heat island effect'?
5. Why are water supplies a problem for large cities?
6. Identify three examples of how elements of the biophysical environment can limit the development of urban areas.
7. Outline how green spaces in urban areas can help to reduce the heat island effect.

Apply your understanding

8. How might rising sea levels, predicted to occur as a result of global warming, affect the place and space of a city such as New York, many populated parts of which are on coastlines or islands?
9. How will the supply of fresh water affect the development of cities in the future?
10. Explain how the development of extensive public transport systems affects the environment.
11. Compare the sustainability of car use with public transport in terms of positive and negative effects on the environment.
12. Explain why urban waste disposal poses a threat to the environment.

Challenge your understanding

13. Suggest ways that the biodiversity of plant species might be increased in an urban area.
14. How might climate change and the impacts of urban areas on the natural environment be interconnected?
15. What might happen if a city is unable to use its water source or its supply runs dry? Propose three strategies that might be used by city authorities to manage such a situation.

To answer questions online and to receive **immediate feedback** and **sample responses** for every question go to your learnON title at www.jacplus.com.au.

8.8 SkillBuilder — Creating and reading pictographs

online only

LEARNING INTENTION

By the end of this subtopic, you will be able to transform a set of data into a pictograph and to analyse pictographs.

The content in this subtopic is a summary of what you will find in the online resource.

8.8.1 Tell me

What is a pictograph?

A pictograph is a graph drawn using pictures to represent numbers, instead of bars or dots that are traditionally used on graphs. Data can be drawn vertically or horizontally. Each picture is given a value.

8.8.2 Show me

How to create a pictograph

Step 1

Decide on a simple picture to represent the data.

Step 2

Determine a number that each picture should represent. Choose a scale that will not require too many pictures for each part of the graph.

Step 3

Draw lines on your page, equal distances apart, to represent each variable for which you have data.

FIGURE 1 Top five countries by population, 2008

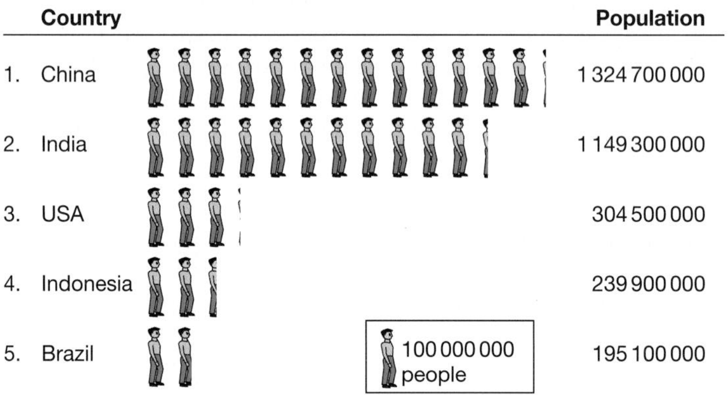

Step 4

Calculate how many pictures you need to represent each number in your data set.

Step 5

Add the drawings to your graph. Create a legend and give your pictograph a clear title.

How to read a pictograph

Read the title and legend to determine the type of data represented by the graph. Count the number of images shown for each category and multiply this by the number each image represents.

8.8.3 Let me do it

learnON

Go to learnON to access the following additional resources to help you build this skill:

- a longer explanation of this skill and its application in Geography (Tell me)
- a video demonstrating the step-by-step process of this skill (Show me)
- an activity and interactivity for you to practise the skill (Let me do it)
- self-marking questions to help you understand and use the skill.

on Resources

eWorkbook	SkillBuilder — Creating and reading pictographs (ewbk-9875)
Video eLesson	SkillBuilder — Creating and reading pictographs (eles-1659)
Interactivity	SkillBuilder — Creating and reading pictographs (int-3155)

8.9 Challenges for sustainable urban environments

LEARNING INTENTION

By the end of this subtopic, you will be able to describe projected urban growth, explain the challenges that this growth presents and recommend sustainable approaches to development.

8.9.1 The influence of technology

Throughout human history, cities have changed as new forms of technology have developed. For instance, high-rise buildings such as skyscrapers could not exist without the modern cement and steel methods of construction and the development of high-speed lifts. What will be the nature of cities as technology progresses, and how can the social, economic and environmental elements of cities develop and be managed in a fair and sustainable manner?

8.9.2 Sustainable urban development

The United Nations (UN) Sustainable Development Goals (SDGs) have been developed as a set of ambitious aims to achieving an improved and sustainable future for everyone on the planet. The SDGs inform bodies that have an interest in urban development, including UN-Habitat, ComHabitat (Commonwealth Habitat), the Cities Alliance and the World Bank. These agencies aim to address the urban challenges of the twenty-first century with a focus on social and economic management criteria. The SDGs are shown in **FIGURE 1**.

FIGURE 1 The 17 Sustainable Development Goals

It has been estimated by UN studies that the global urban population, which is currently around 56 per cent (4.3 billion), will increase to about 5.1 billion in 2030, meaning that over 60 per cent of the world's total population will be living in cities. This increase means that another billion people will need new housing, basic urban infrastructure and services. To achieve this, the equivalent of seven new megacities will need to be created annually (see **FIGURE 2**).

FIGURE 2 Urban population (a) 2018 and (b) projected for 2030

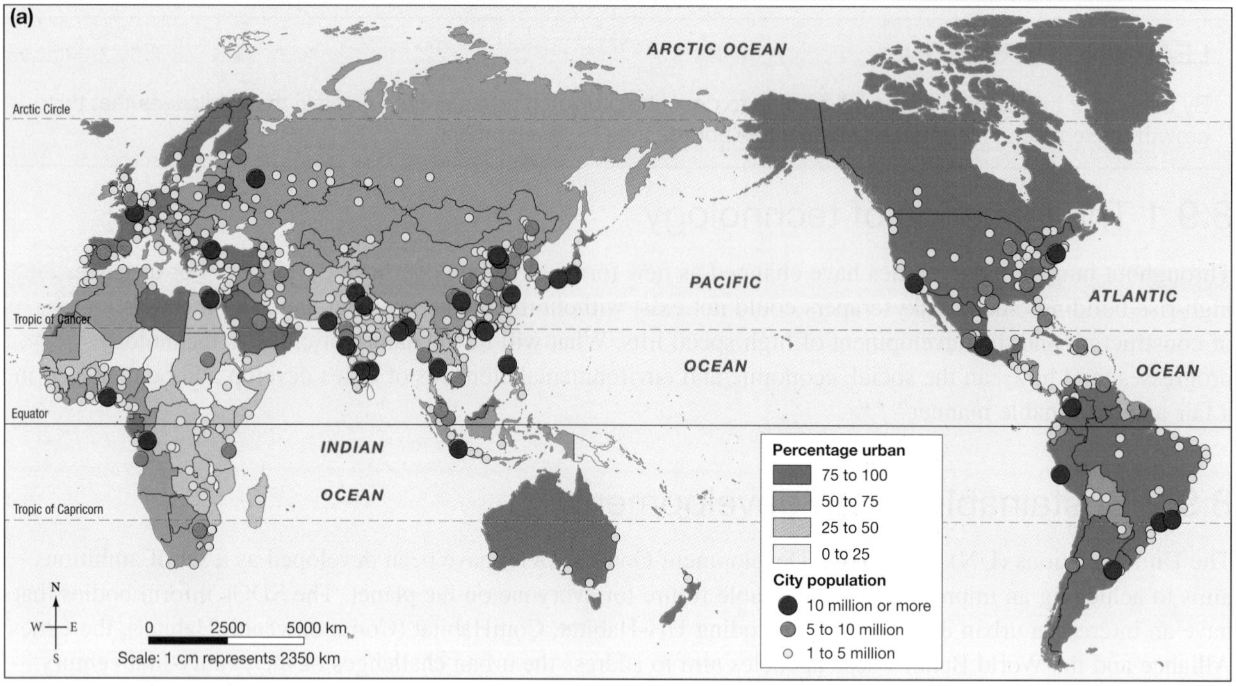

Source: United Nations, Department of Economic and Social Affairs, Population Division 2018. World Urbanization Prospects: The 2018 Revision

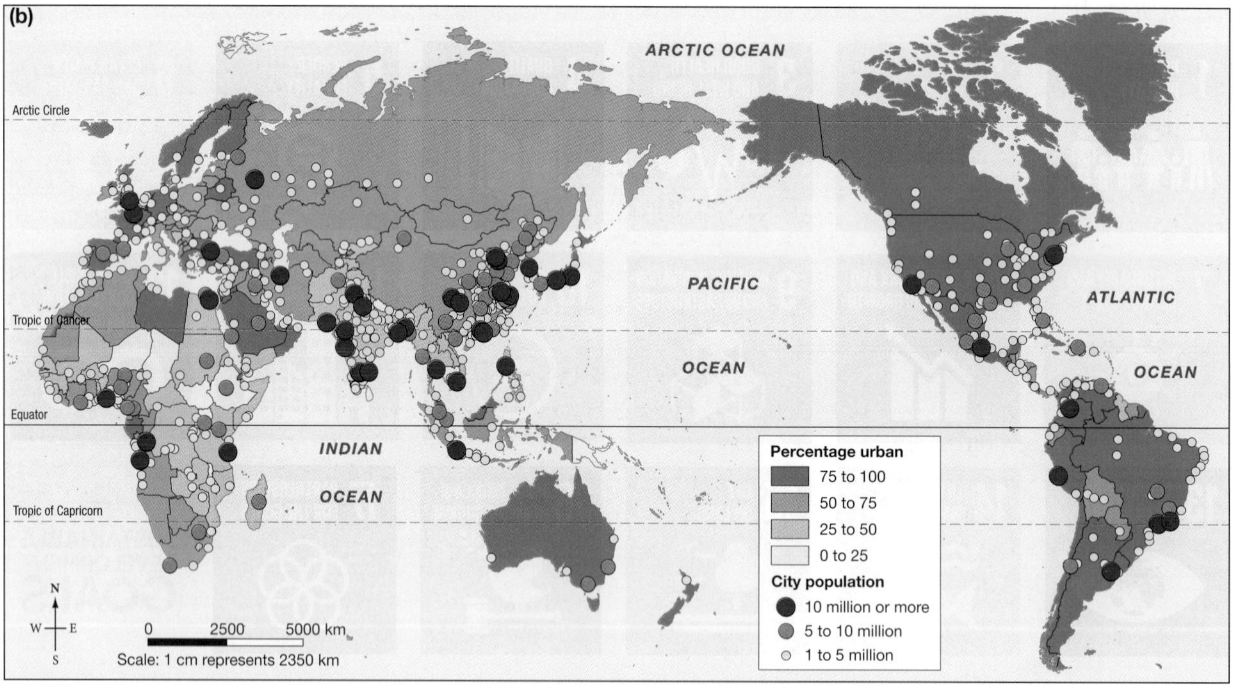

Source: United Nations, Department of Economic and Social Affairs, Population Division 2018. World Urbanization Prospects: The 2018 Revision

8.9.3 Challenges for managing cities

Many cities in the world face environmental, social and economic challenges. Issues such as extensive areas of slum housing and a general lack of infrastructure to support what may be called a socially and economically just lifestyle undermine the sustainability of these environments. Particularly in the poorest or least developed countries, there are significant environmental management issues associated with large cities. Some of these are detailed in **TABLE 1**. Note that there are issues even in cities that could be called wealthy or most developed.

TABLE 1 Urban challenges

Level	Challenges to be addressed to ensure a sustainable urban environment
Least developed countries	• Poverty and inequality • Rapid and chaotic development of slum housing • Increasing demand for housing, urban infrastructure, services and employment • Education and employment needs of the majority population of young people • Shortage of skills in the urban environment sector
Transition countries	• Slow (or even negative) population growth and ageing • Shrinking cities and deteriorating buildings and infrastructure • Urban sprawl and preservation of inner-city heritage buildings • Growing demand for housing and facilities by an emerging wealthy class • Severe environmental pollution from old industries • Rapid growth of vehicle ownership • Financing of local authorities to meet additional responsibilities
Developed countries	• Recent mortgage and housing markets crises • Unemployment and impoverishment due to changing availability of jobs • Large energy use of cities caused by car dependence, huge waste production and urban sprawl • Slow population growth, ageing and shrinking of some cities

Potential management strategies to foster socially, economically and environmentally sustainable urban environments include:

- building energy-efficient houses using energy sources that reduce the **ecological footprint** of cities
- reducing waste by recycling and reusing materials and making grey water tanks compulsory
- improving the efficiency, cost and reliability of public transport systems to reduce reliance on cars, or building cycling and walking paths to encourage passive transport options
- redeveloping to include **medium-density housing** to reduce urban sprawl
- exchanging ideas between governments and planning experts about planning and building policies and best and successful practice in design.

ecological footprint a measure of human demand on the Earth's natural systems in general and ecosystems in particular; the amount of productive land required by each person in the world for food, water, transport, housing, waste management and other purposes

medium-density housing a form of residential development such as detached, semi-attached and multi-unit housing that can range from about 25 to 80 dwellings per hectare

on Resources

eWorkbook	Challenges for sustainable urban environments (ewbk-9879)
Video eLesson	Challenges for sustainable urban environments — Key concepts (eles-5221)
Interactivity	Challenges for sustainable urban environments (int-8650)

8.9 ACTIVITIES

Use the **Ecological footprint** weblink in your online Resources to calculate your ecological footprint. Use the data to create a class average. Construct a column graph to show a comparison between your ecological footprint, the class average and the Australian average.

1. How do the three sets of data compare?
2. How could you reduce your ecological footprint?

8.9 EXERCISE

Learning pathways

■ LEVEL 1	■ LEVEL 2	■ LEVEL 3
1, 2, 7, 10, 14	3, 4, 11, 12, 13	5, 6, 8, 9, 15

Check your understanding

1. Identify two technological achievements that have allowed people to build bigger cities.
2. Name three international organisations that play a role in sustainable urban development.
3. Globally, how many people will need new housing in 2030 based on current predictions of urbanisation?
4. Refer to **FIGURE 1**. What are the SDGs and why were they developed?
5. From **TABLE 1**, identify the urban challenges that relate to life in Australia.
6. Refer to **FIGURES 2(a)** and **2(b)**. Identify two countries that are not expected to have more cities with a population between 1 and 5 million people by 2030.

Apply your understanding

7. Refer to **FIGURE 1**. Which of the SDGs do you think relate to the issue of sustainable urbanisation? Explain your view.
8. How would migration help solve the problems of ageing populations in developed Western cities?
9. Using **TABLE 1**, explain why the urban challenges are different for least developed, transitional and developed countries.
10. List three management strategies for sustaining urban environments and explain the contribution that each of these would make.
11. Which of the SDGs will directly improve social conditions in urban environments? Give reasons for your answer.
12. Which of the SDGs will directly improve environmental conditions in urban environments? Give reasons for your answer.

Challenge your understanding

13. It has been said that if all nations had the same ecological footprint as the developed countries (e.g. the United States, Australia and most European nations), we would need four new worlds the size of planet Earth to accommodate the growth in resource consumption. In what ways can we achieve energy, food and water security with an aim of sustainability into the future?
14. Technology has helped humans build bigger cities. Suggest two technological inventions that would help to solve some of the problems created by big cities. (*Hint:* Consider what problems might need to be solved, and then think of an invention that might solve them.)
15. Cities tend to grow over time, but some countries are creating purpose-built, planned urban environments from the ground up. Is it possible that a planned city could be created that would solve all of the sustainability issues of urbanisation? Suggest which problems might be prevented with careful planning for a new city, and whether there are any issues that are inevitable when there is a large population of people living in the one location.

To answer questions online and to receive **immediate feedback** and **sample responses** for every question go to your learnON title at www.jacplus.com.au.

8.10 Urbanisation in Indonesia

online only

LEARNING INTENTION

By the end of this subtopic, you will be able to describe patterns of urban growth in Indonesia and explain their causes and effects.

The content in this subtopic is a summary of what you will find in the online resource.

Indonesia's population of nearly 273.5 million people (2019) lives on an archipelago of more than 18 000 islands over 1 904 569 square kilometres. However, its population is not evenly distributed. Only about 11 000 of the islands are actually inhabited. Sixty per cent of Indonesia's population is concentrated on only seven per cent of the total land area — on the island of Java.

Like many countries in Asia, Indonesia has experienced rapid urban growth, but this has occurred only relatively recently.

To learn more about urbanisation in Indonesia, go to your learnON resources at www.jacPLUS.com.au.

FIGURE 3 The Jakarta metropolitan area had a population of over 10 million in 2018 and a population density of over 14 000 people per square kilometre. It is the second largest urban area in the world.

Contents

learnON

- 8.10.1 Indonesia's population
- 8.10.2 Causes of urbanisation
- 8.10.3 Consequences of urbanisation

on Resources

eWorkbook	Urbanisation in Indonesia (ewbk-9883)
Video eLesson	Urbanisation in Indonesia — Key concepts (eles-5222)
Interactivity	Urban Indonesia (int-3115)
Google Earth	Jakarta, Indonesia (gogl-0064)
myWorldAtlas	Deepen your understanding of this topic with related case studies and questions. • Urbanisation in Indonesia

8.11 Investigating topographic maps — Jakarta

LEARNING INTENTION

By the end of this subtopic, you will be able to identify land heights and features across Jakarta on a map and predict how these features might influence the direction of urban sprawl and flooding.

8.11.1 Urbanisation in Jakarta

Jakarta is situated on flat lowlands on the northern coast of Indonesia's West Java province. The city has expanded rapidly; in central areas, it has a population density of about 18 500 people per square kilometre.

Jakarta has regularly experienced flooding as a result of a combination of factors including:

- heavy wet-season rainfall
- low relief and land sitting below sea level
- shallow rivers that easily flood
- rubbish deposits in river beds.

FIGURE 1 Areas of Jakarta that experienced severe flooding in January 2014

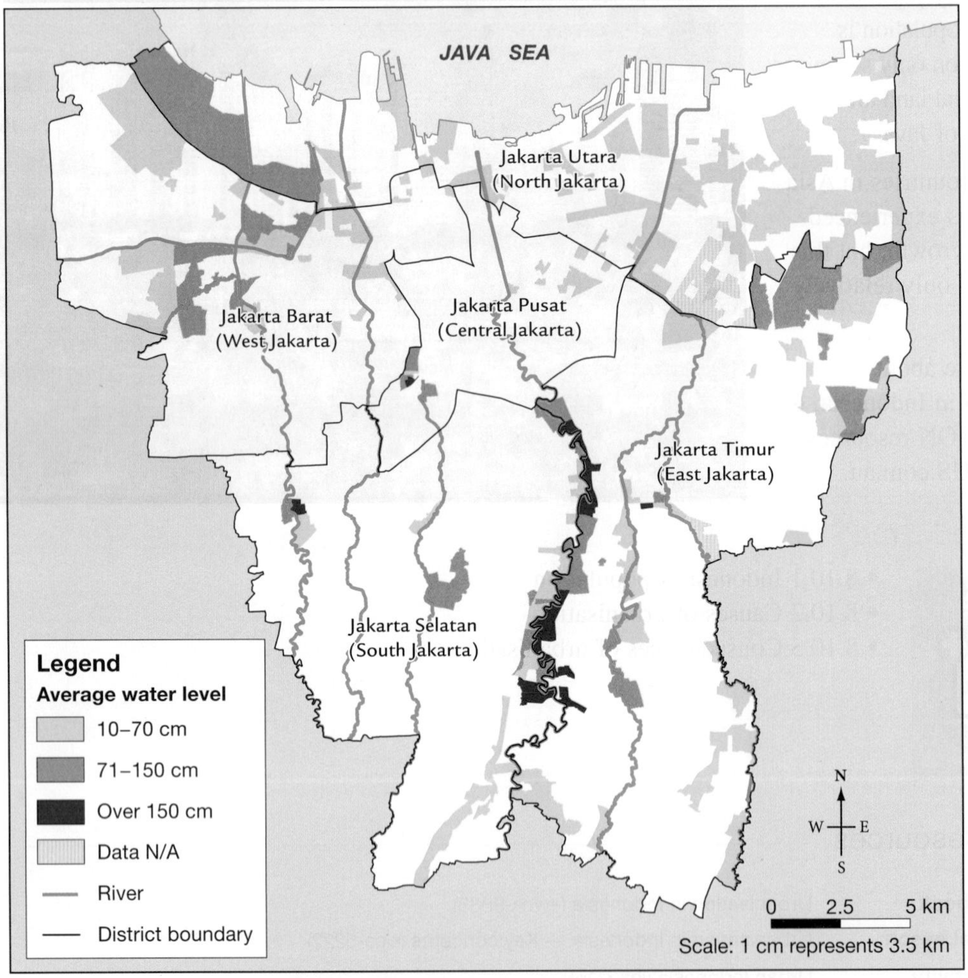

Source: Based on OCHA / ReliefWebSource: ReliefWeb / OCHA Indonesia Jakarta 2014
https://reliefweb.int/sites/reliefweb.int/files/resources/Update on Jakarta Flood as of 21Jan2014-R.pdf

FIGURE 2 Thematic map of land heights in Jakarta

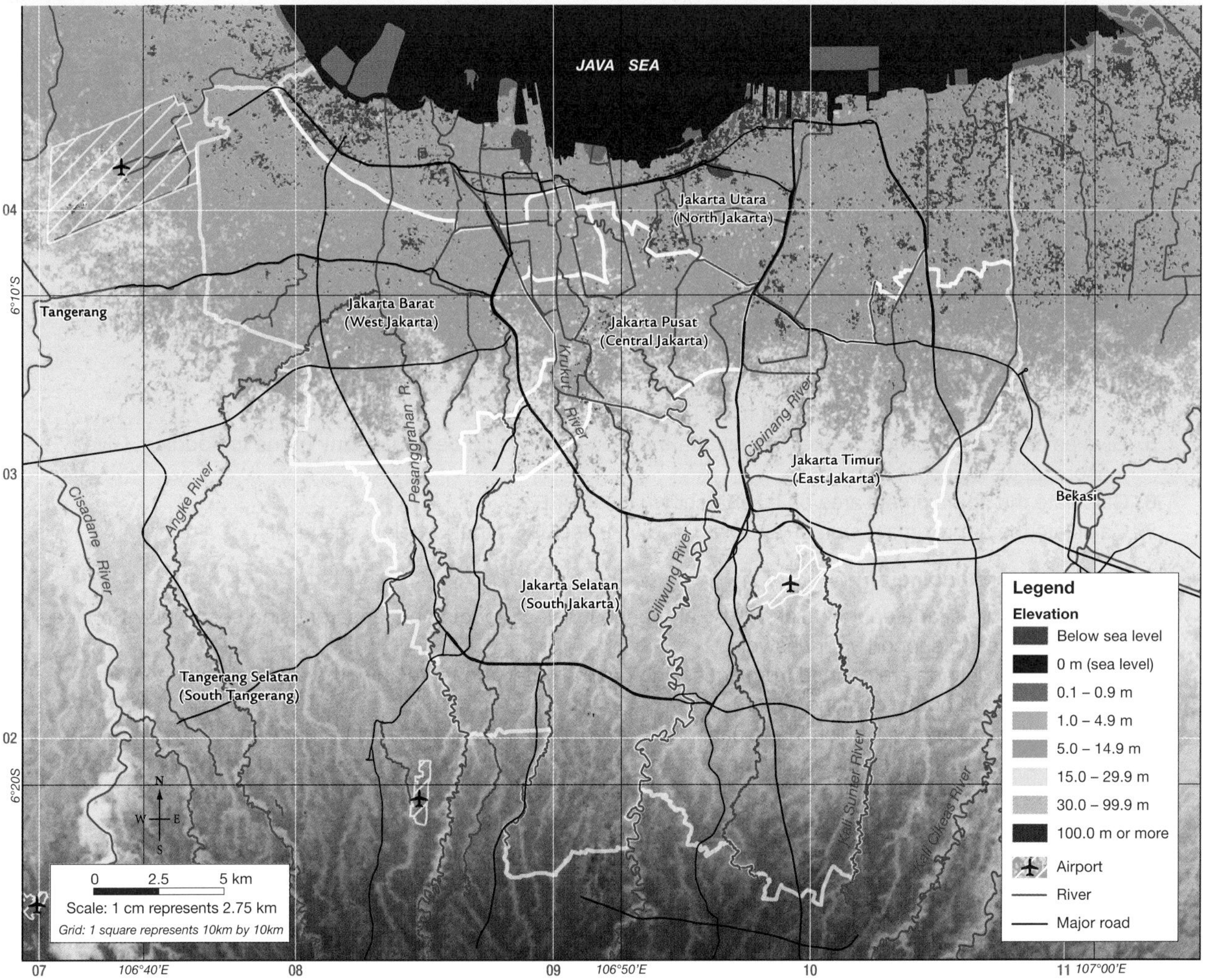

Source: Map data © OpenStreetMap contributors, https://openstreetmap.org. Data is available under the Open Database Licence, https://opendatacommons.org/licenses/odbl/; elevation data sourced from USGS (2013).

on Resources

eWorkbook	Investigating topographic maps — Jakarta (ewbk-9887)
Digital document	Land heights in Jakarta (doc-36319)
Video eLesson	Investigating topographic maps — Jakarta — Key concepts (eles-5223)
Interactivity	Investigating topographic maps — Jakarta (int-8651)
Google Earth	Jakarta, Indonesia (gogl-0064)

8.11 EXERCISE

Learning pathways

■ LEVEL 1	■ LEVEL 2	■ LEVEL 3
3, 4, 5, 10, 12	1, 2, 6, 8, 13	7, 9, 11, 14, 15

Check your understanding

1. Describe the location of Jakarta.
2. What is the approximate population density of central Jakarta?
3. List four reasons why Jakarta regularly floods.
4. Use the map (**FIGURE 2**) to answer the following questions.
   a. Give an approximate latitude and longitude for one of the airports in Jakarta.
   b. Give the direction of West Jakarta from Tangerang.
5. Give an area reference for one part of Jakarta that you think would be particularly prone to flooding. Give reasons to justify your choice.
6. Calculate the approximate area of Jakarta shown in **FIGURE 2**.
7. Calculate the approximate area of landmass shown in **FIGURE 2** that is below sea level.

Apply your understanding

8. Which areas of Jakarta do you think will be at the most risk of flooding because of rising sea levels? Provide area references to support your answer.
9. If residents in Jakarta are banned from accessing groundwater, are there any other natural sources from which they could obtain drinking water?
10. How might disposing of rubbish in the canals and rivers of North Jakarta affect other parts of the city?
11. Determine how far central Jakarta is from the coast. Do you think this area is also at risk of flooding? Give reasons to support your answer.
12. If you were given the task of moving residents whose homes will be underwater in the next twenty years, which areas would be your first priority? Provide area or grid references and reasons for your choice.
13. Rubbish disposal is a significant issue in Jakarta. Explain where the best place for a new rubbish dump would be. Provide area or grid references and reasons for your choice.
14. If you had to build a new sea wall (or increase the height of an existing sea wall), where would you suggest might have the greatest effect in mitigating floods? Provide area or grid references and reasons for your choice.

Challenge your understanding

15. Indonesian governments have been discussing moving Jakarta to a new location on a different island. Consider the logistics of such a move and predict whether it will ever happen. Give reasons for your view, based on the social, environmental and economic costs and benefits.

To answer questions online and to receive **immediate feedback** and **sample responses** for every question go to your learnON title at www.jacplus.com.au.

8.12 Thinking Big research project — Slum improvement proposal

online only

The content in this subtopic is a summary of what you will find in the online resource.

Scenario

It is widely known that the slums of megacities create significant issues that have an impact on human wellbeing. Hence, slums have been identified in the United Nations' Sustainable Development Goals as needing sustainable solutions to improve the social, economic and environmental futures of their inhabitants.

You have been employed by the local council in one of the world's megacities to carry out a study identifying issues associated with life in the city's slums and develop a plan for improving living conditions for slum residents.

Task

With reference to at least three of the 17 SDGs, develop social, economic and environmental plans for sustainable change in your chosen megacity slum.

learnON

You should incorporate your research findings in a proposal to the council of your selected megacity, outlining identified issues and recommended management strategies to improve the livelihood of the people who live in the slum. Your proposal can be presented in written report form or alternatively as a slide deck or multimedia presentation, with comprehensive text information, annotated images, diagrams and maps.

Go to learnON to access the resources you need to complete this research project.

on Resources

ProjectsPLUS Thinking Big research project — Slum improvement proposal (pro-0216)

8.13 Review

8.13.1 Key knowledge summary

8.2 The development of urban environments

- Large urban complexes are a recent phenomenon in the world's history.
- The modern trend is for people to move to urban complexes seeking improvement in lifestyle, but this means having to cope with socioeconomic and environmental challenges in the urban environment, such as congestion, crime, pollution and social isolation.
- Megacities, while offering opportunities for work and access to multiple services, can have issues such as slums and poor waste management.
- Dealing with the impacts of cities on air and water quality is a major issue.
- Urban sprawl in large urban complexes is a major problem for city planners.

8.3 Distribution of urban areas

- Urbanisation is the growth and expansion of urban areas and involves the movement of people from rural to urban areas.
- Patterns of urbanisation across the world are uneven.
- Coastal settlements are often highly urbanised.

8.5 Urban expansion

- Urban infilling and increasing density are two methods of creating more living space in urban areas.
- Expansion into the rural–urban fringe leads to urban sprawl, which has economic and environmental impacts.
- Rapid urban growth in Mumbai has created challenges relating to human wellbeing, urban sprawl, traffic congestion and infrastructure needs.

8.6 Urban decline

- Natural and human-induced changes can lead to processes that build up and lead to the decline of urban complexes.
- Events such as natural and human-induced disasters have large impacts on urban areas.
- The depletion of resources can lead to urban decline.
- De-urbanisation is occurring in some cities around the world.
- Challenges of de-urbanisation include extra financial pressure on the city and issues associated with additional infrastructure

8.7 Impacts of urbanisation on the environment

- Urbanisation has led to significant changes to the natural environment, including changes to the atmosphere, hydrosphere and lithosphere.

8.9 Challenges for sustainable urban environments

- Urban populations are predicted to continue to grow, with megacities being a magnet for rural dwellers.
- Careful management of urban complexes is required so that they may be sustained and offer a good quality of life for their inhabitants.
- The management of sustainable cities must draw inspiration from the Sustainable Development Goals established by the United Nations.

8.10 Urbanisation in Indonesia

- Jakarta is experiencing significant urban growth
- Consequences of urban growth in Jakarta include loss of land, pressure on environment, joblessness and traffic congestion.

8.13.2 Key terms

alpha world city a city generally considered to be important in the global economic system

agglomeration the extended area of a city, including its suburbs and adjoining populated areas

archipelago a chain or group of islands

bankruptcy a legal process that declares that a person cannot pay their debts and allows them to make a fresh start

biophysical environment all elements or features of the natural or physical and the human or urban environment, including the interaction of these elements

conurbations an urban area formed when two or more towns or cities (e.g. Tokyo and Yokohama) spread into and merge with each other

desertification the transformation of land once suitable for agriculture into desert by processes such as climate change or human practices such as deforestation and overgrazing

developing nation a country whose economy is not well developed or diversified, although it may be showing growth in key areas such as agriculture, industries, tourism or telecommunications

ecological footprint a measure of human demand on the Earth's natural systems in general and ecosystems in particular; the amount of productive land required by each person in the world for food, water, transport, housing, waste management and other purposes

economic downturn a recession or downturn in economic activity that includes increased unemployment and decreased consumer spending

foreclosure a legal process that allows a lender to sell a borrower's asset to recover their money if the borrower has stopped making repayments

gross domestic product (GDP) a measurement of the annual value of all the goods and services bought and sold within a country's borders

heat island effect structures in urban environments absorb heat from the sun and raise the temperature of the city environment compared to rural surrounds

human–environment systems thinking using thinking skills such as analysis and evaluation to understand the interaction of the human and biophysical or natural parts of the Earth's environment

industrial effluents waste from manufacturing and other industrial processes

Industrial Revolution the period from the mid-1700s into the 1800s that saw major technological changes in agriculture, manufacturing, mining and transportation, with far-reaching social and economic impacts

infrastructure the facilities, services and installations needed for a society to function, such as transportation and communications systems, water and power lines

JMA Jakarta metropolitan area

medium-density housing a form of residential development such as detached, semi-attached and multi-unit housing that can range from about 25 to 80 dwellings per hectare

megacity a settlement with 10 million or more inhabitants

off-shoring the practice of relocating a business's processes from one country to another, to take advantage of lower costs

pull factor favourable quality or attribute that attracts people to a particular location

push factor unfavourable quality or attribute of a person's current location that drives them to move elsewhere

rural–urban fringe the transition zone where rural (country) and urban (city) areas meet

sanitation services provided to remove waste such as sewage and rubbish

slum an area of a city characterised by poor housing and poverty

subsidence the sinking of land

urban sprawl the spreading of urban developments into areas on the city boundary

urbanisation the social and economic processes whereby an increasing proportion of the population of a country or region live in urban areas

urban infilling the division of larger house sites into multiple sites for new homes

urban environment the humanmade or built structures and spaces in which people live, work and recreate on a day-to-day basis

utilities services provided to a population, such as water, natural gas, electricity and communication facilities

water security the reliable availability of acceptable quality water to sustain a population

water rights refers to the right to use water from a water source such as a river, stream, pond or groundwater source

8.13.3 Reflection

Complete the following to reflect on your learning.

Revisit the inquiry question posed in the Overview:

Why do millions of people choose to live so close to other people in busy urban areas?

1. Now that you have completed this topic, what is your view on the inquiry question above? Discuss with a partner. Has your learning in this topic changed your view? If so, how?
2. Write a paragraph in response to the inquiry question, outlining your views.

Subtopic	Success criteria	●	○	●
8.2	I can define urbanisation.			
	I can describe global patterns of urbanisation.			
8.3	I can identify and describe patterns in where cities and urban areas are located around the world.			
8.4	I can analyse and identify similarities and differences in the data shown in a population profile.			
8.5	I can identify and describe urban expansion.			
	I can explain the consequences of different growth patterns.			
8.6	I can explain how and why urbanised areas decline.			
8.7	I can describe the four spheres that comprise the Earth's biophysical environment.			
	I can explain the ways in which urban environments interconnect with the Earth's biophysical environment.			
8.8	I can transform a set of data into a pictograph.			
	I can analyse pictographs.			
8.9	I can describe projected urban growth.			
	I can explain the challenges that urban growth presents and recommend sustainable approaches to development.			
8.10	I can describe patterns of urban growth in Indonesia and explain their causes and effects.			
8.11	I can locate features of Jakarta on a topographic map.			
	I can predict how the features of Jakarta's topography might influence the direction of urban sprawl and flooding.			

on Resources

eWorkbook   Chapter 8 Extended writing task (ewbk-8550)
Chapter 8 Reflection (ewbk-8549)
Chapter 8 Student learning matrix (ewbk-8548)

Interactivity   Chapter 8 Crossword (int-8652)

ONLINE RESOURCES

on Resources

Below is a full list of **rich resources** available online for this topic. These resources are designed to bring ideas to life, to promote deep and lasting learning and to support the different learning needs of each individual.

eWorkbook

- 8.1 Chapter 8 eWorkbook (ewbk-8100) ☐
- 8.2 The development of urban environments (ewbk-9851) ☐
- 8.3 Distribution of urban areas (ewbk-9855) ☐
- 8.4 SkillBuilder — Comparing population profiles (ewbk-9859) ☐
- 8.5 Urban expansion (ewbk-9863) ☐
- 8.6 Urban decline (ewbk-9867) ☐
- 8.7 Impacts of urbanisation on the environment (ewbk-9871) ☐
- 8.8 SkillBuilder — Creating and reading pictographs (ewbk-9875) ☐
- 8.9 Challenges for sustainable urban environments (ewbk-9879) ☐
- 8.10 Urbanisation in Indonesia (ewbk-9883) ☐
- 8.11 Investigating topographic maps — Jakarta (ewbk-9887) ☐
- 8.13 Chapter 8 Extended writing task (ewbk-8550) ☐
  Chapter 8 Reflection (ewbk-8549) ☐
  Chapter 8 Student learning matrix (ewbk-8548) ☐

Sample responses

- 8.1 Chapter 8 Sample responses (sar-0158) ☐

Digital document

- 8.11 Topographic map of Jakarta (doc-36319) ☐

Video eLessons

- 8.1 Our urban world (eles-1628) ☐
  Causes and consequences of urbanisation — Photo essay (eles-5215) ☐
- 8.2 The development of urban environments — Key concepts (eles-5216) ☐
- 8.3 Distribution of urban areas — Key concepts (eles-5217) ☐
- 8.4 SkillBuilder — Comparing population profiles (eles-1704) ☐
- 8.5 Urban expansion— Key concepts (eles-5218) ☐
- 8.6 Urban decline — Key concepts (eles-5219) ☐
- 8.7 Impacts of urbanisation on the environment — Key concepts (eles-5220) ☐
- 8.8 SkillBuilder — Creating and reading pictographs (eles-1659) ☐
- 8.9 Challenges for sustainable urban environments — Key concepts (eles-5221) ☐
- 8.10 Urbanisation in Indonesia — Key concepts (eles-5222) ☐
- 8.11 Investigating topographic maps — Jakarta — Key concepts (eles-5223) ☐

Interactivities

- 8.2 The growth in urban population over time (int-7857) ☐
  Distribution of the world's population (int-7966) ☐
  Percentage of urban population and urban agglomerations (int-7967) ☐
- 8.3 Distribution of urban areas (int-8646) ☐
- 8.4 SkillBuilder — Comparing population profiles (int-3284) ☐
- 8.5 Urban expansion (int-8647) ☐
- 8.6 Urban decline (int-8648) ☐
- 8.7 Impacts of urbanisation on the environment (int-8649) ☐
- 8.8 SkillBuilder — Creating and reading pictographs (int-3155) ☐
- 8.9 Challenges for sustainable urban environments (int-8650) ☐
- 8.10 Urban Indonesia (int-3315) ☐
- 8.11 Investigating topographic maps — Jakarta (int-8651) ☐
- 8.13 Chapter 8 Crossword (int-8652) ☐

ProjectsPLUS

- 8.12 Thinking Big Research project — Slum improvement proposal (pro-0216) ☐

Weblinks

- 8.4 Population pyramid ☐
  Population profiles ☐
- 8.9 Ecological footprint ☐

Fieldwork

- 8.2 Land use in your area (fdw-0042) ☐
- 8.7 The impact of traffic (fdw-0044) ☐

Google Earth

- 8.10 Jakarta, Indonesia (gogl-0064) ☐

myWorldAtlas

- 8.2 World urbanisation (mwa-4509) ☐
- 8.7 Polluted cities (mwa-7331) ☐
  Mexico City (mwa-4512) ☐
- 8.13 Urbanisation in Indonesia (mwa-7337) ☐

Teacher resources

There are many resources available exclusively for teachers online.

9 Urban settlement patterns

INQUIRY SEQUENCE

To access a pre-test with **immediate feedback** and **sample responses** to every question in this chapter, select your learnON format at www.jacplus.com.au.

9.1 Overview

Numerous **videos** and **interactivities** are embedded just where you need them, at the point of learning, in your learnON title at www.jacplus.com.au. They will help you to learn the content and concepts covered in this topic.

What damage is fast- paced, large- scale urban development creating — to the planet and to its people?

9.1.1 Introduction

Since 2008, more than 50% of the world's population lives in towns and cities. This urban population has grown at the expense of the rural population and is projected to continue growing in the future, reaching 68% by 2050. The fast pace and unplanned nature of this growth has seen the development of large concentrations of people within many countries. Along with the opportunities that urban places offer come many problems. It is a challenge to create sustainable and liveable urban environments that meet the needs of the people living in these places.

STARTER QUESTIONS

1. Would you prefer to live in an urban rather than a rural area? Why?
2. List two positive and two negative things about living in a large city.
3. Study the photograph of the city of Bangkok in **FIGURE 1**.
   a. What urban infrastructure can you see in this image?
   b. In your opinion, what challenges may this city experience?
   c. Suggest three different land use activities examples shown in this image. What evidence in the image supports your suggestion?
4. Watch the **Urban settlement patterns — Photo essay** video in your online Resources. Is this somewhere you'd like to live?

FIGURE 1 Wongwian Yai roundabout in Thonburi, Bangkok, Thailand

on Resources

eWorkbook	Chapter 9 eWorkbook (ewbk-8101)
Video eLessons	Megacities and megaregions (eles-1629) Urban settlement patterns — Photo essay (eles-5224)

9.2 Urban settlement patterns in Australia

LEARNING INTENTION

By the end of this subtopic, you will be able to describe the settlement patterns of Australia's population in the past and present, and explain the reasons for these patterns.

9.2.1 Population distribution in Australia

A settlement refers to a group of people living in one place or location. There may be collections of buildings and transport links within and between settlements that can vary in size, population density, extent, connectivity and economic importance. Many settlements originally grew near features such as water sources, food sources, transport routes, mineral deposits, river crossings and scenic locations. As settlements became more established and supported greater numbers of people, planned settlements with specific land use areas developed. Some examples of Australian settlements and reasons for their establishment are outlined in **FIGURE 1**.

Australians live on the smallest continent and in the sixth largest country on Earth. With a population of almost 25.5 million in 2020 and an area of 7 690 000 square kilometres, our **population density** is 3.2 people per square kilometre. We may think of ourselves as an outback-loving, farming nation, but we mostly live near the coast.

FIGURE 1 Key reasons for the location of settlements in Australia.

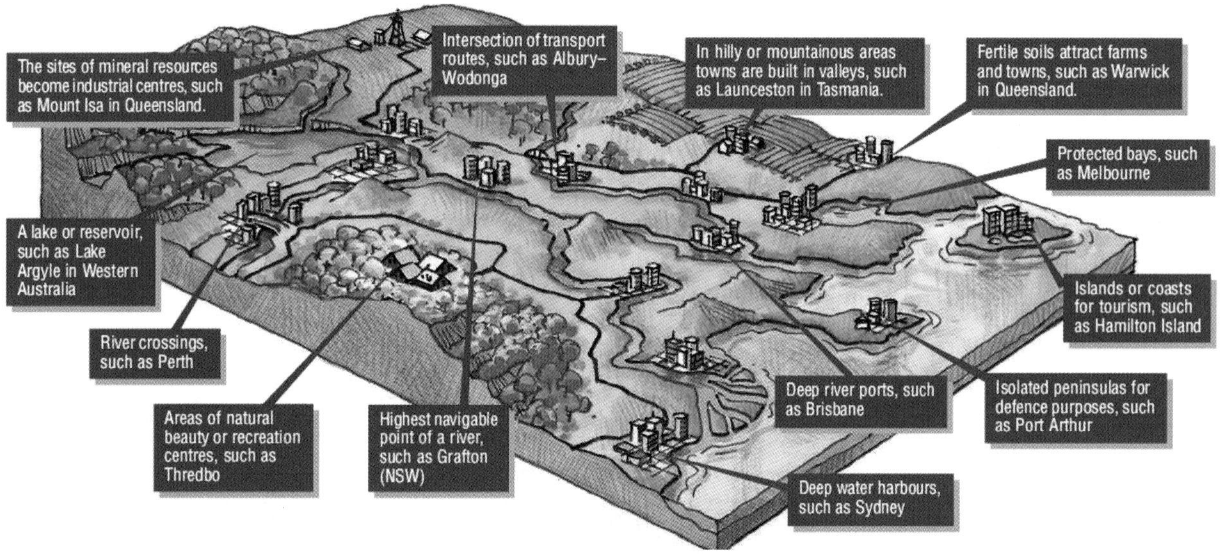
The sites of mineral resources become industrial centres, such as Mount Isa in Queensland.
Intersection of transport routes, such as Albury–Wodonga
In hilly or mountainous areas towns are built in valleys, such as Launceston in Tasmania.
Fertile soils attract farms and towns, such as Warwick in Queensland.
Protected bays, such as Melbourne
A lake or reservoir, such as Lake Argyle in Western Australia
Islands or coasts for tourism, such as Hamilton Island
River crossings, such as Perth
Deep river ports, such as Brisbane
Isolated peninsulas for defence purposes, such as Port Arthur
Areas of natural beauty or recreation centres, such as Thredbo
Highest navigable point of a river, such as Grafton (NSW)
Deep water harbours, such as Sydney

Most Australians currently live within a narrow coastal strip that extends from Brisbane in the north to Adelaide in the south. Seventy-one per cent of Australians live in major cities; in contrast, one in ten people live in small towns of less than 10 000 people. In 2016 there were just over 1000 towns with populations of fewer than 1000. About 85 per cent of people live within 50 kilometres of the coast. Australians love the beach, but is it just a coastal location that can explain this uneven population distribution pattern?

population density the number of people living within one square kilometre of land; it identifies the intensity of land use or how crowded a place is

FIGURE 2 shows the distribution of rainfall within Australia. Comparing **FIGURES 2** and **4**, it is apparent that there is a strong interconnection between the availability of more than 800 millimetres of rainfall per year and population distribution in the east, south-east and south-west of Australia. It would be easy to say that Australians live in places where rainfall is higher, but if you look at these maps carefully there are major exceptions to this spatial pattern. What is the relationship between population distribution and total rainfall in the north of Australia? From **FIGURE 2** we can see that Tully in Queensland has very high levels of rainfall, but there are fewer major cities and large towns on the northern coast than in the south, which received less rainfall. The same can be seen in Tasmania — over 2400 mm of rain falls on average in the west of the state, where there are only a few small towns.

int-8767

FIGURE 2 Distribution of rainfall in Australia.

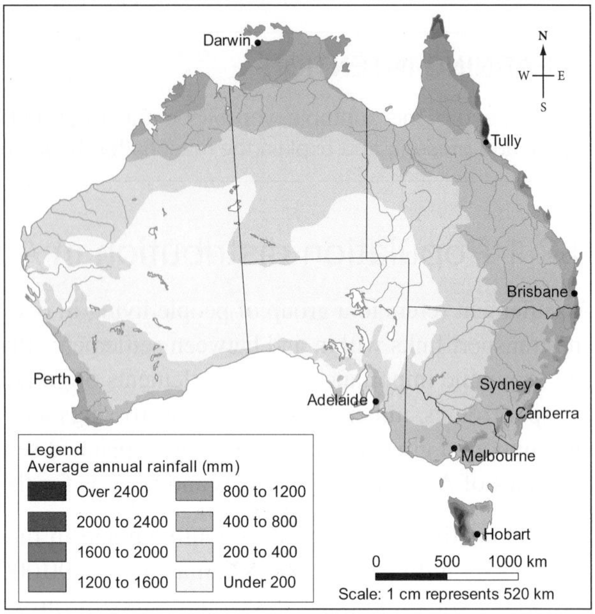

Source: Bureau of Meteorology

Coastal locations and rainfall are not the only reasons Australians live where they do. The availability of mineral resources, irrigation schemes to enhance farm production, and remote and stunning tourist destinations are geographical factors that draw people to live in a particular place.

9.2.2 Comparing population densities

FIGURE 4 shows Australia's population distribution in 2016. To better understand this data, we need to compare Australia's population density with that of other places in the world. This map shows that small areas around the major state capital cities have population densities of over 100 people per square kilometre of land. **TABLE 1** shows that the average population density for Australia is well below the global average and is easily the lowest of any of the permanently inhabited continents.

The population density of Australia is similar to that of Canada (4 people per square kilometre) but much lower than that of New Zealand (15 people per square kilometre), the United States (36 people per square kilometre) or China (148 people per square kilometre). Consider the geographical factors that Australia might share with Canada but not New Zealand, the United States or China that could explain the significant difference between their population densities.

FIGURE 3 Carrieton, South Australia, has a very low population (51) and the area has a population density of about 0.3.

int-8768

FIGURE 4 Distribution of Australia's towns by population size, 2016

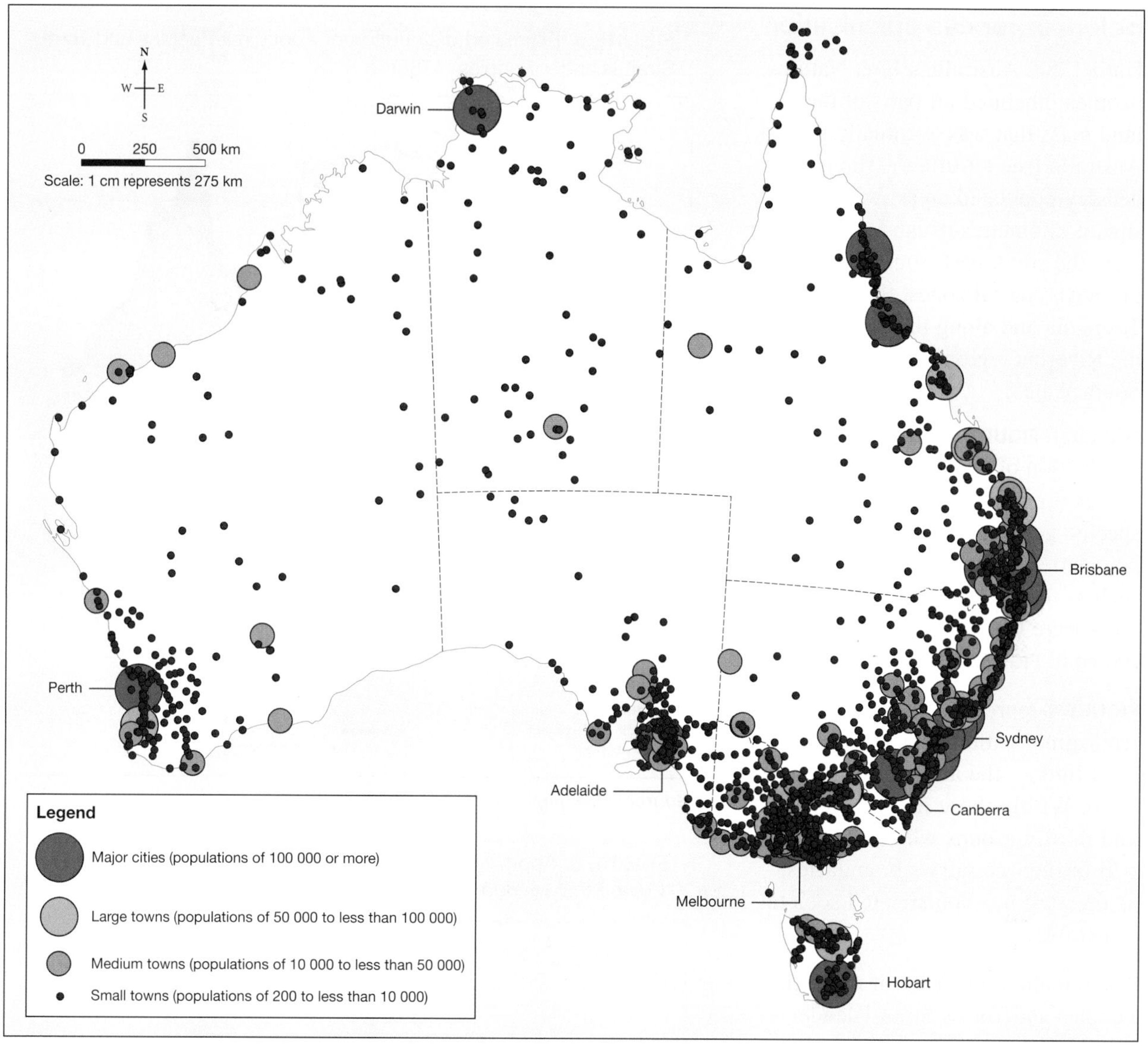

Source: Australian Bureau of Statistics

TABLE 1 The average population density for each continent

Continent	Average population density (people per km^2)
Asia	100
Europe	55
Africa	36
North America	20
South America	32
Australia	3
Antarctica	0.00007

Source: Data obtained from United Nations, Department of Economic and Social Affairs, Population, Division 2015. *World population Prospects: The 2015 Revision*. (rounded to whole figures)

9.2.3 Changing settlement patterns

Before European colonisation

Until 1788, Australia's First Nations peoples inhabited all parts of the land mass that was eventually named Australia (see **FIGURE 5**). The most densely populated areas, with 1–10 square kilometres of land per person, were the south-east, south-west and far north coastal zones, the north of Tasmania and along the major rivers of the Riverina region (south-western New South Wales).

Although **FIGURE 5** estimates the spatial distribution of where communities lived, it does not reflect that groups lived in specific geographical areas that had clear boundaries determined by landforms, such as mountains and rivers. These areas were language-speaking nations as shown in **FIGURE 7** for New South Wales.

FIGURE 7 represents the particular Aboriginal nations that people were born into — the areas in which they lived. Within these nations were clans and family groups who remained within their country's boundaries, as dictated by sophisticated societal structures.

The population density of Aboriginal Peoples and Torres Strait Islander Peoples was highest in places close to coastal and river environments. These places had the best availability of food and other resources. In a location such as Port Jackson, New South Wales, food was abundant and could be cultivated and managed to feed the population all year round. In places where rainfall is less reliable or frequent, such as parts of central Australia, it was harder to cultivate or harvest plentiful food sources in the one place all year or from season to season. When food resources ran low or with changing seasons, communities moved on to another part of Country. In this way, they managed their environment sustainably by not overusing the resources available at any one site.

FIGURE 5 Estimated distribution of Aboriginal Peoples and Torres Strait Islander Peoples (1788)

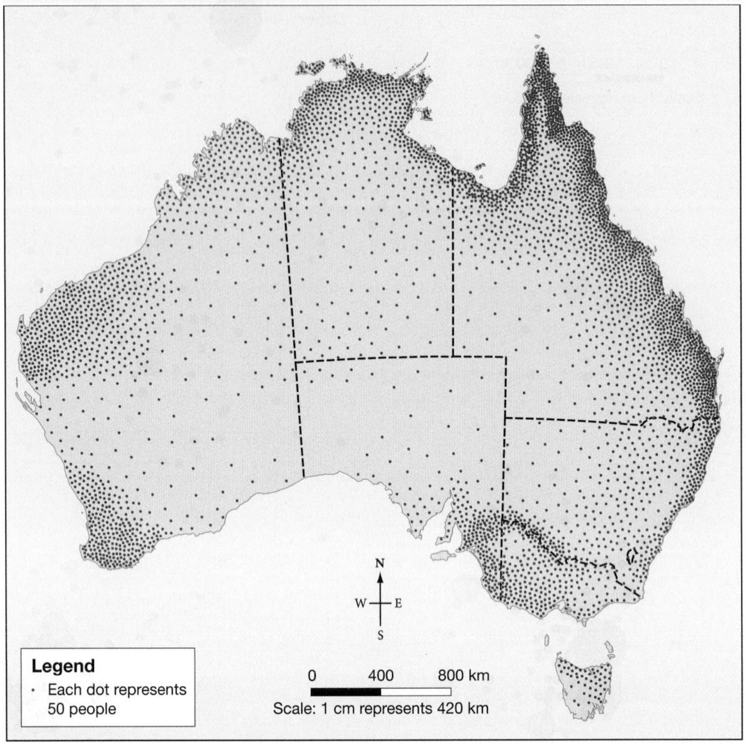

Source: MAPgraphics Pty Ltd, Brisbane

FIGURE 6 Approximate distribution of Aboriginal Peoples and Torres Strait Islander Peoples across Australia today

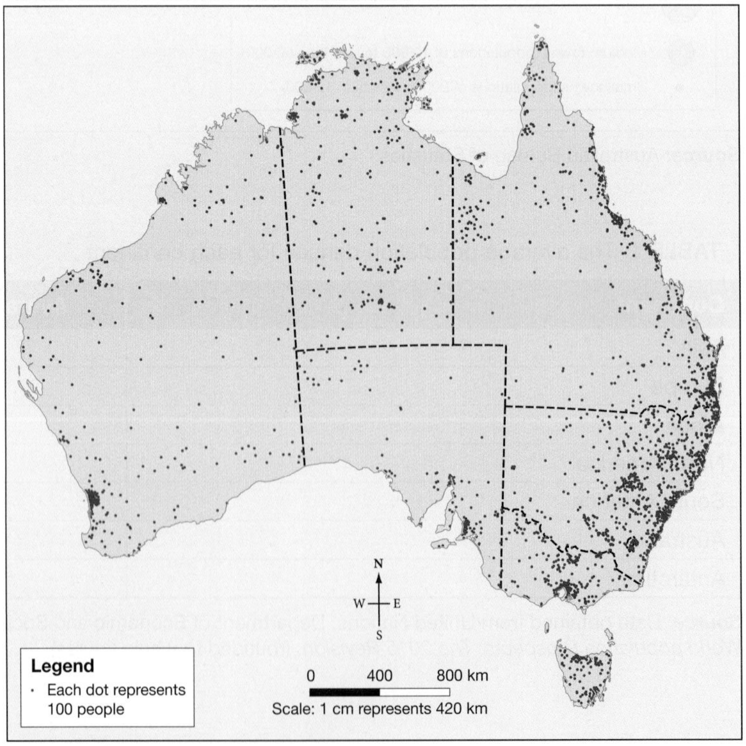

Source: MAPgraphics Pty Ltd, Brisbane

After European colonisation

Current estimates of the population of Australia in 1788 vary widely, from about 350 000 to about 700 000 or higher. Within 50 years of British occupation, this population had been greatly reduced by a range of causes including the introduction of new diseases — accidentally and intentionally and the direct, violent actions of British colonists. The 2016 Australian census recorded 649 171 people who indicated they were of Aboriginal and/or Torres Strait Islander heritage, making up about 2.8 per cent of Australia's population.

Although many Aboriginal Peoples live off-Country, colonisation has not removed Aboriginal nations and their geographical boundaries as identified in **FIGURE 7**. They are ever-present today, and these language-speaking nations are still key factors in the structure of Aboriginal society and people's connection to place today **(FIGURE 6)**.

The Australian environment was managed and altered by Aboriginal Peoples and Torres Strait Islander Peoples over time, but in ways that ensured sustainability of resources and worked with natural systems and cycles. European approaches to land management, however, are vastly different. As a consequence, Australia has changed significantly since 1788. Much land has been cleared for cities, shaped for farms and blasted for mines. By the twenty-first century, little of Australia's environment is not significantly changed by the thousands of years of continual human occupation.

FIGURE 7 Aboriginal nations/language groups in New South Wales and ACT

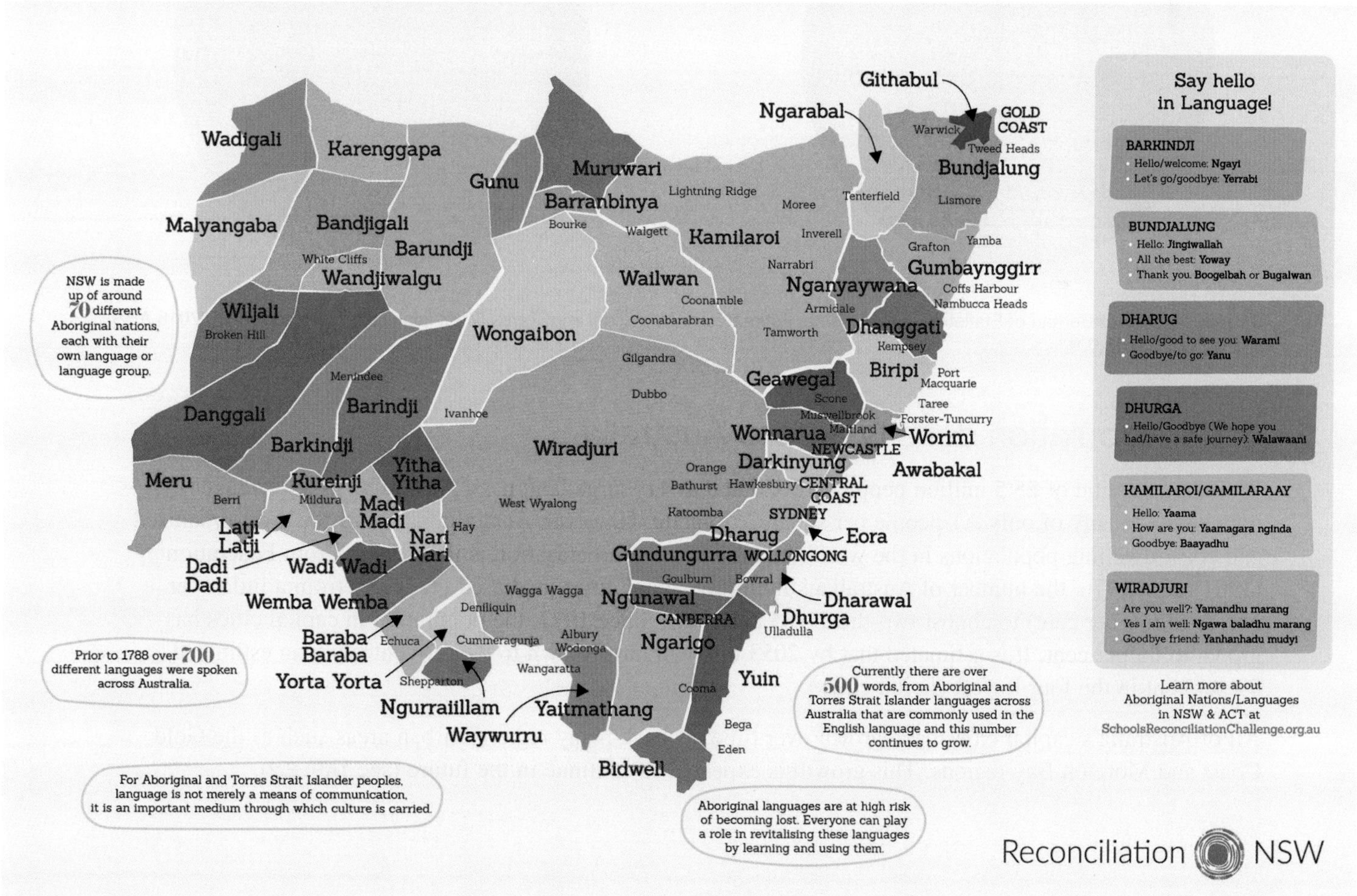

Source: Aboriginal Languages/Nations in NSW & ACT © Reconciliation NSW, www.reconciliationnsw.org.au

The patterns in **FIGURES 5** and **6**, showing the distribution of Aboriginal and Torres Strait Islander populations in 1788 and today, are generally similar. Since before 1788, most of Australia's Aboriginal Peoples and Torres Strait Islander Peoples tended to live in the same relatively small region of this country. As **FIGURE 8** shows, the majority of Aboriginal and Torres Strait Islander Peoples now live in major cities and regional centres.

FIGURE 8 Regional distribution of the Australian population

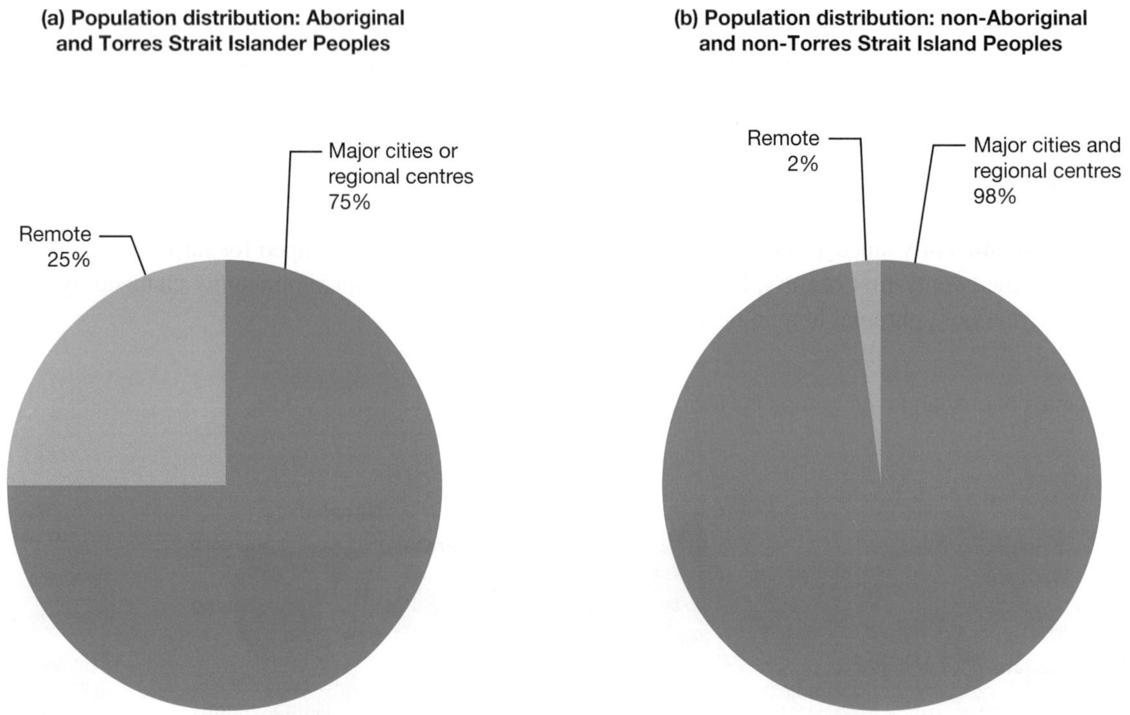

Source: Australian Bureau of Statistics, Experimental Estimates of Aboriginal and Torres Strait Islander Australians, June 2016 ABS cat. no. 3238.0.55.001

9.2.4 Population density across Australia

With a population of 25.5 million people in 2020 and a very large landmass, Australia has an average population density of only 3.1 people per square kilometre. However, Australia is one of the most urbanised and coast-dwelling populations in the world, and the level of urbanisation is increasing. From Federation (1901) until 1976, the number of Australians living in capital cities increased gradually from a little over one-third (36 per cent) to almost two-thirds (65 per cent). Since 1977, the population in capital cities has grown to 66 per cent. It is estimated that by 2053 this will have grown to 72 per cent (with an estimated 89 per cent in the four largest capital cities).

All of Australia's capital cities have grown over time, as have many regional urban areas such as the Gold Coast and Moreton Bay regions. This growth is expected to continue in the future (see **TABLE 2**).

TABLE 2 Australian capital city populations in 2018–19 and projected populations in 2036 and 2066

City	2018–19 population	Projected 2036	Projected 2066
Sydney	5 132 163	7 379 976	11 240 860
Melbourne	5 078 193	7 520 830	12 235 490
Brisbane	2 514 184	3 596 431	5 782 256
Perth	2 059 484	2 798 994	4 330 509
Adelaide	1 345 777	1 605 335	2 068 550
Hobart	246 970	297 085	466 752
Darwin	148 564	195 082	295 458
Total	**16 705 335**	**23 393 733**	**36 419 875**

Source: © Infrastructure Australia 2020

on Resources

eWorkbook Urban settlement patterns in Australia (ewbk-9895)

Video eLesson Urban settlement patterns in Australia — Key concepts (eles-5225)

Interactivity Urban settlement patterns in Australia (int-8653)

myWorldAtlas Deepen your understanding of this topic with related case studies and questions.
- Population of Australia

9.2 ACTIVITIES

1. Use your atlas to identify and list:
   a. geographical landforms or climatic features that are common to Australia and Canada. *Hint:* Look for large regions that have an extreme climate. Explain why.
   b. reasons New Zealand, the United States or China may have a higher population density than Australia. In small groups, discuss the possible reasons and decide which is the most plausible.
2. Consider the issues and problems that increasing city populations will create. Discuss this as a class and construct a consequence chart to summarise all the ideas. What might be some solutions to these issues and problems? Add these to your chart.
3. Refer to **FIGURES 2** and **4** to produce an overlay map that identifies the interconnection between the distribution of population and the distribution of rainfall within Australia.
   a. Describe areas where there are strong similarities between these two features, for example high population distribution and high rainfall, or low population distribution and low rainfall.
   b. Describe places that have a high population distribution but low rainfall or vice versa.
4. Use various theme maps of Australia in your atlas to identify at least four possible place or environmental explanations for the pattern of distribution and density of Australia's population. Discuss your findings with the class or in small groups.
5. Conduct research to find which country in the world has the highest average population density. Find one country with a lower average population density than Australia.
6. Use the **Population density** weblink in your online Resources to investigate your area's population density.
   a. How many people live in your local government area? How much space do they occupy?
   b. Which Aboriginal language group or nation's Country (or Countries) does your local government area include?
   c. Determine what proportion of the population in your local government area are people of Aboriginal and/or Torres Strait Islander heritage.

9.2 EXERCISE

Learning pathways

■ LEVEL 1	■ LEVEL 2	■ LEVEL 3
1, 3, 7, 10, 15	2, 4, 8, 11, 13	5, 6, 9, 12, 14

Check your understanding

1. Define, using your own words, the term *settlement*.
2. Choose two settlements featured in **FIGURE 1** with different locational features. Describe the differences between these two settlements.
3. What is the population density of Australia?
4. What percentage of Australians live in urban areas close to the coast?
5. Use **TABLE 1** to compare the population density of Australia with that of:
   a. North America
   b. Europe
   c. South America.
6. How many Aboriginal people or Torres Strait Islander people:
   a. lived in Australia in 1788
   b. live in Australia today?

Apply your understanding

7. Where do most Australians live? Suggest reasons why.
8. What geographical factors other than rainfall may lead to the uneven distribution of population in Australia?
9. Compare **FIGURES 1**, **2** and **4**.
   a. Identify areas where there are strong similarities between two features (for example high population distribution and high rainfall, or low population distribution and low rainfall).
   b. Explain the interconnection between the two features.
10. Explain why some places with low rainfall still have large populations.
11. Refer to **TABLE 2**. Draw a bar graph to show the predicted change in the populations of Australia's capital cities. What does your graph reveal?
12. Use the statistics in **TABLE 1** to produce a pictograph that illustrates the contrasts between the average population densities for each continent.

Challenge your understanding

13. Write a paragraph to predict the possible change in the distribution of Australia's population over the next 50 years if the following situations occur.
    a. The coastal urban areas become adversely affected by loss of land due to rising sea levels.
    b. A 20-year-long drought occurs in south-eastern Australia.
14. Use information from **FIGURE 2** to predict why, in the future, there may be significant movement of people from the southern states of Australia to places in the tropical north. Your answer must refer to specific information from the map.
15. Refer to **FIGURE 6**. Living so far away from major cities means that 25 per cent of Aboriginal and Torres Strait Islander communities have limited access to many of the services and opportunities that cities offer their residents. Suggest what lifestyle and service difficulties may be associated with living so remotely.

To answer questions online and to receive **immediate feedback** and **sample responses** for every question, go to your learnON title at www.jacplus.com.au.

9.3 Comparing urbanisation — USA and Australia

LEARNING INTENTION

By the end of this subtopic, you will be able to describe the similarities and differences in urbanisation between the United States and Australia, and identify the consequences of urbanisation in both places.

9.3.1 Urbanisation in the United States and Australia

Both the United States and Australia are very large countries that are highly urbanised. In fact, both are among the world's most urbanised nations.

The United States and Australia have some similarities and some differences in terms of how urbanised they are, as revealed in **TABLE 1** and **FIGURE 1**. One of these differences is how urban areas are defined and how the population is measured. This means population data for urban areas can be difficult to compare. United States government data measures a city's population by the number of people living in the urban centre, rather than the greater metropolitan area. This means that data for New York City may state the population as being about 8 million, whereas the population of the New York City's greater metropolitan area (including outer suburbs, which is how Australian data generally shows population) is closer to 19 million. Based on the data **FIGURE 1**, New York City's population could be misinterpreted to be not even twice that of Sydney if the different sources of data collection are not recognised.

FIGURE 1 Population of the top 10 urban settlements in (a) the United States (2018) and (b) Australia (2019)

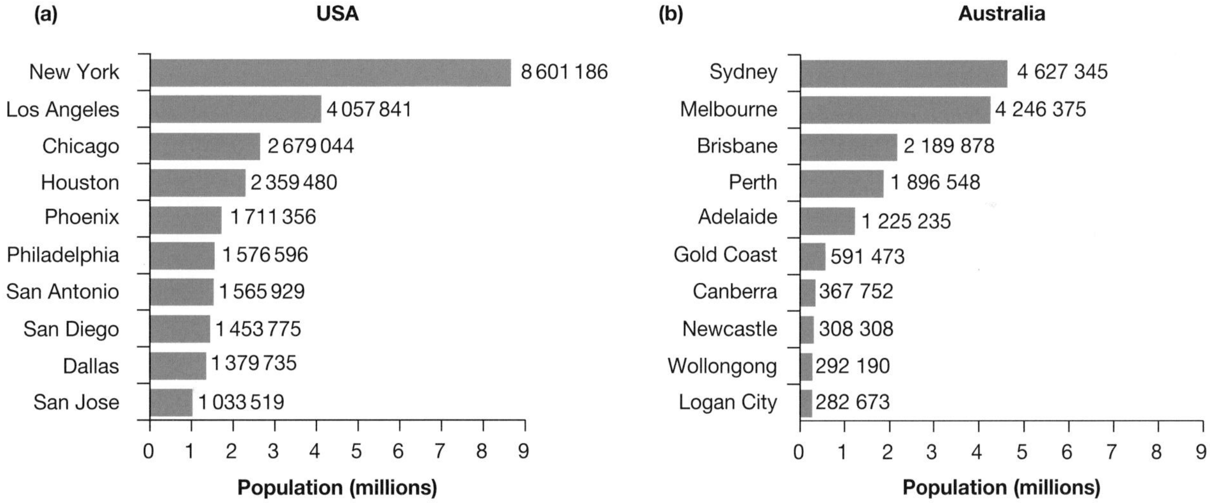

Source: Based on data from US Population Bureau and the ABS

TABLE 1 A comparison of urbanisation in the United States and Australia

	United States	Australia
Population	328 239 523 in 2019	25 364 307 in 2019
UN urban definition	2500+ inhabitants	10 000+ inhabitants
Population distribution	Over 81% live in urban areas.	Over 89% live in urban areas.
People living in large cities	Approximately 1 of every 10 people in the United States live in either the New York or Los Angeles metropolitan areas.	Approximately 4 of every 10 people in Australia live in either Melbourne or Sydney.

Source: United Nations Population Division, World Urbanization prospects, 2018

9.3.2 Causes of urbanisation

The causes of urbanisation are similar for both Australia and the United States. In each case, since the country was founded:

1. fewer people were needed to work in rural areas as technology reduced the demand for labour on farms
2. more jobs and opportunities were available in factories, which were located in urban areas
3. the development of railways allowed goods produced in one city to be transported to rural and urban areas
4. cities could grow and develop thanks to new technologies (steel-framed skyscrapers) and utilities (for example electricity and water supply).

9.3.3 Consequences of urbanisation

Urban sprawl

More affordable personal transport, greater access to internet services and cheaper land values in the urban fringes of US and Australian cities have contributed to high levels of car ownership and urban expansion. As urban places expand outwards, they may attract residents to the outskirts of the city looking for a less congested and more peaceful lifestyle. Transport and technology infrastructure connects them to many of the benefits of city living such as specialist jobs, healthcare, education, entertainment, and shopping opportunities.

Suburban sprawl is a consequence of urbanisation in many US and Australian cities. The fastest-growing city in the US over the past four years has been Phoenix, the state capital of Arizona. **FIGURE 2** shows the expansion of suburban-style housing estates into the countryside around Phoenix.

FIGURE 2 Suburban expansion in Phoenix, Arizona

Conurbations

Sometimes there are so many cities in a particular region that they seem to merge almost into one city as they expand. A conurbation is made up of cities that have grown and merged to form one continuous urban area along corridors of urban growth. Both the United States and Australia have conurbations.

United States

Eleven conurbations have been identified in the United States (see **FIGURE 3**). The major conurbation is in the north-east region. It is often called BosNYWash because it covers the area from Boston in the north, through New York to Washington in the south. This region is home to over 50 million people (17% of the US population) and accounts for 20 per cent of the gross domestic product (GDP) of the United States.

FIGURE 3 Conurbations in the United States

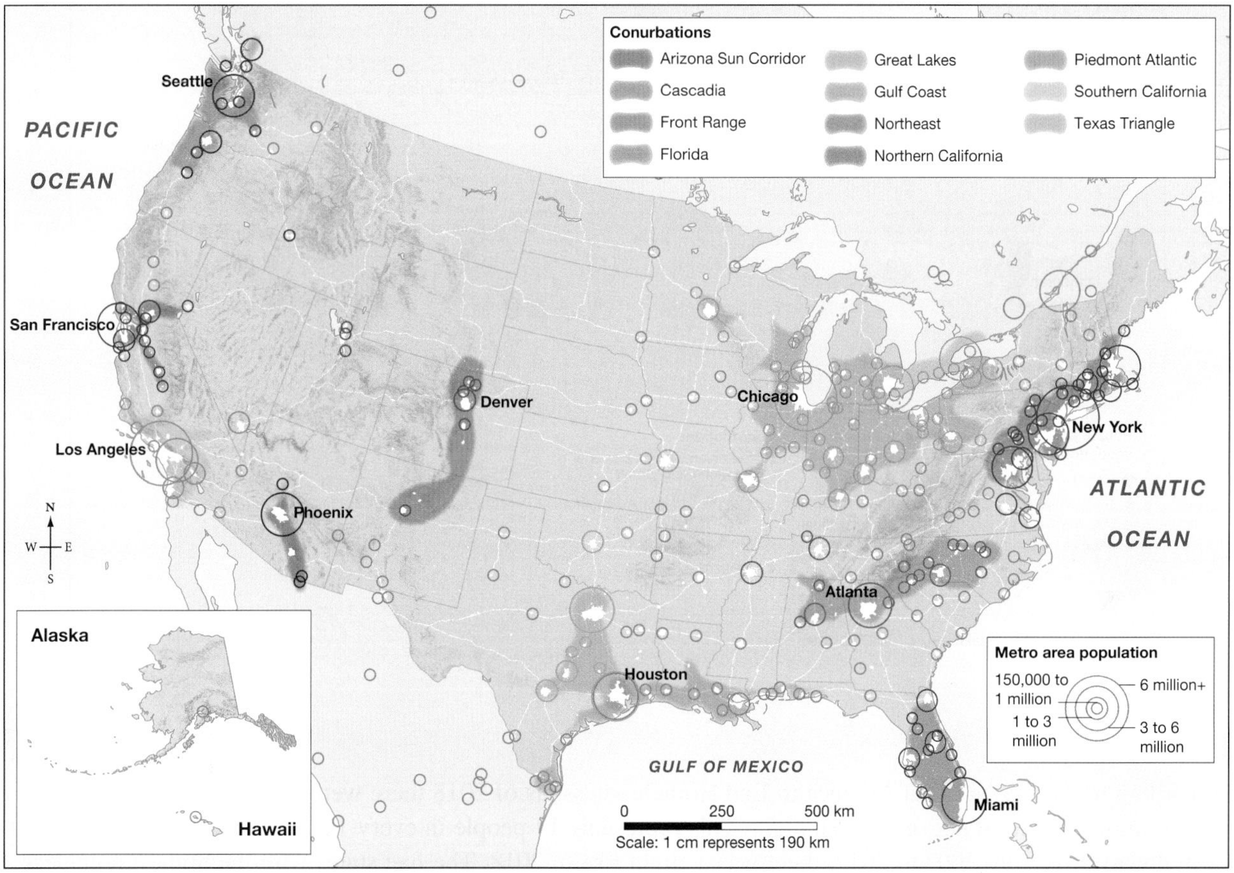

Source: Adapted with permission from Bernard Salt; USGS (2011, 2021).

Australia

Australia, on the other hand, has four conurbations (see **FIGURE 4**). One is in south-east Queensland, one joins Melbourne and Geelong, one is from Perth to Mandurah, and one is from Newcastle to Wollongong. The Newcastle–Wollongong conurbation stretches for over 250 kilometres and is home to almost six million people. The area with the highest growth during 2019 was Camden in Sydney's south-west. It is expected that populations in regional centres of Australia will grow following COVID-19 as residents of densely populated urban places migrate to less densely populated, remotely connected places.

int-8769

FIGURE 4 Australia's population centres and conurbations

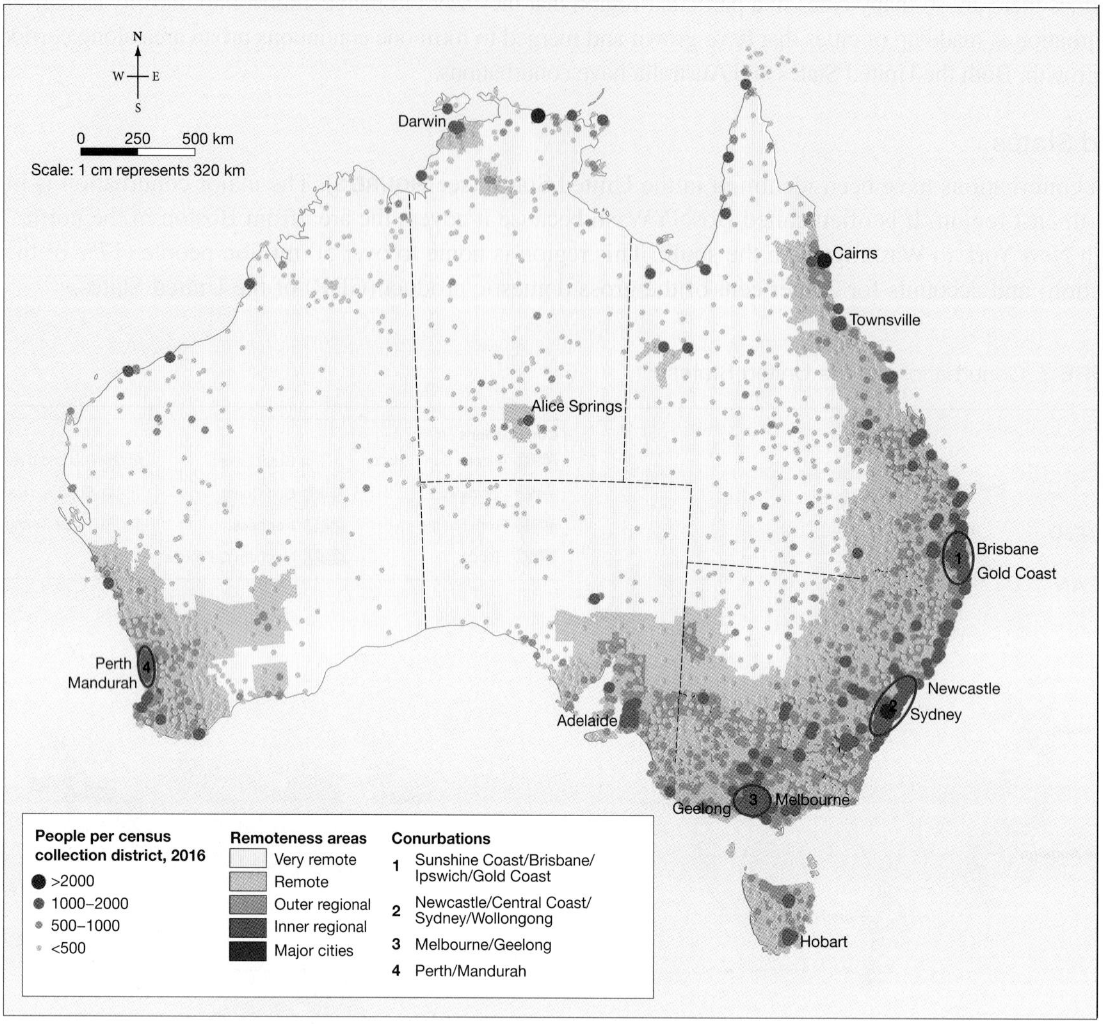

Source: Australian Bureau of Statistics, 2016

Homelessness

According to the US National Alliance to End Homelessness, as of 2018 there were around 553 000 homeless people in the United States on a given night. This represents 17 people in every 10 000. Although the trend had been downwards from 2007 to 2017, there was a slight rise in 2018. The five states with the highest homeless counts in 2018 were California (129 972), New York State (91 897), Florida (31 030), Texas (25 310) and Washington State (22 305).

In comparison, census data shows that the number of homeless people in Australia increased by more than 15 000 (14 per cent) over five years to 2016. According to the Australian Bureau of Statistics, 116 000 people were homeless on census night in 2016, representing 50 homeless people per 10 000. This was an increase of 13.7 per cent from the 2011 census.

Health issues

High population densities in urban areas make it easier for diseases to be transmitted, especially in poor neighbourhoods. The urban poor suffer health issues caused by reduced access to sanitation and hygiene facilities and health care.

Pollution

Air pollution from cars, industry and heating affects people who live in cities. A study in the United States showed that more than 3800 people die prematurely in the Los Angeles Basin and San Joaquin Valley region of Southern California because of air pollution. Generally, Australia has a fairly high level of air quality. Cars and industry are the main factors influencing air quality in urban areas.

9.3.4 Characteristics of cities in the United States

In 1950 there were only two megacities, New York and Tokyo in Japan. In 2015, New York was the sixth-largest city in the world, but by 2025 it is expected to be only sixtieth on this list. There are only 11 states in the United States that are home to more people than New York City.

CASE STUDY

New York City

New York City is located on the eastern Atlantic Ocean at the mouth of the Hudson River. Being located on four islands makes land very scarce and population density very high, on average 10 194 people per square kilometre.

Boroughs

New York City is made up of five boroughs (counties): the Bronx, Brooklyn, Manhattan, Queens and Staten Island. The Bronx is the only part of New York that is connected to the US mainland. Brooklyn is where most New Yorkers live, but Manhattan is the most densely populated borough. Manhattan contains the highest number of skyscrapers and includes Central Park, which is nearly twice as large as the world's second-smallest country, Monaco. Queens is the largest borough in area and is thought to be the place on Earth where the most languages are spoken in one locality. Staten Island has the smallest population of the five boroughs and is home to more Americans of Italian ancestry than any other county in the USA.

FIGURE 5 Aerial view of lower Manhattan

TABLE 2 Population statistics of greater New York City: (a) 2010 census; (b) New York City Planning Population Division, 2019 estimates

Region/borough	2010 Census (a)	2019 NYC PPD (b)
New York State	19 378 102	19 453 561
New York City	8 175 133	8 336 817
Bronx	1 385 108	1 418 207
Brooklyn	2 504 700	2 559 903
Manhattan	1 585 873	1 628 706
Queens	2 230 730	2 253 858
Staten Island	468 730	476 143

Source: Table used with permission of the New York City Department of City Planning. All rights reserved

People

For many years, almost all immigrants came to the United States through New York City — and many of them remained. Many people living in New York are originally from European countries, but there are large numbers from the West Indies, South and Central America, the Middle East and eastern Asia. Around 800 languages are spoken in New York, and around 36 per cent of the city's population were born overseas.

FIGURE 6 Central Park in New York

Recent changes to US immigration policy and COVID-19 restrictions have changed the way New York city operates. Highly concentrated urban places are less popular, with people wanting to minimise their exposure to COVID-19 risk factors. Tighter controls over US government immigration policy have also contributed to a decline in the overseas-born population of New York. Migration flows into the city fell by 46% between 2016 and 2019, and this has impacted the availability of migrant labour, business ownership and the total population size of New York City.

Economy

New York City is a major world centre of trade, commerce and banking, manufacturing, transportation, finance, communications, and cultural and theatrical production. It is the headquarters of the United Nations and a leading seaport. New York is also home to the largest stock exchange in the world.

FIGURE 7 Net international migration, New York City, 2011–2019

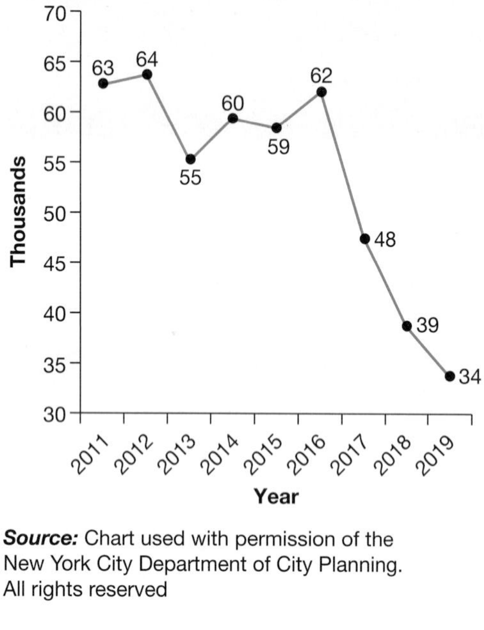

Source: Chart used with permission of the New York City Department of City Planning.

FIGURE 8 Geographical characteristics and famous landmarks of New York City

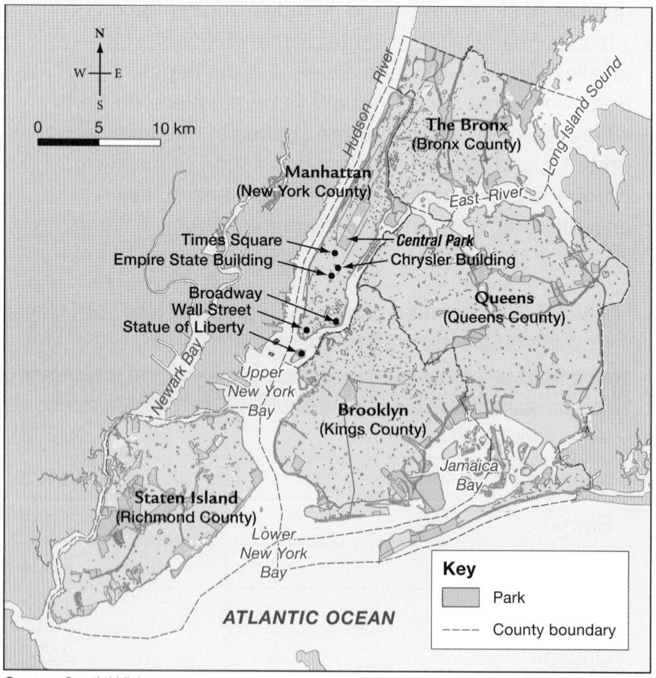

Source: Spatial Vision

FIGURE 9 Satellite image of New York City

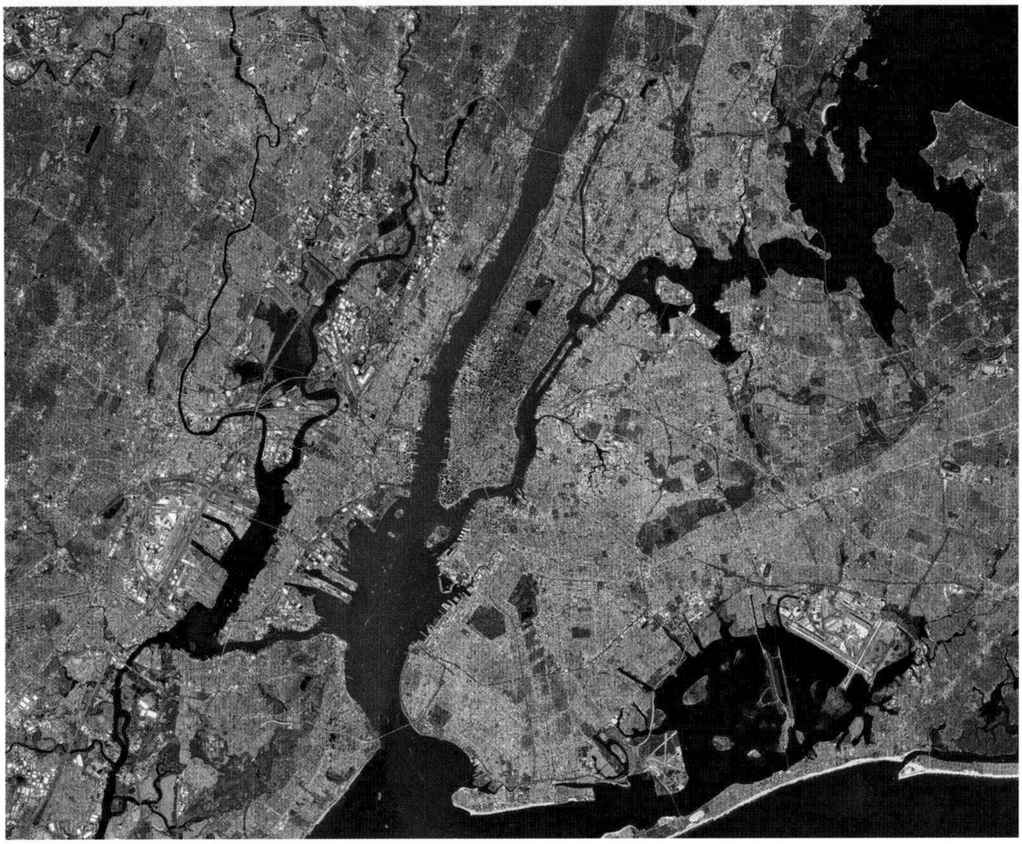

Resources

eWorkbook	Comparing urbanisation — USA and Australia (ewbk-9899)
Video eLesson	Comparing urbanisation — USA and Australia — Key concepts (eles-5226)
Interactivity	Comparing urbanisation — USA and Australia (int-8654)
myWorldAtlas	Deepen your understanding of this topic with related case studies and questions. • Urbanisation in Australia and the USA

9.3 ACTIVITIES

1. Investigate US policy on immigration.
   a. What changes took place to US immigration policy in 2016? Are these policies still in effect? Why were these changes introduced?
   b. What have the benefits to US cities been? What have the costs to US cities been?
   c. Suggest other federal government policies that may change the way a city operates.
2. Find the names of one newspaper, magazine, fashion brand, major bank, art gallery and theatre that are located in New York. Compare your findings with those of other students. Are these well-known businesses? How does this help make New York an important business and cultural centre?
3. Compare two places that have developed rapidly in recent years.
   a. Research the reasons for the rapid development of Phoenix, Arizona between 2016 and 2020. Outline these reasons in a four-sentence written description.
   b. Research the reasons for the rapid development of Camden in Sydney's south-west during 2019. Outline these reasons in a four-sentence written description.
   c. Bring your descriptions together into a coherent paragraph that compares the two places' development.

4. Research and describe the changes in New York due to COVID-19. Prepare a one-page evaluation of the changes to life in New York that considers job opportunities, access to health care, spending levels, housing availability, personal safety and levels of wellbeing.
5. Draw a sketch of **FIGURE 6**. Use a map to help you label Central Park and the Hudson River. In which direction is the photographer facing?

9.3 EXERCISE

Learning pathways

■ LEVEL 1	■ LEVEL 2	■ LEVEL 3
1, 4, 6, 12, 15	2, 5, 8, 11, 14	3, 7, 9, 10, 13

Check your understanding

1. Identify one reason why it can be difficult to compare population data for cities in different countries.
2. Refer to **TABLE 1**. Using data, describe the differences between Australia and the US in terms of the number of people living in cities.
3. Compare the scale of urbanisation in the United States and in Australia.
4. Outline the common factors that have caused urbanisation in Australia and the US.
5. In your own words, define the terms:
   a. *urban sprawl*
   b. *conurbation*.
6. Name the largest conurbation in the United States and in Australia.
7. How does the population of the United States compare to that of Australia? How many times larger (approximately) is one than the other?

Apply your understanding

8. Why do Australia and the US have a different definition of the term *urban*?
9. Refer to **FIGURE 1** and **TABLE 2**. Compare the data given about the 10 largest cities in the United States and in Australia. Explain, using examples, how this data might be misinterpreted.
10. Why might there be more conurbations in the United States than in Australia?
11. Apart from conurbations, what are three consequences of urbanisation?
12. How does Australia compare to the United States in regards to levels of homelessness?

Challenge your understanding

13. The United States was colonised by the British nearly 180 years before Australia. Do you think that Australia will have a population the size of the current US population in 180 years from now? Give reasons for your answer.
14. Suggest one strategy that large cities might use to combat homelessness, especially during difficult times such as during economic downturns or pandemics.
15. Why might a pandemic be particularly dangerous in large cities with high population densities, such as New York? Suggest what the greatest challenges might be and propose how each could be overcome.

To answer questions online and to receive **immediate feedback** and **sample responses** for every question, go to your learnON title at www.jacplus.com.au.

9.4 SkillBuilder — Creating and reading compound bar graphs

online only

LEARNING INTENTION

By the end of this subtopic, you will be able to create and read a compound bar graph.

The content in this subtopic is a summary of what you will find in the online resource.

9.4.1 Tell me

What are compound bar graphs?

A compound bar graph is a bar or series of bars divided into sections to provide detail for a total figure. These bars can be drawn vertically or horizontally. The height or length of each section represents a percentage, with the total length of the bar representing 100 per cent.

9.4.2 Show me

How to create and interpret a compound bar graph

Step 1

To complete a compound bar graph you must have a set of data that totals 100 per cent, with information as to how that total is made up.

Step 2

Decide on a horizontal axis and vertical axis for the bar graph. The easiest length to work with is 10 centimetres, so that each millimetre represents 1 per cent, or 10 millimetres represents 10 per cent. Draw your vertical axis 10 centimetres long. Add a scale alongside the axis (see **FIGURE 1**).

FIGURE 1 Compound bar graph illustrating 2018 world population estimates

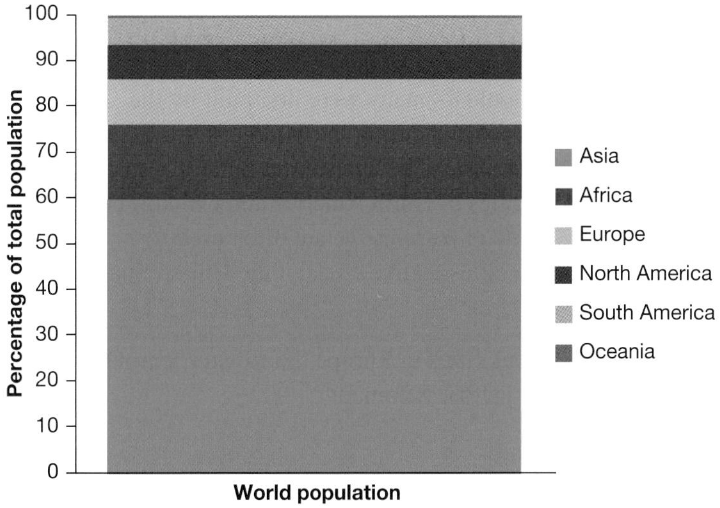

Step 3

Since this is a compound graph, all numbers compound, or add onto one another. Mark on your graph the length of the section of bar representing your first piece of data as a percentage. Add the next percentage to the percentage for the first piece of data. In **FIGURE 1**, 16.7 per cent has been added to the previous 59.8 per cent for a total of 76.5 per cent. Draw a line where this percentage is represented on your bar. Shade the segment in a different colour; add this colour to the key.

Step 5

Complete the graphing, colouring and key. Add a title and state the source of your data under the graph.

Step 6

To write a description of the information, begin with a comment on the most obvious feature — the colour that fills the largest section of the bar. Consider each of the other coloured sections of the compound bar and comment on how these relate to one another.

9.4.3 Let me do it

learnON

Go to learnON to access the following additional resources to help you build this skill:

- a longer explanation of this skill and its application in Geography (Tell me)
- a video demonstrating the step-by-step process of this skill (Show me)
- an activity and interactivity for you to practise the skill (Let me do it)
- self-marking questions to help you understand and use the skill.

on Resources

eWorkbook	SkillBuilder — Creating and reading compound bar graphs (ewbk-9903)
Video eLesson	SkillBuilder — Creating and reading compound bar graphs (eles-1705)
Interactivity	SkillBuilder — Creating and reading compound bar graphs (int-3285)

9.5 Characteristics of cities in Europe

online only

LEARNING INTENTION

By the end of this subtopic, you will be able to describe the characteristics of cities located in Europe and explain the factors that make these cities unique.

The content in this subtopic is a summary of what you will find in the online resource.

European cities are old — many were first built by the Romans, and most existed during the Middle Ages. They are often smaller in scale and have shorter buildings than the huge modern cities of North America and China. The terms *romantic, chic* or *picturesque* are often used to describe European cities, unlike cities of the United States or Australia.

To learn more about cities in Europe, go to your learnON resources at www.jacPLUS.com.au.

FIGURE 3 The medieval quarter of central Saint-Emilion, Gironde, Aquitaine, France

Contents

learnON

- 9.5.1 Typical features of European cities
- 9.5.2 Consequences of urbanisation in Europe

on Resources

eWorkbook	Characteristics of cities in Europe (ewbk-9907)
Video eLesson	Characteristics of cities in Europe — Key concepts (eles-5227)
Interactivity	Characteristics of cities in Europe (int-8655)

9.6 SkillBuilder — Constructing and describing isoline maps

online only

LEARNING INTENTION

By the end of this subtopic, you will be able to construct and analyse the information presented in an isoline map.

The content in this subtopic is a summary of what you will find in the online resource.

9.6.1 Tell me

What is an isoline map?

An isoline map shows lines that join all the places with the same value. Isoline maps show gradual change in one type of data over a continuous area. **FIGURE 2** shows a completed coloured isoline map.

FIGURE 2 A coloured isoline map showing travel time to Copenhagen airport by car

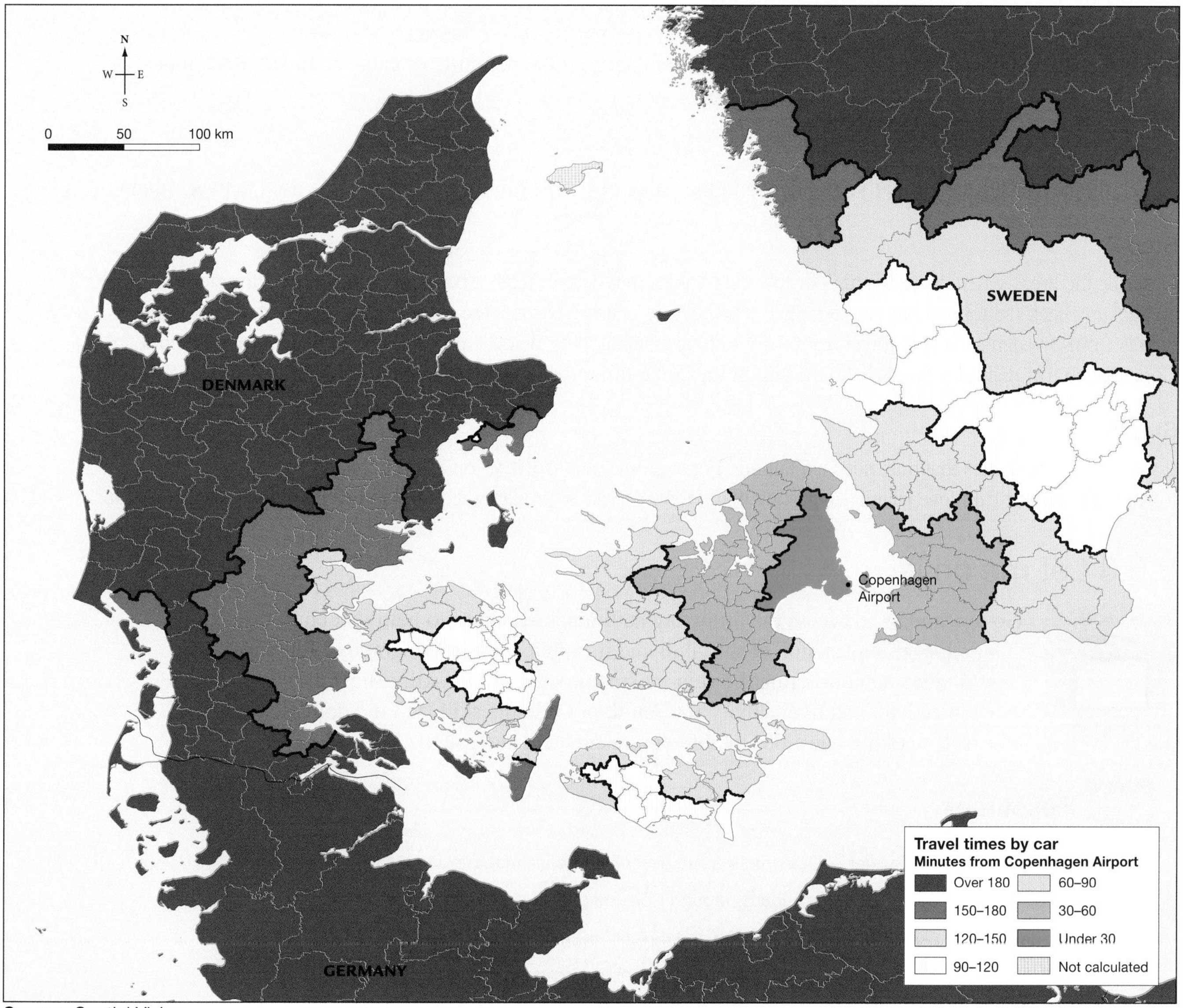

Source: Spatial Vision

9.6.2 Show me

How to construct an isoline map

Step 1

Select a set of data to map, and plot the relevant figure at each of the places listed.

Step 2

Select a value to use for intervals within the data set. In **FIGURE 2**, the interval was set at 30 minutes. Draw lines (isolines) joining places of the same value; in this example, that would be 30 minutes, 60 minutes, 90 minutes and so on.

Step 3

Create a legend for your map with a colour system that indicates a gradation of colour, where the lowest data is the lightest shade and the highest data is the darkest. Shade between the isolines according to the legend.

Step 4

Apply BOLTSS to your map.

Interpreting an isoline map

Step 5

Make sure you know what feature is being mapped by checking the map title or caption. In **FIGURE 2**, this feature is travel time by car to Copenhagen airport.

Step 6

Check the key or legend so that you understand the value of each isoline and the intervals used between them.

Step 7

Describe the areas where there are high or low data values that help to form a pattern. You may need to refer to an atlas to check the topography and establish whether any country borders are involved. In **FIGURE 2**, some of the data is from Sweden as people travel from there to Copenhagen, their nearest airport. People living west of central Copenhagen travel similar distances to the airport in similar times (60–90 minutes) to those living in southern Sweden.

Step 8

Look for any anomalies that may need explaining. For example, in **FIGURE 2** you can see that it takes 150–180 minutes to get to the airport from one island, suggesting that a ferry service is probably required to reach the road system by car.

9.6.3 Let me do it

learnON

Go to learnON to access the following additional resources to help you build this skill:

- a longer explanation of this skill and its application in Geography (Tell me)
- a video demonstrating the step-by-step process of this skill (Show me)
- an activity and interactivity for you to practise the skill (Let me do it)
- self-marking questions to help you understand and use the skill.

on Resources

eWorkbook	SkillBuilder — Constructing and describing isoline maps (ewbk-9915)
Digital document	Base map of the suburbs around Copenhagen city centre (doc-36410)
Video eLesson	SkillBuilder — Constructing and describing isoline maps (eles-1737)
Interactivity	SkillBuilder — Constructing and describing isoline maps (int-3355)

9.7 Characteristics of cities in South America — São Paulo

online only

LEARNING INTENTION

By the end of this subtopic, you will be able to describe general settlement patterns in South America and discuss the causes and consequences of urbanisation with specific examples from São Paulo, Brazil.

The content in this subtopic is a summary of what you will find in the online resource.

São Paulo is located on the south-eastern coast of Brazil in South America. It is the most populated city in Brazil. From 1950 to 1980, São Paulo's population quadrupled from two million to more than eight million. In 2020, the population of the greater metropolitan area was estimated to be 22 043 000.

To learn more about urbanisation in São Paulo, go to your learnON resources at www.jacPLUS.com.au.

FIGURE 3 São Paulo, Brazil

Contents

learnON

- 9.7.1 Urbanisation in SouthAmerica
- 9.7.2 Case study: São Paulo

FIGURE 7 The meeting of the Tietê and Pinheiros rivers in São Paulo

on Resources

eWorkbook	Characteristics of cities in South America — São Paulo (ewbk-9911)
Video eLesson	Characteristics of cities in South America — São Paulo — Key concepts (eles-5228)
Interactivity	Characteristics of cities in South America — São Paulo (int-8656)

9.8 Consequences of urban concentration

LEARNING INTENTION

By the end of this subtopic, you will be able to evaluate the impacts of urban concentrations on cities around the world.

9.8.1 Impacts of settlement patterns

There are many reasons for the growth of urban centres and the patterns of settlement that develop. Urban concentrations can impact the characteristics, liveability and sustainability of a city in a number of ways — both positive and negative. Across the world, cities have also implemented a range of strategies to lessen the negative and foster the positive impacts of large urban settlements.

9.8.2 Urban sprawl

More land is needed when cities expand and this results in the greatest change — from agricultural to urban land. This has been called urban sprawl (see subtopic 9.2).

Historically, urban areas were settled where the land was flat, the water and soil were good, and the climate was temperate — in other words, where good farmland is located. When cities spread, the sprawl takes over arable land (land able to be farmed for crops). Urban sprawl has long-term effects, as it is very difficult to bring the soil back to its former state once the predominant land use has been for buildings.

TABLE 1 Features and consequences of urban sprawl

<table>
<tr><th colspan="3">What is urban sprawl?
The city physically grows outwards to accommodate larger numbers of people.</th></tr>
<tr><td rowspan="2"></td><th>Evidence of change</th><th>Urban challenges</th></tr>
<tr><td>• Land situated on the outskirts of an urban settlement is rezoned for housing.
• Widespread land clearing occurs.
• Infrastructure is built to support the new suburban growth and to link it to more established parts of the city (e.g. road networks, internet, water, electricity, schools, shops).
• High demand for services occurs in a short space of time.
• Changes occur to the population structure.</td><td>• Loss of wildlife habitat or farming land
• Threats to vulnerable plant and animal species
• Delays in constructing services for new residents e.g. schools, health, shops
• Long distance commutes and infrequent transport services
• Slow development of sense of community amongst the residents
• Air quality issues with a greater reliance on cars</td></tr>
</table>

9.8.3 Land reclamation

Some of the world's cities have limited space on which to rebuild and redevelop infrastructure capable of supporting the needs of growing populations. One solution that several wealthier cities are turning to is **land reclamation**. This involves extracting or importing sand or seabed deposits to infill low-lying land. A dredging ship is used to bring the deposits to the surface, as shown in **FIGURE 2**, and the land is built up to create a more level and useable piece of land for construction.

FIGURE 3 shows how land reclamation has been used in Dubai to create an industrial complex on artificial islands reclaimed from the sea. These facilities provide job opportunities and advanced port handling facilities for future trade opportunities. Sand mining is leading to the destruction of many wetland environments around the world and its continued use for land reclamation may not be sustainable.

land reclamation expansion onto land created by humans

FIGURE 1 Dredging ships being used to create an artificial island.

FIGURE 2 Land reclamation has been used in Dubai to build industrial facilities.

TABLE 2 Features and consequences of land reclamation

What is land reclamation? Expansion onto land created by humans to accommodate larger numbers of people		
	Evidence of change	**Urban challenges**
	• Low-lying land is filled with river or seabed sediments to create a flat, useable space for buildings. • Seabed sediments is removed in a process called dredging. • The shoreline may develop an artificial shape with straight lines. • Historical images show the urban space as a water body or wetlands. • Sand is being exported into the country to support the land reclamation projects.	• Land may subside over time if infilling is poorly planned • Changes to natural river catchment or coastal processes that may cause erosion or deposition • Disturbance to marine and wetland habitats and loss of access • Loss of water views • Changes to land use for existing residents

9.8.4 Urban consolidation

Urban planners may redevelop centrally located or accessible parts of a city to increase population density and create more housing opportunities for a growing urban population. Suburbs on the rural–urban fringe of large settlements often lack the infrastructure of established suburbs. Electricity, schools, hospitals, roads and water supplies need to be established. Urban planners may adopt **urban consolidation** strategies in established places so the existing infrastructure can be upgraded.

urban consolidation to develop and upgrade existing urban space

FIGURE 3 In the late 1960s, in parts of inner Melbourne, blocks of homes considered 'slums' — like these surviving workers' cottages (a) — were demolished to build high-rise public housing towers, like these shown (b) dotting Fitzroy and Collingwood.

TABLE 3 Features and consequences of urban consolidation

What is urban consolidation? Developing and upgrading existing urban space to accommodate larger numbers of people		
	Evidence of change	**Urban challenges**
	• Low- to medium-level buildings are replaced with high-rise buildings. • Suburbs have higher population density. • People live in apartment style homes in confined spaces. • More congested transport routes and higher volumes of traffic occur during peak hour.	• Higher population density • Loss of privacy • Personal security issues • Limited access to green space • Shadowing of neighbouring areas • Traffic congestion • Waste management • Noise pollution

9.8.5 Housing inequality

Growing urban settlements may be overwhelmed by the large number of people moving into the city in search of employment and housing. Many people end up living in insecure housing arrangements or homelessness. In cities with very large populations and limited government control of housing standards, slum settlements may develop in the less desirable areas of the city, such as land subject to flooding or very steep land. **FIGURE 4** shows slum settlements along the shoreline in Mumbai. People with limited resources are forced into these poorly located and poorly resourced settlements.

FIGURE 4 Housing inequality in Mumbai along the low-lying shoreline

TABLE 4 Features and consequences of housing inequality

What is housing inequality? Growth of poorly resourced areas of a city to accommodate larger numbers of people		
	Evidence of change	**Urban challenges**
	• High rates of in-migration into the city from rural areas creates a lack of housing options for new arrivals to the city. • Slum settlements form in poorly resourced parts of the city, situated in less desirable locations (e.g. flood prone areas, on steep hillsides). • High rates of participation in the informal sector of the economy occur (i.e. in unregulated jobs).	• Poorly resourced local government • Poorly maintained water, sanitation and garbage services • Lack of job opportunities • Overcrowding issues • Poor air quality and water quality • Adequate supply of clean water into the future • Personal security issues • Lacking a legal right to occupy space in the city by slum dwellers

9.8.6 Urban renewal

Singapore is a city state that currently supports a population of 5.69 million people (2020) and a population density of 7952.8 people per square kilometre. The city embarked on a plan of **urban renewal** and land reclamation to address its need for more land and more efficient use of existing land resources. Urban policies led to an increase in green space, the development of a well-coordinated public transport network, forced relocation of street vendors and pollution control measures. To create more land, land reclamation projects have taken place, and to improve air quality within the city vertical walls and water-sensitive buildings have been built. These efforts have led to a city that better meets the needs of Singapore's population.

urban renewal repurposing land and improving services to better meet the needs of more people

FIGURE 5 Urban renewal projects have revitalised Singapore harbour, providing important open, public green space and private developments

FIGURE 6 The Oasia Hotel complex includes open-air terraces and plants to help with ventilation.

TABLE 5 Features and consequences of urban renewal

What is urban renewal? Repurposing land and improving services to better meet the needs of larger numbers of people		
	Evidence of change	**Urban challenges**
	• More energy-efficient design is introduced. • New services are provided and the land use planning better meets the needs of the new population. • Local government invests in improvements to previously rundown areas.	• Reliable and cost-effective energy options for large concentrations of people • Increase in land and housing prices making housing less affordable for low-income residents that previously lived in these areas of the city • Loss of community as residents are forced to re-locate

9.8.7 Consequences of a highly urbanised Australia

The amount of productive land needed on average by each person (in the world or in a country, city or suburb, for example) for food, water, transport, housing and waste management is known as an **ecological footprint**. It is measured in hectares per person per year. In 2016, the World Wide Fund for Nature (WWF) reported that the average global ecological footprint was 2.8 hectares per person. In 2017, Australia had an ecological footprint of 7.3 hectares per person. The United States had an ecological footprint of 8.2 hectares per person in 2017.

Melbourne's growth has resulted in many new suburbs and extensive growth into and over food-growing areas, particularly in the west and south-east of the CBD (see **FIGURE 8**). Sydney, Perth and Brisbane have also spread into distant, previously agricultural areas.

ecological footprint the amount of productive land needed on average by each person for food, water, transport, housing and waste management

Many of Australia's cities have been called 'car cities' due to their reliance on cars and road networks for transport. This has an impact on distances and commuting times for people travelling to and from workplaces.

TABLE 6 Ecological footprints of Australian capital cities

City	Ecological footprint value*
Perth	7.66
Canberra	7.09
Darwin	7.06
Brisbane	6.87
Sydney	6.82
Adelaide	6.72
Melbourne	6.33
Hobart	5.50

*hectares/person/year
Source: *Canberra's Ecological Footprint: what does it mean?* The Office of the Commissioner for Sustainability and Environment, ACT by Joy Murray & Christopher Dey, The University of Sydney, 2011.

FIGURE 7 Hobart has the lowest ecological footprint of any of Australia's capital cities.

FIGURE 8 Melbourne's urban growth over time

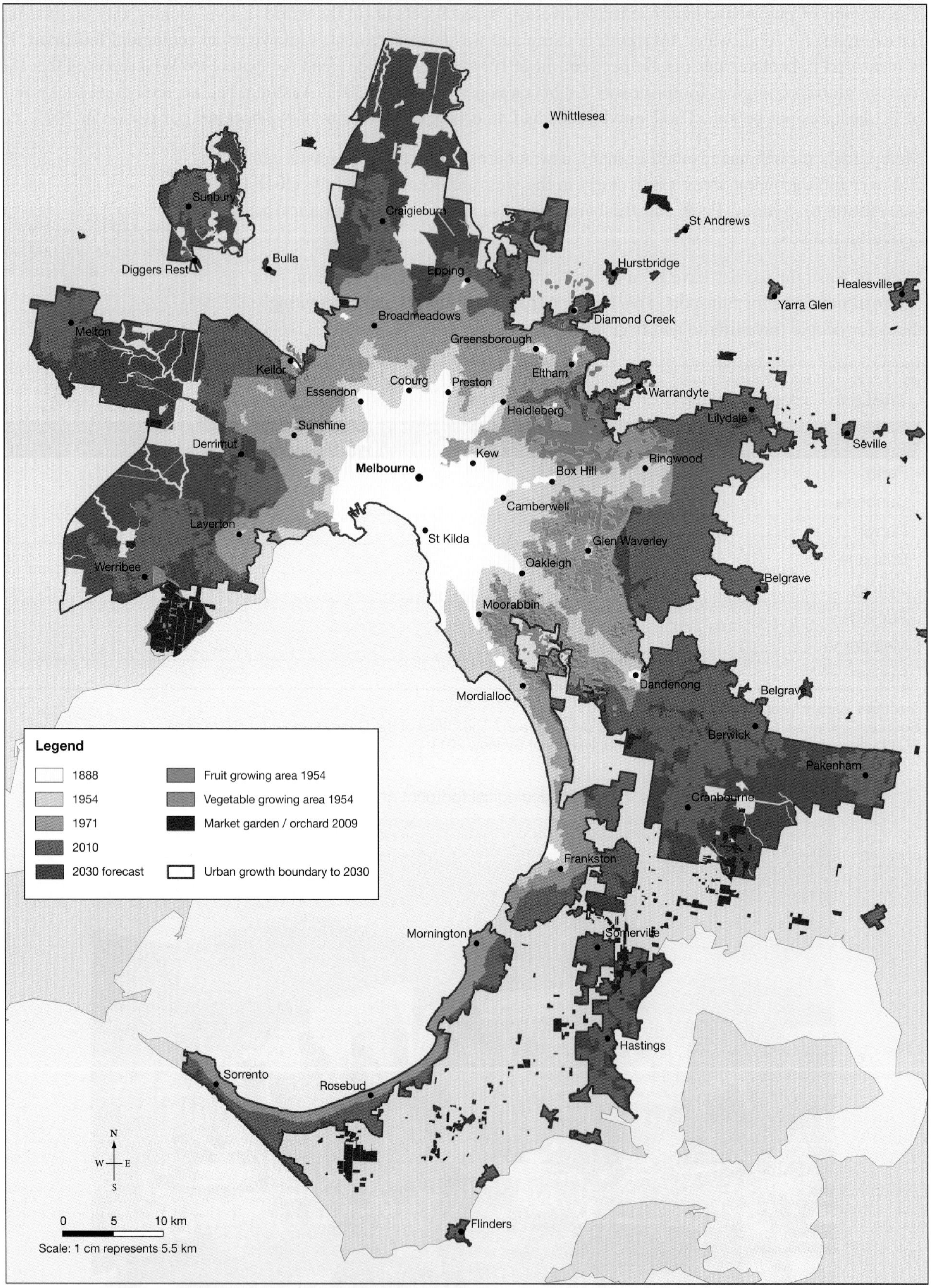

Source: Various Victorian planning studies and current land use mapping. Map produced by Spatial Vision 2019.

Resources

eWorkbook	Consequences of urban concentration (ewbk-9919)
Video eLesson	Consequences of urban concentration — Key concepts (eles-5229)
Interactivity	Consequences of urban concentration (int-8657)

9.8 ACTIVITY

Use your atlas or online research to find an urban growth map for an Australian capital city. Describe the change that has taken place over time. Using this map and a physical map of the city's state or territory, predict where future growth might occur. Justify your responses.

9.8 EXERCISE

Learning pathways

■ LEVEL 1	■ LEVEL 2	■ LEVEL 3
1, 2, 7, 12, 13	3, 4, 8, 10, 14	5, 6, 9, 11, 15

Check your understanding

1. List the five consequences of urbanisation listed in this topic.
2. Identify the terms defined by each of the following:
   a. ____________________: creating new land using sand to build up low-lying land
   b. ____________________: when a city's populated areas expand outwards
   c. ____________________: using land for a new purpose
   d. ____________________: upgrading existing spaces to house more people
   e. ____________________: when areas with few resources grow
3. What is an ecological footprint?
4. What are three disadvantages of urban sprawl?
5. Describe the growth of Melbourne over time.
6. Describe two ways that urban sprawl can affect traffic congestion.

Apply your understanding

7. Refer to **TABLE 6**.
   a. How does the ecological footprint data compare for Australian cities?
   b. How do these figures compare with the average global ecological footprint?
8. Compare two of the consequences of urbanisation presented in this topic.
9. Choose one example of a large urban settlement and identify the processes outlined in this topic that are taking place there.
10. What impact might the growth of urban areas have on food production areas?
11. What is the most difficult challenge resulting from housing inequality in a city? Give reasons for your answer.
12. How can land reclamation damage the natural features in or near a city?

Challenge your understanding

13. Generally, do you think cities in Australia will continue to expand? Give reasons for your view.
14. Based on **FIGURE 8**, do you expect that the market garden/orchard areas near Melbourne will still be used for food production in 2030?
15. Australia's ecological footprint is one of the highest in the world. Suggest two ways that we could manage our urban environments to reduce our footprint.

To answer questions online and to receive **immediate feedback** and **sample responses** for every question, go to your learnON title at www.jacplus.com.au.

9.9 Creating sustainable and liveable cities

LEARNING INTENTION

By the end of this subtopic, you will be able to explain sustainable urban design factors and apply them to an urban settlement.

9.9.1 Sustainable urban solutions

To be sustainable, cities need to develop so that they meet present needs and leave sufficient resources for future generations to meet their needs.

A sustainable city, or eco-city, is a city designed to reduce its environmental impact by minimising energy use, water use and waste production (including heat), and reducing air and water pollution.

Every city in the world experiences some type of problem that needs to be overcome — inadequate housing, urban sprawl, air and/or water pollution and waste disposal are just a few. Solutions to city problems have a better chance of succeeding if:

- responsibility is shared between governments, communities and citizens
- communities are involved in projects and decision making.

food miles the distance food is transported from the time it is produced until it reaches the consumer

9.9.2 Sustainable urban projects

Urban greening program, Sri Lanka

Producing food in cities provides people with an income and improves local environments, as well as reducing the distance that food must travel to a consumer — '**food miles**'. With support from the Department of Agriculture, the Department of Education and the Youth Services Council, three city councils in Sri Lanka developed a program of community environmental management that led to the creation of 300 home gardens and 100 home-composting programs. It also helped organise and empower community groups, and the idea has now spread to many other municipalities in the country (**FIGURE 1**).

FIGURE 1 Urban greening, Sri Lanka

Beekeeping in urban areas

A worldwide movement of urban beekeeping has had beekeepers in partnership with businesses and property owners in major cities to place beehives on rooftops. The movement makes a strong connection between urban areas and food supply. This is happening in cities such as London, New York, San Francisco, Paris, Berlin and Toronto. In London, for example, high-end department store Fortnum & Mason keeps bee hives on the roof of its building to harvest the honey to sell in store (see **FIGURE 2**). In Australia, there is a growing number of hives on city rooftops in Melbourne, Sydney and Brisbane.

FIGURE 2 Fortnum & Mason bee hives, London

Solar panels

Vatican City, Italy

Vatican City is the world's smallest independent state and is hoping to become the first solar-powered nation in the world. It plans to create Europe's largest solar power plant, which will provide enough energy to power all of the state's 40 000 households. The roof of the Paul VI Hall is now covered in photoelectric cells (see **FIGURE 3**).

Ota, Japan

Ota is located 80 kilometres north-west of Tokyo and is one of Japan's sunniest locations. Through investment by the local government, Ota is one of Japan's first solar cities — three-quarters of the town's homes are covered by solar panels that have been distributed free of charge.

FIGURE 3 Solar panels on the roof of Paul VI Hall, Vatican City

FIGURE 4 Ota, Japan — solar panels are visible on most of the houses.

Waste incineration in Vienna

The Spittelau waste incineration and heat generation plant is part of a hard-waste management system in Vienna, Austria. This plant became the first in the world to burn waste that cannot be recycled and use the energy generated by the plant in a heating network. The plant burns more cleanly and produces more heat and energy than many other waste generation plants, making the design attractive to many urban communities. Each year, waste is turned into heat and electricity and supplied to 190 000 homes and 4200 public buildings, including Vienna's largest hospital. Landfill waste has been reduced by 60 per cent in the city.

FIGURE 5 Spittelau waste treatment plant in Vienna, Austria. This power station burns waste, thus reducing landfill, to produce heat that is supplied to thousands of buildings.

The Loading Dock, Baltimore

The Loading Dock is an organisation based in Baltimore, Maryland, in the United States, that recycles building material that was destined for landfill. The material is reused to help develop affordable housing while preserving the urban environment. The organisation works with non-profit housing groups, environmental organisations, local government, building contractors, manufacturers and distributors, and uses human resources from within the community, improving living conditions for families, neighbourhoods and communities.

Since 1984, The Loading Dock has saved low-income housing and community projects more than US$16.5 million and has rescued over 33 000 tons of building materials from landfill. There has been interest in the project from 3000 other cities within the United States and in Mexico, the Caribbean, Hungary, Germany and five countries in Africa. All these projects will have a positive impact on people's lives and the urban environment.

fdw-0042

FOCUS ON FIELDWORK

Cultural diversity in the built environment

Sustainability projects are more likely to succeed if there is community involvement — if people feel like they have a sense of belonging to and responsibility for where they live. But how do you know who lives in your community to get them involved? Census data and other statistics can provide some information about who lives in a place and what their heritage is, but understanding a community's needs means understanding its cultural diversity and the heritage of the area.

Learn how to assess evidence of cultural diversity in the built environment using the **Cultural diversity** fieldwork activity in your learnON Fieldwork Resources.

on Resources

eWorkbook	Creating sustainable and liveable cities (ewbk-9923)
Video eLesson	Creating sustainable and liveable cities — Key concepts (eles-5230)
Interactivity	Creating sustainable and liveable cities (int-8658)

9.9 ACTIVITIES

1. Work in pairs to identify one urban problem and design a sustainable program that would help to improve the condition. You will need to conduct some research to find similar problems and ways in which they have been tackled. Your program should include responses to the following.
   - What is the urban problem? Include statistics (graphs or tables).
   - Where is the problem located? Describe the location and include city/state/country map(s).
   - What are the aims of your project? Describe what you hope to achieve.
   - How will you achieve your aims? Describe your program or idea.
   - Which individuals or groups are to be involved?
   - What results would reflect success for your project?

   Present your program to the class with a slide deck or multimedia presentation, a panel discussion or other format of your choice. Alternatively, you could share your programs through a class blog or wiki.
2. Choose one of the sustainable projects featured in this topic. What are the benefits of this project to the local residents of the city? What are the benefits of this project to the urban environment? Discuss how these projects might help to enhance the sustainability of your community.

9.9 EXERCISE

Learning pathways

■ LEVEL 1	■ LEVEL 2	■ LEVEL 3
1, 2, 7, 9, 13	3, 4, 10, 12, 14	5, 6, 8, 11, 15

Check your understanding

1. In your own words define the term *sustainable*.
2. The distance that food must travel to reach the consumer is known as ______________.
3. Which city was the first to burn waste that cannot be recycled to generate heat and energy?
4. Name the worldwide movement that makes a strong connection between urban areas and food supply.
5. Describe what make a city sustainable.
6. What sustainability benefits do urban beehives create?

Apply your understanding

7. Explain what the term *food miles* means using an example of your choice.
8. The projects described in this section have been completed on a local scale. Why do you think this is the case? Do you think any of these sustainable city projects would work on a suburb-wide or city-wide scale? Why or why not?
9. Explain why Ota in Japan is described as a solar city.
10. Vatican City is aiming to be the first solar-powered nation in the world.
    a. What characteristics of Vatican City make this an achievable goal?
    b. Explain two factors that might contribute to the success or failure of such a program in Australia?
11. Explain why The Loading Dock in Baltimore in the United States can be considered a sustainable project.
12. Who is responsible for ensuring that cities are as sustainable as possible?

Challenge your understanding

13. Suggest one way you could reduce the food miles of the meals you eat at home.
14. Many people in urban areas rent, rather than own, their homes. That means they may not be able to install solar panels, bee hives or garden beds — particularly in apartments. Propose one strategy that a family living in a rented apartment with no outdoor space could put in place to reduce their ecological footprint.
15. Do you think Australia should burn its rubbish to produce heat or electricity in the future? Give reasons for your view.

To answer questions online and to receive **immediate feedback** and **sample responses** for every question, go to your learnON title at www.jacplus.com.au.

9.10 Investigating topographic maps — Examining the city of São Paulo

LEARNING INTENTION

By the end of this subtopic, you will be able to identify key features and landforms in the Sao Paulo area from a topographic map, and suggest how these features affect the liveability and sustainability of the city.

9.10.1 The biggest city in the southern hemisphere

São Paulo is a large, vibrant urban centre located in the South American country of Brazil. It is considered to be the most populated city in the southern hemisphere and the wealthiest state of Brazil. São Paulo is very multicultural, and many parts of the city reflect the rich cultural heritage of Italian, Portuguese, Spanish, and Japanese influences. Japan Town is a popular tourist destination offering traditional Japanese food and markets in streets decorated with distinctive architecture.

São Paulo is located in the south east of Brazil on elevated land more than 800 metres in altitude. **FIGURE 1** shows the sprawling urban spread of São Paulo and nearby Santos, South America's largest port area. Despite the extensive growth of the city, the last area of coastal rainforest has been preserved in the Parque Trianon.

FIGURE 1 The built up-area can clearly be seen in this satellite image of São Paulo.

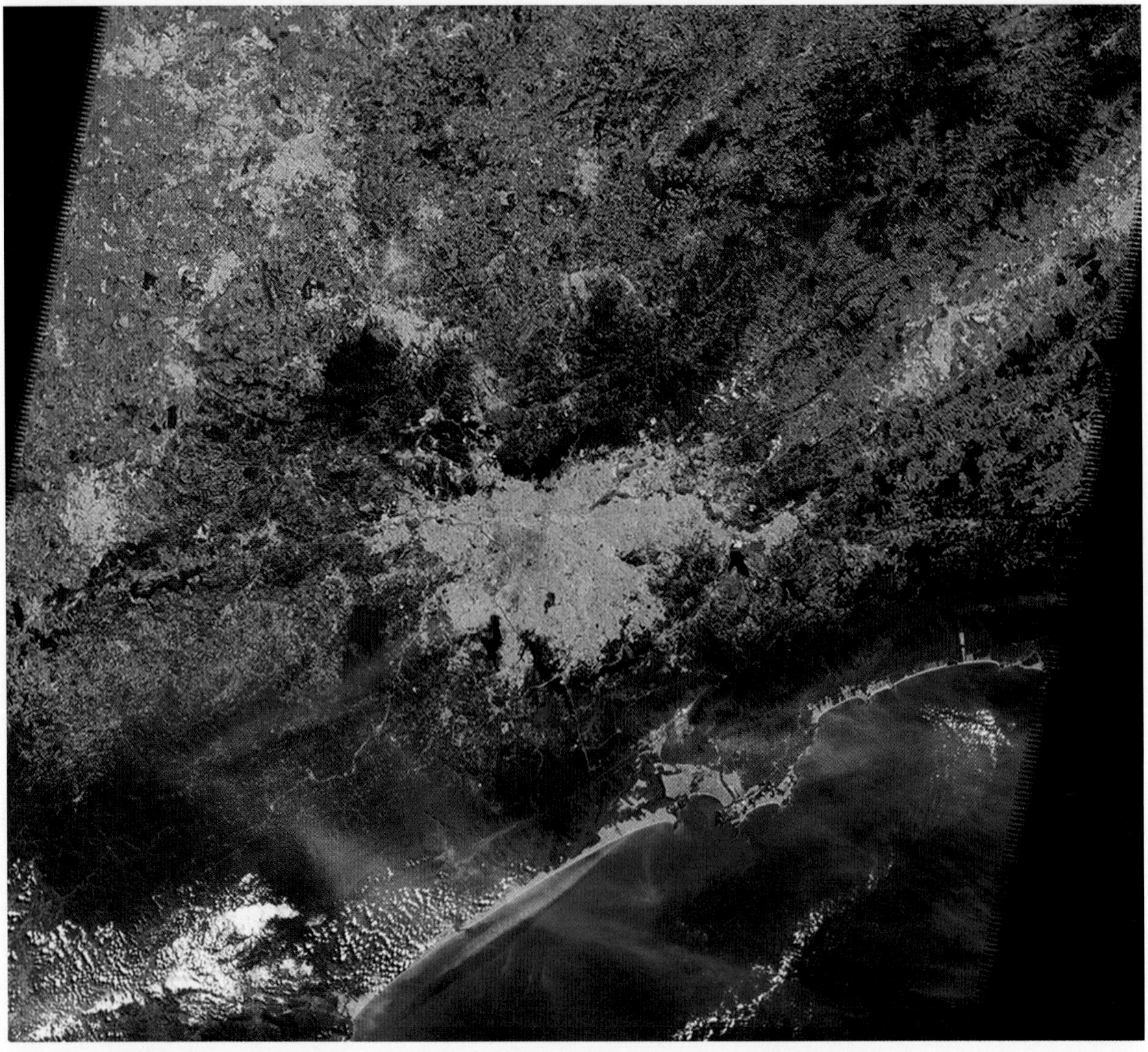

FIGURE 2 A topographic map of São Paulo, Brazil

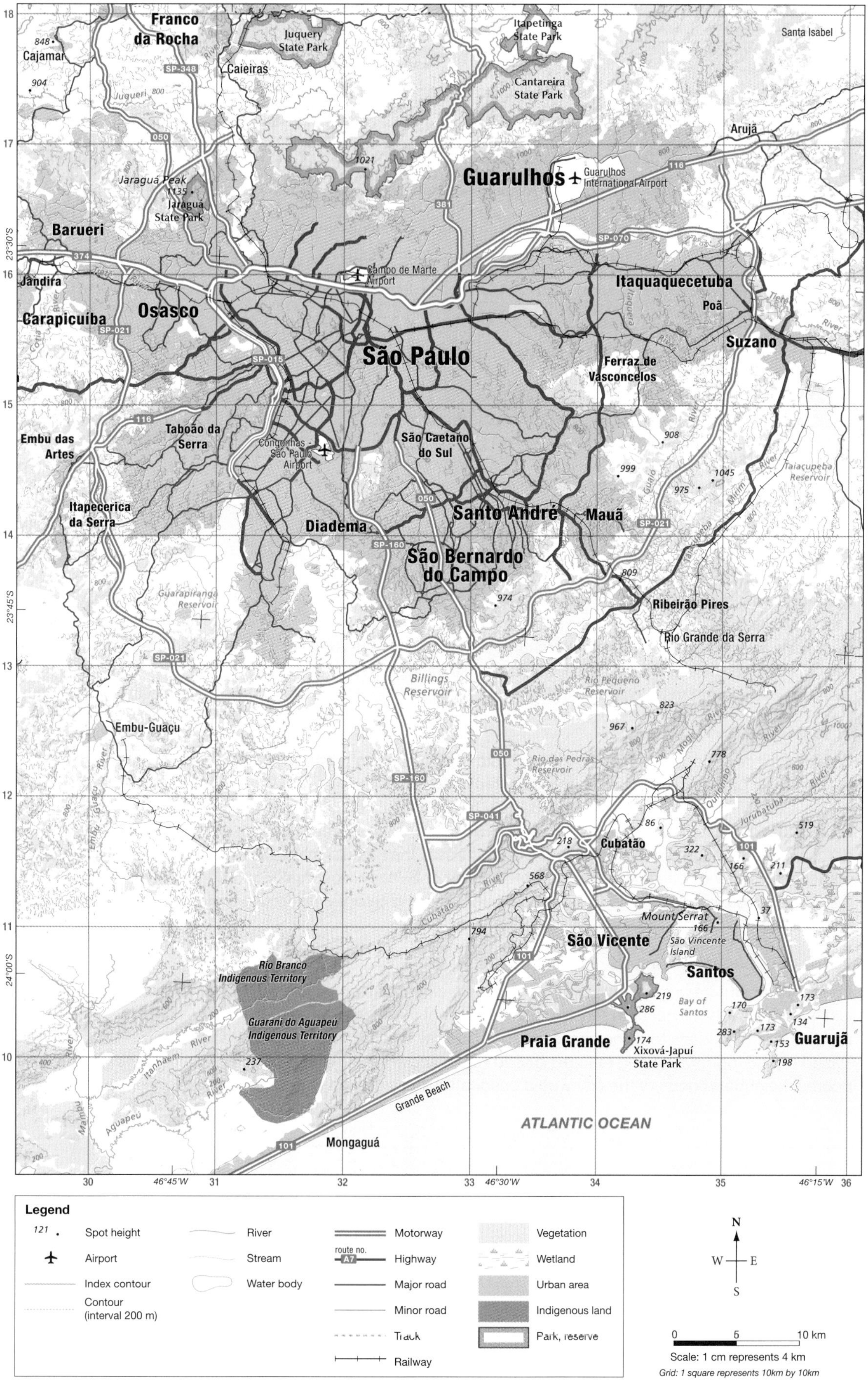

Source: Map data © OpenStreetMap contributors, https://openstreetmap.org. Data is available under the Open Database Licence, https://opendatacommons.org/licenses/odbl/; elevation data sourced from USGS.

Resources

eWorkbook	Investigating topographic maps — Examining the city of São Paulo (ewbk-9927)
Digital document	Topographic map of São Paulo (doc-36320)
Video eLesson	Investigating topographic maps — Examining the city of São Paulo — Key concepts (eles-5231)
Interactivity	Investigating topographic maps — Examining the city of São Paulo (int-8659)
Google Earth	São Paulo (gogl-0139)

9.10 EXERCISE

Learning pathways

■ LEVEL 1	■ LEVEL 2	■ LEVEL 3
1, 2, 3, 9, 13	4, 5, 6, 10, 14	7, 8, 11, 12, 15

Check your understanding

1. Give the area reference for Guarulhos International airport.
2. Give the area references for four state parks shown on the map.
3. Give the direction of Santos from the central part of São Paulo.
4. Record the highest and lowest elevations shown.
5. Calculate the total area of São Paulo's urban area.
6. Name one river that flows from the plateau to the sea and one that flows inland.
7. Describe where the steepest land is located on this map.
8. Describe the rail and road routes from the coast to São Paulo. How can you explain the pattern shown?

Apply your understanding

9. What evidence supports the fact that São Paulo is located on a plateau?
10. What are the physical limitations to the growth of São Paulo?
11. Which areas of São Paulo would be most likely to flood? Provide evidence to support your answer.
12. Study **FIGURE 1**. Describe the location of São Paulo and draw a sketch of the satellite image showing the area of the city. If Melbourne and Sydney were grid-shaped, they would measure 40 and 33 square kilometres respectively. Now calculate the area that Melbourne or Sydney would take up and draw this over São Paulo. Compare the size of São Paulo with that of Melbourne and Sydney. Write two statements to describe the differences.

Challenge your understanding

13. Predict where future urban growth will occur by shading areas on your map. Make notes on your map to justify why growth will occur in these locations and not in others.
14. If global warming leads to rising sea levels, predict how this might affect São Paulo.
15. Based on the topography and features of São Paulo shown in **FIGURE 2**, suggest where you expect the poorest neighbourhoods of the city would be located. Provide grid or area references to support your answer.

To answer questions online and to receive **immediate feedback** and **sample responses** for every question, go to your learnON title at www.jacplus.com.au.

9.11 Thinking Big research project — One day in Jakarta, one day in New York City

online only

The content in this subtopic is a summary of what you will find in the online resource.

Scenario

Jakarta and Greater New York City are both megacities — they have a population of over 10 million people. What is it like to live in these megacities, located in very different parts of the world?

Task

learnON

Your task is to plan an itinerary for someone visiting each of these cities for just one day. What places can they visit that will provide them with an experience of the life of these two cities? At the end of the day, the visitor should have an understanding about the population characteristics, culture and environmental challenges of each city.

Go to learnON to access the resources you need to complete this research project.

on Resources

ProjectsPLUS Thinking Big Research project — One day in Jakarta, one day in New York City (pro-0174)

9.12 Review

9.12.1 Key knowledge summary

9.2 Urban settlement patterns

- Australia is the sixth-largest continent and has a population density of 3.2 people per square kilometres.
- In Australia, 85 per cent of people live within 50 kilometres of the coast.
- There is a strong interconnection between the availability of more than 800 millimetres of rainfall per year and population distribution in the east, south-east and south-west of Australia.
- Before European colonisation, the most densely populated areas in Australia, with 1–10 square kilometres of land per person, were the south-east, south-west and far north coastal zones, the north of Tasmania and along the major rivers of the Riverina region

9.3 Comparing urbanisation — USA and Australia

- Causes of urbanisation are similar for both Australia and the United States.
- A conurbation is made up of cities that have grown and merged to form one continuous urban area along corridors of urban growth.
- The fastest-growing city in the US over the past four years has been Phoenix in the state of Arizona.
- The area with the highest growth in Australia during 2019 was Camden in Sydney's south-west.
- Recent changes to US immigration policy and COVID-19 restrictions have changed the way New York City operates. Migration flows into the city fell by 46% between 2016 and 2019 and this has affected the availability of migrant labour, business ownership and the total population size of New York City.

9.5 Characteristics of cities in Europe

- Most European cities became cities 700 to 1000 years ago.
- A vibrant main square is a feature of European cities from Spain to Sweden and from England to Greece. The square was usually the site of a marketplace in medieval times, as well as being the communal and cultural centre of the city.
- In European cities, the tallest building is often a church.
- To try to solve some transport problems, most European cities are encouraging people to walk, cycle and use public transport within the city.

9.7 Characteristics of cities in South America— São Paulo

- São Paulo is the most populated city in Brazil. It is located on a plateau on the south-eastern coast of Brazil, in South America.
- An average of 36 per cent of people in South America live in favelas.
- São Paulo has grown rapidly and in an unplanned manner, leaving little space for highways and parks. Six million cars contribute to crippling traffic congestion and choking levels of air pollution in the city.
- Floods are common in São Paulo because there are very few green spaces to soak up the water.

9.8 Consequences of urban concentration

- Many of Australia's cities have been called 'car cities' due to their reliance on cars and road networks.
- Some of the consequences of urban concentration include urban sprawl, land reclamation, urban consolidation, housing inequalities and urban renewal.
- The amount of productive land needed on average by each person (in the world or in a country, city or suburb, for example) for food, water, transport, housing and waste management is known as an ecological footprint.

9.9 Creating sustainable and liveable cities

- A sustainable city, or eco-city, is a city designed to reduce its environmental impact by minimising energy use, water use and waste production (including heat), and reducing air and water pollution.
- Producing food in cities provides people with an income and improves local environments, as well as reducing the distance that food must travel to a consumer is known as 'food miles'.

9.12.2 Key terms

ecological footprint the amount of productive land needed on average by each person for food, water, transport, housing and waste management

food miles the distance food is transported from the time it is produced until it reaches the consumer

land reclamation expansion onto land created by humans

metropolitan region an urban area that consists of the inner urban zone and the surrounding built-up area and outer commuter zones of a city

population density the number of people living within one square kilometre of land; it identifies the intensity of land use or how crowded a place is

urban consolidation to develop and upgrade existing urban space

urban renewal repurposing land and improving services to better meet the needs of more people

9.12.3 Reflection

Complete the following to reflect on your learning.

Revisit the inquiry question posed in the Overview:

What damage is fast-paced, large-scale urban development creating — to the planet and to its people?

1. Now that you have completed this chapter, what is your view on the inquiry question above? Discuss with a partner. Has your learning in this chapter changed your view? If so, how?
2. Write a paragraph in response to the inquiry question, outlining your views.

Subtopic	Success criteria	●	○	◐
9.2	I can describe the settlement patterns of Australia's population (past and present).			
	I can explain the reasons for Australia's settlement patterns.			
9.3	I can describe the similarities and differences in urbanisation between the United States and Australia.			
	I can identify the consequences of urbanisation in both places.			
9.4	I can create a compound graph from data.			
	I can analyse the data in compound graphs.			
9.5	I can describe the characteristics of cities located in Europe.			
	I can explain and give examples of the factors that make these cities unique.			
9.6	I can describe the characteristics of cities located in South America, providing examples related to São Paulo, Brazil.			
9.8	I can evaluate the impacts of urban concentrations on cities around the world.			
9.9	I can explain sustainable urban design factors and apply them to an urban settlement.			
9.10	I can identify key features and landforms in the São Paulo area from a topographic map.			

on Resources

eWorkbook Chapter 9 Extended writing task (ewbk-8554)
Chapter 9 Reflection (ewbk-8553)
Chapter 9 Student learning matrix (ewbk-8552)

Interactivity Chapter 9 Crossword (int-8660)

ONLINE RESOURCES

on Resources

Below is a full list of **rich resources** available online for this topic. These resources are designed to bring ideas to life, to promote deep and lasting learning and to support the different learning needs of each individual.

eWorkbook

- 9.1 Chapter 9 eWorkbook (ewbk-8101) ☐
- 9.2 Urban settlement patterns in Australia (ewbk-9895) ☐
- 9.3 Comparing urbanisation — USA and Australia (ewbk-9899) ☐
- 9.4 SkillBuilder — Creating and reading compound bar graphs (ewbk-9903) ☐
- 9.5 Characteristics of cities in Europe (ewbk-9907) ☐
- 9.6 SkillBuilder — Constructing and describing isoline maps (ewbk-9915) ☐
- 9.7 Characteristics of cities in South America — São Paulo (ewbk-9911) ☐
- 9.8 Consequences of urban concentration (ewbk-9919) ☐
- 9.9 Creating sustainable and liveable cities (ewbk-9923) ☐
- 9.10 Investigating topographic maps — Examining the city of São Paulo (ewbk-9927) ☐
- 9.12 Chapter 9 Extended writing task (ewbk-8554) ☐
  Chapter 9 Reflection (ewbk-8553) ☐
  Chapter 9 Student learning matrix (ewbk-8552) ☐

Sample responses

- 9.1 Chapter 9 Sample responses (sar-0159) ☐

Digital documents

- 9.6 Base map of the suburbs around Copenhagen city centre (doc-36410)
- 9.10 Topographic map of São Paulo (doc-36320) ☐

Video eLessons

- 9.1 Megacities and megaregions (eles-1629) ☐
  Urban settlement patterns — Photo essay (eles-5224)
- 9.2 Urban settlement patterns — Key concepts (eles-5225) ☐
- 9.3 Comparing urbanisation — USA and Australia — Key concepts (eles-5226) ☐
- 9.4 SkillBuilder — Creating and reading compound bar graphs (eles-1733) ☐
- 9.5 Characteristics of cities in Europe — Key concepts (eles-5227) ☐
- 9.6 SkillBuilder — Constructing and describing isoline maps ☐
- 9.7 Characteristics of cities in South America — São Paulo — Key concepts (eles-5228) ☐
- 9.8 Consequences of urban concentration — Key concepts (eles-5229) ☐
- 9.9 Creating sustainable and liveable cities — Key concepts (eles-5230) ☐
- 9.10 Investigating topographic maps — Examining the city of São Paulo — Key concepts (eles-5231) ☐

Interactivities

- 9.2 Distribution of rainfall in Australia (int-8767) ☐
  Distribution of Australia's towns by population size, 2016 (int-8768) ☐
  Urban settlement patterns (int-8653) ☐
- 9.3 Australia's population centres and conurbations (int-8769) ☐
  Comparing urbanisation — USA and Australia (int-8654) ☐
- 9.4 SkillBuilder — Creating and reading compound bar graphs (int-3285) ☐
- 9.5 Characteristics of cities in Europe (int-8655) ☐
- 9.6 SkillBuilder — Constructing and describing isoline maps (int-3355) ☐
- 9.7 Characteristics of cities in South America — São Paulo (int-8656) ☐
- 9.8 Consequences of urban concentration (int-8657) ☐
- 9.9 Creating sustainable and liveable cities (int-8658) ☐
- 9.10 Investigating topographic maps — Examining the city of São Paulo (int-8659) ☐
- 9.12 Chapter 9 Crossword (int-8660) ☐

ProjectsPLUS

- 9.11 Thinking Big Research project — One day in Jakarta, one day in New York City (pro-0174) ☐

Weblinks

- 9.2 Population density (web-6377) ☐
- 9.5 European traffic (web-3905) ☐
  European parking (web-3904) ☐
- 9.8 São Paulo (web-3838) ☐

Fieldwork

- 9.9 Cultural diversity in the built environment (fdw-0042) ☐

Google Earth

- 9.10 São Paulo ☐

myWorldAtlas

- 9.2 Population of Australia (mwa-4505) ☐
- 9.3 Urbanisation in Australia and the USA (mwa-7338) ☐

Teacher resources

There are many resources available exclusively for teachers online.

10 Internal and international migration

INQUIRY SEQUENCE

To access a pre-test with **immediate feedback** and **sample responses** to every question in this chapter, select your learnON format at www.jacplus.com.au.

10.1 Overview

Numerous **videos** and **interactivities** are embedded just where you need them, at the point of learning, in your learnON title at www.jacplus.com.au. They will help you to learn the content and concepts covered in this topic.

Australia boasts of having 'boundless plains to share'. Where do our migrants come from and what benefits do they bring?

10.1.1 Introduction

The movement of people from one place to another place temporarily or permanently with an intention to live is known as migration. This topic will investigate reasons people migrate to Australia and other countries. We will look at why people move within Australia and other countries and the effects these population movements have on the places of origin and the destination.

FIGURE 1 Border wall between USA and Mexico, El Paso, Texas, USA, November 2019

STARTER QUESTIONS

1. People migrate for many reasons. Brainstorm a list of reasons for migration.
2. List two positive and two negative things about living where you live.
3. Study the photograph of the US–Mexico border wall in **FIGURE 1**.
   a. What do you know about this border wall?
   b. In your opinion, should a country restrict the movement of people into and out of their country? Why?
   c. Outline one reason to support the construction of this wall and one reason against:
      - from a US point of view
      - from a Mexican point of view.
4. Watch the **Internal and international migration — Photo essay** video in your online Resources. If you could migrate to anywhere in the world, where would you choose?

on Resources

eWorkbook	Chapter 10 eWorkbook (ewbk-8102)
Video eLessons	Migration (eles-5350) Internal and international migration — Photo essay (eles-5286)

10.2 Reasons people move in Australia

LEARNING INTENTION

By the end of this subtopic, you will be able to outline the main reasons for internal migration within Australia.

10.2.1 Understanding internal migration

People move for many reasons. It is believed the average Australian will live in 11 houses during their lifetime. For some people, this figure will be much higher.

This movement within a country is known as **internal migration**. In Australia it may take place within or between states or territories, or people may choose to move within a city to another suburb more suited to their needs.

FIGURE 1 Key reasons for migration within Australia

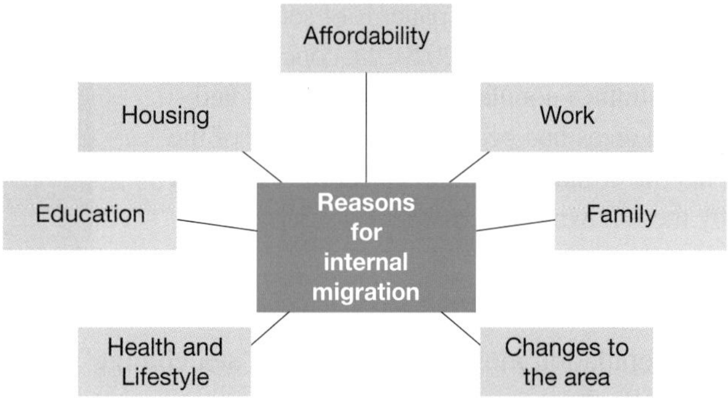

Australia is among the most mobile societies in the world, with 39 per cent of people changing their address from 2011 to 2016. For comparison, across the globe, on average 7.9 per cent of people move domestically each year, while 21 per cent move at least once every five years. Most of the moves in Australia were limited to local areas, especially within capital cities. About 4.4 per cent of these moves involved a change of state or territory.

One method used to measure internal migration is to investigate the total number of interstate migrations. During 2020, state and territory border restrictions were put in place to manage the spread of COVID-19 within the Australian population. Internal migration plans changed when these regulations were enforced by state and territory governments. **FIGURE 2** shows that the intensity of internal migration has been declining over time. During the COVID-19 pandemic period, rates of internal migration dropped sharply.

internal migration the movement of people temporarily or permanently within parts of a country

FIGURE 2 Interstate migration rates 1972 to 2018 for several countries

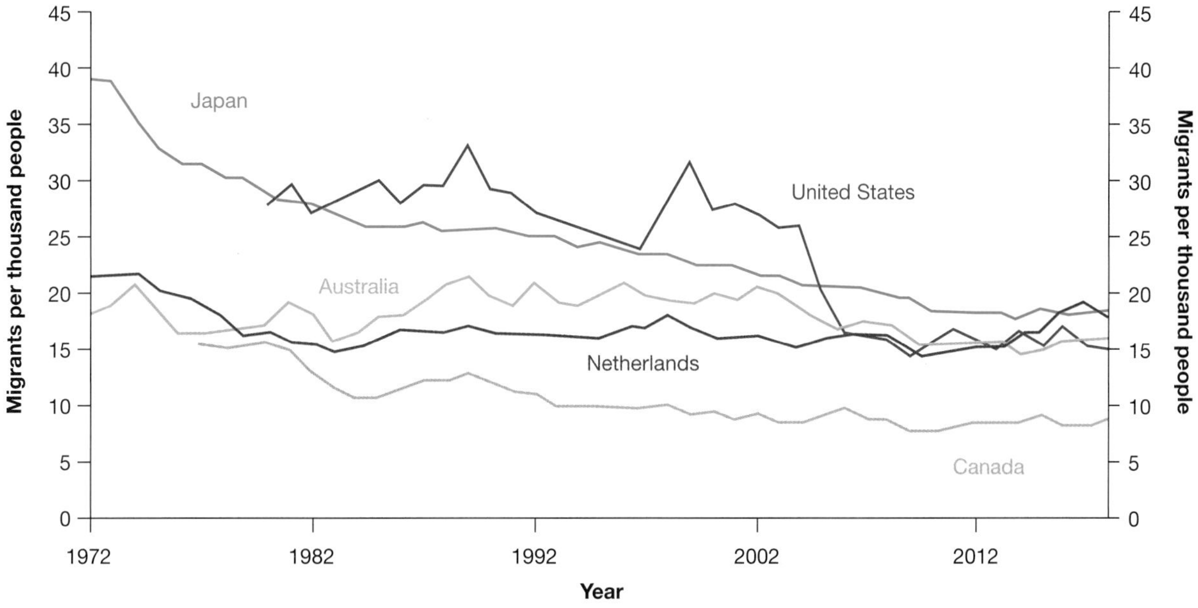

Source: Australian Government Centre for Population, 2020

10.2.2 Internal migration processes

Australia's population pattern shows that the south-east coastline of Australia is home to the majority of the population, who live in modern, urbanised and industrial locations. In contrast, the north-western areas of Australia are sparsely populated regions that are remote but rich in resources. The concentration of Australia's population is in the south-eastern parts of the country (refer to **FIGURE 4** in section 9.3.3 to see Australia's patterns of settlement). Until the pandemic in 2020, the concentration of Australia's population in capital cities and coastal areas had been increasing. Some of the patterns of internal migration can be explained by the following processes.

FIGURE 3 The Central Coast of New South Wales offers coastal living within an hour's drive north of Sydney.

Sea change or tree change

The population movement caused by '**sea change**' or '**tree change**' is a national issue affecting coastal and forested mountain communities in every state in Australia. The movement involves people who are searching for a more peaceful or meaningful existence, who want to know their neighbours and have plenty of time to relax. Local communities in high- growth coastal and mountain areas often cannot afford the services and increased infrastructure, such as roads, water and sewerage, that a larger population requires. Geelong, Wollongong, Cairns and the Gold Coast are all popular places for sea changers to settle.

sea change movement of people from major cities to live near the coast to achieve a change of lifestyle

tree change movement of people from major cities to live near forests to achieve a change of lifestyle

One initiative promoted by several regional New South Wales cities is the Evocities campaign, which encouraged people from capital cities to move to regional cities such as Dubbo, Bathurst, Albury and Tamworth. Incentives to relocate businesses and families have led to population increases in some of these regional cites since the campaign began in 2010.

Not every sea changer loves their new life, and many return to the city. Factors such as distance from family, friends, cultural activities and various professional or health services may pull people back to their previous city residences.

FIGURE 4 Mittagong offers a rural lifestyle at about an hour's drive south of Sydney.

Fly-in, fly-out (FIFO) workers

Employment opportunities have grown within the mining industry in places such as the Pilbara. However, local towns do not have the infrastructure, such as water, power and other services, to support a large population increase. Rental payments for homes can be as high as $3000 per week. One way to attract workers to these regions is to have a **fly-in, fly-out (FIFO)** workforce.

FIGURE 5 Miners in the Pilbara region in Western Australia departing after a two-week shift

FIFO workers are not actually 'settlers', as they choose not to live where they work. Some mine workers from the Pilbara live in Perth or even Bali, and commute to their workplace on a weekly, fortnightly or longer-term basis.

Permanent residents of these remote towns do not always benefit from FIFO workforces. In addition to pushing up many costs of accommodation, FIFO workers change the nature of the town by choosing not to make it their home. By not living locally, most of their wages leave the region rather than being invested in local businesses and services.

Those people that did migrate for employment opportunities generally moved away from Australia's capital cities, with Sydney and Melbourne experiencing the greatest population loss during and after this time.

Seasonal agricultural workers

Many jobs in rural areas are seasonal. For example, the picking and pruning of grapes and fruit trees requires a large workforce for only a few months each year. Many children born in rural areas leave their homes and move to the city for education, employment or a more exciting lifestyle than the one they knew in the country. This means that there are not enough agricultural workers to cover the seasonal activities, as many of these jobs were typically filled by young locals.

fly-in, fly-out (FIFO) system in which workers fly to work in places such as remote mines and after a week or more fly back to their home elsewhere

FIGURE 6 Temporary workers harvesting wine grapes in South Australia

fdw-0033

FOCUS ON FIELDWORK

Internal migration

How much do you know about where your friends have lived in their life? Are there people in your class who have lived interstate or overseas? Sometimes, people might not feel comfortable telling you in person about their family's migration, but they might be willing to share if you are collecting anonymous data from lots of different people.

Learn how to conduct research that respects your subjects' privacy using the **Internal migration** fieldwork activity in your learnON Fieldwork Resources.

on Resources

eWorkbook Reasons people move in Australia (ewbk-9935)

Video eLesson Reasons people move in Australia — Key concepts (eles-5287)

Interactivity Reasons people move in Australia (int-8702)

10.2 ACTIVITIES

1. Select one regional city listed on the **Evocities** weblink in your online Resources.
   a. Compare the benefits of moving from your current home location to this evocity for:
      - a young family
      - an elderly person.
   b. What may be some of the disadvantages of this move for each age group?
   c. Suggest ways these evocities could attract greater numbers of residents.
2. Investigate current work opportunities for seasonal agricultural workers in Australia.
   a. Suggest two ways the arrival of seasonal workers may change the way a regional town operates during a fruit-picking season.
   b. Research the problems that growers and workers have faced during the COVID-19 pandemic because of state border closures, including any strategies state or federal governments have put in place to help farmers and workers.

FIGURE 7 Tamworth, NSW, from Oxley lookout

10.2 EXERCISE

Learning pathways

■ LEVEL 1	■ LEVEL 2	■ LEVEL 3
1, 6, 8, 11, 14	2, 3, 7, 10, 13	4, 5, 9, 12, 15

Check your understanding

1. In your own words define the term internal migration.
2. Describe what makes someone a FIFO worker.
3. Outline the reasons why people may choose to relocate to a different place within Australia.
4. Describe the general trends in internal migration in Australia during 2020.
5. Refer to **FIGURE 2**. Outline two reasons why the rate of interstate migration may have been declining globally before 2020.
6. Complete the following sentence:
   A ________________________________ moves from a major city to live by the sea and ______________________________ moves to live near the forest or bush.

Apply your understanding

7. Use your own examples to explain the differences between a sea change and a tree change.
8. Consider what being a FIFO worker might be like.
   a. Suggest two personal challenges a FIFO worker may experience.
   b. What would be two benefits of being a FIFO worker?
9. A more recent population migration is towards high-rise apartment living in the centre of major cities. How might this trend impact on these new residents and the sustainability of the environment their migration is creating? Use examples to justify your stance.
10. Explain one way in which the following factors might influence a person's decision to move to the city from a rural area, or to move to a rural area from the city.
    a. Housing affordability
    b. Work opportunities
    c. Health and wellbeing
    d. Educational opportunities
11. Explain why permanent residents in remote communities might not be in favour of FIFO workers being flown in to work in local mining operations.
12. Many people seek a sea change because they believe it will offer them time to relax with their families.
    a. What features of city living are they trying to avoid by moving to the coast?
    b. What features of coastal living are they assuming will help them to achieve this aim?
    c. Evaluate this reason for moving. What other factors might play a more significant role in achieving this aim than living in a coastal location?

Challenge your understanding

13. Predict the most popular destinations for internal migrants in Australia.
14. It is common in regional and rural Australia for young people to move away from home for university or work. What factors might you need to take into consideration if you were thinking about moving from your home town/suburb to attend university or start a job in another state?
15. What impacts or restrictions might the COVID-19 border closures have had on FIFO workers? Imagine you are a representative of a large mining company that relies on FIFO workers from interstate. Suggest two strategies your company could you put in place to ensure your workers could continue to work.

To answer questions online and to receive **immediate feedback** and **sample responses** for every question, go to your learnON title and www.jacplus.com.au.

10.3 Internal migration trends in Australia

LEARNING INTENTION

By the end of this subtopic, you will be able to describe the main trends in temporary and permanent internal migration within Australia.

10.3.1 Movement of people within Australia

In Australia and other parts of the world, younger people are typically the most mobile group. Australia's internal migration pattern peaks around age 25 years and then declines through older working ages, with a small increase taking place during the retirement years. **FIGURE 1** shows that this pattern coincides with significant life events such as entering the labour force and starting a family.

FIGURE 1 Reasons for internal migration in Australia at different stages of life

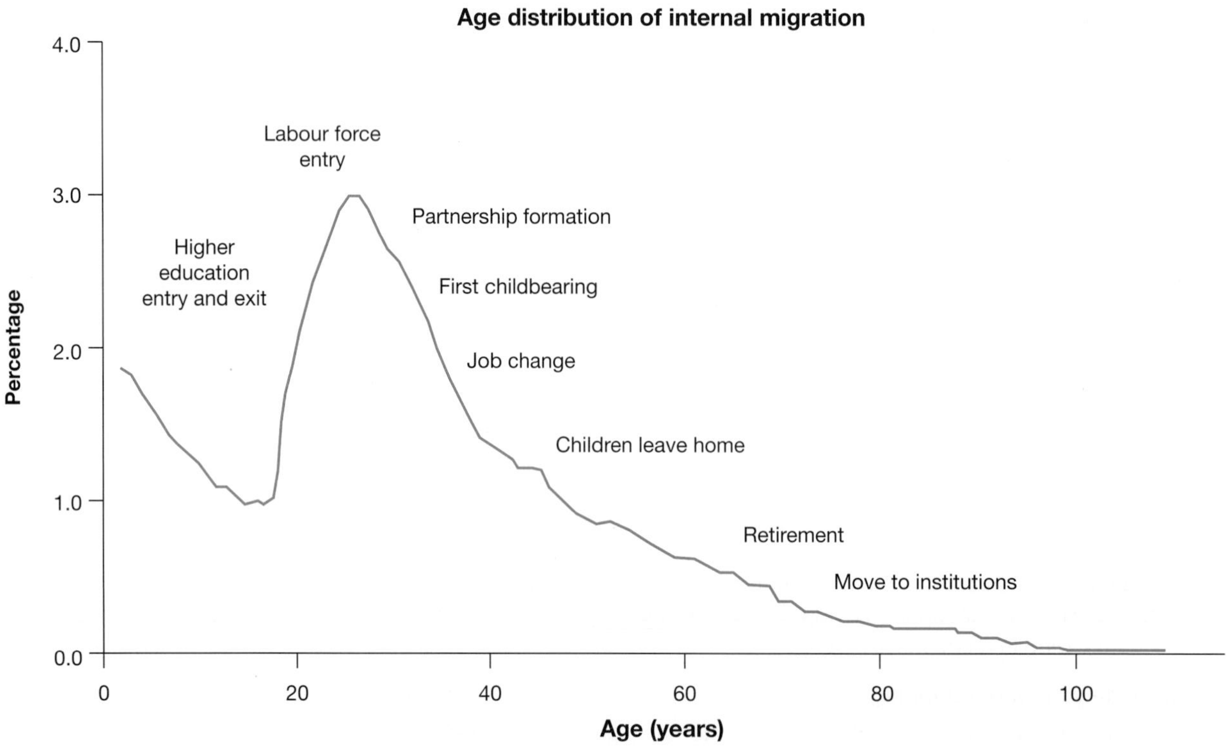

Source: Centre for Population, https://population.gov.au/docs/why_do_people_move_understanding_internal_migration_in_australia.pdf

The motivation to move changes as people pass through different life stages. Young people have different priorities to people in their older years looking to retire, or couples wanting to raise families. Young people seeking education or work opportunities tend to be more mobile and less restricted by home ownership and family commitments than people in their older years. The patterns of internal migration reflect these different priorities. Australia's Centre for Population reports that overseas migrants and people who rent their homes are more mobile than Australian-born people and homeowners.

10.3.2 Movement patterns over time

The major movements of Australians since 1788 are shown in **FIGURE 2**. Some of the factors that have contributed to where people have settled over time are water availability, the shape and productivity of the land, climatic factors, land ownership, accessibility and government policy. Advances in technology have opened more areas to settlement over time.

FIGURE 2 Australia's population movements over time

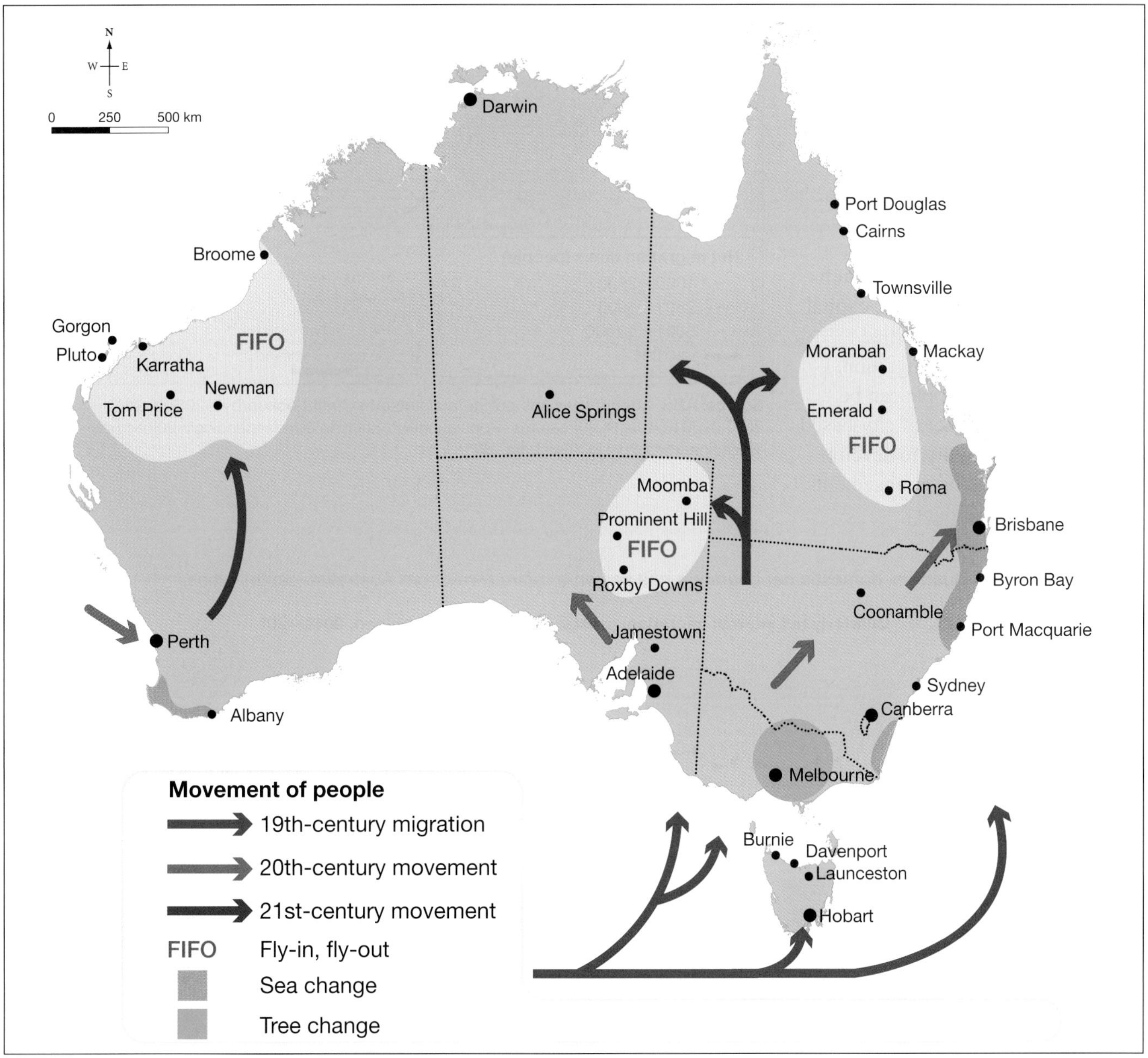

Source: Bernard Salt, KPMG Demographics, 2001.

The 2016 census map (**FIGURE 3**) shows the flow of population movements between states and territories over the period 2011–2016. The largest flow of people was away from New South Wales. Many factors contributed to this flow, including lifestyle choices, job opportunities and rising house prices in major cities.

FIGURE 3 Flow map of interstate movements of people between 2011 and 2016

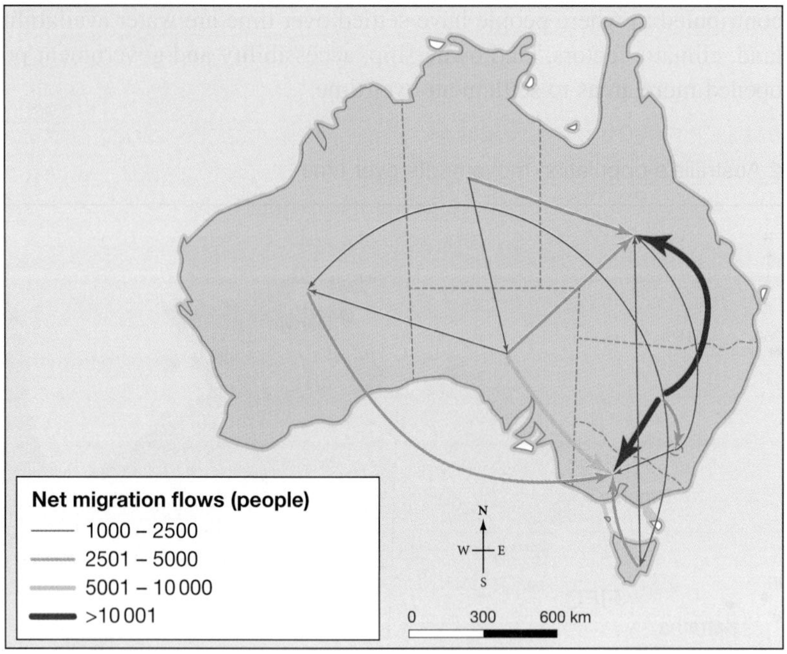

Source: ABS, https://www.abs.gov.au/ausstats/abs@.nsf/Lookup/by%20Subject/2071.0~2016~Main%20Features~Population%20Shift:%20Understanding%20Internal%20Migration%20in%20Australia~69

10.3.3 Movement patterns before and during COVID-19

Before the COVID-19 pandemic, there was a movement of people away from Sydney and New South Wales. During the pandemic, capital cities saw a significant decline in the rates of internal migration. This movement of people can be attributed to the impact of city-based lockdowns, opportunity to develop more flexible working arrangements and economic uncertainty.

FIGURE 4 Quarterly domestic net migration has shown a move away from Australian capital cities.

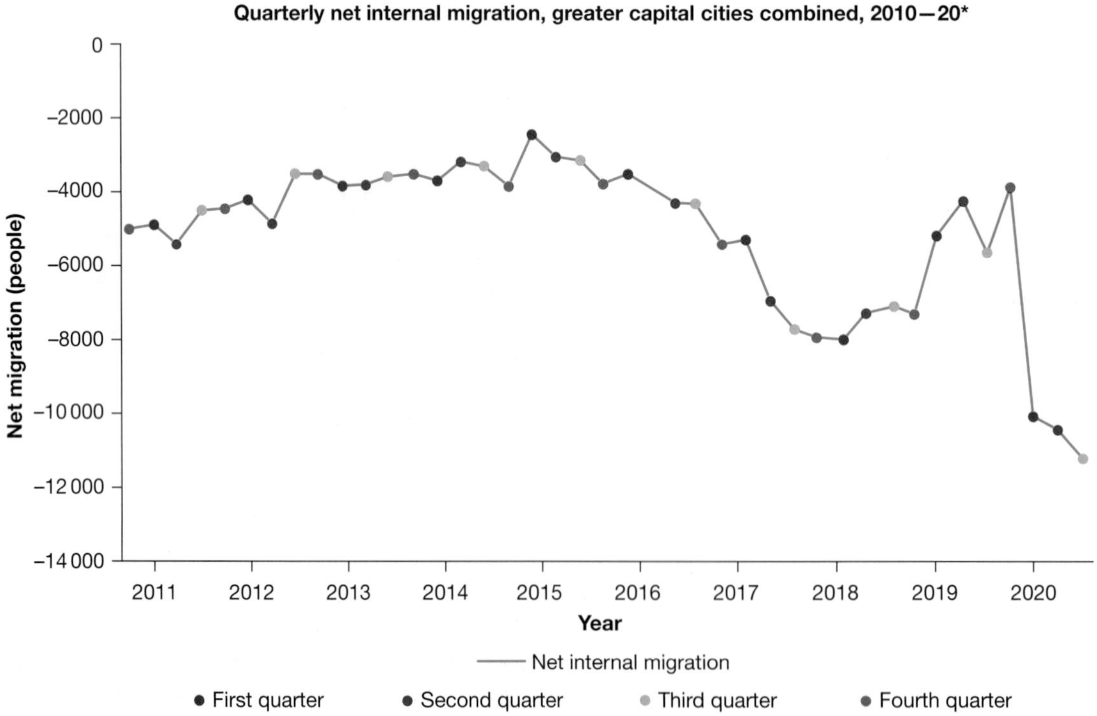

Source: ABS, https://www.abs.gov.au/statistics/people/population/regional-internal-migration-estimates-provisional/sep-2020#interstate-migration

* From fourth quarter of 2010 until third quarter of 2020

In February 2021, the ABS released provisional data that showed a significant net loss of people from both Sydney and Melbourne during the pandemic — the largest in 20 years. In contrast, cities that were less affected by lockdowns and community transmission of the virus recorded net population increases (see **TABLE 1**).

TABLE 1 Internal migration across Australia's capital cities.

	September 2019 quarter			June 2020 quarter			September 2020 quarter		
	Arrivals	**Departures**	**Net**	**Arrivals**	**Departures**	**Net**	**Arrivals**	**Departures**	**Net**
Sydney	16633	22848	−6215	16839	23217	−6378	14634	22416	−7782
Melbourne	19992	20977	−985	17023	25017	−7994	14405	21850	−7445
Brisbane	20579	16892	3687	20325	17136	3189	18743	15528	3215
Adelaide	6140	7233	−1093	6704	6885	−181	5827	6161	−334
Perth	9808	9643	165	9411	9003	408	9167	7779	1388
Hobart	1749	1934	−185	1846	1888	−42	1555	1717	−162
Darwin	2391	2974	−583	2567	2769	−202	2488	2480	8
Canberra	4568	4990	−422	4723	4480	243	4127	4262	-135
Total	46762	52393	−5631	47453	58410	−10957	41840	53087	−11247

Source: ABS, https://www.abs.gov.au/statistics/people/population/regional-internal-migration-estimates-provisional/sep-2020

Across New South Wales, the ABS identified the following migration patterns.

- There was a net loss of 4100 people from New South Wales to other states in the September 2020 quarter, 4000 in the June quarter and 4600 in the September 2019 quarter.
- Of people moving to New South Wales, the most came from Victoria (+500).
- Of people moving from New South Wales, the most went to Queensland (-4000).

on Resources

eWorkbook Internal migration in Australia (ewbk-9939)

Video eLesson Internal migration in Australia — Key concepts (eles-5288)

Interactivity Internal migration in Australia (int-8703)

10.3 ACTIVITIES

1. Investigate the FIFO settlements shown in **FIGURE 2**.
   a. What is being mined at this location?
   b. Suggest reasons why the workers don't live in the FIFO settlement permanently.
2. Using the **Population data** weblink in your online Resources, select your Local Government Area and suburb name.
   a. Record the net internal migration value and the population growth for your suburb.
   b. What changes have you noticed in your local area in the past three years?
   c. Research your area online to find evidence of these changes, for example images, media articles, new building projects.
3. Explore the internal migration data for the periods before and during the COVID-19 pandemic using the **Internal migration (COVID-19)** weblink in your online Resources.
   a. Analyse the data for one state (individually or in small groups). For each of the three tables in the ABS data set, complete the following calculations and create a table to summarise your findings for each.
   b. What age groups of people were most and least likely to depart in each of the three quarters?
   c. What age groups of people were most and least likely to arrive in each of the three quarters?

d. What percentage change occurred in the net migration from September 2019 to September 2020 for:
   - the whole state
   - the state's capital city
   - the rest of the state (i.e. excluding the capital city)?

e. Discuss your findings as a class or with others who investigated the changes in other states. How might some of your findings be interconnected? What conclusions can you draw about internal migration during the pandemic?

10.3 EXERCISE

Learning pathways

■ LEVEL 1	■ LEVEL 2	■ LEVEL 3
1, 2, 7, 11, 13	3, 6, 8, 9, 14	4, 5, 10, 12, 15

Check your understanding

1. List five life events shown in **FIGURE 1** that commonly result in people moving to another place.
2. Outline two reasons why young people are more likely to move house than older people.
3. Use **FIGURE 2** to answer the following questions.
   a. In which century would people have been more likely to move from southern Western Australia to northern Western Australia?
   b. In which century was the earlier migration to Tasmania from Europe?
   c. Which states are shown as having large 'sea change' populations?
4. Use **FIGURE 3** to answer the following questions.
   a. Which state had the largest net flow of migration into the state between 2011 and 2016?
   b. Which state had the largest net flow of migration out of the state between 2011 and 2016?
5. Describe how COVID-19 affected internal migration in Australia. Use data in your answer.

Apply your understanding

6. Are people more like to move when they get their first job, or when they change jobs later in life? Explain why this might be the case.
7. What reasons might account for the large movement of people in **FIGURE 3** away from New South Wales?
8. Explain why people might be more likely to relocate within Australia before they turn 30 than after.
9. Explain the interconnection between levels of home ownership and internal migration rates.
10. Based on **FIGURE 4**, write a short paragraph to compare the rates of internal net migration away from capital cities in September 2010 and September 2020.
11. How did lockdowns during COVID-19 affect internal migration? Use data in your answer.
12. Choose two cities from **TABLE 1**. Write a paragraph comparing their net migration rates between September 2019 and September 2020. Use data in your answer.

Challenge your understanding

13. If you had to move to another location within New South Wales, which location would you choose and why?
14. Suggest why there were no significant movements of people to the northern states of Australia in the nineteenth century.
15. **FIGURE 3** shows movement of people for the first part of the twenty-first century. Predict what the pattern might show at the end of the twenty-first century. Give reasons to support your view.

To answer questions online and to receive **immediate feedback** and **sample responses** for every question, go to your learnON title and www.jacplus.com.au.

10.4 SkillBuilder — Drawing a line graph using Excel

online only

LEARNING INTENTION

By the end of this subtopic, you will be able to use Excel to draw line graphs.

The content in this subtopic is a summary of what you will find in the online resource.

10.4.1 Tell me

What is a line graph?

A line graph is a clear method of displaying information so it can be easily understood. It is best used to show changes in data over time. A line graph can be drawn by hand. In this SkillBuilder, you will develop your skills in constructing a line graph using Excel.

10.4.2 Show me

How to draw a line graph using Excel

Step 1

Enter the data into a worksheet. Put time (hours, days, months or years) in column A and the other variable in column B. Do not leave blank rows or columns. If there is more than one set of data, list the second data set in column C and so on.

FIGURE 3 The required data (all values in the example shown here) is selected.

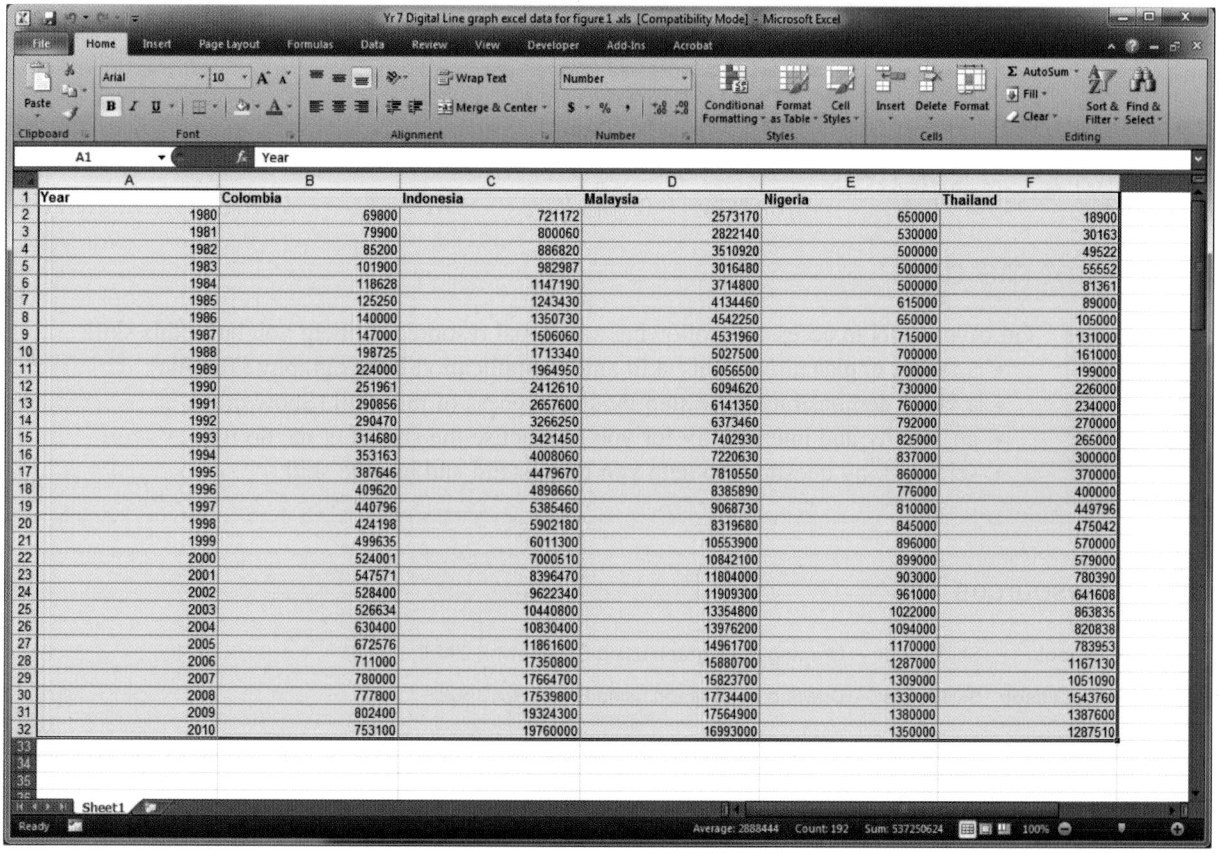

Year	Colombia	Indonesia	Malaysia	Nigeria	Thailand
1980	69800	721172	2573170	650000	18900
1981	79900	800060	2822140	530000	30163
1982	85200	886820	3510920	500000	49522
1983	101900	982987	3016480	500000	55552
1984	118628	1147190	3714800	500000	81361
1985	125250	1243430	4134460	615000	89000
1986	140000	1350730	4542250	650000	105000
1987	147000	1506060	4531960	715000	131000
1988	198725	1713340	5027500	700000	161000
1989	224000	1964950	6056500	700000	199000
1990	251961	2412610	6094620	730000	226000
1991	290856	2657600	6141350	760000	234000
1992	290470	3266250	6373460	792000	270000
1993	314680	3421450	7402930	825000	265000
1994	353163	4008060	7220630	837000	300000
1995	387646	4479670	7810550	860000	370000
1996	409620	4898660	8385890	776000	400000
1997	440796	5385460	9068730	810000	449796
1998	424198	5902180	8319680	845000	475042
1999	499635	6011300	10553900	896000	570000
2000	524001	7000510	10842100	899000	579000
2001	547571	8396470	11804000	903000	780390
2002	528400	9622340	11909300	961000	641608
2003	526634	10440800	13354800	1022000	863835
2004	630400	10830400	13976200	1094000	820838
2005	672576	11861600	14961700	1170000	783953
2006	711000	17350800	15880700	1287000	1167130
2007	780000	17664700	15823700	1309000	1051090
2008	777800	17539800	17734400	1330000	1543760
2009	802400	19324300	17564900	1380000	1387600
2010	753100	19760000	16993000	1350000	1287510

Step 2

Select to highlight the cells containing the data to be included in your line graph. *Note:* Make sure you select any column and row headings that you want included in the graph.

Step 3

In the 'Insert' menu, select 'Charts'; click on the 'Scattergraph' category and select 'Scatter with Straight Lines'. Click on the 'Layout' tab within the 'Chart Tools' section. Select 'Axis Titles' and enter the axis names. Select 'Chart Title' and choose the option 'Above Chart'. Add your title.

Step 4

Select the 'Design' tab within the 'Chart Tools' section. Click on the 'Move Chart' button to place your chart on a new page within your spreadsheet.

Step 5

Add the source of the data in a text box. Select the chart. Click on the 'Insert' tab and select 'Text Box'. Format your box and text to a suitable size and style. Move the text box to an area where it does not interfere with the reading of the graph.

FIGURE 1 Production of palm oil for the top five producers (1980–2010)

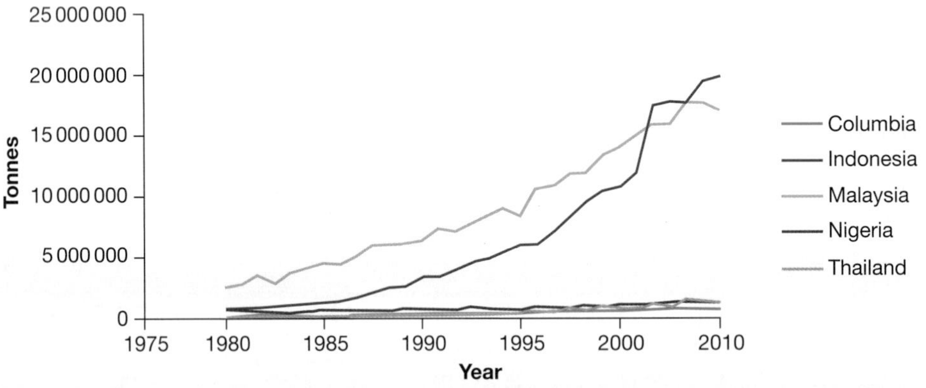

Source: Food and Agriculture Organization of the United Nations, 2012, FAOSTAT, http: //faostat3.fao.org/home//index.html

10.4.3 Let me do it

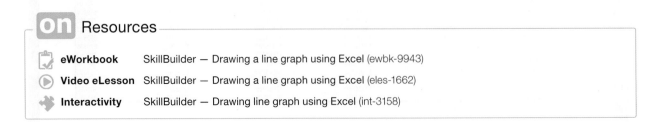

Go to learnON to access the following additional resources to help you build this skill:

- a longer explanation of this skill and its application in Geography (Tell me)
- a video demonstrating the step-by-step process of this skill (Show me)
- an activity and interactivity for you to practise the skill (Let me do it)
- self-marking questions to help you understand and use the skill.

on Resources

eWorkbook SkillBuilder — Drawing a line graph using Excel (ewbk-9943)

Video eLesson SkillBuilder — Drawing a line graph using Excel (eles-1662)

Interactivity SkillBuilder — Drawing line graph using Excel (int-3158)

10.5 Effects of internal migration in Australia

LEARNING INTENTION

By the end of this subtopic, you will be able to outline the main effects of internal migration on places of origin and destinations within Australia.

10.5.1 Consequences for places of origin

As people migrate within Australia, they collectively have an impact on the area that they have migrated from, known as the origin or source, and the area to which they migrate, called the destination. The effects of this movement of people are largely felt across the economic and social spheres of the place of origin.

For smaller towns, changes in the town can have negative flow-on effects for the population. For example, rural or remote communities can find it difficult to attract new medical staff. Many rural towns in Australia do not have a local GP and rely on visiting doctors' clinics or have to travel to the nearest regional centre for medical care.

If a major employer within a small town goes out of business or relocates, the loss of the employment opportunities may lead to many people leaving the town at short notice This is potentially damaging for the local economy and can begin a cycle of further business closures and job losses.

Most internal migrants within Australia are young adults who do not have home ownership commitments. There has been an increasing number of people under 35 years moving away from the larger cities such as Sydney towards regional and other smaller urban centres in recent years. Many are moving to take advantage of less congested and more affordable lifestyle opportunities, as well as new work opportunities. **FIGURE 1** shows a loss of 25 555 people from Sydney and an influx of 15 914 people to Brisbane during 2018–19 due to internal migration.

FIGURE 1 Rates of internal migration, 2018–19

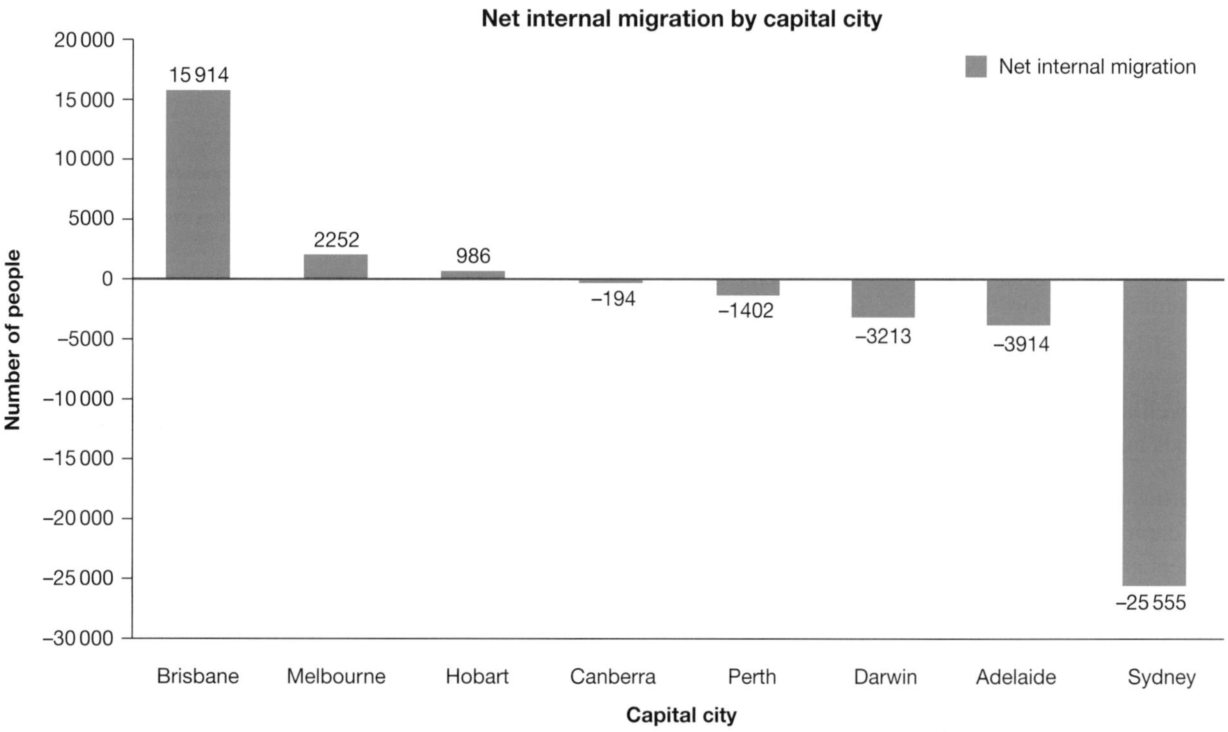

Source: ABS, https://www.abs.gov.au/statistics/people/population/migration-australia/latest-release

The movement of young people out of Sydney not only reduces population size; it also represents a loss of skilled workers and spending for the local economy. Large cities grow in response to the needs of the current and future predicted populations. Plans for a more populated and economically stronger Sydney have resulted in construction projects across the city and upgrades to transport, industry, and housing infrastructure. The Greater Sydney Region Plan (**FIGURE 2**) is designed to address future concerns by creating three cities within the larger Sydney urban area and developing transport corridors to facilitate movement between special zones called strategic centres.

FIGURE 2 Outline of the Greater Sydney Region Plan, 2056

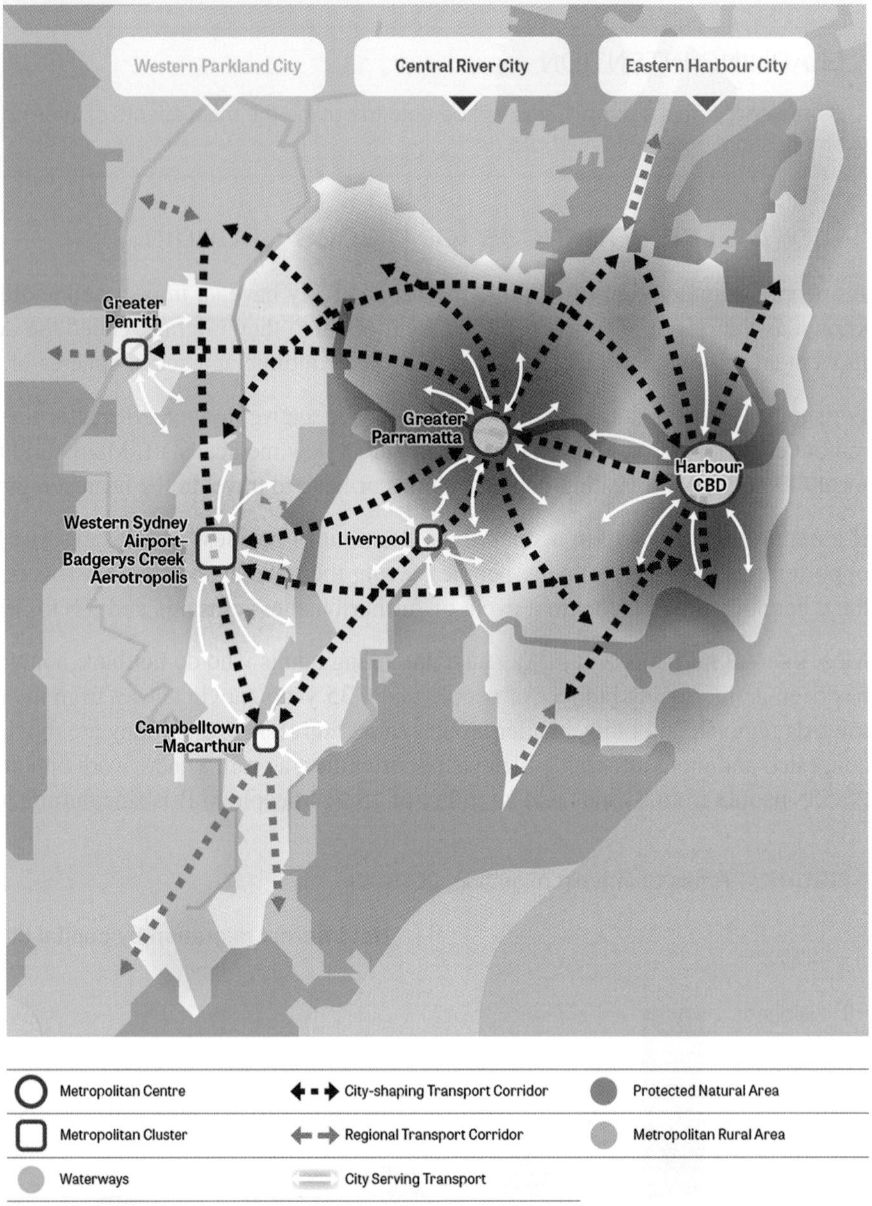

Source: © State of New South Wales through the Greater Sydney Commission

Generally, housing is more affordable on the outer edges of a city; however, jobs are often unevenly distributed across the city. This creates challenges for people who need to commute or travel to other areas in a city to work. In Sydney, for example, the average commuting time is 82 minutes, and 31 per cent of commuters report times of over two hours. New construction projects also generate traffic and congestion issues that extend commuting times. This impacts the wellbeing of residents whose lengthy travel times detract from the time they have available to spend with friends and family or for recreational activities.

As cities expand, the cost of renting or buying a home becomes less affordable. The average Australian living in a capital city spends between 20 per cent and 30 per cent of their weekly income on rent. Young first home-owners and renters in Sydney are finding cheaper options outside the built-up areas or choosing to migrate to other places with more affordable housing. This also changes the social makeup of an area, as young people and families leave behind older people and those who cannot move or cannot find work elsewhere.

The lockdown of concentrated areas and workplaces during the COVID-19 pandemic forced a change to the way many people work. Home internet access and work-from-home strategies put in place by businesses and other organisations allowed people to conduct work remotely — many for the first time. This encouraged more people to reconsider the necessity of living in a large city for work and to think about making a sea change or tree change in search of a better work–life balance and a cheaper cost of living.

10.5.2 Consequences for the destination

Destinations experience economic, environmental and social impacts from migration to their area. The movement of people into an area will have a positive effect on the local economy as spending levels increase, boosting work opportunities and demand for services. Many new arrivals enjoy the shorter commute times, cheaper housing options and improved work–life balance that moving to quieter regional areas provide. Regional areas also promote a greater sense of community than larger cities and a more relaxed lifestyle.

FIGURE 3 Wivenhoe Dam supplies much of south-eastern Queensland with water but dropped below half capacity in 2020.

Population growth can benefit the place that migrants move to, but it may also be a challenge. The growth in population allows urban planners to invest in infrastructure to ensure continued health, education and job opportunities, and to ensure the population can access these services. With a growing reliance on 'remote' workplaces, the construction and maintenance of reliable high-speed internet connections will continue to be a priority for communities. Funding such projects may need to be given priority over other much-needed services that local people have been waiting a long time for if migrants are going to move to the area.

Environmental consequences of population growth may include the additional demand for fresh water from an insecure water source, and the effects of land clearing on the soil and habitats. Many regional locations face seasonal water shortages and drought, and are forced to upgrade their water supply when the population grows.

FIGURE 4 Land clearing for a housing development at Strathnairn, a new suburb in the north of Canberra

Housing developers release home and land packages within growing regional centres. Many of the housing estates are constructed on the outskirts of the urban centre and require land clearing and levelling. The removal of the natural vegetation disrupts animal and plant habitats and exposes the soil to erosion. Many housing sites use impermeable paving and bitumen surfaces that prevent runoff from entering and nourishing the soil.

If the growing population exceeds the opportunities provided in smaller settlements, then the local community may face challenges such as higher housing costs, unemployment and homelessness. For example, when rates of migration from capital cities rose during 2020, the increase in demand for housing in rural and coastal towns led to price rises for both rental homes and properties for sale. This caused problems for long-term residents, who faced rent rises and greater competition for properties, often from people who had sold or plan to sell properties in the city for far more than the value of properties in regional areas.

Moving to another place does not work out for some people, who find the slower pace of life, fewer community resources and infrastructure and further distance from family a challenge. They may choose to move again, either back to their previous community or to a place they believe will be more suitable.

on Resources

eWorkbook	Effects of internal migration in Australia (ewbk-9947)
Video eLesson	Effects of internal migration in Australia — Key concepts (eles-5289)
Interactivity	Effects of internal migration in Australia (int-8704)

10.5 ACTIVITIES

1. Use the **ABS QuickStats** weblink in your online Resources to investigate the household median weekly income and median rent around Australia. (Use the Census QuickStats to compare data for the same year.)
   a. Choose five places across Australia to investigate, with a mix of large cities, regional centres and smaller towns. Create a graph to show the wages and rent data for each place.
   b. Which location is the most affordable?
   c. Which location is the least affordable?
   d. What factors might account for the differences between the places you chose?
2. The southern and western regions of Sydney are some of the city's most rapidly expanding areas. Investigate the changes that are planned for southern and western Sydney using the **Greater Sydney Region Plan** weblink in your online Resources.
   a. What plans are outlined to promote jobs growth? (You will find this in the Productivity menu.)
   b. What plans are outlined to improve transport options? (You will find this in the Infrastructure and collaboration menu.)
   c. What plans are outlined to improve housing options? (You will find this in the Liveability menu.)
   d. What plans are outlined to reduce pollution and protect green space? (You will find this in the Sustainability menu.)
   e. As a class or in small groups, discuss these plans. Map the benefits and challenges that they present for residents of different ages, cultural backgrounds, incomes and any other key categories you can identify.

10.5 EXERCISE

Learning pathways

■ LEVEL 1	■ LEVEL 2	■ LEVEL 3
1, 2, 7, 8, 13	3, 4, 9, 10, 14	5, 6, 11, 12, 15

Check your understanding

1. List three reasons why a young person living in an urban centre may move to another location.
2. Outline three advantages for the destination of more people moving into the town.
3. What are two potential problems faced by small towns when people leave?
4. What is the main purpose of creating transport corridors across Sydney?

5. When cities and towns expand, does the cost of renting or buying a home increase or decrease? Outline why.
6. Describe one way that the COVID-19 pandemic led to more people migrating to regional and coastal areas.

Apply your understanding

7. What two main problems is the 'three cities' aspect of the Greater Sydney plan expected to solve?
8. Explain why young adults are more likely to migrate within Australia.
9. Suggest two challenges that a person moving into a smaller regional town may have in adjusting to the new lifestyle.
10. Explain how the COVID-19 pandemic affected internal migration in Australia. Use data in your answer.
11. What is the interconnection between long urban commuting times and internal migration patterns?
12. Explain the environmental impacts of increased migration to regional centres.

Challenge your understanding

13. How might communities safeguard the homes and wellbeing of vulnerable people, such as those on low incomes or who are unwell, if their town becomes a popular place for tree changers or sea changers.
14. What would the main social challenges be for people who move to rural communities from the city?
15. Predict how the Greater Sydney Region Plan will affect net migration from Sydney.

To answer questions online and to receive **immediate feedback** and **sample responses** for every question, go to your learnON title and www.jacplus.com.au

10.6 Case study — Internal migration in China

online only

LEARNING INTENTION

By the end of this subtopic, you will be able to describe the patterns of internal movement in China and explain reasons for these movements.

The content in this subtopic is a summary of what you will find in the online resource.

China has been experiencing a changing population distribution. The country's urban population became larger than that of rural areas for the first time in its history in 2012, as rural people moved to towns and cities to seek better living standards. China has become the world's largest urban nation.

To learn more about patterns of internal migration in China, go to your learnON resources at www.jacPLUS.com.au.

FIGURE 3 In 2019, Shanghai's population was estimated to be 24.28 million.

Contents

learnON

on Resources

eWorkbook	Case study — Internal migration in China (ewbk-9951)
Video eLesson	Case study — Internal migration in China — Key concepts (eles-5290)
Interactivities	Rural–urban migration in China (int-3116) Case study — Internal migration in China (int-8705)
myWorldAtlas	Deepen your understanding of this topic with related case studies and questions. • China

10.7 SkillBuilder — Constructing a land use map

online only

LEARNING INTENTION

By the end of this subtopic, you will be able to construct a basic land use map and explain why they are used.

The content in this subtopic is a summary of what you will find in the online resource.

10.7.1 Tell me

What is a basic sketch map?

A land use map shows simplified information about the uses made of an area of land (**FIGURE 1**). In a built environment, a land use map may show a shopping centre, a local shopping strip or the types of houses in a street. In a rural environment, a land use map may show vegetation types or agricultural activities.

10.7.2 Show me

How to construct a land use map

Step 1

To complete a land use map from an aerial photograph or map, or during fieldwork, you must determine the area to be mapped and acquire or create a base map of that area. Orientate the base map and the aerial photograph or, if on fieldwork, orientate yourself with the base map.

Step 2

Create a key/legend that you will use for the colouring of your map. Include building types, services and activity areas. Colour code these in the legend. These have then been simplified into broader activity categories in **FIGURE 2b**.

Step 3

Identify a starting point to colour your base map according to your pre-determined key.

Step 4

When the map is complete, ensure that the key is attached. Check that the BOLTSS are complete. The map would look like **FIGURE 2b**.

FIGURE 1 Land use maps are often found to help people navigate public areas, such as historic towns.

FIGURE 2(b) Land use map of Blue Lake Shopping Centre.

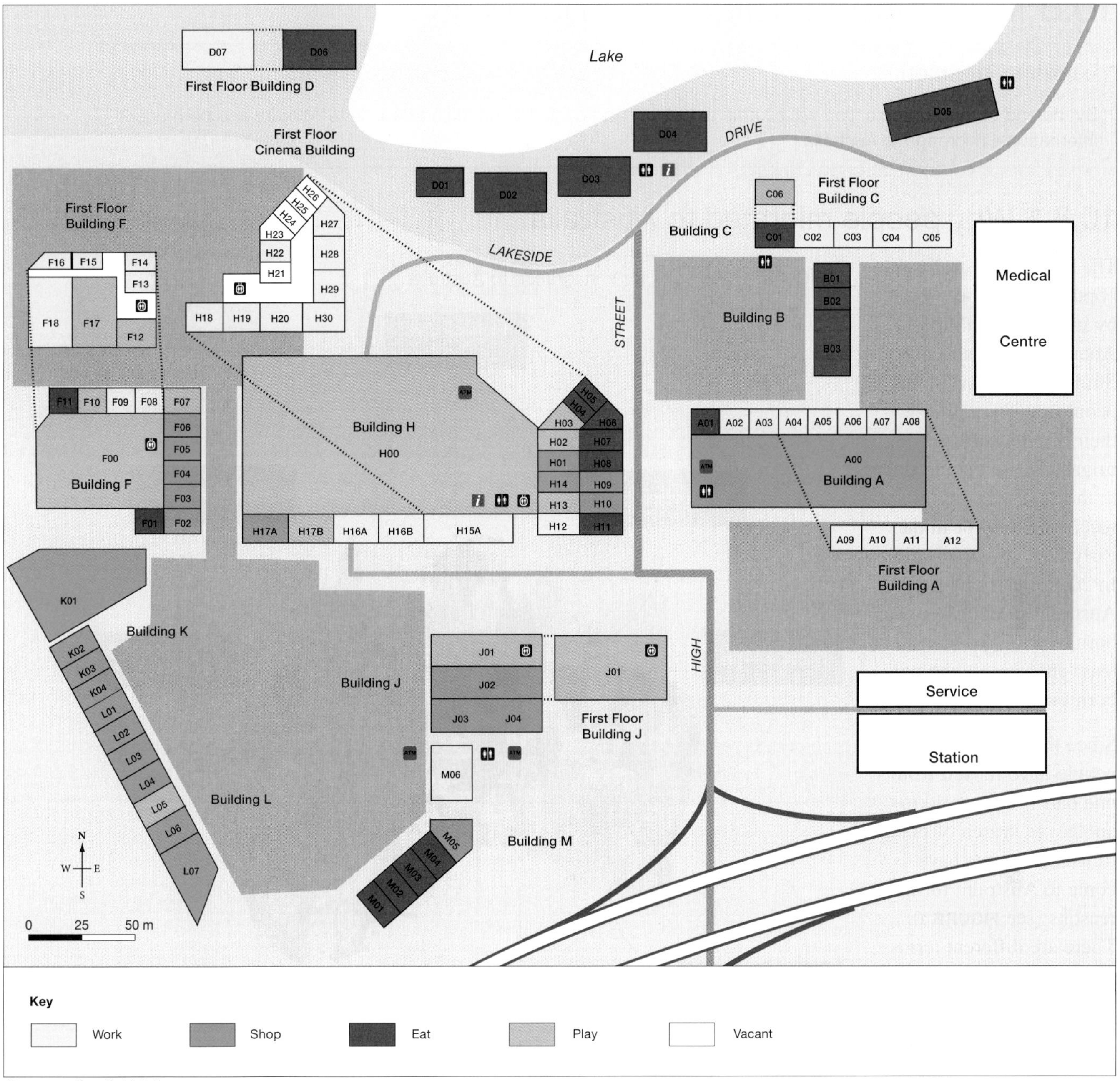

Source: Spatial Vision

10.7.3 Let me do it

learnON

Go to learnON to access the following additional resources to help you build this skill:

- a longer explanation of this skill and its application in Geography (Tell me)
- a video demonstrating the step-by-step process of this skill (Show me)
- an activity and interactivity for you to practise the skill (Let me do it)
- self-marking questions to help you understand and use the skill.

on Resources

eWorkbook	SkillBuilder — Constructing a land use map (ewbk-9955)
Video eLesson	SkillBuilder — Constructing a land use map (eles-1755)
Interactivity	SkillBuilder — Constructing a land use map (int-3373)

10.8 International migration patterns

Learning intention

By the end of this subtopic, you will be able to identify and describe the main trends in temporary and permanent international migration to Australia.

10.8.1 Why people migrated to Australia

The modern Australian population has been built by migration. All non-Aboriginal and non-Torres Strait Islander Australian people can trace their origins back to **migrants** — at some stage in the past, whether in recent memory or in the early days of colonisation. In 2016, nearly half of Australia's population was born overseas or had at least one parent who was born overseas.

Since the earliest times, people have moved from one part of the world to another in search of places to live. Migrants have come to Australia for many reasons (see **FIGURE 1**). There are different terms to describe migrants, depending on whether they are leaving or arriving in a country. People who come to Australia are immigrating (or are **immigrants**) to Australia. People who leave Australia areemigrating (or **emigrants**) from Australia.

FIGURE 1 Reasons for immigration to Australia

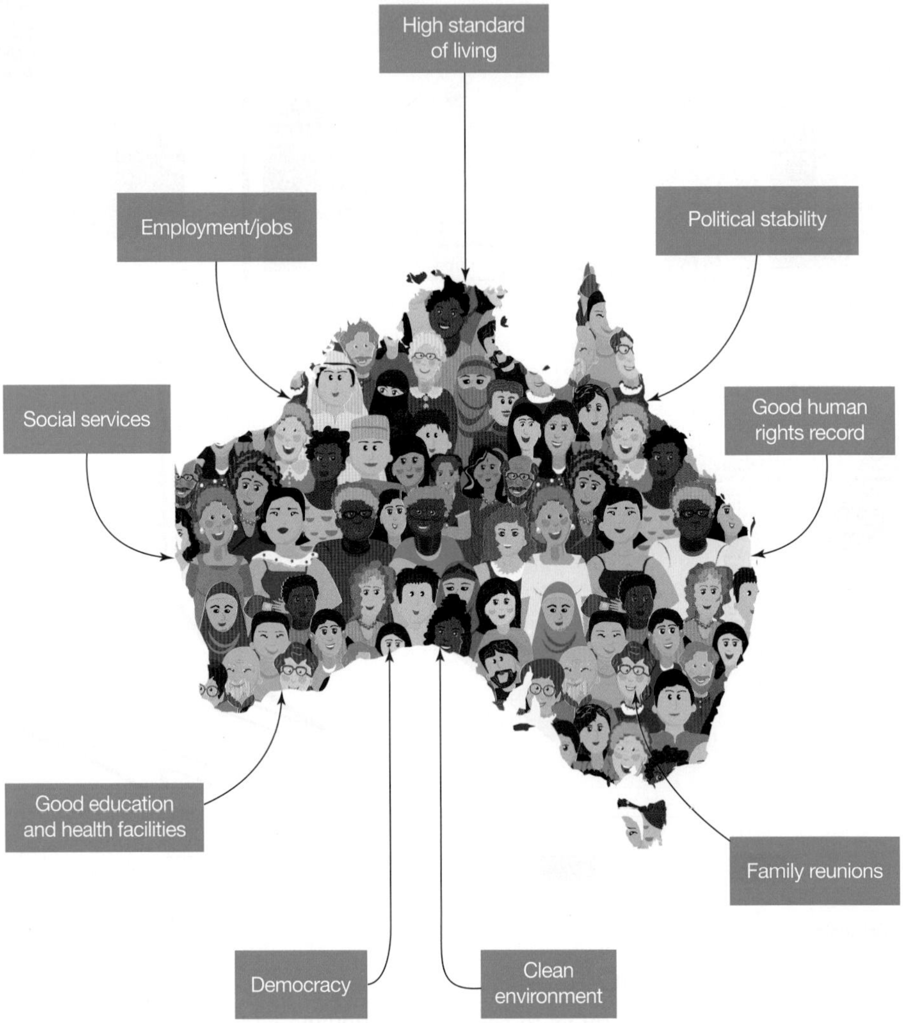

10.8.2 Where our migrants come from

Between 1851 and 1861 over 600 000 people came to Australia. Although the majority were from Britain and Ireland, 60 000 came from Continental Europe, 42 000 from China, 10 000 from the United States and just over 5000 from New Zealand and the South Pacific. However, since 1975, the country has attracted more immigrants from Asia (see **FIGURE 3** and **TABLE 2**). Despite this, the most common ancestries today are still English, Australian, Irish, and Scottish (see **TABLE 1**).

migrant a person who moves from living in one location to another

immigrant a person who migrates to a place

emigrant a person who migrates from a place

FIGURE 2 Origin of Australia's migrants, 1949–1959

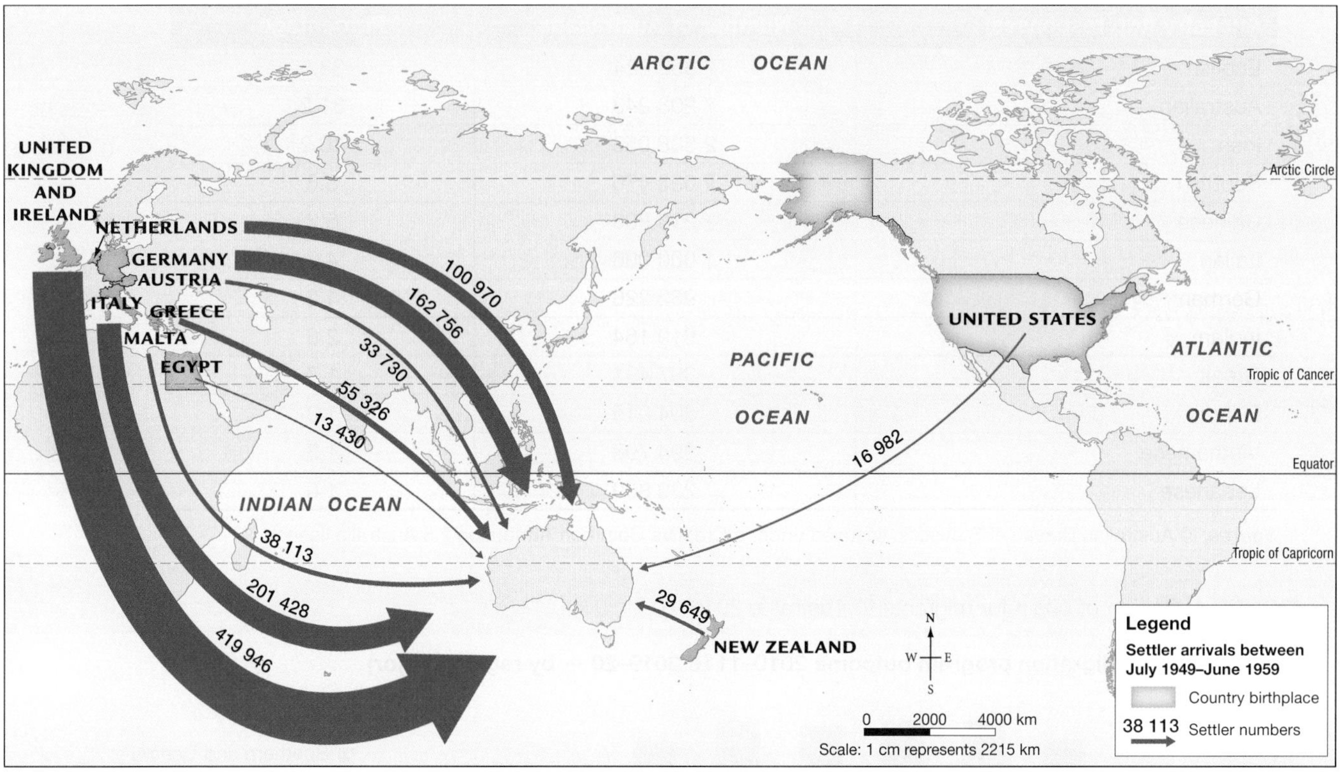

Source: Map drawn by Spatial Vision

FIGURE 3 Origin of Australia's migrants, 2015–16

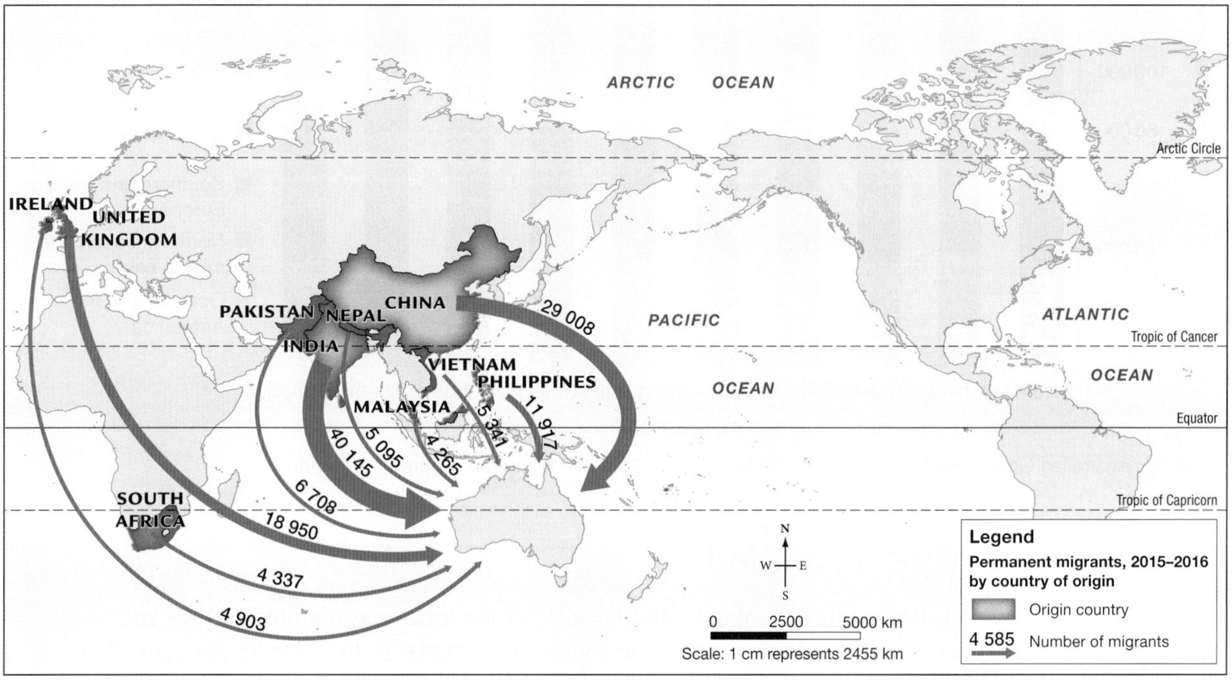

Source: Australian Government, Department of Home Affairs

TABLE 1 Ancestry by birthplace of parents, 2016

Ancestry (top responses)	Number of Australians	Percentage
English	7 852 224	33.6
Australian	7 298 243	31.2
Irish	2 388 058	10.2
Scottish	2 023 470	8.6
Chinese	1 213 903	5.2
Italian	1 000 006	4.3
German	982 226	4.2
Indian	619 164	2.6
Greek	397 431	1.7
Filipino	304 015	1.3
Vietnamese	294 798	1.3
Lebanese	230 869	1.0

Source: © Australian Bureau of Statistics, licensed under a Creative Common Attribution 2.5 Australia licence

FIGURE 4 Region of origin for migrants to Australia, 2010–20

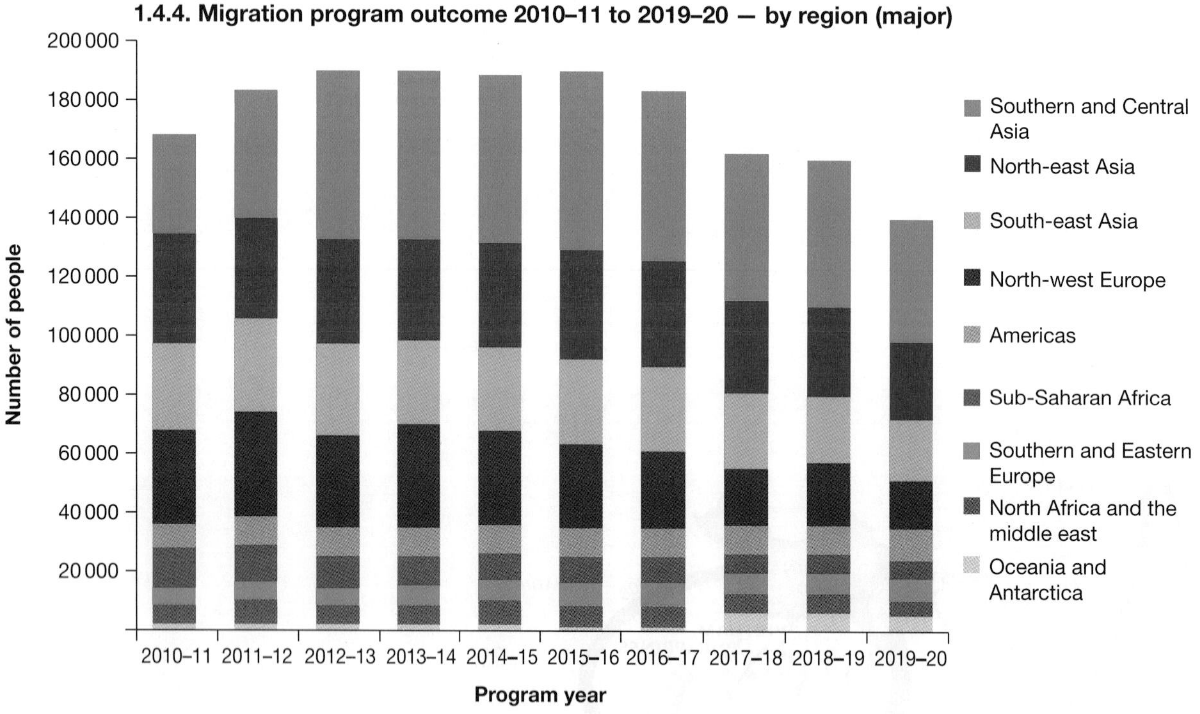

Source: Australian Government, Department of Home Affairs, 2019–20 Migration Programme Report

10.8.3 Where migrants settled

When they arrive, migrants tend to live in capital cities because of the greater availability of jobs and to be near family members, friends and people from the same country (see **TABLE 2**). In 2016, 83 per cent of the overseas-born population in Australia lived in capital cities. About one-third of the population in our large cities was born overseas.

Overseas-born migrants who arrived in the past 20 years are more likely to live in a capital city than those who arrived before 1992 (85 per cent compared to 79 per cent). Migrants from certain countries tend to be attracted to certain Australian states or territories more than others (see **TABLE 3**).

TABLE 2 Top 10 birthplaces of Australians, 2016

Country of birth	Number of people	Percentage of state population	Percentage of state population living in capital city
United Kingdom	1087749	4.6%	5.0
New Zealand	518466	2.2%	2.3
China	509563	2.2%	3.1
India	455388	1.9%	2.7
Philippines	232397	1.0%	1.2
Vietnam	219349	0.9%	1.4
Italy	174051	0.7%	1.0
South Africa	162450	0.7%	0.8
Malaysia	138371	0.6%	0.8
Sri Lanka	109841	0.5%	0.7

Source: © Australian Bureau of Statistics

TABLE 3 Top four countries of birth by state or territory ('000), 2016

ACT	NSW	NT	Qld	SA	TAS	VIC	WA
England (13.3)	China (256.0)	England (6.7)	New Zealand (200.4)	England (103.7)	England (20.5)	England (192.7)	England 213.9
China (11.9)	England (250.7)	Philippines (7.0)	England (219.9)	India (29.0)	New Zealand (5.4)	India (182.8)	New Zealand (87.4)
India (10.9)	India (153.8)	New Zealand (5.6)	India (53.1)	China (26.8)	China (3.3)	China (176.6)	India (53.4)
New Zealand (5.0)	New Zealand (127.9)	India (4.2)	China (51.6)	Italy (20.2)	India (2.1)	New Zealand (102.7)	Philippines (33.4)

Source: © Australian Bureau of Statistics

Examples include the following:

- In 2016, Western Australia had the highest proportion of residents that were born in England of any state or territory (8.4%), more than twice the Australian proportion of 4.1 per cent.
- Western Australia recorded the highest proportion of the population born overseas, at 35 per cent (895 400 persons).
- Victoria recorded the second highest proportion, with 30.7 per cent of its residents born overseas (1 892 500 persons).
- Queensland had the highest proportion of the population that were born in New Zealand (4.5%).
- New South Wales had a higher proportion of residents born in China (3.3%) and South Korea (0.8%) than any other state or territory.
- Victoria had the highest proportions of residents born in India (3.0%), Vietnam (1.5%), Italy (1.3%), Sri Lanka (1.0%) and Greece (0.9%).
- The Northern Territory had the highest proportion of people born in the Philippines (2.8%).

Not only have immigrants tended to settle in larger cities, they have also settled in particular suburbs and regions within the capital cities. For example, many migrants have settled in inner Sydney, especially in the western suburbs (see **FIGURE 5**).

FIGURE 5 Distribution of new overseas migrants to Sydney, 2017

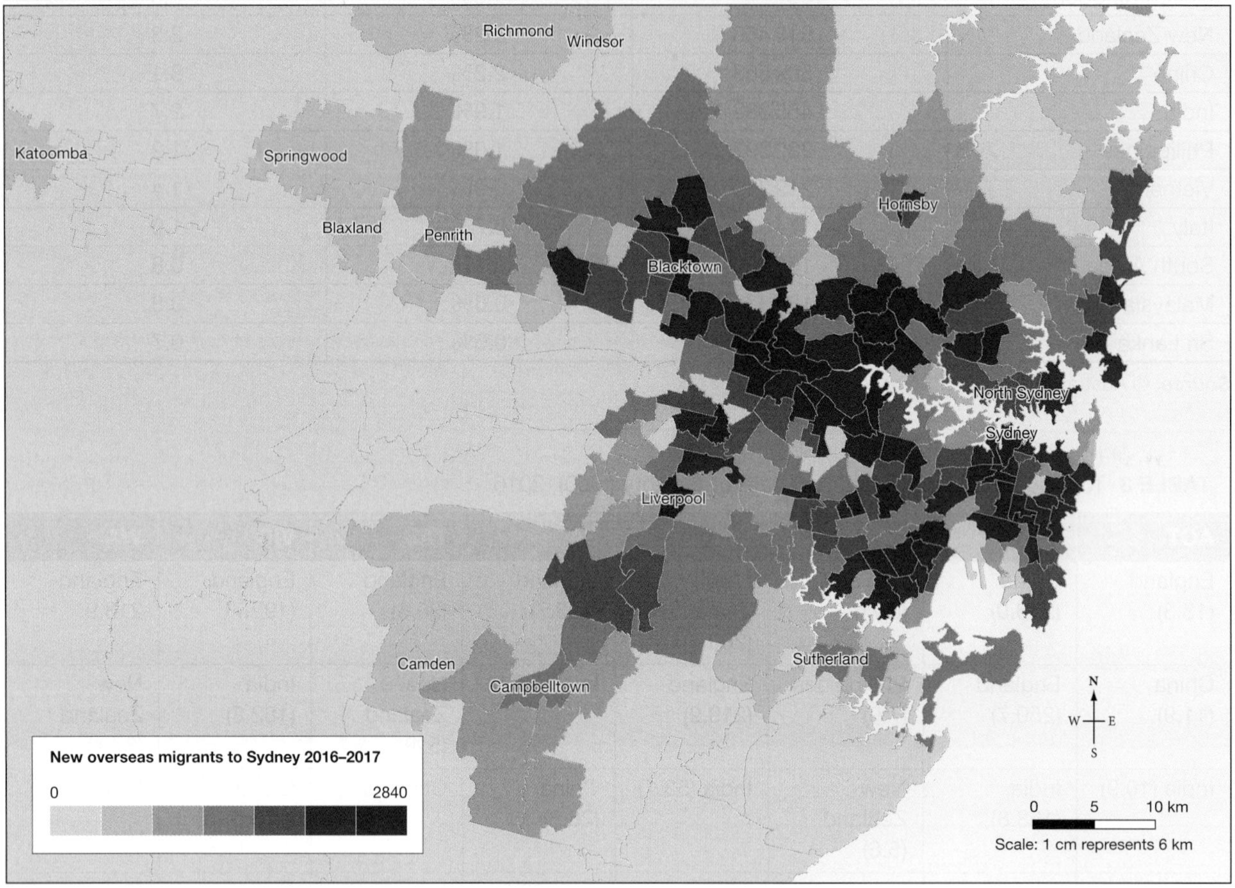

Source: Australian Government, Department of Home Affairs

on Resources

eWorkbook	International migration patterns (ewbk-9959)
Video eLesson	International migration patterns — Key concepts (eles-5291)
Interactivity	International migration patterns (int-8706)

10.8 ACTIVITY

Work in groups to survey various family members or members of the public to find out their ancestry. Compare your findings with **TABLE 1**.

10.8 EXERCISE

Learning pathways

■ LEVEL 1	■ LEVEL 2	■ LEVEL 3
1, 4, 8, 12, 13	3, 5, 6, 11, 14	2, 7, 9, 10, 15

Check your understanding

1. Define the term migrant in your own words.
2. Outline the key difference between an *immigrant* and an *emigrant*.
3. Using statistics, describe how migration has influenced Australia's growth.
4. Compare **FIGURES 2** and **3**. Describe how the origins of our migrants have changed since 1949.
5. Refer to **TABLE 3** and **FIGURE 5**. Describe how the distribution of the areas of settlement by migrants varies within Australia.
6. Using **FIGURE 5**, describe the pattern of where overseas migrants live in Sydney.
7. Using **FIGURE 4**, describe how the regions of origin for migrants to Australia have changed over time.

Apply your understanding

8. What do you consider to be the main reasons for why people would migrate to Australia?
9. Why might people who immigrate to Australia choose to live in particular suburbs? Provide two reasons why this might be the case.
10. Explain why migrants to Australia tend to live in capital cities.
11. Analyse the reasons people migrate to Australia shown in **FIGURE 1**. Are they mainly social environmental or economic reasons? Justify your answer.
12. What cultural benefits does Australia receive from its multicultural population? Choose one benefit and explain how it enriches Australian society.

Challenge your understanding

13. Predict which country of birth (apart from Australia) will be the highest at the next Australian Census. Give reasons for your answer.
14. For most Australian states and territories, the top four countries of birth are the same countries: China, England, New Zealand and India. Suggest reasons that might account for the different countries in the top four for the Northern Territory, South Australia and Western Australia.
15. Consider **FIGURES 3** and **4**. Predict the top ten countries immigrants to Australia will come from in the 2051 census.

FIGURE 6 British family (Peter and Sylvia Harding and their 13 children) boarding a liner at Southampton to emigrate to Australia, 1968

To answer questions online and to receive **immediate feedback** and **sample responses** for every question, go to your learnON title and www.jacplus.com.au.

10.9 Effects of international migration

LEARNING INTENTION

By the end of this subtopic, you will be able to outline the benefits of international migration to Australia.

10.9.1 Population changes

Migration has helped increase Australia's population. The increase in population from only seven million at the end of World War II to more than triple that now was caused by both the arrival of migrants and increased birth rates since then (see **FIGURE 1**).

Migrants to Australia have contributed to our society, culture and prosperity. Many communities hold festivals and cultural events where we can all share and enjoy the foods, languages, music, customs, art and dance.

Australian society is made up of people from many different backgrounds and origins. We have come from more than 200 countries to live here. Therefore, we are a very multicultural society — one that needs to respect and support each other's differences, and the rights of everyone to have their own culture, language and religion.

FIGURE 1 Australia's population growth, 1900–2020

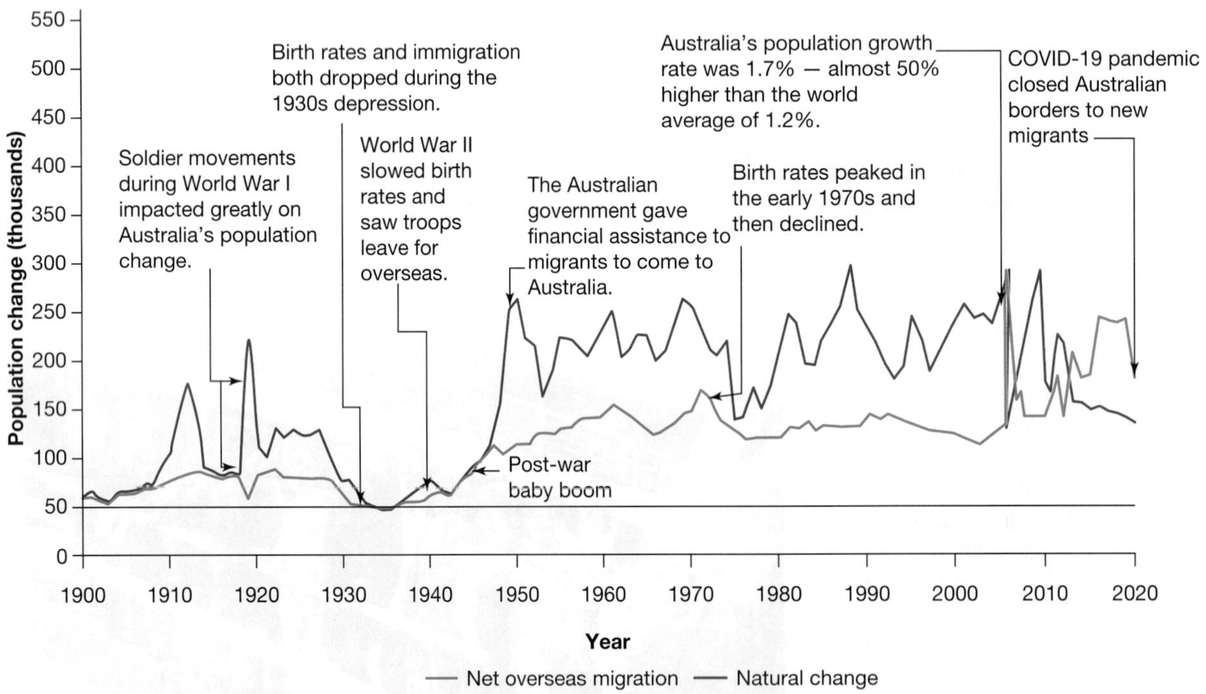

Source: ABS, Annual national population change data, 2020

Note: Data is reported for financial years (July–June) so the 2020 data is from July 2019–June 2020

Parts of regional Australia rely on international migration to drive population growth. Regional cities such as Toowoomba and Logan in Queensland; Wagga Wagga, Wollongong and Coffs Harbour in New South Wales; and Shepparton in Victoria have settled many refugees. Specialist services are provided to help new arrivals to adapt to Australian society.

Overseas migrants add to population growth and help to even out the population structure. Many of the migrants are aged between 20 and 34, so they add a youthful contribution to Australia's ageing population structure. This has positive social and economic effects on the economy.

10.9.2 Economic benefits

An increased population also means a greater demand for goods and services, which stimulates the economy. Migrants need food, housing, education and health services, and their taxes and spending allow businesses to expand. Apart from labour and capital (money), migrants also bring many skills to Australia (see **FIGURE 2**). Some migration programs target skilled migrants able to fill skill shortages in our economy.

FIGURE 2 Types of migrants to Australia, 1991–2020

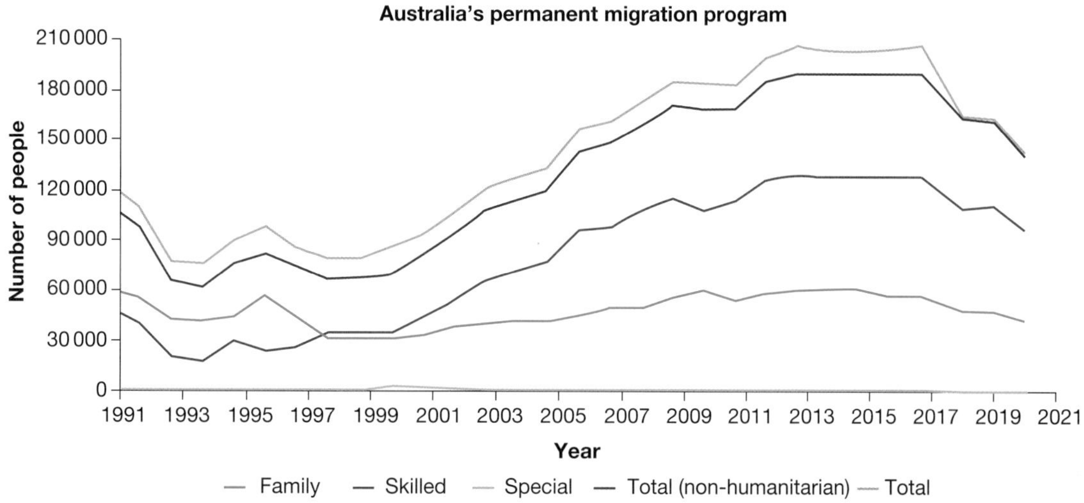

Note: Data is reported for financial years (July–June), so the 2020 data is from July 2019–June 2020.
Source: Australian Government, Department of Home Affairs, 2019–20 Migration Programme Report

Migrants generate more in taxes than they consume in benefits and government goods and services. As a result, migrants contribute more financially than they take from society.

During the COVID-19 pandemic, farmers faced crop losses as international border restrictions and quarantine measures left them without fruit pickers for the harvest. The government has negotiated with workers from nine Pacific countries and Timor-Leste to help fill labour shortages.

CASE STUDY

Pacific island workers arrive in Western Australia to combat labour shortages on farms

Extract from ABC news 9 January 2021

'Back home in Vanuatu we haven't got many jobs because of this crisis … so I must come here to work, to earn and to send money back home.'

Saving grace for grape grower

The arrival of the Vanuatu workers came at a critical time for Fruitico, the state's largest table grape producer, with the start of harvest just weeks away.

Up to 800 000 boxes of grapes will be picked at the farm from next month but without the Vanuatu workers doubling the company's workforce, assistant farm manager Kevin Dell'Agostino said they might not have made it to harvest.

Up to 800 000 boxes of grapes will be picked at Fruitico from February.

'The fruit, we probably wouldn't get there to pick it, so we'd be leaving it on the vine,' he said.

'It's [also] all the work that comes before that … We have just been through our leafing period and if that doesn't get done, the quality of the fruit is poor.'

Mr Dell'Agostino said many locals had turned their back on the work.

Source:https://www.abc.net.au/news/rural/2021-01-09/vanuatu-workers-arrive-wa-farm-labour-shortage/13043304

In the past, people argued that immigrants put pressures on Australia's environment and resources by increasing our population and the need for water, energy and other requirements. However, today many people believe that Australia's environmental problems are not caused by migration and population increase, but by inadequate planning and management.

10.9.3 Cultural benefits

Results from the 2016 census revealed that 33.6% of New South Wales residents and 27.3% of Victorian residents were born overseas. Cultural groups within an area shape and generate change within communities. Interconnections develop between the residents and their homelands, as shops and services cater to local tastes and preferences.

Community groups provide social opportunities for migrants to engage in and to contribute to Australian society. Multicultural festivals and events such as Harmony Day (21 March) aim to promote respect, a sense of belonging and a celebration of Australia's diversity (**FIGURE 3**). Urban art in **FIGURE 3d** celebrates cultural diversity in an inner-city Melbourne community.

FIGURE 3 Celebrating multiculturalism with the wider community helps to promote a sense of belonging and respect, and to allow people to share their traditions and culture. (a) Chinese New Year Lantern Festival, Sydney (b) Holi celebrations, Brisbane (c) Eid Festival, Melbourne (d) A commissioned mural on public housing towers in Collingwood, Melbourne, featuring local residents

10.9.4 The future of migration to Australia

Since 1995, the Australian government has been working to encourage new migrants to settle in regional and rural Australia. The Regional Sponsored Migration Scheme (RSMS) allows employers in areas of Australia that are regional, remote or have low population growth to sponsor employees to work with them in those regions (see **FIGURE 4**). This takes the pressure off large cities and also provides regional employers with skilled workers. As we have seen, it has always been the case that most immigrants settle first in our cities, especially the state capitals. In 2017–18, 101 255 migrants arrived in Australia and of these, 6637 settled in regional Australia. There are many regional locations that want to attract migrants.

FIGURE 4 Migrants arriving under the Regional Sponsored Migration Scheme, 2006–2020.

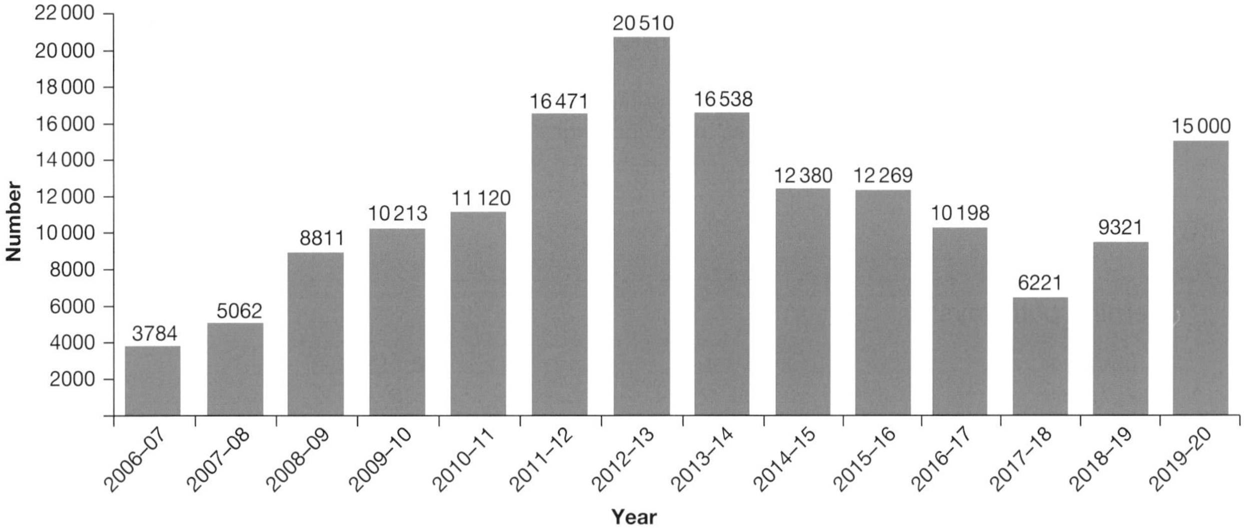

Note: A new employer-sponsored regional work visa was introduced in Nov 2019, which provided visas for a further 8372 people in the 2019–20 period.

Source: Australian Government, Department of Home Affairs, 2019–20 Migration Programme Report

FIGURE 5 Celebrations of multiculturalism across Australia: (a) Australia Day march, Adelaide, 2020; (b) Indigenous Art and Cultures Festival, Cairns, 2016; (c) Lunar New Year Festival, Perth, 2020

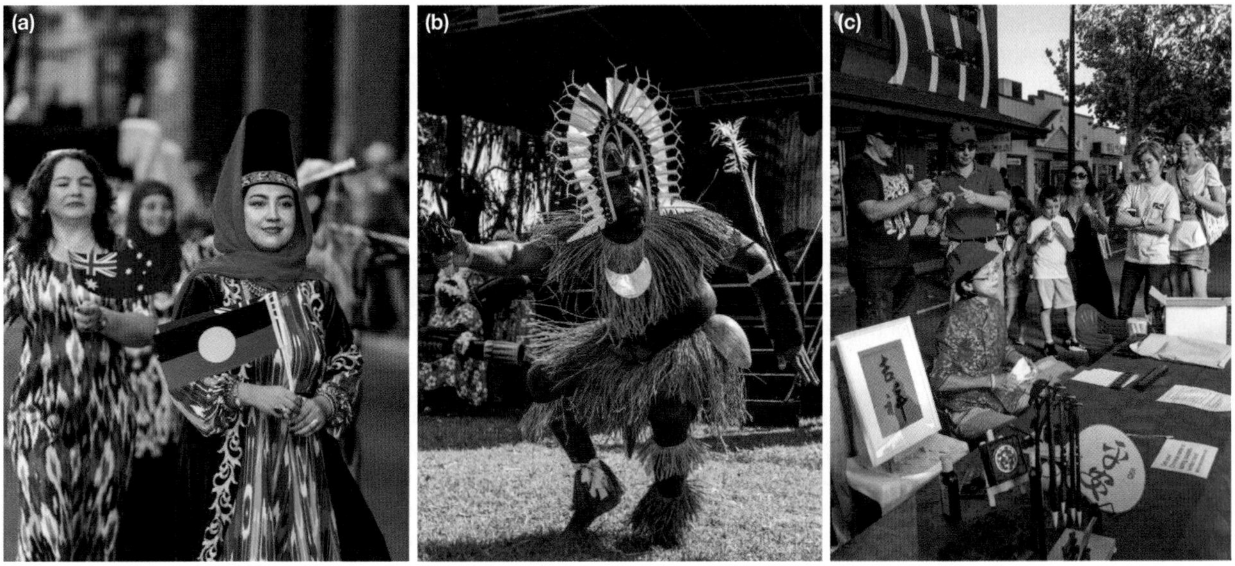

Resources

eWorkbook — Effects of international migration (ewbk-9963)
Video eLesson — Effects of international migration — Key concepts (eles-5292)
Interactivity — Effects of international migration (int-8707)

myWorldAtlas — Deepen your understanding of this topic with related case studies and questions.
- International migration and Australian cities

10.9 ACTIVITIES

1. Research the cultures of your local area. Create a calendar of cultural events taking place throughout the year for display in the classroom.
2. What challenges did the Australian economy face during the COVID-19 lockdown? How would open international borders have helped to address these challenges?

10.9 EXERCISE

Learning pathways

■ LEVEL 1	■ LEVEL 2	■ LEVEL 3
1, 2, 7, 8, 13	3, 4, 9, 10, 14	5, 6, 11, 12, 15

Check your understanding

1. Describe the features of Australia that make it a multicultural society.
2. Describe two factors that affected migration numbers to Australia in the twentieth century.
3. Which type of migrant visa leads to the highest number of immigrants to Australia each year?
4. Describe the general pattern for each of the following types of migration to Australia between 1991 and 2020.
   a. Skilled migrant
   b. Family
   c. Total numbers
5. Refer to **FIGURE 4**. Describe how the number of migrants coming into Australia under the Regional Sponsored Migration Scheme has changed between 2006 and 2020.
6. Why did the Australian government allow workers from Vanuatu and other Pacific Island nations to come to Australia to work during the COVID-19 pandemic?

Apply your understanding

7. What do you believe are the two main benefits of migration to Australia? Give reasons for your answer.
8. What impact did World War II have on Australia's birth rate and why?
9. In which year was Australia's birth rate almost 50 per cent more than the world average? Suggest why.
10. Why are skilled migrants being encouraged to move to regional Australia?
11. How do cultural festivals help to promote a sense of belonging in Australia?
12. Evaluate the importance of migration to Australia's population growth.

Challenge your understanding

13. What types of skills would you place at the top of the list for skilled migrants to Australia? Justify your answer.
14. Some rural communities have used creative strategies to attract specific types of professional (such as doctors) to their area, including posting musical video advertisements involving the whole community online. Suggest one other creative strategy country towns and regional areas might use to attract migrants to their area.
15. How would you explain the cultural makeup of your community to someone who had never been there?

To answer questions online and to receive **immediate feedback** and **sample responses** for every question, go to your learnON title and www.jacplus.com.au.

10.10 Investigating topographic maps — Urbanisation in Albury–Wodonga

LEARNING INTENTION

By the end of this subtopic, you will be able to define decentralisation and describe how this is occurring in Albury–Wodonga, providing evidence from a topographic map.

10.10.1 Decentralisation

Most of Australia's urbanisation has occurred in our large coastal capital cities. However, there are also some large inland regional centres that have experienced urban growth. Albury–Wodonga is located on the border between New South Wales and Victoria border. The two locations share a 'twin city' relationship. Albury, 580 kilometres from Sydney, is located on the northern bank of the Murray River, and Wodonga, 300 kilometres from Melbourne, is located on the southern bank.

To try to halt the spread of Australia's large coastal cities, the federal government developed a policy in the early 1970s to encourage people to move from large cities, such as Sydney and Melbourne, to regional centres such as Albury–Wodonga. This process is called decentralisation. Industries and government departments were encouraged to locate in Albury–Wodonga, which attracted people to live there.

anabranch section of a river or stream that diverts from the main channel

FIGURE 1 An **anabranch** of the Murray River, which separates Albury and Wodonga

Resources

eWorkbook	Investigating topographic maps — Urbanisation in Albury–Wodonga (ewbk-9967)
Digital document	Topographic map of Albury–Wodonga (doc-36321)
Video eLesson	Investigating topographic maps — Urbanisation in Albury–Wodonga — Key concepts (eles-5293)
Interactivity	Investigating topographic maps — Urbanisation in Albury–Wodonga (int-8708)
Google Earth	Albury–Wodonga (gogl-0140)

FIGURE 2 Topographic map extract of Albury–Wodonga

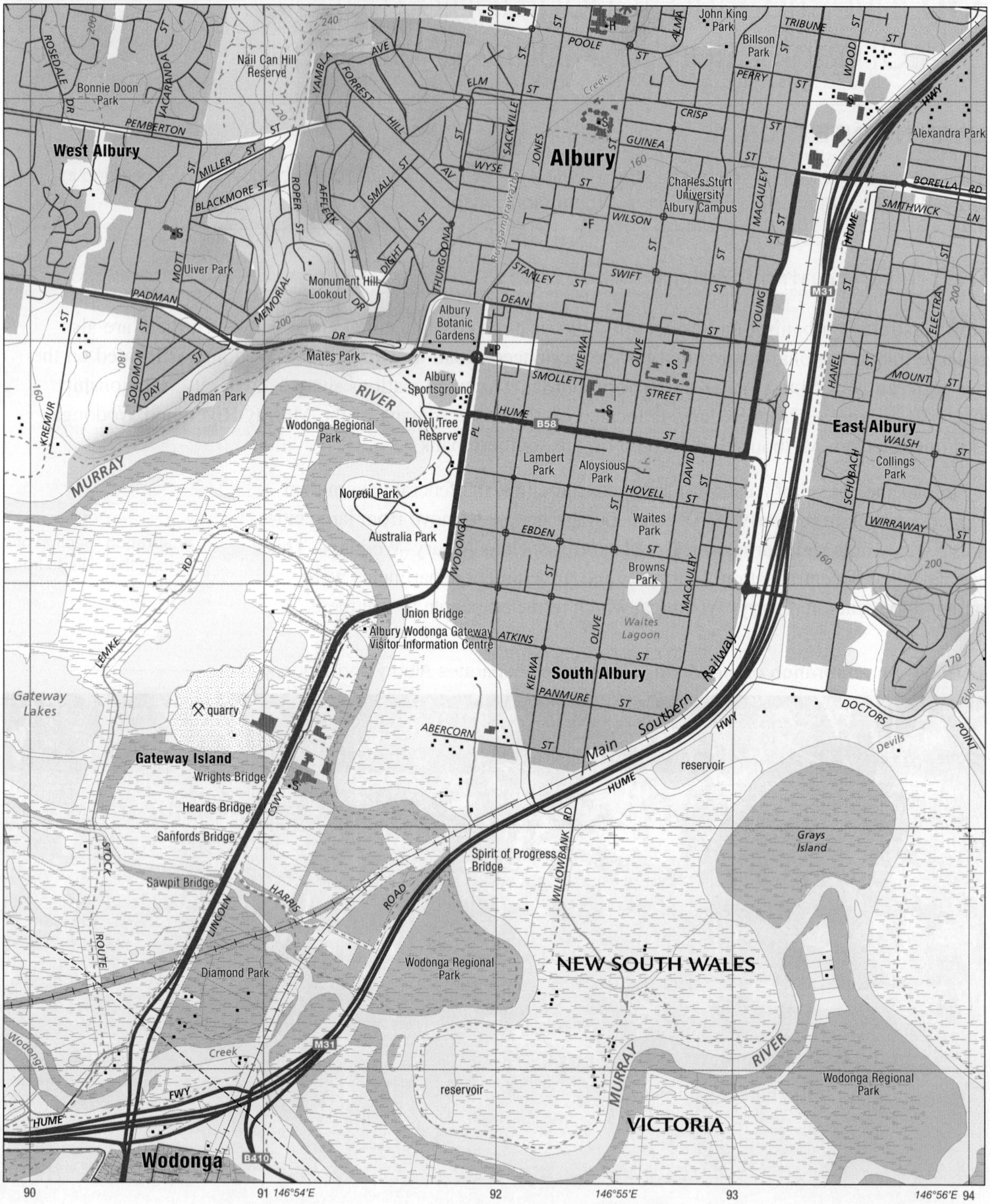

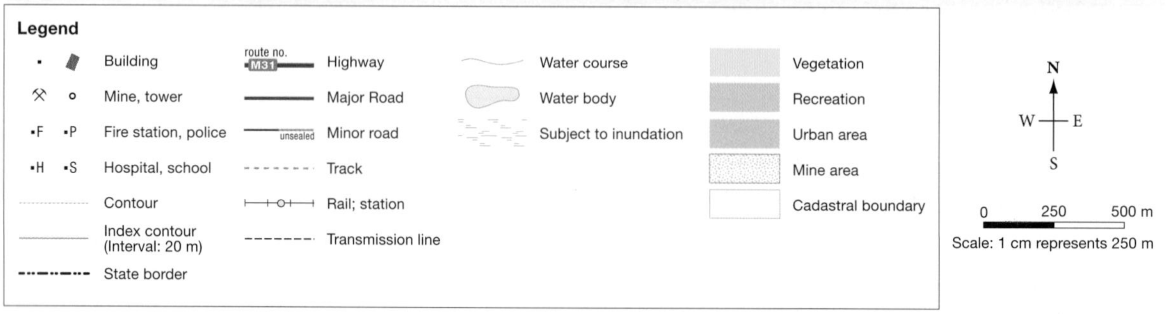

Source: Data based on Spatial Datamart, Sourced on 16 December 2020, https://services.land.vic.gov.au/SpatialDatamart/ and © State of Victoria (Department of Environment, Land, Water and Planning)

10.10 EXERCISE

Learning pathways

■ LEVEL 1	■ LEVEL 2	■ LEVEL 3
1, 4, 6, 9	2, 5, 8	3, 7, 10

Check your understanding

1. Identify the freeway that passes through both Albury and Wodonga.
2. What is located at the following points?
   a. GR905065
   b. GR927076
   c. GR940080
   d. GR913073
   e. GR920050
   f. GR932067
3. What is the approximate distance from the Albury Visitor Information Centre to Albury Botanic Gardens?
4. What do you notice about the number of parks on the eastern side of Albury compared to the number of parks on the western side?
5. In which direction does Albury seem to be expanding?
6. Is the state boundary between New South Wales and Victoria in the middle, on the Victorian side or on the New South Wales side of the Murray River?

Apply your understanding

7. What evidence is there on the map in **FIGURE 2** that Albury and Wodonga will physically remain two separate cities in the future and not merge into one city?
8. Identify four features shown in **FIGURE 2** that might attract people to the town. Provide grid references and an explanation of the feature's appeal.
9. Which parts of Albury would be the most in danger of inundation (flooding)? Identify an area and justify your choice.

Challenge your understanding

10. Considering the location of the river and surrounding wetlands and swamp, propose one strategy that could be used to change the landscape so that new developments could be built without the city expanding further north, south, east or west.

FIGURE 3 Albury, New South Wales, (a) 2018 and (b) 1960

To answer questions online and to receive **immediate feedback** and **sample responses** for every question, go to your learnON title at www.jacplus.com.au.

10.11 Thinking Big Research project — Multicultural Australia photo essay

online only

The content in this subtopic is a summary of what you will find in the online resource.

Scenario

Australia is celebrating a new national holiday — Multicultural Australia Day — to acknowledge the fact that Australia is made up of people from many backgrounds and origins. You are entering the inaugural photo essay competition, which aims to show aspects of Australia's rich multicultural heritage.

Task

learnON

Go to learnON to access the resources you need to complete this research project.

Create a photo essay — a story told through a series of photographs with some accompanying text. The purpose of this photo essay is to inform people of the rich and diverse cultures that make up Australian society. Your submission must reflect and sensitively include:

- your local area's Aboriginal Peoples' cultures
- the cultures of the current top five migrant groups in Australia.
- cultures of migrant heritage present at your school.

on Resources

ProjectsPLUS Thinking Big research project — Multicultural Australia photo essay (pro-0173)

10.12 Review

10.12.1 Key knowledge summary

10.2 Reasons people move in Australia

- Australia is among the most mobile societies globally.
- The eastern seaboard is the most attractive place for Australians to move.
- Internal migration is driven by the processes of sea change, tree change, and movement of fly-in-fly-out workers and seasonal farm workers across Australia.
- Border restrictions and economic instability due to the COVID-19 pandemic reduced interstate movements of people throughout 2020.

10.3 Internal migration patterns

- Young adults are the most mobile group of people.
- Rates of internal migration change with age and life events.
- Overseas migrants and home renters are more mobile than Australian-born people and homeowners.

10.5 Effects of internal migration

- Internal migration results in the loss of skilled and young people for the source location.
- Population growth boosts spending in the local economy and generates a greater demand for goods and services in the destination location.
- Population growth may put pressure on vulnerable water and land resources.

10.6 Case study — Internal migration in China

- China's population is rapidly becoming more urban.
- Large numbers of surplus agricultural workers move to urban areas to find work

10.8 International migration patterns

- Overseas migrants settle in Australia for many reasons
- The most common ancestries today are still English, Australian, Irish and Scottish.
- Cultural group clusters have developed in places across Australia and within large cities.
- When migrants arrive, they tend to live in capital cities because of the greater availability of jobs and to be near family members, friends and people from the same country.

10.9 Effects of international migration on Australia

- Australia is a migrant nation with people from many nations.
- There are many ancestries that make up the Australian population.
- Migrants predominantly settle in large cities, especially capital cities.

10.10 Investigating topographic maps: Urbanisation in Albury–Wodonga

- Decentralisation is occurring in Albury–Wodonga.
- Urban growth is occurring because of decentralisation.

10.12.2 Key terms

anabranch section of a river or stream that diverts from the main channel
emigrant a person who migrates from a place
fly-in, fly-out (FIFO) system in which workers fly to work in places such as remote mines and after a week or more fly back to their home elsewhere
immigrant a person who migrates to a place
internal migration the movement of people temporarily or permanently within parts of a country
migrant a person who moves from living in one location to another
sea change movement of people from major cities to live near the coast to achieve a change of lifestyle
tree change movement of people from major cities to live near forests to achieve a change of lifestyle

10.12.3 Reflection

Complete the following to reflect on your learning.

Revisit the inquiry question posed in the Overview:

Australia boasts of having 'boundless plains to share'. Where do our migrants come from and what benefits do they bring?

1. Now that you have completed this chapter, what is your view on the inquiry question above? Discuss with a partner. Has your learning in this chapter changed your view? If so, how?
2. Write a paragraph in response to the inquiry question, outlining your views.

Subtopic	Success criteria			
10.2	I can outline the main reasons for internal migration within Australia.			
10.3	I can describe the main trends in temporary and permanent internal migration within Australia.			
10.4	I can use Excel to draw line graphs.			
10.5	I can outline the main effects of internal migration on places of origin and destination within Australia.			
10.6	I can describe the patterns of movement in China and explain reasons for these movements.			
10.7	I can construct a land use map and describe why they are used.			
10.8	I can identify and describe the main trends in temporary and permanent international migration to Australia.			
10.9	I can outline the effects of international migration to Australia.			
10.10	I can define decentralisation and describe how this is occurring in Albury–Wodonga, providing evidence from a topographic map.			

on Resources

eWorkbook Chapter 10 Extended writing task (ewbk-8558)
Chapter 10 Reflection (ewbk-8557)
Chapter 10 Student learning matrix (ewbk-8556)

Interactivity Chapter 10 Crossword (int-8709)

ONLINE RESOURCES

on Resources

Below is a full list of **rich resources** available online for this topic. These resources are designed to bring ideas to life, to promote deep and lasting learning and to support the different learning needs of each individual.

eWorkbook

- 10.1 Chapter 10 eWorkbook (ewbk-8102) ☐
- 10.2 Reasons people move in Australia (ewbk-9935) ☐
- 10.3 Internal migration in Australia (ewbk-9939) ☐
- 10.4 SkillBuilder — Drawing a line graph using Excel (ewbk-9943) ☐
- 10.5 Effects of internal migration in Australia (ewbk-9947) ☐
- 10.6 Case study — Internal migration in China (ewbk-9951) ☐
- 10.7 SkillBuilder — Constructing a land use map (ewbk-9955) ☐
- 10.8 International migration patterns (ewbk-9959) ☐
- 10.9 Effects of international migration (ewbk-9963) ☐
- 10.10 Investigating topographic maps — Urbanisation in Albury–Wodonga (ewbk-9967) ☐
- 10.12 Chapter 10 Extended writing task (ewbk-8558) ☐
  - Chapter 10 Reflection (ewbk-8557) ☐
  - Chapter 10 Student learning matrix (ewbk-8556) ☐

Sample responses

- 10.1 Chapter 10 Sample responses (sar-0160) ☐

Digital document

- 10.10 Topographic map of Albury–Wodonga (doc-36321) ☐

Video eLessons

- 10.1 Migration (eles-5350) ☐
  - Internal and international migration — Photo essay (eles-5286) ☐
- 10.2 Reasons people move in Australia — Key concepts (eles-5287) ☐
- 10.3 Internal migration in Australia — Key concepts (eles-5288) ☐
- 10.4 SkillBuilder — Drawing a line graph using Excel (eles-1662) ☐
- 10.5 Effects of internal migration in Australia — Key concepts (eles-5289) ☐
- 10.6 Case study — Internal migration in China — Key concepts (eles-5290) ☐
- 10.7 SkillBuilder — Constructing a land use map (eles-1755) ☐
- 10.8 International migration patterns — Key concepts (eles-5291) ☐
- 10.9 Effects of international migration — Key concepts (eles-5292) ☐
- 10.10 Investigating topographic maps — Urbanisation in Albury–Wodonga — Key concepts (eles-5293) ☐

Interactivities

- 10.2 Reasons people move in Australia (int-8702) ☐
- 10.3 Internal migration in Australia (int-8703) ☐
- 10.4 SkillBuilder — Drawing a line graph using Excel (int-3158) ☐
- 10.5 Effects of internal migration in Australia (int-8704) ☐
- 10.6 Rural–urban migration in China (int-3116) ☐
  - Case study — Internal migration in China (int-8705) ☐
- 10.7 SkillBuilder— Constructing a land use map (int-3373) ☐
- 10.8 International migration patterns (int-8706) ☐
- 10.9 Effects of international migration (int-8707) ☐
- 10.10 Investigating topographic maps — Urbanisation in Albury–Wodonga (int-8708) ☐
- 10.12 Chapter 10 Crossword (int-8709) ☐

ProjectsPLUS

- 10.11 Thinking Big research project — Multicultural Australia photo essay (pro-0173) ☐

Weblinks

- 10.2 Small cities tool (web-6384) ☐
  - Evocities (web-2844)
- 10.3 Population data (web-6385) ☐
  - Internal migration (COVID-19) (web-6386) ☐
- 10.5 ABS QuickStats (web-2731) ☐
  - Greater Sydney Region plan (web-6387) ☐
- 10.6 China's urban growth (web-0099) ☐
  - China's population (web-6392) ☐

Fieldwork

- 10.2 Internal migration (fdw-0033) ☐

Google Earth

- 10.6 Shanghai (gogl-0062) ☐
- 10.10 Albury–Wodonga (gogl-0140) ☐

myWorldAtlas

- 10.6 China (mwa-7336) ☐
- 10.9 International migration and Australian cities (mwa-7339) ☐

Teacher resources

There are many resources available exclusively for teachers online.

11 Australia's urban future

INQUIRY SEQUENCE

To access a pre-test with **immediate feedback** and **sample responses** to every question in this chapter, select your learnON format at www.jacplus.com.au.

11.1 Overview

Numerous **videos** and **interactivities** are embedded just where you need them, at the point of learning, in your learnON title at www.jacplus.com.au. They will help you to learn the content and concepts covered in this topic.

Can Australia live in and grow its urban areas without making things worse for the future?

11.1.1 Introduction

Sustainability means meeting our own current needs while still ensuring that future generations can do the same. To make this happen, human and natural systems must work together without depleting our resources. Ultimately, sustainability is about improving the quality of life for all — socially, economically, and environmentally — both now and in the future. As the Prince of Wales said in his speech to the Copenhagen climate summit in 2010, 'As our planet's life-support system begins to fail and our very survival as a species is brought into question, remember that our children and grandchildren will ask not what our generation said, but what it did.'

STARTER QUESTIONS

1. From what you already know, do you think Australia's urban communities need to be planned more carefully?
2. What do you think the terms *harmony* and *balance in nature* mean?
3. How can the principles of harmony and balance be applied to human environments such as cities?
4. Why is it important to think about and plan for the future of our cities?
5. In what ways does the way we live today affect how people will live in the future?
6. Watch the **Australia's urban future — Photo essay** video in your online Resources. To what extent is your home sustainably designed?

FIGURE 1 The heliostat at One Central Park, Sydney, reflects sunlight onto the ground and plants below.

on Resources

eWorkbook	Chapter 11 eWorkbook (ewbk-8103)
Video eLessons	Sustainable cities (eles-3495) Australia's urban future — Photo essay (eles-5294)

11.2 Australia's projected population

LEARNING INTENTION

By the end of this subtopic, you will be able to outline projected population growth of urban Australia and Sydney.

11.2.1 The future of Australia

Australia's population will continue to grow and change. In particular, it will become more urban and its composition will age. Population increases provide many challenges to sustainability. These include loss of biodiversity, limits on water supply, more greenhouse gas emissions and threats to food security (access to enough healthy food). Our cities experience more traffic congestion and there are problems with housing availability and **affordability**. Access to services, infrastructure and green space are also limited for some people in our communities.

To handle these many challenges, we must plan effectively for an increased population by building communities that can accommodate future changes. This will build communities in which all Australians live and prosper.

FIGURE 1 Change in Australian urban and rural populations over time (projected)

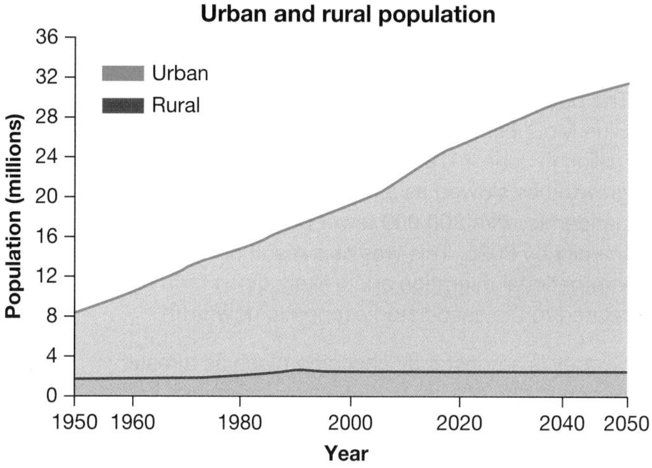

Source: Australian Bureau of Statistics

Approximately 93 per cent of Australia's growing population will be living in urban areas by 2050. However, some urban residents still make a tree change or a sea change and relocate to rural areas or the coast to improve their **quality of life**. The population in rural communities is generally stable or decreasing, as many young people leave in search of jobs and study opportunities. Some rural communities manage to keep their populations stable by shifting their employment focus from manufacturing to services; by utilising better internet connections, to allow people to work remotely from their office; or by improving public transport links. This became more common during and after the COVID-19 pandemic. Being required to work from home led to an increase in numbers of people moving to rural and coastal areas.

11.2.2 Implications for future growth and sustainability

Forecasts for growth of Australia's population have a range of implications for sustainability. Sustainability means development that meets the needs of the present population without endangering the capacity of future generations to meet their needs. Indicators of sustainability in urban areas include air and water quality, biodiversity, integration of green building initiatives, health and wellbeing measures, employment rates, transport infrastructure and access to employment.

As Australia's population grows, this will affect how Australian cities will grow and how sustainable they will be. Issues of sustainability include access to water, affordability of food and the distance food travels to get on the plate, loss of habitat areas and species diversity, and greenhouse gas emissions. Planning for Australia's urban future involves strategically planning for equitable and affordable access to services and infrastructure, in order to develop resilient communities that can cope with and manage changes in the future.

affordability the quality of being affordable — priced so that people can buy an item without inconvenience

quality of life your personal satisfaction (or dissatisfaction) with the conditions under which you live

Population growth in urban areas has many implications beyond the urban area itself. These include loss of agricultural land, habitat areas and open space, as well as increased pressure on transport infrastructure, resulting in heavy flows of commuter traffic and traffic congestion.

Sustainable planning for future growth includes strategies for:

- providing accessible, efficient and cost-effective public transport
- upgrading and more efficient use of existing transport infrastructure, and providing additional infrastructure
- planning that takes into consideration equitable access and reduction in carbon emissions
- providing green and public space
- creating and supporting employment centres.

CASE STUDY

Sydney's recent population trend

The population of Greater Sydney (including the Blue Mountains and Central Coast) reached 5.3 million in June 2019. Since that time, population growth has slowed as a result of the COVID-19 pandemic, with 200 000 fewer people expected in the city by 2025. This was as a result of reduced international migration and a likely dip in birth rates related to the associated economic downturn.

Sydney is a constantly changing place as a result of population growth and changing demands on parts of the city. A range of strategies are used throughout parts of Sydney to create liveable urban places that are economically, socially and environmentally sustainable. These include reducing commuting times, reducing energy consumption, consolidation, and shared public natural spaces. Individuals and communities can actively contribute to a more sustainable urban future.

FIGURE 2 Population growth in Sydney slowed during the COVID-19 pandemic as people moved out of urban areas.

FOCUS ON FIELDWORK

fdw-0045

Capturing and annotating photographs

One way to gather information on a theme for inquiry is to take photographs. To add context to the images and explain any links to geographical content, it is useful to fully annotate photographs.

Learn how to refine your eye for taking and annotating your fieldwork photos using the **Capturing and annotating photographs** fieldwork activity in your learnON Fieldwork Resources.

on Resources

eWorkbook	Australia's projected population (ewbk-9975)
Video eLesson	Australia's projected population — Key concepts (eles-5295)
Interactivity	Australia's projected population (int-8710)

11.2 ACTIVITIES

1. Growing communities create growing problems. For example, social problems may include poverty, chronic unemployment, welfare dependence, drug and alcohol abuse, crime and homelessness. Working in small groups, brainstorm some of the impacts that growing communities may have on:
   a. the environment
   b. the economy.
   Collect your ideas together as a class.
2. Conduct your own research on population growth in Sydney. Create a summary including the location of the highest growth areas and the impact of migration on growth in Sydney. Use data tables, column graphs and/or line graphs to present your information.
3. A hectare is equivalent to 10 000 square metres or about 2.5 acres. In urban Australia, most houses were traditionally built on quarter-acre blocks (about 12 house blocks per hectare). How does your area compare?

11.2 EXERCISE

Learning pathways

■ LEVEL 1	■ LEVEL 2	■ LEVEL 3
2, 3, 9, 12, 14	1, 4, 8, 11, 13	5, 6, 7, 10, 15

Check your understanding

1. In your own words, write a definition of affordability.
2. Describe Australia's and Sydney's expected population growth.
3. Describe the differences between urban and rural living.
4. Identify four areas that need careful planning for growing cities to be sustainable.
5. Outline how each of the following are sustainability issues.
   a. Access to water
   b. Affordability of food
   c. Loss of habitat areas
   d. Greenhouse gas emissions
6. What are three key services that rural towns can improve to keep their populations stable?

Apply your understanding

7. Explain how Australia's urban population growth will create challenges for creating a sustainable urban future.
8. Explain how COVID-19 is likely to affect population growth rates in Australia.
9. How important is careful planning in creating a sustainable urban future in Australia? Provide reasons to support your response.
10. Examine **FIGURE 1**. Explain the trends evident in the graph.
11. Why is an ageing population a challenge to sustainability?
12. Account for Sydney's slowed population growth in 2020.

Challenge your understanding

13. In cities, we must face the challenges and opportunities of productivity, sustainability and liveability. If we address one goal, we can have an impact, either positively or negatively, on others. This demonstrates interconnection. For example, efficient public transport can fix congestion and improve access to jobs and opportunity (productivity). It can also reduce greenhouse gas emissions (sustainability) and improve access to education, health and recreational facilities (liveability).
    a. Using the example of the National Broadband Network, how might productivity, sustainability and liveability be affected?
    b. Classify the impacts you have listed as positive or negative.
14. Young people leave rural areas in search of employment and education. Describe the factors that could contribute to you leaving the area where you live.
15. What will be the biggest sustainability challenge for Australia in the twenty-first century? Justify your decision.

To answer questions online and to receive **immediate feedback** and **sample responses** for every question, go to your learnON title at www.jacplus.com.au.

11.3 Designing sustainable cities

LEARNING INTENTION

By the end of this subtopic, you will be able to discuss the implications of population growth on the sustainability of places.

11.3.1 Challenges of sustainability

Our cities are facing an important challenge. Some predict that Australia's population will reach 42 million by 2050. If this is the case, then our cities will need to adapt and become more efficient to allow us to maintain our current quality of life.

Sustainable communities share a common purpose of building places where people enjoy good health and a high quality of life. A sustainable community can thrive without damaging the land, water, air, and natural and cultural resources that support them, and ensures that future generations have the chance to do the same. The basic **infrastructure** should be designed to minimise consumption, waste, pollution and the production of greenhouse gases. Sustainable urban areas strike a delicate but achievable balance between economic, environmental and social factors. We can address the challenges and opportunities for sustainable communities at two different scales: neighbourhood level and city level.

A sustainable city is one that has a small **ecological footprint**. The ecological footprint of a city is the surface area required to supply a city with food and other resources and to absorb its wastes. At the same time, a sustainable city is improving its quality of life in health, housing, work opportunities and liveability.

infrastructure the facilities, services and installations needed for a society to function, such as transportation and communications systems, water and power lines

ecological footprint the amount of productive land needed on average by each person in a selected area for food, water, transport, housing and waste management

FIGURE 1 City-level scale means considering the greater urban area. Perth, Western Australia

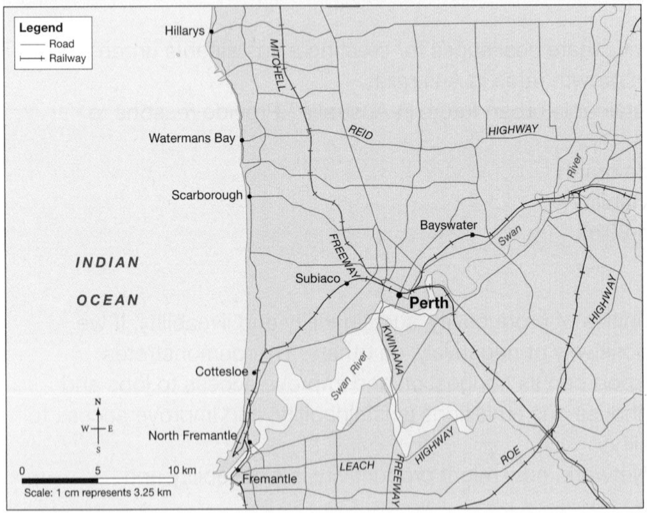

Source: Spatial Vision

FIGURE 2 A neighbourhood level means considering a small part of the larger city. Perth, Western Australia

FIGURE 3 City Farm, a neighbourhood community farm on a reclaimed industrial site in East Perth, Western Australia

Ways to improve sustainability at the neighbourhood scale include:
- reducing the ecological footprint
- protecting the natural environment
- increasing community wellbeing and pride in the local area
- changing behaviour patterns by providing better local options
- encouraging compact or dense living
- providing easy access to work, play and schools.

Ways to improve sustainability at the city scale include:
- building strong central activity areas (either one major hub or a number of specified activity areas)
- reducing traffic congestion
- protecting natural systems
- avoiding suburban sprawl and reducing inefficient land use
- distributing infrastructure and transport networks equally and efficiently to provide accessible, cheap transportation options
- promoting inclusive (accessible for all people) planning and urban design
- providing better access to healthy lifestyles (e.g. cycle and walking paths)
- improving air quality and waste management
- using stormwater more efficiently
- increasing access to parks and green spaces
- reducing car dependency and increasing walkability
- promoting green space and recreational areas
- demonstrating a high mix of uses (e.g. commercial, residential and recreational).

on Resources

eWorkbook	Designing sustainable cities (ewbk-9979)
Video eLesson	Designing sustainable cities — Key concepts (eles-5296)
Interactivity	Designing sustainable cities (int-8711)

11.3 ACTIVITY

Consider the areas listed in which a neighbourhood can become more sustainable. Create a table and detail the ways in which your own suburb, area or neighbourhood is meeting these aims. Add another column and use the internet to research how your local council is trying to make your suburb more sustainable. Conclude by writing a few sentences to answer the following questions.

a. Is my neighbourhood sustainable?
b. How will liveability be improved?
c. What needs to change in order to make it even more sustainable?

11.3 EXERCISE

Learning pathways

■ LEVEL 1	■ LEVEL 2	■ LEVEL 3
1, 2, 8, 9, 13	3, 6, 7, 10, 14	4, 5, 11, 12, 15

Check your understanding

1. Complete the following sentence: Some organisations have projected that Australia's population will reach ____ million by ____.
2. Identify the two main aims of a sustainable community.
3. Identify the two scales at which we can work to improve the sustainability of our communities.
4. Describe what a sustainable neighbourhood might look like.
5. Define the term *ecological footprint* in your own words.
6. Outline four changes that could be made to improve sustainability at a neighbourhood scale or city scale.

Apply your understanding

7. Explain how an ecological footprint is measured.
8. List and explain the differences between local and city scales of sustainability action.
9. Explain how each of the following might improve people's quality of life.
   a. Better access to cycling paths
   b. More green space
   c. Protecting the natural environment
   d. Reducing traffic congestion
10. Which of the ways to improve sustainability at the neighbourhood scale do you think would contribute the most to the environmental sustainability of a place? Give reasons for your answer.
11. Which of the ways to improve sustainability at the city scale do you think would contribute the most to peoples' quality of life? Give reasons for your answer.
12. Explain what the phrase 'demonstrating a high mix of uses' means and how this might contribute to the sustainability of a place.

Challenge your understanding

13. Suppose that you have unlimited funds to invest in the sustainability of your neighbourhood. What would your first improvement be and why?
14. Propose one change that would improve the sustainability of where you live.
15. Think about the biggest city or urban area you have ever visited (or seen online in Street view).
    a. What conclusions can you draw about that place's main sustainability challenge?
    b. Choose one problem that you noticed and suggest how it might be better managed.

To answer questions online and to receive **immediate feedback** and **sample responses** for every question, go to your learnON title at www.jacplus.com.au.

11.4 SkillBuilder — Describing change over time

online only

LEARNING INTENTION

By the end of this subtopic, you will be able to identify and describe change over time on a map showing movement timeframes.

The content in this subtopic is a summary of what you will find in the online resource.

11.4.1 Tell me

What is a description of change over time?

A description of change over time is a verbal or written description of how far a feature moves, or how much it alters, over an extended time period. A description of change over time is used to show us the distance that a feature has moved or the extent to which it has altered, and to alert us to the possible impacts over a wider region. For example, the intensity of earthquake tremors indicates that energy has moved across a region.

11.4.2 Show me

How to describe change over time

Step 1

Read the map title, legend, and any captions.

Step 2

Study the movement lines across the map and relate these to places, either by name, latitude and longitude, or direction from other places.

Step 3

Begin with an opening statement that generalises about what has been mapped. For example, 'In **FIGURE 1**, the Peru 2007 magnitude 8.0 earthquake was monitored, and a warning was sent across the Pacific based on the timeframes of movement of the tsunami's energy.'

Step 4

Provide some specific statements about places impacted close to the time of origin of the event. For example, within three hours the wave energy would have reached the Galapagos Islands, a distance of 1500 kilometres.

FIGURE 1 Tsunami mapping from Peru, 2007, showing the magnitude 8.0 earthquake that occurred on 15 August. SEAFRAME stations on Pacific islands detected a tsunami.

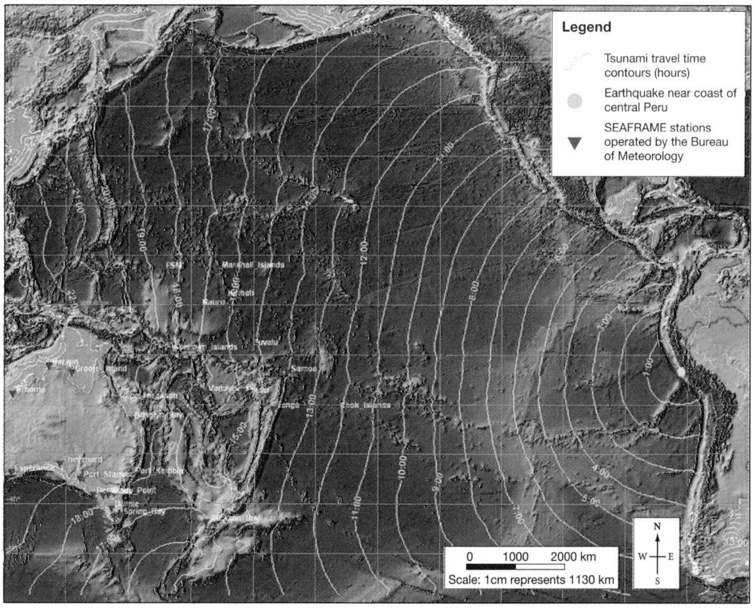

Source: © Bureau of Meteorology

Step 5

From the timeframes discussed in Step 4, infer what impact the event will have on people and places at different times.

Step 6

Conclude your analysis with an overall statement about the level or magnitude of the event.

11.4.3 Let me do it

learnON

Go to learnON to access the following additional resources to help you build this skill:
- a longer explanation of this skill and its application in Geography (Tell me)
- a video demonstrating the step-by-step process of this skill (Show me)
- an activity and interactivity for you to practise the skill (Let me do it)
- self-marking questions to help you understand and use the skill.

on Resources

eWorkbook	SkillBuilder — Describing change over time (ewbk-9983)
Video eLesson	SkillBuilder — Describing change over time (eles-1753)
Interactivity	SkillBuilder — Describing change over time (int-3371)

11.5 Sustainable cities in Australia

LEARNING INTENTION

By completing this subtopic, you will be able to discuss the implications of population growth on the sustainability of places.

11.5.1 Measuring city sustainability

Australian cities often perform well in worldwide rankings of liveability. Liveability is an assessment of the quality of life in a particular place—living in comfortable conditions in a pleasant location. But being liveable is not the same as being sustainable, which involves living in a way that sustains the environment and conserves resources into the future.

What makes a city sustainable? In 2010, the Australian Conservation Foundation conducted a study to measure the sustainability of Australia's 20 largest cities. The indicators measured were a combination of:
- environment—air quality, ecological footprint, water, green building and biodiversity
- quality of life—health, transport, wellbeing, population density and employment
- resilience (the ability of a city to cope with future change): climate change, public participation, education, household repayments and food production.

FIGURE 1 Darwin was ranked the most sustainable city in Australia in 2010.

The results showed that Darwin was the most sustainable city in Australia in 2010. It performed best in terms of the economic indicators, such as low unemployment rates and residents paying relatively low proportions of their income for loan repayments. Darwin's ecological footprint was poor (at 7.06 hectares/person/year) but the city rated well for other environmental factors such as biodiversity and air quality. **FIGURE 2** shows the ranking for other cities.

FIGURE 2 A sustainability ranking of Australia's cities

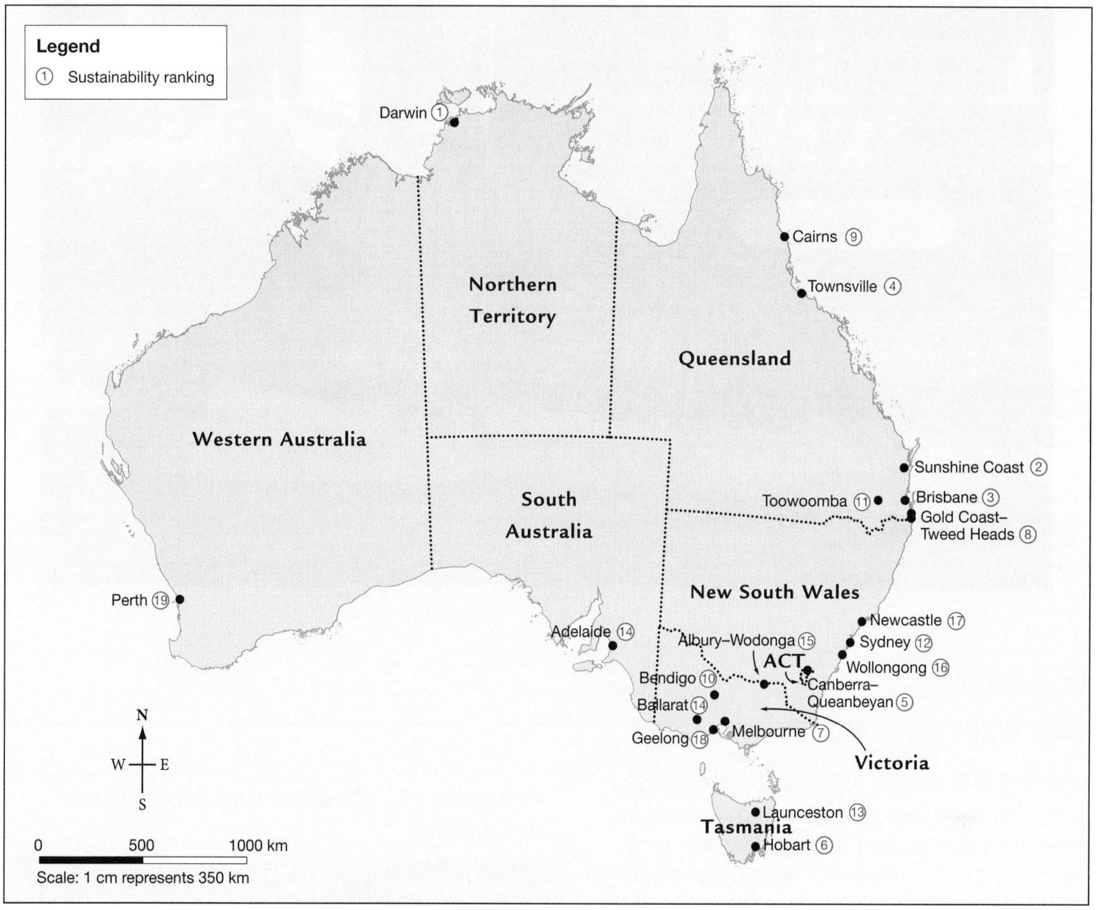

Source: Spatial Vision

11.5.2 Local urban communities

In most cities, it is action at a local community scale that can make the most difference in improving city sustainability. State governments and local councils have responsibility for improving complex infrastructure (for example transport and water supply) for whole cities, but change at a local level can have positive results.

Sustainable communities in cities may have features in common. Generally, sustainable communities:

- are friendly and social
- consume less energy and water and produce less waste
- have **medium**- to **high-density** rather than **low-density housing**
- are within walking distance of some public facilities and have excellent public transport links for longer trips
- include public places that people can walk or cycle to
- have good landscaping, such as rain gardens that allow water to stay in the soil and increase oxygen supply, and gaps in kerbs to allow water to flow into gardens
- have dwellings that have been built to a budget to make them affordable.

medium-density housing residential developments with around 20–50 dwellings per hectare

high-density housing residential developments with more than 50 dwellings per hectare

low-density housing residential developments with around 12–15 dwellings per hectare; usually located in outer suburbs

FIGURE 3 Sustainable urban planning strategies in a densely populated urban area

CASE STUDY 1

Christie Walk, Adelaide

Christie Walk is located in Adelaide, South Australia. It is a small urban village of 27 dwellings located on a quarter of an acre of land. The site is within easy walking distance of Adelaide's markets, parklands and CBD, which means car use is reduced. Around 40 people live at Christie Walk, ranging in age from very young children to people over 80 years old.

A number of principles were used in the design of Christie Walk:

- building for low energy demand (passive heating and cooling; natural lighting and sealed double glazing in all windows and glass doors)
- maximising the use of renewable/solar-based energy sources (photovoltaic cells on the roof) and minimising the use of non-renewable energy sources
- capturing and using storm water in large underground rainwater tanks and recycling waste-water
- creating healthy gardens and maximising the biodiversity of local flora and fauna (the gardens also produce herbs, vegetables and fruit)
- avoiding the use of products that damage human health
- minimising the use of non-recyclable materials.

FIGURE 4 One of the sustainable buildings in Christie Walk

FIGURE 5 A plan of Christie Walk in Adelaide

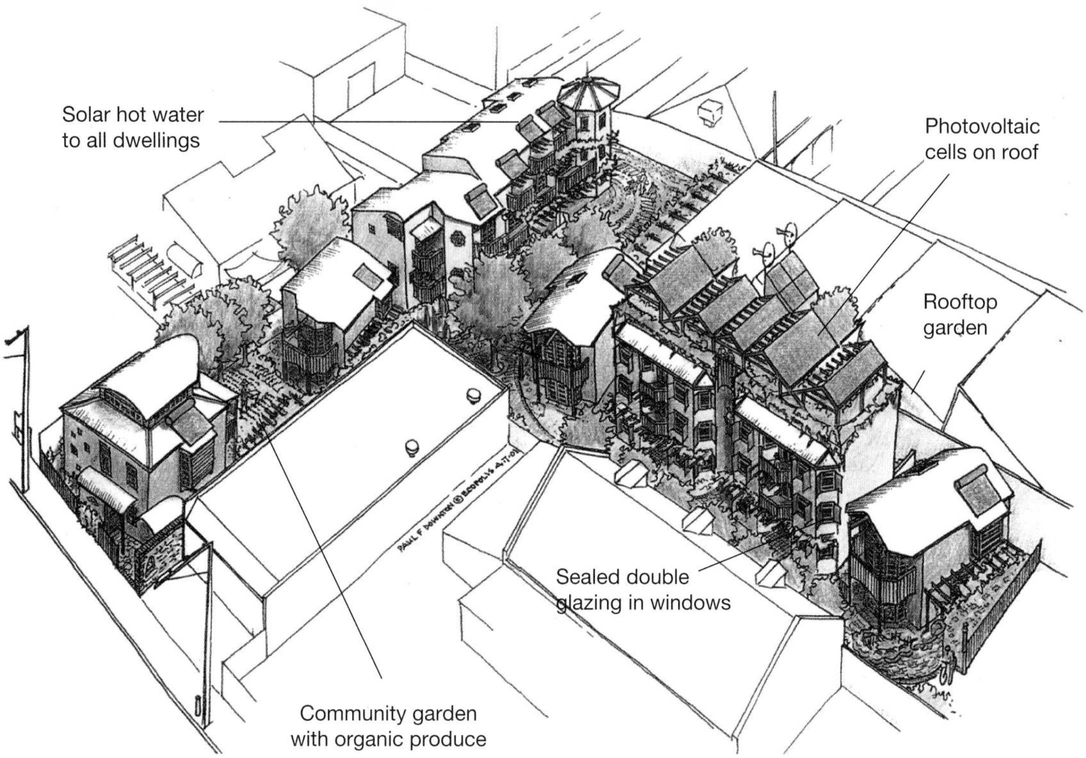

FIGURE 6 Rooftop gardens provide good insulation, protecting the buildings below from the hot sun in summer. In winter, they keep warmth from escaping from the building below.

11.5.3 Sustainability in Sydney

The City of Sydney Council has developed the Sustainable Sydney 2030 Community Strategic Plan, which is a set of goals to guide the community in becoming as connected, global and green as possible. The city is planning on becoming internationally recognised for its outstanding environmental initiatives and green industries. The City of Sydney is constructing green infrastructure to reduce water, waste water and energy demands. It is also developing new housing with transport, facilities and open space integrated in the planning process. Innovation, new technologies, creativity and collaboration make Sydney an attractive place for business activities and social and cultural activities.

CASE STUDY 2

One Central Park

Central Park is a development on the site of an old brewery in Chippendale, Sydney. It was designed by architects from Sydney, London, Paris and Copenhagen as a residential, business and retail development, and was built with sustainability in mind (**FIGURE 7**). The development was hailed as an innovative next step in sustainable design at the time of its construction, and it has won a variety of awards for its construction, design and sustainability features.

FIGURE 7 Vertical gardens and heliostat

Sustainability features of the development

Ninety-three per cent of the demolition waste from the site was recycled.

The development has an on-site water recycling plant. The recycled water comes from a variety of sources such as rainwater collected from the roofs, groundwater and sewage water. This water is used for such things as toilets, washing machines and irrigating the rooftop gardens and green walls.

The design of several of the buildings incorporates huge vertical gardens. Frames of growing plants were constructed around the walls and roofs of the buildings. As the plants grow, the gardens on the walls and roofs will grow into each other to cover the buildings. A water recycling system has been installed to water the plants. A heliostat (a suspended platform, see **FIGURE 7**) reflects light onto the plants and ground below.

Green walls and roofs allow cities to manage their storm water better. They soak up and filter water, which is then stored in the leaves of plants. The increase in vegetation provides shading, increased **evapotranspiration** and the possibility of more rain. The vegetation will also absorb some greenhouse gases and serve as a form of **carbon sequestration**, reducing levels of carbon dioxide and nitrogen oxides associated with **photochemical smog**.

Green walls and roofs can improve human health, wellbeing and levels of happiness and improve social interactions within the community. There is also an aesthetics issue — they can be a thing of beauty in built-up city areas.

As well as a water recycling plant, Central Park has its own low-carbon natural gas power plant, which produces thermal energy for the residents and businesses at the park. This plant is estimated to be twice as energy efficient as an energy plant powered by coal.

evapotranspiration the process by which water is transferred to the atmosphere from surfaces such as the soil and plants

carbon sequestration capturing and storing carbon dioxide from the air

photochemical smog smog that is produced by a reaction between chemicals in the air and ultraviolet light from the sun

Finally, Central Park has a 6400-square-metre park at its centre, an area that is mainly grassed to allow for recreational activities. It has shade and barbecue areas for all the residents to enjoy.

FIGURE 8 (a) Central communal green space and (b) vertical gardens

fdw-0046

FOCUS ON FIELDWORK
Sustainable streetscapes

Are there examples of sustainable urban planning or sustainability projects in your school or neighbourhood? Even though there may be programs in place, it can be difficult to assess their effectiveness in the short term, but we can measure the areas and extents of programs to see if there is anything that has been overlooked.

Learn how to conduct a detailed sustainability survey of a place near you using the **Sustainable streetscapes** fieldwork activity in your learnON Fieldwork Resources.

on Resources

eWorkbook	Sustainable cities in Australia (ewbk-9987)
Video eLesson	Sustainable cities in Australia — Key concepts (eles-5297)
Interactivity	Sustainable cities in Australia (int-8712)

11.5 ACTIVITIES

1. Work in groups of three. Use the **Sustainable Cities Awards** weblink in your online Resources to learn about projects that have won Sustainable Cities Awards in Australia. Each group member should read about three awards and summarise the projects to the others. Using a diamond ranking chart, rank the projects from most to least important for sustainability. Write the name or description of the best project in the top space of the diamond and the least sustainable at the bottom. Add the other projects to the chart after your group has discussed and agreed on the ranking.
2. Research and list the similarities and differences between the Woolloongabba neighbourhood plan in Brisbane and the Christie Walk development in Adelaide.

3. Use ideas from this section and further research to design a small sustainable urban neighbourhood. You may choose to work in groups or individually. You may like to use photographs of examples you find in your city/town or on the internet to draw your plan. Alternatively, make a video of some examples and incorporate them into your design. Justify the inclusion of all the features you choose by annotating the plan or writing some notes to explain your choices.
4. Research Central Park in central Sydney or another sustainable building project. Explore some of the ways in which the building works, not only in conserving resources but also in improving the wellbeing of its workers.
   a. Identify the features that are most interesting to you. Why? How does it work?
   b. Could this feature, or any others, be applied to your home? Would any of the features be more suited to larger buildings such as your school?
   c. Would you like to work in a building like this? Why or why not? Justify your explanation.

11.5 EXERCISE

Learning pathways

■ LEVEL 1	■ LEVEL 2	■ LEVEL 3
2, 3, 6, 7, 13	1, 4, 8, 9, 14	5, 10, 11, 12, 15

Check your understanding

1. Describe the difference between a liveable city and a sustainable city.
2. Define the following terms.
   a. Low-density housing
   b. Medium-density housing
   c. High density housing
3. Which Australian city was considered Australia's most sustainable in 2010?
4. What were the three indicators used to measure the sustainability of Australian cities in the Australian Conservation Foundation's study of sustainability?
5. Describe the advantages and disadvantages of low-, medium- and high-density housing. Outline three ways that local communities can improve the sustainability of their area.

Apply your understanding

6. Explain how the Sustainable Sydney 2030 Community Strategic Plan can help Sydney City Council in their planning for a sustainable urban future in Sydney.
7. How do different housing densities impact on the sustainability of urban places?
8. Outline the features of Central Park that contribute to sustainability. Explain how these ideas could be used more widely to contribute to the sustainability of Sydney.
9. Identify two sustainability principles that were used in the design of Christie Walk in Adelaide. Analyse the images and plans of the village (**FIGURES 4, 5** and **6**) and explain how these principles can be seen in the design. Give specific examples to support your explanation.
10. Explain how allowing water to run off a footpath into a roadside garden makes a community more sustainable.
11. Explain how vertical gardens contribute to:
    a. the sustainability of a building
    b. the wellbeing of the local community.
12. Explain the purpose of a heliostat.

Challenge your understanding

13. Which development would you rather live in: Christie Walk or Central Park? Provide reasons for your choice.
14. Propose your own set of five key criteria that are essential for a building to be considered sustainable.
15. Should the technology and sustainability features of Christie Walk and Central Park be subsidised by the government so that they are affordable for all homes? Give reasons for your view.

To answer questions online and to receive **immediate feedback** and **sample responses** for every question, go to your learnON title at www.jacplus.com.au.

11.6 Managing the suburbs

LEARNING INTENTION

By the end of this subtopic, you will be able to explain strategies to create sustainable suburban places.

11.6.1 Living on the urban fringes

There is much at stake on the rural–urban fringe, with the conflict between farming and urban residential development reaching a critical point on the outskirts of Australia's cities. Australia is the driest inhabited continent on Earth, and just six per cent of its total area (45 million hectares) is **arable** land. The areas targeted by our state governments for residential development continue to expand. When some of our most fertile farmland is lost to urban sprawl, we reduce our productive capacity. Is this a recipe for sustainability?

On the edge of many Australian cities, new homes are being built as part of planned developments. These were previously wildlife habitats and productive farmland on the urban fringe and are now known as greenfield sites. Accompanying these housing developments are plans for kindergartens, schools, parks, pools, cafés and shopping centres (often called amenities and facilities).

FIGURE 1 The history of Sydney's urban sprawl

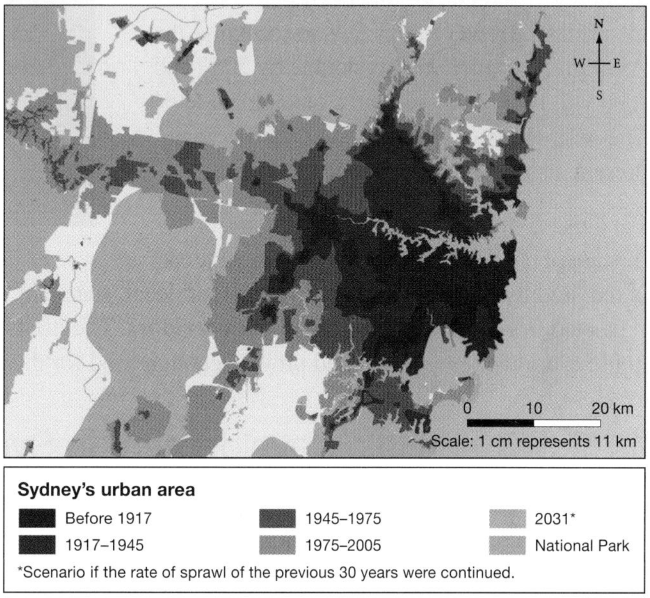

Source: Spatial Vision

Having an 'affordable lifestyle' is the main attraction for people who purchase these brand new homes. They like the idea of joining a community and having the feeling of safety in their newly established neighbourhood.

Most new houses on the rural–urban fringe are bought by young first-home buyers, who are attracted by cheaper housing and greener surroundings. Generally the residents of these fringe households feel that the benefits of their location outweigh the poor public transport provisions and long journeys to work and activities — trips that are usually made in a car.

11.6.2 Feeding our growing cities

Market gardens have traditionally provided much of a growing city's food needs, supplying produce to central fruit and vegetable markets. These 'urban farms' were located on fertile land within a city's boundaries but close to its edge, with a water source nearby and often on floodplains. They have been in existence in and around Australia's major cities since the 1800s, and some (such as La Perouse market gardens in Sydney) are now listed on the National Trust heritage garden register.

Fifty per cent of Victoria's fresh vegetable production still occurs in and around Melbourne, on farms such as those at Werribee South and Bacchus Marsh. More than 60 per cent of Sydney's fresh produce is grown close to the city, with the bulk of it coming from commercial gardens such as those in Bilpin, Marsden Park and Liverpool.

arable describes land that is suitable for growing crops

These farms are important because:

- they provide us with nutritious food that does not have to be transported very far
- they provide local employment
- they preserve a mix of different land uses in and around our cities.

Currently, we can obtain our food from almost anywhere because we have modern transportation (such as trucks and planes), better storage technology (refrigeration and ripening techniques) and cheap sources (not necessarily the closest). However, this fails to recognise that Australia's population may double by 2050 and food will become more scarce on a global level. The destruction of our local food sources may present a significant risk to our food security.

FIGURE 2 The suburbs of Maitland, expanding across former farms

Land use zoning is generally the responsibility of state planning departments, but cooperation is required by all three levels of government: local, state and federal. We need to ensure that our **green wedges** are protected from becoming **development corridors**. The needs on both sides of the argument are valid. How can we house a growing population and provide enough food for them? Can we do both?

11.6.3 Greenfield development in Sydney

Greenfield developments occur when undeveloped land is released by the government for housing developments. In Sydney, greenfield development areas are generally more readily available in outer suburban areas. Of Sydney's greenfield stocks that are zoned for residential development, most are located in west and south-west Sydney. As a result, the largest increases in population in Sydney are expected to occur in Liverpool (for example Carnes Hill), Camden (for example Oran Park) and Wollondilly. Both the south-west and north-west corridors are expected to experience significant growth, as well as outer suburban areas such as the Central Coast.

As part of the South West Growth Area, new communities are being developed in Leppington North, East Leppington, Austral, Edmondson Park, Catherine Field, South Creek West, Lowes Creek Maryland, Oran Park and Turner Road. There will also be new suburbs developed close to the new Western Sydney Airport and Badgerys Creek as part of the Western Sydney Employment Area. This development has been facilitated by major transport infrastructure projects such as the South West Rail Link and the upgrade of Camden Valley Way, Bringelly Road and Northern Road.

It is expected that most land currently available for greenfield development will have been developed by 2036.

green wedges land in urbanised areas that is protected from development and kept as farming, parkland or green space

development corridor area set aside for urban growth or development

greenfield developments when undeveloped land is released by the government for housing developments

on Resources

eWorkbook	Managing the suburbs (ewbk-9991)
Video eLesson	Managing the suburbs — Key concepts (eles-5298)
Interactivity	Managing the suburbs (int-8713)

11.6 ACTIVITIES

1. Investigate rates of suburban growth.
   a. Sydney's growth has occurred around a series of urban precincts. New suburbs include Ropes Crossing, Jordan Springs and The Ponds. Use Google Earth to locate and 'placemark' each of these places. Use the ruler tool to measure the distance from each suburb to Sydney's CBD.
   b. Complete the same task to identify growth areas in regional centres, for example Maitland, Dubbo, Wollongong, Bathurst, Albury and Wagga Wagga.
   c. Are there any patterns or differences in the extent, direction and rates of growth?
2. The demands of housing and agriculture on land are two of the biggest dilemmas of the twenty-first century. A growing population needs to be housed, but it also needs to be fed, and the cost of relying on imported food can be very high. Set up a debate with your classmates on the following statement: 'Green belts close to the city should be preserved and protected.' The affirmative team will argue for this, while the negative team will argue that green belts should be removed and used for new housing developments.
3. Research some companies that sell house and land packages in your area. What are some of the marketing messages that are used to sell the properties? Do you think they are able to deliver on their promises?
4. Many new homes on the urban fringe are built with six-star or seven-star energy efficiency. What does this mean? Should there be a requirement that new developments are energy efficient?

11.6 EXERCISE

Learning pathways

■ LEVEL 1	■ LEVEL 2	■ LEVEL 3
1, 3, 7, 11, 15	2, 4, 9, 12, 14	5, 6, 8, 10, 13

Check your understanding

1. List the general groups involved in the conflict over our rural–urban spaces.
2. Refer to **FIGURE 1**. Describe how Sydney's urban sprawl has changed in direction and pace over time.
3. Why are many homes in outer suburbs bought by first home buyers?
4. Describe why it is important for people to have rural spaces, such as market gardens, close to the city.
5. Define the term *greenfield development*.
6. Outline two reasons why modern transport made it easier for urban areas to expand.

Apply your understanding

7. Explain why it is important for cities to have access to more land for urban development.
8. Analyse why south-west Sydney is likely to experience significant growth in the future. What are the implications of this for Sydney's sustainability?
9. Most land currently available for greenfield development will have been developed by 2036. What are the implications of this for Greater Sydney's:
   a. population
   b. ecological footprint
   c. traffic congestion
   d. public transport
   e. food security?
10. Explain the interconnection between suburban growth and food security.
11. How does urban sprawl affect biodiversity?
12. What compromises might each of the following have to make in order to ensure our suburbs develop sustainably? List three for each.
    a. Home owners on the rural-urban fringe
    b. Local councils in growth corridors
    c. Developers
    d. Market gardeners

Challenge your understanding

13. Should our governments be allowed to regulate how and where suburbs grow (for example limiting growth or infrastructure in favour of protecting local farms or wildlife habitat) if it reduces the lifestyle benefits for the residents?
14. How can we house a growing population and provide enough food for them? Suggest a strategy that might help to solve this problem.
15. How might people who live in the inner urban areas of a city be affected by suburban sprawl?

To answer questions online and to receive **immediate feedback** and **sample responses** for every question, go to your learnON title at www.jacplus.com.au.

11.7 Living vertically

online only

LEARNING INTENTION

By the end of this subtopic, you will be able to explain strategies to create sustainable urban places.

The content in this subtopic is a summary of what you will find in the online resource.

Australian cities are experiencing an apartment revolution. More people are choosing to live near the centre of cities in high-rise apartments rather than in houses on big suburban blocks. Families and individuals are moving to the inner city for a variety of reasons, such as seeking to make a smaller ecological footprint or avoiding long commutes to school, work and shops.

To learn more about the sustainability benefits of vertical living, go to your learnON resources at www.jacPLUS.com.au.

FIGURE 3 The library courtyard at Green Square

Contents

learnON

- 11.7.1 Higher-density living, smaller households
- 11.7.2 Going green
- 11.7.3 Priority Precincts and increased density

on Resources

eWorkbook	Living vertically (ewbk-9995)
Video eLesson	Living vertically — Key concepts (eles-5299)
Interactivity	Living vertically (int-8714)

11.8 Urban renewal

LEARNING INTENTION

By completing this subtopic, you will be able to explain and evaluate strategies to create more sustainable places through urban renewal.

11.8.1 Understanding urban renewal

Urban renewal is the redevelopment of an area of a large city. It can involve redeveloping obsolete housing, industrial or commercial areas. The purpose of urban renewal may be to address social problems, haphazard planning or traffic congestion, and/or to improve quality of life for residents. Urban renewal can include reusing buildings for a new purpose, clearing structures that are deteriorated, and rehabilitation and conservation of heritage buildings.

FIGURE 1 Honeysuckle, on the waterfront in Newcastle, redeveloped into public space

Sites considered appropriate for urban renewal will often already have access to infrastructure, as it was developed for the previous users of the site. This includes access to water, electricity, gas and sewerage. Often there are already established transport links to these locations that developed through the history of the site. New infrastructure costs are minimised.

The benefits of urban renewal are that it can spur economic growth, improve the urban environment, remove unsightly or obsolete buildings, create jobs and improve the quality of life for the local community. Buildings that are considered historically or architecturally significant can be repurposed and preserved through adaptive reuse.

urban renewal the redevelopment or improvement of disused or underused urban areas

FIGURE 2 (a) Honeysuckle boardwalk with the path of the old railway, 2020, and (b) Honeysuckle Railway Station, 1910

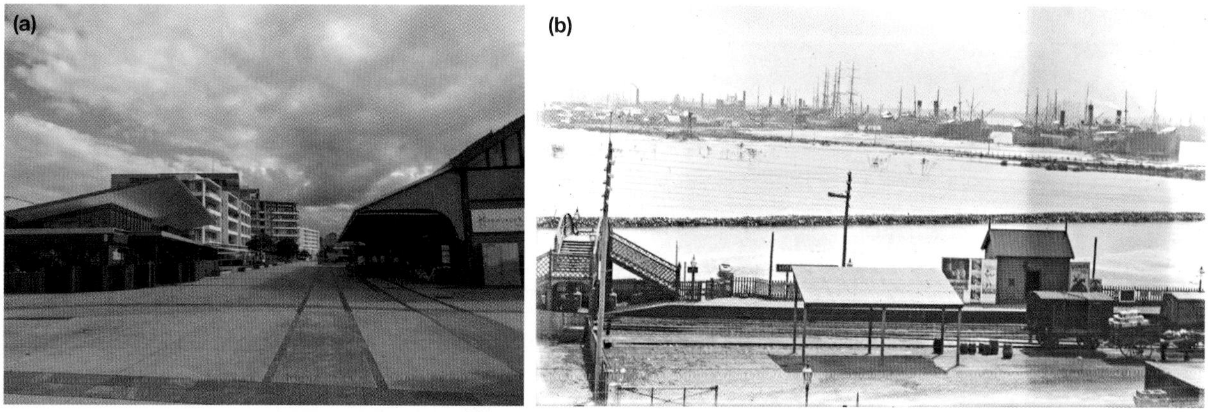

11.8.2 Types of urban renewal

Brownfield development

Brownfield developments involve using land that is disused or derelict, and was generally previously used for industrial uses (**FIGURE 3**), and as a result may be contaminated with waste or pollutants. A significant factor in the development of a brownfield site is the extent and type of contamination onsite. Contamination can limit the potential use of the site, and the cost of cleaning up any dangerous pollutants or waste can be very high.

FIGURE 3 A former brewery transformed into a high-density apartment complex in inner-city East Melbourne

Greyfield developments

Greyfield developments involve land that is no longer being used, or only being partly used because the site is not profitable due to economic obsolescence (that is, its previous use did not make or stopped making the owners/users money). Examples of kinds of land bought for greyfield developments include old retail areas or shopping centres that have closed, as well as old office buildings. Greyfield developments are less likely to be contaminated than brownfield developments, and tend to be cheaper and easier to redevelop. Like brownfield developments, greyfield developments also have existing infrastructure such as water, sewerage, and transportation links.

Bluefield developments

Bluefield developments are those that take place on land next to a waterfront, for example land along a harbour, river, ocean or lake. These areas can be subject to flooding and can have issues with stormwater, but are otherwise very valuable because of their proximity to the water (**FIGURE 4**). Like brownfield and greyfield developments, bluefield developments often have the benefit of existing infrastructure such as water, sewerage, and transportation links.

brownfield developments urban development projects on land that was previously used for industrial purposes

greyfield developments urban development projects on land that is being underused or not used

bluefield developments urban development projects on land that is next to the water (e.g. harbour, river, lake, ocean)

FIGURE 4 (a) and (b) Riverside, East Perth: 40 hectares of land on the Swan River being developed to include a mix of commercial and residential use

11.8.3 Urban renewal in Sydney

Sydney's Inner West is still experiencing deindustrialisation as industrial land users continue to move further west. Zoning for high-density residential developments has sped up increases in land values of industrial properties in Inner West suburbs. As a result some of the last remnants of the suburbs' blue-collar, industrial working-class history are being redeveloped. Old waterfront industrial sites such as Rozelle Bay and White Bay have already been rezoned as part of the Bays Precinct urban renewal initiative. Recent rezoning for high-density residential housing in suburbs such as Marrickville and Dulwich Hill will see a decline in small industries in coming years.

Several of the case studies used throughout this chapter are examples of urban renewal through brownfield developments, such as Green Square and Central Park. Other examples include Barangaroo and Homebush Bay.

CASE STUDY

Bays Precinct, Sydney

The Bays Precinct includes White Bay and White Bay Power Station, Glebe Island, Rozelle Rail Yards, Rozelle Bay and Bays Waterways, Bays Market District, Bays Waterfront Promenade and Wentworth Park. Sections of the Bays Precinct are two kilometres west of the Sydney CBD. It includes 95 hectares of prime waterfront land, most of which is government owned, and 5.5 kilometres of harbourfront.

FIGURE 5 (a) and (b) Bays Precinct, Sydney

Some of the land is currently used for maritime, port and commercial uses. Glebe Island currently receives shipments of gypsum, sugar, salt and cement. White Island is currently a cruise ship terminal. The Sydney Fish Market is currently located at Blackwattle Bay. Rozelle Bay is used by a range of maritime industries and the Sydney Superyacht Marina. Some areas, such as the White Bay Power Station and Rozelle Rail Yards, have not been in use for some years, and several of the sites contain buildings that are derelict or underutilised. The Bays Precinct also has a number of significant heritage buildings, such as the grain silos and White Bay Power Station. Much of the land in the Bays Precinct, including harbour sites, cannot currently be accessed by the public.

The urban renewal plan involves repurposing the White Bay Power Station, repurposing Glebe Island Bridge, introducing the Bays Waterfront Promenade at Blackwattle Bay on the old site of the Sydney Fish Markets, redeveloping the Sydney Fish Markets into the Bays Market Precinct on the previous Hanson Concrete site at Blackwattle Bay, and making better use of Wentworth Park as part of the precinct.

The urban renewal of the Bays Precinct will incorporate water quality improvements to address marine ecology and water quality issues in Sydney Harbour. It is planned to incorporate aspects of energy production within some of the structures and an increase in parkland, nature reserves and public access to the harbour.

When completed, the renewal project will incorporate community food markets and restaurants, and encourage temporary activation (short-term cultural and recreational uses) such as pop-ups and mobile food trucks. It will be an economic hub for the fresh food industry. The project will create new jobs, provide workspaces for knowledge-intensive sectors, support new enterprises and support the maritime industries.

FIGURE 6 Bays Precinct, Sydney

on Resources

eWorkbook	Urban renewal (ewbk-9999)
Video eLesson	Urban renewal — Key concepts (int-5300)
Interactivity	Urban renewal (int-8715)

11.8 ACTIVITY

Research the Bays Precinct, Sydney, and the Riverside development in East Perth. Begin with the **Bays Precinct** and **Riverside** weblinks listed in your online Resources. You could also complete this activity by comparing one or more developments that are planned for your area.

- Outline the main features and land uses of each development. Include details of housing capacity, types of housing, commercial land use, and community and recreational land use.
- What do the developments have in common?
- How are they different?
- Consider the timelines and other specific details that have been released to the public about the developments. How far are they along in the process? When are they due for completion?
- Have the projects been embraced by their local communities, or have there been objections and protests?

As a class or as an individual, evaluate how successful you think each of these projects will be in achieving its aims.

11.8 EXERCISE

Learning pathways

■ LEVEL 1	■ LEVEL 2	■ LEVEL 3
1, 2, 7, 10, 15	3, 4, 8, 9, 14	5, 6, 11, 12, 13

Check your understanding

1. What is *urban renewal*?
2. Why is the Honeysuckle area of Newcastle an example of urban renewal?
3. List three benefits of urban renewal for the community.
4. Define the following terms in your own words.
   a. Brownfield development
   b. Greyfield development
   c. Bluefield development
5. What does the term *deindustrialisation* mean?
6. What advantages do developers receive from brownfield and greyfield developments that they might not receive with bluefield developments?

Apply your understanding

7. Examine **FIGURE 4**. Explain what kind of development is being created in Riverside in East Perth.
8. What kind of development is the Bays Precinct project? Provide examples to justify your decision.
9. What are the advantages of urban renewal projects? Explain, in detail, how each of the following will benefit and provide examples based on one of the case studies you have examined on urban renewal.
   a. The community
   b. The environment
   c. The developers
10. What are the potential dangers of redeveloping former industrial sites?
11. How does urban renewal encourage economic growth in an area?
12. Which type of development would require the most remediation (removing contamination and fixing ready for building)? Provide reasons for your answer.

Challenge your understanding

13. Predict whether the rezoning of land for high-density housing in inner Sydney will continue to push up land values.
14. Suggest three criteria for assessing the community benefits of an urban renewal project.
15. Examine **FIGURES 4** and **7**. Predict whether local residents in the areas surrounding East Perth might object to this kind of development. Explain why.

FIGURE 7 East Perth and the Swan River

To answer questions online and to receive **immediate feedback** and **sample responses** for every question, go to your learnON title at www.jacplus.com.au.

11.9 Transport infrastructure

online only

LEARNING INTENTION

By the end of this subtopic, you will be able to explain transport strategies that create more sustainable urban places.

The content in this subtopic is a summary of what you will find in the online resource.

Australians who live in cities are experiencing longer commuting times than ever before, and this is only going to get worse. A growing population will mean an increase in cars — unless we start to tackle the problem from a sustainable perspective. Transport is one of the largest sources of greenhouse gas emissions in Australia (34 per cent), with passenger cars contributing more greenhouse gases than any other part of the transport sector.

To learn more about the sustainability benefits of better public transport, go to your learnON resources at www.jacPLUS.com.au.

Contents

learnON

FIGURE 1 Greenhouse gas emissions from different forms of transport

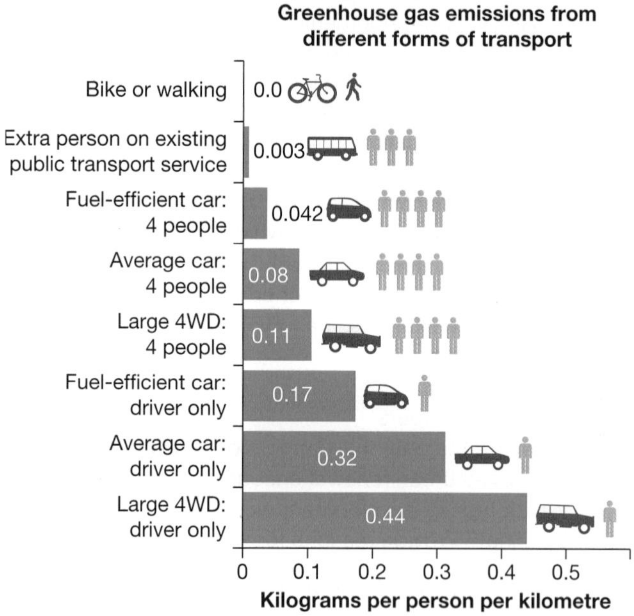

Source: Material/information courtesy of Department of Climate Change and Energy Efficiency

FIGURE 3 The light rail interchange at Dulwich Hill

Resources

eWorkbook	Transport infrastructure (ewbk-10003)
Video eLesson	Transport infrastructure — Key concepts (eles-5301)
Interactivity	Transport infrastructure (int-8716)

11.10 SkillBuilder — Writing a submission

online only

LEARNING INTENTION

By the end of this subtopic, you will be able to write a structured submission to an organisation or government body that clearly identifies a problem and suggests what action should be taken to resolve the problem.

The content in this subtopic is a summary of what you will find in the online resource.

11.10.1 Tell me

What is a formal submission?

A formal submission is a written communication (letter or email) to an organisation or government body that requests that a specific action takes place or expresses an opinion on an issue. Submissions from the public are often called for by parliamentary committees or local councils when they are investigating community issues.

11.10.2 Show me

Step 1

Identify the specific issue(s) you wish to have addressed and why they are a problem. Examples of issues that someone might want addressed are the removal of trees in a local park or the extension of transport infrastructure.

Step 2

Research the issue to clearly identify specific problems that need addressing and possible solutions to those problems. Use the information provided by the government (the **Making a submission** weblink in your online Resources) to help you collect the right information.

Step 3

Identify the people or organisations that have responsibility for the issue (or may have influence in addressing the issue). Monitor public information channels to find out if there are opportunities for public submissions.

Step 4

Write your submission, following any guidelines the organisation has provided. Complete several drafts and edit your final version to create the best submission you can.

Step 5

Send your submission. Members of Parliament and local councils generally have websites and social media accounts that provide contact details. These details usually include email and postal addresses.

11.10.3 Let me do it

learnON

Go to learnON to access the following additional resources to help you build this skill:
- a longer explanation of this skill and its application in Geography (Tell me)
- a video demonstrating the step-by-step process of this skill (Show me)
- an activity and interactivity for you to practise the skill (Let me do it)
- self-marking questions to help you understand and use the skill.

on Resources

eWorkbook	SkillBuilder — Writing a submission (ewbk-10007)
Video eLesson	SkillBuilder — Writing a submission (eles-5302)
Interactivity	SkillBuilder — Writing a submission (int-8717)

11.11 Individual actions to enhance sustainability

LEARNING INTENTION

By the end of this subtopic, you will be able to explain ways that individuals can contribute to a more sustainable future.

11.11.1 Consumer decisions

Individuals' personal actions can contribute to a more sustainable urban future and facilitate a gradual move towards policy development and institutional change to address sustainability.

Responsible consumption and production

An important aspect of creating a sustainable urban future is the decisions consumers make when they purchase goods and services. This can range from the types of shops people visit to the shares people choose to buy. In relation to food, people may choose to purchase ethically sourced and cruelty-free products, and produce that has been grown locally in urban farms.

Free-range chickens and eggs and grass-fed beef are examples of products that people who consume meat can buy to reduce their environmental impact. Vegetarianism and veganism are choices some people make to reduce their impact on the environment even further. The suburb of Newtown is known colloquially as the vegetarian capital of Sydney and is home to Australia's first vegetarian butcher, a vegan gelateria, and a vegan pizzeria. The Red Lion Hotel in Rozelle sells only vegan meals and wines. In Glebe, The Cruelty Free Shop sells vegan cleaning products, pet supplies and snacks.

FIGURE 1 Sourcing ethically raised and organically grown food can reduce your environmental impact.

FIGURE 2 Installing solar panels decreases a home's reliance on power generated from fossil fuels.

Energy usage

Individuals can change their habits to reduce their energy usage. For example, they can install energy-efficient light bulbs, purchase products with high energy ratings, switch appliances off at the power point, hang clothes on a line instead of using a dryer, and use a gas hot water system instead of an electric one. They can change their home energy infrastructure by installing rooftop solar panels or purchasing energy from a renewable energy supplier.

11.11.2 Active citizenship

Biodiversity

To encourage native bird species, residents can plant native plants in their gardens. Providing a pond can encourage frogs, and beekeeping is becoming increasingly popular.

Individuals can become involved in citizen science projects. These projects can involve individuals collecting data about their local environment or contributing to local community projects.

Sydney residents can remove the grass on **council verges** and replace it with native shrubs and grasses. This action is dependent on local council regulations. Some parts of Marrickville, such as Tamar Street and Neville Street, are part of the Marrickville Sustainable Streets Program and have converted their verges into native plantings. Other examples of streets that have embraced verge gardening are McKye Street, Waverton; Wilga Avenue, Dulwich Hill; and Browns Avenue, Marrickville.

council verges the land between the footpath and the road

Return and Earn vending machines 'reverse' vending machines where people can return recyclable bottle and cartons for a refund (part of the NSW government scheme launched in 2017)

Waste

The mantra 'reduce, reuse and recycle' encourages individuals to rethink and manage how they dispose of their waste. Managing waste effectively reduces the raw materials needed to make products and reduces the amount of rubbish that ends up in landfill. **Return and Earn vending machines** have monetised recycling to encourage more people to actively recycle, offering a 10 cent refund on eligible containers.

Rotting food waste contributes methane gas into the atmosphere. Composting and worm farms can process this food waste so that it can be productive waste, adding nutrients to soil to help grow more food.

FIGURE 3 Careful separation of waste products plays a key role in effective waste management.

FIGURE 4 Return and Earn station, Narrabri, New South Wales

11.11.3 Transport decisions

Commuting

Reducing car dependence and using alternate forms of travel is one way to be more sustainable. However, even though Sydney has an extensive public network incorporating trains, buses, trams and ferries, some areas of the city are better serviced by public transport than others. According to the Australian Bureau of Statistics, in the 2016 Census 22.1 per cent of people in the City of Sydney travelled to work using a car, 35.5 per cent used public transport, and 27 per cent of people rode a bike or walked. In 2016 in Greater Sydney (incorporating the City of Sydney local government area as well as several other surrounding local government areas) 52.7 per cent of residents drove to work, 3.9 per cent travelled by car as a passenger. 16.9 per cent travelled by train, 6.1 per cent travelled by bus, 0.4 per cent travelled by ferry or tram, 0.7 per cent cycled, and 4.0 per cent walked to work.

FIGURE 5 (a) Quakers Hill and (b) Redfern Metro stations, Sydney

11.11.4 What you can do

We can all seek to enjoy a quality of life that does not damage the environment. Although you might feel powerless, in the next decade you will be making your own contribution to society and thinking about what kind of world you would like to grow old in. What is your personal sustainability plan? Ultimately, if you want to improve your quality of life and the environment, make your choices sustainable ones. You could get involved by:

- riding or walking to school each day
- establishing an eco-classroom at your school
- learning more about the connections between local Aboriginal Cultural knowledges about reducing your impact and managing the land
- installing solar hot water or solar panels at your residence
- growing your own food or buying locally grown food.

on Resources

eWorkbook	Individual actions to enhance sustainability (ewbk-10011)
Video eLesson	Individual actions to enhance sustainability — Key concepts (eles-5303)
Interactivity	Individual actions to enhance sustainability (int-8718)

11.11 ACTIVITIES

1. Develop a sustainability action plan that includes one action that you could take every day to reduce your impact on the environment for each of the following key areas.
   - Being a responsible consumer
   - Reducing your energy use
   - Encouraging biodiversity
   - Managing your rubbish
   - Reducing car use
2. In many areas of Sydney, people are allowed to plant their verges with native plants to encourage biodiversity. Investigate how your local council encourages people to foster biodiversity in the street outside their homes, or by the side of roads in rural areas.

11.11 EXERCISE

Learning pathways

■ LEVEL 1	■ LEVEL 2	■ LEVEL 3
2, 4, 9, 10, 13	1, 3, 7, 8, 14	5, 6, 11, 12, 15

Check your understanding

1. What makes an energy company a renewable energy supplier?
2. Describe two ways that people can make responsible choices in the supermarket.
3. Describe two ways that people can reduce their energy use at home.
4. Why shouldn't you put food scraps in your rubbish?
5. What proportion of people in the City of Sydney travel to work:
   a. by car
   b. by bike
   c. by train
   d. by bus
   e. by ferry or tram/light rail?
6. What does the term *car dependence* mean?

Apply your understanding

7. Explain why vegan and vegetarian diets have less impact on the environment.
8. Why is separating your rubbish from your recycling an important way to reduce your impact on the environment?
9. Evaluate the benefits of the Return and Earn vending machine scheme. Do you think the benefits will outweigh any inconvenience people might feel by having to collect and return their recycling?
10. If you are not able to walk or cycle to school, what is the next best environmentally friendly option?
11. Explain how understanding local Aboriginal knowledge about your area can help to reduce your impact on the environment.
12. Compare the data about how people travel to work for the City of Sydney and the Greater Sydney area. What conclusions can you draw?

Challenge your understanding

13. Predict whether meat-free food options will become more or less popular in the future.
14. Suggest reasons why some urban areas are not as well serviced by public transport as other areas in the same city. What factors might contribute to this imbalance?
15. Some people living in remote or rural communities have little choice but to drive long distances to regional centres to buy most of their food and other household needs. If you were (or are) in this situation, describe how you might organise your shopping trips or deliveries to minimise their impact on the environment.

To answer questions online and to receive **immediate feedback** and **sample responses** for every question, go to your learnON title at www.jacplus.com.au.

11.12 Community actions to enhance sustainability

LEARNING INTENTION

By the end of this subtopic, you will be able to describe ways that community groups contribute to a sustainable future.

11.12.1 Community contributions

Communities and organisations are working with governments, businesses and individuals to respond to global challenges such as climate change. There are many measures in place to improve transport and mobility, develop effective use of our land, and plan and develop appropriate policies.

Communities maintain and improve infrastructure and open spaces, and can help us work at the neighbourhood level to build a more sustainable community. An example of this is the Sustainability Street program run by many councils, in which residents are encouraged to work together with their neighbours on improving local liveability. Residents taking part in this program might establish community gardens or purchase solar systems in bulk. Some examples of communities working with governments to improve liveability and sustainability are shown in **FIGURES 1, 2** and **3**.

FIGURE 1 (a) and (b) The Highline, New York City: a disused railway converted into an elevated public linear park

11.12.2 Types of community groups

Community gardens

Community gardens allow local residents who don't have access to suitable land to participate with other members of the community to plant vegetables and ornamental plants in a shared space. Community members can access food grown locally, make connections with other locals and connect with nature. Examples of community gardens in Sydney are Taringa Street Community Gardens in Ashfield, Thornton Community Garden in Penrith, Curly Community Garden in North Curl Curl, Carss Park Community Garden and Coogee Community Garden.

Biodiversity volunteers

A range of different community and volunteer groups contribute to environmental sustainability. Bushcare programs are run by community volunteers and involve weeding, reinforcing sites and planting native species. Examples of volunteer groups include the Inner West Microbat Monitors, the Tempe Birdos, the Mudcrabs Cooks River Eco Volunteers, the GreenWay Birdos, Rozelle Landcare, Friends of Orphan School Creek and the Glebe Society Blue Wren Group.

CASE STUDY 1

The Mudcrabs

The Mudcrabs Cooks River Eco Volunteers are a local community group actively contributing to the environmental sustainability of Sydney. They care for the Cooks River and its foreshore by restoring bush along the river bank, removing weeds and collecting rubbish. In the past 15 years, the group has removed more than 10 000 bags of rubbish from the river and 2500 bags of weeds from its banks, and has planted more than 12 000 native trees.

Due to the activities of the Mudcrabs, combined with remedial work completed by the local council and a number of other community volunteer groups, the state of the Cooks River has been considerably improved.

FIGURE 2 The Cooks River bank and foreshore

11.12.3 Community groups that contribute to social sustainability

Community welfare volunteers

Community members who volunteer to help those who are socially disadvantaged are contributing to **social sustainability**. Some organisations specialise in working with specific groups of people such as children, families who have experienced trauma, refugees, the homeless, those with disabilities or those requiring mental health support.

Volunteer organisations include Lifeline, Red Cross, the St Vincent de Paul Society, Barnados Australia, Mission Australia, the Benevolent Society, Mission Australia, the Asylum Seeker Resource Centre, the House with No Steps and the Fred Hollows Foundation. There are also many small-scale charities that only operate within specific suburbs, such as the Reverend Bill Crews Foundation in Ashfield

Resident Action Groups

Resident Action Groups (RAGs) are a form of social movement that tend to be localised and focused on single issues. Unlike large-scale social movements, RAGs have limited aims, and sometimes they can be interpreted as having NIMBY ('not in my backyard') motives. However, they can also be effective at prompting change on a broader scale, for example when RAGs from neighbouring communities start to campaign for change across a broader area.

Recent transport infrastructure development and proposals for high-density housing throughout the Inner West of Sydney have created an increase in the number of RAGs and concentrated the patterns of RAGs around development sites. There are currently a large number of these groups in the Inner West of Sydney protesting and lobbying against WestConnex and increased development. Examples include Rozelle Against WestConnex, Save Dully and the Newtown WestConnex Action Group.

social sustainability ensuring that all people in a community or society have their needs met and have equal access to resources, facilities and services

Resident Action Groups (RAGs) community groups formed by residents from a specific area to campaign for or against a specific local issue

CASE STUDY 2

Westconnex Action Group and Rozelle Against Westconnex

The WestConnex Action Group was formed in 2015 in response to the original plans for the WestConnex development. It is made up of residents from western, south-western and inner western suburbs in opposition to the WestConnex development.

FIGURE 3 Protest banner across Darling Street, Rozelle, 2018

As WestConnex construction has continued, various resident action groups in local communities around Haberfield, Ashfield and Newtown have formed, and some have become part of the larger WestConnex Action Group. The Rozelle Against WestConnex group is an example of this. The Rozelle Against WestConnex group lobbies against WestConnex in general, but more specifically the Rozelle Interchange in the vicinity of the Rozelle Goods Yard, as well as the tunnels running below Denison and Darling Streets. Construction of the interchange and tunnels will involve the demolition of homes and businesses, and the creation of 12-metre-high unfiltered smoke stacks.

FIGURE 4 'Thank you for using WestConnex' road sign, Homebush West, 2021

CASE STUDY 3

Save Dully, Ashbury Community Action Group and Canterbury Racecourse Action Group

Resident Action Groups are common in areas experiencing great change, for example when the change occurring is going to create a place that is very different from the existing place. This is common in older suburbs, where residents may value the traditional appearance of the suburb. Examples are Dulwich Hill and Ashbury, which both contain buildings with heritage architecture, particularly Federation architecture.

FIGURE 5 The 'Save Dully' RAG campaigns against local developments that they feel are inappropriate for their community.

The Save Dulwich Hill Community Group, or Save Dully, promotes issues related to the redevelopment of the suburb and lobbies the government to preserve its heritage. Dulwich Hill experienced growth in the late 1800s following the introduction of the tram line, and although some areas along main roads have had some higher-density rezoning, many of the houses and blocks remain largely unchanged. The Sydenham to Bankstown Urban Renewal Strategy encompasses the suburb of Dulwich Hill, and as a result rezoning for higher density and redevelopment of older buildings will occur. Save Dully is lobbying to ensure that the historic and diverse nature of Dulwich Hill is preserved.

Similarly, the Ashbury Community Action Group lobbies the government over concerns about overdevelopment. Parts of Ashbury are protected as a Heritage Conservation Area, which protects homes and attempts to preserve the unique character of the suburb. The conservation area is adjacent to some large areas of land that are prime for redevelopment, including a large former industrial site. Proposals from a number of property development companies have fanned concerns about the impacts of future developments on the special character of the area.

11.12.4 Community groups that contribute to economic sustainability

Many local communities have organisations and governance structures to promote economic development and sustainability. A chamber of commerce or business alliance is a group made up of business owners, employees and representative of other facilities such as education facilities. Such groups aim to improve the sustainability of their businesses and foster positive relationships between businesses to be able to take advantage of business opportunities as they arise. Participation generally requires a paid membership. Examples in Sydney include the Liverpool Chamber of Commerce, the Hornsby Chamber of Commerce, the Parramatta Chamber of Commerce and the Surry Hills Business Alliance.

FIGURE 6 In Brunswick, Victoria, an old garbage dump was converted into the Centre for Education and Research into Environmental Strategies (CERES), comprising a community garden, a café, an organic food store, and a resource and environment education centre.

11.12.5 The role of governments

Managing and planning Australia's future urban areas will take the efforts of many. As citizens of Australia and the world, we must be prepared to make significant changes to the way we live if we wish to enjoy a good quality of life in the future. Sustainability and liveability must be on the agenda for governments, communities and individuals.

Governments can commit to sustainability in a number of ways. They may offer incentives such as **rebates** on solar panels or water-efficient showerheads. They can fund research into sustainable technologies. Governments can adopt strict planning regulations and well-defined urban growth boundaries. They can set clear policies on levels of air quality, business sustainability, and the construction or **retrofitting** for sustainability of 'green' buildings. They can develop land use and management plans that encourage sustainability and biodiversity. This also includes the regular assessment of vulnerable areas and updating of plans to manage their care.

rebate a partial refund on something that has been bought or paid for

retrofitting adding a component or accessory to something that did not have it when it was originally built or manufactured

on Resources

eWorkbook	Community actions to enhance sustainability (ewbk-10015)
Video eLesson	Community actions to enhance sustainability — Key concepts (eles-5304)
Interactivity	Community actions to enhance sustainability (int-8719)

11.12 ACTIVITIES

1. Conduct a class debate on the following statement: 'The sustainability of cities is more reliant on the role of individuals and volunteers than on governments.'
2. Research a Bushcare, Coastcare or biodiversity group that operates in your local area.
   a. Identify the purpose of the group.
   b. Describe the actions they have taken and what they have achieved.
   c. Assess the contribution they have made to sustainability of the local area.

11.12 EXERCISE

Learning pathways

■ LEVEL 1	■ LEVEL 2	■ LEVEL 3
1, 2, 5, 10, 14	3, 4, 8, 9, 13	6, 7, 11, 12, 15

Check your understanding

1. What is the purpose of a community garden?
2. What do biodiversity volunteers hope to achieve in urban areas?
3. Outline what the term *social sustainability* means.
4. Outline what the term *economic sustainability* means.
5. Give examples of each of the following at a local scale.
   a. An environmental sustainability project
   b. An organisation working to improve social sustainability
   c. An organisation that promotes economic sustainability
6. Identify two ways governments can make changes to create a more sustainable urban future.
7. List three examples of each of the following.
   a. Resident action groups
   b. Biodiversity volunteer groups
   c. Community welfare volunteer groups.

Apply your understanding

8. Assess the role of resident action groups in addressing sustainability in their local area.
9. Welfare groups such as Barnados and the St Vincent de Paul Society play an important role in addressing the social and economic needs of members of the community. Discuss whether you think these groups should be necessary in a sustainable urban place.
10. Why might people who have space to grow their own vegetables at home still want to be members of a community garden? Give reasons for your answer, explaining what other benefits they might receive other than fresh food.
11. What changes might make your local community more socially sustainable?
12. What does the term *NIMBY* mean? In your answer, explain what the letters stand for and the connotations of calling someone a NIMBY.

Challenge your understanding

13. Predict whether community action groups will play more or less of a role in urban areas in the future. Give reasons to support your opinion.
14. Suggest one strategy that a community group might take to convey their message to local councils and state governments. Write a paragraph to describe how this strategy might be put in place and why it might or might not achieve the desired effect.
15. How can local decision makers (for example, council and state governments) determine whether the views of a RAG represents the whole community? Suggest two ways they could get an objective understanding of the whole community's views on an issue.

FIGURE 7 Community groups use a range of strategies to make their views known.

To answer questions online and to receive **immediate feedback** and **sample responses** for every question, go to your learnON title at www.jacplus.com.au.

11.13 Investigating topographic maps — Urban sprawl in Narre Warren

LEARNING INTENTION

By the end of this subtopic, you will be able to identify and describe how a suburban area has grown over time, and explain what some of the consequences of this growth might be.

11.13.1 Urban sprawl in Narre Warren

Narre Warren is a suburb located 42 kilometres south-east of Melbourne's city centre. It is an area of flat plains and undulating hills that has previously been used mainly for agricultural and pastoral purposes (see **FIGURE 1**).

FIGURE 1 Topographic map extract of Narre Warren, 1966

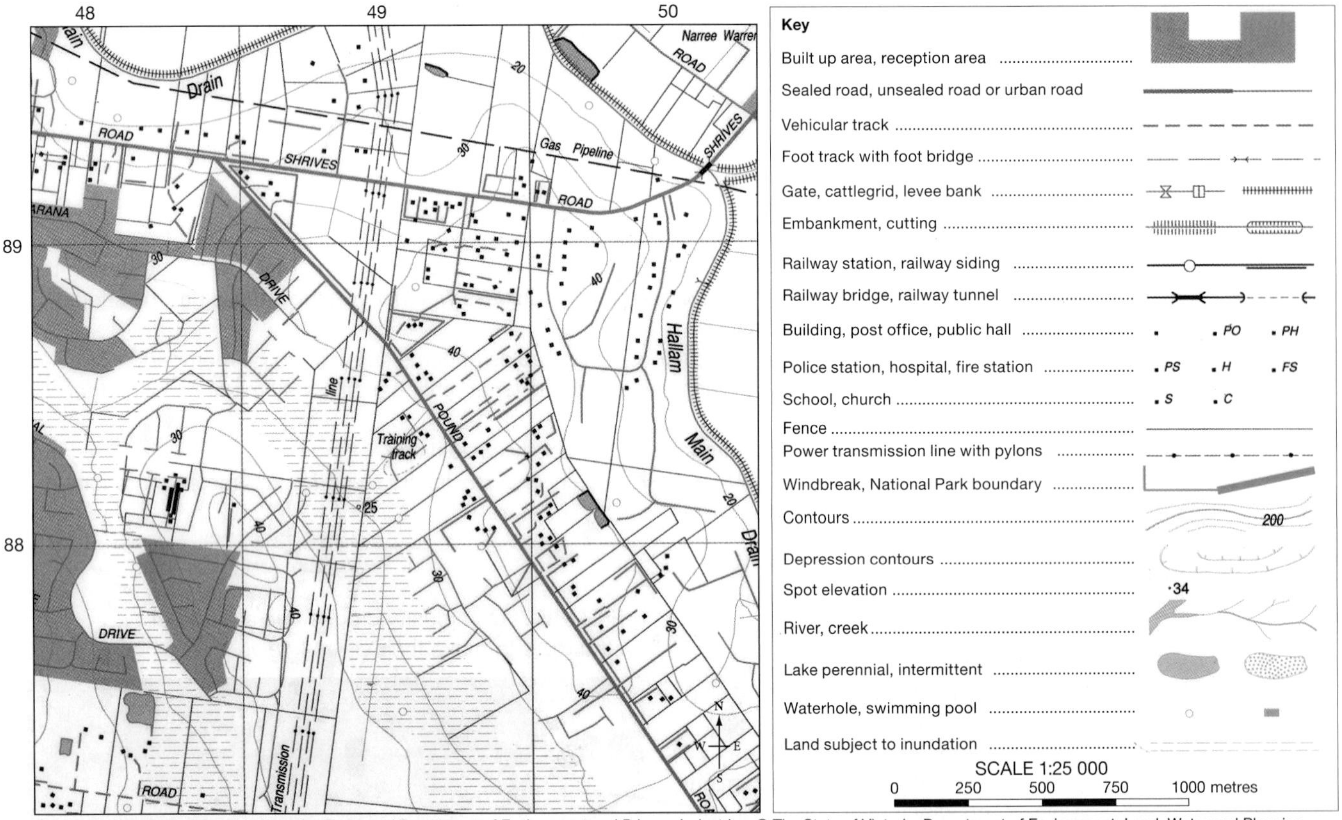

Source: © Vicmap Topographic Mapping Program / Department of Environment and Primary Industries. © The State of Victoria, Department of Environment, Land, Water and Planning, 2016. This publication may be of assistance to you, but the State of Victoria and its employees do not guarantee that the publication is without flaw of any kind or is wholly appropriate for your particular purposes and therefore disclaims all liability for any error, loss or other consequence which may arise from you relying on any information in this publication.

At the time of the 2016 census, Narre Warren had a population of 26 621 with a median age of 35 years. There were 7204 families living in 9485 private dwellings, and an average of two motor vehicles per family. In 2021, the City of Casey estimated the population to be 28 529.

Since 1966, the area has undergone significant change. Its close proximity to main arterial roads (the Monash Freeway and Princes Highway) and the Casey commercial area makes it a perfect location for medium- and high-density housing. It has drainage ponds and constructed wetlands to alleviate problems associated with the floodway zones and provide visual attractions.

FIGURE 2 Topographic map extract of Narre Warren, 2020

Source: © Vicmap Topographic Mapping Program / Department of Environment and Primary Industries. © The State of Victoria, Department of Environment, Land, Water and Planning, 2016. This publication may be of assistance to you, but the State of Victoria and its employees do not guarantee that the publication is without flaw of any kind or is wholly appropriate for your particular purposes and therefore disclaims all liability for any error, loss or other consequence which may arise from you relying on any information in this publication.

Resources

eWorkbook	Investigating topographic maps — Urban sprawl in Narre Warren (ewbk-10019)
Digital documents	Topographic map of Narre Warren, 1966 (doc-20457) Topographic map of Narre Warren, 2020 (doc-36322)
Video eLesson	Investigating topographic maps — Urban sprawl in Narre Warren — Key concepts (eles-5319)
Interactivitie	Investigating topographic maps — Urban sprawl in Narre Warren, 2020 (int-8610)
Google Earth	Narre Warren (gogl-0141)

11.13 EXERCISE

Learning pathways

■ LEVEL 1	■ LEVEL 2	■ LEVEL 3
1, 2, 3, 9	4, 5, 8	6, 7, 10

Check your understanding

1. Identify four different land uses in Narre Warren in 1966.
2. Identify four land uses that were present in Narre Warren in 1966 but not in 2020.
3. Narre Warren is a suburb located 42 kilometres ____________________ of Melbourne's city centre.
4. What was the main land use in this area in 1966?
5. What was the main land use for the area in 2020?
6. The area at GR485880 is subject to flooding (inundation). Use evidence from the map to suggest two reasons why it is flood prone.
7. List and give grid references for any new forms of infrastructure established. Consider schools, shopping centres, parks and transport.

FIGURE 3 Housing estate in Narre Warren South

Apply your understanding

8. What evidence is there on the map to suggest that this area is part of the rural–urban fringe? Provide grid or area references to support your answer.
9. Suggest one human and one environmental factor that make Narre Warren suitable for a housing estate.
10. How have planners used this flood-prone land when designing this housing estate? Use **FIGURE 3** as a starting point, then examine **FIGURE 2** to find similar types of land use.

To answer questions online and to receive **immediate feedback** and **sample responses** for every question, go to your learnON title at www.jacplus.com.au.

11.14 Thinking Big research project — Celebrating diversity

online only

The content in this subtopic is a summary of what you will find in the online resource.

Scenario

A cohesive society is defined by an OECD report as one that 'fights exclusion and marginalisation, creates a sense of belonging and promotes trust within communities'. But what creates a cohesive society in which people trust and accept each other? How can we create a community where everyone feels they can thrive, knowing their culture, beliefs and traditions are respected by their community? How can we create urban spaces where everyone has access to what they need to live a sustainable life: economically, socially and environmentally?

Task

learnON

Investigate the main factors that create a cohesive society. Using these factors as a base, design and conduct a survey to determine the level of social cohesion at your school, with a specific focus on determining how the diversity of your school could be celebrated to improve everyone's sense of belonging. Based on your results, you will then plan a celebration of your school community's diversity.

Go to learnON to access the resources you need to complete this research project.

on Resources

ProjectsPLUS Thinking Big research project — Celebrating diversity (pro-0254)

11.15 Review

11.15.1 Key knowledge summary

11.2 Australia's projected population

- Australia is urbanising at a rapid rate as people move away from rural and regional centres, especially on the east coast in Sydney, Melbourne and south-east Queensland.
- Continued population growth and urbanisation have implications for the sustainability of our urban places, including access to water, affordability of food and the distance food travels to get on the plate, loss of habitat areas and species diversity, and greenhouse gas emissions.

11.3 Designing sustainable cities

- To allow our quality of life to match this growth, we will need to develop more sustainable urban areas and infrastructure design to minimise consumption, waste, pollution and the production of greenhouse gases, and strike the balance between the needs of the environment, society and the economy.
- Sustainable communities must have a small ecological footprint. Communities can reduce their footprint by growing their own food, reducing their waste output, and promoting sustainable living among residents and visitors.

11.5 Sustainable cities in Australia

- Indicators of sustainability include the environment (air quality, ecological footprint, water, green building and biodiversity), quality of life (health, transport, wellbeing, population density and employment) and resilience (the ability of a city to cope with climate change, public participation, education, household repayments and food production)
- State governments and local councils have responsibility for improving complex infrastructure (for example transport and water supply) for whole cities.
- Action at a local community scale can be important in improving city sustainability.
- The City of Sydney Council has developed the Sustainable Sydney 2030 Community Strategic Plan, which is a set of goals to guide the community in becoming as connected, global and green as possible.

11.6 Managing the suburbs

- Urban sprawl is a significant challenge in Australia's major cities.
- Sprawl is threatening these cities' ability to provide food for their expanding populations.
- Urban expansion is also threatening the ecologically valuable green spaces that surround our cities.
- Greenfield developments occur when undeveloped land is released by the government for housing developments. Of Sydney's greenfield stocks that are zoned for residential development, most are located in west and south-west Sydney.

11.7 Living vertically

- The construction of smaller apartments in Sydney, Melbourne and Brisbane is occurring at a rapid rate.
- Medium- and high-density housing can offer the best benefits of sustainable living.
- Australians living in urban areas need to work together to improve their quality of life now and in the future.
- Green Square is a suburb in Sydney with substantial high rise housing that has received the 6 Star Green Star Communities rating from the Green Building Council of Australia.

11.8 Urban renewal

- Urban renewal is the redevelopment of an area of a large city.
- Brownfield developments involve using land that is disused or derelict, was generally previously used for industrial uses, and as a result may be contaminated.
- Greyfield developments involve land that is underused, such as abandoned retail malls and commercial sites.
- Bluefield developments are built on land next to a waterfront, for example land along a harbour, river, ocean or lake.

11.9 Transport infrastructure

- Road transport produces approximately 18 per cent of greenhouse gas emissions in Australia. The movement of goods around our cities also increases traffic congestion.
- Public transport infrastructure in Australian cities and the use of a private vehicle is still the most common method of travelling to work in Sydney, Melbourne and Brisbane. Public transport interconnections are vital as many commuters use more than one form of public transport on their way to work, school or university.
- Traffic congestion issues can be improved by increasing public transport infrastructure expenditure, reducing costs and changing commuter behaviour.
- Improving local road networks does not always have a positive impact on traffic congestion.

11.11 Individual actions to enhance sustainability

- Individuals' personal actions can contribute to a more sustainable urban future and facilitate a gradual move towards policy development and institutional change to address sustainability
- An important aspect of creating a sustainable urban future is individual action such as the decisions consumers make when they purchase goods and services; reducing energy usage and waste; and being active citizens by becoming involved in citizen science projects, encouraging wildlife in their backyards and planting native plants.
- Reducing car dependence and using alternate forms of travel is one way to be more sustainable.

11.12 Community actions to enhance sustainability

- Community gardens allow local residents to plant vegetables and ornamental plants in a shared space.
- A range of different community and volunteer groups contribute to environmental sustainability.
- Community members who volunteer to help those who are socially disadvantaged are contributing to social sustainability.
- Resident Action Groups are a form of social movement that tends to be localised and focused on a single issue. They can at times bring about change at a small scale to improve sustainability.

11.15.2 Key terms

affordability the quality of being affordable — priced so that people can buy an item without inconvenience

arable describes land that is suitable for growing crops

bi-articulated bus an extension of an articulated bus, with three passenger sections instead of two

bluefield developments urban development projects on land that is next to the water (e.g. harbour, river, lake, ocean)

brownfield developments urban development projects on land that was previously used for industrial purposes

carbon sequestration capturing and storing carbon dioxide from the air

council verges the land between the footpath and the road

development corridor area set aside for urban growth or development

ecological footprint the amount of productive land needed on average by each person in a selected area for food, water, transport, housing and waste management

evapotranspiration the process by which water is transferred to the atmosphere from surfaces such as the soil and plants

family household two or more persons, one of whom is at least 15 years of age, who are related by blood, marriage (registered or de facto), adoption, step-relationship or fostering

green wedges land in urbanised areas that is protected from development and kept as farming, parkland or green space

greenfield developments when undeveloped land is released by the government for housing developments

greyfield developments urban development projects on land that is being underused or not used

high-density housing residential developments with more than 50 dwellings per hectare

incentive something that motivates or encourages a person to do something

infrastructure the facilities, services and installations needed for a society to function, such as transportation and communications systems, water and power lines

investment an item that is purchased or has money dedicated to it with the hope that it will generate income or be worth more in the future

low-density housing residential developments with around 12–15 dwellings per hectare; usually located in outer suburbs

medium-density housing residential developments with around 20–50 dwellings per hectare

photochemical smog smog that is produced by a reaction between chemicals in the air and ultraviolet light from the sun

quality of life your personal satisfaction (or dissatisfaction) with the conditions under which you live

▶

rebate a partial refund on something that has been bought or paid for

retrofitting adding a component or accessory to something that did not have it when it was originally built or manufactured

Return and Earn vending machines 'reverse' vending machines where people can return recyclable bottle and cartons for a refund (part of the NSW government scheme launched in 2017)

Resident Action Groups (RAGs) community groups formed by residents from a specific area to campaign for or against a specific local issue

social sustainability ensuring that all people in a community or society have their needs met and have equal access to resources, facilities and services

urban sprawl the spreading of urban areas into surrounding rural areas to accommodate an expanding population

urban renewal the redevelopment or improvement of disused or underused urban areas

11.15.3 Reflection

Complete the following to reflect on your learning.

Revisit the inquiry question posed in the Overview:

Can Australia live in and grow its urban areas without making things worse for the future?

1. Now that you have completed this chapter, what is your view on the question? Discuss with a partner. Has your learning in this chapter changed your view? If so, how?
2. Write a paragraph in response to the inquiry question, outlining your views.

Subtopic	Success criteria			
11.2	I can outline projected population growth of Australia and Sydney.			
11.3	I can discuss the implications of population growth on the sustainability of urban places.			
11.4	I can analyse change over time to create a sustainable urban place.			
11.5	I can discuss the implications of population growth on the sustainability of places.			
11.6	I can explain strategies to create more sustainable urban places.			
11.7	I can explain how density management strategies can create more sustainable urban places.			
11.8	I can explain and evaluate urban renewal strategies that aim to create sustainable urban places.			
11.9	I can explain transport strategies that create more sustainable urban places.			
11.10	I can write a structured submission to an organisation or government body that clearly identifies a problem and suggests what action should be taken to resolve the problem			
11.11	I can describe ways individuals can contribute to a sustainable future.			
11.12	I can describe and provide examples of ways community groups can contribute to a sustainable future.			
11.13	I can describe how an urban area has changed and provide examples from a topographic map.			

Resources

eWorkbook Chapter 11 Extended writing task (ewbk-8562)
Chapter 11 Reflection (ewbk-8561)
Chapter 11 Student learning matrix (ewbk-8560)

Interactivity Chapter 11 Crossword (int-8721)

FIGURE 1 Aerial view of Wollongong, New South Wales

ONLINE RESOURCES

on Resources

Below is a full list of **rich resources** available online for this topic. These resources are designed to bring ideas to life, to promote deep and lasting learning and to support the different learning needs of each individual.

eWorkbook

- **11.1** Chapter 11 eWorkbook (ewbk-8103) ☐
- **11.2** Australia's projected population (ewbk-9975) ☐
- **11.3** Designing sustainable cities (ewbk-9979) ☐
- **11.4** SkillBuilder — Describing change over time (ewbk-9983) ☐
- **11.5** Sustainable cities in Australia (ewbk-9987) ☐
- **11.6** Managing the suburbs (ewbk-9991) ☐
- **11.7** Living vertically (ewbk-9995) ☐
- **11.8** Urban renewal (ewbk-9999) ☐
- **11.9** Transport infrastructure (ewbk-10003) ☐
- **11.10** SkillBuilder — Writing a submission (ewbk-10007) ☐
- **11.11** Individual actions to enhance sustainability (ewbk-10011) ☐
- **11.12** Community actions to enhance sustainability (ewbk-10015) ☐
- **11.13** Investigating topographic maps — Urban sprawl in Narre Warren (ewbk-10019) ☐
- **11.15** Chapter 11 Extended writing task (ewbk-8562) ☐
  Chapter 11 Reflection (ewbk-8561) ☐
  Chapter 11 Student learning matrix (ewbk-8560) ☐

Sample response

- **11.1** Chapter 11 Sample responses (sar-0161) ☐

Digital documents

- **11.13** Topographic map of Narre Warren, 1966 (doc-20457) ☐
  Topographic map of Narre Warren, 2020 (doc-36322) ☐

Video eLessons

- **11.1** Sustainable cities (eles-3495) ☐
  Australia's urban future — Photo essay (eles-5294) ☐
- **11.2** Australia's projected population — Key concepts (eles-5295) ☐
- **11.3** Designing sustainable cities — Key concepts (eles-5296) ☐
- **11.4** SkillBuilder — Describing change over time (eles-1753) ☐
- **11.5** Sustainable cities in Australia — Key concepts (eles-5297) ☐
- **11.6** Managing the suburbs — Key concepts (eles-5298) ☐
- **11.7** Living vertically — Key concepts (eles-5299) ☐
- **11.8** Urban renewal— Key concepts (eles-5300) ☐
- **11.9** Transport infrastructure — Key concepts (eles-5301) ☐
- **11.10** SkillBuilder — Writing a submission (eles-5302) ☐
- **11.11** Individual actions to enhance sustainability — Key concepts (eles-5303) ☐
- **11.12** Community actions to enhance sustainability — Key concepts (eles-5304) ☐
- **11.13** Investigating topographic maps — Urban sprawl in Narre Warren — Key concepts (eles-5305) ☐

Interactivities

- **11.2** Australia's projected population (int-8710) ☐
- **11.3** Designing sustainable cities (int-8711) ☐
- **11.4** SkillBuilder — Describing change over time (int-3371) ☐
- **11.5** Sustainable cities in Australia (int-8712) ☐
- **11.6** Managing the suburbs (int-8713) ☐
- **11.7** Living vertically (int-8714) ☐
- **11.8** Urban renewal (int-8715) ☐
- **11.9** Transport infrastructure (int-8716) ☐
- **11.10** SkillBuilder — Writing a submission (int-8717) ☐
- **11.11** Individual actions to enhance sustainability (int-8718) ☐
- **11.12** Community actions to enhance sustainability (int-8719) ☐
- **11.13** Investigating topographic maps — Urban sprawl in Narre Warren, 2020 (int-8610) ☐
- **11.15** Chapter 11 Crossword (int-8721) ☐

Fieldwork

- **11.2** Capturing and annotating photographs (fdw-0045) ☐
- **11.5** Sustainable streetscapes (fdw-0046) ☐

Weblinks

- **11.4** Oil spill (web-2341) ☐
- **11.5** Sustainable Cities Awards (web-3840) ☐
- **11.8** Bays Precinct (web-6388) ☐
  Riverside (web-6389) ☐
- **11.9** Sydney Metro (web-6390) ☐
  Future Transport 2056 (web-6391) ☐
- **11.10** Making a submission (Parliament of NSW) (web-5282) ☐
  Making a submission to a Senate Committee (Australian Parliament House) (web-5283) ☐
  Petition to Council or a Committee submission (web-5356) ☐

ProjectsPLUS

- **11.14** Thinking Big Research project — Celebrating diversity (pro-0254) ☐

Google Earth

- **11.13** Narre Warren (gogl-0141) ☐

Teacher resources

There are many resources available exclusively for teachers online.

12 Geographical inquiry — Investigating Asian megacities

12.1 Overview

LEARNING INTENTION

By the end of this subtopic, you will be able to explain key liveability characteristics for an Asian megacity.

12.1.1 Scenario and task

Task: Create a website designed to inform the residents of an Asian megacity about its characteristics.

The latest liveability report for Asian megacities has been released, and residents are concerned. Populations are increasing by between one and five per cent every year, putting city infrastructure under extreme pressure.

City authorities have commissioned your team to put together a website increasing awareness of the characteristics of an Asian megacity and informing residents of current and newly proposed sustainable development planning initiatives.

Your task

Your team has been put in charge of creating a website designed to inform the residents of an Asian megacity about its characteristics. Each city will be different depending on its location, wealth or poverty, size and climate. Your investigations need to ensure that the audience can gain a comprehensive understanding of both population characteristics and city characteristics, and that any urban problems are presented. A key feature of your website will be to cover any urban solutions and innovations that are currently being implemented in your megacity.

Shinagawa Station, Tokyo, Japan

12.2 Inquiry process

12.2.1 Process

Open the ProjectsPLUS application for this chapter located in your online Resources. Watch the introductory video lesson and then click the 'Start Project' button and set up your project group. You can complete this project individually or invite members of your class to form a group. Save your settings and the project will be launched.

- **Planning:** You will need to research the characteristics of your chosen Asian megacity. Navigate to your Research Forum. Research topics that have been loaded in the system to provide a framework for your research include location and city characteristics (main economy, tourism, culture); population characteristics (migrants and migration, languages, religion); and urban problems, solutions and innovations. Choose a number of these topics to include in your website and ensure you add your own. Divide the research tasks among the members of your group.

12.2.2 Collecting and recording data

Begin by discussing with your group what you might already know about your chosen Asian megacity. Then discuss the information you will be looking for and where you might find it. To discover extra information about life in your megacity, find at least three sources other than the textbook. At least one of these should be an offline source such as a book or an encyclopaedia. Remember that you will need to choose specific keywords to enter into your search engine to find other data. The weblinks in your Media Centre will help you get started. You can view and comment on other group members' articles and rate the information they have entered.

12.2.3 Processing and analysing your information and data

You now need to decide what information to include in your website. Maps to show location, graphs, tables and lists to illustrate data, and images and photos with annotations (descriptive notes) should all be included. Each of these should also have a written description. You should make sure that you have addressed each of the following points:

- Describe the pattern of distribution on each of the maps or satellite images you have drawn or collected.
- What are the main characteristics of your city?
- How has your city changed over time? Is information available on how it is predicted to change in the future?
- For what reasons are people attracted to move to this city?
- What are the main problems in this city? Are any solutions being introduced to try to overcome these problems?

Visit your Media Centre and download the website model and website planning template to help you build your website. Your Media Centre also includes images and audio files to help bring your site to life.

Use the website planning template to create design specifications for your site. You should have a home page and at least three link pages per topic. You might want to insert features such as 'Amazing facts' and 'Did you know?' into your interactive website. Remember the three-click rule in web design — you should be able to get anywhere in a website (including back to the homepage) with a maximum of three clicks.

12.2.4 Communicating your findings

Use website-building software to build your website. Remember that less is more with website design. Your mission is to inform people about your Asian megacity in an informative and engaging way. You want people to take the time to read your entire website.

12.3 Review

12.3.1 Reflecting on your work

Think back over how well you worked with your partner or group on the various tasks for this inquiry. Determine strengths and weaknesses, and recommend changes you would make if you were to repeat the exercise. Identify one area where you were pleased with your performance, and one area where you would like to improve. Write two sentences outlining how you might be able to do this.

Print out your Research Report from ProjectsPLUS and hand it in with your website and reflection notes.

The financial district of Shanghai, China

on Resources

ProjectsPLUS Geographical inquiry — Investigating Asian megacities (pro-0146)

UNIT

3 Environmental change and management

13 Introducing environmental change and management

INQUIRY SEQUENCE

To access a pre-test with **immediate feedback** and **sample responses** to every question in this chapter, select your learnON format at www.jacplus.com.au.

13.1 Overview

Numerous **videos** and **interactivities** are embedded just where you need them, at the point of learning, in your learnON title at www.jacplus.com.au. They will help you to learn the content and concepts covered in this topic.

The Earth is our home and provides us with everything we need to live. What are we doing to it in return?

13.1.1 Introduction

Environments change as a result of a range of natural and human-induced factors. Natural changes to ecosystems can be severe and drastic, resulting in an ecosystem having to adapt or change in a short period of time in order to survive. More often, natural changes occur over an extended period, allowing ecosystems to gradually change to suit changing conditions. Examples of natural changes to ecosystems include drought, flood, fire, volcanic eruptions, storm surges, cyclones, changes in climate, movements of species, and adaptation to cope with the way that other organisms have changed. There are also more gradual, natural changes to ecosystems.

FIGURE 1 Birling Gap on the southern English coast, near Eastbourne

Across the world there are also many environmental changes that have been caused by humans, such as pollution, land degradation and impacts on aquatic environments. People have different points of view, or worldviews, on many of these changes. Climate change is a major environmental change as it affects all aspects of the environment, such as our land, inland water resources, coastal and marine environments. It is vital that we respond intelligently to and effectively manage all future environmental changes.

STARTER QUESTIONS

1. The environment supports all life on Earth — humans, plants and animals. As a class, brainstorm examples of environmental changes people have caused, and discuss where these are occurring.
2. Choose one environmental change from question 1 and discuss the various viewpoints different people, groups or organisations have about it.
3. Brainstorm specific examples of environmental changes people have caused that have been positive, and that have come about by people deliberately and efficiently managing the change.
4. Watch the **Introducing environmental change and management — Photo essay** video in your online Resources. What are some of the ways that the environment where you live is under threat, and how does your community act to protect it?

on Resources

eWorkbook Chapter 13 eWorkbook (ewbk-8105)

Video eLessons What are we doing? (eles-1707)
Introducing environmental change and management — Photo essay (eles-5232)

13.2 Environmental functioning

LEARNING INTENTION

By the end of this subtopic, you will be able to describe how natural environments function.

13.2.1 The four spheres

The four spheres of the environment are the atmosphere, lithosphere, hydrosphere and biosphere. The atmosphere is the layer of gases enveloping the Earth. The lithosphere is all the rocks, soils and crust on the Earth's surface. The hydrosphere is all the water on Earth, including oceans, lakes, rivers and glaciers. The biosphere is all living things on Earth, such as plants and animals.

FIGURE 1 The interaction of the four spheres of the environment

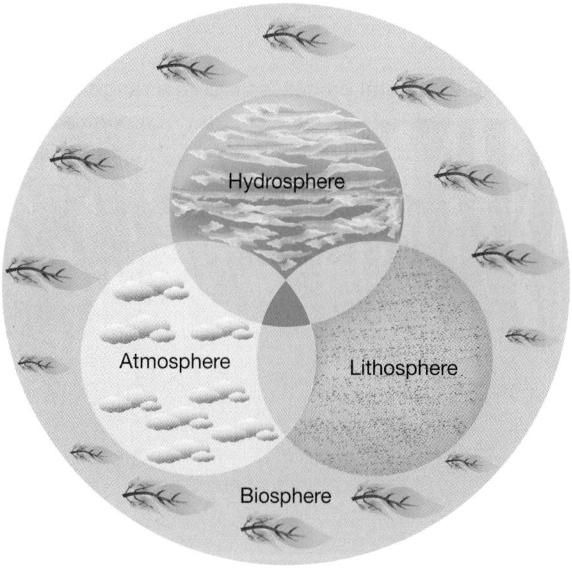

13.2.2 Geographical processes

Places and environments are formed and transformed by a range of natural **geographic processes** and influences related to each sphere.

The carbon cycle is an important part of this formation and transformation across all four spheres of the environment. It is the process of reusing or moving carbon from one sphere to another.

TABLE 1 Geographical processes that form and transform environments

Biospheric processes	Lithospheric processes	Hydrospheric processes	Atmospheric processes
Carbon cycle			
Oxygen cycle Food chains Photosynthesis Evolution Extinction Migration	Erosion Weathering Nitrogen and phosphorus cycle Tectonic processes	Precipitation Infiltration Run-off Evaporation Transpiration Condensation Transportation Deposition	Absorption (light) Reflection (light) Scattering (light) Aeolian (winds) Transportation Deposition

geographic processes the physical forces that form and transform our world

fdw-0034

FOCUS ON FIELDWORK

Ecological footprint

Fieldwork requires us to be able to collect objective data, draw conclusions on the likely consequences of what the data shows, and suggest ways to solve any negative consequences. We all have an impact on the planet, but many people are not aware of exactly what level of impact they are having.

Learn some strategies to help you draw conclusions and propose solutions for problems using the **Ecological footprint** fieldwork activity in your online Resources.

FIGURE 2 Natural processes create a range of unique environments.

on Resources

eWorkbook	Environmental functioning (ewbk-10029)
Video eLesson	Environmental functioning — Key concepts (eles-5233)
Interactivity	Environmental functioning (int-8662)

13.2 ACTIVITIES

1. Investigate one of the natural geographical processes listed in **TABLE 1.** Describe the process and explain the role of each sphere in that process.
2. Recreate the Venn diagram shown in **FIGURE 1** using images from your local area. (You could take the photos yourself or use images you find online.) For an extra challenge, hand-draw or digitally create a version that could be used as a logo for a company involved in environmental sustainability.

13.2 EXERCISE

Learning pathways

■ LEVEL 1	■ LEVEL 2	■ LEVEL 3
1, 3, 7, 12, 15	2, 4, 8, 10, 14	5, 6, 9, 11, 13

Check your understanding

1. Complete the following sentence:
   ________________ processes are responsible for forming and transforming environments.
2. Identify each of the following as a biospheric, lithospheric, hydrospheric or atmospheric process.
   a. Reflection: ____________ process
   b. Photosynthesis: ____________ process
   c. Weathering: ____________ process
   d. Condensation: ____________ process
3. Complete the following sentence:
   The ____________ cycle is a component of all four geographical processes.
4. Define and give two examples of the features of the following.
   a. Atmosphere
   b. Lithosphere
   c. Hydrosphere
   d. Biosphere
5. List three geographical processes that form or transform each of the following.
   a. Atmosphere
   b. Lithosphere
   c. Hydrosphere
   d. Biosphere
6. Identify one geographical process occurring in each of the images in **FIGURE 2**.

Apply your understanding

7. Construct a diagram of a natural environment. Label key features of the environment and, in brackets, state whether each feature is part of the hydrosphere, lithosphere, atmosphere or biosphere.
8. Explain the interaction between the four spheres of the environment.
9. How might evolution work as a biospheric process?
10. How quickly do tectonic processes form and transform environments? Give reasons for your decision.
11. Complete a flow chart to show a food chain that begins and ends with a plant but has a human in the chain.
12. Examine the photographs in **FIGURE 2**.
    a. Identify any geographical processes occurring in any of these photographs.
    b. Select a geographical process in one of the photographs and explain how you were able to identify the process.

Challenge your understanding

13. Predict how increased levels of carbon in the atmosphere might affect other parts of the carbon cycle.
14. Predict the possible rate and extent of the impact of evolution as a biospheric process.
15. Suggest how the extinction of one species might affect each of the four spheres of the environment.

To answer questions online and to receive **immediate feedback** and **sample responses** for every question, go to your learnON title at www.jacplus.com.au.

13.3 Lithospheric processes

LEARNING INTENTION

By the end of this subtopic, you will be able to describe the lithospheric process that enable environments to function.

13.3.1 Understanding the lithosphere

The **lithosphere** is all the rocks, soils and crust on the Earth's surface. Processes related to the lithosphere work over very varied degrees of intensity and speed. For example, a volcano may erupt very violently, creating significant change in a matter of minutes, whereas tectonic movement creates change comparatively slowly, forcing the crust into fold mountains over millions of year.

13.3.2 Erosion and weathering

Erosion is the wearing away of earth by wind, water or ice. Moving water and rain carry away soil and rock fragments. Waves crash against shorelines and move sand. Wind can carry lighter sediments, such as dust, sand and ash, away from their source. In dry areas with high winds, these materials can also blast against rocks, intensifying erosion. Moving glaciers can also carry away sediments and large rocks. As they move, they rub against the ground, further eroding soils and breaking boulders.

Weathering is the physical and chemical disintegration of rocks and minerals. It occurs through physical processes such as water freezing and expanding, seeds germinating and cracking rocks, and when rocks are exposed and expand. Chemical weathering occurs when minerals react with oxygen and form oxides, acids dissolve minerals in rocks, and when rocks expand as their minerals combine with water.

lithosphere all the rocks, soils and crust on the Earth's surface

erosion the wearing down of rocks and soils on the Earth's surface by the action of water, ice, wind, waves, glaciers and other processes

weathering the breaking down of rock through the action of wind and water and the effects of climate, mainly by water freezing and cooling as a result of temperature change

FIGURE 1 Geographical processes form rocks and landforms, such as erosion from (a) rivers or (b) glaciers.

13.3.3 Tectonic processes

The Earth's crust is divided into tectonic plates, which float around on top of the semi-molten rock of the mantle. Where plates have collided, huge mountain ranges have been formed. Along plate boundaries, volcanoes and earthquakes are common. Fold mountains, such as the Himalayas, form when the Earth's plates crunch into each other and layers of the crust are pushed up into loops and bumps (see **FIGURE 2**).

FIGURE 2 Formation of fold mountains

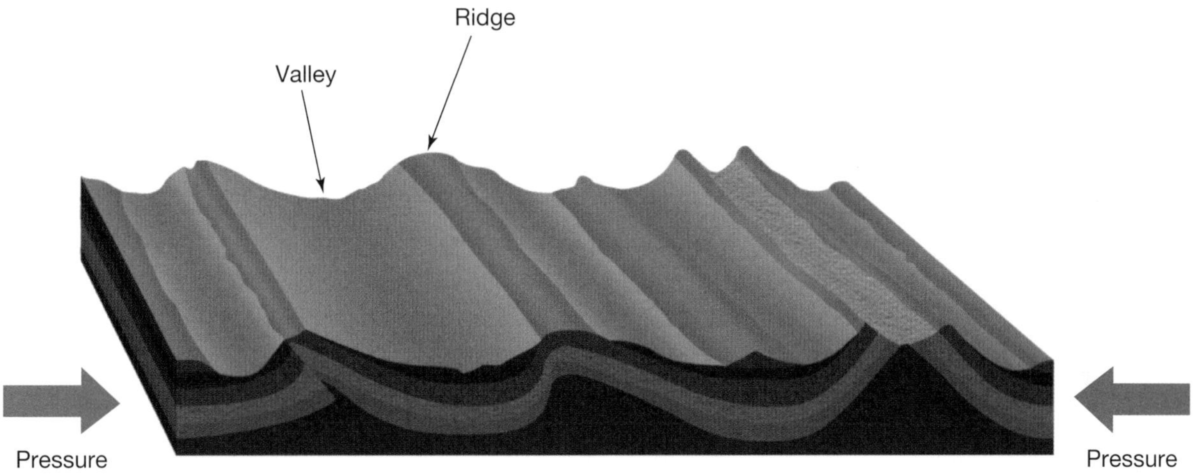

Fault mountains are made when part of the crust is forced up or collapses between two cracks in a plate. These cracks are called faults (see **FIGURE 3**).

Earthquakes occur as a result of movement between sections of the Earth's crust. They commonly occur along fault lines and along plate boundaries. Sections of tectonic plates can be forced upwards, exposing new sections. Volcanoes involve the process of molten rock from the mantle being forced up onto the Earth's surface, building continents.

FIGURE 3 Fault mountains are created by pressure on the Earth's plates from both sides.

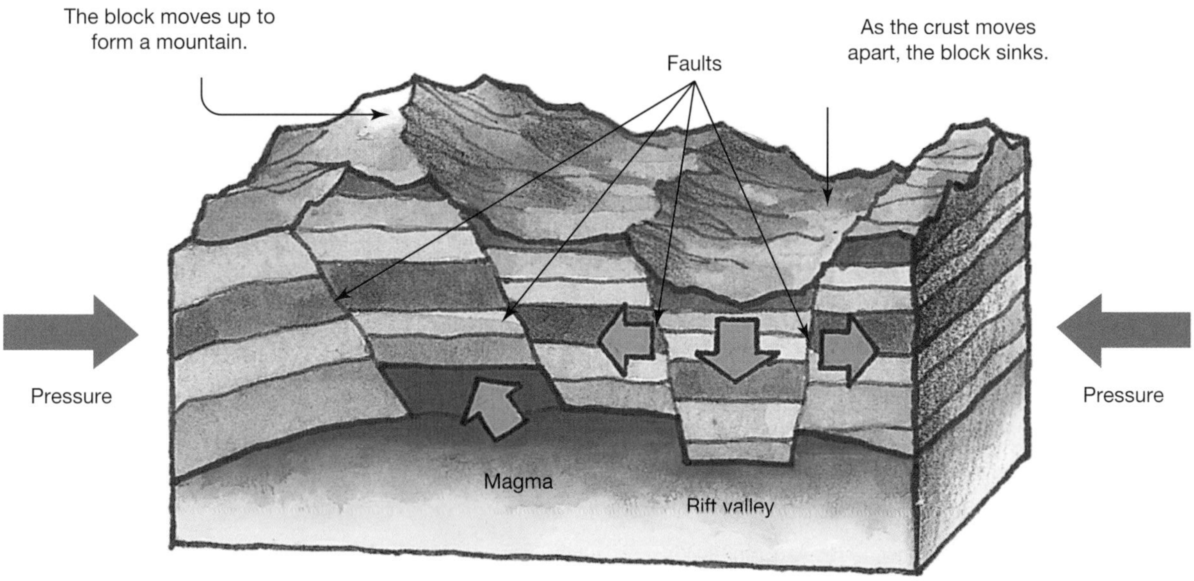

13.3.4 The nitrogen and phosphorus cycles

Living things need nitrogen for growth. The basic nitrogen cycle begins when green plants take in chemicals such as nitrogen and phosphorus from the soil. Plants use these chemicals to build proteins for growth. When plants die, decomposers such as bacteria and fungi break down the proteins into ammonium compounds. Animals also eat the plants, and animal waste and dead animal carcasses are broken down into ammonium compounds. Bacteria in the soil convert the ammonium back into nitrates.

FIGURE 4 The nitrogen cycle

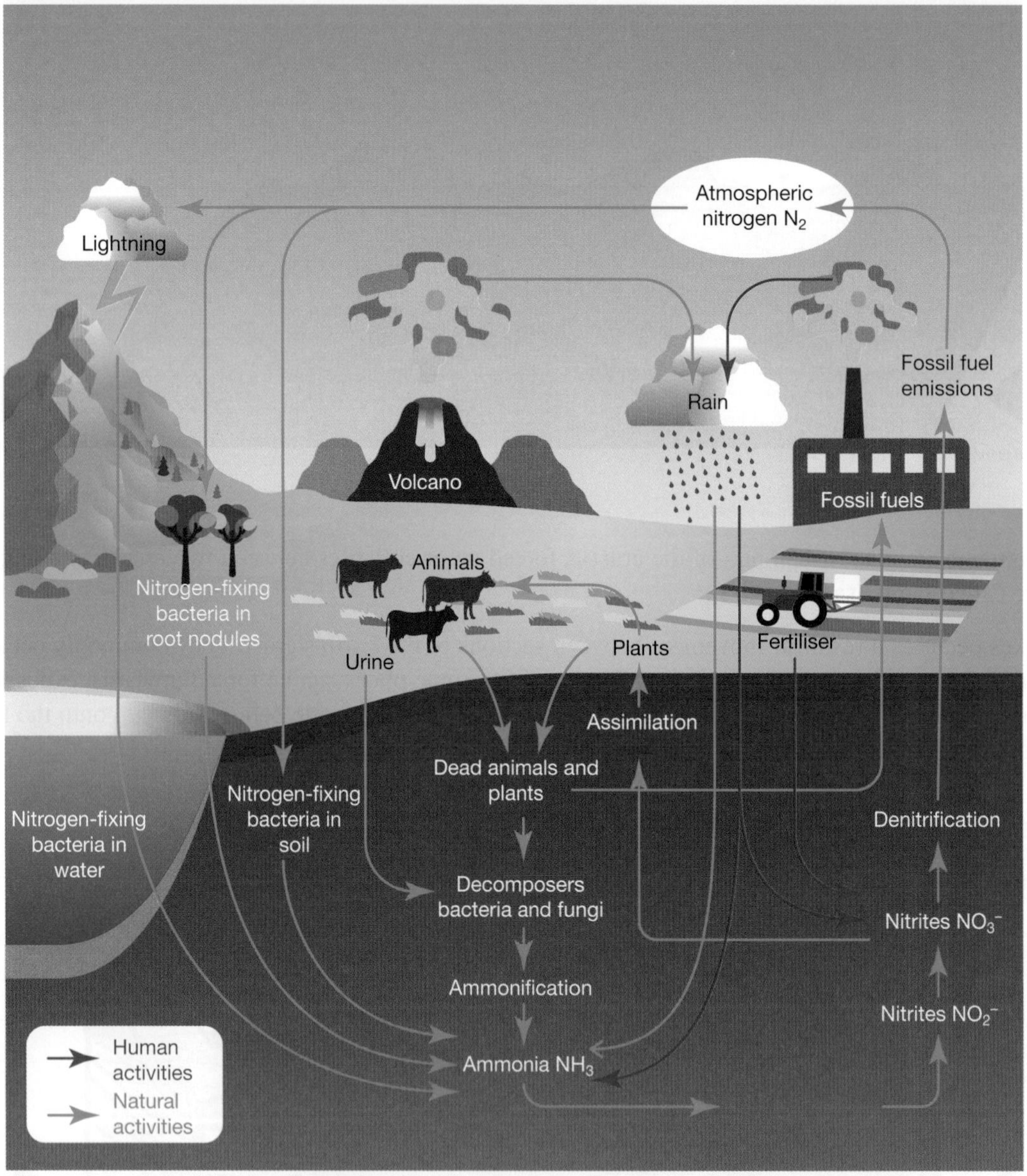

Resources

- **eWorkbook** Lithospheric processes (ewbk-10033)
- **Video eLesson** Lithospheric processes — Key concepts (eles-5234)
- **Interactivity** Lithospheric processes (int-8663)

13.3 ACTIVITIES

1. Research the process of plate tectonics.
   a. With the aid of diagrams, describe how this process occurs.
   b. Describe one landform that has been created by plate tectonics. Include images and diagrams to support your answer.
   c. Discuss one source of evidence that supports the theory of plate tectonics.
2. Find an example of a volcano or earthquake that has transformed an environment.
   a. On a world map, show the location of the volcano or earthquake.
   b. Describe the details of the volcanic eruption or earthquake (date, severity).
   c. Explain how the volcanic eruption or earthquake has transformed an environment.

13.3 EXERCISE

Learning pathways

■ LEVEL 1	■ LEVEL 2	■ LEVEL 3
1, 2, 7, 12, 13	3, 5, 8, 11, 15	4, 6, 9, 10, 14

Check your understanding

1. Which of the following are examples of lithospheric processes? Select all possible answers: oxygen cycle, erosion, photosynthesis, weathering, tectonic processes.
2. Complete the following sentence:
   ________________ is the wearing away of earth by wind, water or ice, and ________________ is the physical or chemical breakdown of rocks and minerals.
3. List the different ways that sediments can be moved from one location to another.
4. Are volcanoes and earthquakes more common at the centre of floating tectonic plates or the edges?
5. Complete the following sentence:
   In relation to tectonic plates, when part of the crust is forced up or collapses between two cracks in a plate, ________________ mountains are formed.
6. Describe the difference between erosion and weathering.
7. Define the term *tectonic plate*.
8. Match the steps of the nitrogen and phosphorus cycle with the correct number so that the process is in order.

1	Decomposers break down the proteins into ammonium compounds.
2	Plants use chemicals to build proteins for growth.
3	Bacteria in soil converts ammonium back into nitrates.
4	Plants die.
5	Green plants take in chemicals (nitrogen and phosphorus) from the soil.

Apply your understanding

9. Construct an annotated diagram of the nitrogen cycle.
10. Identify and explain the relationship between tectonic plates and the formation of volcanoes and earthquakes.
11. Explain how earthquakes and volcanoes can transform environments.
12. What is the difference between physical weathering and chemical weathering?

Challenge your understanding

13. Climate change is expected to increase the rate at which glaciers melt. How might this affect the process of erosion in glacial areas?
14. Predict whether you would see more rapid change from chemical weathering in urban areas with a lot of heavy industry, or in a national park covered with dense bushland. Give reasons for your answer.
15. Suggest two ways that a hot-burning bushfire might have an impact on the nitrogen cycle. Consider the role of plants and animals in the cycle.

To answer questions online and to receive **immediate feedback** and **sample responses** for every question, go to your learnON title at www.jacplus.com.au.

13.4 Biospheric processes

LEARNING INTENTION

By completing this subtopic, you will be able to describe the biospheric processes that enable environments to function.

13.4.1 Understanding the biosphere

The biosphere is all the living things on Earth. Processes related to the biosphere are central in the creation of environments. These processes affect the type of vegetation in an environment, how large and how densely the vegetation grows, the quality of the soil, the levels of oxygen and moisture in the air, and rates of erosion. Some of the processes related to the biosphere include the carbon cycle, the oxygen cycle, food chains and food webs, photosynthesis, evolution, and population fluctuations and movements.

13.4.2 The carbon cycle

Carbon is one of the most basic elements that make up all living things. The carbon cycle is a process that involves all the spheres. It is the process where carbon is transferred through the environment. Carbon dioxide exists in the air and is used by plants to photosynthesise and make food. Oxygen is released into the air as a by-product (plant respiration). Animals eat plants and then use carbon from plants for energy and growth. Animals produce waste, breathing out carbon dioxide (animal respiration), and decompose when they die. Decomposers, such as bacteria and fungi, feed on the dead matter and release carbon dioxide into the air as they respire. Factories and petrol-using vehicles also release carbon compounds into the air.

Carbon is also absorbed into the ocean as carbon dioxide in a process called ocean uptake. Carbon dioxide, when dissolved in salt water, separates into carbon and oxygen. Increased levels of carbon in the ocean can increase its acidity or chemical balance.

FIGURE 1 The carbon cycle

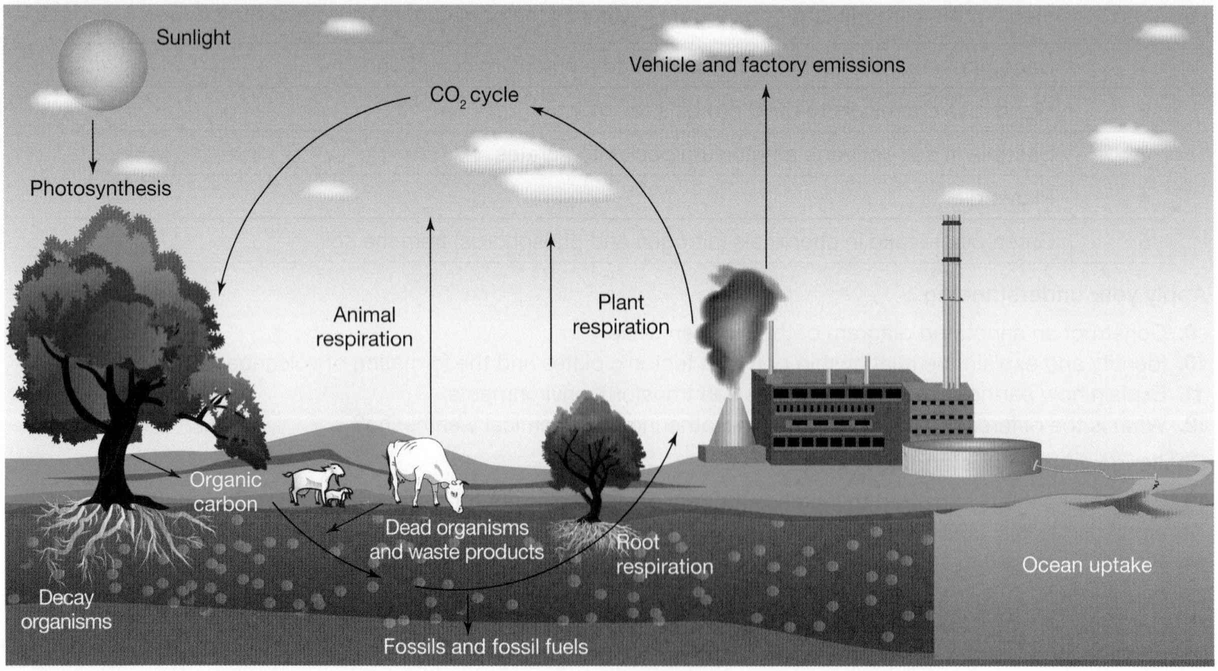

13.4.3 The oxygen cycle

The oxygen cycle begins with the oxygen that exists in the air. Animals obtain oxygen by breathing, and plants produce oxygen and release it through their pores.

13.4.4 Food chains

A food chain is a series of organisms, each eating or decomposing the preceding one. A food web is a more complicated branching diagram that shows the feeding relationships of all living things in an ecosystem or particular area. In nearly all ecosystems, the source of energy is the Sun (hydrothermal vents provide energy in some rare ecosystems). Every food chain begins with a **producer** organism, a plant that can photosynthesise. **Consumers** are organisms that eat other organisms. **Decomposers** break down dead organisms. Energy is transferred from one animal to another through food chains and food webs, and flows in the direction of the arrows on the diagram. Some energy is lost at each step in the food chain as heat.

FIGURE 2 A desert food web

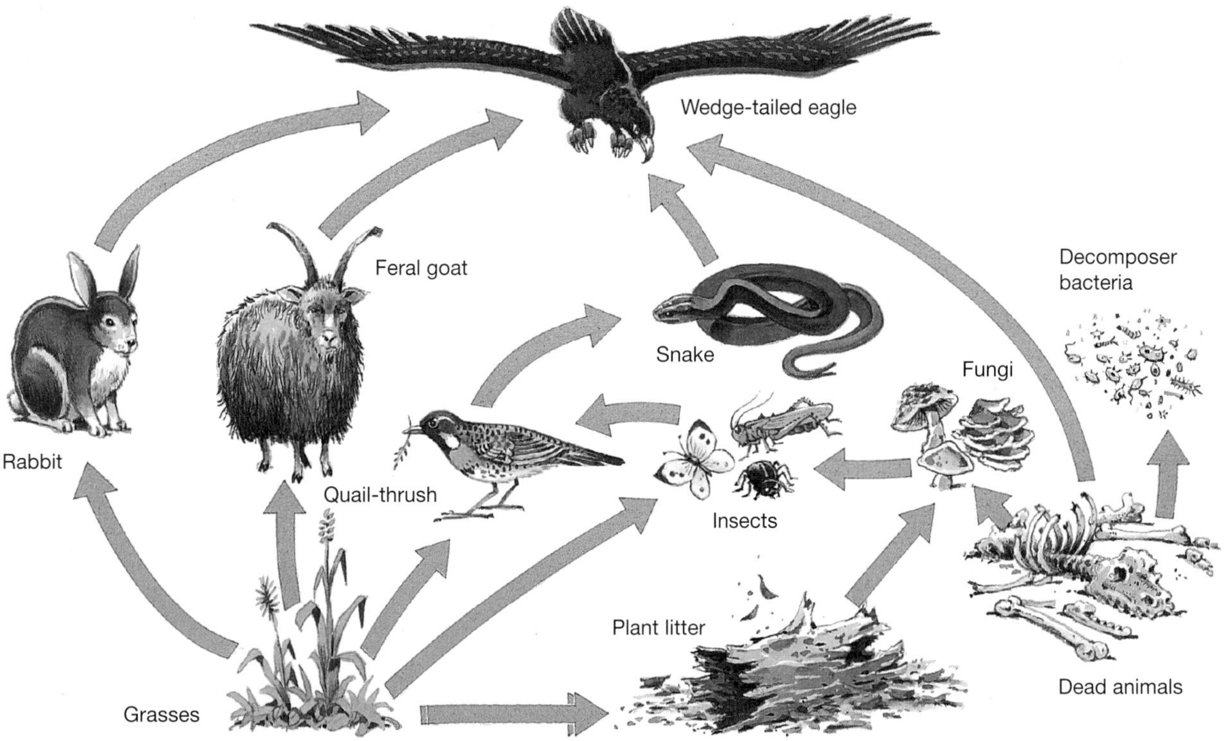

13.4.5 Photosynthesis

Most plants make food in their leaves, which contain a special green pigment, or colouring, called **chlorophyll**. Chlorophyll uses energy absorbed from sunlight to convert carbon dioxide from the air and water from the ground into a form of sugar called glucose.

producer a plant that can photosynthesise

consumers organisms that eat other organisms

decomposers organisms that break down dead organisms

chlorophyll green pigment in plants that uses energy from the Sun to transform carbon-dioxide into glucose

13.4.6 Evolution

Evolution is the long-term process where species have changed and developed from earlier forms of species to become more suited to particular environments. Survival of the fittest means that organisms that are most suited to a particular environment are more likely to live a long life and reproduce. Organisms that are not suited to a particular environment are more likely to be killed, be eaten or starve, and are less likely to reproduce. In this way, organisms with the most suitable genes pass them on to future generations. Each successive generation contains a greater number of individuals with suitable genes. As time goes on, the species becomes more and more specialised.

FIGURE 3 Plant species in the Daintree rainforest have adapted to suit the specific climatic and soil conditions of the region.

13.4.7 Population fluctuations and movements

Changes in the size and distribution of plant and animal species impacts greatly on the functioning of environments. Changes to climate, natural hazards and human interactions are some of the factors that affect population numbers.

Many species of animals migrate from place to place. Some migrations are seasonal, where animals move looking for the best food sources, to avoid extreme heat or cold, and to find suitable breeding grounds. The migrations of some species don't follow any particular pattern, while others simply move on when they have exhausted the food source. Animals that live in mountainous regions move to higher or lower regions to avoid snowfalls. If an environment becomes inhospitable, organisms may permanently migrate from an area.

FIGURE 4 Crabs moving en masse

Extinction is the death of all individuals of a species. This occurs when a species can't adapt to the changes in an environment. Humans have greatly accelerated the number of extinctions through modifications of ecosystems.

extinction when all individuals of a species have died

on Resources

eWorkbook	Biospheric processes (ewbk-10037)
Video eLesson	Biospheric processes — Key concepts (eles-5235)
Interactivity	Biospheric processes (int-8664)

13.4 ACTIVITIES

1. Using the internet, investigate one species that is well adapted to its environment.
   a. Describe the unique conditions of the environment (for example climate, soils and vegetation).
   b. Describe the characteristics this species has that enable it to live and flourish in this environment.
   c. Discuss how a similar organism might have different characteristics if it lived in a different environment.
   d. Describe a related species that lives in a different environment. Outline the differences between the two species.
2. Investigate one migratory species.
   a. Why does the species migrate?
   b. What migration patterns does it follow?

13.4 EXERCISE

Learning pathways

■ LEVEL 1	■ LEVEL 2	■ LEVEL 3
2, 4, 10, 11, 14	1, 6, 8, 9, 13	3, 5, 7, 12, 15

Check your understanding

1. Which of the following is *not* a process related to the biosphere: population movements, food chains, the oxygen cycle, photosynthesis, or the nitrogen and phosphorus cycle?
2. The source of carbon dioxide for plants to photosynthesise is ______________.
3. Construct a simple flow chart showing the relationship between producers, decomposers, consumers and the Sun in a simple food chain.
4. Complete the following sentence:
   A food ____________________ shows the relationships of all living things in an ecosystem; a food ____________________ is a series of organisms, each eating or decomposing the preceding one.
5. What is the difference between a food chain and a food web?
6. Match each component of a food chain to its role.

Role	Component
Breaks down dead organisms	Producer
Source of energy	Consumer
Can photosynthesise	Decomposer
Eats other organisms	Sun

Apply your understanding

7. Compare and contrast the carbon cycle and the oxygen cycle.
8. Explain the flow of energy through food chains and food webs.
9. Explain the process of evolution.
10. How might the migration of a species transform an environment?

11. Analyse the following statement, and explain whether it is true or false. 'Evolution is the process where many species of animals move from place to place to find the best food sources.'
12. Evolution is the long-term process where species have changed and developed from earlier forms of species. How does the concept of 'survival of the fittest' relate to evolution?

Challenge your understanding

13. What might happen to a population if their environment becomes inhospitable?
14. Can evolution occur over just a few generations of a species? Give reasons for your answer.
15. Predict how increased levels of carbon in the atmosphere might affect other parts of the biosphere.

To answer questions online and to receive **immediate feedback** and **sample responses** for every question, go to your learnON title at www.jacplus.com.au.

13.5 Hydrological and atmospheric processes

13.5.1 Understanding the hydrosphere

The hydrosphere is water in all its forms, including rain, ice, sleet, and vapour. Processes and features of the hydrosphere are referred to as hydrospheric. The atmosphere is the gaseous layer surrounding the Earth. Processes and features of the atmosphere are referred to as atmospheric. Many of the geographic processes that form and transform environments involve an interaction between the hydrosphere and atmosphere.

FIGURE 1 Solar radiation from the Sun is concentrated and dispersed at the equator.

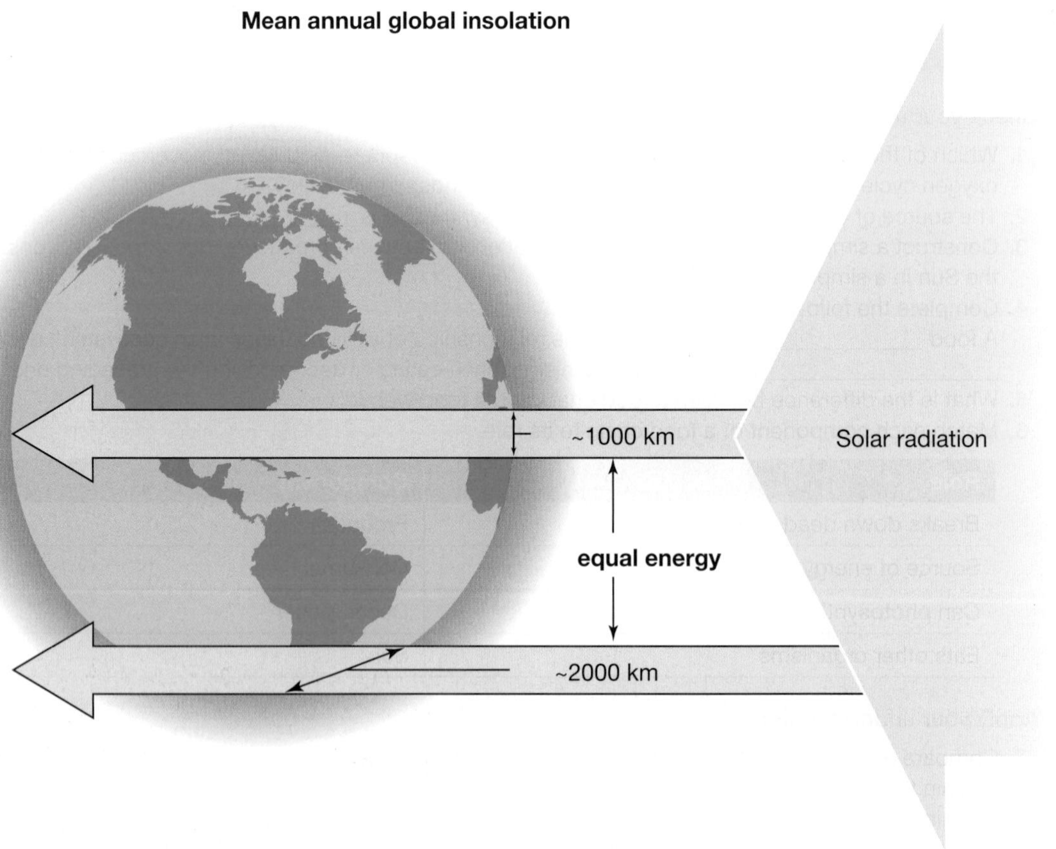

13.5.2 The water cycle

The water cycle relates to all four of the spheres, but specifically connects the hydrosphere and atmosphere. The water cycle is the continuous movement of water. It begins in water storage areas such as the oceans, lakes and rivers. Water then evaporates from water storage areas into the air or is transpired from trees. In the atmosphere, the water vapour begins to form clouds; when it condenses, it falls to the ground as **precipitation**. Some water will run over the surface of the ground, collecting in low areas or joining rivers, lakes and oceans. Some water will soak into the ground and join groundwater. The processes associated with the water cycle are precipitation, **infiltration**, **run-off**, **evaporation**, **transpiration**, **evapotranspiration** and **condensation**.

precipitation occurs when water droplets or ice crystals become too heavy to be suspended in the air and fall to Earth as rain, snow, sleet or hail

infiltration water that is absorbed into the ground, flows downward and collects above an impermeable layer or rock

run-off water that is unable to be absorbed into the ground, flows over its surface and collects in nearby waterways or reaches stormwater drains

evaporation when water contained in water bodies is heated by the Sun and the liquid changes into a gaseous state and rises into the atmosphere

transpiration when water contained in plants is heated and changes from a liquid into a gaseous state and rises into the atmosphere

evapotranspiration a process in which liquid water is evaporated from soil, trees, and the ocean surface, and transferred into the atmosphere

condensation when water in the atmosphere cools and changes from a gaseous state into a liquid state. This occurs when the water vapour clusters around a solid particle (such as dust)

13.5.3 Heating and cooling

Solar radiation is the heat that is received from the Sun's rays. The atmosphere plays an important role in distributing this heat around the planet.

In some locations there is more heat received from the Sun than is reflected by the Earth. These locations are mainly in the tropics. In contrast, in polar regions and at high altitudes, less heat is received from the Sun than is reflected by the Earth. The atmosphere scatters the Sun's rays, helping to distribute heat, but most excess of heat that results in the tropics is transferred to the poles and high altitudes by air movements.

The atmosphere absorbs some incoming radiation, which helps to balance temperatures overnight when no direct radiation is received. Some radiation is reflected, which helps to regulate temperatures in areas of extreme heat.

FIGURE 2 Oceans also play an important role in absorbing heat.

Resources

eWorkbook	Hydrological and atmospheric processes (ewbk-10041)
Video eLesson	Hydrological and atmospheric processes — Key concepts (eles-5236)
Interactivity	Hydrological and atmospheric processes (int-8665)

13.5 ACTIVITY

Using the internet, investigate the water cycle and explain the role it plays in forming and transforming environments.

13.5 EXERCISE

Learning pathways

■ LEVEL 1	■ LEVEL 2	■ LEVEL 3
1, 3, 4, 9, 13	2, 6, 7, 10, 14	5, 8, 11, 12, 15

Check your understanding

1. Complete the following sentence:
   The ____________ is the gaseous layer surrounding the Earth, and the ____________ is water in all its forms.
2. Define the term *water cycle*.
3. Complete the following sentence:
   In the atmosphere, when water vapour begins to form clouds and condenses, it falls to the ground as ____________.
4. Describe how water moves through the water cycle.
5. Choose one of the processes associated with the water cycle and create a diagram that explains the process.
6. Describe how the water cycle links the hydrosphere and the atmosphere.
7. Describe the role of the biosphere and lithosphere in the water cycle.

Apply your understanding

8. Does the water cycle relate to all four environmental spheres? Give reasons for your answer.
9. Explain how the atmosphere distributes heat around the globe.
10. Explain the difference between evaporation, transpiration and evapotranspiration.
11. How does the water cycle connect the hydrosphere and the atmosphere?
12. What is **FIGURE 1** demonstrating in relation to the distribution of solar radiation?
13. How is the distribution of heat from the Sun different between tropic regions and polar regions?

Challenge your understanding

14. How might oceans contribute to absorbing heat from the Sun? Suggest how the process might work.
15. How might climate change alter the process of solar radiation being dispersed around the globe?

To answer questions online and to receive **immediate feedback** and **sample responses** for every question, go to your learnON title at www.jacplus.com.au.

13.6 Human-induced change

LEARNING INTENTION

By completing this subtopic you will be able to outline and provide examples of how and why humans change the environment.

13.6.1 Human causes of environmental change

Environments can change because of natural or human-induced factors. Humans have the ability to simplify natural ecosystems in order to grow food, build habitats and remove or extract resources. Unwanted species are removed and other species are provided with an environment made favourable for their survival by human intervention. Human-induced change can be intentional, inadvertent or through negligence.

Humans have changed environments since prehistoric times. Early humans domesticated animals, hunted and undertook basic irrigation. As the size of the world's population has increased over time, so too has the demand on resources to provide for it.

FIGURE 1 Terraced farmlands, Lavaux, Switzerland

13.6.2 Consequences of human-induced environmental change

Large-scale agriculture has changed environments through the clearing of land to introduce monocultures of crops, the introduction of livestock and subsequent overgrazing, the removal of native species, and ongoing activities such as ploughing and the use of pesticides, insecticides and fertilisers to maximise harvests.

Urbanisation and urban growth have resulted in the replacement of natural environments with roads, buildings and manicured parklands. In many places waterways have been straightened or covered, and riverbanks have been cemented.

Industrial land uses have resulted in the leaking of toxic substances. Mining has changed vast areas of land and left sinkholes, contaminated surface water, chemical leakage, mine dumps and tailing dams.

Over a longer period of time, human-induced environmental change can have long-term and sometimes irreversible impacts. Consequences can include salinisation and soil waterlogging, compaction and erosion, pollution, habitat loss, species loss, introduction of exotic species and reduced biodiversity.

13.6.3 Scales of human impact

Change can occur at different levels or scales. On a map we use a scale to give an idea of the size or focus of the map, and to allow us to take accurate measurements. Issues or environmental changes are also examined at a range of scales. Most commonly we examine change at a local, regional, national or global scale. Examples include climate change at the global scale, cyclones in the Great Barrier Reef and deforestation in the Amazon at the regional scale, and land degradation on a single farm at the local scale.

FIGURE 2 Like many other rivers around the world, the Arno River in Florence, Italy has been modified from a natural river system to a system dominated by human development.

CASE STUDY

Environmental change as a result of energy production in Australia

Black coal energy production in Australia increased by 2 percent in 2017–2018. This was despite an increase in renewable energy production of 1 per cent in 2017–2018 and a decrease in brown coal production by 2 per cent. Renewable energy sources accounted for just over 2 per cent of total energy production in 2017–2018. Many of the natural resources that are extracted in Australia are exported for consumption in other countries.

FIGURE 3 Australian electricity generation — fuel mix

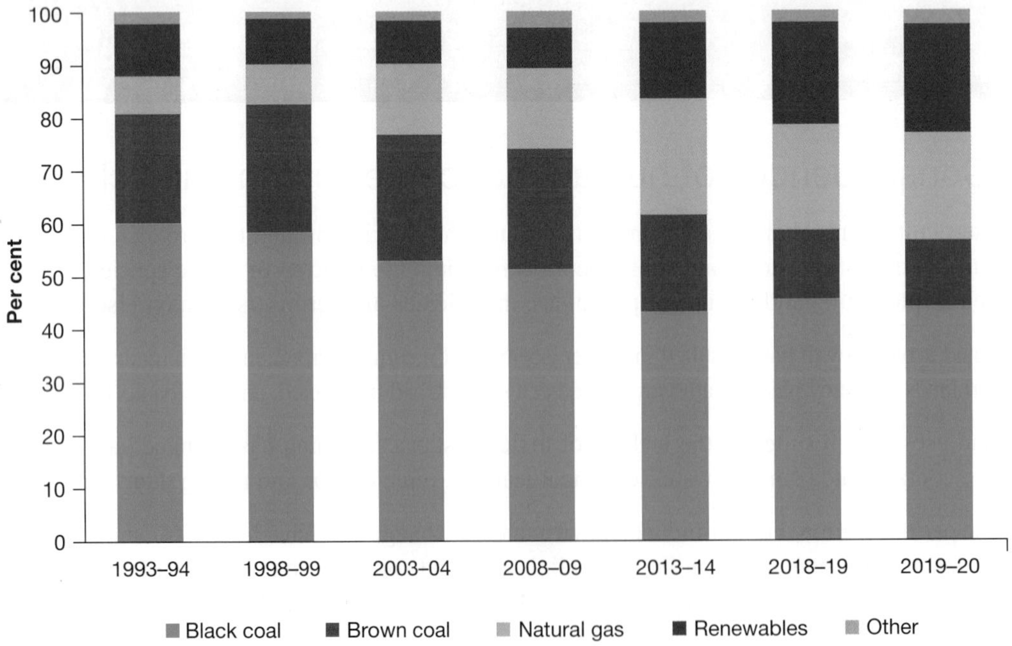

Source: Australian Government, Department of Industry, Science, Energy and Resources, 2020, https://www.energy.gov.au/data/australian-electricity-generation-fuel-mix

FIGURE 4 Energy resources in Australia

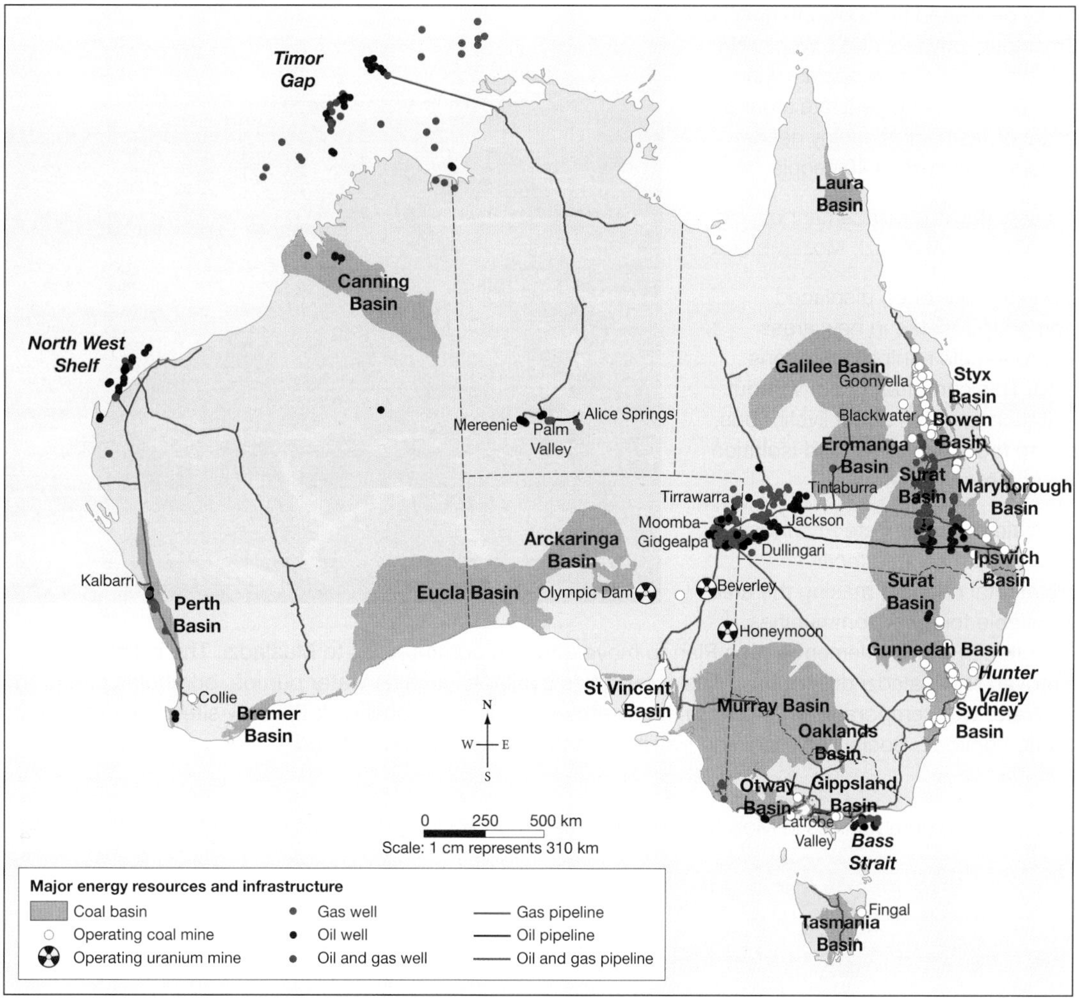

Source: Spatial Vision

Australians consume more resources per person than most other nations.

Over the last 30 years, Australia's energy consumption from all forms of fuel has increased by over 200 per cent, with most of it coming from non-renewable energy sources.

Australia is fortunate to have large reserves of coal, natural gas and uranium, but the use of these non-renewable energy sources has significant environmental impact. There is a concern about how Australia will satisfy its energy needs in the future.

FIGURE 5 A Glencore Xstrata coal mine in Queensland

Regional/state case study: wind farms

Many locations throughout Australia have strong winds that can be used to produce clean, sustainable energy. Wind farms require wind speeds of 15 to 90 kilometres per hour to drive the turbines, so operators look for locations where wind speeds are generally high. Woolnorth in Tasmania has 62 wind turbines, each 60 metres high, which take the force of the Roaring Forties, the prevailing westerly wind in southern latitudes. The electricity generated at Woolnorth goes to the Smithtown power station. Woolnorth Renewables Wind Farms operate wind farms in Tasmania that generate around 9 percent of Tasmania's energy needs and employ in excess of 40 people.

FIGURE 6 Wind turbines in Tasmania

Local case study: displacement of communities in Mualadzi, Mozambique

Communities are often displaced and forced to resettle in new areas when large-scale mining operations develop. This can often occur without consultation and can affect livelihoods, access to food and water, and isolation for the communities affected.

Large-scale mining in Tete Province of Mozambique caused local environmental change, making the area uninhabitable for local communities. This resulted in a Resettlement Action Plan to move affected communities to Mualadzi. The initial families who were moved to Mualadzi did not have basic amenities available such as water pumps, boreholes or storage tanks. Water was temporarily trucked into the resettlement. The soil quality at the new site also made food production difficult. Social networks and livelihood patterns were also fractured.

FIGURE 7 Some communities in Mozambique are forced to travel long distances in order to access drinking water.

Resources

eWorkbook	Human-induced change (ewbk-10053)
Video eLesson	Human-induced change — Key concepts (eles-5237)
Interactivity	Human-induced change (int-8666)

13.6 ACTIVITIES

1. Investigate one human-induced environmental change. Create a multimedia presentation that includes the following information:
   a. Describe the biophysical processes that relate to the environment.
   b. Examine the causes of the environmental change.
   c. Explain the short- and long-term consequences of the change.
   d. Suggest possible management strategies to address the environmental change.
2. Using Google Earth and the internet, examine an example of environmental change at a regional scale. Create a Google Tour that examines evidence of the environmental change over various locations.

13.6 EXERCISE

Learning pathways

■ LEVEL 1	■ LEVEL 2	■ LEVEL 3
3, 4, 7, 10, 13	1, 2, 11, 9, 15	5, 6, 8, 12, 14

Check your understanding

1. Define the term *sustainability*.
2. Using examples, describe the two categories of causes of environmental change.
3. As the world's population has increased, the demand on the environment has ____________.
4. Determine whether the following are examples of natural change or human-induced change:
   a. introducing livestock
   b. monocultures
   c. species migration
   d. erosion.
5. Are natural changes are always slow and gradual? Give examples to support your answer.
6. Identify and describe two long-term consequences of human-induced environmental change.

Apply your understanding

7. Explain the difference between natural environmental changes and human-induced environmental changes.
8. Explain how population growth and environmental change have been interconnected in the past.
9. Do you think that human-induced environmental changes have long-term consequences? Give examples to support your view.
10. Provide an example of an environmental change at a global scale, a national scale and a local scale.
11. Identify and explain two environmental consequences of Australia's reliance on non-renewable energy.
12. Analyse and respond to the following statement: 'Human-induced environmental change can have long-term impacts, but nothing is irreversible.'

Challenge your understanding

13. Construct criteria to help you assess the environmental sustainability of a place.
14. Propose a range of strategies to improve sustainability in your local community.
15. Describe whether you think population growth continues to be a major factor in environmental change. Provide reasons how and/or why.

To answer questions online and to receive **immediate feedback** and **sample responses** for every question, go to your learnON title at www.jacplus.com.au.

13.7 Climate change

LEARNING INTENTION

By the end of this subtopic, you will be able to explain the processes through which the Earth's atmosphere is heated and describe the role of human activity in the enhanced greenhouse effect.

13.7.1 Climate change and global warming

The world's climate has been changing for millions of years, but recently the concentration of greenhouse gases in the atmosphere has increased, leading to **global warming**. In particular, it is believed that burning fossil fuels such as coal and oil has led to what is known as the **enhanced greenhouse effect**, which is heating the Earth and its atmosphere. Despite ongoing debate about the nature and extent of **climate change**, the majority of the scientific community agrees that global warming and climate change exist and will result in ongoing changes to world weather patterns and, in the longer term, climates from the equatorial to polar regions. The wider consequences of global warming will also lead to environmental change across a wide range of biophysical systems (see **FIGURE 1**).

global warming increased ability of the Earth's atmosphere to trap heat

enhanced greenhouse effect the observable trend of rising world atmospheric temperatures over the past century, particularly during the last couple of decades

climate change any change in climate over time, whether due to natural processes or human activities

FIGURE 1 Consequences of changes in the global climate

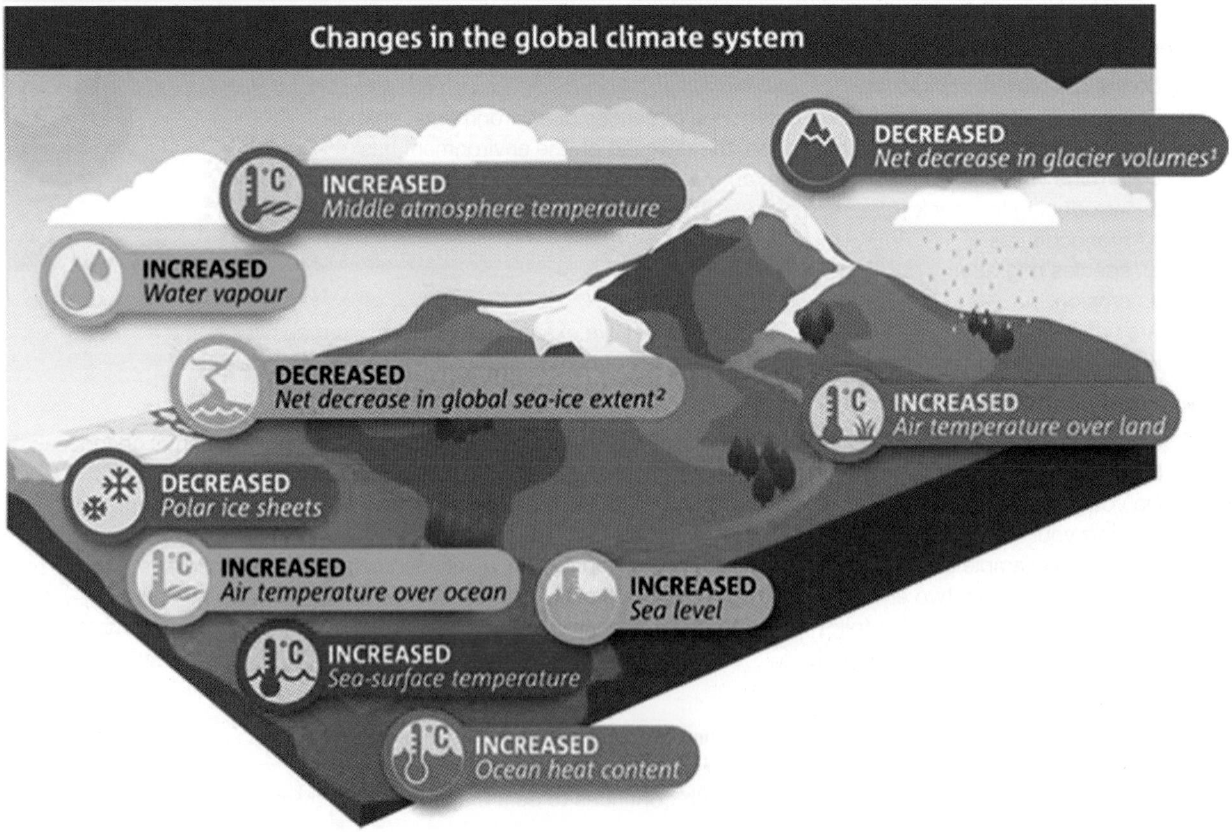

Source: Bureau of Meteorology and CSIRO.

Climate, which can be defined as the long-term weather patterns of a particular area, is highly variable over the Earth's surface. As such, climates in the tropics contrast markedly with climates near the poles. Climate also varies over extensive periods of time, and scientists have described these changes — which date back millions of years, long before the emergence of the human species — as warm periods and ice ages. Currently the Earth is in a warm period, having moved out of ice age conditions as recently as 6000 years ago. Today we realise that human activity is increasing the rate of global warming leading to climate change, particularly in the past few hundred years, and this can have serious consequences for the planet (see **FIGURE 2**).

FIGURE 2 Global temperature change to 2017

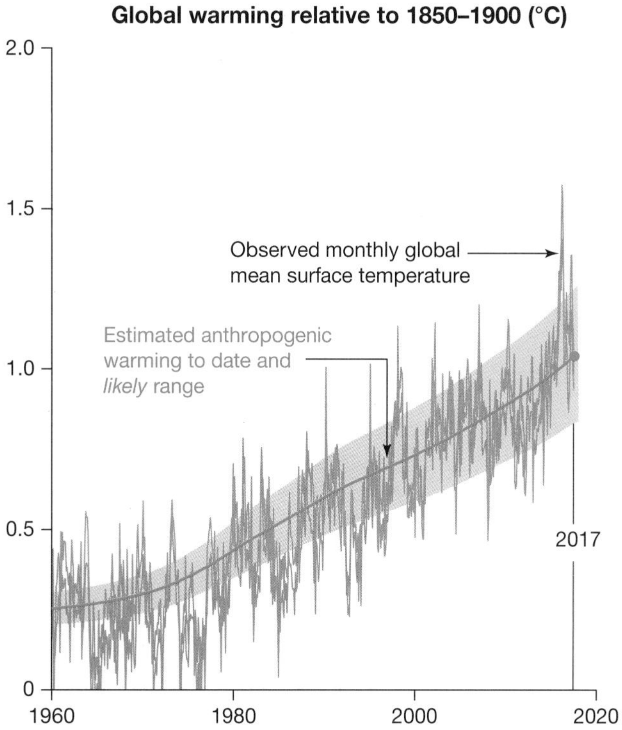

Source: IPCC, Special Report 2018: Global Warming of 1.5 °C. Summary for Policymakers, figure SPM.1 a), p. 6.

13.7.2 Human activity and the enhanced greenhouse effect

The greenhouse effect is the mechanism where solar energy is trapped by water vapour and gases in the atmosphere, heating the atmosphere and helping to retain this heat, as in a glasshouse (see **FIGURE 3**). The three most important gases responsible for the greenhouse effect are carbon dioxide, nitrous oxide and methane. Without this greenhouse effect, the atmosphere would be much cooler and ice age conditions would prevail over the planet, making life as we know it impossible.

Changes in the balance of the greenhouse gases are a natural event, leading to the different climatic conditions on the planet as experienced over geological time. The issue today is how much impact human activity is having on the natural cycle of events and how this activity is leading to climate change and global warming.

The term 'enhanced greenhouse effect' has been developed to show that heating of the atmosphere is moving at a rate that is above what could be expected by natural processes of change (see **FIGURE 4**). Recent research by government and non-government organisations has indicated that all parts of the world are vulnerable to the impacts of the enhanced greenhouse effect and associated climate change. Six key risks that have been identified in Australia alone include higher temperatures, sea-level rise, heavier rainfall, greater wildfire risk, less snow cover, reduced run-off over southern and eastern Australia, and more intense tropical cyclones and storm surges along the coast.

FIGURE 3 How the greenhouse effect works

FIGURE 4 The enhanced greenhouse effect

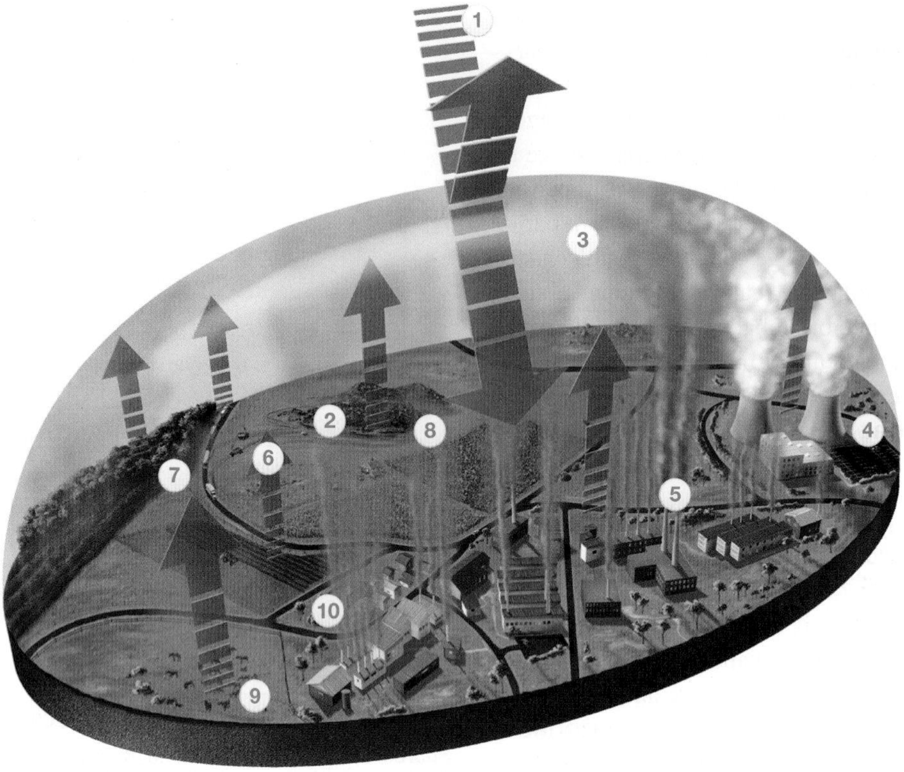

1 Heat from the Sun

2 Heat trapped by greenhouse gases

3 Heat radiating back into space

4 Greenhouse gases produced by power stations burning fossil fuels

5 Greenhouse gases produced by industry burning fossil fuels

6 Greenhouse gases produced by transport burning fossil fuels

7 Greenhouse gases released by logging forests and clearing land

8 Methane escaping from waste dumps

9 Methane from ruminant (cud-chewing) livestock, e.g. cattle, sheep

10 Nitrous oxide released from fertilisers and by burning fossil fuels

13.7.3 The impact of climate change

Climate change is a global phenomenon. The greenhouse gases produced in one country spread through the atmosphere and affect other countries. Action by only a few countries to reduce greenhouse gases will therefore have little impact — it requires international cooperation, especially by the largest polluters.

Since the 1990s, countries have met at the United Nations Intergovernmental Panel on Climate Change (IPCC) conferences and agreed to take steps to reduce emissions of greenhouse gases. An early conference developed the Kyoto Protocol, an agreement that sets targets to limit greenhouse gas emissions, and 128 countries have agreed to this protocol. Further conferences in 2009 in Copenhagen, Denmark, in 2010 in Cancun, Mexico, and in 2015 in Paris, France, led to an important new direction, with all countries agreeing to contain global warming within 2 °C.

This means that emissions of CO_2, which were at 395 parts per million (ppm) in 2013, must be kept below 550 ppm to reach this target. If no actions (mitigation measures) are taken, temperatures could increase by 5 °C, as shown in **FIGURE 5**. To date, 192 of the world's 195 countries have signed the Kyoto Protocol; however, close to half have modified their commitment to reach targets for greenhouse emission reductions set for 2020. The United States has signed the Protocol but has not ratified emission targets and Canada has withdrawn from the Protocol.

FIGURE 5 Global average temperature outcomes for three emissions cases, 1990–2100*

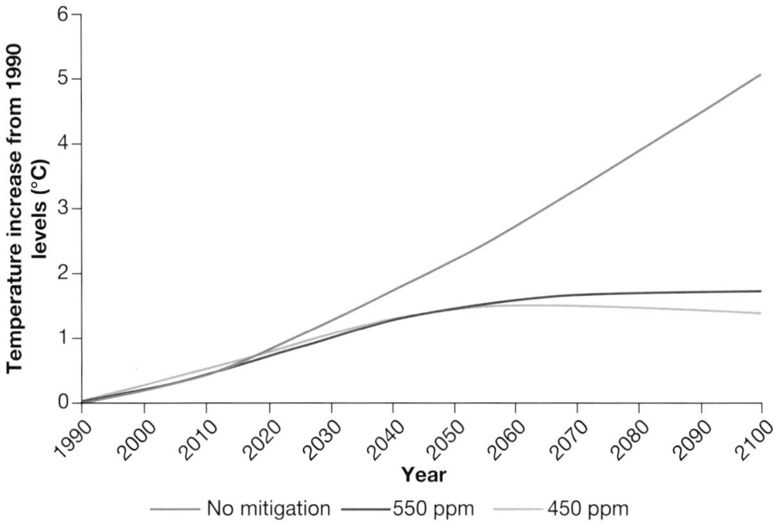

*projected

Source: The Garnaut Climate Change Review 2008, p. 88.

To meet the greenhouse gas emissions targets defined by these agreements, countries must make changes that reduce their level of emissions. They can also meet the targets in two other ways.

1. A country can carry out projects in other countries that reduce greenhouse gas emissions and offset these reductions against their own target.
2. Companies can buy and sell the right to emit carbon gases. For example, a major polluter, such as a coal power station, is allowed to emit a certain volume of greenhouse gases. If it is energy efficient and emits less than its limit, it gains **carbon credits**. It has the right to sell these credits to another company that is having difficulty reducing its emissions. Companies can also gain credits by investing in projects that reduce greenhouse gases (such as renewable energy), improve energy efficiency or act as carbon sinks (such as tree planting and underground storage of CO_2).

Australia's response

The 2011 Garnaut Report and the findings of the 2018 IPCC state that it is in Australia's national interest to do its fair share in a global effort to mitigate climate change. Australia agreed to reduce greenhouse gas emissions by 26–28 per cent by 2030 (from 2005 levels) as part of the Paris Agreement of 2015. In 2019, the Australian government announced the Climate Solutions Package, a $3.5 billion investment to deliver on Australia's 2030 Paris climate commitments.

In 2018, the United Nations Environment Programme Emissions Gap Report indicated that Australia was not on track to achieve its greenhouse gas emissions reduction goal of 26–28 per cent but would achieve or come close to some goals set in Cancun for 2020. The report noted, however, that Australia's climate policy had not improved since 2017.

The 2020 Climate Change Performance Index (Germanwatch, NewClimate Institute and Climate Action Network) rated Australia very low, with only five countries performing worse overall on their key indicators: greenhouse gas emissions, renewable energy, energy use and climate policy. Australia was rated last, with a score of 0, on its climate policy. Examples cited of why Australia achieved such a low ranking in that category were a lack of explanation as to how the government would achieve their emissions reductions targets, approving expansion of fossil fuel production and mining operations, and not attending the UN Climate Action Summit in September 2020.

carbon credit a tradable certificate representing the right of a company to emit one metric tonne of carbon dioxide into the atmosphere

Resources

	eWorkbook	Climate change (ewbk-8161)
	Video eLesson	Climate change — Key concepts (eles-5238)
	Interactivity	The enhanced greenhouse effect (int-7951)

myWorldAtlas

Deepen your understanding of this topic with related case studies and questions.

- **Causes of climate change**
- **Larsen Ice Shelf break-up**
- **Impacts on polar bears**
- **Climate change and Australia**
- **Global warming and Antarctica**

13.7 ACTIVITIES

1. In groups, prepare a report that explains how the enhanced greenhouse effect operates, based on the information in **FIGURE 4**. You may wish to also carry out further research. Prepare a presentation for the class that includes your suggestions about what we can do to reduce the impacts of the enhanced greenhouse effect.
2. Some people claim that climate change is not human-induced, or that climate change is not occurring differently to in the past. Research the issue, making a list of the different ideas or opinions you can find expressed in the Australian media. In your notes include details of who expressed each view, what qualifications they have, how they support their conclusions (research, citing other people) and what sources they refer to. Using the steps outlined in 13.9 SkillBuilder: Evaluating alternative responses, objectively evaluate the ideas you find.
3. Investigate and evaluate the implication from UN reports and global surveys that Australia's climate change policy is not working.

13.7 EXERCISE

Learning pathways

■ LEVEL 1	■ LEVEL 2	■ LEVEL 3
1, 3, 9, 10, 14	2, 5, 7, 8, 13	4, 6, 11, 12, 15

Check your understanding

1. Complete the following sentences:
   ________________ refers to an ongoing process in the Earth's atmosphere trending through ice ages and warm periods over the geological time scale. ________________ is a warm period in which rising temperatures of the atmosphere impact on land and ocean water temperatures and lead to a change in climate over time.
2. What is the enhanced greenhouse effect?
3. List three atmospheric gases that are responsible for the greenhouse effect.
4. What changes have occurred to the Earth's climate over geological time?
5. Consider **FIGURE 2**.
   a. What is the approximate total temperature increase shown between 1960 and 2017?
   b. Describe the general trend of the graph.
6. What are six key risks of climate change for Australia?

Apply your understanding

7. Analyse and discuss the following statement: 'If there was no greenhouse effect, the Earth would be a much cooler place and life as we know it would still be sustainable.'
8. Why would sea levels be much lower in an ice age period?
9. What role do trees play in the carbon cycle and in controlling the level of greenhouse gases?
10. Without the greenhouse effect, would there be sufficient heat trapped inside the Earth's atmosphere to sustain human, animal and plant life?

11. Is human life on Earth possible without the greenhouse effect? Explain why or why not.
12. How will climate change affect the functioning of the four spheres of the environment? Provide one example for each sphere.

Challenge your understanding

13. What impacts will global warming, and in particular higher water temperatures, have on a marine ecosystem such as the Great Barrier Reef? Consider the impact in ten, fifty and one hundred years if nothing is done to stop rising sea temperatures.
14. Based on **FIGURE 2**, predict what the average surface temperature will be in 2100.
15. Human activity has warmed the climate. Do you think human activity can also stop, slow or reverse this impact? Give reasons for your view.

To answer questions online and to receive **immediate feedback** and **sample responses** for every question, go to your learnON title at www.jacplus.com.au.

13.8 Worldviews and environmental management

online only

LEARNING INTENTION

By the end of this subtopic, you will be able to assess how people's worldviews affect their attitudes to environmental management.

The content in this subtopic is a summary of what you will find in the online resource.

People have different perceptions or views about how the world works and how they fit into the world. These are known as environmental worldviews. An environmental worldview helps a person make decisions about how he or she will behave towards their environment, and environmental ethics will determine their beliefs about right or wrong behaviour.

FIGURE 2 Ecocentric and egocentric worldviews are at opposing ends of the continuum.

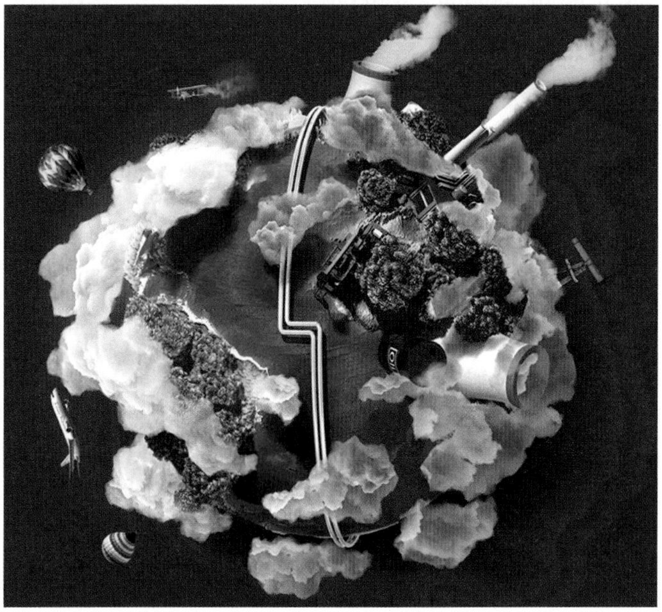

To learn more about environmental world views, go to your learnON resources at www.jacPLUS.com.au.

Contents

learnON

- 13.8.1 Humans and the environment
- 13.8.2 Environmental worldviews

Resources

eWorkbook	Worldviews and environmental management (ewbk-10049))
Video eLesson	Worldviews and environmental management — Key concepts (eles-5239)
Interactivity	Worldviews and environmental management (int-8667)

13.9 SkillBuilder — Evaluating alternative responses

online only

LEARNING INTENTION

By the end of this subtopic, you will be able to examine a geographical issue from a range of perspectives and evaluate alternative responses.

The content in this subtopic is a summary of what you will find in the online resource.

13.9.1 Tell me

What are alternative responses?

Alternative responses are a range of different ideas/opinions on an issue. You may or may not agree with the alternative responses, but evaluating ideas effectively requires weighing up and interpreting your research to reach a judgement or a decision based on the information.

13.9.2 Show me

How to evaluate alternative responses

Step 1

Read through all the data, seek clarification of ideas, and develop a viewpoint on the information.

Step 2

Divide a page into two columns and head the columns with one of the following heading pairs — whichever is relevant to your issue: Advantages/Disadvantages, Positives/Negatives, Strengths/Weaknesses, Costs/Benefits. In each column, list the information from the data that you believe is important in determining your viewpoint on the issue. Consider a range of perspectives: economic, environmental, social justice, historical, political, technological and sustainable.

TABLE 1 Table of alternative responses

Alternative responses	Advantages	Disadvantages
1. Allow tourism to develop without restraint.		
2. Restrict tourist numbers to the island.		
3. Restrict tourist numbers only in the peak season.		
4. Ban tourists from the island.		
5. Introduce tighter rules on tourist movements on the island.		

Step 3

When the columns are complete, consider which column outweighs the others and why. Are there more points in one column than another? Are some arguments stronger than others? Use the answers to these questions to shape your opinion and help you decide which responses are better than others.

13.9.3 Let me do it

learnON

Go to learnON to access the following additional resources to help you build this skill:

- a longer explanation of this skill and its application in Geography (Tell me)
- a video demonstrating the step-by-step process of this skill (Show me)
- an activity and interactivity for you to practise the skill (Let me do it)
- self-marking questions to help you understand and use the skill

on Resources

eWorkbook SkillBuilder — Evaluating alternative responses (ewbk-10045)

Video eLesson SkillBuilder — Evaluating alternative responses (eles-1744)

Interactivity SkillBuilder — Evaluating alternative responses (int-3362)

13.10 Sustainable management

LEARNING INTENTION

By the end of this subtopic, you will be able to outline the interconnections essential for sustainable management.

13.10.1 Environmental sustainability

Sustainability refers to something's ability to continue into the future. It means living within the resources of the planet without damaging the environment now or in the future. Sustainability refers to taking the long-term view of how our actions affect future generations and making sure we don't cause pollution or deplete resources at rates faster than the Earth is able to renew them.

Types of sustainability include economic and political, social and cultural, and environmental. All of the different types of sustainability are important in a functioning society. However, environmental sustainability is the most relevant when investigating the effectiveness of management strategies. Environmental sustainability emphasises the importance of ecosystems and environments to be able to continue to function effectively into the future.

Environmental sustainability is based on the idea of preserving the Earth's capacity to support human life and maintain the four S functions: source, sink, service and spiritual.

- The **source function** refers to the capacity of the environment to provide us with materials we rely on such as timber, water, and soil.
- The **sink function** refers to the ability of the environment to remove and breakdown waste.
- The **service function** refers to the processes that occur that enable our existence, such as pollinating food crops and stabilising the climate.
- The **spiritual function** refers to how environments can provide us with psychological benefits or spiritual connections.

Effective management of environments will preserve the processes within them and their ability to perform the four functions. In turn this will ensure the wellbeing of people.

FIGURE 1 The four functions associated with environmental sustainability

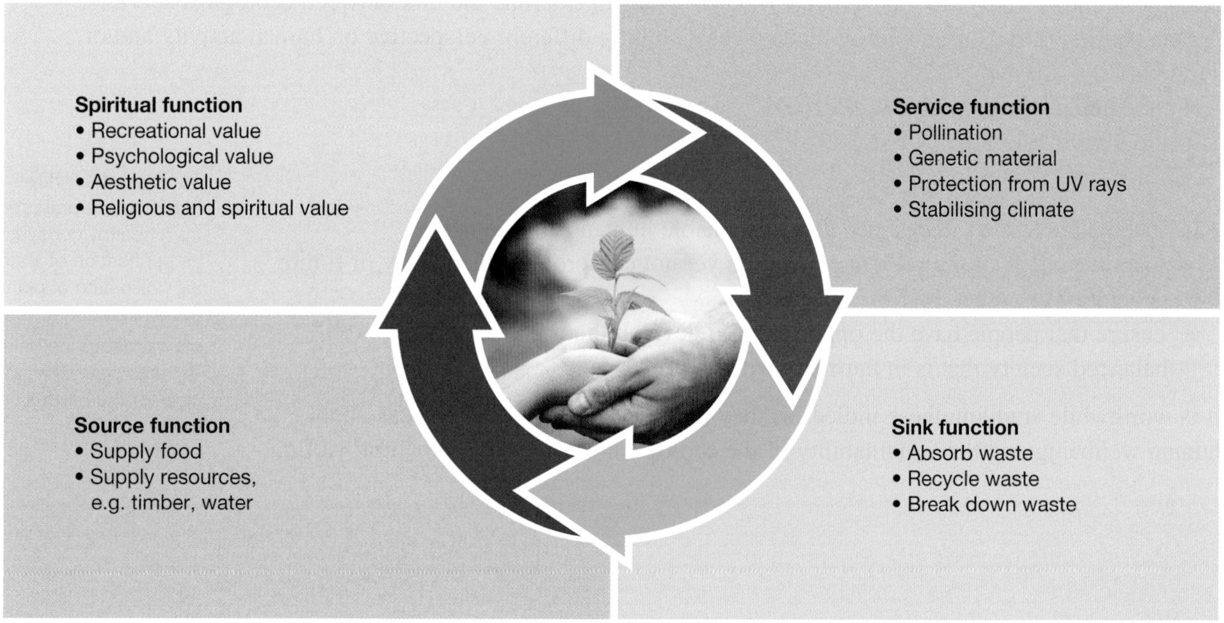

13.10.2 Ecological services

Another view of the relationship between the environment and people is one of an ecological service or 'what nature provides for humanity'.

Ecological services can be thought of as biological and physical processes that occur in natural or semi-natural ecosystems and maintain the habitability and livelihood of people on the planet. These services are shown in **FIGURE 2**.

Understanding the interconnection between ecological services and human action is important as it can lead to more sustainable practices. The idea of ecological management takes an Earth-centred environmental worldview, promoting **stewardship** or custodial management. This view considers caring for the land and the ecological services it provides as paramount. By applying this Earth-centred viewpoint to human uses and management of the environment, future options for human wellbeing will be sustainable. The question is: how do we evaluate human impacts on the environment and what management strategies can be implemented to reverse damage and create a sustainable future? As such, we need to consider the costs and benefits, or more simply, the advantages and disadvantages of changes we make to the environment, as there will be consequences in terms of economic viability and social justice.

FIGURE 2 Ecological services model

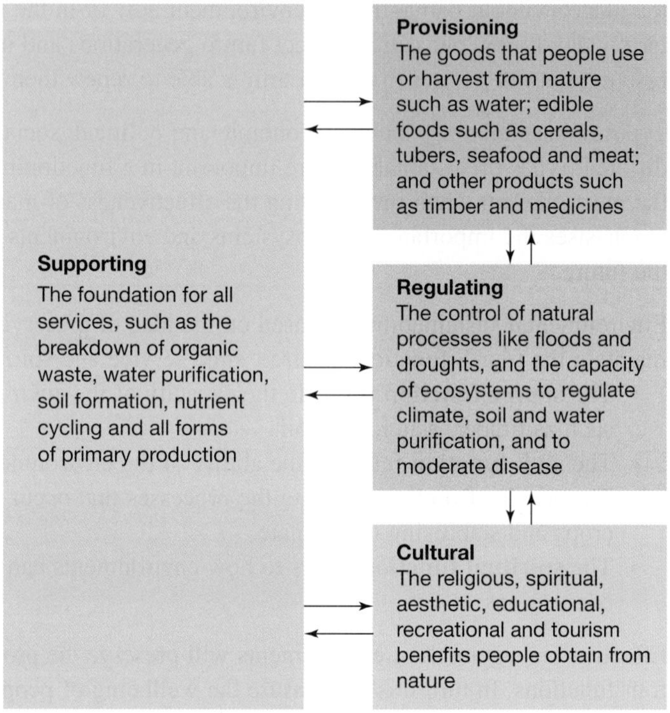
Supporting
The foundation for all services, such as the breakdown of organic waste, water purification, soil formation, nutrient cycling and all forms of primary production
Provisioning
The goods that people use or harvest from nature such as water; edible foods such as cereals, tubers, seafood and meat; and other products such as timber and medicines
Regulating
The control of natural processes like floods and droughts, and the capacity of ecosystems to regulate climate, soil and water purification, and to moderate disease
Cultural
The religious, spiritual, aesthetic, educational, recreational and tourism benefits people obtain from nature

13.10.3 Measuring sustainability

A range of indices have been developed in recent years to examine the link between ecological services, human wellbeing and sustainability. Each gives a slightly different perspective on human activity and/or sustainability. These include:

- the Human Development Index (HDI)
- the Sustainable Society Index (SSI)
- the Happy Planet Index (HPI).

The Sustainable Society Index says that sustainable human action must:

- meet the needs of the present generation yet not compromise the ability of future generations to meet their own needs
- ensure that people have the opportunity to develop themselves in a free, well-balanced society that is in harmony with nature.

It is worthwhile studying these indices as they put forward many sound ideas about human wellbeing and the sustainability of the ecological services of the natural world.

ecological services any beneficial natural process arising from healthy ecosystems, such as purification of water and air, pollination of plants and decomposition of waste

stewardship the belief that humans have a responsibility to care for the Earth to protect its future

The Sustainable Society Index gives values to 21 factors across a range of social, political, economic and environmental considerations. For example, Australia rates highly in clean air and sufficient food but lower in renewable energy and consumption. In another example, Burundi rates extremely low in organic farming, genuine savings, GDP, good governance and population growth, but highly in greenhouse gases, energy use and renewable water resources.

FIGURE 3 Australia's situation based on the Sustainable Society Index

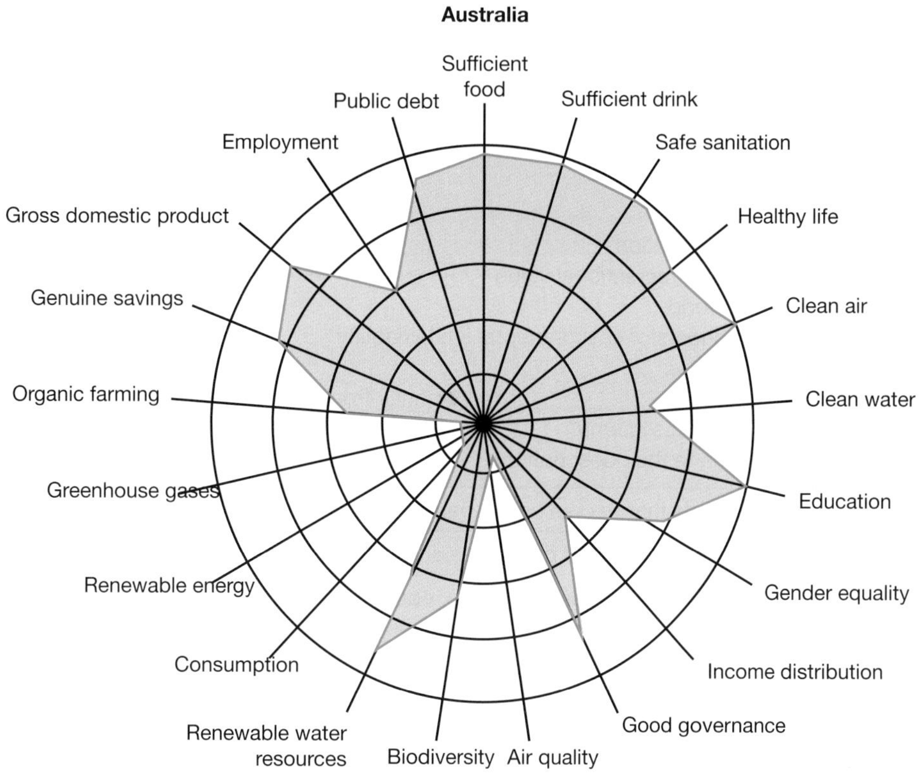

Source: © Sustainable Society Foundation

Resources

eWorkbook	Sustainable management (ewbk-10061)
Video eLesson	Sustainable management — Key concepts (eles-5240)
Interactivity	Sustainable management (int-8668)

13.10 ACTIVITIES

1. Use the **Forest depletion** weblink in your online Resources to explore information on this topic.
   a. What aspects of sustainability and the concept of stewardship can you draw from this information?
   b. Make a list of nations that have an unsustainable level of forest depletion.
2. Mahatma Gandhi once said, 'The Earth offers enough for everyone's needs, not for everyone's greed.' Provide an argument with an Earth-centred viewpoint about this quote and then a counter-argument based on a human-centred viewpoint. Ensure that your arguments are logical, clearly expressed and supported by evidence. Discuss and compare your arguments with a partner.

13.10 EXERCISE

Learning pathways

■ LEVEL 1	■ LEVEL 2	■ LEVEL 3
3, 5, 8, 10, 13	4, 6, 7, 12, 14	1, 2, 9, 11, 15

Check your understanding

1. Describe how environmental sustainability is different to other types of sustainability.
2. Which type of sustainability is the most relevant when investigating the effectiveness of management strategies?
3. Sustainability takes many forms. Identify to which sustainability function each of the following actions belongs.
   a. Absorb and recycle waste
   b. Pollination and climate stabilisation
   c. Religious, psychological and recreational value
   d. Supplying resources and food
4. Outline the four main components of environmental sustainability.
5. Complete the following sentence:
   Ecological services are the ________________ to humanity from the resources and processes that are supplied by ______________ ecosystems.
6. Match the ecological services to their descriptions.

Supporting	Goods that people use or harvest from nature
Provisioning	Control of natural processes and capacity to regulate climate, soil and water
Regulating	Religious, spiritual, aesthetic, educational, recreational and tourism benefits derived from nature
Cultural	The foundation of all services

Apply your understanding

7. Why are different measures used to evaluate sustainability?
8. Explain the main criteria that the Sustainable Society Index uses to qualify sustainable human action.
9. Discuss the following statement: In terms of the SSI, Australia rates highly in renewable energy and consumption.
10. Divide the various factors shown around the circle in **FIGURE 3** into the categories human wellbeing, environmental wellbeing and economic wellbeing.
11. Choose three variables identified in **FIGURE 3** as impacting on Australia's sustainability rating. Explain how these three factors are interconnected.
12. What factors does Australia need to change to be a more sustainable nation? Consider environmental, social and economic criteria from the Sustainable Society Index to inform your recommendations.

Challenge your understanding

13. Suggest reasons why Australia rates poorly in some of the factors shown in **FIGURE 3**.
14. The ecological services model considers what nature provides for humanity. Discuss what humanity gives back to nature.
15. Anthropocentric models of environmental sustainability are based on the idea of preserving the Earth's capacity to support human life. Construct a model of sustainability that takes a strong anthropocentric worldview.

To answer questions online and to receive **immediate feedback** and **sample responses** for every question, go to your learnON title at www.jacplus.com.au.

13.11 Ecological footprint and biocapacity

LEARNING INTENTION

By the end of this subtopic, you will be able to explain and outline the interconnection between ecological footprint and biocapacity.

13.11.1 Understanding the ecological footprint

The **ecological footprint** is one means of measuring human demand for ecological services. The footprint takes into account the regenerative capacities of biomes and ecosystems, which are described as the Earth's **biocapacity**. The footprint is given as a number, in hectares of productive land and sea area, by measuring a total of six factors, as shown in **FIGURE 1**. The ecological footprint is a useful indicator of environmental sustainability.

ecological footprint a measure of human demand on the Earth's natural systems in general and ecosystems in particular

biocapacity the capacity of a biome or ecosystem to generate a renewable and ongoing supply of resources and to process or absorb its wastes

FIGURE 1 Measuring the Earth's ecological footprint

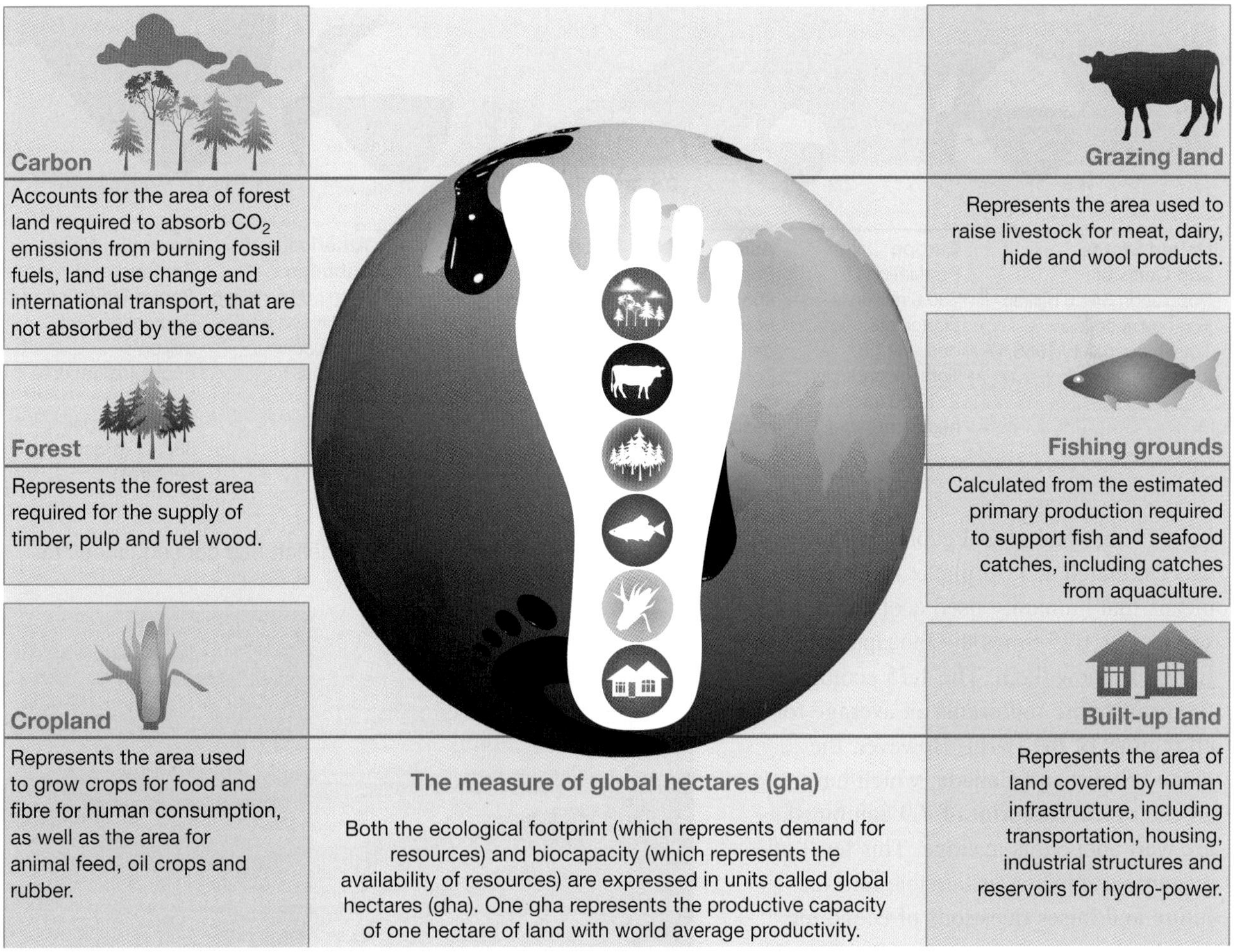
Carbon
Accounts for the area of forest land required to absorb CO_2 emissions from burning fossil fuels, land use change and international transport, that are not absorbed by the oceans.
Forest
Represents the forest area required for the supply of timber, pulp and fuel wood.
Cropland
Represents the area used to grow crops for food and fibre for human consumption, as well as the area for animal feed, oil crops and rubber.
Grazing land
Represents the area used to raise livestock for meat, dairy, hide and wool products.
Fishing grounds
Calculated from the estimated primary production required to support fish and seafood catches, including catches from aquaculture.
Built-up land
Represents the area of land covered by human infrastructure, including transportation, housing, industrial structures and reservoirs for hydro-power.
The measure of global hectares (gha)
Both the ecological footprint (which represents demand for resources) and biocapacity (which represents the availability of resources) are expressed in units called global hectares (gha). One gha represents the productive capacity of one hectare of land with world average productivity.

FIGURE 2 compares the ecological footprint with biocapacity. The elephants represent each region's footprint (per capita) and the balancing balls represent the size of the region's biocapacity (per capita). The dark green background represents the gross footprint of regions that exceed their biocapacity, and the light blue background represents those regions that use less than their biocapacity.

FIGURE 2 Biocapacity and ecological footprint

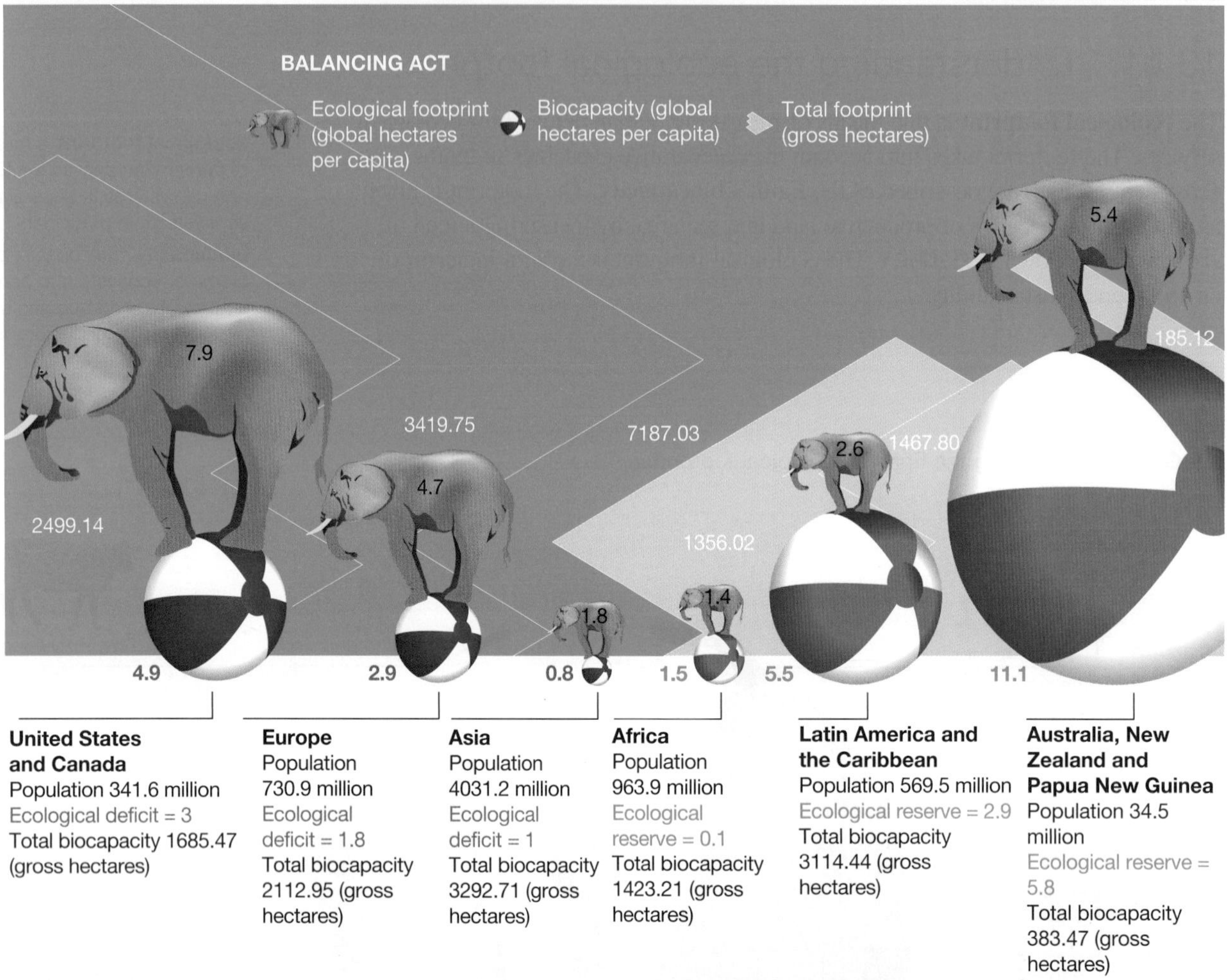

In 2019 the total global ecological footprint was estimated at 1.75 planet Earths, which means that humanity used ecological services at 1.75 times the biocapacity of the Earth to renew them. The 1.75 ecological footprint figure represents an average for all regions of the Earth. However, the United States and Canada, which have an ecological footprint of 7.9 combined, are well above this average. This level of resource use is not sustainable into the future and raises questions of economic viability, environmental benefit and social justice.

FIGURE 3 Continued ecological deficit has consequences for food and water security

int-8669

FIGURE 4 Ecological debt map

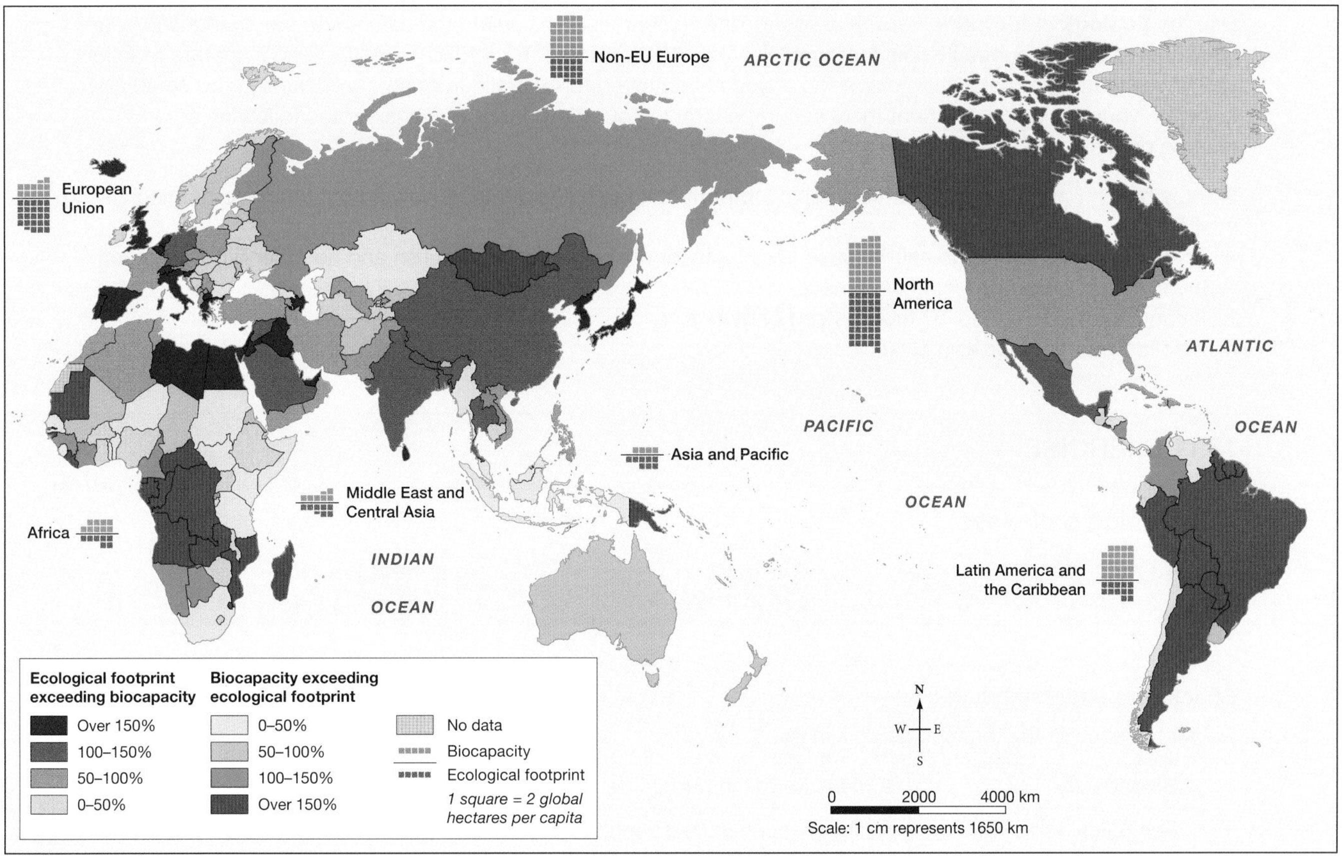

Source: Global Footprint Network.

fdw-0035

FOCUS ON FIELDWORK

Perception study

Throughout our lives, culture and experience shape our worldviews, which in turn influence our perceptions or views about places and regions. How do you know whether your perceptions match reality, or whether the public has a realistic view of an environmental problem or event? Being able to separate facts from opinion is an important part of analysing field work data.

Learn some strategies to help you assess and compare people's opinions of environmental issues using the **Perception study** fieldwork activity in your online Resources.

on Resources

eWorkbook	Ecological footprint and biocapacity (ewbk-10065)
Video eLesson	Ecological footprint and biocapacity — Key concepts (eles-5241)
Interactivity	Ecological footprint and biocapacity (int-8669)

13.11 ACTIVITY

Use the **Ecological footprint** weblink in your online Resources to calculate the difference that changes in your life would make to your ecological footprint. Use the 'ADD DETAILS TO IMPROVE ACCURACY' feature to tailor your investigation to make it more specific. If you have moved to Australia from another country, you could also compare your ecological footprint there in comparison to your footprint now. Consider the following as starting points.

By how many 'Earths' would your ecological footprint change if you did the following but changed nothing else in your life?

- If you ate more, less or no animal-based products (or ate only chicken and fish and no other types of meat)?
- If you 'often' carpooled instead of 'never'?
- If you travelled to school by train instead of by bus?
  Discuss your findings as a class.

13.11 EXERCISE

Learning pathways

■ LEVEL 1	■ LEVEL 2	■ LEVEL 3
1, 3, 4, 9, 14	5, 6, 7, 8, 13	2, 10, 11, 12, 15

Check your understanding

1. Match each of the terms to its definition.

Biocapacity	A measure of the human demand for Earth's natural resources
Ecological footprint	Ability to regenerate resources
Ecological service	The benefits humanity gains from nature

2. Both biocapacity and ecological footprint are measured in gha. What does one gha represent?
3. Complete the following sentences:
   a. Generally, countries with large populations and/or wealthy countries tend to have ______________ ecological footprints and ______________ biocapacities.
   b. To reduce its ecological footprint, a country should aim to ______________ its use of ecological services and ______________ its biocapacity.
4. Use **FIGURE 1** to summarise how an ecological footprint is measured.
5. Identify two regions that use fewer resources than their biocapacity.
6. Identify two regions that use more resources than their biocapacity.

Apply your understanding

7. Examine **FIGURE 2**. Explain what is being demonstrated by this diagram.
8. Explain how a region can exceed their biocapacity.
9. What reasons can you suggest for the very high environmental or ecological footprint for the United States and Canada?
10. How might the three regions with the dark green very high gross footprint improve their biocapacity?
11. Explain the difference between an ecological footprint and biocapacity.
12. Why is Australia in such a good position in terms of ecological footprint compared to biocapacity?

Challenge your understanding

13. Choose one of the regions shown to be in ecological reserve in **FIGURE 2**. Suggest why this might be the case, and predict whether this will change over the next 50 years and why.
14. Propose one new government policy that might help to improve Australia's ecological footprint.
15. Consider Australia's ecological footprint and biocapacity. To what extent might Australia be considered to be in harmony with nature?

To answer questions online and to receive **immediate feedback** and **sample responses** for every question, go to your learnON title at www.jacplus.com.au.

13.12 SkillBuilder — Drawing a futures wheel

online only

LEARNING INTENTION

By the end of this subtopic, you will be able to present geographical information diagrammatically, and predict outcomes in a futures wheel.

The content in this subtopic is a summary of what you will find in the online resource.

13.12.1 Tell me

What is a futures wheel?

A futures wheel is a series of bubbles or concentric rings with words written inside each to show the increasing impact of change. It helps show the consequences of change.

FIGURE 1 Possible responses by the ski and alpine resort industry to climate change

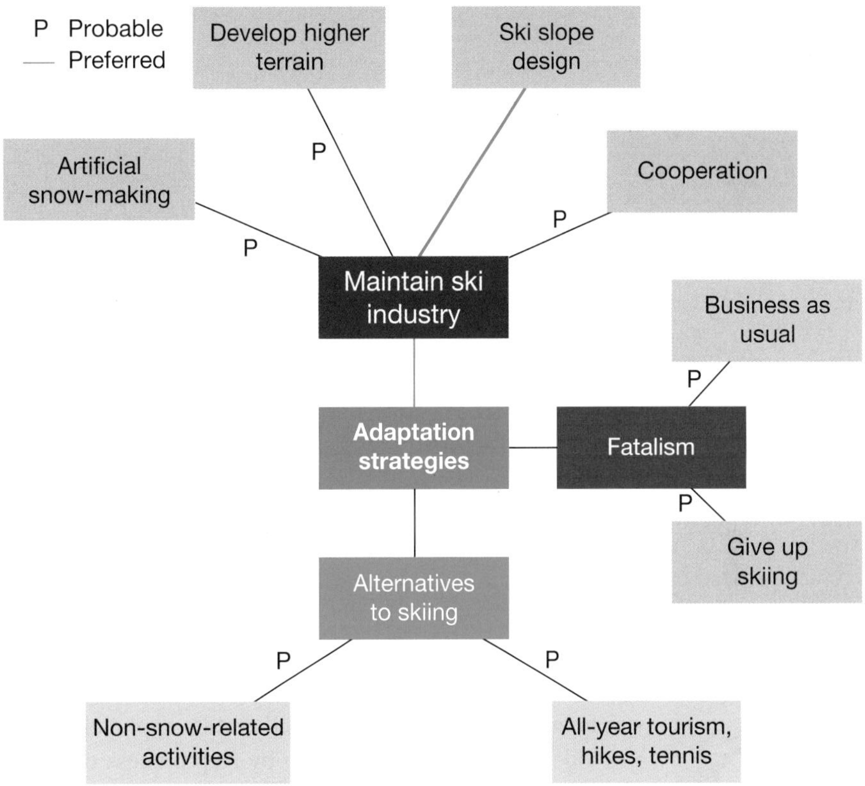

13.12.2 Show me

How to draw a futures wheel

Step 1

Draw a number of concentric rings — four is a good starting point. Make sure the inner circle is big enough to write in.

Step 2

In the inner circle or bubble, write the issue.

Step 3

In the first ring out from the centre, write what you think are the most possible consequences. Draw a square block around these possible ideas.

Step 4

In the next layer out, take each of the consequences from the previous ring and think of two or more impacts that this change would imply. These thoughts are those that you see as most probable — a view of things that could happen. Label each of these ideas with a P.

Step 5

Continue adding rings of ideas. The outer ring will have a whole range of ideas, whereas the rings closer to the centre of the wheel will have fewer ideas.

Step 6

You may notice that there are interconnections between ideas. If you can see a link, you should draw a line between the interconnecting components.

Consider the different connections that you have made between ideas. Find a route that you consider as the preferred option — a view that you see as most desirable. Colour this route in some way to show the thread of ideas. Can you justify your choice?

Step 7

Give your futures wheel a title.

FIGURE 2 Examples of futures wheels

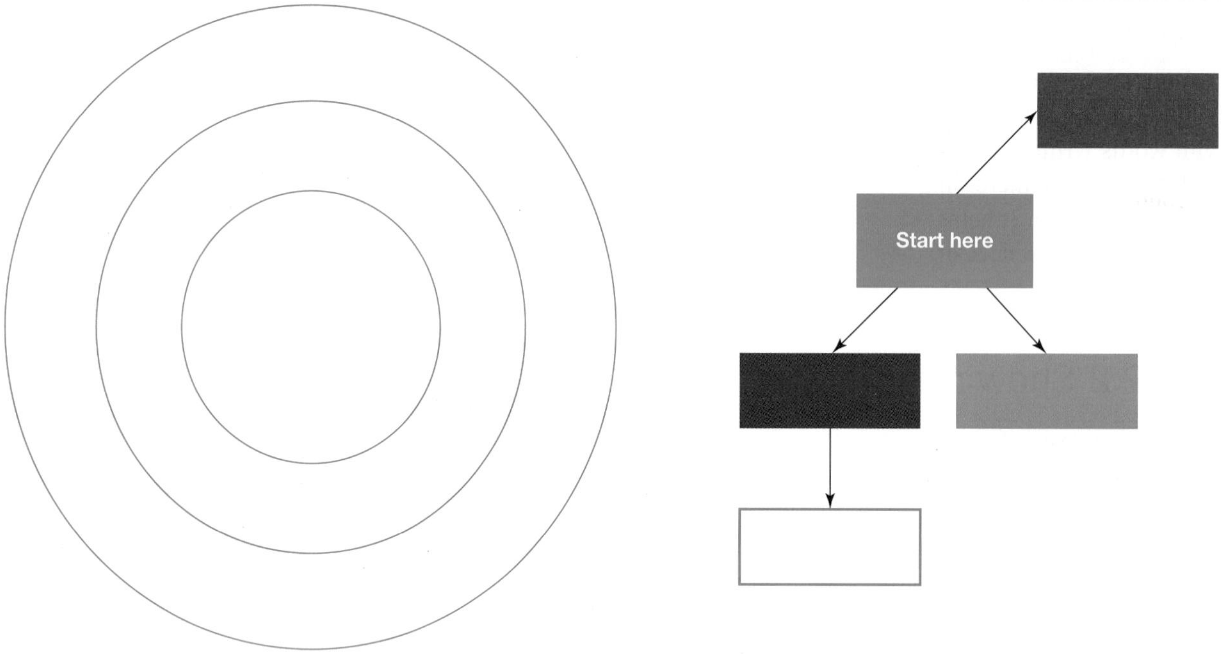

13.12.3 Let me do it

learnON

Go to learnON to access the following additional resources to help you build this skill:

- a longer explanation of this skill and its application in Geography (Tell me)
- a video demonstrating the step-by-step process of this skill (Show me)
- an activity and interactivity for you to practise the skill (Let me do it)
- self-marking questions to help you understand and use the skill.

on Resources

eWorkbook	SkillBuilder — Drawing a futures wheel (ewbk-10057)
Video eLesson	SkillBuilder — Drawing a futures wheel (eles-1745)
Interactivity	SkillBuilder — Drawing a futures wheel (int-3363)

13.13 Investigating topographic maps — Environmental change in Jindabyne

LEARNING INTENTION

By the end of this subtopic, you will be able to describe potential effects of environmental change in Jindabyne and provide examples from a topographic map to support your ideas.

13.13.1 Environmental change in Jindabyne

Jindabyne was originally located under the current location of Lake Jindabyne. It was moved in the 1960s when the Snowy River was dammed as part of the Snowy Mountains Scheme.

This alpine region is popular for its mountain peaks, which are used for snow activities such as skiing, snowboarding and tobogganing. It is a holiday destination that is popular as a base for skiers visiting ski resorts, such as Perisher and Thredbo, but also offers a range of other nature-based activities year round. The increasing numbers of tourists is another environmental change that could impact on the fragility and value of the alpine environment. In addition, natural snowfalls in the region are likely to be affected by warmer conditions associated with climate change. This could have flow-on impacts for businesses and communities in the area.

FIGURE 1 An oblique aerial view of Jindabyne

on Resources

eWorkbook	Investigating topographic maps — Environmental change in Jindabyne (ewbk-10069)
Digital document	Topographic map of Jindabyne (doc-36367)
Video eLesson	Investigating topographic maps — Environmental change in Jindabyne — Key concepts (eles-5242)
Interactivity	Investigating topographic maps — Environmental change in Jindabyne (int-8670)
Google Earth	Jindabyne (gogl-0102)

FIGURE 2 Topographic map of Jindabyne

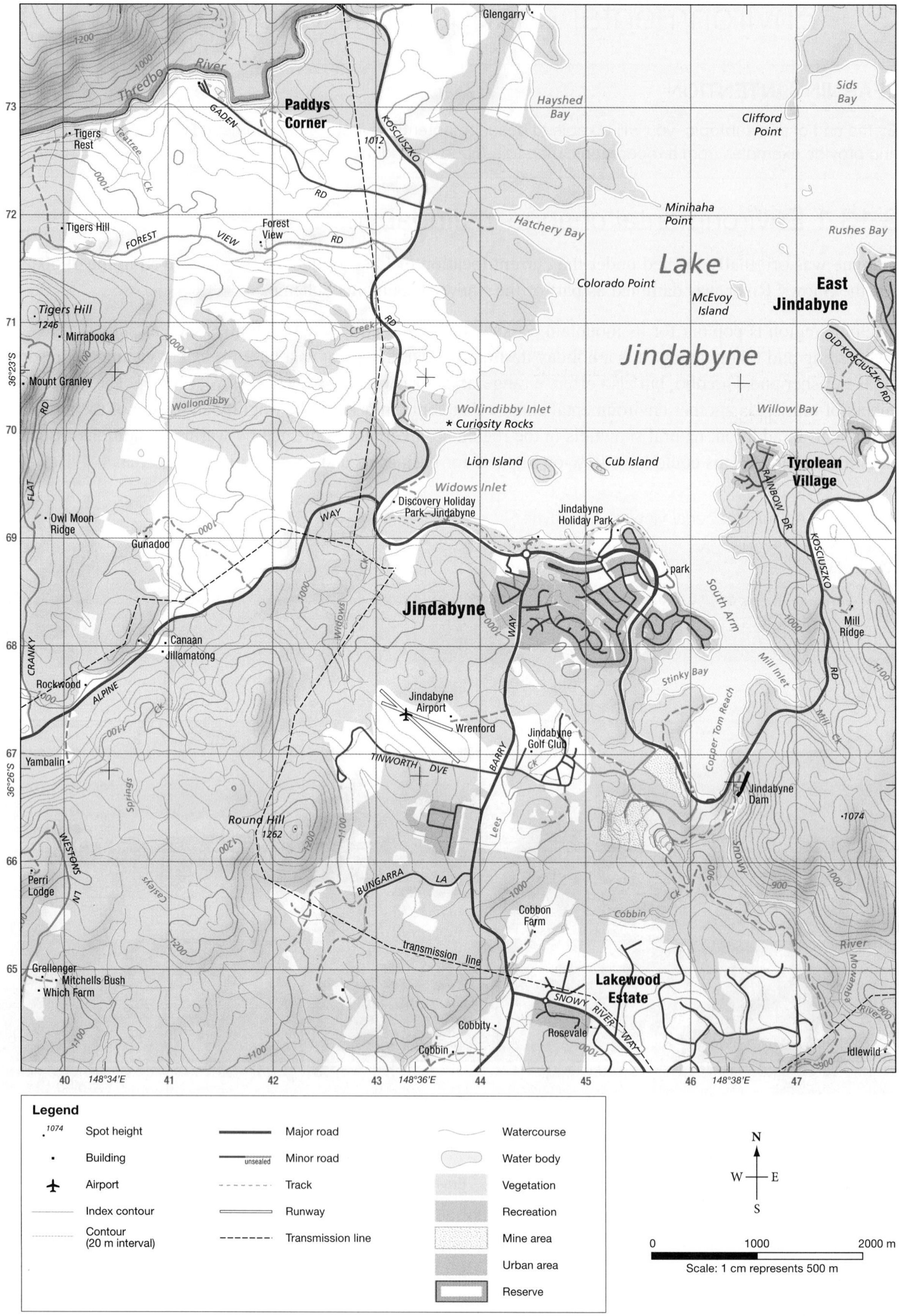

Source: Spatial Vision

13.13 EXERCISE

Learning pathways

■ LEVEL 1	■ LEVEL 2	■ LEVEL 3
1, 2, 8	3, 4, 6, 9	5, 7, 10

Check your understanding

1. Identify the grid references of the following.
   a. Cobbon Farm
   b. Lakewood Estate
   c. Mill Ridge
2. Write the scale of the map as a ratio and a sentence, and draw it as a linear scale.
3. Identify the direction, bearing and straight line distance from:
   a. Owl Moon Ridge to Grellenger
   b. Jindabyne Dam to Jindabyne Golf Club
   c. Jindabyne Airport to Tigers Rest
4. Identify the altitude of each of the following.
   a. Tigers Hill
   b. Forest View
   c. Jindabyne Golf Club
5. Calculate the local relief in:
   a. AR4668
   b. AR4770
   c. AR4365.
6. Identify and describe examples of the following landforms. Use area references to identify the locations in your descriptions.
   a. A saddle
   b. A plateau
   c. A valley

Apply your understanding

7. Examine the features of the map and **FIGURE 1**. Consider an environmental change that you could study in this location.
   a. Identify an aim for research.
   b. Suggest some fieldwork methods you could use to observe, measure, collect and analyse data related to this environmental change.
   c. Identify any fieldwork instruments that you might use in your investigation.
   d. Suggest environmental management strategies that could be implemented to address this change.
8. Outline your route, including distances and directions of travel, if you were staying on the unsealed part of Rainbow Drive in Tyrolean Village overlooking Willow Bay and wanted to go fishing in the Thredbo River.
   a. Plan the most direct route.
   b. Plan the route that might offer the most interesting scenery. Explain your choices.
9. If you were planning a bushwalk to a high point in this area, which of the peaks shown on this map would you decide to climb? Give reasons for your choice, considering how easy or challenging the walk might be for someone of your fitness level and experience, and the benefits you might receive from climbing it.

Challenge your understanding

10. Based on the direction and features of the land, do you think the Old Kosciusko Road that ends in East Jindabyne used to meet up with the route of the current Kosciusko Road? Give reasons for your answer using specific data from the map.

To answer questions online and to receive **immediate feedback** and **sample responses** for every question, go to your learnON title at www.jacplus.com.au.

13.14 Thinking Big research project — Wacky weather presentation

online only

The content in this subtopic is a summary of what you will find in the online resource.

Scenario

As the regular weather presenter for an evening news program, you have been asked by the producer to compile a segment on the history of extreme weather events in Australia, to outline the link between these events and climate change, and to respond to claims made by 'climate change sceptics' that climate change is not occurring.

Task

learnON

For your segment, you will need to create a slide deck or multimedia presentation and an accompanying speech that:

- addresses the link between extreme weather events and climate change
- outlines what Australia is doing to tackle global warming and climate change
- considers and responds to the views of climate-change sceptics.

Go to learnON to access the resources you need to complete this research project.

Resources

ProjectsPLUS Thinking Big research project — Wacky weather presentation (pro-0211)

13.15 Review

13.15.1 Key knowledge summary

13.2 Environmental functioning

- The four spheres of the environment are the atmosphere, lithosphere, hydrosphere and biosphere.
- The spheres interact through geographical processes to form and transform environments.

13.3 Lithospheric processes

- The lithosphere is all the rocks, soils and crust on the Earth's surface.
- Erosion is the wearing away of earth by wind, water or ice. Weathering is the physical and chemical disintegration of rocks and minerals.
- Tectonic processes have transformed environments throughout history. These include volcanic eruptions and earthquakes and the formation of fold and fault mountains.
- The nitrogen and phosphorus cycle moves nutrients through the lithosphere, allowing plants to grow and organisms to decompose.

13.4 Biospheric processes

- The biosphere is all the living things on Earth.
- Carbon is transferred through the environment by the carbon cycle. Carbon dioxide in the air is used by plants to photosynthesise, and animals eat plants and release carbon into the atmosphere when they respire or when they decompose after death.
- The oxygen cycle involves plants producing oxygen and animals obtaining this oxygen by breathing.
- A food chain is a series of organisms, each eating or decomposing the preceding one. Food webs are more complicated and involve the combination of a number of food chains.

13.5 Hydrological and atmospheric processes

- The hydrosphere is water in all its forms including rain, ice, sleet and vapour. Processes and features of the hydrosphere are referred to as hydrospheric.
- The water cycle is the continuous movement of water and includes the processes of precipitation, infiltration, run-off, evaporation, transpiration, evapotranspiration and condensation.
- The atmosphere is the gaseous layer surrounding the Earth. Processes and features of the atmosphere (such as the effects of solar radiation) are referred to as atmospheric. Solar radiation is the heat that is received from the Sun's rays.

13.6 Human-induced change

- Humans modify ecosystems to grow food, build habitats and extract resources.
- Large-scale agriculture has introduced monocultures, removed native species, introduced livestock and resulted in overgrazing.
- Urbanisation replaces natural environments with roads, buildings, straightened waterways and parkland.

13.7 Climate change

- The greenhouse effect has been altered by human activities, such that an enhanced greenhouse effect is warming the Earth's atmosphere and oceans.
- Increased use of renewable energy sources for power generation will reduce the level of global warming.

13.9 Worldviews and environmental management

- Environmental worldviews are the varying beliefs about how the world works and how humans fit within its cycles. Environmental worldviews shape the assumptions and values that guide an individual's actions towards the environment and its management.

13.10 Sustainable management

- Sustainability refers to living within the resources of the planet without damaging the environment in a way that threatens our ability to continue living this way in the future. Environmental sustainability aims to maintain the four functions of the environment: the source, sink, service, and spiritual functions.

- The Sustainable Society Index says that sustainable human action must meet the needs of the present generation without compromising the ability of future generations to meet their own needs.

13.11 Ecological footprint and biocapacity

- The ecological footprint is a means of measuring human demand for ecological services and takes into account the regenerative capacities of biomes and ecosystems.
- Biocapacity is the capacity of a biome or ecosystem to generate a renewable and ongoing supply of resources and to process or absorb the wastes.

13.15.2 Key terms

anthropocentric the belief that humans needs and wants are more important than protecting the environment

biocapacity the capacity of a biome or ecosystem to generate a renewable and ongoing supply of resources and to process or absorb its wastes

biocentric the belief that humans should minimise their impact on the Earth to ensure it is able to sustain life for all environments and species

carbon credit a tradable certificate representing the right of a company to emit one metric tonne of carbon dioxide into the atmosphere

chlorophyll green pigment in plants that uses energy from the Sun to transform carbon-dioxide into glucose

climate change any change in climate over time, whether due to natural processes or human activities

condensation when water in the atmosphere cools and changes from a gaseous state into a liquid state. This occurs when the water vapour clusters around a solid particle (such as dust)

consumers organisms that eat other organisms

decomposers organisms that break down dead organisms

ecocentric the belief that protecting the environment should come before all other needs and wants

ecological footprint a measure of human demand on the Earth's natural systems in general and ecosystems in particular

ecological services any beneficial natural process arising from healthy ecosystems, such as purification of water and air, pollination of plants and decomposition of waste

enhanced greenhouse effect the observable trend of rising world atmospheric temperatures over the past century, particularly during the last couple of decades

environmental ethics an individual's beliefs about what is right or wrong behaviour in relation to the Earth and its environments and communities

environmental worldview varying viewpoints of how the world works and where people fit into the world. The worldview will form the assumptions and values that guide an individual's actions towards the environment.

erosion the wearing down of rocks and soils on the Earth's surface by the action of water, ice, wind, waves, glaciers and other processes

evaporation when water contained in water bodies is heated by the Sun and the liquid changes into a gaseous state and rises into the atmosphere

evapotranspiration a process in which liquid water is evaporated from soil, trees, and the ocean surface, and transferred into the atmosphere

extinction when all individuals of a species have died

geographic processes the physical forces that form and transform our world

global warming increased ability of the Earth's atmosphere to trap heat

infiltration water that is absorbed into the ground, flows downward and collects above an impermeable layer or rock

lithosphere all the rocks, soils and crust on the Earth's surface

precipitation occurs when water droplets or ice crystals become too heavy to be suspended in the air and fall to Earth as rain, snow, sleet or hail

producer a plant that can photosynthesise

run-off water that is unable to be absorbed into the ground, flows over its surface and collects in nearby waterways or reaches stormwater drains

stewardship the belief that humans have a responsibility to care for the Earth to protect its future

transpiration when water contained in plants is heated and changes from a liquid into a gaseous state and rises into the atmosphere

weathering the breaking down of rock through the action of wind and water and the effects of climate, mainly by water freezing and cooling as a result of temperature change

13.15.3 Reflection

Complete the following to reflect on your learning.

Revisit the inquiry question posed in the Overview:

Why is an understanding of environmental processes and interconnections essential for sustainable management of environments?

1. Now that you have completed the topic, choose a particular environment and outline the environmental processes that enable that environment to function. Describe how the processes are interconnected.
2. Write an extended response.
   Explain how an understanding of environmental processes and interconnections essential for sustainable management of environments.

Subtopic	Success criteria	●	●	●
13.2	I can describe how environments function.			
13.3	I can describe the lithospheric processes that enable environments to function.			
13.4	I can describe the biospheric processes that enable environments to function.			
13.5	I can describe the atmospheric and hydrological processes that enable environments to function.			
13.6	I can describe how humans have changed the environment.			
13.7	I can explain how human activity has altered the Earth's climate.			
13.8	I can explain how people's worldviews might affect their attitudes to environmental management.			
13.9	I can evaluate different points of view on an issue.			
13.10	I can define the term sustainability and outline how it is measured.			
	I can understand interconnections essential for sustainable management.			
13.11	I can explain the concept of an ecological footprint and explain how this is connected to biocapacity.			
13.12	I can present geographical information diagrammatically and predict outcomes in a futures wheel.			
13.13	I can use maps and stimulus material to assess environmental change in Jindabyne.			

on Resources

eWorkbook Chapter 13 Student learning matrix (ewbk-8568)
Chapter 13 Reflection (ewbk-8569)
Chapter 13 Extended writing task (ewbk-8570)

Interactivity Chapter 13 Crossword (int-8671)

ONLINE RESOURCES

on Resources

Below is a full list of **rich resources** available online for this topic. These resources are designed to bring ideas to life, to promote deep and lasting learning and to support the different learning needs of each individual.

eWorkbook

- **13.1** Chapter 13 eWorkbook (ewbk-8105) ☐
- **13.2** Environmental functioning (ewbk-10029) ☐
- **13.3** Lithospheric processes (ewbk-10033) ☐
- **13.4** Biospheric processes (ewbk-10037) ☐
- **13.5** Hydrological and atmospheric processes (ewbk-10041) ☐
- **13.6** Human-induced change (ewbk-10053) ☐
- **13.7** Climate change (ewbk-8161) ☐
- **13.8** Worldviews and environmental management (ewbk-10049) ☐
- **13.9** SkillBuilder — Evaluating alternative responses (ewbk-10045) ☐
- **13.10** Sustainable management (ewbk-10061) ☐
- **13.11** Ecological footprint and biocapacity (ewbk-10065) ☐
- **13.12** SkillBuilder — Drawing a futures wheel (ewbk-10057) ☐
- **13.13** Investigating topographic maps — Environmental change in Jindabyne (ewbk-10069) ☐
- **13.15** Chapter 13 Student learning matrix (ewbk-8568) ☐
  Chapter 13 Reflection (ewbk-8569) ☐
  Chapter 13 Extended writing task (ewbk-8570) ☐

Sample responses

- **13.1** Chapter 13 Sample responses (sar-0162) ☐

Digital document

- **13.13** Topographic map of Jindabyne (doc-36367) ☐

Video eLessons

- **13.1** What are we doing? (eles-1707)
  Introducing environmental change and management — Photo essay (eles-5232) ☐
- **13.2** Environmental functioning — Key concepts (eles-5233) ☐
- **13.3** Lithospheric processes — Key concepts (eles-5234) ☐
- **13.4** Biospheric processes — Key concepts (eles-5235) ☐
- **13.5** Hydrological and atmospheric processes — Key concepts (eles-5236) ☐
- **13.6** Human-induced change — Key concepts (eles-5237) ☐
- **13.7** Climate change — Key concepts (eles-5238) ☐
- **13.8** Worldviews and environmental management — Key concepts (eles-5239) ☐
- **13.9** SkillBuilder — Evaluating alternative responses (eles-1744)
- **13.10** Sustainable management — Key concepts (eles-5240) ☐
- **13.11** Ecological footprint and biocapacity — Key concepts (eles-5241) ☐
- **13.12** SkillBuilder — Drawing a futures wheel (eles-1745)
- **13.13** Investigating topographic maps — Environmental change in Jindabyne — Key concepts (eles-5242) ☐

Interactivities

- **13.2** Environmental functioning (int-8662) ☐
- **13.3** Lithospheric processes (int-8663) ☐
- **13.4** Biospheric processes (int-8664) ☐
- **13.5** Hydrological and atmospheric processes (int-8665) ☐
- **13.6** Human-induced change (int-8666) ☐
- **13.7** The enhanced greenhouse effect (int-7951) ☐
- **13.8** Worldviews and environmental management (int-8667) ☐
- **13.9** SkillBuilder — Evaluating alternative responses (int-3362)
- **13.10** Sustainable management (int-8668) ☐
- **13.11** Ecological footprint and biocapacity (int-8669) ☐
- **13.12** SkillBuilder — Drawing a futures wheel (int-3363)
- **13.13** Investigating topographic maps — Environmental change in Jindabyne (int-8670) ☐
- **13.15** Chapter 13 Crossword (int-8671) ☐

ProjectsPLUS

- **13.14** Thinking Big research project — Wacky weather presentation (pro-0211) ☐

Weblinks

- **13.10** Forest depletion (web-2319) ☐
- **13.11** Ecological footprint (web-1199) ☐
- **13.12** Bubbl.us (web-2315) ☐

Fieldwork

- **13.2** Ecological footprint (fdw-0034) ☐
- **13.11** Perception study (fdw-0035) ☐

Google Earth

- **13.13** Jindabyne (gogl-0102) ☐

myWorldAtlas

- **13.7** Causes of climate change (mwa-4519) ☐
  Larsen Ice Shelf break-up (mwa-4520) ☐
  Impacts on polar bears (mwa-4521) ☐
  Climate change and Australia (mwa-4522) ☐
  Global warming and Antarctica (mwa-4523) ☐

Teacher resources

There are many resources available exclusively for teachers online.

14 Land environments under threat

INQUIRY SEQUENCE

To access a pre-test with **immediate feedback** and **sample responses** to every question in this chapter, select your learnON format at www.jacplus.com.au.

14.1 Overview

Numerous **videos** and **interactivities** are embedded just where you need them, at the point of learning, in your learnON title at www.jacplus.com.au. They will help you to learn the content and concepts covered in this topic.

From housing to food production, we use land for many different things. What impact are we having on this important resource?

14.1.1 Introduction

Land is one of our most valuable resources. Left alone it exists in a state of balance, and if managed wisely it can continue to do so. However, the land is under increasing pressure as a direct result of population growth — agriculture, mining and the expansion of settlements — all of which have the potential to interfere with natural processes.

STARTER QUESTIONS

1. What do you think is meant by the term *natural balance*?
2. Copy the following table.

How we use the land	Sustainable land use	Cause of land degradation

   In column 1, list the ways in which we use the land. Then use columns 2 and 3 to record whether you think each use is a sustainable land use.
3. Briefly explain how human activity can have a negative impact on natural processes.
4. Watch the **Land environments under threat — Photo essay** video in your online Resources. Think about all the places you have been in last week. How much of the land have you seen? How often have you walked on soil rather than concrete or some other form of land covering? Do you feel like you have a strong connection with the land, or just the things humans have built on it?

on Resources

eWorkbook	Chapter 14 eWorkbook (ewbk-8106)
Video eLessons	Wasting our land (eles-1708) Land environments under threat — Photo essay (eles-5243)

14.2 The causes and impacts of land degradation

LEARNING INTENTION

By the end of this subtopic, you will be able to explain what land degradation is and describe its causes and impacts.

14.2.1 Explaining land degradation

Land degradation is the process that reduces the land's capacity to produce crops, support natural vegetation and provide fodder for livestock. Land degradation causes physical, chemical and biological changes; the natural environment deteriorates and the landscape undergoes a dramatic change (see **FIGURE 1**). Common causes of land degradation include soil erosion, increased salinity, pollution and desertification.

FIGURE 1 Land degradation causes physical, chemical and biological changes to the natural environment.

14.2.2 Causes of land degradation

Land can be degraded in many ways, but most of the causes can be traced back to the influences of human activity on the natural environment. **FIGURE 2** outlines these activities and their impacts.

int-5591

FIGURE 2 Why land degrades

- **A** When land is cleared or overgrazed, it becomes vulnerable to erosion by wind and water. The nutrient-rich soil is either washed or blown away, reducing the quality and quantity of crop yields. Dust storms result and sediment is transported to rivers, where it can smother marine species.
- **B** Introduced species such as rabbits eat grass, shrubs and young trees (saplings) down to the soil, thus exposing it to erosion. Their burrows increase erosion as they destabilise the soil. Rabbits also compete with native animals for food and burrows.
- **C** Tourism encourages the clearing of sand dunes for high-density housing and mountain slopes for ski runs, leaving the surface exposed to erosion.
- **D** Overgrazing leads to nutrient-rich soil being washed or blown away. Animals with hard hooves such as sheep and cattle trample vegetation and compact the soil, making it increasingly difficult for native species to grow. This leads to increased run-off after heavy rain.
- **E** Climate change will affect land degradation in the future. Higher sea levels will flood low-lying coastal areas. Expanding cities, removal of vegetation and use of concrete reduce the ability of the land to absorb moisture. This not only increases erosion, but can reduce the amount of rainfall in an area.
- **F** Urban communities produce large quantities of waste, which is deposited in landfills. Much of the rubbish remains toxic or, in the case of plastic bags, takes hundreds of years to break down. Liquid and solid waste seeps into groundwater and runs off into rivers and eventually into the sea, killing marine species.
- **G** Introduced plant species such as blackberries and Paterson's Curse (Salvation Jane) choke the landscape and compete with native vegetation. Their dense groundcover prevents light from reaching the soil.
- **H** Salinity occurs naturally in areas where there is low rainfall and high evaporation and where the land was below sea level millions of years ago. Salinity is also caused by excess irrigation and clearing natural vegetation. In some cases the watertable rises, bringing salt to the surface.

Factors that contribute to land degradation

- Poor management leads to the loss of nutrients vital for plant growth.
- Removal of vegetation makes the land vulnerable to erosion by wind and water.
- When urban development encroaches on agricultural land, vegetation is removed and the waste generated is disposed of in landfill.
- Poor agricultural practices, especially related to irrigation and the use of chemical fertilisers, can lead to the soil becoming saline or acidic.

FIGURES 3 and **4** show that agricultural activities and overgrazing combined account for more than 50 per cent of land degradation in the Asia–Pacific region and globally.

FIGURE 3 Causes of land degradation in the Asia–Pacific region; Australia is ranked fifth in clearing of native vegetation.

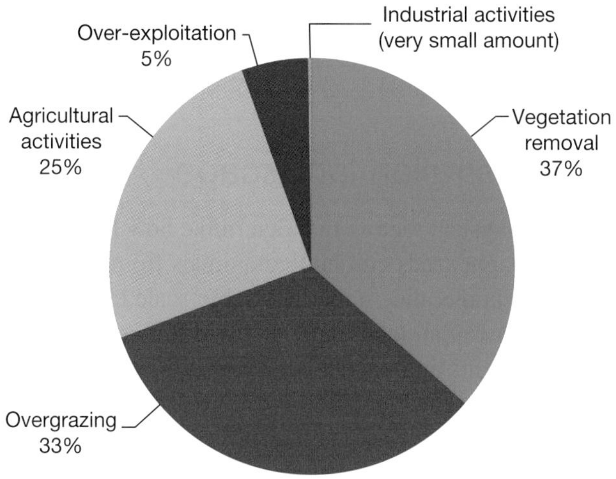

Source: FAO

FIGURE 4 Main causes of land degradation globally

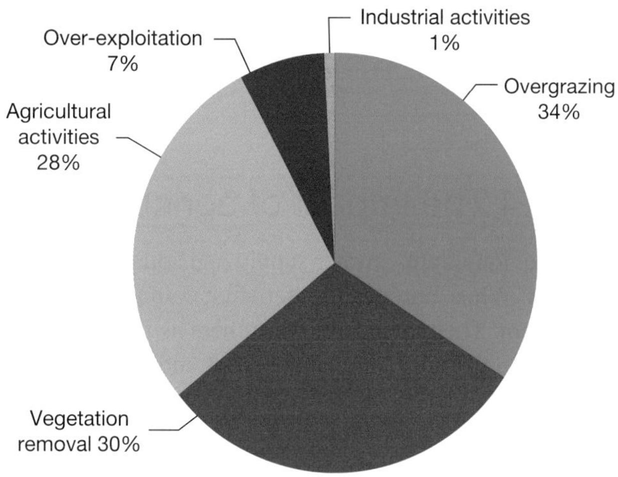

Source: FAO

14.2.3 The effects of land degradation

Even small changes can have dramatic effects on the land. The shortcut students take from the oval to the classroom can soon reduce a grassy area to dust. Drought can quickly reduce the productivity of an area used for farming. A farmer who neglects the land after one growing season may still be able to raise a good crop the following season, but if the land is neglected year after year it will eventually become unproductive.

The effects of land degradation are far-reaching. Farmland productivity diminishes and yields drop because the soil becomes exhausted through overuse or deforestation (see **FIGURE 5**). Costs increase as the land requires more treatment with fertiliser, and **topsoil** and nutrients in the soil need to be replaced. Valuable topsoil is often washed away into rivers and out to sea. Nutrients cause foul-smelling blue–green **algal blooms** that choke waterways. These blooms decrease water quality, poison fish and pose a direct threat to other aquatic life.

When rain falls on a well-vegetated hillside, the water is absorbed by plant roots and held in the soil. However, if the vegetation is removed, there is nothing to stabilise the soil and hold it together, especially when it rains. Rills and gullies form (see 13.7 Climate change) where the unprotected soil is washed away, and landslides may occur. **FIGURE 5** shows the impacts on the landscape of deforestation.

topsoil the top layers of soil that contain the nutrients necessary for healthy plant growth

algal bloom rapid growth of algae caused by high levels of nutrients (particularly phosphates and nitrates) in water

int-7952

FIGURE 5 (a) and (b) Land clearing and deforestation leave the land vulnerable to erosion.

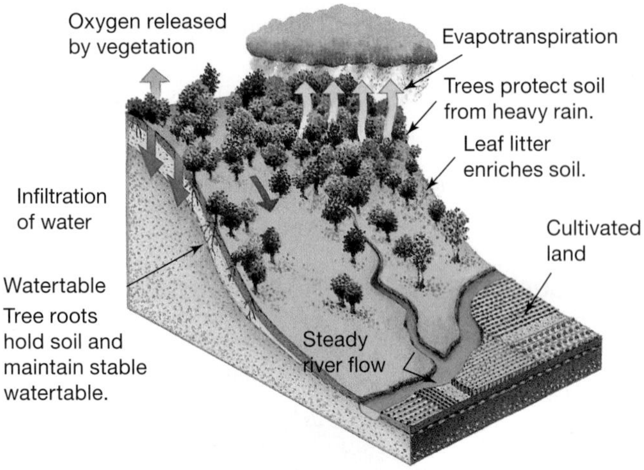

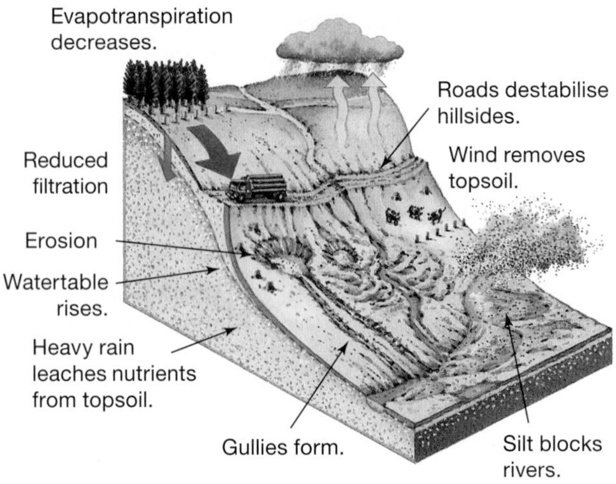

14.2.4 The impact of agriculture on the Australian landscape

Climate, topography, water supply and soil quality are the major physical factors that determine how land can be used. When European settlers first colonised Australia they brought seeds and hoofed animals from Europe with them. They intended to farm here as they had always done at home; they undertook large-scale clearing of trees and shrubs and planted crops and pasture. However, the Australian landscape is very different to the land they had left behind. Australia's soils are naturally low in nutrients and have a poor structure. Much of the vegetation is shallow-rooted and easily disturbed when the land is ploughed and made ready for cultivation. Even in areas where the soil is fertile, over-irrigation and deforestation can raise the **watertable** and bring salt to the surface, decreasing soil fertility. Australia also has variable rainfall, and drought can last for years. This leaves the earth dry, parched, barren and unproductive. Floods can wash away a farmer's livelihood and leave the land flooded.

watertable upper level of groundwater; the level below which the earth is saturated with water

FIGURE 6 A former freshwater lake affected by salinity. The high salt levels have killed the native eucalypts; the smaller plants are more salt tolerant.

Kangaroo farming

Early European settlers knew they had to be self-sufficient, for their own survival and that of the new colony, but Australia's early economic growth and development also depended on agriculture. They farmed soil that was often hard, stony and exposed to a variety of climatic extremes. Overgrazing by heavy, hard-hoofed animals such as sheep and cattle increased the rate of land degradation, especially in arid and semi-arid regions. Kangaroo farming has been presented as an alternative sustainable solution to this problem.

Those in favour of kangaroo farming claim it would be more environmentally friendly as kangaroos are not hard-hoofed; issues such as soil compaction and vegetation trampling would be reduced. There could also be human health benefits, as kangaroo meat contains less fat and fewer calories than both lamb and beef. Those against the idea argue that because of various species characteristics, kangaroo farming is not commercially viable in the long term (see **FIGURE 7**).

FIGURE 7 Comparing commercial viability of kangaroo farming with sheep farming

- Young dependent on mother for 14 months
- Cannot be sold live
- One-off use (meat and skin)
- 18 months before meat can be harvested
- A 60-kilogram kangaroo yields 6 kilograms of prime meat; the rest is suitable only as pet food
- Can meet only 0.5 per cent of current needs

- Young dependent on mother for a few months
- Can be sold live
- Multiple uses (wool, meat and skin)
- Breed from 12 months; multiple births possible
- Meat can be harvested from 3–6 months
- Yields 20 kilograms of prime meat
- Easier to herd and care for

14.2.5 Spatial distribution of land degradation

In 1961 globally, there were 0.37 hectares of arable (productive) land available to grow food for every man, woman and child. By 2019, this figure had fallen to 0.19 hectares. **TABLE 1** shows changes to arable land availability over time for Australia and the world. These changes are due to factors such as population growth, urban sprawl, land degradation and climate change. In Australia, approximately two-thirds of the land used for agricultural production is degraded. **FIGURES 8** and **9** show the severity of land degradation in Australia and globally.

TABLE 1 Arable land per person over time

	Arable land per person (hectares)	
Year	Global	Australia
1961	0.37	2.88
1976	0.29	3.0
1991	0.23	2.64
2019	0.19	1.90

Source: World Bank data

According to the United Nations, around 42 per cent of the world's poorest people live on the most degraded lands. Areas where the land degradation is most rapid are also those where population growth is greatest. In sub-Saharan Africa, for example, the annual rate of population growth is 2.7 per cent annually — significantly higher than the world average of 1.15 per cent. Africa is one of the most vulnerable continents, having lost 65 per cent of its arable lands and 25 per cent of its overall land area to desert. Experts estimate that 10 million hectares of land needs to be rehabilitated each year to reverse the current trends in land degradation.

In 1950, about 65 per cent of the world's population lived in developing nations; however, this figure is expected to rise to 85 per cent by the year 2030. These people are dependent on the most fragile environment for their survival.

FIGURE 8 Severity of soil degradation in Australia

int-7953

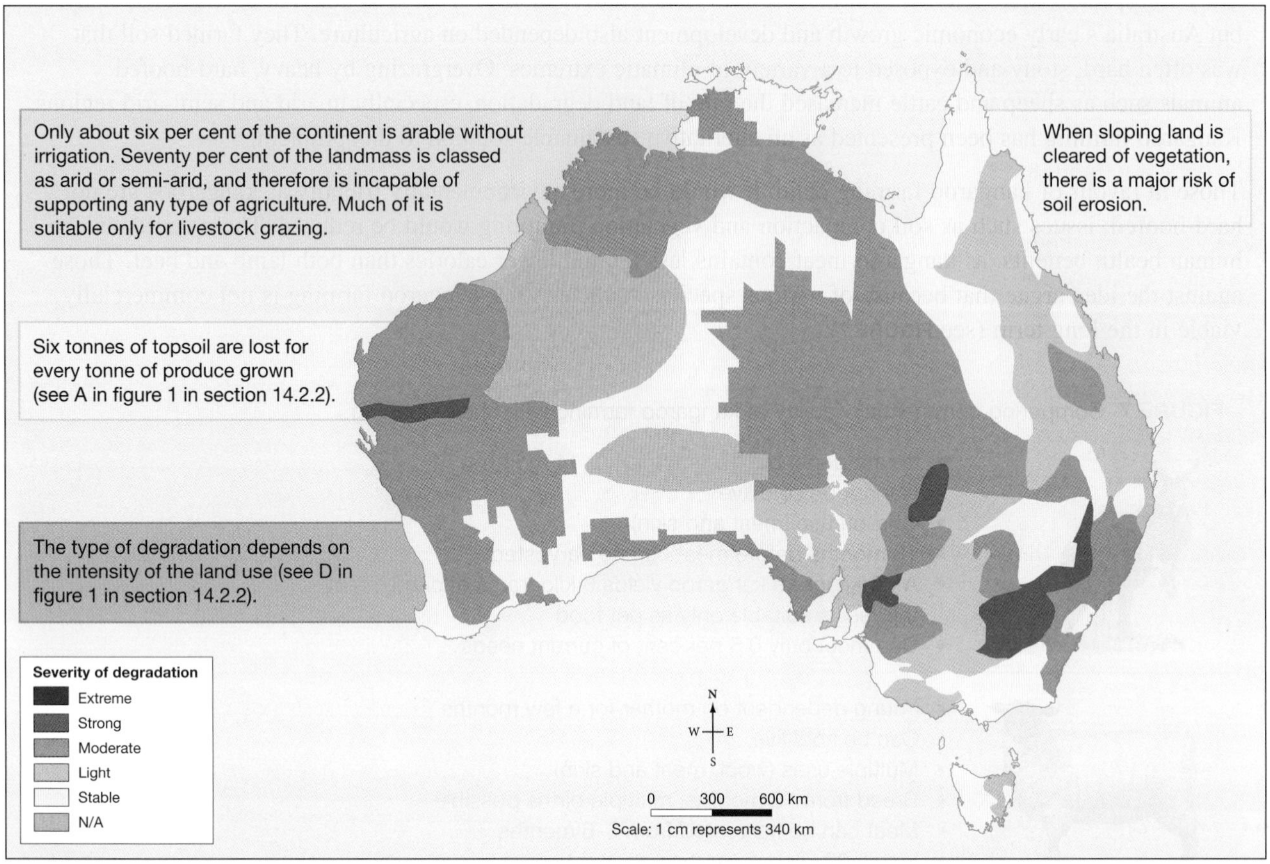

Source: ISRIC (1991)

FIGURE 9 Soil degradation is a global problem affecting every permanently inhabited continent.

int-5597

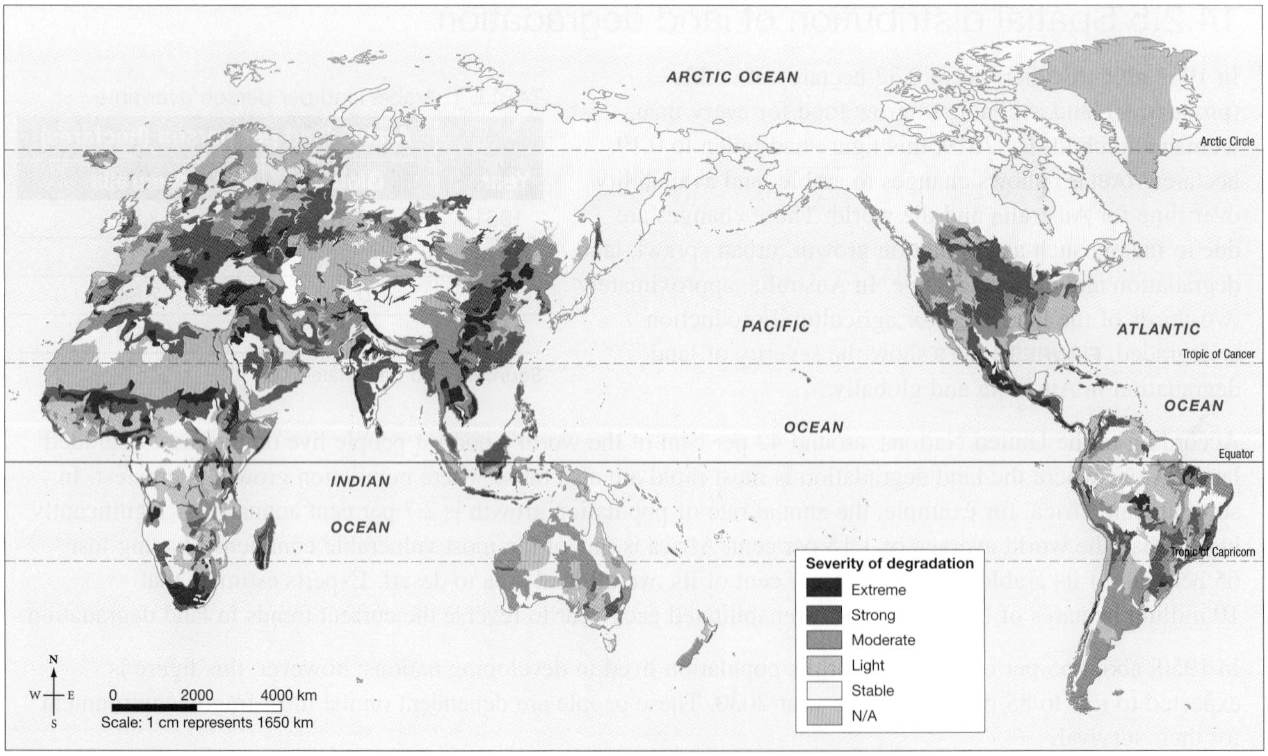

Source: ISRIC (1991)

14.2.6 Challenges to food production

Today, there are more than three times the number of people living on the Earth than in 1950: over 7.8 billion, compared with 2.5 billion. Our primary energy use is five times higher and our use of fertilisers has increased eightfold. In addition, the amount of nitrogen pumped into our oceans has quadrupled. While the global population is increasing, the land upon which food is grown to feed this population is degrading.

Everyone on Earth relies on the land. Apart from providing us with a place to live, the land also provides most of our food and products such as oil and timber. Since 1990, the amount of land irrigated for agriculture has doubled and agricultural production has trebled. With the world's population expected to reach 9.8 billion by 2050 and 11.2 billion by the end of the century, the land and its resources will be placed under even more pressure.

Globally, 75 per cent of the Earth's total land area is classed as degraded, and around 60 per cent of this degraded land is used for agricultural production. In 2019 alone, 24 billion tonnes of fertile soil was lost worldwide, with the worst affected areas being in sub-Saharan Africa. Global food production is already being undermined by land degradation and shortages of both farmland and water resources, making feeding the world's rising population even more daunting.

Land degradation is a global problem. If the current trends continue, our ability to feed a growing world population will be threatened. Although we have a better understanding of factors that contribute to land degradation, the challenge is to manage the land sustainably for the future and reverse the trends.

FIGURE 10 Sustainable land use will be important for ensuring the health of the land while increasing our capacity to feed the Earth's growing population.

Resources

eWorkbook The causes and impacts of land degradation (ewbk-10077)

Video eLesson The causes and impacts of land degradation — Key concepts (eles-5244)

Interactivities Destroying the land (int-3289)
Why land degrades (int-5591)
Soil degradation — a global problem affecting every permanently inhabited continent (int-5597)
Effects of land clearing and deforestation (int-7952)
Severity of soil degradation in Australia (int-7953)

14.2 ACTIVITIES

1. Investigate a particular type of land degradation and produce an annotated visual display to show the parts of Australia affected by it. Cover major contributing factors and possible management strategies. Add an inset diagram that examines a particular place, scale and rate of change associated with this type of degradation. Include your own recommendations for sustainable use of the environment.
2. a. In small groups, prepare a fold-out educational pamphlet outlining the damage caused by the waste produced by urban communities each year. Make sure you clearly outline the interconnection between human activity and environmental harm. Devise a strategy to reduce this waste and estimate the difference this would make to the amount of waste generated.
   b. In your group, evaluate your own and others' contributions to this group task, critiquing roles including leadership, and provide useful feedback to your peers. Evaluate your group's task achievement and make recommendations for improvements in relation to team goals.
3. Create an overlay theme map. Prepare a base map that shows the extent of land degradation around the world. Prepare an overlay map showing land use. Annotate your overlay with any similarities and differences between the two maps.
4. a. Investigate alternatives to traditional livestock farming of sheep and cattle, such as kangaroos or emus. Use the information presented in this subtopic as a starting point. Present a reasoned argument for or against this type of farming as a sustainable alternative.
   b. Evaluate emotional responses and the management of emotions in terms of this type of farming.
5. Create a presentation showing the different ways people use and manage the land in another country.
   a. Design a suitable symbol for land degradation and use this to highlight any uses you think might result in land degradation.
   b. Add annotations to explain how activities might degrade environments, and the scale of this change.
   c. Suggest a possible sustainable solution for each type of degradation identified.

14.2 EXERCISE

Learning pathways

■ LEVEL 1	■ LEVEL 2	■ LEVEL 3
1, 6, 8, 10, 13	2, 4, 7, 9, 14	3, 5, 11, 12, 15

Check your understanding

1. List the different ways in which the land can become degraded.
2. Outline the impact of land degradation on water resources.
3. Calculate the percentage by which arable land has decreased globally and in Australia since 1960.
4. Describe in your own words what land degradation is.
5. Outline why some people consider kangaroo farming to be one way of reducing land degradation in Australia.
6. Complete the follow sentences that summarise the land degradation in Australia:
   Only about _____ per cent of the continent is arable _____ irrigation. _____ per cent of the landmass is classed as _____or _____, and therefore is incapable of supporting any type of agriculture. Much of it is suitable only for _____ grazing.

Apply your understanding

7. Explain why land degradation is a current geographical issue.
8. Why were European farming methods unsuitable for the Australian environment?
9. Do you think land degradation is happening on a small or large scale? Explain.
10. Study **FIGURE 6**.
    a. In your own words, describe the damage that has occurred to the environment.
    b. Suggest how these changes have come about.
    c. How would you try to restore these places and manage their resources in a sustainable manner?
11. Analyse the economic benefits of farming sheep and kangaroos. Provide a reasoned argument supporting the benefits of one over the other that considers both environmental and economic factors.
12. Explain the interconnection between food security and land degradation.

Challenge your understanding

13. Land degradation is often the result of many little actions and events, the effects of which interact and build up over time. Identify some things that you do, consciously or unconsciously, that might be contributing to land degradation where you live. Explain the impact of these actions.
14. Describe an area or place that is near where you live, that you have visited recently or you have heard about in the media, and that you think is degraded. Give reasons for your choice and suggest how and why you think this degradation came about.
15. Predict the scale and severity of land degradation in Australia in 50 years. Give reasons for your prediction.

To answer questions online and to receive **immediate feedback** and **sample responses** for every question, go to your learnON title at www.jacplus.com.au.

14.3 SkillBuilder — Interpreting a complex block diagram

online only

LEARNING INTENTION

By the end of this subtopic, you will be able to describe the features and processes, and explain the interconnection between features and spaces in a complex block diagram.

The content in this subtopic is a summary of what you will find in the online resource.

14.3.1 Tell me

What is a complex block diagram?

A complex block diagram is a diagram that is made to appear three-dimensional. It shows a great deal of information about a number of aspects of a topic or location.

14.3.2 Show me

How to interpret a complex block diagram

Step 1

Read the title and identify the topic or location being studied.

Step 2

Examine the diagram, carefully reading any labelling that explains the topic being covered. Look for arrows or icons that indicate movement, change or processes.

Step 3

Check that you know what all of the terms in the diagram mean. Use a geographic dictionary or online research to understand the terms.

Step 4

Write a short paragraph explaining the processes at work in the diagram.

FIGURE 1 Saltbush Farm, land audit, 2012. Saltbush Farm is in the catchment of the Naangi River, a tributary of the Murray River.

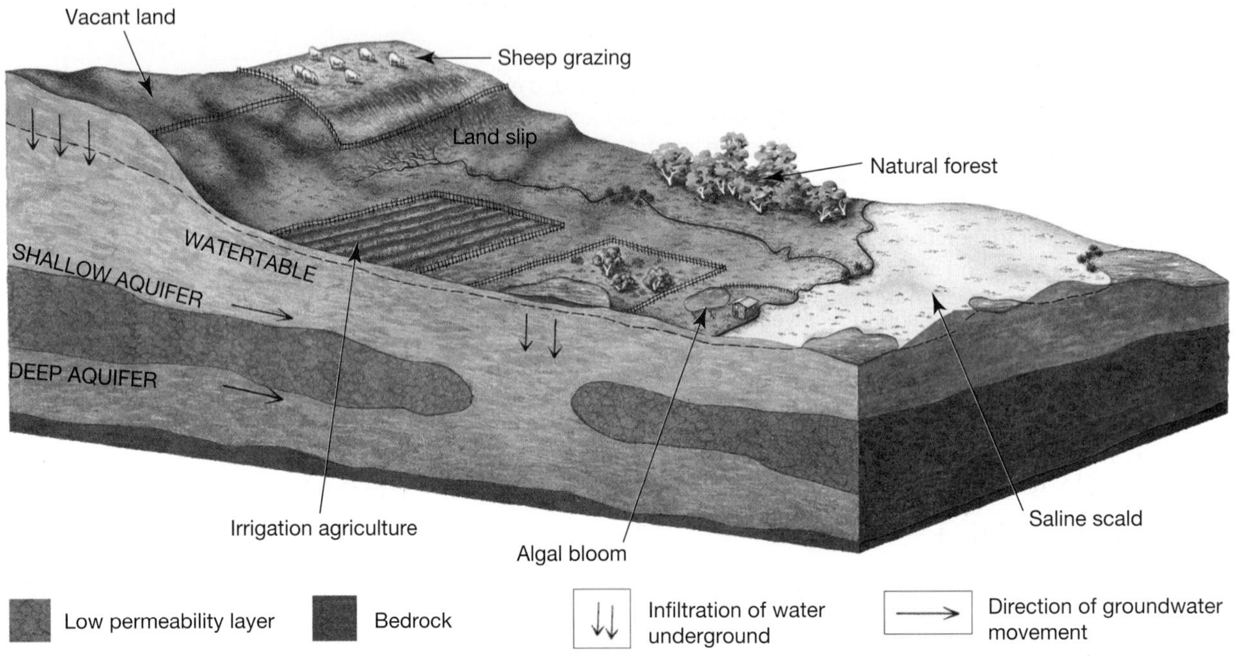

14.3.3 Let me do it

learnON

Go to learnON to access the following additional resources to help you build this skill:

- a longer explanation of this skill and its application in Geography (Tell me)
- a video demonstrating the step-by-step process of this skill (Show me)
- an activity and interactivity for you to practise the skill (Let me do it)
- self-marking questions to help you understand and use the skill.

on Resources

eWorkbook SkillBuilder — Interpreting a complex block diagram (ewbk-10081)

Video eLesson SkillBuilder — Interpreting a complex block diagram (eles-1746)

Interactivity SkillBuilder — Interpreting a complex block diagram (int-3364)

14.4 Managing land degradation

LEARNING INTENTION

By the end of this subtopic, you will be able to identify the causes and likely consequences of environmental change, and identify and explain solutions to an environmental issue.

14.4.1 The development and importance of soil

Soil is a mixture of broken-down rock particles, living organisms and **humus**. Over time, as surface rock breaks down through the process of **weathering** and mixes with organic material, a thin layer of soil develops and plants are able to take root (see **FIGURE 1**). These plants then attract animals and insects, and when these die their dead bodies decay, making the soil rich and thick.

Soil formation is a complex process brought about by the combination of time, climate, landscape and the availability of organic material. In some areas it takes hundreds of years to develop, while in others soil can form in a few decades.

The land is one of our most valuable resources. We depend on it for food, shelter, fibres and the oxygen we breathe. Yet the demands of an ever-increasing population place great pressure on it. To meet our needs, swamps and coastal marshes have been drained, vegetation removed and minerals extracted from the ground. Large-scale clearing and poor agricultural practices have left the land vulnerable. Although erosion is a natural process, farming land clearing and the construction of roads and buildings can accelerate the process.

humus decaying organic matter that is rich in nutrients needed for plant growth

weathering the breaking down of rocks

FIGURE 1 Wild flowers taking root in cracks in the rocks

14.4.2 Types of erosion

Sheet erosion

Sheet erosion (see **FIGURE 2**) occurs when water flowing as a flat sheet flows smoothly over a surface, removing a large, thin layer of topsoil. Sheet erosion might happen down a bare slope. It occurs when the amount of water is greater than the soil's ability to absorb it.

Strategies to combat this form of erosion include planting slopes with vegetation and adding **mulch** to the exposed soil so that it can absorb greater volumes of water. Another solution is to 'terrace' the landscape — to form the land into a series of steps rather than a steep slope.

FIGURE 2 Sheet erosion

Rill erosion

Rill erosion (see **FIGURE 3**) often accompanies sheet erosion, occurring where rapidly flowing sheets of water start to concentrate in small channels (or rills). These channels, less than 30 centimetres deep, are often seen in open agricultural areas. With successive downpours, rills can become deeper and wider, as fast-flowing water scours out and carries away more soil.

Strategies to combat rill erosion include tilling the soil (turning it over before planting crops) to slow the development of the rills. Building contours in the soil and planting a covering of grass can help slow the flow of water and hold the soil in place.

FIGURE 3 Rill erosion

Gully erosion

Gully erosion (see **FIGURE 4**) often starts as rill erosion. Over time, one or more rills may deepen and widen as successive flows of water carve deeper into the soil. Gully erosion may also start when a small opening in the surface such as a rabbit burrow or a pothole is opened up over time. Soil is often washed into rivers, dams and reservoirs, muddying the water and killing marine species. Large gullies need bridges or ramps to allow vehicles and livestock to cross.

Strategies to combat gully erosion largely involve stopping large water flows reaching the area at risk, through measures such as planting vegetation or crops to soak up the water. Other strategies include building diversion banks to channel the water away from the area, and constructing dams.

FIGURE 4 Gully erosion

mulch organic matter such as grass clippings

Tunnel erosion

Sometimes water will flow under the soil's surface — for example under dead tree roots or through rabbit burrows — carving out an underground passage or tunnel (see **FIGURE 5**). The roof of the tunnel may be thin and collapse under the weight of livestock or agricultural machinery. When these tunnels collapse they create a pothole or gully.

Strategies to combat tunnel erosion include planting vegetation both to absorb excess water and to break up its flow. Sometimes major earthworks are needed to repack the soil in badly affected areas.

FIGURE 5 Tunnel erosion

Wind erosion

When the surface of the land is bare of vegetation, the wind can pick up fine soil particles and blow them away (see **FIGURE 6**). It is more common during periods of drought or if the land has been overgrazed. The soil can be transported large distances and deposited in urban areas.

Strategies to combat wind erosion include planting bare areas with vegetation, mulching, planting wind breaks and avoiding overgrazing.

FIGURE 6 Wind erosion results when wind picks up and carries away fine soil particles. Did you know that soil from China has been deposited in the United States?

CASE STUDY

Managing land degradation in Costerfield, Victoria

Costerfield is around 100 kilometres north of Melbourne (see **FIGURE 7**). The landscape is characterised by gentle slopes with undulating (wavelike) pastures. It has an average annual rainfall of around 575 millimetres, but due to relatively high average evaporation rates, the climate is described as semi-arid or **Mediterranean**. Summers are hot and dry, and winters cool and wet. The area is also subject to climate extremes, so heavy rain, drought, frost and dust storms are not uncommon. Costerfield once had a dense covering of trees, predominantly eucalyptus. Native grasses also dominated the area. Soils in the area are generally considered to have a low **carrying capacity** for livestock because of poor fertility levels. Bushfires were a constant threat throughout the nineteenth century.

FIGURE 7 Costerfield lies approximately 100 kilometres north of Melbourne.

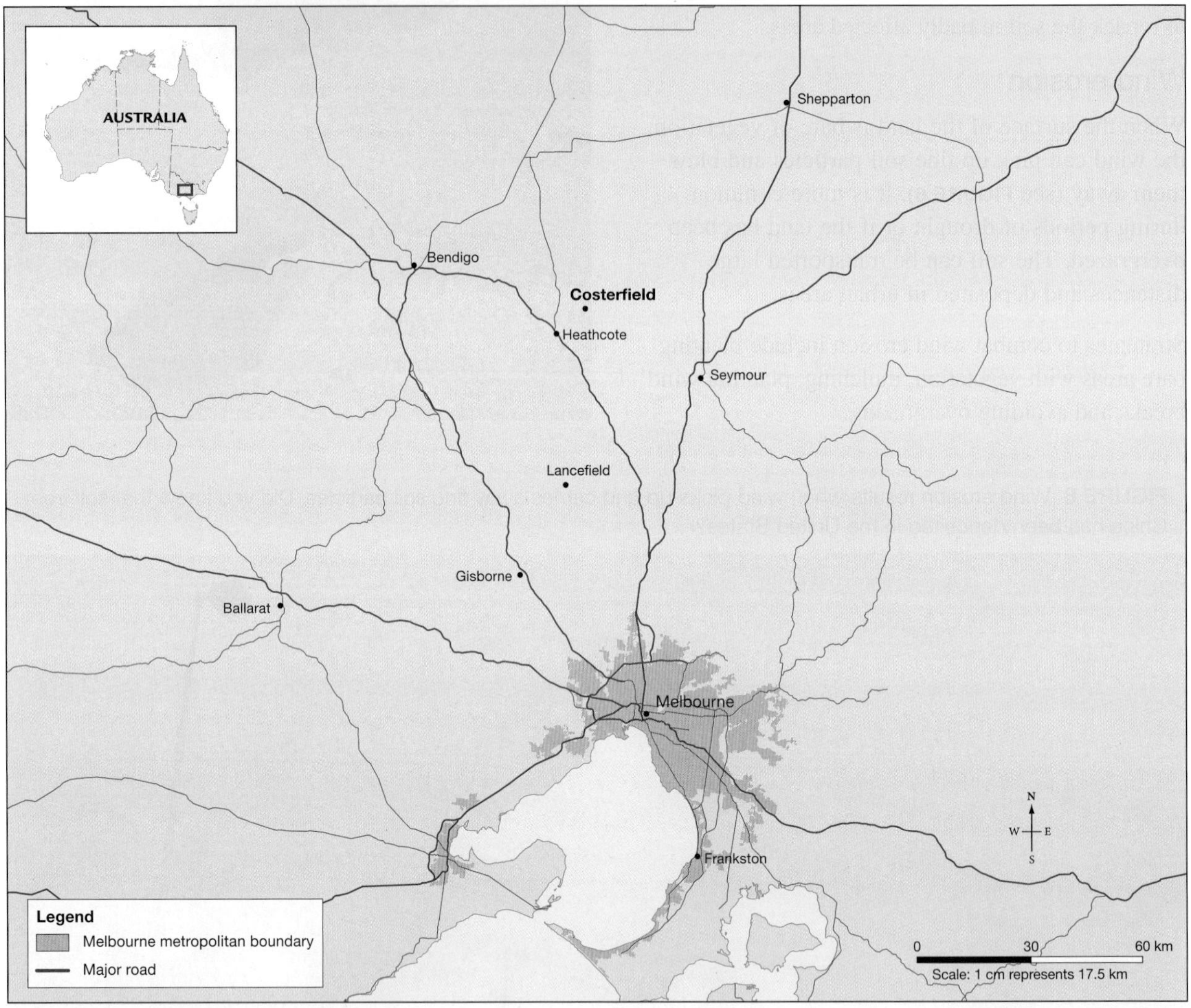

Source: © The State of Victoria, Department of Environment and Primary Industries 2013.

European settlements in the area

The first **pastoral run** was established in 1835, and the land was extensively cleared. European settlers introduced sheep, cattle, rabbits and foxes soon afterwards. Sheep grazing soon became the dominant activity in the region. The lack of native vegetation cover allowed rainfall to flow across the surface, eroding soil and making the run-off **turbid**. It also allowed rainfall to infiltrate the subsoil, leading to sheet, rill, gully and tunnel erosion (see **FIGURES 8** and **9**). The problem was further exacerbated by the rabbit population. Subdivision plans were drawn up in the early 1850s; however, in 1852 gold was discovered at McIvor Creek, which led to an influx of up to 40 000 gold prospectors in the region, causing the land to become even more degraded.

Mediterranean (climate) characterised by hot, dry summers and cool, wet winters

carrying capacity the ability of the land to support livestock

pastoral run an area or tract of land for grazing livestock

turbid water that contains sediment and is cloudy rather than clear

FIGURE 8 A dead tree stump or old fence post can allow water to infiltrate the soil.

Source: © State of Victoria Department of Environment and Primary Industries 2013. Reproduced with permission. Photograph by Stuart Boucher.

Following the gold rushes and into the twentieth century, the area around Costerfield was largely used for grazing livestock, predominantly sheep. However, there is evidence that both horses and cattle were raised in the region on a much smaller scale.

FIGURE 9 Notice that tunnel erosion forms where the surface of the land is bare. What do you think is likely to happen next?

Source: Stuart Boucher/Dept of Primary Industries Vic

Strategies to tackle erosion

It is much easier to prevent gully erosion than control it once it has developed. Without intervention, gullies can continue to become larger and larger. A number of measures were introduced by local Landcare groups to tackle the issues in the Costerfield area.

The gullies were stabilised by constructing banks, gully check dams (see **FIGURE 10**) and terracing, all aimed at reducing and redirecting run-off. Other strategies included:

- re-establishing ground cover, especially plants and grasses that are native to the region (see **FIGURE 11**)
- rabbit eradication programs to control the population and reduce burrowing activities that can create access points for run-off to enter the subsoil and promote development of new gullies. In addition, they protect the newly sown grasses from being eaten by the rabbits
- introduction of chemicals such as lime and gypsum to improve soil structure and pH levels to assist in the revegetation process
- protection of revegetated areas by preventing access, especially by livestock, during the restoration process.

FIGURE 10 Check dam construction water quality. Outlet pipes allow water to be redirected and control the flow of water across the landscape.

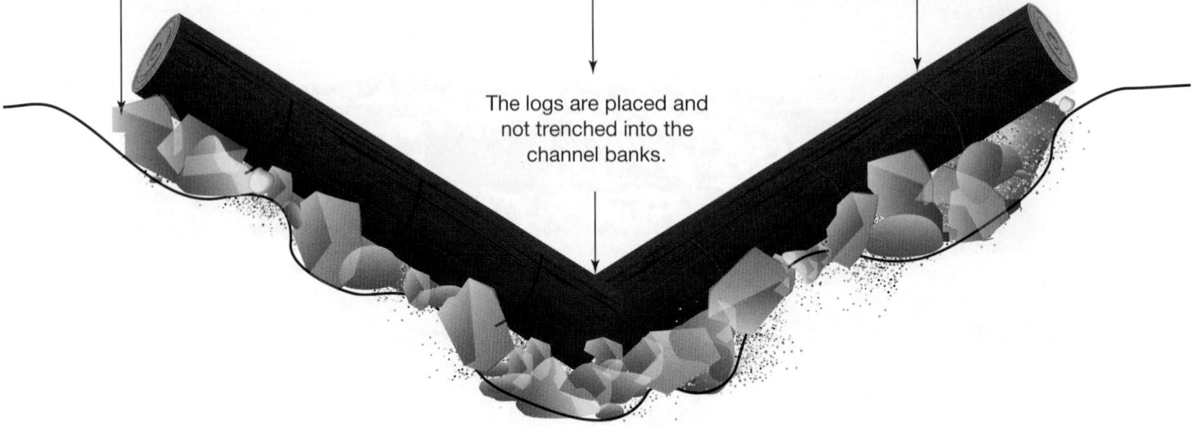
Cross-section view
Rocks, small branches, woody debris and needles are used to fill the gaps between the log and the ground.
Flow
Logs are anchored into place with a 1.2-metre-long reinforcing bar that is bent to conform to the logs and held into place using 5.1-centimetre fencing staples.
The logs are placed and not trenched into the channel banks.

Permanent check dams may be constructed using logs or stone. Sometimes they are lined to prevent seepage into the ground so that the water is trapped and can be used for irrigation purposes. They also trap sediment and prevent it being washed into waterways. Additionally, some are designed to trap nutrients and so help maintain water quality. Outlet pipes allow water to be redirected and control the flow of water across the landscape.

FIGURE 11 Revegetated area near Costerfield

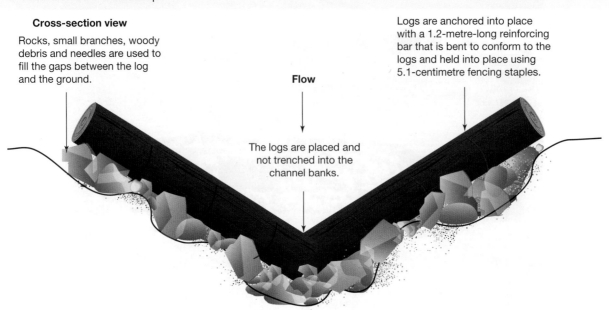

Source: Stuart Boucher/Dept of Primary Industries Vic

Resources

eWorkbook	Managing land degradation (ewbk-10085)
Video eLesson	Managing land degradation — Key concepts (eles-5245)
Interactivity	Down in the dirt (int-3290)

14.4 ACTIVITIES

1. Refer to **FIGURES 1** to **6**. With the aid of a flow diagram, show the interconnection between sheet, rill and gully erosion. Use the captions and the questions that appear with each image to help you.
2. Working with a partner, use the internet to investigate an international environment, such as the Dust Bowl in the United States or the Yellow River in China, that has been degraded because of soil erosion.
   a. Annotate a sketch of this environment to explain what has happened to the area. Include an inset sketch map that shows the location of this place, and describe the scale and rate of change.
   b. Swap your eroded environment sketch with another pair. For the other pair's sketch, devise a series of management strategies to rehabilitate the environment and allow it to be used in a sustainable manner. Swap back and add their strategies to your annotated sketch
3. Use your atlas to find a map showing vegetation in Victoria. Explain why wind erosion is more common in north-west Victoria than south-east Victoria.

14.4 EXERCISE

Learning pathways

■ LEVEL 1	■ LEVEL 2	■ LEVEL 3
1, 2, 7, 11, 13	4, 5, 8, 9, 14	3, 6, 10, 12, 15

Check your understanding

1. Describe each of the following types of erosion.
   a. Sheet
   b. Rill
   c. Gully
   d. Tunnel
   e. Wind
2. Identify the types of soil erosion that occurred in Costerfield.
3. Describe what is meant by the phrase 'making the run-off turbid'.
4. Define the term *pastoral run.*
5. Outline the steps you would take to construct a check dam.
6. Describe the features of a Mediterranean climate.

Apply your understanding

7. In your own words, explain how the use of a check dam might reduce the development of gullies.
8. Explain the interconnection between the removal of native vegetation and land degradation.
9. How would the arrival of gold prospectors increase land degradation? Explain the type of erosion that would most likely occur and the activities relating to gold mining that might cause these changes to develop.
10. Explain the interconnection between rill erosion and gully erosion.
11. An important part of any land management program is the control of introduced species such as rabbits. Explain why rabbits are a problem in areas where the land has been degraded.
12. Explain why it is easier to prevent than repair gully erosion.

Challenge your understanding

13. Suggest why soil erosion in all its forms is such a significant cause of land degradation.
14. Predict which of the strategies implemented at Costerfield would be the most successful in combating the erosion.
15. Would eradicating rabbits from Australia make a significant change in levels of land degradation? Give reasons for your prediction.

To answer questions online and to receive **immediate feedback** and **sample responses** for every question, go to your learnON title at www.jacplus.com.au.

14.5 Desertification

LEARNING INTENTION

By the end of this subtopic, you will be able to explain the causes of desertification and the areas most at risk, and discuss and evaluate strategies to deal with desertification.

14.5.1 Regions at risk of desertification

As the Earth's population increases, more pressure is placed on the land to provide both food and shelter. In many parts of the world this has meant that land has been overused and become exhausted. This is especially true in **dryland** regions. Many of these areas have become so degraded that they are at risk of being turned into desert, placing the survival and livelihood of the people who depend on them in jeopardy.

The United Nations estimates that approximately 41 per cent of the Earth's land surface is at risk of turning into desert. This is a process known as **desertification**, an extreme form of land degradation that affects arid, semi-arid and dry sub-humid areas of the Earth. These dryland regions often border existing deserts but, unlike deserts, they support population and agriculture. Drylands are fragile environments that degrade rapidly when the land is not carefully managed.

The areas affected are home to more than two billion people in 168 countries. **FIGURE 2** shows the Earth's desert areas and those places most at risk from desertification.

dryland ecosystems characterised by a lack of water. They include cultivated lands, scrublands, shrublands, grasslands, savannas and semi-deserts. The lack of water constrains the production of crops, wood and other ecosystem services.

desertification the transformation of land once suitable for agriculture into desert by processes such as climate change or human practices such as deforestation and overgrazing

FIGURE 1 Trees planted on the sand dunes of the Gobi Desert in Inner Mongolia to reduce desertification

Estimates predict that by 2025, without intervention, two-thirds of the arable lands in Africa will be lost, along with one-third in Asia and one-fifth in China. Based on current trends, Bangladesh will have no fertile soil available in 50 years. Desertification is a global issue as it is present on all continents in both developed and developing economies.

FIGURE 2 The drylands of the Earth are spreading. It is estimated that 12 million hectares of productive land (an area almost three times the size of Switzerland) is lost annually due to desertification.

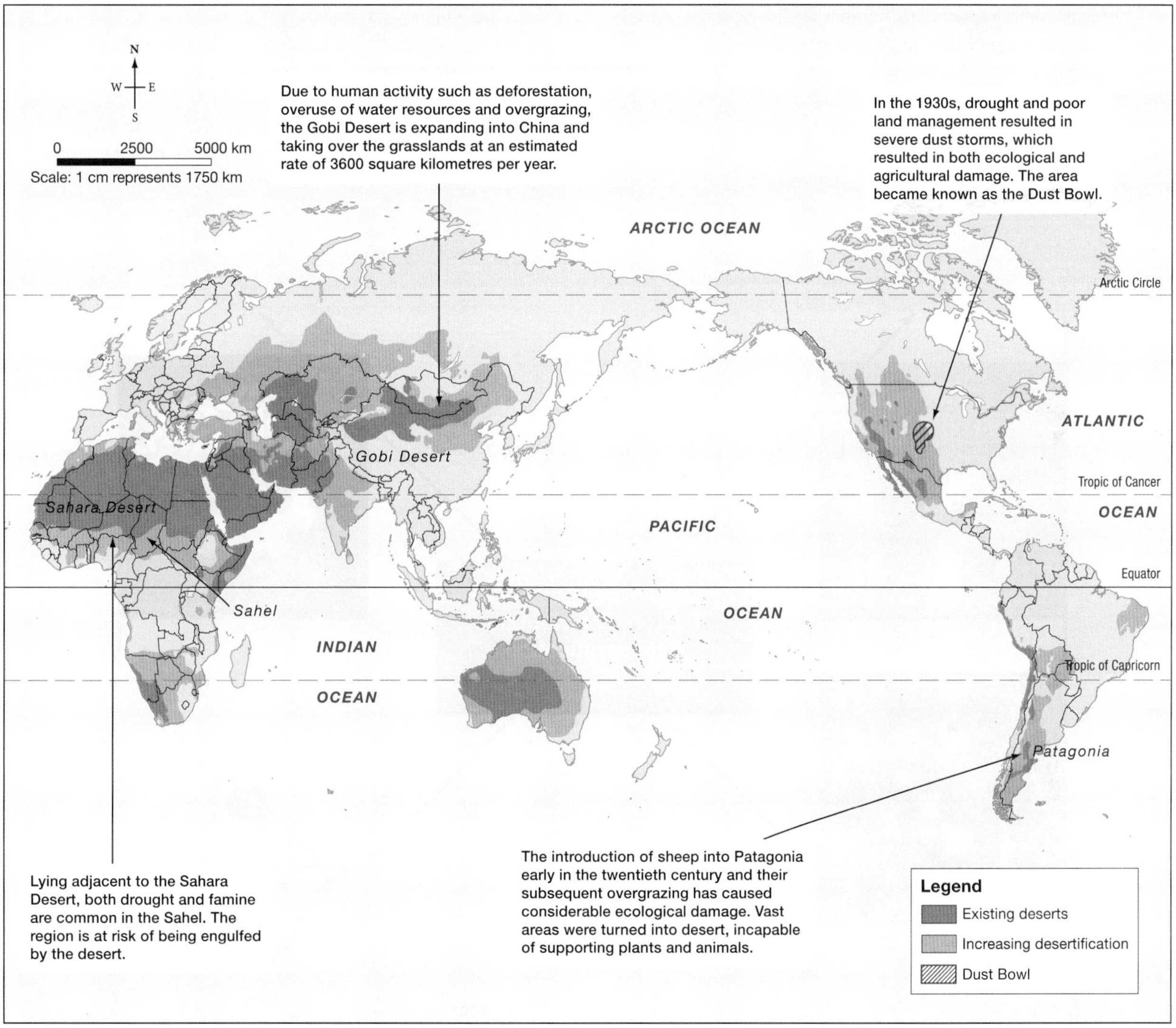

Source: UNEP World Conservation Monitoring Centre

14.5.2 Causes of desertification

Desertification is largely a result of human-induced environmental change, caused by the complex interconnection of environmental, political, cultural and economic factors. It generally arises from the poor management of dryland environments. Increasing populations, the demand for more agricultural production and overuse of the soil degrades the land to the extent that once productive places turn into wastelands (see **FIGURE 3**).

FIGURE 3 Factors contributing to desertification

Economic factors
- Overgrazing — desire to increase stock numbers to increase income leads to more animals than the land can cope with
- Overcropping to produce more food but not allowing time for soil to rest between crops
- Intensive farming depletes nutrients in the soil.
- Crops not suited to the environment require irrigation, using valuable, scarce water resources.
- Switching from cultivation to grazing, where more money can be made
- Clearing trees for sale as fuel wood and for construction

Environmental factors
- Low rainfall, frequent droughts and high evaporation dry out soil.
- Overgrazing leads to loss of vegetation and compacts the soil.
- Lack of vegetation exposes soil to evaporation and increases wind erosion.
- Often drylands are located in the rain shadow of mountain ranges and so experience lower rainfall and dry winds.
- Often poor quality marginal lands are used (e.g. steep slopes), which are not suitable for agriculture.

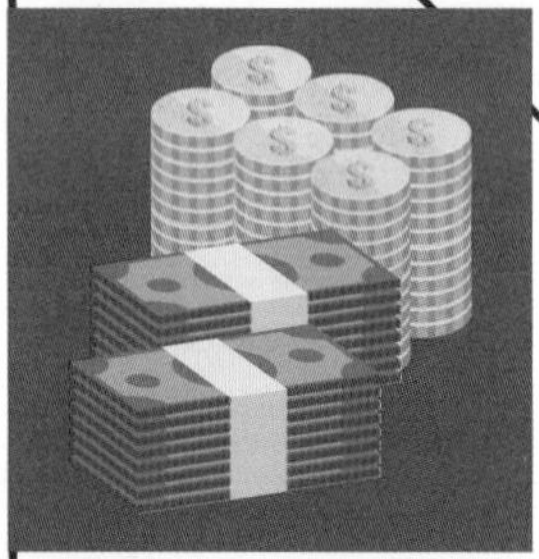

Factors contributing to desertification

Social factors
- Increase in population creates increased demand for more food.
- Lack of infrastructure, skills and knowledge to prevent land degradation
- Poor farming techniques
- Often wealth is measured by the number of livestock owned, so people breed larger herds.
- Wood is often the main source of fuel for cooking and heating, leading to large-scale deforestation.
- During periods of drought, increasingly marginal land is farmed to produce food. The tilled soil may be blown away, making the effects of the drought worse.

Political factors
- Control of political borders, conflict and expansion of agricultural and urban areas reduces the range of nomadic pastoralists, increasing the pressure on remaining grasslands.
- Governments and aid agencies often construct permanent water wells for nomadic grazing; however, these tend to promote increased herd sizes, which create land degradation around the wells.
- Governments encourage multinational agricultural companies to farm more intensively in order to alleviate poverty and create employment opportunities. In dryland regions, this approach may not be sustainable over time.

14.5.3 The impacts of desertification

Currently, the world loses approximately 12 million hectares of land annually, an area almost three times the size of Switzerland, enough to grow 20 million tonnes of grain. The cost to global economies is estimated to be $490 billion per annum.

Desertification brings about environmental change as the loss of topsoil and protective vegetation enables desert sand dunes to migrate and smother former farmland (see **FIGURE 4**).

FIGURE 4 Fence drowned by a huge sand dune in the United Arab Emirates

Desertification also affects the wellbeing of over one billion people in the world. While poverty can contribute to desertification, it is also a consequence of it, as poverty forces people to over-exploit the land, which can then accelerate land degradation. It can also increase the risk of food insecurity as food production decreases. As the land fails, social and cultural networks become lost, as whole villages can effectively be abandoned when people leave farming in search of employment in urban areas.

14.5.4 Tackling the problem

Desertification, climate change and the loss of biodiversity were identified as the greatest challenges to sustainable development during the 1992 Rio Earth Summit. As a result of this, the United Nations developed the United Nations Convention to Combat Desertification (UNCCD), an agreement supported by 197 parties with the aims of:

- improving living conditions for people living in drylands
- maintaining and restoring land and soil productivity
- reducing the impacts of drought.

This worldview encourages cooperation in exchanging knowledge and technology between developed and developing countries and promotes the idea of a 'bottom-up' approach to a problem. This means encouraging and supporting people to develop their own solutions rather than a government-led 'top-down' approach.

CASE STUDY 1

Combating desertification in China

China is one of the countries most severely affected by desertification; it affects over 30 per cent of the total land area (approximately 3327 million km^2) there and has a negative impact on 400 million people (see **FIGURE 6**).

In the wake of rapid population growth (from 550 million in 1950 to almost 1.44 billion in 2020), the demand for food, fuel, construction timber and livestock feed surged. With a viewpoint of 'growth at all costs', more farmland was opened up on desert fringes and the number of livestock increased exponentially. This expansion was done without any consideration of the environmental impacts. Thus, human activities in the form of inappropriate land use have magnified the problem of desertification (see **FIGURE 5**).

FIGURE 5 Causes of desertification in China

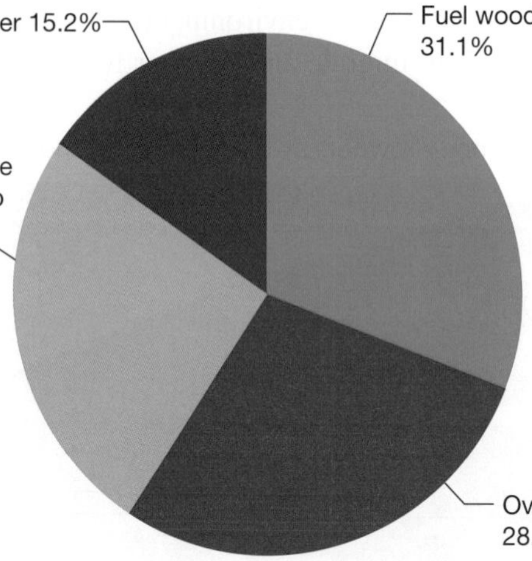

Source: FAO

int-7954

FIGURE 6 Vulnerability to desertification in China

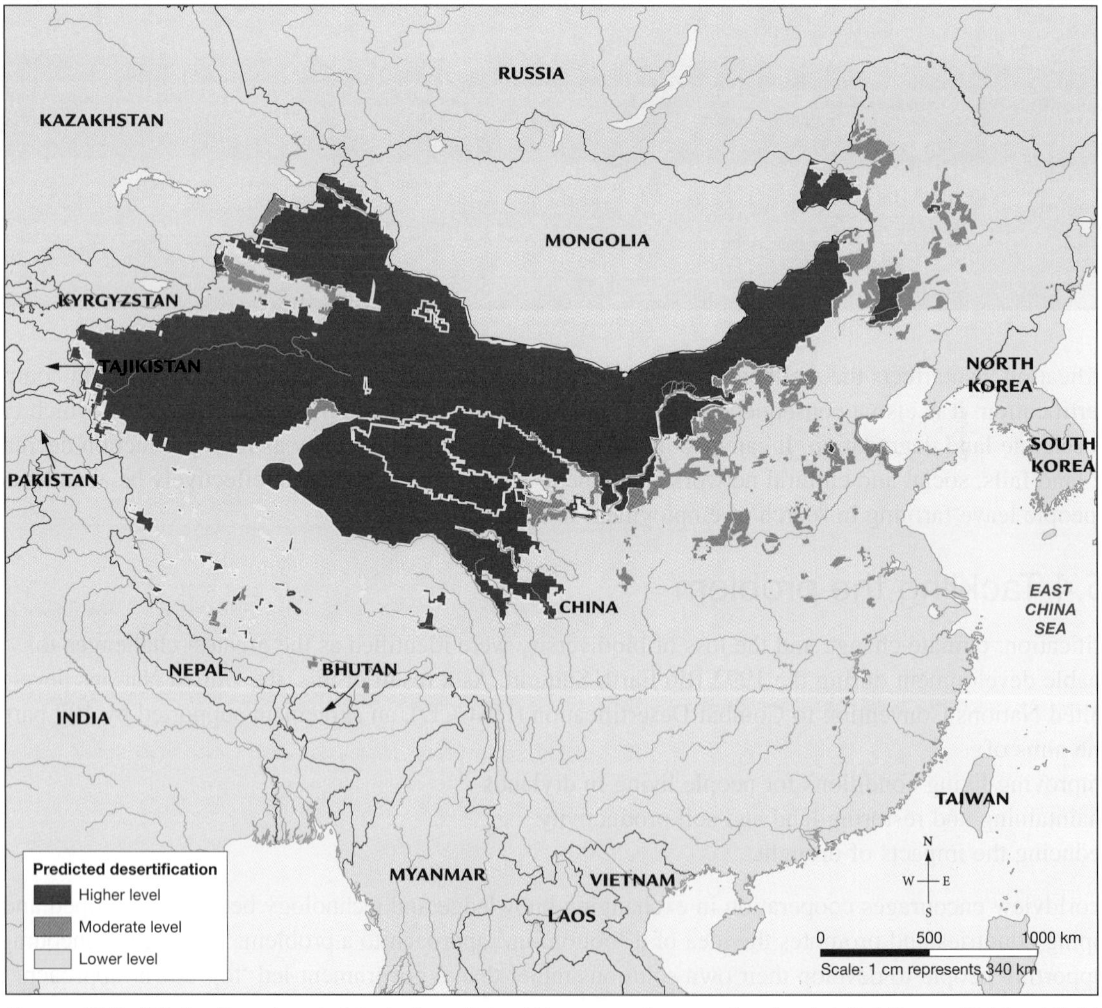

Source: United States Department of Agriculture, Natural Resources Conservation Service, Soil Survey Division, World Soil Resources; Paul Reich, Geographer. 1998. Global Desertification Vulnerability Map. Washington, D.C.

Each year, the Gobi Desert in Mongolia swallows 360 000 hectares of grasslands, and dust storms remove 2000 square kilometres of topsoil. Sand and dust from the desert regions of China is carried eastwards by the prevailing winds, choking the city of Beijing: destroying crops, closing airports and creating a surge in respiratory ailments. The sand storms, or 'Yellow Dragon' as they were traditionally called, continue their journey and affect international communities in South Korea, Japan, Russia and even the United States (see **FIGURE 7**).

The Chinese government has been working relentlessly on the problem, implementing a range of schemes, two of which are described in section 13.6.

China currently spends $5 billion each year on combating desertification. The aim is to reclaim all the treatable land area by 2050. Already China has slowed the rate of desertification by more than one-third of 1999 levels, making it a world leader in this field.

FIGURE 7 A dust storm from China affecting neighbouring countries

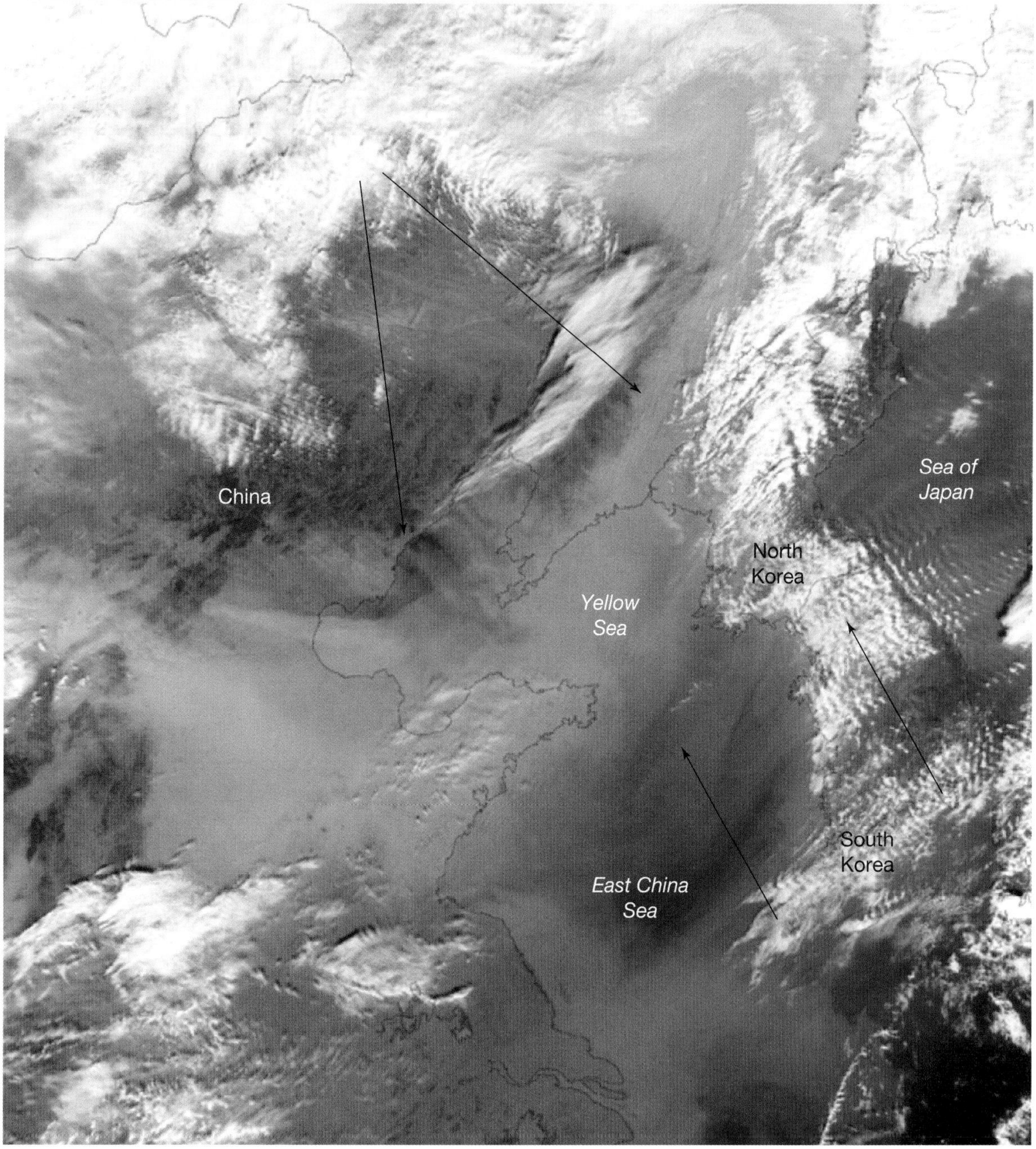

CASE STUDY 2

The Great Green Wall of China

To halt the spread of deserts and reduce the impacts of climate change, the Chinese government embarked on a plan to create the Great Green Wall of China. Green walls are ambitious initiatives designed to act as a barrier against desert winds and prevent desertification. Both the **Sahel** region in Africa and China have embarked on massive replanting projects that are expected to reduce erosion, enhance biodiversity, provide new grazing lands, boost agriculture and provide employment (see **FIGURE 8**).

FIGURE 8 The Great Green Wall of China: an ambitious attempt to stop the advance of desert sands from the Gobi Desert

The Chinese government envisaged a 4800-kilometre series of forest strips spanning the country from east to west, and stretching 1500 kilometres from the southern edge of the Gobi Desert, to protect valuable farmland and waterways against wind erosion. To make this target a reality, every citizen over the age of 11 was expected to plant at least three saplings each year. Since the start of the millennium, Chinese citizens have planted over 66 billion trees. By 2016, the Chinese State Forestry Commission had succeeded in creating almost 30 million hectares of forest, with the goal of reaching a national forest cover of 42 per cent by 2050. However, an early study done by geographers at the University of Alabama noted that 'the reforestation efforts have done little to abate China's great yellow dust storms' (see **TABLE 1**).

Sahel a semi-arid region in sub-Saharan Africa. It is a transition zone between the Sahara Desert to the north and the wetter tropical regions to the south. It stretches across the continent, west from Senegal to Ethiopia in the east, crossing 11 borders.

TABLE 1 Impacts of planting green walls

Environmental benefits	Environmental drawbacks
• Mass planting of fast-growing species (known as monoculture) helps to slow desertification. • Trees act as windbreak and reduce erosion. • There is a long-term possibility of harvesting trees as a commercial wood crop or for pulp and paper. • Growth of trees acts as a carbon store, reducing greenhouse gases.	• Monoculture reduces biodiversity and provides poor habitat for endangered native animal and bird species. • Monoculture is highly susceptible to disease. A pest can wipe out an entire plantation, ruining decades of work. • Many tree species chosen were not native and after initial growth soon died. In some places up to 85 per cent of the plantings failed. • Initial rapid growth of trees used a lot of soil moisture and lowered watertables. • Trees out-competed native grasses, which have a more extensive root system for holding soil. • Plantations generate less leaf litter than native forests, so less nutrients are entering the soil.

Although China reports an overall increase in forested areas, from 5 per cent to 12 per cent, Greenpeace reports that only 2 per cent of China's original vegetation remains. Many of the trees planted have a lifespan of only 40 years. Nevertheless, in areas where the local community is prepared to care for the newly planted trees, the spread of the desert appears to have been halted, with the area of land affected by desertification shrinking by almost 2000 square kilometres annually. Sandstorms are now also reported to have decreased by 20 per cent.

As one Chinese ecologist, Jian Gaoming, has stated, there is a need for 'nurturing the land by the land itself'. This is an Earth-centred approach to the problem of desertification. He noted through his research in Inner Mongolia that native grasslands will restore themselves in as little as two years, if protected from grazing animals and human activities.

Restoring grasslands

It is estimated that 80 per cent of China's natural grasslands (42 per cent of its land area) are degraded as a result of overgrazing. A wide range of rehabilitation programs are being introduced. These include:

FIGURE 9 Nomadic grazing on grasslands in Mongolia.

- moving people. In places especially at risk of desertification, people are being resettled to prevent further damage and halt the spread of the desert. Relentless sandstorms threaten the traditional lifestyles and farming practices of nomads in both Tibet and Mongolia. Failing crops and a lack of pasture for grazing livestock are forcing them to join other climate refugees and move into new settlements (see **FIGURE 9**).
- changing land use. Land use is converted from grazing to tree crops and forests, with farmers receiving compensation for the loss of stock and income.
- total grazing bans. Between 2005 and 2010, a total ban was placed on animal grazing on 7 million hectares of land (an area twice the size of Germany). This was part of a larger plan to restore more than 660 million hectares of grasslands at an estimated cost of approximately A$4 billion. This has meant that more than 20 million animals had to be farmed indoors and hand fed. In test projects, after three years of grazing bans the vegetation rate increased from 20 per cent to over 60 per cent, and local sand storms have reduced. The grazing ban has since been extended, with farmers paid a subsidy to safeguard their livelihood. Additional bans were also put in place banning hunting and declaring some areas national parks. Money received by the traditional nomadic herders has been used to leave the land.

on Resources

eWorkbook	Desertification (ewbk-10093)
Video eLesson	Desertification — Key concepts (eles-5246)
Interactivities	Vulnerability to desertification in China (int-7954) Desertification (int-8672)
myWorldAtlas	Deepen your understanding of this topic with related case studies and questions. • **Desertification in Mauritania**

14.5 ACTIVITIES

1. Investigate one of the causes of desertification outlined in **FIGURE 3** and write a news report that explains the interconnection between this factor and environmental change. In your report include the following:
   a. a description of the impact of this factor over space
   b. an example of a place that has been changed as a result of this factor
   c. the scale of this change
   d. a strategy for the sustainable management of the environment to combat this factor.
2. Use the **Great Green Wall of China (1)** and **(2)** weblinks in your online Resources to find out more about the Great Green Wall of China. With a partner, discuss the actions taken. Do you think the plan will succeed? Why or why not?

14.5 EXERCISE

Learning pathways

■ LEVEL 1	■ LEVEL 2	■ LEVEL 3
1, 2, 8, 9, 15	3, 4, 6, 10, 11, 13	5, 7, 12, 14

Check your understanding

1. Define the term *monoculture*.
2. List two factors that contribute to desertification for each of the following categories.
   a. Economic
   b. Environmental
   c. Political
   d. Social
3. What is the biggest human-induced cause of desertification in China?
4. What is the difference between deserts and drylands?
5. Why are drylands especially vulnerable to desertification?
6. Refer to **FIGURE 2**. Describe the distribution of those places in the world most at risk of desertification.
7. Identify an economic, social and environmental impact of desertification in China.

Apply your understanding

8. Explain in your own words what is meant by desertification and why it is a global issue.
9. Explain the difference between a 'top-down' and a 'bottom-up' approach to resolving a problem.
10. How effective do you think a top-down approach can be in combating desertification in China? How effective do you think a bottom-up approach might be? Explain your view.
11. Evaluate, according to environmental, social and economic impacts, the effectiveness of:
    a. green walls
    b. grazing bans for combating desertification in China.
12. Explain why the Sahel is under threat from desertification.

Challenge your understanding

13. How would you envisage the issue of desertification in China in the year 2050? Give reasons for your answer.
14. The Chinese ecologist Jian Gaoming's viewpoint on managing desertification is 'nurturing the land by the land itself'. How does this Earth-centred viewpoint compare to the Green Wall scheme, which is a human-centred viewpoint? Which do you think is the more sustainable approach? Outline and justify your views.
15. Consider the perspectives of the nomadic herders who can no longer live their traditional lifestyles. What might their response be to the strategies being employed by the Chinese government to protect against desertification? Consider their view from political, economic, social and environmental perspectives.

To answer questions online and to receive **immediate feedback** and **sample responses** for every question, go to your learnON title at www.jacplus.com.au.

14.6 Environmental change and salinity

LEARNING INTENTION

By the end of this subtopic, you will be able to:
- explain the causes of salinity
- describe why it is a serious environmental issue
- identify possible solutions to issues related to salinity.

14.6.1 Understanding salinity

Salinity is not a new problem. In fact, it was an environmental issue for the earliest civilisations some 6000 years ago. Historical records indicate that the Sumerians, who farmed the land between the Tigris and Euphrates rivers in the area known as Mesopotamia, ruined their land as a result of their poorly managed irrigation practices.

FIGURE 1 Salinity distribution

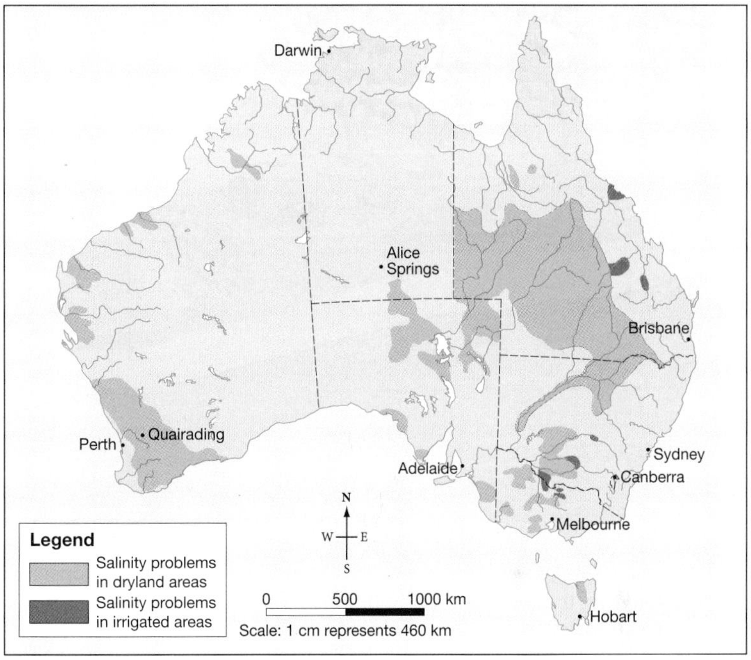

Source: Spatial Vision

Salt has become a major contributor to land degradation in Australia. Rising from below the land surface, it is destroying native vegetation and threatening the livelihood of many Australians. As plants die as a result of salinity, other problems emerge: the soil no longer has a protective cover of vegetation, which means it is more easily blown away or eroded. Some 140 million years ago, parts of the Australian continent were covered by shallow seas and saltwater lakes. The salt stores from these waters have lain dormant below the surface of the land, much of them in the **groundwater**. In addition, salt continues to be deposited on the land's surface by rain and winds blowing in from the oceans, and by the weathering of mineral-carrying rocks.

Australia's native vegetation had built up some tolerance to the salt levels in the soil. The deep-rooted vegetation also soaked up water in the soil before it could seep down into the groundwater. This meant that the watertable stayed at a fairly constant level, and that the concentrated salt stores stayed where they were. This natural balance changed with the arrival of European settlers. The farming and land-clearing practices they introduced were, and still are according to many experts, unsuited to Australia's generally harsh, dry climate, as well as to its geological history.

Salt has now become a serious problem. There are two ways in which the soil can become too salty: dryland salinity and irrigation salinity. The areas in Australia affected by salinity are shown in **FIGURE 1**.

salinity an excess of salt in soil or water, making it less useful for agriculture

groundwater water held underground within water-bearing rocks or aquifers

14.6.2 Dryland salinity

Dryland salinity occurs in areas that are not irrigated. When settlers cleared the land, they replaced deep-rooted native vegetation with crop and pasture plants. These plants generally have shorter roots and cannot soak up as much rainfall as native vegetation. Excess moisture seeped down into the groundwater, raising the watertable and bringing concentrated saline water into direct contact with plant roots (see **FIGURE 2**). Vegetation, even salt-tolerant plants, started dying as the salt concentrations rose. Once the vegetation dies off, the soil is left bare and is prone to erosion. Often layers of salt, known as **salt scald**, are visible on the surface of the land.

FIGURE 2 The effects of a rising watertable

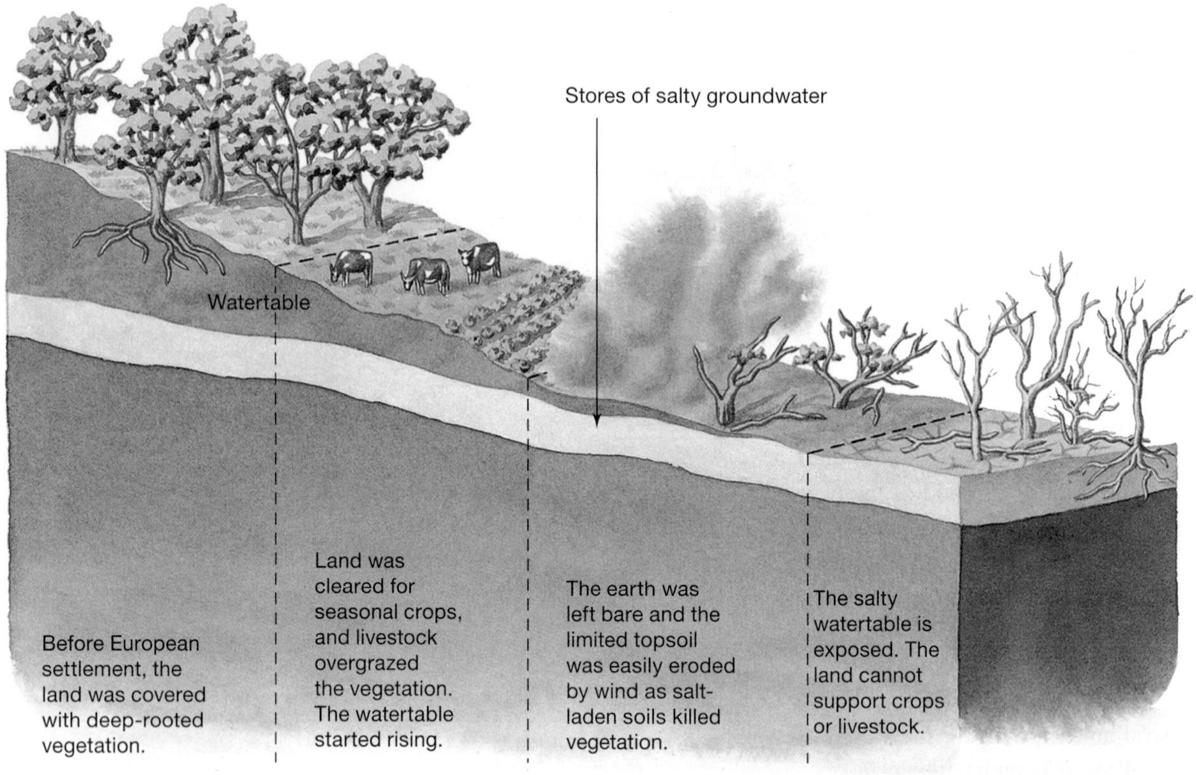
Stores of salty groundwater
Watertable
Before European settlement, the land was covered with deep-rooted vegetation.
Land was cleared for seasonal crops, and livestock overgrazed the vegetation. The watertable started rising.
The earth was left bare and the limited topsoil was easily eroded by wind as salt-laden soils killed vegetation.
The salty watertable is exposed. The land cannot support crops or livestock.

14.6.3 Irrigation salinity

One-third of the world's food is produced on irrigated land. Irrigation salinity occurs in irrigated regions and is a direct result of overwatering. When more water is applied to crops or pasture plants than they can soak up, the excess water seeps down through the soil into the groundwater, causing the salty watertable to rise to the surface. Some of this salt is washed into rivers, either as run-off or groundwater seepage, and transported to other places.

How much land around the world is affected by salinity?

- Africa: 2 per cent of Africa's landmass
- China: 21 per cent of arid lands or around 30 million hectares of land
- Western Europe: 10 per cent of the land area
- United States: Land across 17 states
- South America: Most countries have areas of land affected.
- Australia: 2.5 million hectares
- Worldwide: Estimated that 10 million hectares of arable land succumb to the effects of irrigation-related salinity each year and that, without intervention, the affected area might triple by 2050.

salt scald the visible presence of salt crystals on the surface of the land, giving it a crust-like appearance

14.6.4 Salinity management strategies

Many programs are in place to identify and monitor problem areas. Action being taken includes:

- changing irrigation practices to reduce overwatering
- planting deep-rooted native trees and shrubs in open areas
- developing new crops that are more salt tolerant, such as new strains of wheat
- replacing introduced pasture grasses with native vegetation such as saltbush (see **FIGURE 3**)
- using satellite technology to map areas at risk to enable early intervention.

FIGURE 3 Native plants such as saltbush help solve the problem of salinity on Australian grazing lands.

CASE STUDY 1

Salinity in the Murray–Darling Basin

The Murray–Darling Basin is Australia's largest **drainage area**. Extending across parts of four states and the entire Australian Capital Territory, it contains the country's three longest rivers: the Murray (2508 kilometres), the Darling (2740 kilometres when including its three main tributaries) and the Murrumbidgee (1690 kilometres). It is also one of the country's most significant agricultural regions, producing close to 45 per cent of the nation's food. Because it receives very little rainfall, the area depends heavily on irrigation. In fact, 70 per cent of Australia's irrigation occurs here.

FIGURE 4 The earliest signs of salinity: the watertable has risen, bringing salt to the root zone.

Over time, human activities in the Murray–Darling Basin have increasingly threatened the basin's **ecology** (see **FIGURE 4**). These activities have included introducing non-native plant and animal species, changing the natural flow of the river for irrigation purposes, clearing the land and over-watering crops. It is the last two activities that have particularly contributed to the region's salinity. (Although, in 1829 when explorer Charles Sturt discovered the Darling River during the dry season, he observed that the water was too salty to drink.) It has been estimated that by 2050, 1.3 million hectares (or 93 per cent) of land in the region could be salt affected.

drainage area (or basin) an area drained by a river and its tributaries

ecology the environment as it relates to living organisms

Tackling the problems

Over the years, a range of strategies have been investigated to better manage the Murray–Darling Basin and reduce salinity problems. Strategies have included the development of action plans such as revegetation programs and educational programs. In 2008, the Commonwealth Government took control of the region to allow for the implementation of a comprehensive management strategy that would provide for the needs of all states and also be environmentally sustainable. The Murray–Darling Basin Authority (MDBA) was created to oversee the entire project.

As part of the measures to control salinity, salt interception schemes have been established along the Murray River (see **FIGURE 5**). Collectively, they remove 500 000 tonnes of salt annually from groundwater and drainage basins (see **FIGURE 6**). Before these schemes were established, the Murray River carried huge amounts of salt — for example 250 tonnes per day and 100 tonnes a day, respectively, past Woolpunda and Waikerie, between Renmark and Morgan in South Australia. Recent surveys show that salinity levels have decreased to less than 10 tonnes a day in each area.

FIGURE 5 Salt interception in the Murray–Darling Basin

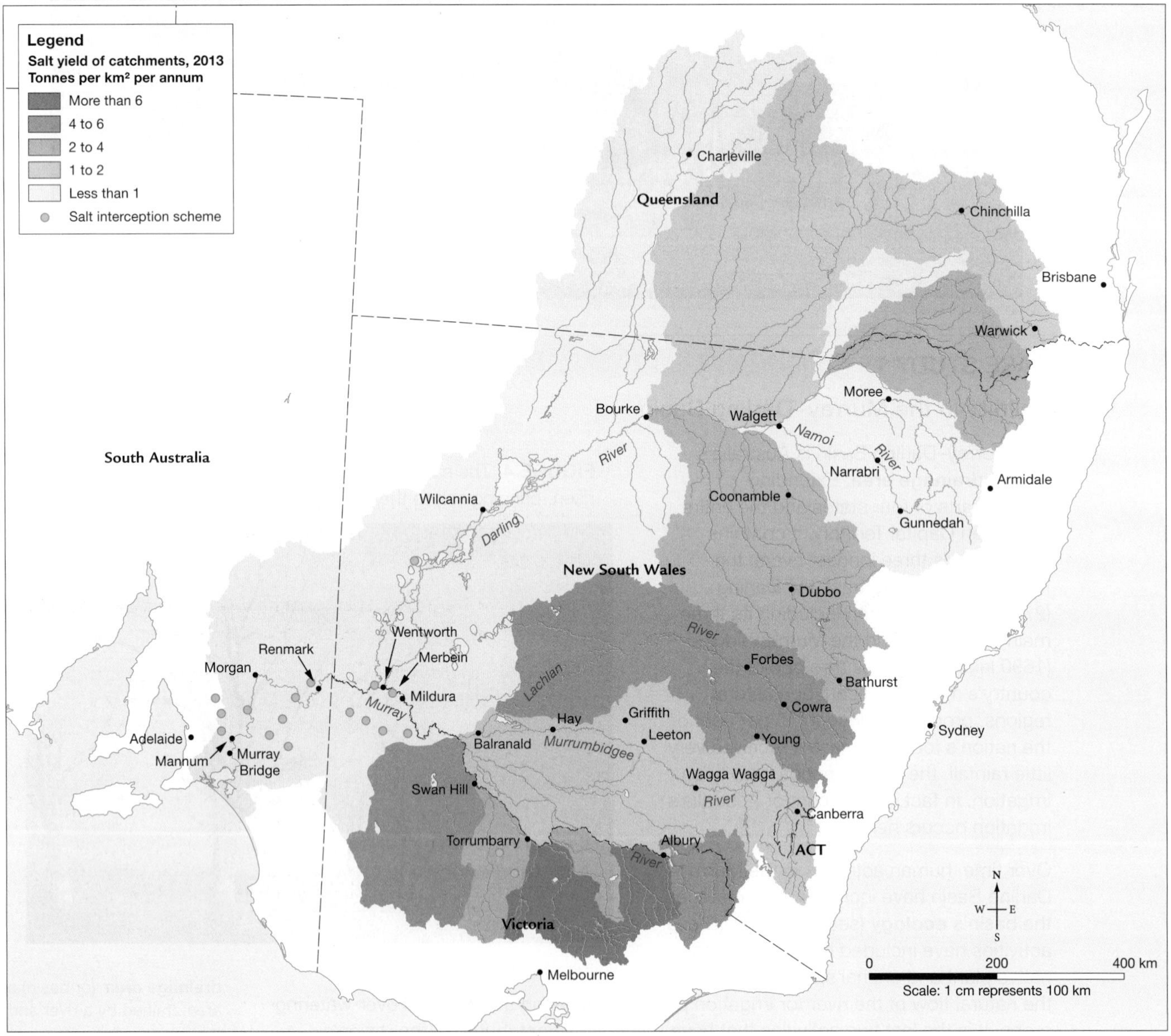

Source: © Commonwealth of Australia Geoscience Australia 2013. Map by Spatial Vision.

FIGURE 6 Salt harvested from evaporation ponds is exported all around the world.

Murray–Darling Basin Royal Commission 2018

In 2018, a report by the ABC program Four Corners levelled serious allegations of inadequate management, negligence and water theft in the Murray–Darling Basin. Following this, the South Australian Government ordered a **Royal Commission** into the management of the Basin.

The Royal Commission found that the existing management strategies were in need of a complete overhaul to ensure that more water was diverted from irrigation and back into the environment.

The report was also highly critical of the Menindee Lakes project (in NSW), which involved shrinking and emptying the lakes more often to save water from evaporation — an action that failed to take into account the impact not only on the environment when river flow stopped, but also on those further downstream in Victoria and South Australia.

The report argued the aim of the MDBA should be to share limited water resources and ensure that the needs of the environment, agriculture, Indigenous communities and 2.6 million people who depend on the Murray–Darling River system to supply their water were considered — especially those who live downstream and all the way to the mouth of the river where it empties into the sea. The sustainable management of this vital water resource remains an ongoing challenge.

Royal Commission a public judicial inquiry into an important issue, with powers to make recommendations to government

CASE STUDY 2

Vietnam — adapting to salinity issues

Vietnam's Mekong Delta region (see **FIGURE 7**) is a major exporter of both rice and shrimp. Drought and the early arrival of the dry season, which is being attributed to climate change, is allowing sea water to encroach on valuable farming land.

FIGURE 7 The Mekong Delta

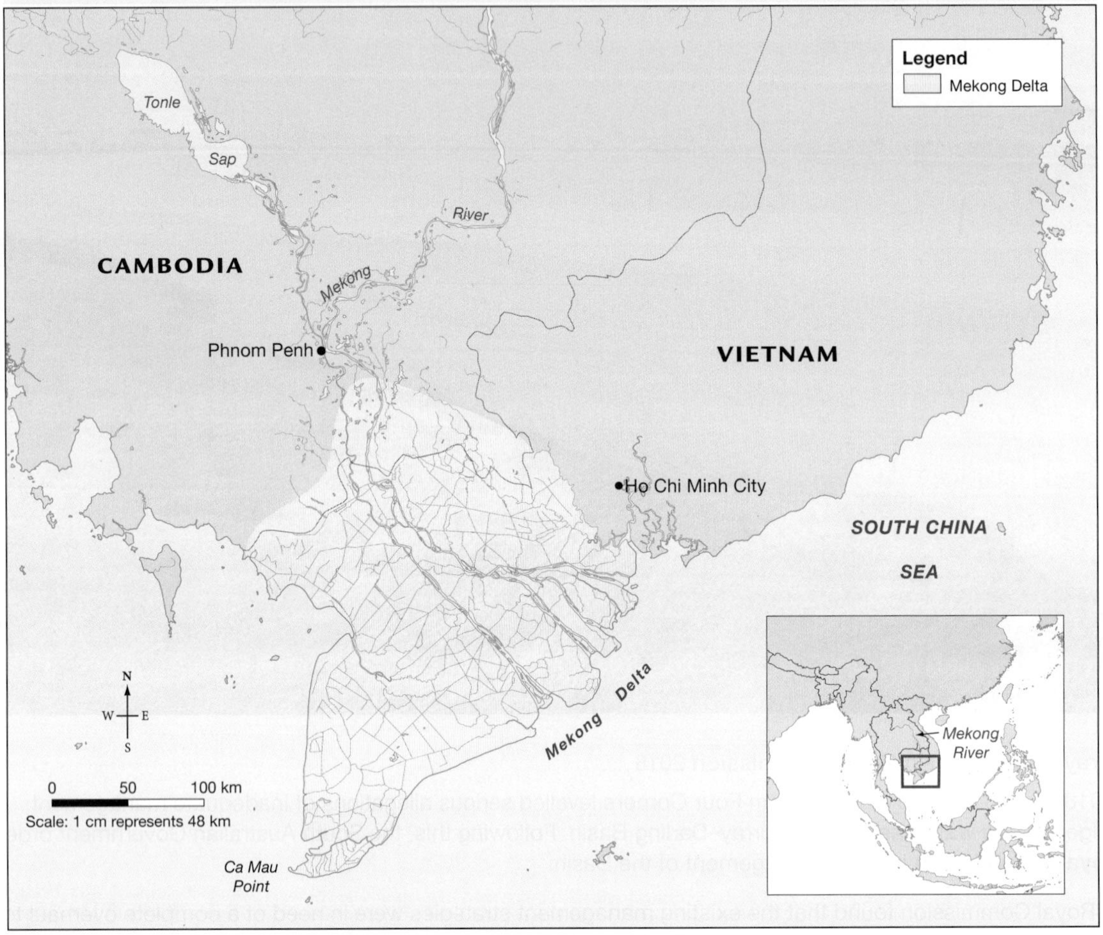

Source: Made with Natural Earth; Vector Map Level 0 (Digital Chart of the World). Map by Spatial Vision.

Although rice is a water-thirsty crop, it does not like to be completely submerged for its entire growing season. Rice grows best in fertile soils where there is an abundant supply of water that can be controlled throughout the growing season, but it can adapt to a variety of growing conditions.

Scientists are now developing strains of rice that are not only salt-resistant but can also withstand being submerged in water for almost three weeks, whereas traditional strains die within a week of being flooded and fully submerged.

Additionally, in some regions, farmers are making use of the **brackish water** that results from periods of saltwater intrusion. Although brackish water is not suited to rice farming, it is ideal for cultivating shrimp. With the onset of the monsoon season, the farmers rely on the heavy rains to flush out the salt water and allow them to plant their rice crops.

brackish water water that contains more salt than fresh water but not as much as sea water

FIGURE 8 Dry rice husks suitable only for poultry are the result of brackish water from the sea flowing inland.

on Resources

eWorkbook	Environmental change and salinity (ewbk-10097)
Video eLesson	Environmental change and salinity — Key concepts (eles-5247)
Interactivities	A pinch of salt (int-3291) Environmental change and salinity (int-8673)
myWorldAtlas	Deepen your understanding of this topic with related case studies and questions. • **Salinity in the Murray–Darling Basin**

14.6 ACTIVITIES

1. Investigate the history of agriculture in an ancient civilisation, such as Mesopotamia.
   a. Include a sketch map of the area. Annotate this map to show how the region was affected and why.
   b. What lessons might modern farmers learn from ancient practices?
2. In groups, investigate a method of combating salinity and sustainable practices that will improve the productivity of agricultural land. Before you begin, decide as a class which groups will cover dryland salinity and which will focus on irrigation salinity. Present your findings as a news report.
3. Find out the total land area of Australia and of the world. If areas affected by irrigation salinity are expected to triple by 2050, estimate the proportion of land that will be affected on a national and global scale. Use your findings as the basis for writing a letter to the editor, urging governments to take action and halt this trend.

4. In groups, investigate the Mekong River, its delta and its importance to Vietnam.
   a. Prepare a report on how the river is used and the issue of land degradation. In your report make reference to the scale of the problem and the rate at which change is occurring.
   b. What strategies have been suggested or used to deal with this issue? Are these strategies a sustainable option for caring for the environment? Why, or why not?
5. Investigate land use in the Murray–Darling Basin and explain the interconnection between land use and salinity in this region.
6. With a partner, investigate how saltbush may help to reduce salinity. Create an annotated diagram of a saltbush plant and its root system to explain this role.

14.6 EXERCISE

Learning pathways

■ LEVEL 1	■ LEVEL 2	■ LEVEL 3
1, 3, 7, 8, 14	2, 4, 9, 10, 13	5, 6, 11, 12, 15

Check your understanding

1. Define the term *drainage area* in your own words.
2. What is the difference between dryland and irrigation salinity?
3. Which general area of the Murray–Darling Basin is most affected by salinity: the north-east, the western region or the south-east? Describe the scale of the problem.
4. Why is the Murray–Darling Basin a significant part of the Australian environment?
5. Refer to **FIGURE 1**. Describe the distribution of dryland salinity areas in Australia.
6. What factors have contributed to the degradation of the Murray–Darling Basin's land and water resources since the arrival of Europeans?
7. What is 'salt scald'?

Apply your understanding

8. Why would planting deep-rooted trees help solve the problem of salinity?
9. What actions could an irrigation farmer take to reduce the risk of salinity?
10. Explain the interconnection between soil salinity and land degradation.
11. With the aid of a diagram, explain what a delta is and why it is important.
12. Refer to **FIGURE 5**.
    a. What is the aim of salt interception schemes? Explain.
    b. Do you think this is a sustainable management strategy? Explain.
    c. Discuss the impact of this scheme on river ecosystems.
    d. What do you think happens to the salt that is extracted?
    e. Do you think a similar scheme could be developed for the Mekong Delta? Justify your point of view.
13. With the aid of a Venn diagram, compare salinity issues that exist in the Murray–Darling Basin and Vietnam's Mekong Delta. Include references to the scale of the issue and the rate of change.

Challenge your understanding

14. How do you think the salinity problems in the Murray–Darling basin could be resolved?
15. Irrigation salinity is caused by overwatering. Is solving the problems of irrigation salinity as simple as banning irrigation or drastically reducing water allocations for irrigation? Suggest what other impacts such measures might have.
16. Predict whether climate change will affect salinity in the areas of Australia that are already most affected by the problem. Give reasons for your answer.

To answer questions online and to receive **immediate feedback** and **sample responses** for every question, go to your learnON title at www.jacplus.com.au.

14.7 Introduced species and land degradation

online only

LEARNING INTENTION

By the end of this subtopic, you will be able to explain the term *invasive species* and why they pose a threat to the Australian environment, and identify and explain possible strategies that can be used to control or eradicate invasive species.

The content in this subtopic is a summary of what you will find in the online resource.

An invasive species (sometimes referred to as an exotic species) is any plant or animal species that colonises areas outside its normal range and becomes a pest. Such species take over the environment at the expense of those that occur naturally in the region, generally causing damage to native habitats and degrading the landscape. Invasive species are a major cause of land degradation. Often introduced for a specific reason, they can soon take over the environment, threatening indigenous plant and animal species and taking over what was once valuable farming land.

To learn more about the impacts of introduced species on land environments, go to your learnON resources at www.jacPLUS.com.au.

FIGURE 1 Goats were a food source for early settlers and sailors.

Contents

learnON

- 14.7.1 Understanding invasive species
- 14.7.2 Introduced animals
- 14.7.3 Introduced plants
- 14.7.4 Controlling invasive species

on Resources

eWorkbook	Introduced species and land degradation (ewbk-10101)
Video eLesson	Introduced species and land degradation — Key concepts (eles-5248)
Interactivity	Introduced species and land degradation (int-8674)
myWorldAtlas	Deepen your understanding of this topic with related case studies and questions. • **Introduced species in Australia**

14.8 Native species and environmental change

online only

LEARNING INTENTION

By the end of this subtopic, you will be able to explain the threats to biodiversity from introduced species and environmental change.

The content in this subtopic is a summary of what you will find in the online resource.

Many native Australian species have been threatened by the spread of urban settlements, human actions, and introduced plants and animals. As human populations expand, the natural ranges of animals such as koalas, kangaroos and wallabies are diminished. The same changes are occurring around the world, having a direct impact on species' ability to survive and the environments they live in.

To learn more about protecting native species and their habitat, go to your learnON resources at www.jacPLUS.com.au.

FIGURE 1 A koala drinking from a dog's water bowl in a suburban backyard

Contents

learnON

- 14.8.1 Protecting biodiversity
- 14.8.2 Culling Australian native animals
- 14.8.3 The African elephant
- 14.8.4 Possums

on Resources

eWorkbook	Native species and environmental change (ewbk-10105)
Video eLesson	Native species and environmental change — Key concepts (eles-5249)
Interactivity	Native species and environmental change (int-8675)

14.9 SkillBuilder — Writing a fieldwork report as an annotated visual display (AVD)

online only

LEARNING INTENTION

By the end of this subtopic, you will be able to prepare and write a fieldwork report as an annotated visual display (AVD).

The content in this subtopic is a summary of what you will find in the online resource.

14.9.1 Tell me

What is a fieldwork report?

A fieldwork report helps you process all the information that you have gathered during fieldwork. You sort your data, create tables and graphs, and select images. You interpret the data as text or annotated images to convey your ideas. To convey your ideas, you synthesise, or pull together, all the data in a logical presentation.

14.9.2 Show me

How to create an AVD

Step 1

Determine a simple, short and concise title for your fieldwork study.

FIGURE 1 An annotated visual display (AVD) completed from secondary sources

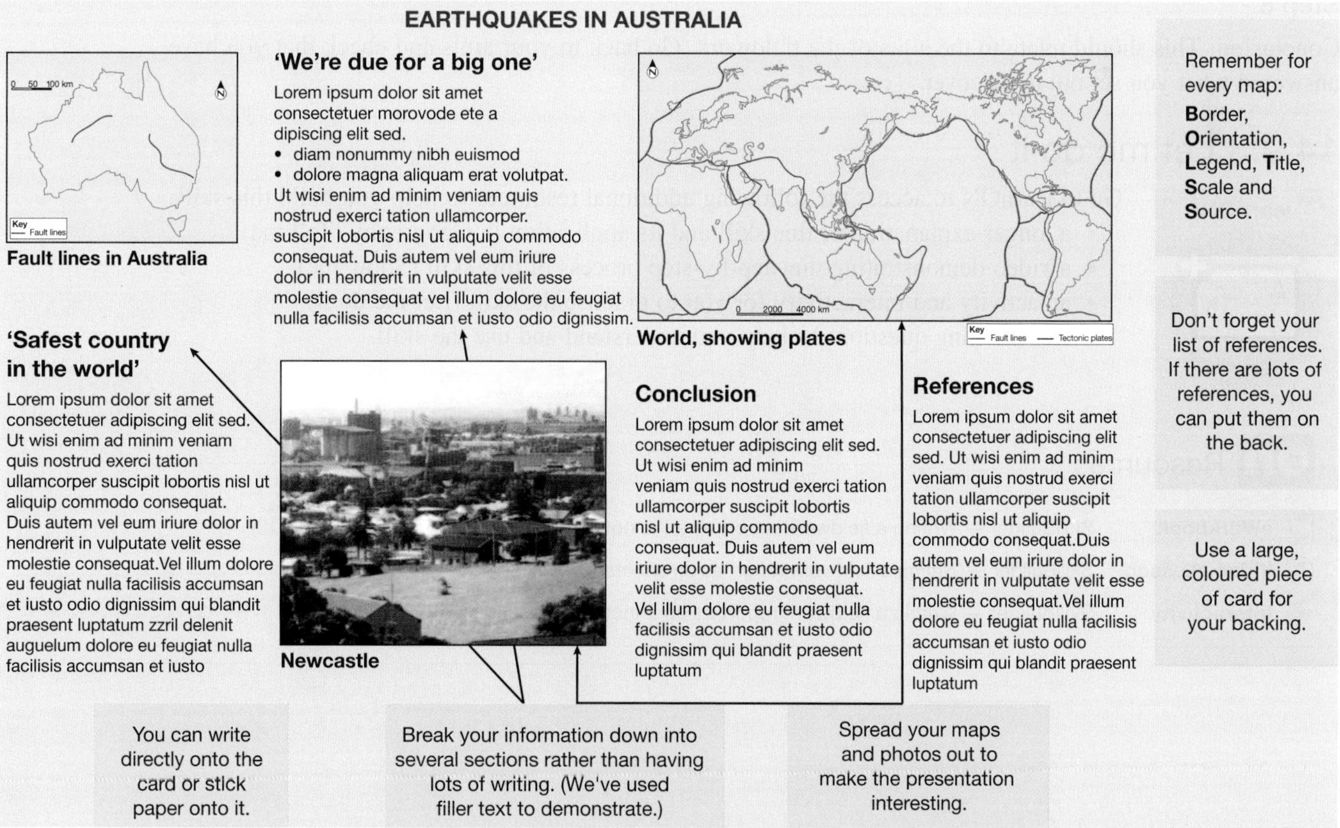
EARTHQUAKES IN AUSTRALIA
0 50 100 km
Key
Fault lines
Fault lines in Australia
'We're due for a big one'
Lorem ipsum dolor sit amet consectetuer morovode ete a dipiscing elit sed.
• diam nonummy nibh euismod
• dolore magna aliquam erat volutpat.
Ut wisi enim ad minim veniam quis nostrud exerci tation ullamcorper. suscipit lobortis nisl ut aliquip commodo consequat. Duis autem vel eum iriure dolor in hendrerit in vulputate velit esse molestie consequat vel illum dolore eu feugiat nulla facilisis accumsan et iusto odio dignissim.
0 2000 4000 km
Key
Fault lines
Tectonic plates
World, showing plates
Remember for every map:
Border,
Orientation,
Legend, Title,
Scale and
Source.
'Safest country in the world'
Lorem ipsum dolor sit amet consectetuer adipiscing elit sed. Ut wisi enim ad minim veniam quis nostrud exerci tation ullamcorper suscipit lobortis nisl ut aliquip commodo consequat. Duis autem vel eum iriure dolor in hendrerit in vulputate velit esse molestie consequat.Vel illum dolore eu feugiat nulla facilisis accumsan et iusto odio dignissim qui blandit praesent luptatum zzril delenit auguelum dolore eu feugiat nulla facilisis accumsan et iusto
Newcastle
Conclusion
Lorem ipsum dolor sit amet consectetuer adipiscing elit sed. Ut wisi enim ad minim veniam quis nostrud exerci tation ullamcorper suscipit lobortis nisl ut aliquip commodo consequat. Duis autem vel eum iriure dolor in hendrerit in vulputate velit esse molestie consequat. Vel illum dolore eu feugiat nulla facilisis accumsan et iusto odio dignissim qui blandit praesent luptatum
References
Lorem ipsum dolor sit amet consectetuer adipiscing elit sed. Ut wisi enim ad minim veniam quis nostrud exerci tation ullamcorper suscipit lobortis nisl ut aliquip commodo consequat.Duis autem vel eum iriure dolor in hendrerit in vulputate velit esse molestie consequat.Vel illum dolore eu feugiat nulla facilisis accumsan et iusto odio dignissim qui blandit praesent luptatum
Don't forget your list of references. If there are lots of references, you can put them on the back.
Use a large, coloured piece of card for your backing.
You can write directly onto the card or stick paper onto it.
Break your information down into several sections rather than having lots of writing. (We've used filler text to demonstrate.)
Spread your maps and photos out to make the presentation interesting.

Step 2

On a separate sheet of paper, sketch a layout for your work.

Step 3

Introduction: This should be short and should state clearly the aims of the fieldwork and the location of the investigation, shown as a map.

Step 4

Method: State where you went within the broader location; what information you gathered; the methods you used to gather information; and why you collected that information.

Step 5

Findings: This is the main focus of the report, in which you present the information that you gathered in the field. This must have a clear structure that guides the reader through the development of the ideas. Look for interconnections between the data and set out the information in an organised manner.

Step 6

Secondary sources: This information is not the focus of your work and must only supplement your fieldwork findings. Therefore, it must be very brief.

Step 7

Include a statement about the limitations and successes of the fieldwork. The limitations should cover anything that went wrong or ways in which the fieldwork could be improved. The successes should include new things learned and any interest that you may have gained from the investigation, particularly if you want to recommend active citizenship.

Step 8

Conclusion: This should relate to the aims of the fieldwork. Go back to your aims and check that you have answered what you set out to discover.

14.9.3 Let me do it

learnON

Go to learnON to access the following additional resources to help you build this skill:

- a longer explanation of this skill and its application in Geography (Tell me)
- a video demonstrating the step-by-step process of this skill (Show me)
- an activity and interactivity for you to practise the skill (Let me do it)
- self-marking questions to help you understand and use the skill.

on Resources

eWorkbook	SkillBuilder — Writing a fieldwork report as an annotated visual display (AVD) (ewbk-10113)
Video eLesson	SkillBuilder — Writing a fieldwork report as an annotated visual display (AVD) (eles-1747)
Interactivity	SkillBuilder — Writing a fieldwork report as an annotated visual display (AVD) (int-3365)

14.10 Aboriginal Peoples' land management

LEARNING INTENTION

By the end of this subtopic, you will be able to explain the traditional sustainable land management techniques used by some Aboriginal Peoples.

14.10.1 Land management technologies

Before the arrival of Europeans, Aboriginal Peoples had their own systems of land management, each developed to suit their own Country. Land management practices that maintained Country were intrinsically linked to Aboriginal knowledges, along with roles and responsibilities that were determined by a sophisticated societal structure based on kinship. For example, grasslands were managed through the use of fire, which encouraged plant regrowth and attracted a variety of animals. Management strategies were largely governed by the seasons and the local cycles of growth and weather, with each change dictating a change in the use of the land and its management. The use of cultural burning practices enabled Aboriginal Peoples to shape the land to best suit their needs while also maintaining sustainable and healthy environments.

Environmental change is not new. First Nations communities around the world, including Aboriginal Peoples and Torres Strait Islander Peoples, managed their environments carefully. In **FIGURE 1** you can see how fire was used to manage the landscape by some Aboriginal communities. It is interesting to note that the use of fire in

FIGURE 1 Cultural land management practices

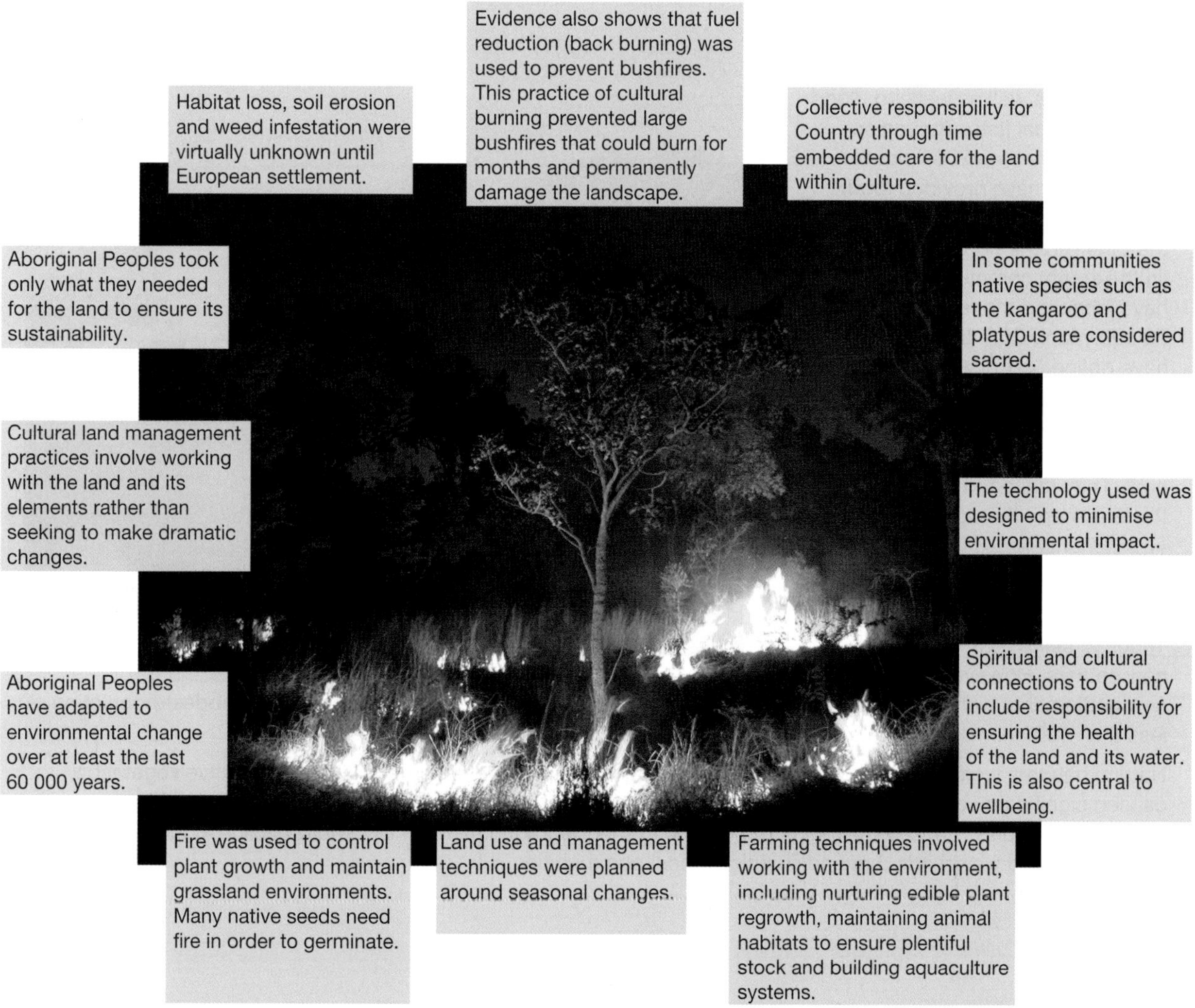
Evidence also shows that fuel reduction (back burning) was used to prevent bushfires. This practice of cultural burning prevented large bushfires that could burn for months and permanently damage the landscape.
Habitat loss, soil erosion and weed infestation were virtually unknown until European settlement.
Collective responsibility for Country through time embedded care for the land within Culture.
Aboriginal Peoples took only what they needed for the land to ensure its sustainability.
In some communities native species such as the kangaroo and platypus are considered sacred.
Cultural land management practices involve working with the land and its elements rather than seeking to make dramatic changes.
The technology used was designed to minimise environmental impact.
Aboriginal Peoples have adapted to environmental change over at least the last 60 000 years.
Spiritual and cultural connections to Country include responsibility for ensuring the health of the land and its water. This is also central to wellbeing.
Fire was used to control plant growth and maintain grassland environments. Many native seeds need fire in order to germinate.
Land use and management techniques were planned around seasonal changes.
Farming techniques involved working with the environment, including nurturing edible plant regrowth, maintaining animal habitats to ensure plentiful stock and building aquaculture systems.

this way for more than 60 000 years had a significant impact on the species of plants that thrive in Australia today. Those plants that adapted to the use of fire thrived, for example eucalypts. Using fire was just one of a variety of strategies employed by Aboriginal and Torres Strait Islander Peoples to ensure the land was used in a sustainable manner.

14.10.2 Season-based land management

Aboriginal Peoples were very careful in maintaining Country and ensuring its resources remained plentiful through seasonal migration. This land management practice was heavily influenced by deep knowledge of the environments that they lived in and were culturally responsible for. Detailed knowledges of weather systems and what these specific times supplied in terms of resources allowed Aboriginal Peoples to move about their Country with specific purpose at various times of the year, thus contributing to a sustainable and thriving society. For example, the Yolngu People, who live in north-east Arnhem Land, knew that each of their five seasons brought different opportunities for food collection and cultivation. They hunted and fished for particular species and harvested bulbs, fruits and other edible vegetation at different times of the year. Different Peoples' and Cultures' calendars varied, but whatever the location, seasonal events provided information about what to do. For instance, the arrival of march flies might signal the time to collect crocodile eggs and bush honey.

CASE STUDY 1

Managing Kakadu Wetlands

Kakadu is a kaleidoscope of both cultural and ecological biodiversity. The landscape varies from savannah and woodlands to escarpments and ridges as well as wetland, flood plains and tidal flats. The region includes more than 2000 plant species that have provided food, medicine and weaving materials for the Bininj (in the north) and Mungguy (in the south) communities that have inhabited the region for some 50 000 years. Over this time they have defined six distinct seasons, all signalled by subtle changes in the weather patterns that mark the transition from one season to the next. They managed and maintained the landscape through the use of fire.

FIGURE 2 Wetlands after removal of buffalo and before burning

The arrival of European settlers saw a massive change in the region. Buffalo were introduced in the early to mid 1800s to serve as a food supply for new settlers. However, once these new settlements were abandoned in the mid 1900s the buffalo population expanded from a modest population of less than 100 animals to more than 350 000. The impact on local habitats was extreme.

Natural habitats were devastated. The now feral buffalo took over wetland areas, disturbed native vegetation, caused significant soil erosion and changed the characteristics of the region's floodplains. Saltwater intrusion of freshwater wetlands caused the region to become further degraded, leading to a rapid decline in the flora and fauna, including waterbirds that had sustained local communities.

In the late twentieth century, a massive culling program commenced to remove the feral buffalo and allow the region to regenerate. However, an invasive native plant species that had once been the main food source of the buffalo spread unchecked. It choked the wetlands and prevented waterbirds from feeding and recolonising the region.

The CSIRO undertook extensive research into the Bininj's and Mungguy's sustainable land management practices. A joint management initiative was introduced into the area. At the heart of the initiative was traditional fire management.

The results have been dramatic. The wetlands are once again home to a rich assortment of flora and fauna. The project provides an internationally recognised example of sustainable land management using the practices developed and perfected by traditional owners over thousands of years.

FIGURE 3 Wetlands after burning

14.10.3 Using traditional knowledges today

Traditional knowledges that protect and manage land and sea environments are being used more often in conjunction with conservation techniques that developed from European traditions and science. For example, the number of Indigenous Protected Areas (IPAs) across Australia is increasing, as is the number of Ranger programs. In 2020, there were 78 dedicated IPAs over 74 million hectares across the country. An IPA is a sea or land area that is protected and conserved under the management of local Aboriginal and/or Torres Strait Islander Peoples, which helps to maintain continuing cultural connection with Country and preservation of important sites with traditional land management technologies. Indigenous Ranger programs employ Aboriginal Peoples and Torres Strait Islander Peoples in conservation programs to protect Country. The programs aim to combine contemporary and traditional land management knowledges to protect vulnerable sea and land environments and threatened species, and to reduce the impact of introduced species.

The diverse and exceptional knowledges that Aboriginal Peoples have of Country and how to care for it have resulted in increasing respect today in relation to sustainably managing the land. Cultural practices that successfully maintained Country for tens of thousands of years are now being recognised for their dynamic ability to counteract the impact of exploited lands in today's world. Further, the value of native plants and their uses are being respected, with many Aboriginal businesses and communities at the forefront of introducing and advocating native plants as both a source of food and healing. This is directly aligned with knowledge systems that hold a deep understanding and respect of Country and the environment.

CASE STUDY 2

Land management and climate change

The Conversation, January 31, 2020, Zena Cumpston (Research Fellow, University of Melbourne)

https://theconversation.com/to-address-the-ecological-crisis-aboriginal-peoples-must-be-restored-as-custodians-of-country-108594

To address the ecological crisis, Aboriginal peoples must be restored as custodians of Country

In the wake of devastating bushfires across the country, and with the prospect of losing a billion animals and some entire species, transformational change is required in the way we interact with this land.

Australia's First Peoples have honed and employed holistic land management practices for thousand of generations. These practices are embedded in all aspects of our culture. They are so effective, so perfectly suited to this harshest of continents, that we are the oldest living culture in the world today.

A reintroduction of traditional land management is essential if we want to address the ecological crisis we now face.

Not just 'consultants'

For a little over 200 years, Country in Australia has been predominantly managed without empowering or reflecting Aboriginal and Torres Strait Islander peoples' cultural practices, voices or aspirations.

To meaningfully engage First Nations communities' ways of knowing and interacting with Country, they need to cease being 'informants', 'actors' and 'consultants' which, at best, marginally inform ecological and agricultural imperatives.

The machine of colonisation continues to restrict our involvement in decision-making processes at every level. There are very few areas in Australia where Traditional Owners have succeeded in not only gaining back large land holdings, but also enjoy any real power to significantly maintain and nurture Country.

An example of this can be seen on my own Barkandji Country where in 2015, after 18 years of fighting, Barkandji people were recognised as the Traditional Owners of one of the largest areas ever before granted in a Native Title determination.

And yet, our Barka (the Darling River), our Mother, is now dying. It is poisonous and foul with algae, bone dry in many areas, with millions of fish washing up dead.

The devastation was caused by the gross mismanagement of this precious river by those in power — a destruction wrought through greed. Rights to land, with no rights to water, is a poignant example of our continued disempowerment in managing and caring for our lands in line with cultural obligations.

Our many thousands of generations of careful observations (science) and effective management and custodianship, must see us empowered to lead decision-making. Our community leaders must not only be given a seat at the table, they should set the menu too.

Different mob, different knowledge

Our mobs are extremely diverse, as are our land management practices. But some overarching beliefs sit at the core of our culture, and are important to understand.

First Peoples have a relationship with Country that is loving, reciprocal and engaged. This 'kincentric' relationship includes custodianship obligations — often lacking within non-Indigenous views of Country. Instead of being seen as kin — something to be cared for, listened to, deeply respected and nurtured — Country is seen by many non-Indigenous people as a resource to be exploited and controlled.

Our custodianship of Country, our Law and our vast ecological knowledges are all attached to a place. For each area in Australia, the mob belonging to that place must be engaged, and empowered to speak for that Country.

It's time to stop seeing Aboriginal ecological knowledges as something which can exist separately from the people who are its custodians. Our vast knowledges are embedded in our communities, and always have been.

Aboriginal knowledges aren't lost

When it comes to Aboriginal agricultural and land management practices there is still so much to uncover, adopt and reinvigorate. And there are still many who do not believe in our expertise in this area.

Too many ignorantly perceive our knowledges as lost, or call for elders to hand over their knowledges as a matter of urgency, unaware that our communities still practice intricate systems of sharing knowledge across generations.

The belief that our knowledges are lost harks back to early 'scientific' theories which emerged around the time of colonisation, when we were considered an inferior race which would soon die out.

Our knowledges are not lost. We are very much still here, still a living culture. But many of our practices and systems need more resources to reinvigorate them.

The extraordinary lifetime work of ethnobiologist Dr Beth Gott to reawaken Aboriginal plant knowledge is a brilliant example of this reinvigoration.

Dr Gott took a truly collaborative, respectful and empowering approach to working with Aboriginal communities. This enabled a safe space for Elders and communities to share and create a significant archive of their unparalleled knowledge of the medicinal, nutritional and cultural uses of Indigenous plants in south-eastern Australia.

Agriculture and fire

With temperatures rising, many of our food systems will fail. Introduced grain crops we rely heavily upon may not cope with the fluctuations predicted.

Traditional crops endemic to Australia such as native millet (panicum) and kangaroo grass will perhaps again become staple food sources.

As explored by Uncle Bruce Pascoe in Dark Emu, Australian crops are the most nutrient-rich and sustainable crops that can be grown here, requiring little water and no fertilisers. First Nations communities domesticated these crops over thousands of generations, and hold the best knowledge of how to grow them.

Cultural fire management practices are integral to our agricultural practices and are medicine for Country. Their continued reinvigoration will undoubtedly prove an important aspect in land management, protection and healing for all communities.

The recent horrifying and unprecedented bushfires traumatised and distressed all Australians. The loss of life, both people and animals, and the devastation wrought on Country triggered many calls for Aboriginal management systems to be more meaningfully incorporated.

Empowering and resourcing First Nations peoples' ecological knowledges would help address the effects of climate change on the land, through practices of care and custodianship. But it must not perpetuate well-established systems of exploitation. It must happen in true partnership.

Enacting healing

Finally, making Indigenous cultural practices central to Australia's ecological management could be vital to the process of 'truth-telling'.

Truth-telling here means acknowledging the complexity and richness of our culture, acknowledging the science we have developed over many many millennia to care for Country, and challenging still-embedded narratives which deny our diversity, our agency and most damaging, our sovereignty.

Truth-telling could not only bring long overdue public recognition of atrocities suffered and their continuing legacies, but could also finally dispense with the lie of peaceful settlement. The psychosis of denial impoverishes us all.

A process to enact a healing would begin a path to enlightened acceptance of our systems of management, opening up new possibilities for coming together to heal and enact vital reparations for both people and Country. Empower us and our active custodianship of Country and you empower yourselves.

As long as Aboriginal and Torres Strait Islander peoples and communities continue to be disenfranchised with our sovereignty denied, as long as we are excluded from leadership roles in meeting the challenges of climate change, we all stand to lose so much more than we can imagine.

---

on Resources

eWorkbook	Aboriginal Peoples' land management (ewbk-10109)
Video eLesson	Aboriginal Peoples' land management — Key concepts (eles-5250)
Interactivity	Aboriginal Peoples' land management (int-8676)
myWorldAtlas	Deepen your understanding of this topic with related case studies and questions. • **Indigenous Australians — Caring for Country**

14.10 ACTIVITIES

1. With the aid of diagrams, explain how land-use practices have changed over time in your area before and after European colonisation. Make sure you include references to practices that promoted sustainable use of the environment. Include links to how these changes would have resulted in salinity and degraded the environment.
2. Use the **Indigenous Weather Knowledge** weblink in your online Resources to explore the seasonal calendars of Aboriginal Peoples and Torres Strait Islander Peoples. Discuss how each community's understanding of the seasons informed their land management strategies.

14.10 EXERCISE

Learning pathways

■ LEVEL 1	■ LEVEL 2	■ LEVEL 3
1, 6, 7, 9, 13	2, 3, 4, 10, 14	5, 8, 11, 12, 15

Check your understanding

1. Why wasn't land degradation an issue before European settlers arrived?
2. Explain what determined Aboriginal Peoples' land management practices before European colonisation.
3. Give one example that demonstrates an interconnection between Aboriginal Peoples' cultural beliefs about the land and sustainable management of the environment.
4. Explain why the use of fire is a significant cultural practice for Aboriginal Peoples.
5. Explain why water buffalo are considered a significant threat to the Kakadu wetlands.
6. Identify two arguments for Aboriginal Peoples' increased control over land management in Australia raised in Case study 2.
7. What is an IPA?
8. What is the purpose of the Indigenous Ranger program?

Apply your understanding

9. Study **FIGURES** 3 and 4. Describe the environment shown in **FIGURE** 3 and the changes that have occurred in **FIGURE** 4.
10. What steps that might have been taken to turn the landscape shown in **FIGURE** 3 to the one shown in **FIGURE** 4?
11. Do you think the land management practices used in Kakadu could be used in other parts of Australia? Give reasons for your answer.
12. It has been suggested that the four seasons currently used in Australia do not adequately reflect the changing nature of our seasons. Do you agree or disagree with this suggestion? Give reasons for your opinion based upon the area you live in.

Challenge your understanding

13. Create a list of questions that you could ask Elders or traditional custodians about the best ways to protect the Country where you live.
14. Construct an argument to persuade farmers in your area to work with local Aboriginal elders in caring for the land. Justify your point of view.
15. Explore the idea that Aboriginal land management strategies might help to combat the affects of climate change.

To answer questions online and to receive **immediate feedback** and **sample responses** for every question, go to your learnON title at www.jacplus.com.au.

14.11 Investigating topographic maps — Managing land degradation in the Parwan Valley

LEARNING INTENTION

By the end of this subtopic, you will be able to describe the extend of land degradation in the Parwan Valley

14.11.1 Parwan Valley

The Parwan Valley, situated 60 kilometres west of Melbourne, is an area that is experiencing significant land degradation due to human interactions since European settlement. This region was one of the first areas of settlement in Victoria and was used for dairy farming. The land was cleared by the farmers, which led to the erosion of the poor and unstable soil.

Gully erosion and tunnel erosion have occurred throughout the White Elephant Range within the Parwan Valley, providing a habitat for an exploding wild rabbit population, which in turn caused further soil erosion. Since the 1940s, the area has been the focus of regeneration projects including the control of rabbit populations in efforts to rebuild and sustain the landscape.

FIGURE 1 Gully erosion

FIGURE 2 Topographic map of the Parwan Valley

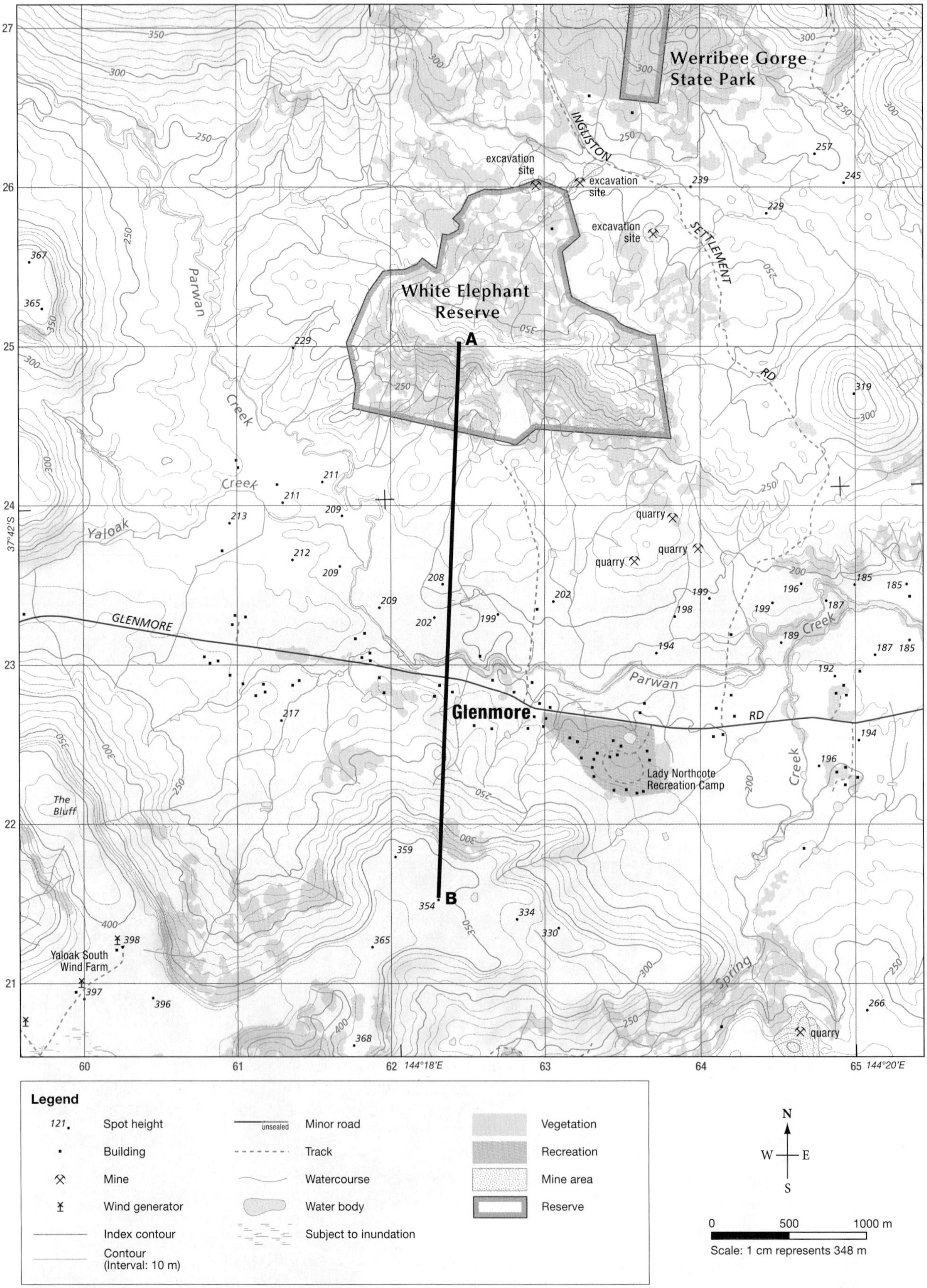

Source: Data based on Vicmap, Department of Environment, Land, Water and Planning.

Resources

eWorkbook	Investigating topographic maps — Managing land degradation in the Parwan Valley (ewbk-10089)
Digital document	Topographic map of the Parwan Valley (doc-36368)
Video eLesson	Investigating topographic maps — Managing land degradation in the Parwan Valley — Key concepts (eles-5251)
Interactivity	Investigating topographic maps — Managing land degradation in the Parwan Valley (int-8677)
Google Earth	Parwan Valley (gogl-0103)

14.11 EXERCISE

Learning pathways

■ LEVEL 1	■ LEVEL 2	■ LEVEL 3
1, 2, 5	3, 7, 8	4, 6, 9, 10

Study the topographic map of Parwan Valley (**FIGURE 2**) to answer the following questions.

Check your understanding

1. What is the elevation (height above sea level) for the following locations?
   a. GR625251
   b. GR623215
   c. GR603213
   d. GR613236
   e. GR645234
2. What is the name of the area found in AR6225?
3. Describe the topography of the Parwan Valley.
4. Describe the built environment on the map and its location in relation to the topography.
5. Identify examples of the following land features, giving a grid reference for each.
   a. Saddle
   b. Round hill
   c. Ridge
   d. Valley

Apply your understanding

6. Would some parts of the built environment have had a greater influence on erosion in the area than others? Explain why.
7. Draw a cross-section from point A to point B on the map.
8. Calculate the gradient from point A to Glenmore.
9. Explain how the topography might influence land degradation in the area.

Challenge your understanding

10. Construct a proposal to help regenerate the area by suggesting strategies that could be implemented in the valley and justifying why they would work. Some strategies may be more effective, expensive or long-term than others.
    Consider strategies such as:
    - placing straw bales and rock filters on top of the land to reduce the speed of water running over the land, thereby reducing erosion.
    - planting trees and shrubs
    - attaching strong filter fabric wire fences to slow the flow of water
    - spreading nets or mats over the soil surface to prevent erosion
    - reducing rabbit numbers.

To answer questions online and to receive **immediate feedback** and **sample responses** for every question, go to your learnON title at www.jacplus.com.au.

14.12 Thinking Big research project — Invasive species Wanted poster

online only

The content in this subtopic is a summary of what you will find in the online resource.

Scenario

The time has come to act, to eradicate invasive species before they cause further damage to the environment! The Department of the Environment and Energy's 'Invasive species' webpage provides information about the issues relating to invasive species, and strategies to deal with these pests.

The following species are of particular concern:

- Brumbies
- Cane toads
- Dromedary camels
- European carp
- European rabbits
- Feral cats
- Red foxes
- Water buffaloes

As an environmental crusader, you want to present an idea to the Department: to publish a series of Wanted! posters on its website to raise community awareness and, in particular, to educate upper primary and lower secondary school students about the damage that is caused by invasive species. You will start by creating one poster to help you pitch your idea to the Department.

Task

Select one of the species listed in the **Scenario** section and create a Wanted! poster suitable to be featured on the website of the Department of Water and Environmental Regulation. You should include the following on your poster:

learnON

- the name of the species
- when it first arrived in Australia and why it is a problem
- what has been/is being done to try and solve this problem
- map showing the species distribution throughout Australia
- what people should do if they see this species.

Ensure that on your poster you also explain what is meant by the terms *exotic* and *invasive species*.

Go to learnON to access the resources you need to complete this research project.

on Resources

ProjectsPLUS Thinking Big research project — Invasive species Wanted poster (pro-0212)

14.13 Review

14.13.1 Key knowledge summary

14.2 The causes and impacts of land degradation

- Land degradation is a complex issue; however, most of the causes of this degradation are human induced.
- As the population of the Earth increases, the land is under more pressure to accommodate the population.
- The land becomes degraded when we alter its natural state, through vegetation removal and introduced species, predominantly to expand our cities and increase our agricultural land.
- Developing nations where the population is growing fastest (especially those in sub-Saharan Africa) are most at risk; when the land becomes degraded, they lack the resources to deal with the issue.
- Land degradation presents challenges to future food production.

14.4 Managing land degradation

- A critical issue in land degradation is the loss of fertile soil. This soil has taken decades and, in some areas, thousands of years to develop. However, clearing the land leaves it vulnerable to erosion.
- Erosion of the topsoil may lead to rill, gully and tunnel erosion that affects the capacity of the land to support vegetation. Strong winds can pick up the soil and carry it large distances.
- The only way to repair the damage is through revegetation and programs designed to stabilise the soil.
- Costerfield is an example of how poor land management can have a devastating impact on the environment. However, it is also an example of how improving land management strategies can restore the land, enabling it to be used in a sustainable way.

14.5 Desertification: the drylands are spreading

- Experts estimate that 41 per cent of the Earth's surface is at risk of turning into desert, largely due to poor management of the semi-arid lands on the margins of deserts.
- Desertification is a human-induced problem. Population growth and the need to increase food production place drylands under increasing pressure. Once the land has been overrun by the desert, it is difficult to reclaim.
- China is attempting to halt the spread of the Gobi Desert by planting a massive green wall along the southern border of the desert. A similar project is taking place in the Sahel in Africa to halt the spread of the Sahara Desert.

14.6 Environmental change and salinity

- Salt occurs naturally in the environment; in Australia, it lies below the surface in the groundwater.
- Poor irrigation practices and the removal of deep-rooted vegetation have seen this salt rise to the surface, reducing the fertility of the land.
- Salinity is a major issue in the Murray–Darling Basin and in other parts of the world such as in the Mekong Delta.
- In order for issues related to salinity to be addressed, it is essential to consider the needs of all stakeholders, especially the needs of the environment.

14.7 Introduced species and land degradation

- Introduced species can have a devastating impact on the environment.
- Over time, these species can escape and take over the landscape. Goats, for example, have no natural predators, compete with native animals for food, damage the soil and overgraze the land. Introduced plants such as Paterson's Curse choke out natural vegetation and prevent sunlight from reaching the soil.
- Introducing goats to areas that are infested with invasive plant species can assist in reclaiming the land and keeping this pest at bay.
- Other species such as foxes and rabbits have been more difficult to control. However, Maremma dogs have been used to protect colonies of penguins from foxes.

14.8 Native species and environmental change

- Native species can also cause significant environmental change — koalas can literally eat themselves out of house and home. Koalas have, however, lost much of their original habitat, so issues related to food supply are largely caused by fragmented habitats; that is, they have nowhere to go.
- Elephants, because of their size, can cause damage to the environment. However, they are also an ecological species in that they maintain their ecosystem. Elephant numbers are declining in Africa, placing these ecosystems at risk.

14.9 Aboriginal Peoples' land management

- Aboriginal Peoples' land management involved the use of fire, which maintained a grassland environment, encouraged new vegetation growth and assisted in managing food sources.
- Aboriginal Peoples also had their own unique calendars linked to the environmental and seasonal changes that they observed. These changes governed their land management and, therefore, the food they ate throughout the year.
- Management practices were designed to ensure that there would always be a plentiful supply of food from year to year and the land would be used sustainably.

14.13 2 Key terms

algal bloom rapid growth of algae caused by high levels of nutrients (particularly phosphates and nitrates) in water
biodiversity the variety of plant and animal life within an area
brackish water that contains more salt than fresh water but not as much as sea water
carrying capacity the ability of the land to support livestock
chlamydia a sexually transmitted disease infecting koalas
cull selective reduction of a species by killing a number of animals
desertification the transformation of land once suitable for agriculture into desert by processes such as climate change or human practices such as deforestation and overgrazing
drainage area (or basin) an area drained by a river and its tributaries
dryland ecosystems characterised by a lack of water. They include cultivated lands, scrublands, shrublands, grasslands, savannas and semi-deserts. The lack of water constrains the production of crops, wood and other ecosystem services.
ecology the environment as it relates to living organisms
invasive species any plant or animal species that colonises areas outside its normal range and becomes a pest
groundwater water held underground within water-bearing rocks or aquifers
humus decaying organic matter that is rich in nutrients needed for plant growth
invasive plant species commonly referred to as weeds; any plant species that dominates an area outside its normal region and requires action to control its spread
Mediterranean (climate) characterised by hot, dry summers and cool, wet winters
mulch organic matter such as grass clippings
national park an area set aside for the purpose of conservation
pastoral run an area or tract of land for grazing livestock
ringbark removing the bark from a tree in a ring that goes all the way around the trunk. The tree usually dies because the nutrient-carrying layer is destroyed in the process.
Royal Commission a public judicial inquiry into an important issue, with powers to make recommendations to government
Sahel a semi-arid region in sub-Saharan Africa. It is a transition zone between the Sahara Desert to the north and the wetter tropical regions to the south. It stretches across the continent, west from Senegal to Ethiopia in the east, crossing 11 borders.
salinity an excess of salt in soil or water, making it less useful for agriculture
salt scald the visible presence of salt crystals on the surface of the land, giving it a crust-like appearance
topsoil the top layers of soil that contain the nutrients necessary for healthy plant growth
turbid water that contains sediment and is cloudy rather than clear
watertable upper level of groundwater; the level below which the earth is saturated with water
weathering the breaking down of rocks
weed any plant species that dominates an area outside its normal region and requires action to control its spread

14.13.3 Reflection

Complete the following to reflect on your learning.

Revisit the inquiry question posed in the Overview:

From housing to food production, we use land for many different things. What impact are we having on this important resource?

1. Now that you have completed this topic, what is your view on the question? Discuss with a partner. Has your learning in this topic changed your view? If so, how?
2. Write a paragraph in response to the inquiry question, outlining your views.

Subtopic	Success criteria	●	◐	●
14.2	I can describe what land degradation is.			
	I can describe the causes of land degradation.			
	I can describe the impacts of land degradation.			
14.3	I can interpret a complex block diagram.			
14.4	I can identify the causes and likely consequences of environmental change.			
	I can identify and explain solutions to an environmental issue.			
14.5	I can define desertification.			
	I can explain the causes of desertification.			
	I can evaluate strategies to manage desertification.			
14.6	I can describe the causes of salinity.			
	I can identify and explain solutions to salinity.			
14.7	I can define the term *invasive species*.			
	I can explain why invasive species are a threat to the Australian environment.			
	I can identify and explain solutions to control or eradicate invasive species.			
14.8	I can explain how native species create environmental change.			
	I can explain how native species threaten biodiversity.			
14.9	I can recall all of the elements required in an AVD.			
14.10	I can explain some of the land management practices used by Aboriginal Peoples.			
14.11	I can describe the causes and extent of land degradation in the Parwan Valley.			

on Resources

eWorkbook Chapter 14 Student learning matrix (ewbk-8572)
Chapter 14 Reflection (ewbk-8573)
Chapter 14 Extended writing task (ewbk-8574)

Interactivity Chapter 14 Crossword (int-8723)

ONLINE RESOURCES

on Resources

Below is a full list of **rich resources** available online for this topic. These resources are designed to bring ideas to life, to promote deep and lasting learning and to support the different learning needs of each individual.

eWorkbook

- 14.1 Chapter 14 eWorkbook (ewbk-8103) ☐
- 14.2 The causes and impacts of land degradation (ewbk-10077) ☐
- 14.3 SkillBuilder — Interpreting a complex block diagram (ewbk-10081) ☐
- 14.4 Managing land degradation (ewbk-10085) ☐
- 14.5 Desertification (ewbk-10093) ☐
- 14.6 Environmental change and salinity (ewbk-10097) ☐
- 14.7 Introduced species and land degradation (ewbk-10101) ☐
- 14.8 Native species and environmental change (ewbk-10105) ☐
- 14.9 SkillBuilder — Writing a fieldwork report as an annotated visual display (AVD) (ewbk-10113) ☐
- 14.10 Aboriginal Peoples' land management (ewbk-10109) ☐
- 14.11 Investigating topographic maps — Managing land degradation in the Parwan Valley (ewbk-10089) ☐
- 14.13 Chapter 14 Student learning matrix (ewbk-8572) ☐
- Chapter 14 Reflection (ewbk-8573) ☐
- Chapter 14 Extended writing task (ewbk-8574) ☐

Sample responses

- 14.1 Chapter 14 Sample responses (sar-0163) ☐

Digital document

- 14.11 Topographic map of the Parwan Valley (doc-36368) ☐

Video eLessons

- 14.1 Wasting our land (eles-1708) ☐
- Land environments under threat — Photo essay (eles-5243) ☐
- 14.2 The causes and impacts of land degradation — Key concepts (eles-5244) ☐
- 14.3 SkillBuilder — Interpreting a complex block diagram (eles-1746) ☐
- 14.4 Managing land degradation — Key concepts (eles-5245) ☐
- 14.5 Desertification — Key concepts (eles-5246) ☐
- 14.6 Environmental change and salinity — Key concepts (eles-5247) ☐
- 14.7 Introduced species and land degradation — Key concepts (eles-5248) ☐
- 14.8 Native species and environmental change — Key concepts (eles-5249) ☐
- 14.9 SkillBuilder — Writing a fieldwork report as an annotated visual display (AVD) (int-1747) ☐
- 14.10 Aboriginal Peoples' land management — Key concepts (eles-5250) ☐
- 14.11 Investigating topographic maps — Managing land degradation in the Parwan Valley — Key concepts (eles-5251) ☐

Interactivities

- 14.2 Destroying the land (int-3289) ☐
- Why land degrades (int-5591) ☐
- Soil degradation — a global problem affecting every permanently inhabited continent (int-5597) ☐
- Land clearing and deforestation leave the land vulnerable to erosion (int-7952) ☐
- Severity of soil degradation in Australia (int-7953) ☐
- 14.3 SkillBuilder — Interpreting a complex block diagram (int-3364) ☐
- 14.4 Down in the dirt (int-3290) ☐
- 14.5 Desertification (int-8672) ☐
- 14.6 Environmental change and salinity (int-8673) ☐
- A pinch of salt (int-3291) ☐
- 14.7 Introduced species and land degradation (int-8674) ☐
- 14.8 Native species and environmental change (int-8675) ☐
- 14.9 SkillBuilder — Writing a fieldwork report as an annotated visual display (AVD) (int- 3365) ☐
- 14.10 Aboriginal Peoples' land management (int-8676) ☐
- 14.11 Investigating topographic maps — Managing land degradation in the Parwan Valley (int-8677) ☐
- 14.13 Chapter 14 Crossword (int-8723) ☐

ProjectsPLUS

- 14.12 Thinking Big research project — Invasive species Wanted poster (pro-0212) ☐

Weblinks

- 14.5 Great Green Wall of China (1) (web-3342) ☐
- Great Green Wall of China (2) (web-3343) ☐
- 14.7 Weed species (web-3347) ☐
- 14.10 Indigenous Weather Knowledge (web-6397) ☐

Google Earth

- 14.11 Parwan Valley (gogl-0103)

myWorldAtlas

- 14.5 Desertification in Mauritania (mwa-7349) ☐
- 14.6 Salinity in the Murray–Darling Basin (mwa-7350) ☐
- 14.7 Introduced species in Australia (mwa-4487) ☐
- 14.10 Indigenous Australians — Caring for Country (mwa-7348) ☐

Teacher resources

There are many resources available exclusively for teachers online.

15 Inland water management

INQUIRY SEQUENCE

To access a pre-test with **immediate feedback** and **sample responses** to every question in this chapter, select your learnON format at www.jacplus.com.au.

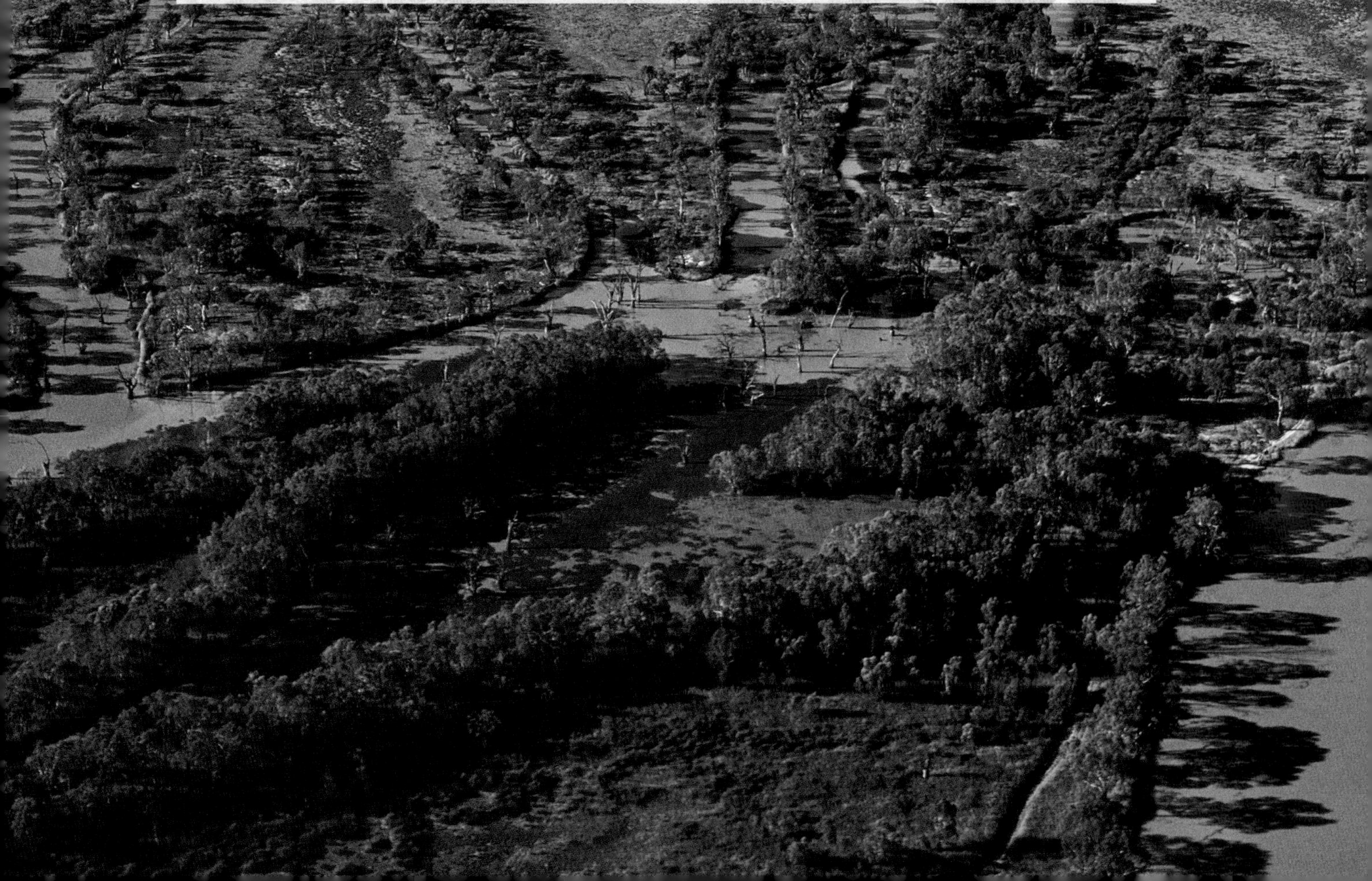

15.1 Overview

Numerous **videos** and **interactivities** are embedded just where you need them, at the point of learning, in your learnON title at www.jacplus.com.au. They will help you to learn the content and concepts covered in this topic.

Humans would find life very hard without healthy inland water sources. Are we being careful with how we use and change them?

15.1.1 Introduction

Water makes life on Earth possible, and rivers are like blood running through the veins of a body. Over time we have dammed, diverted and drained water, and this has brought about significant environmental change. Careful stewardship of these resources will provide a health insurance policy for a sustainable future.

Inland waters are important sources of water for both environments and people.

FIGURE 1 Hume Weir on Lake Hume at the start of the Murray River

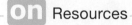

STARTER QUESTIONS

1. How many different types of freshwater bodies can you think of within 100 kilometres of where you live?
2. Where does your fresh water come from? Name and describe the location of the freshwater bodies that supply your house. You may need to refer to an atlas.
3. If you didn't have shops, supermarkets, water taps and pipes where you live, what water sources would you get your daily fresh water from?
4. Watch the **Inland water management — Photo essay** video in your online Resources. What is your experience of rivers in Australia?

on Resources

eWorkbook	Chapter 15 eWorkbook (ewbk-8107)
Video eLessons	Drained away (eles-1709) Inland water management — Photo essay (eles-5252)

15.2 Inland water

LEARNING INTENTION

By the end of this subtopic, you will be able to describe the different types of water bodies that make up inland water, explain their importance and identify how human activities threaten them.

15.2.1 Understanding inland water

Have you ever stopped to think that the water flowing down a river or rippling across a lake is providing us with a life support system? The rivers, lakes and **wetlands** that make up our inland water are important for supplying water for our domestic, agricultural, industrial and recreational use. They provide important habitats for a wide range of terrestrial and aquatic life.

Inland water systems cover a wide range of landforms and environments, such as lakes, rivers, floodplains and wetlands. The water systems may be **perennial** or **ephemeral**, flowing (such as rivers), or standing water (such as lakes) (see **FIGURE 1**). There are interconnections between surface water and **groundwater**, and between inland and coastal waters. Inland water is an important link in the water cycle, as water evaporates from its surface into the atmosphere. In return, rainfall can be stored in rivers and lakes, or soak through the soil layers to become groundwater.

wetland an area covered by water permanently, seasonally or ephemerally. They include fresh, salt and brackish waters such as rivers, lakes, rice paddies and areas of marine water, the depth of which at low tide does not exceed 6 metres.

perennial describes a stream or river that flows permanently

ephemeral describes a stream or river that flows only occasionally, usually after heavy rain

groundwater water held underground within water-bearing rocks or aquifers

Why is inland water important?

Inland water provides both the environment and people with fresh water, food and habitats. It provides environmental services; for example, it can filter pollutants, store floodwater and even reduce the impacts of climate change. The economic value of these services cannot easily be measured. Their importance, however, can be taken for granted and may not be appreciated until the services are lost or degraded.

15.2.2 Threats to inland water

Inland water is extremely vulnerable to change. It has been estimated that in the last century over 50 per cent of inland water (excluding lakes and rivers) has been lost in North America, Europe and Australia. Those systems remaining are often polluted and reduced in size. The loss is largely a result of human activity. **TABLE 1** illustrates some of the reasons for changes to inland water systems and their possible impacts on the environment and people. As water is such a valuable resource, much of our inland waterways have been dammed, diverted or drained to meet the needs of people.

FIGURE 1 The Parana River floodplain in northern Argentina shows a variety of different types of inland water.

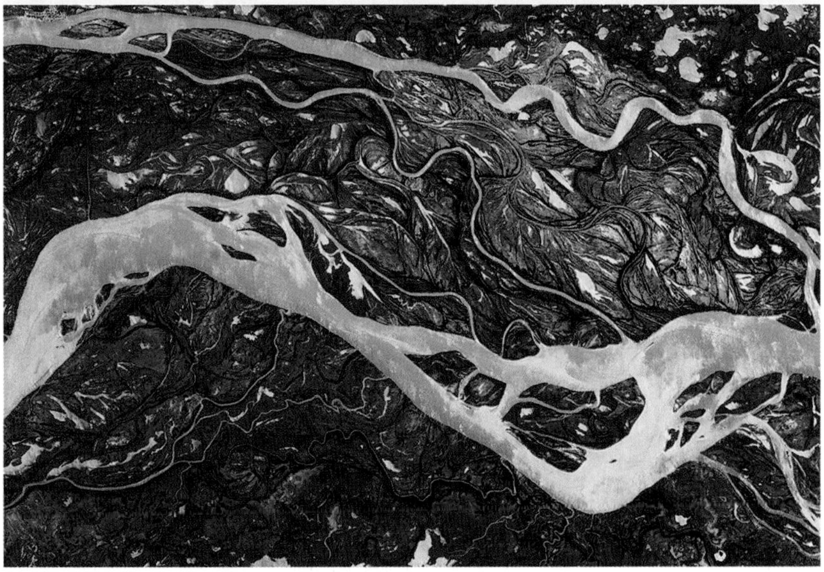

TABLE 1 Threats to inland water

Causes of change	Environmental functions threatened	Impacts of change
Increasing population and increasing demand for water across space	• Most services (e.g. fresh water, food and biodiversity) • Regulatory features such as recharging groundwater and filtering pollutants	• Increased withdrawal of water for human and agricultural use • Large-scale draining of wetlands to create farmland
Construction of infrastructure including dams, weirs and levee banks, and diverting water to other drainage basins	• Services supporting the quality and quantity of water • Biodiversity, habitat, river flow and river landforms	• Changes to the amount and timing of river flow. The transportation of sediment can be blocked and dams can restrict fish movements.
Changing land use (e.g. draining of wetlands, urban development on floodplains)	• Holding back floodwaters and filtering pollutants • Habitats and biodiversity	• Alters run-off and infiltration patterns • Increased risk of erosion and flood
Excessive water removal for irrigation	• Reduced water quantity and quality • Less water available for groundwater supply	• Reduced water and food security • Loss of habitat and biodiversity in water bodies
Discharge of pollutants into water or on to land	• Change in water quality, habitat • Pollution of groundwater	• Decline in water quality for domestic and agricultural use • Changes ecology of water systems

FIGURE 2 Wetlands are an example of inland water systems that are vulnerable to human-induced damage.

fdw-0037

FOCUS ON FIELDWORK

Features of your local catchment

In Australia, most people can turn on a tap in their homes for potable (drinkable) water. It's something that many of us take for granted, especially if we live in urban areas.

What we rarely think about is the work that went into ensuring that water was piped to our homes. Do you know how much human intervention it takes to get the water from your catchment to you so that it is safe to drink and plentiful? How did people change the landscape to collect and move water?

Learn more about how people modified the environment in your local area to bring you water using the **Features of your local catchment** activity in your learnON Fieldwork Resources.

on Resources

eWorkbook	Inland water (ewbk-10121)
Video eLesson	Inland water — Key concepts (eles-5253)
Interactivity	Inland water (int-8678)

infrastructure the facilities, services and installations needed for a society to function, such as transportation and communications systems, water and power lines

15.2 ACTIVITIES

1. Make a simplified sketch of **FIGURE 1** and clearly label an example of each of the following features: *main channel*, *tributary*, *anabranch*, *meander*, *oxbow lake (or billabong)*, *floodplain*. A dictionary may help you define the terms.
2. Make a list of as many inland water sources you can name, and classify them according to whether they are surface or underground, natural or constructed by people. (Note that some can be both natural and constructed.) Collect your lists together as a class and map them.
3. Match the following terms with their correct definition in the table below: *main channel*, *tributary*, *anabranch*, *meander*, *oxbow lake (or billabong)*, *floodplain*.

Term	Definition
	A smaller stream that flows into a larger stream
	A bend in the river
	An area of relatively flat, fertile land on either side of a rive
	A main river
	A cut-off meander bend
	Where a river branches off and joins back into itself

15.2 EXERCISE

Learning pathways

■ LEVEL 1	■ LEVEL 2	■ LEVEL 3
2, 3, 4, 9, 13	1, 5, 7, 8, 10, 14	6, 11, 12, 15

Check your understanding

1. What is the difference between a perennial and an ephemeral river?
2. What is a wetland?
3. Define the term 'groundwater' in your own words.
4. Look at the many inland water storage features below and classify them according to whether they are surface or underground, natural or human-made. (Note that some can be both natural and human-made.)
   a. Dam/reservoir
   b. River/creek/stream
   c. Wetland/swap
   d. Irrigation channel
   e. Groundwater/aquifer
   f. Well
   g. Lake
5. Identify two short-term and two long-term examples of human-induced changes that could have an impact on the wetland.
6. Identify two reasons why a wetland, such as that shown in **FIGURE 2**, might be drained.

Apply your understanding

7. Refer to **FIGURE 1**. The brown shading visible in the water and on the land represents the river's muddy sediment. This is material such as sand and silt carried and deposited by a river.
   a. Where has this sediment come from?
   b. How does the sediment get onto the floodplain?
8. Explain how groundwater is part of the water cycle.
9. Refer to **TABLE 1**. Indicate whether each of the following statements is true or false.
   a. Large-scale draining of wetlands will not affect groundwater.
   b. The spread of settlement over a floodplain will alter the amount of water available to replenish groundwater.
   c. Habitat destruction can occur with both draining of wetlands and construction of dams.
   d. Water that is diverted from one drainage basin to another is lost to the water cycle.
10. What are two short-term and two long-term examples of human-induced changes that could have an impact on the wetland in **FIGURE 2**?
11. Refer to **FIGURE 1**. Imagine that the Parana River flooded, and the floodwaters have now subsided. Would the floodplain look the same as it does in this image? Explain your answer.
12. The Parana River is 4880 kilometres long, making it the second longest river in South America. The river flows from the south-east central plateau of Brazil south to Argentina. **FIGURE 1** is a small section of this river. What evidence is there to suggest that this river frequently floods?

Challenge your understanding

13. Are there justifiable reasons to drain a wetland? Give reasons to justify your answer.
14. If the river shown in **FIGURE 1** was dammed upstream, what changes are likely to happen to the sediment carried and to the floodplain?
15. Which of the threats to inland water listed in **TABLE 1** do you think poses the greatest threat to Australia's water security? Give reasons to justify your answer.

To answer questions online and to receive **immediate feedback** and **sample responses** for every question, go to your learnON title at www.jacplus.com.au.

15.3 Damming rivers

LEARNING INTENTION

By the end of this subtopic, you will be able to explain why dams are used and the positive and negative effects of dams.

15.3.1 The purpose of dams

Are dams marvellous feats of modern engineering or environmental nightmares? Without them, we would not have a dependable supply of water or electricity, nor would we feel relatively safe from floods. For many decades, dams have been seen as symbols of a country's progress and economic development. But increasingly, the social, economic and environmental costs are emerging.

A reliable water supply has always been critical for human survival and settlement. The global demand for water has increased by 600 per cent in the last century — more than twice the rate of population growth. If this rate continues, global water demand will exceed supply by 2030. Water is also unevenly distributed across the globe. Some places suffer from regular droughts, while others experience massive floods. As a result, people have learned to store, release and transfer water to meet their water, energy and transport needs. This could be in the form of a small-scale farm dam or a large-scale multi-purpose project such as the Snowy River Scheme. Constructing dams is one of the most important contributors to environmental change in river basins. Globally, over 60 per cent of the world's major rivers are controlled by dams.

FIGURE 1 shows the degree of **river fragmentation**, or interruption, in the world's major drainage basins. River fragmentation is an indicator of the degree to which rivers have been modified by humans. Highly affected rivers have less than 25 per cent of their main channel remaining without dams, and/or the annual flow pattern has changed substantially. Unaffected rivers may have dams only on tributaries but not the main channel, and their discharge has changed by less than 2 per cent. Today, 48 per cent of the world's river volume is moderately to severely affected by dams.

river fragmentation the interruption of a river's natural flow by dams, withdrawals or transfers

int-7956

FIGURE 1 Degree of river fragmentation in the world's major drainage basins

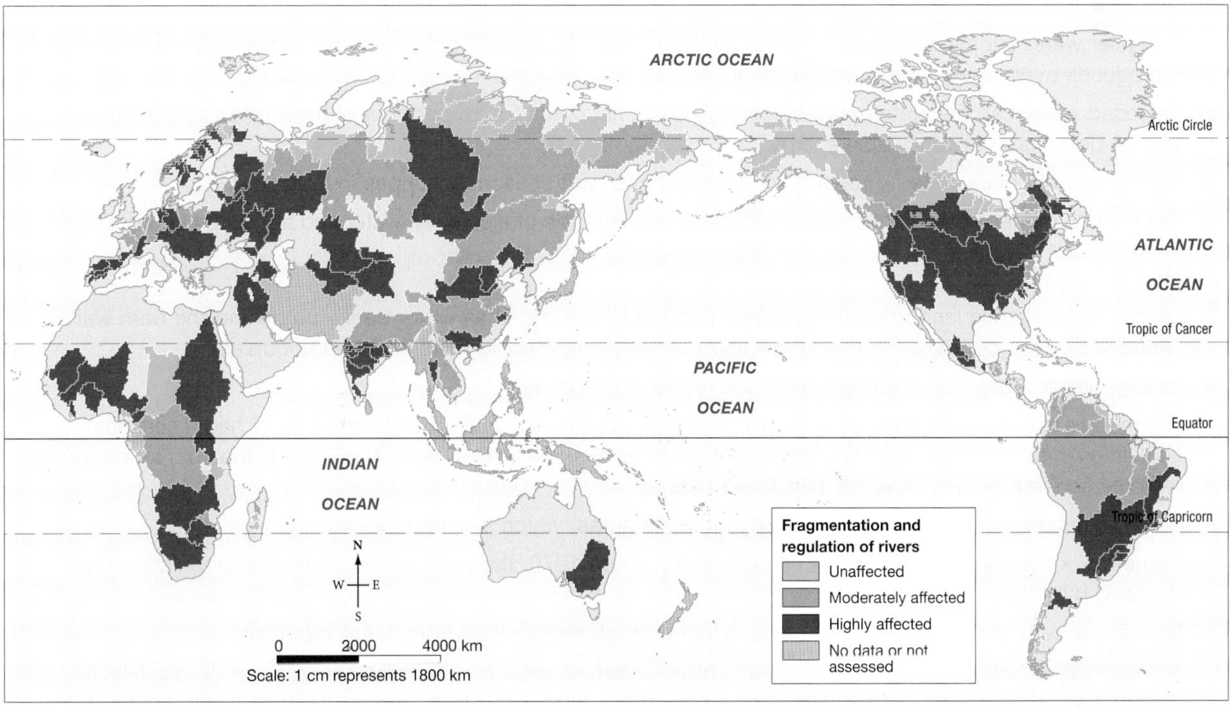

Source: Made with Natural Earth. University of New Hampshire UNH/Global Runoff Data Centre GRDC http://www.grdc.sr.unh.edu/ Map by Spatial Vision

Dams, **reservoirs** and **weirs** have been constructed to improve human wellbeing by providing reliable water sources for agricultural, domestic and industrial use. Dams can also provide flood protection and generate electricity.

However, while there are many benefits, large-scale or mega dams bring significant changes to the environment and surrounding communities, both positive and negative, as shown in **FIGURE 2**.

reservoir large natural or artificial lake used to store water, created behind a barrier or dam wall

weir wall or dam built across a river channel to raise the level of water behind. This can then be used for gravity-fed irrigation.

int-5592

FIGURE 2 The advantages and disadvantages of large-scale dams

Positive changes

1. A regular water supply allows for irrigation farming. Only 20 per cent of the world's arable land is irrigated, but it produces over 40 per cent of crop output.
2. Released water can generate hydro-electricity, which accounts for 16 per cent of the world's total electricity and 71 per cent of all renewable energy.
3. Dams can hold back water to reduce flooding and even out seasonal changes in river flow.
4. Income can be generated from tourism, recreation and the sale of electricity, water and agricultural products.

Negative changes

5. Large areas of fertile land upstream become flooded or inundated as water backs up behind the dam wall. Alluvium or silt is deposited in the calm water that previously would have enriched floodplains.
6. Initially, flooded vegetation rots and releases greenhouse gases.
7. The release of cold water from dams creates thermal pollution. Originally the Colorado River had a seasonal fluctuation in temperature of 27 °C. Today, temperatures average 8 °C all year. The water is too cold for native fish reproduction, but is ideal for some introduced species.
8. Some dams are constructed in tectonically unstable areas, which are prone to earthquakes.
9. Dams block the natural migration of fish upstream. Since 1970, the world's freshwater fish population has declined by 80 per cent.
10. Over 7 per cent of the world's fresh water is lost through evaporation from water storages.
11. A conservative estimate has stated that dams have negatively affected 472 million people worldwide. Tens of millions have been relocated from dam sites, while other communities both upstream and downstream have lost their livelihoods or had their land flooded.

CASE STUDY 1

Dammed disasters: The Samarco Mine Dam

Samarco is a Brazilian mining company jointly owned by Vale and BHP Billiton. On 5 November 2015 a dam owned by Samarco, containing by-products of iron mining, collapsed.

The dam, located in the Bento Rodrigues subdistrict of Mariana, Brazil (see **FIGURE 3**), sent up to 60 million cubic metres of water, mud and iron-ore by-products down the mountainside, causing significant flooding and killing 16 people (see **FIGURE 4**).

FIGURE 3 Map of Bento Rodrigues

Source: www.theguardian.com

FIGURE 4 Bento Rodrigues after the dam burst

The toxic sludge, containing substances such as mercury and arsenic, flowed into the Rio Doce (the Doce River) and travelled more than 500 kilometres to the Atlantic Ocean (see **FIGURE 5**). The pollution has killed thousands of fish, cut off drinking water to a quarter of a million people and significantly altered marine food chains along the coast of Brazil and into the Southern Atlantic Ocean.

FIGURE 5 Satellite images showing Bento Rodrigues: (a) shows the landscape before the dam burst and (b) shows the toxic sludge spreading across the landscape.

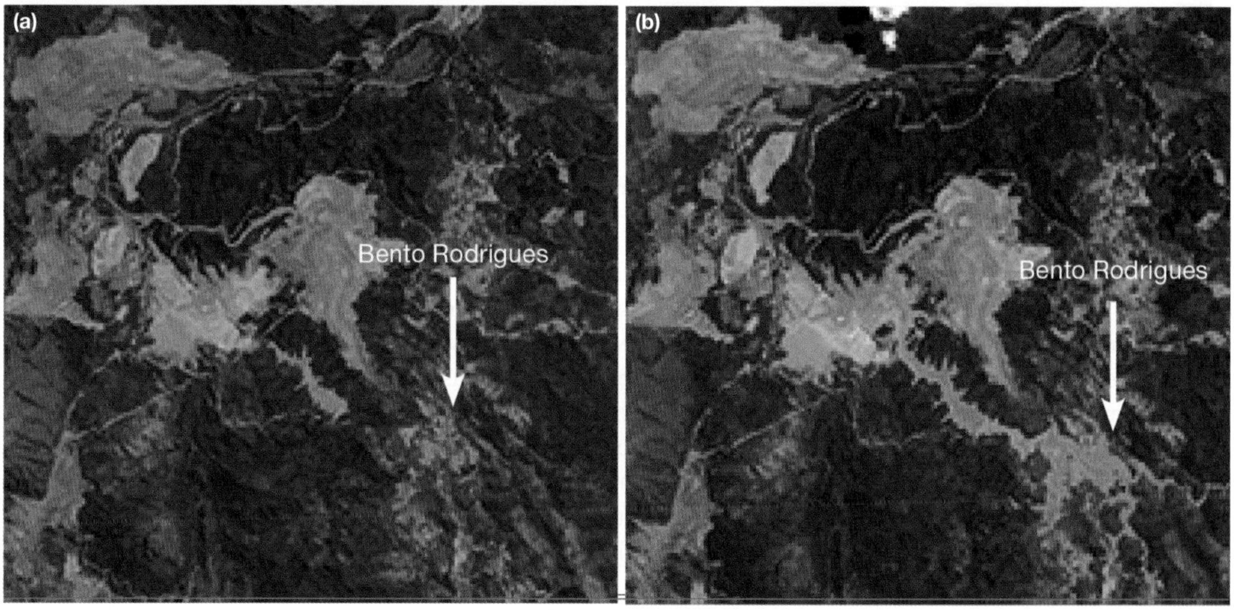

15.3.2 Why rivers flow

Traditionally, water flowing out to sea was seen as a waste. If it could be stored, then it could be used. Little thought was given to the health of the river and the importance of keeping water in a stream. Governments around the world have favoured damming rivers to make use of water resources.

Mega dams have been linked to economic development and improvement in living standards. Only in recent times have people questioned the real cost of these schemes — environmentally, economically and socially. **FIGURE 6** shows the number of downstream communities in each country that have the potential to be affected by the construction of mega dams.

int-5599

FIGURE 6 Distribution of downstream communities affected by large dams

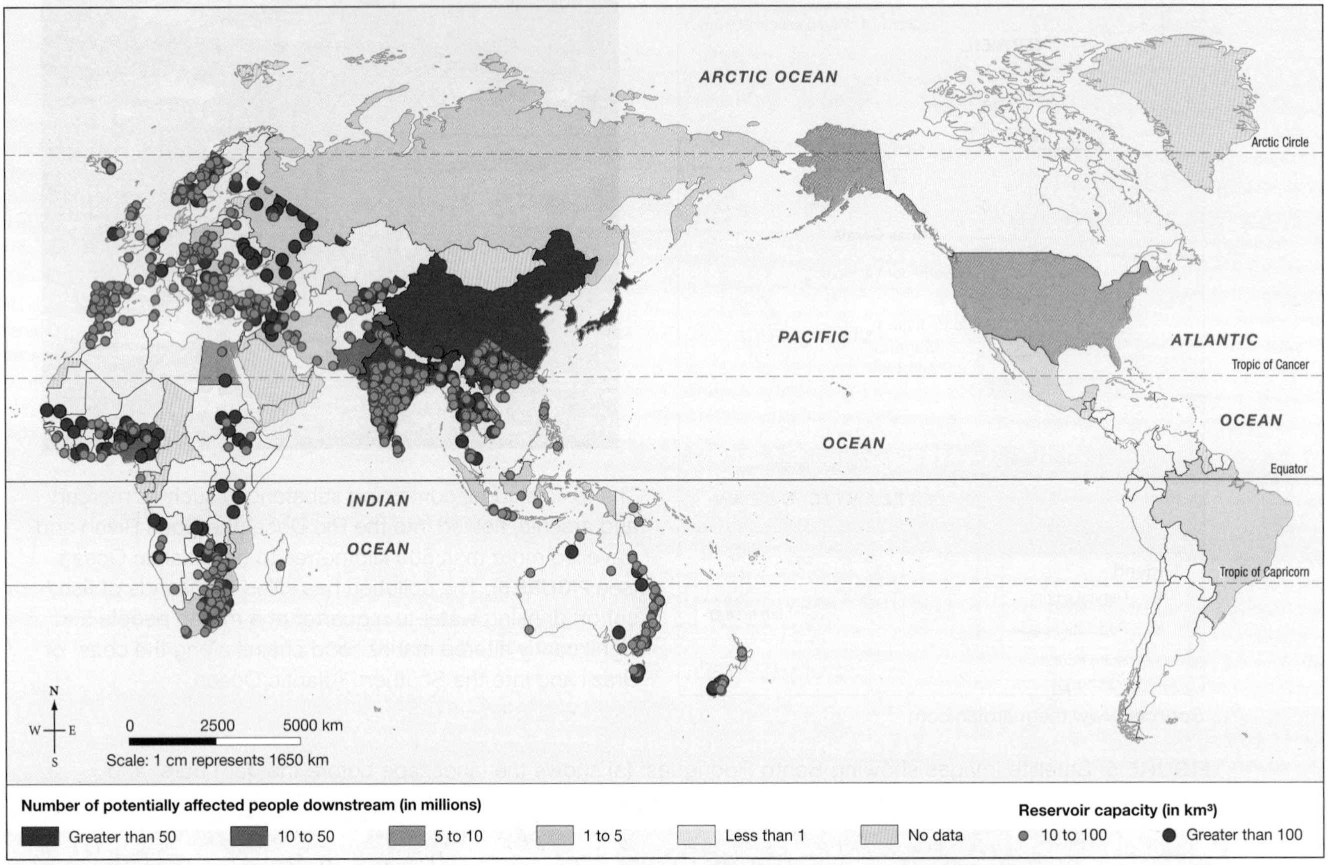

Source: Lehner et al.: High resolution mapping of the world's reservoirs and dams for sustainable river flow management. Frontiers in Ecology and the Environment. GWSP Digital Water Atlas (2008). Map 81: GRanD Database (V1.0). Available online at http://atlas.gwsp.org.

There is also the concern that multi-purpose dams have conflicting aims. To generate hydro-electricity you need to release a large volume of stored water. To provide **flood mitigation** you need to keep water levels low. To use water for irrigation you need a large store.

More than one billion people worldwide lack access to a decent water supply, yet it has been estimated that only 1 per cent of current water use could supply 40 litres of water per person per day, if the water was properly managed. The problem is not so much the quantity or distribution of water resources but the mismanagement of it. During the twentieth century, over $2 trillion was spent on constructing more than 50 000 dams. The emphasis now is to switch from *controlling* river flow to *adapting to* river flow — in other words, shifting from a human-centred to an earth-centred approach. This means building small-scale projects that promote social and environmental sustainability. In many regions of the world, there are ongoing community protests against the need for mega dams in preference to smaller schemes that benefit local people directly.

flood mitigation managing the effects of floods rather than trying to prevent them altogether

15.3.3 Challenges to dams

Across the globe, from Africa to Asia to South America, there has been a growing movement of community and environmental groups challenging the construction of mega dams in terms of location, sustainability and the potential social, economic and environmental impacts. Organisations such as International Rivers work with local groups to help restore justice to dam-affected communities, find better alternatives and promote the restoration of rivers through better dam management.

CASE STUDY 2

Belo Monte Dam, Brazil

For over 20 years, there has been an ongoing protest over the construction of the Belo Monte Dam in Brazil. The original design called for five huge dams on the Xingu River, but after large-scale local and international protests by indigenous groups and environmentalists, the scheme was scaled back to one large dam — the world's third largest (see **FIGURE 7**).

int-7958

FIGURE 7 (a) Location of the Belo Monte Dam and (b) the related environmental changes

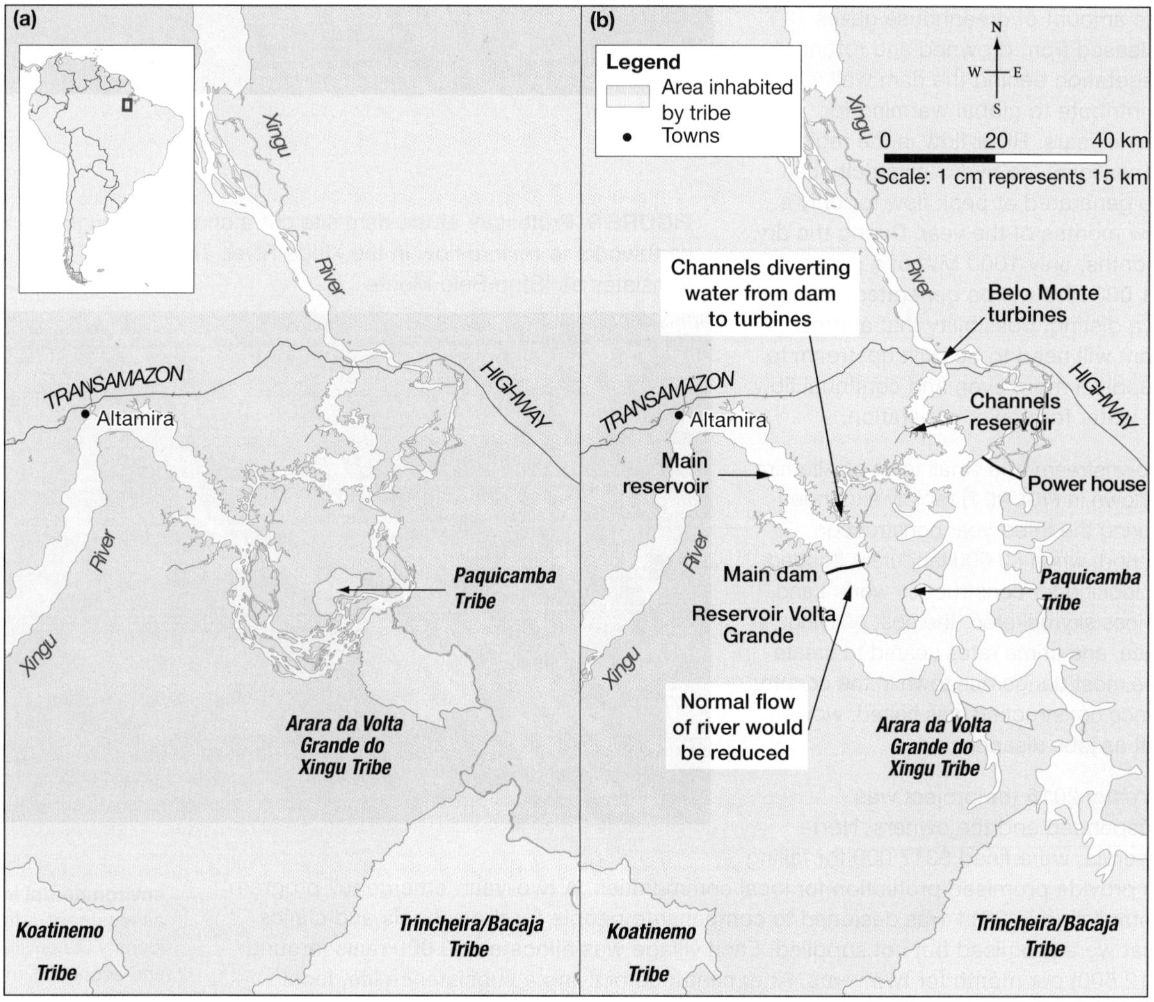

Source: Spatial Vision

Belo Monte was designed to divert more than 80 per cent of the flow of the Xingu River, drying out over 100 kilometres of river, known as the Big Bend (see **FIGURE 8**). As a result, over 516 km^2 of rainforest was flooded, and between 20 000 and 40 000 indigenous people were displaced from their homelands.

Construction was delayed and battles fought in court over the legality of the **environmental impact assessment**, which was done after work had already started on the project. For the indigenous people, diverting water from the river channel meant a reduction in fish populations. Additionally, because there were few roads in the region, river trading was essential, but has now been reduced. The loss of rainforest, lowering of watertables and drying out of soils are further predicted impacts. Traditional livelihoods and cultures based on small-scale fishing, floodplain farming and forest management have been threatened.

FIGURE 8 First stages of the construction of the Belo Monte Dam

Although the government has claimed the dam will provide **green energy**, the amount of greenhouse gases released from drowned and rotting vegetation behind the dam wall will contribute to global warming for some years. River flow in the region is seasonal, so hydro-electricity can be generated at peak flow for only a few months of the year. During the dry months, only 1000 MW of a potential 11 000 MW will be generated. There is a distinct possibility that another dam will need to be built upstream to supply a more even and continual flow of water for power generation.

FIGURE 9 Protesters at the dam site cut a channel through earthworks to restore flow in the Xingu River. The wording translates as 'Stop Belo Monte'.

Downstream, the small town of Altamira (shown in **FIGURE 7**) rapidly expanded during the three-year construction period, when 60 000 labourers flooded in looking for construction work. Land prices skyrocketed, the cost of living rose, and crime rates soared to create the most dangerous town in the country. Once construction was halted, workers left as jobs disappeared.

In early 2016 the project was suspended and the owners, Norte Energia, were fined $317 000 for failing to provide promised protection for local communities. A two-year 'emergency program' established in 2011 was designed to compensate people for the schools and clinics that were promised but not supplied. Each village was allocated 30 000 reais (around $12 500) per month for two years. After centuries of living a subsistence life, local tribes were introduced to the modern world. Fishing and hunting was replaced with supermarket fast food, alcohol and sweets. Motorbikes and outboard motors replaced canoes, and the role of the tribal Elders was pushed aside by younger people who could speak Portuguese with the construction workers. Traditional social activities were replaced by televisions, and plastic and other garbage accumulated as no-one knew what to do with it.

environmental impact assessment a tool used to identify the environmental, social and economic impacts, both positive and negative, of a project prior to decision-making and construction

green energy sustainable or alternative energy (e.g. wind, solar and tidal)

For nearly three decades the Jurana tribe have fought the dam construction and much of their traditional lifestyle activities have been replaced by meetings with government and company officials, environmental activists and journalists. Attitudes towards the native communities have changed. As one dam employee noted, 'In the old days you just gave the Indians a mirror and they were happy. Now they want iPads and four-wheel drives'.

Scientists are now questioning whether large-scale infrastructure projects can balance economic benefits with environmental and social costs. With the increasing threat of climate change and recent drought, which has reduced flow along the Xingu River, the Belo Monte dam may never meet its promised economic or energy-producing goals.

In 2018, the Brazilian government announced that they would cease constructing mega dams in the Amazon. Brazil has the potential to generate 50 gigawatts of energy by 2050 if they built all the dams under design, but 77 per cent of these would to some extent impact indigenous land or federally protected areas. It appears that the ongoing resistance of indigenous peoples and environmentalists, combined with other political and economic influences, has led to a hard-won change in policy.

fdw-0038

FOCUS ON FIELDWORK

Impacts of weirs and walls

Visit a weir, floodgate or dam wall in your local catchment to investigate the effects it has on the water quality. Make visual observations and take measurements of the water temperature, turbidity and pH at different points along the waterway, including inside the barrier and in the main water channel. Tabulate these results and the time of day at which they were taken.

Compare the results at each location and make suggestions as to how the water quality is being affected at each location.

Learn more about investigating the impacts of barriers in waterways using the **Impacts of weirs and walls** activity in your learnON Fieldwork Resources.

on Resources

eWorkbook	Damming rivers (ewbk-10125)
Video eLesson	Damming rivers — Key concepts (eles-5254)
Interactivities	Dam it (int-3292) The advantages and disadvantages of large-scale dams (int-5592) Distribution of downstream communities affected by large dams (int-5599) Degree of river fragmentation in the world's major drainage basins (int-7956) Location of the Belo Monte Dam and the related environmental changes (int-7958)

15.3 ACTIVITIES

1. Research one other controversial dam site around the world and compare it with the Belo Monte dam in terms of (a) size, (b) purpose and (c) impacts.
2. Does a large company such as Norte Energia have obligations to the people dislocated by such a large-scale scheme? Before deciding, carefully consider the consequences of the company being deemed responsible or not responsible.

15.3 EXERCISE

Learning pathways

■ LEVEL 1	■ LEVEL 2	■ LEVEL 3
1, 4, 5, 7, 8, 13	2, 9, 10, 11, 14	3, 6, 12, 15

Check your understanding

1. What human activities are responsible for changing or fragmenting rivers?
2. Using **FIGURE 1**, describe the location of places with rivers that are largely unaffected by river fragmentation.
3. Traditionally, why has water flowing out to sea been considered a waste? Is this a human-centred or Earth-centred viewpoint?
4. What is the primary aim of an environmental impact assessment?
5. Where are the world's largest (over 100 km^3) dams?
6. Refer to **FIGURES 7(a)** and **7(b)** to describe the environmental changes brought to the Xingu River by the dam.
7. Using information from **FIGURE 2** and the text in this subtopic, construct a table with the following headings to classify the impacts of dam building.

Positive effects for people	Negative effects for people
Positive effects for the environment	**Negative effects for the environment**

Apply your understanding

8. Would people and environments upstream of large dams be affected by the dams? Explain how.
9. Why are large-scale dam projects often seen as indicators of progress?
10. 'The positive impacts of large dam-building projects on people outweigh the negative impacts on the environment.' Do you agree or disagree with this statement? Give reasons for your point of view.
11. What makes a place suitable for a large dam? Consider landforms, climate, soil and rock type.
12. Evaluate the positive and negative changes shown in **FIGURE 2**. Which has the most significant positive impact, and which has the most negative? Justify your choice with reasons.

Challenge your understanding

13. Which countries in the world have the greatest number of people affected by large dams? Suggest a reason why.
14. Is there a sustainable future for mega dam projects such as the Belo Monte Dam? Justify your answer.
15. Predict whether Australia will need to build more or fewer dams in the future. Consider factors such as population projections, climate change and the quality of our current water sources in your answer.

To answer questions online and to receive **immediate feedback** and **sample responses** for every question, go to your learnON title at www.jacplus.com.au.

15.4 Alternatives to dams

LEARNING INTENTION

By the end of this subtopic, you will be able to describe some alternatives to dams.

15.4.1 Save water

Agriculture

Globally, more than 70 per cent of fresh water is used for agriculture. Irrigation is often very inefficient, with over half of the water applied not actually reaching the plants. High rates of evaporation and leaking infrastructure waste water. Often governments subsidise and encourage farmers to grow water-thirsty crops, such as cotton, in semi-arid regions. Poorly designed and managed irrigation schemes can become unsustainable if they develop waterlogging and salinity problems.

Vast water savings could be made by improving irrigation methods, switching to less water-consuming crops and taking poor quality land out of production. If the amount of water consumed by irrigation was reduced by 10 per cent, water available for domestic use could double across the globe.

Urban use

It is estimated that as much as 40 per cent of water is wasted in urban areas just through leaking pipes and taps. Savings can be made by:

- reducing leaking pipes and improving water delivery infrastructure
- encouraging the use of water- and energy-efficient appliances and fixtures
- changing the pricing of water to a 'the more you use, the more you pay' system
- offering incentives to industry to reduce water waste and recycle
- harvesting rainwater, collecting rainwater off roofs, recycling domestic wastewater and other efficiency schemes.

Small-scale solutions

It has been estimated that it would cost $9 billion a year between now and 2025 to provide all of the world's people with adequate water and sanitation using small-scale technologies. This amount is only one-third of current spending in developed nations on water and sanitation. It is the equivalent of nine day's defence spending by the United States of America. Rather than one large, expensive dam, smaller projects that benefit local communities can be more desirable. These are often constructed and maintained by people who benefit directly from control over their own resources, at a minimal cost (see **FIGURE 1**).

FIGURE 1 Research in India has shown that 10 micro dams with one-hectare catchments will store more water than one dam of 10 hectares.

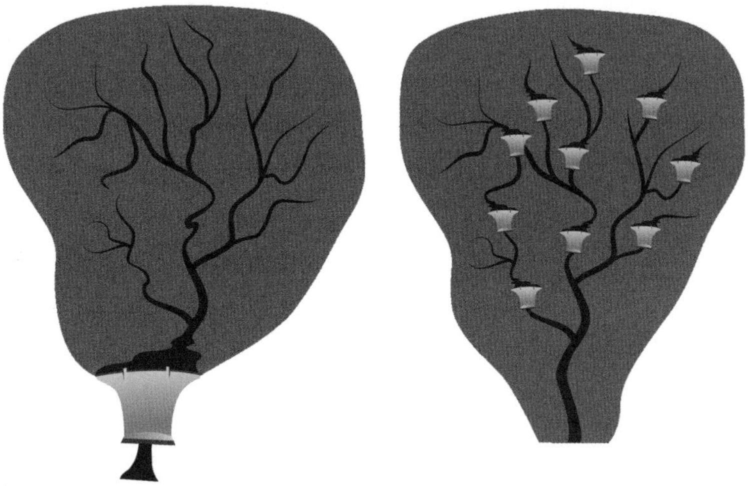

15.4.2 Other water collection methods

As many countries are actually running out of suitable places to locate large dams, alternatives need to be found. **Rainwater harvesting** schemes such as illustrated in **FIGURE 2** can be used for storing water. **Micro hydro-dams** (see **FIGURE 3**) can be used for generating electricity. Both of these schemes are easier and cheaper to build than large dams, and have lower environmental impacts.

rainwater harvesting the accumulation and storage of rainwater for reuse before it soaks into underground aquifers

micro hydro-dams produce hydro-electric power on a scale serving a small community (less than 10 MW). They usually require minimal construction and have very little environmental impact.

FIGURE 2 Two methods for water harvesting: (a) rainwater tank and (b) groundwater recharging

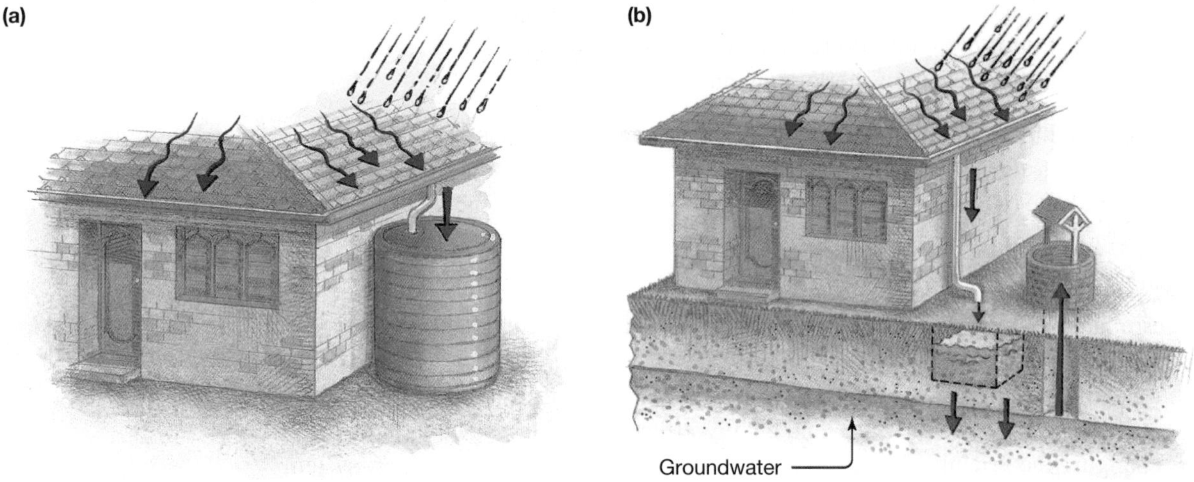

FIGURE 3 Water collected from a stream uphill rushes down the pipe and drives a small turbine in the hut to generate electricity for a local community in the Philippines.

CASE STUDY

Rajasthan, India

The state of Rajasthan is located in the arid north-west of India (see **FIGURE 4**). The region has only 1 per cent of the country's surface water and a population growth rate of 21 per cent (compared to Australia's 1.5 per cent). The largest state in India faces both water scarcity and frequent droughts. Continual pumping of groundwater has seen underground water supplies dropping.

FIGURE 4 Distribution of rainfall in India. The state of Rajasthan is highlighted.

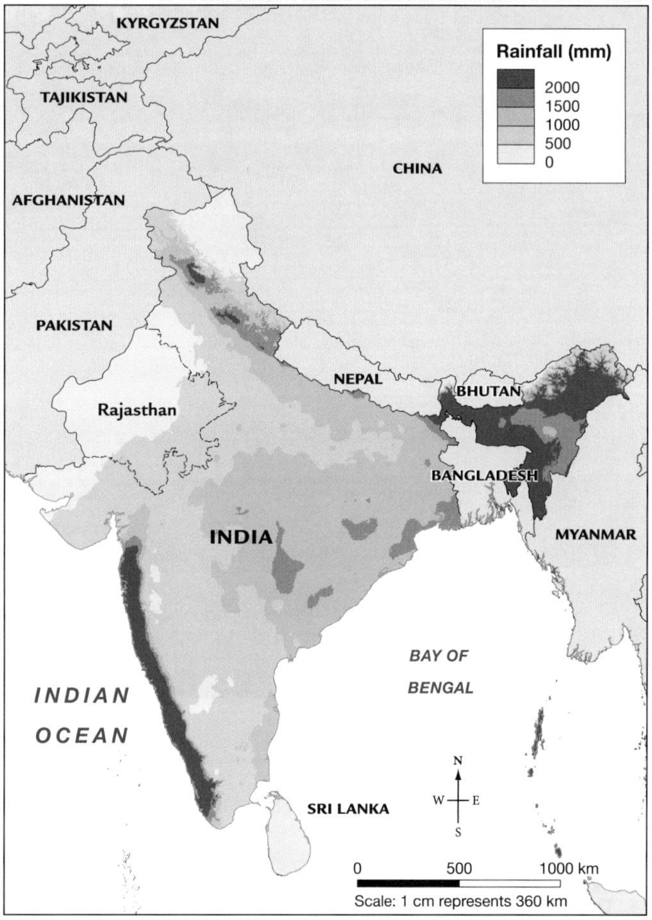

Source: World Climate - http://www.worldclim.org/; map by Spatial Vision

Traditionally, forests, grasslands and animals were considered property to be shared by all, and were carefully managed by a strict set of rules by local communities. These resources were used sustainably to ensure continual regeneration of plants and trees to enable farming to continue each year. By t he mid twentieth century, government initiatives had taken control of local resources and promoted excessive mining and logging in the area. Large-scale deforestation resulted in severe land degradation, which increased the frequency of flash floods and droughts. There was little motivation for villages to maintain traditional water systems, or johads, and so there was a gradual decline in people's economic and social wellbeing.

Tarun Bharat Sangh (TBS) is an aid agency that was established in the mid 1980s. It set about trying to re-establish traditional water management practices. It focused its attention on the construction and repair of nearly 10 000 johads in over 1000 villages. Johads are often small dirt embankments that collect rainwater and allow it to soak into the soil and recharge groundwater aquifers (see **FIGURE 5**).

Another johad design features small concrete dams across gullies that would seasonally flood, trapping the water and allowing it to infiltrate. Water stored in aquifers can later be withdrawn when needed via wells. The benefits have been remarkable and the estimated cost calculated to be an average of US$2 or 100 rupees per person. This is compared to over 10 000 rupees per head for water supplied from the Narmada River Dam Project.

What have been the benefits?

Environmental benefits

- Groundwater has risen by six metres.
- Five rivers that flowed only after the monsoon season now flow all year (fed by **base flow**).
- Revegetation schemes have increased forest cover by 38 per cent, which helps improve the soil's ability to hold water and reduce evaporation and erosion.

base flow water entering a stream from groundwater seepage, usually through the banks and bed of the stream

FIGURE 5 A johad or traditional small water harvesting dam in India

Social benefits

- More than 700 000 people across Rajasthan have benefited from improved access to water for household and farming use.
- There has been a revival of traditional cultural practices in constructing and maintaining johads.
- The role of the village council (Gram Sabha) is promoted for encouraging community participation and social justice.
- With a more reliable water supply communities became more economically viable.

FIGURE 6 Young women taking cattle to drink at a johad

Resources

- **eWorkbook** Alternatives to dams (ewbk-10129)
- **Video eLesson** Alternative to dams — Key concepts (eles-5255)
- **Interactivity** Alternatives to dams (int-8679)

15.4 ACTIVITIES

1. Complete the **Alternatives to dams** worksheet to learn more about traditional water harvesting schemes.
2. Investigate the different methods of irrigating crops, such as flood, furrow and drip irrigation.
   a. What are the advantages and disadvantages of each in terms of water use and waste?
   b. Which irrigation method would:
      i. be the most economically viable
      ii. have the most environmental benefit?

15.4 EXERCISE

Learning pathways

■ LEVEL 1	■ LEVEL 2	■ LEVEL 3
1, 2, 7, 8, 13	3, 4, 9, 10, 14	5, 6, 11, 12, 15

Check your understanding

1. Define the term *rainwater harvesting*.
2. What is a micro hydro-dam?
3. Explain how water can be wasted through poor farming methods.
4. List two advantages and two disadvantages of micro hydro-dams (shown in **FIGURE 1**).
5. Refer to **FIGURE 4**. Describe the distribution of rainfall in Rajasthan.
6. Suggest one environmental, one social and one political factor that have contributed to the decline in water availability in Rajasthan.

Apply your understanding

7. Explain what a johad does and how it works.
8. What have been the benefits of revegetation schemes around the villages restoring johads?
9. Study the information in **FIGURE 4**. Explain why Rajasthan has water issues. Use data in your answer.
10. Have small-scale water management schemes in Rajasthan been successful? Why or why not?
11. Are micro dams better alternatives to large dams?
12. Which of the methods of saving water in urban environments do you think would be the most effective in Australia? Give reasons for your answer.

Challenge your understanding

13. The two goals of sustainable water management are to reduce the demand for water, and to use existing water more efficiently. Propose two methods that your family could use to meet these goals.
14. Some places in India can receive up to 2500 mm of rainfall per year, but this can all fall in 100 hours. Suggest possible consequences of this for local communities.
15. Do you think the johad method of water harvesting could be used in other places around the world? Give reasons for your answer.

To answer questions online and to receive **immediate feedback** and **sample responses** for every question, go to your learnON title at www.jacplus.com.au.

15.5 SkillBuilder — Creating a fishbone diagram

online only

LEARNING INTENTION

By the end of this subtopic, you will be able to construct a fishbone diagram to represent information examining key points of a particular issue.

The content in this subtopic is a summary of what you will find in the online resource.

15.5.1 Tell me

What is a fishbone diagram?

A fishbone diagram is a graphic representation of the causes of a particular effect. Fishbone diagrams are useful for visualising a problem (or effect) and showing the causes of that problem. Bones above and below the central line are used to identify causes, while the 'head' of the diagram gives the problem or effect. Each major category of cause then has a number of causes and even sub-causes. These are all linked to convey the interconnection of ideas.

15.5.2 Show me

How to develop a fishbone diagram

Step 1

Determine the problem to be considered — this becomes your 'effect'. Place the effect in the head of the fishbone diagram.

FIGURE 2 Fishbone diagram template

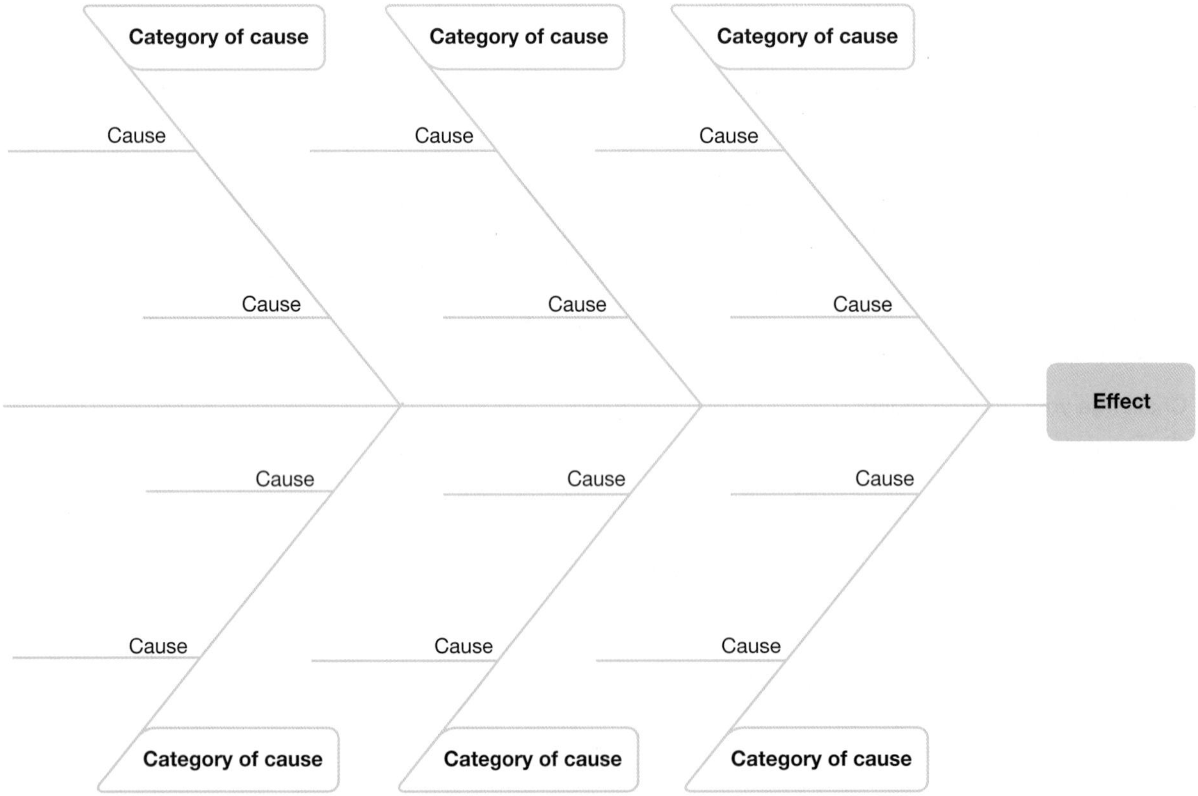

Step 2

Consider all the possible causes of the problem and decide what major categories these fall into. Then decide which of these categories are the most significant, and place them in the category of causes boxes closest to the fish head; place the least important categories of causes close to the fish tail.

Step 3

For each category of causes, now brainstorm a number of causes within that category. Keep asking

'Why is this a problem?' or 'Why does this happen?'

Step 4

Consider another category and its related causes. Complete the bones of the fish with all your ideas.

Step 5

Now in the tail of the fish you can draw your conclusion.

FIGURE 5 Conclusion shown on fishbone diagram

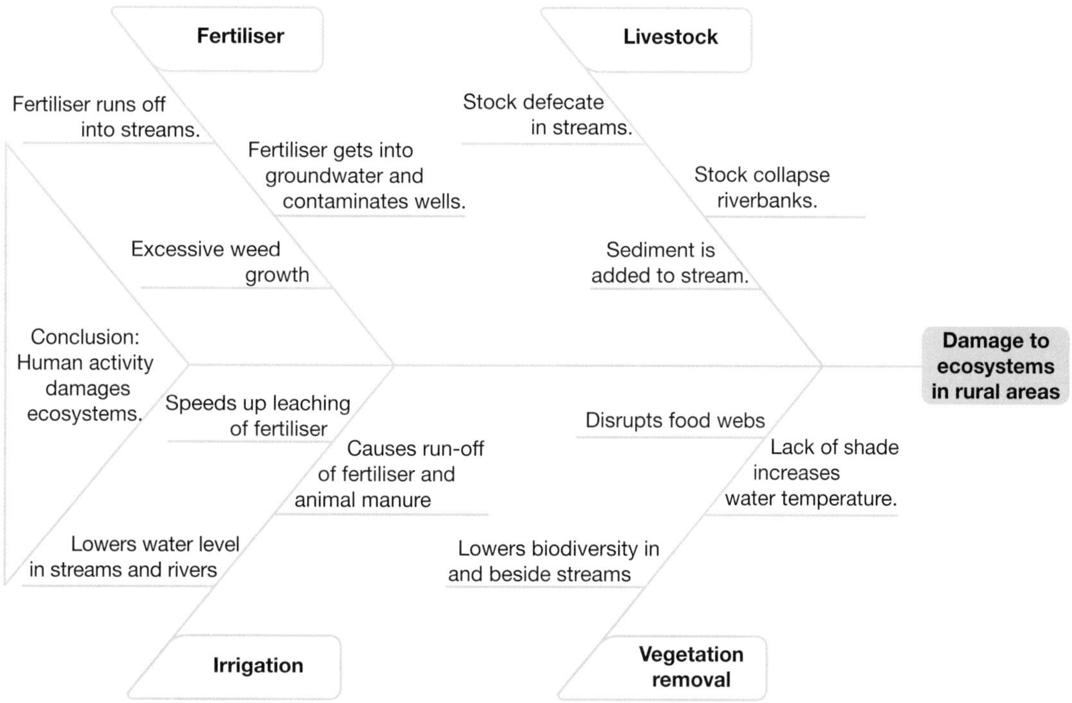

15.5.3 Let me do it

learnON

Go to learnON to access the following additional resources to help you build this skill:

- a longer explanation of this skill and its application in Geography (Tell me)
- a video demonstrating the step-by-step process of this skill (Show me)
- an activity and interactivity for you to practise the skill (Let me do it)
- self-marking questions to help you understand and use the skill.

on Resources

eWorkbook SkillBuilder — Creating a fishbone diagram (ewbk-10133)

Video eLesson SkillBuilder — Creating a fishbone diagram (eles-1740)

Interactivity SkillBuilder — Creating a fishbone diagram (int-3366)

15.6 Using our groundwater reserves

LEARNING INTENTION

By the end of this subtopic, you will be able to describe how groundwater is used and its importance. You will also be able to explain how groundwater use is an issue in China and the strategies used to fix it.

15.6.1 Understanding groundwater

Of all the fresh water in the world not locked up in ice sheets and glaciers, less than 1 per cent is available for us to use — and most of that is groundwater. More than two billion people use groundwater, making it the single most used natural resource in the world. It is also the most reliable of all water sources. Fresh water stored deep underground is essential for life on Earth.

Groundwater is one of the invisible parts of the water cycle, as it lies beneath our feet. Rainfall that does not run off the surface or fill rivers, lakes and oceans will gradually seep into the ground. **FIGURE 1** shows where groundwater is stored in porous rock layers called aquifers. Water is able to move through these aquifers and can be stored for thousands of years. Unlike most other natural resources, groundwater is found everywhere throughout the world.

FIGURE 1 Diagram showing groundwater

15.6.2 Advantages of using groundwater

Since the mid-twentieth century, advances in drilling and pumping technology have provided people with an alternative to surface water for meeting increasing water demands. Groundwater has many advantages:

- It can be cleaner than surface water.
- It is less subject to seasonal variation and there is less waste through evaporation.
- It requires less and cheaper infrastructure for pumping as opposed to dam construction.
- It has enabled large-scale irrigated farming to take place.
- In arid and semi-arid places, groundwater has become a more reliable water supply, which has led to improved water and food security.

If groundwater is removed unsustainably, that is, at a rate that is greater than is being replenished naturally by rainfall, run-off or underground flow, then watertables drop and it becomes harder and more expensive to pump. In areas of low rainfall, there is very little **recharge** of groundwater, so it may take thousands of years to replace. Over-extraction of groundwater can result in wells running dry, less water seeping into rivers, and even land **subsidence** or sinking. **FIGURE 2** identifies those places in the world most at risk of groundwater depletion. Many of these are important food bowls for the world.

recharge the process by which groundwater is replenished by the slow movement of water down through soil and rock layers

subsidence the gradual sinking of landforms to a lower level as a result of earth movements, mining operations or over-withdrawal of water

FIGURE 2 The world's use of groundwater

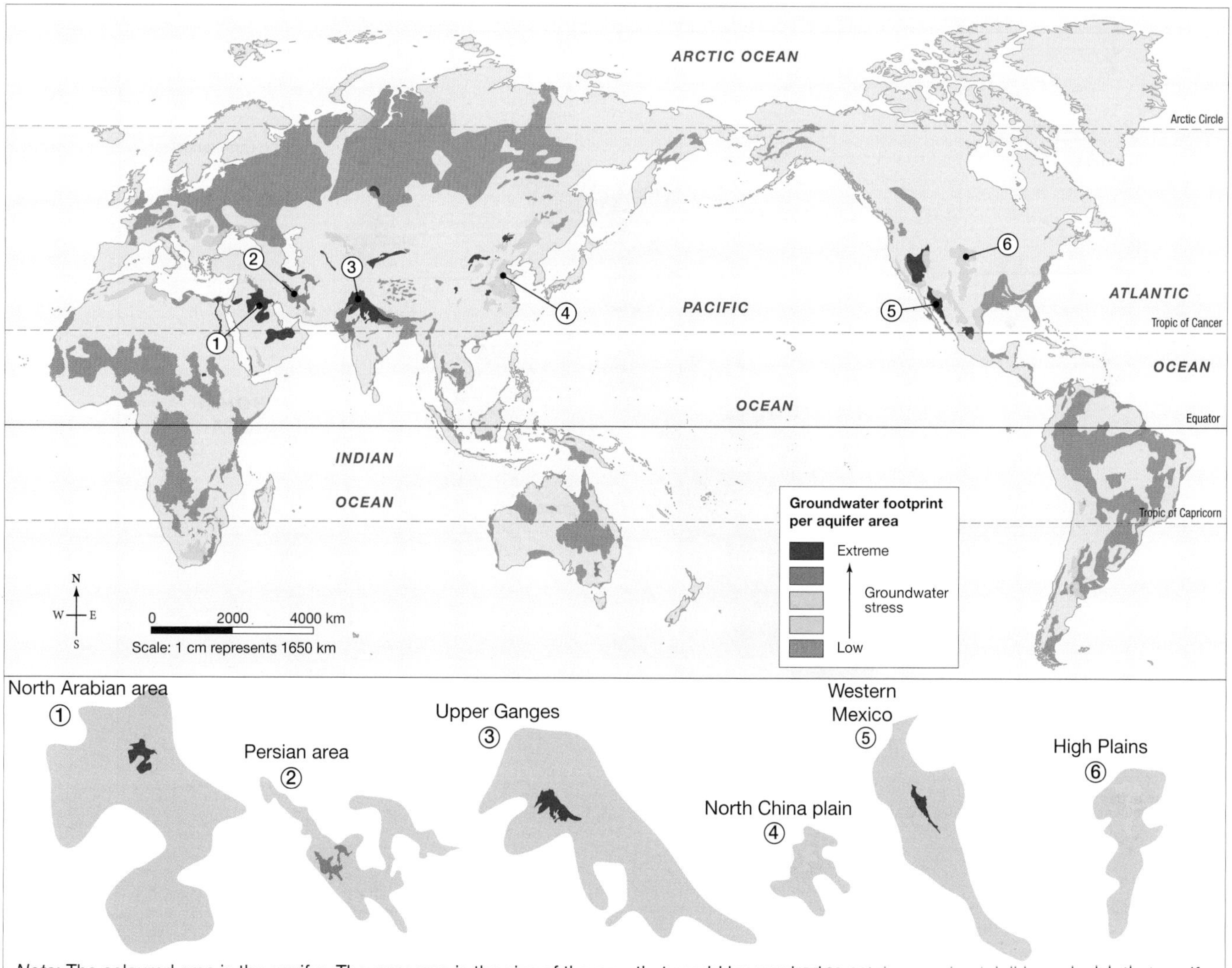

Note: The coloured area is the aquifer. The grey area is the size of the area that would be required to catch enough raintall to replenish that aquifer.

Source: BGR & UNESCO 2008: Groundwater Resources of the World 1 : 25 000 000. Hannover, Paris.

15.6.3 Improving our use of groundwater

In the past, we had limited knowledge of the interconnection between groundwater and surface water. As agriculture is the biggest user of groundwater, any improved efficiencies in water use can reduce the demand to pump more water. Improved irrigation methods and the reuse of treated effluent water could reduce our unsustainable use of groundwater. Many countries share aquifers, so pumping in one place can affect water supplies in another. There is a need for more international cooperation and management of the aquifer as a single shared resource.

CASE STUDY

Groundwater in China

What do you do if you don't have access to a reliable water source? You do what hundreds of millions of people do around the world every day. You dig for it. Beneath our feet lie vast quantities of fresh water that may have taken thousands of years to slowly work its way deep into rock layers. Since ancient times, people have used groundwater to provide for their water needs.

Rapid growth in both population and irrigated agriculture, combined with increasing demand for water, has seen the increased pumping of groundwater in northern China. As a consequence, the water table around Beijing has been dropping by 5 metres per year. Groundwater supplies more than 70 per cent of the water needs for over 100 million people living on the North China Plain (see **FIGURE 3**).

FIGURE 3 Decline in the aquifer under the North China Plain

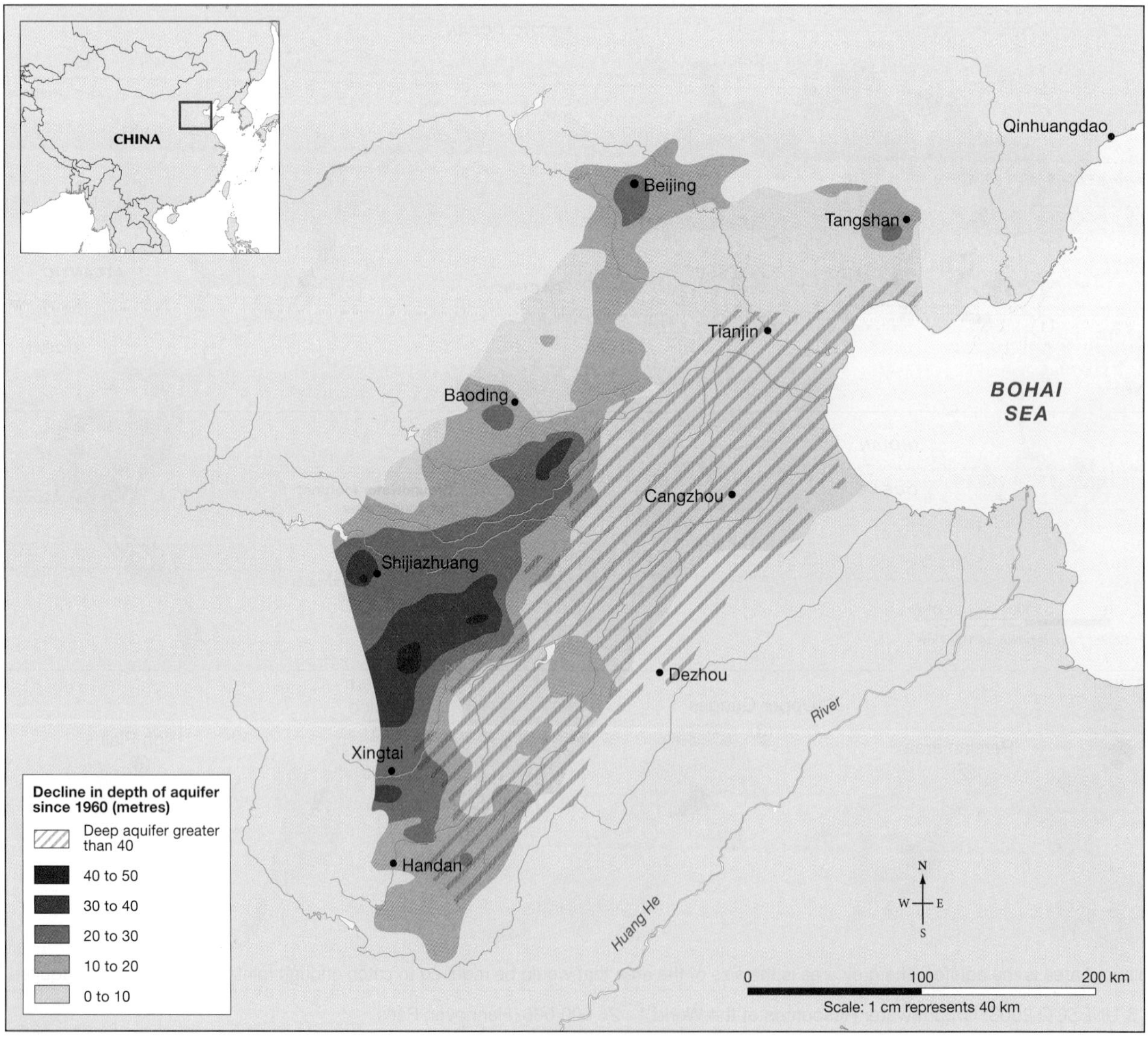

Source: UNEP Global Environmental Alert Service GEAS. Map by Spatial Vision.

The northern regions of China receive only 20 per cent of the country's rainfall. The southern regions, home to about half the population, receive the other 80 per cent. Eleven provinces in the north have less than 100 cubic metres of water per person per year, which officially classifies these places as being 'water stressed'. Eight other provinces in the north have less than 500 cubic metres of water per person per year. To date, management of water resources has been poor and unsustainable. The emphasis has always been on meeting an increasing demand for water using large-scale engineering 'fixes', rather than looking at ways of using water more efficiently, slowing down demand by increasing the cost, reducing irrigation wastage and improving the catchment areas to recharge the aquifers.

After extensive flooding in the 1960s, the government set about building dams and canals to reduce flood impacts and provide water to rapidly growing cities. Farmers were encouraged to increase grain production by drawing on groundwater to irrigate a second crop each year. As cities continued to expand, they too began to pump unsustainable amounts of groundwater for domestic and industrial use. Scientists now also believe that climate change has reduced rainfall in the region, which will only make the situation worse.

The South–North Water Transfer Project

In 2002, an ambitious 50-year project was started to effectively 're-plumb' the country: the South–North Water Transfer Project. At an estimated cost of US$62 billion, over 44.8 billion cubic metres of water per year will be diverted north from the Yangtze River via three canals into the Huang He River Basin in the north of the country (see **FIGURE 4**). Before the transfer project development, the Yangtze River, on average, released 960 billion cubic litres of fresh water into the sea each year. Construction of the central and eastern sections has been completed, but the western route is still being planned. The completed central section now supplies 73 per cent

FIGURE 4 South–North Water Transfer Project for transfer of water from the Yangzte River in the south to the Huang He River in the north

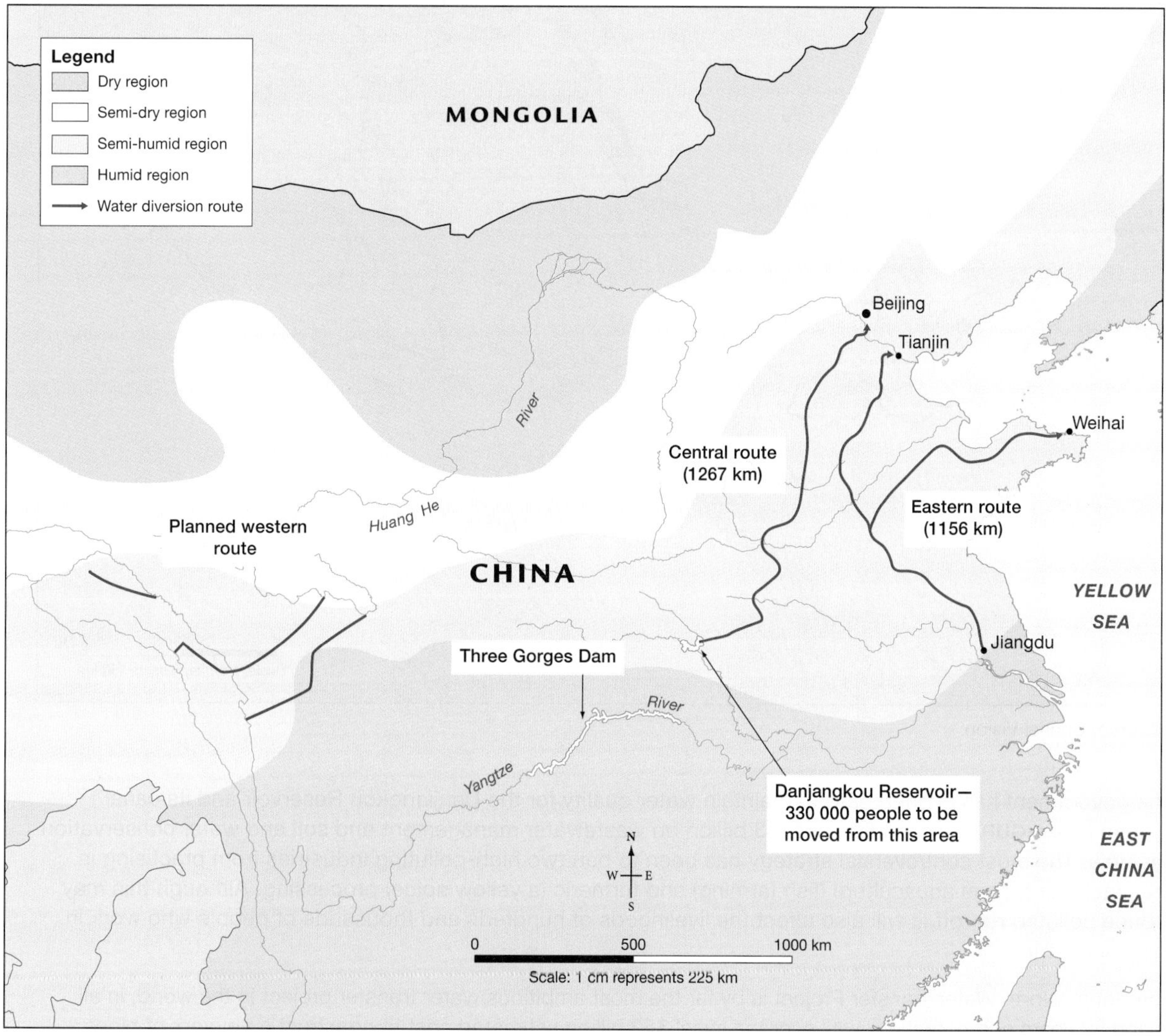

Source: BBC News, http://news.bbc.co.uk/2/hi/8545321.stm

of the Beijing's tap water, to provide for its population of 21.5 million people. The water transfer has reduced the exploitation of groundwater by 800 million cubic metres. As the extra surface water filters into the ground, the watertable has started to rise, with levels increasing by around half a metre.

The project's water will largely go to expanding industries and cities such as Beijing and Tianjin. Little of the water will be directed towards food production. Irrespective of the cost and relocation of hundreds of thousands of people, the biggest ongoing concerns of the scheme will be about water quality. The Huang He River collects over 4.29 billion tonnes of waste and sewage each year (see **FIGURE 5**), and over 40 per cent of China's total waste water is dumped in the Yangtze River (see **FIGURE 6**). These figures are likely to increase as more and more industry moves close to new water sources. With less water to flow downstream, there will be less water available to dilute the polluted water. This will affect river environments and it is possible that the water reaching the north will be too contaminated for human or even agricultural use.

FIGURE 5 Levels of pollution along the Huang He River

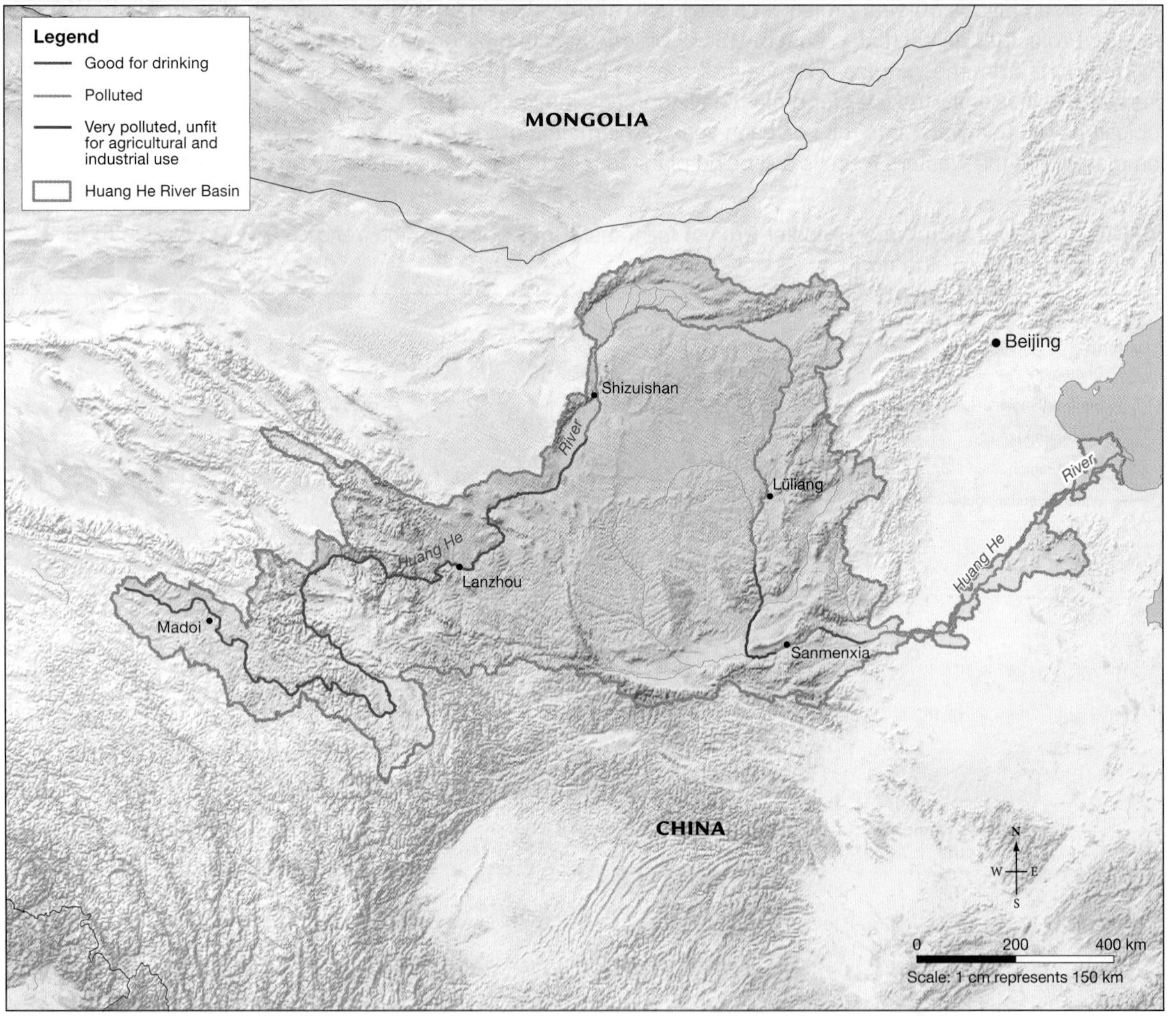

Source: Spatial Vision.

The government has pushed hard to maintain water quality for the Danjiangkou Reservoir and its canal system (see **FIGURE 4**), spending over $3 billion on wastewater management and soil and water conservation systems. The most controversial strategy has been to ban two high-polluting industries from practising in the catchment: cage aquaculture (fish farming) and turmeric (a yellow spice) processing. Although this may reduce polluted run-off, it will also affect the livelihoods of hundreds and thousands of people who work in these industries.

The South–North Water Transfer Project is by far the most ambitious water transfer project in the world, in all taking 50 years and potentially well over the initial $62 billion estimated cost to construct a network of pipes, canals and tunnels that would stretch in a straight line from Melbourne to Fiji. Is it sustainable? Beijing consumes more than 3.6 billion cubic metres of water each year, supplied partially by its own traditional surface and groundwater sources and, increasingly, by the transfer scheme. Predictions are that its needs will soon outstrip the new scheme's capacity to supply. Scientists are already questioning the 'big scheme' approach rather than the use of water recycling, desalination and harvesting more rainwater as more environmentally friendly and sustainable methods of supplying water.

FIGURE 6 Polluted water flows into the Yangtze River.

Resources

eWorkbook	Using groundwater reserves (ewbk-10137)
Video eLesson	Using groundwater reserves — Key concepts (eles-5256)
Interactivity	That sinking feeling (int-3293)

15.6 ACTIVITY

Using an atlas, find a map of world food production and compare this with any three places from **FIGURE 2**. Conduct research to determine the following:

a. What types of food are produced in those regions of the world where watertables are severely depleted?
b. What are the future implications for sustainable food production in these regions?

15.6 EXERCISE

Learning pathways

■ LEVEL 1	■ LEVEL 2	■ LEVEL 3
1, 3, 6, 7, 13	2, 5, 8, 9, 14	4, 10, 11, 12, 15

Check your understanding

1. What is a watertable?
2. What are the advantages and disadvantages of using groundwater for domestic and agricultural purposes?
3. What is the difference between groundwater and the watertable?
4. Describe how water can move vertically and horizontally through the ground.
5. Refer to **FIGURE 2**. Describe the location of places in the world that have the highest groundwater stress.
6. How does groundwater recharge?

Apply your understanding

7. What is the interconnection between atmospheric water, surface water and groundwater?
8. Looking at **FIGURE 2**, compare the scale of the selected aquifers with the scale of the area needed to recharge them.
9. Using **FIGURE 3**, describe the location of the North China Plain. Use distance, direction and place names in your answer.
10. Refer to **FIGURE 4**. Suggest a reason why northern China uses groundwater to supply over 70 per cent of its water needs.

11. What might be the possible source for the pollution you can see entering the river in **FIGURE 6**?
12. Explain the two most significant advantages of using groundwater for China.

Challenge your understanding

13. Is the South–North Water Transfer Project sustainable? Is a human-centred rather than earth-centred viewpoint the best option for water management in northern China? Write a paragraph outlining your views.
14. There is often talk about transferring water from the wetter regions of northern Australia to the water-hungry regions further south. What would you need to know before planning a project similar to the one in China? Thinking geographically, write a list of 10 questions to consider before designing such a project.
15. Who owns groundwater? How can we manage the resource sustainably? Write a paragraph expressing your viewpoint.

To answer questions online and to receive **immediate feedback** and **sample responses** for every question, go to your learnON title at www.jacplus.com.au.

15.7 The impacts of drainage and diversion

LEARNING INTENTION

By the end of this subtopic, you will be able to explain the importance of wetlands and describe the environmental changes brought about by draining and diverting water.

15.7.1 Understanding wetlands

Often referred to as the area where 'earth and water meet', wetlands are one of the most important and valuable biomes in the world. Wetlands are areas that are covered by water permanently, seasonally or ephemerally, and can include fresh, salty and brackish waters. They include such things as ponds, bogs, swamps, marshes, rice paddies and coastal lagoons. Wetlands are intricately connected to other elements in the landscape, especially rivers and floodplains as water, nutrients and sediments move between them (see **FIGURE 1**).

FIGURE 1 Interconnections between the river and wetlands on the floodplain

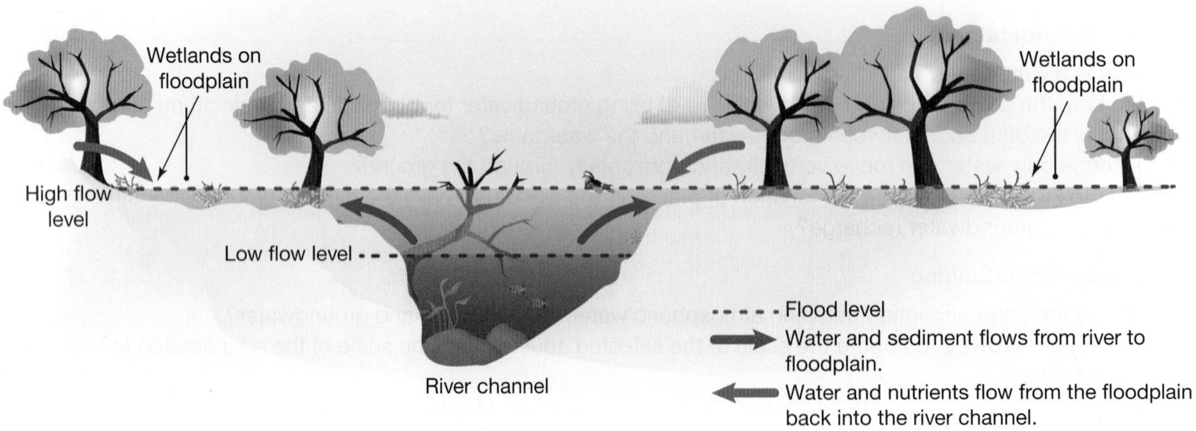

The importance of wetlands

Wetlands perform many important functions. They purify water; for example, much of Melbourne's sewage water is filtered through a series of lagoons and wetlands at the Western Treatment Plant in Werribee, producing high-quality recycled water. Wetlands located along river floodplains reduce the impact and speed of floods by holding vast quantities of flood water and then slowly releasing it back into the river system. Water in wetlands also infiltrates the soil, recharging groundwater reserves. In addition, wetlands provide habitat and breeding grounds for 40 per cent of the world's species, such as aquatic fish, insects, reptiles and birds. Globally, more than one billion people rely on wetlands for a living, for their water and food supply, and for tourism and recreation (see **FIGURE 2**).

FIGURE 2 A wetland in Queensland. What features in this image are typical of a wetland?

15.7.2 Threats to wetlands

The degradation and loss of wetlands and the species that inhabit them have been more rapid than any other ecosystem; in fact, wetlands are disappearing three times faster than forests. It has been estimated that between 1970 and 2015, the world lost 35 per cent of its wetlands. Competition from other land uses and increasing populations have contributed to the decline.

- Agricultural expansion is the largest contributor to wetland loss and degradation globally. Farming often requires the draining of wetlands to create more land, reducing biodiversity. In addition, the water run-off from agriculture is often polluted with fertilisers and pesticides, and increased pumping from aquifers depletes groundwater resources.
- Dams alter seasonal floods and block supply of sediment and nutrients onto the floodplain and deltas. Often, damming means little water and sediment reaches the mouths and deltas of large rivers.
- Loss of wetlands affects populations and the migratory patterns of birds and fish. The introduction of invasive species results in changed ecosystems and loss of biodiversity. For example, 70 per cent of amphibian species are affected by habitat loss.
- Clearing for urban growth, industry, roads and other land uses replaces wetlands with hard **impervious** surfaces. This reduces infiltration and leads to polluted run-off and increased impacts of flooding.
- Although wetlands can naturally filter many pollutants, excessive amounts of fertilisers and sewage causes algal blooms and **eutrophication**, depriving aquatic plants and animals of light and oxygen.
- Climate change is expected to increase the rate of wetland degradation and loss.

Wetlands, like all other water resources, are prone to over-exploitation and need to be managed carefully to ensure sustainable use.

15.7.3 Water diversion

Because populations and water sources are distributed unevenly, we often need to transfer or divert large amounts of water. This means piping or pumping water from one drainage basin to another. For example, in Australia, water from the Snowy River is diverted into the Murray and Murrumbidgee rivers. Diverting water can alleviate water shortages and allows for the development of irrigation and the production of hydro-electricity. Diversions, however, are not always the most sustainable use of water resources.

Many of the world's greatest lakes are shrinking, and large rivers such as the Colorado, Rio Grande, Indus, Ganges, Nile and Murray discharge very little water into the sea for months and even years at a time. Up to one-third of the world's major rivers and lakes are drying up, and the groundwater wells for three billion people are being affected. The overuse and diversion of water is largely to blame.

impervious a rock layer that does not allow water to move through it due to a lack of cracks and fissures

eutrophication a process where water bodies receive excess nutrients that stimulate excessive plant growth

CASE STUDY

Lake Urmia

The largest lake in the Middle East and one of the largest salt lakes in the world is drying up. Since the 1970s, Lake Urmia in northern Iran has shrunk by nearly 90 per cent. In 1999, the lake's volume was 30 billion cubic metres; by 2018, this had reduced to 2 billion cubic metres, exposing extensive areas of salt flats (see **FIGURE 3**). After unprecedented torrential rains in 2019, the depth of Lake Urmia increased by 62 cm and the area grew by almost 1000 square kilometres.

The lake was declared a Wetland of International Importance by the Ramsar Convention in 1971 and a UNESCO Biosphere Reserve in 1976. The lake and its surrounding wetlands serve as a seasonal habitat and feeding ground for migratory birds that feed on the lake's shrimp. This shrimp is the only thing, other than plankton, that can live in the salty water.

FIGURE 3 Lake Urmia (a) in 1998, (b) in 2016, (c) in February 2019 and (d) in April 2019

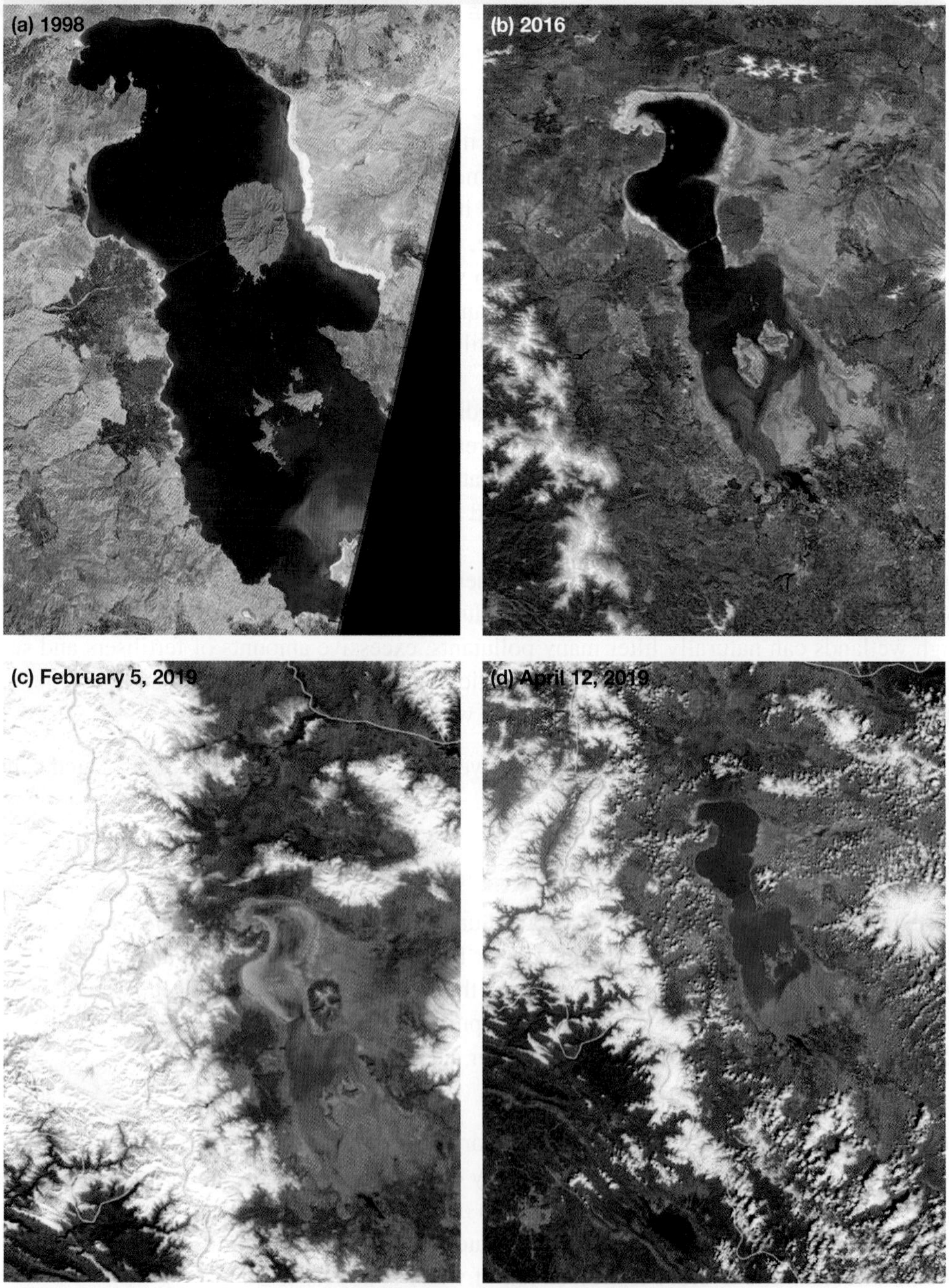

Lake Urmia is a **terminal lake**; the rivers that flow into the lake (some permanent and some ephemeral) bring naturally occurring salts. Because of the arid climate, high evaporation causes salt crystals to build up around the shoreline. **FIGURE 4** shows the rapid decline in the surface area of Lake Urmia from 2006 to 2013.

Why is the lake drying up?

A combination of environmental, economic and social factors has been blamed for the large-scale changes in Lake Urmia. Prolonged drought and the illegal withdrawal of water by farmers who do not pay or who take more than their allocation are minor contributors to the problem.

Researchers have found that 60 per cent of the decline can be attributed to climate changes (increased frequency of drought and higher temperatures), and 40 per cent of the decline relates to water diversions and the increased demand for water in the region as the population has risen and land use for farming has tripled. The result is a form of 'socioeconomic drought' — a human-induced drought caused when the demand for water is greater than the available supply.

Impacts of this drought include:

- increased salinity of the shallow lake due to high evaporation (rates of between 600 mm and 1000 mm per year) and reduced freshwater flowing in via rivers (salt levels have increased from 160 g/L to 330 g/L)
- collapse of the lake's ecosystem and food chain (salt levels over 320 g/L are fatal to the shrimp that form the basis of its food chain)
- loss of habitat as surrounding wetlands dry up, which then reduces tourism to view wetland wildlife
- over 400 km^2 of exposed lakebed around its shores is nothing but salty deserts, unable to support native vegetation or food crops
- salt storms occurring, as wind blows salt and dust from the exposed, dry lakebed; the storms damage crops and are also a potential health hazard
- less water available to produce food crops.

FIGURE 4 Surface area of Lake Urmia

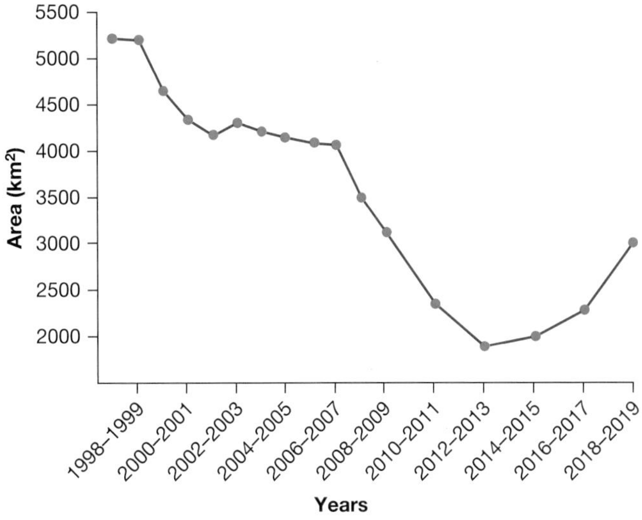

Source: United Nations Environment Programme UNEP

FIGURE 5 Distribution of dams, existing and under construction, in the lake's catchment area. This level of diversion is unsustainable.

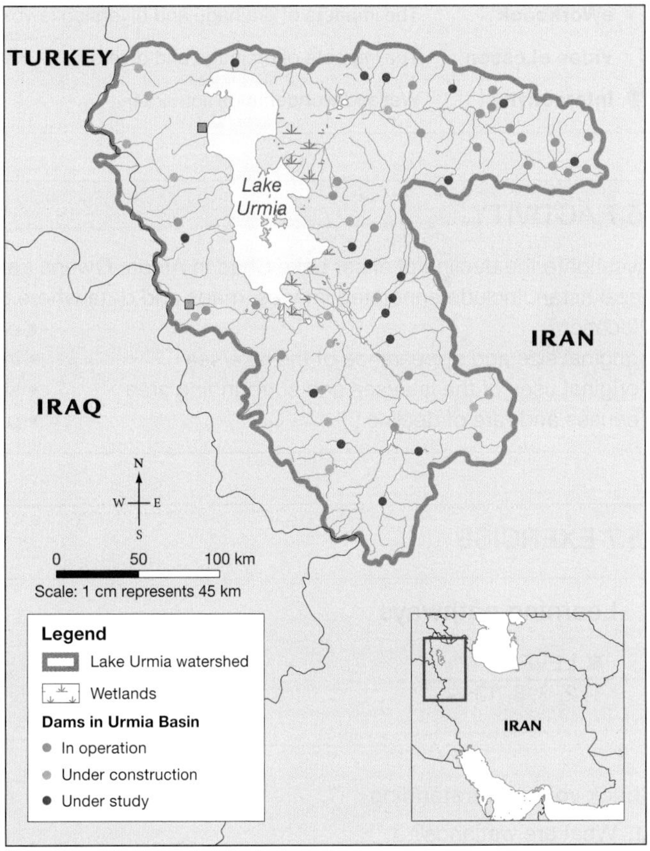

Source: United Nations Environment Programme. Vector Map Level 0 Digital Chart of the World.

The lake is divided in two by a causeway and bridge constructed to improve access across the lake (the bridge can be seen in **FIGURE 3b**). However, there is concern that the nearly 1.5-km-long bridge does not allow sufficient mixing of water between the north and south sections of the lake. The bridge, completed in 2008, is already rusting as a result of the highly saline water.

terminal lake a lake where the water does not drain into a river or sea. Water can leave only through evaporation, which can increase salt levels in arid regions. Also known as an endorheic lake.

The current situation

Essentially, more water is required to flow into the lake to increase the water level, dilute the salt and maintain an ecological balance. However, with higher than normal rainfall, the lake is beginning to see the return of animals including brine shrimp and water birds.

In 2017, the government pledged an annual budget of US$460 million to help restore the lake and its surrounding wetlands. However, only about US$5 million has been made available. Other programs in place include:

- a water transfer scheme moving water from the Little Zab basin through tunnels and channels
- engineering works to help clear sediment clogging many of the rivers that feed into the lake
- releasing water from dams to flow into the lake
- constructing 13 treatment plants in the region to treat wastewater from urban areas and deliver it to the lake
- reducing consumption of potable water by 30 per cent by 2021 and to use desalinated water to meet 30 per cent of the water demands in South Iran
- trials of planting vegetation to reduce wind speed and salt storms
- promoting water-saving farming techniques as 85 per cent of the water in the region is used for agriculture.

In recent years, water levels have increased but further progress is limited by a lack of funding. In addition, the lake can only support 300 000 hectares of farmland, whereas farming currently uses 680 000 hectares. Some progress has been made in changing farming practices, with a 35 per cent reduction in water use by local farmers and a 40 per cent reduction of pesticide use.

---

on Resources

eWorkbook	The impacts of drainage and diversion (ewbk-10141)
Video eLesson	The impacts of drainage and diversion — Key concepts (eles-5257)
Interactivity	Wetland wonderlands (int-3294)

15.7 ACTIVITY

Investigate the decline of either Lake Chad in Africa, Owens Lake in the United States or the Aral Sea in Kazakhstan. Include annotated images, maps and data where possible. Use the following as a research guide:

- location
- original size and appearance of the lake/sea
- original uses of the lake/sea and surrounding area
- causes and rate of decline
- changes that have taken place
- impacts on people and the environment
- strategies in place to reverse damage
- possible solutions.

15.7 EXERCISE

Learning pathways

■ LEVEL 1	■ LEVEL 2	■ LEVEL 3
1, 2, 3, 8, 13	4, 5, 9, 12, 15	6, 7, 10, 11, 14

Check your understanding

1. What are wetlands?
2. In what ways are wetlands important?
3. Outline three major threats to wetlands.
4. Why is it sometimes necessary to divert water?
5. Lake Urmia is considered a terminal lake. What does this mean?
6. Outline two ways that higher than normal rainfall affected Lake Urmia.
7. What is the interconnection between rivers and wetlands?

Apply your understanding

8. Explain eutrophication, its causes, and the impact it has on wetland biomes.
9. Discuss the advantages and disadvantages of diverting water from one drainage basin to another.
10. Examine **FIGURES 3a** and **3b**. Compare the appearance of Lake Urmia in 1998 and 2016.
11. Refer to **FIGURE 4**. Describe, with the use of data, the changes in the surface area of Lake Urmia.
12. What do you consider to be the worst threat to wetland environments? Give reasons for your answer.

Challenge your understanding

13. Do you think there is a future for Lake Urmia? How successful will the restoration project be? Explain your view.
14. Suggest how urban developments and infrastructure might be made less damaging to wetland environments. Propose a regulation that might be introduced to allow for greater infiltration.
15. Predict how climate change will affect wetland degradation, giving reasons for your view.

To answer questions online and to receive **immediate feedback** and **sample responses** for every question, go to your learnON title at www.jacplus.com.au.

15.8 Case study — Managing the Murray–Darling

online only

LEARNING INTENTION

By the end of this subtopic, you will be able to explain how the water resources of the Murray–Darling Basin are managed to cater for agricultural, domestic, industrial, recreational and environmental needs, and evaluate strategies to improve the basin's health.

The content in this subtopic is a summary of what you will find in the online resource.

The rivers, lakes and wetlands of the Murray–Darling Basin make it Australia's most important inland water body. Decades of continually diverting water from its rivers and prolonged periods of drought have brought significant changes to the rivers, surrounding floodplains and surrounding wetlands.

To learn more about the human impact on the Murray–Darling Basin and the strategies that are being used to manage its system, go to your learnON resources at www.jacPLUS.com.au.

Contents

learnON

Resources

eWorkbook	Case study — Managing the Murray–Darling (ewbk-10153)
Video eLesson	Case study — Managing the Murray–Darling – Key concepts (eles-5258)
Interactivity	Case study — Managing the Murray–Darling (int-8680)
myWorldAtlas	Deepen your understanding of this topic with related case studies and questions. • **Murray–Darling Basin**

15.9 SkillBuilder — Reading topographic maps at an advanced level

online only

LEARNING INTENTION

By the end of this subtopic, you will be able to calculate gradient, relief and area on a topographic map.

The content in this subtopic is a summary of what you will find in the online resource.

15.9.1 Tell me

What is reading a topographic map at an advanced level?

Topographic maps are more than just contour maps showing the height and shape of the land. They also include local relief and gradients and allow us to calculate the size of various areas. Reading this information requires more advanced skills.

15.9.2 Show me

How to read a topographic map at an advanced level

Step 1

Revise your skills: check the legend symbols, determine the map scale and check your grid reference skills. Cast your eye over the map and make interpretations of the area. What are the obvious features of this map?

Step 2

Discovering the local relief of the area is best done using spot heights. Spot heights are used to indicate the highest or lowest point, but can also be given across a map when the land is flat and few contours appear.

Calculate the local relief within the region. Local relief is the measure of the difference in height between the highest and lowest points within a relatively small area.

Step 3

Gradient is the measurement of the steepness of the land between two places. To calculate the gradient, you need to divide the vertical interval or 'rise' by the horizontal distance or 'run'.

Step 4

To calculate the area of an unusual shape on the map, a scaled grid can be placed over the map and the parts can be added up to give a squared area. Draw a grid on a piece of tracing paper.
Lay the tracing paper over the mapped area and mark the squares that are complete in the mapped area.
Now mark the incomplete squares and count them as half squares. That is, halve the number of incomplete squares. Add up the number of markings from steps 7 and 8 to obtain the total size.

15.9.3 Let me do it

learnON

Go to learnON to access the following additional resources to help you build this skill:
- a longer explanation of this skill and its application in Geography (Tell me)
- a video demonstrating the step-by-step process of this skill (Show me)
- an activity and interactivity for you to practise the skill (Let me do it)
- self-marking questions to help you understand and use the skill.

on Resources

eWorkbook	SkillBuilder — Reading topographic maps at an advanced level (ewbk-10149)
Digital documents	Topographic map of Wentworth, New South Wales (doc-11569) Topographic map of Berri, South Australia (doc-11570)
Video eLesson	SkillBuilder — Reading topographic maps at an advanced level (eles-1749)
Interactivity	SkillBuilder — Reading topographic maps at an advanced level (int-3367)

15.10 Investigating topographic maps — Wetlands along the Murray River

LEARNING INTENTION

By the end of this subtopic, you will be able to explain the importance of managing wetlands.

15.10.1 Tar-Ru Wetlands

The Tar-Ru Wetlands within the Carrs, Capitts and Bunberoo Creeks are located on the floodplains within Locks 9 and 8 of the Murray River and Frenchmans Creek, about 45 kilometres west of Wentworth, New South Wales. Over time, the natural water flows of the wetlands have been interrupted by river regulation methods such as weirs and levee banks. Managing the health of the Murray Darling Basin wetlands is essential to the functioning of the entire ecosystem, as wetlands provide an abundance of food and support the majority of native fish at some point in their lifecycle.

FIGURE 1 Southern Bell frog

In April 2016, a project was initiated to spread nearly one billion litres of environmental water across five sites on Tar-Ru Lands. This included 60 hectares of wetlands between the Carrs, Capitts and Bunberoo Creek systems. This project was expected to increase the biodiversity of native waterbirds, the population of the Southern Bell frog and improve the health of River Red gums; it saw immediate results. In the weeks after the watering of the wetlands, the number of observed bird species more than doubled and the number of individual birds increased by 250 per cent, as well as sightings of two species of frogs, new green vegetation and new aquatic vegetation increasing tenfold. The Tar-Ru Wetland is managed by the Tar-Ru Aboriginal Land Managers, who will assist in maintaining the health of the aquatic habitat and ongoing environmental monitoring.

FIGURE 2 Topographic map of Tar-Ru Wetlands, west of Wentworth, New South Wales

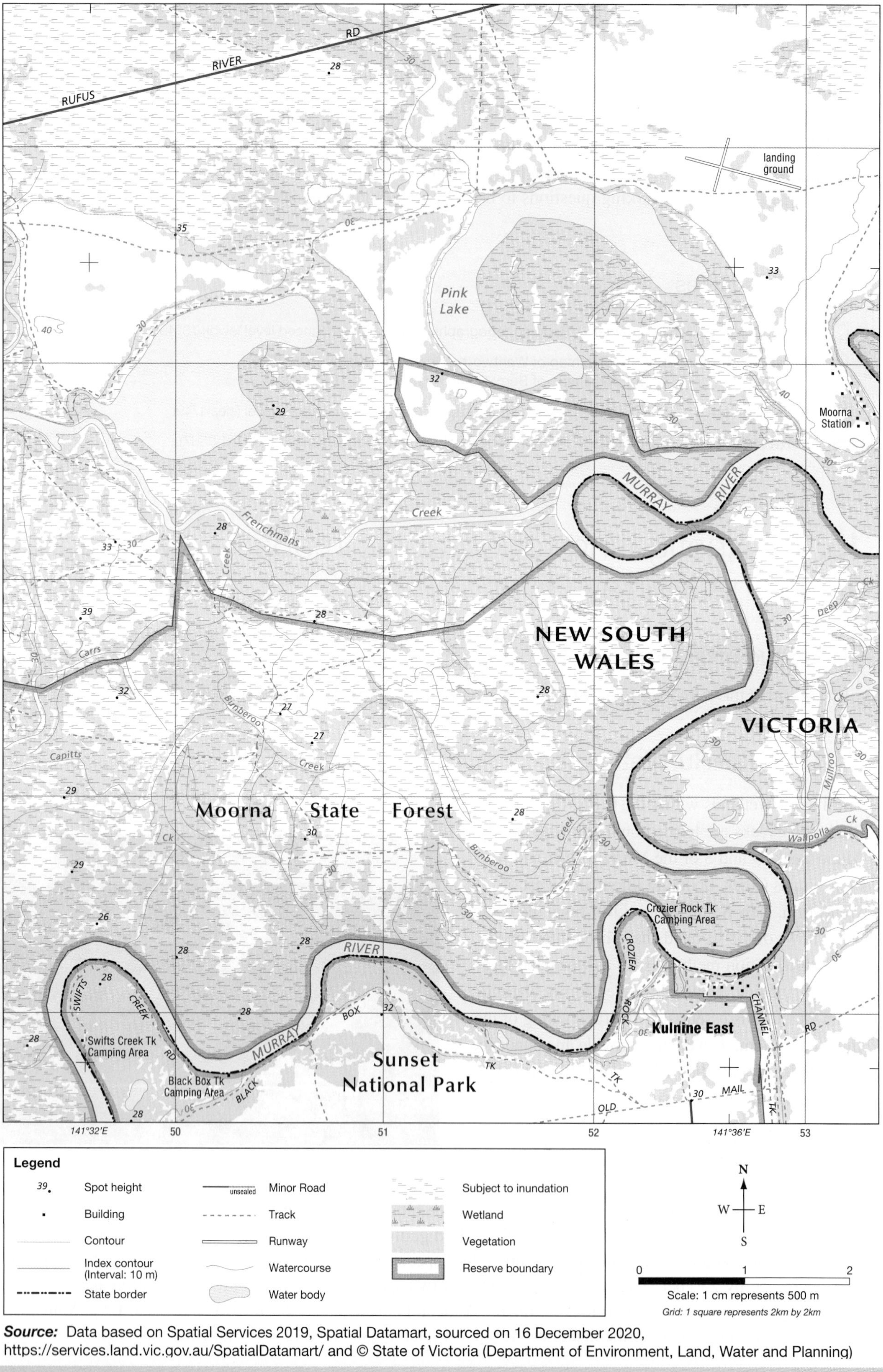

Source: Data based on Spatial Services 2019, Spatial Datamart, sourced on 16 December 2020, https://services.land.vic.gov.au/SpatialDatamart/ and © State of Victoria (Department of Environment, Land, Water and Planning)

Resources

eWorkbook	Investigating topographic maps — Wetlands along the Murray River (ewbk-10145)
Digital documents	Topographic map of the Tar-Ru Wetlands (doc-36369)
Video eLesson	Investigating topographic maps — Wetlands along the Murray River (eles-5259)
Interactivity	Investigating topographic maps — Wetlands along the Murray River (int-8681)
Google Earth	Tar-Ru Wetlands (gogl-0142)

15.10 EXERCISE

Learning pathways

■ LEVEL 1	■ LEVEL 2	■ LEVEL 3
3, 5, 6, 9	1, 2, 7, 11	4, 8, 10, 12

Check your understanding

1. Identify and give the grid reference for the following in **FIGURE 2**.
   a. Crozier Rock Track Camping Area
   b. the aircraft landing ground
   c. a spot height over 30 m in Victoria
   d. the eastern most point of Pink Lake
2. What is the relief of the area shown?
3. Mark the locations of the following water-management constructions on your map, and add an appropriate symbol to the legend.
   a. Lock 9 (GR524163)
   b. weir/bridge (GR493196)
4. Give the longitude and latitude of the following (to the nearest minute).
   a. Kulnine East settlement
   b. the most easterly point of Pink Lake
   c. Moorna Station
5. How long is the section of Rufus River Road show in **FIGURE 2**?
6. Calculate the area of the land shown on **FIGURE 2** that is north of Rufus River Road.
7. Calculate the area of the land shown on **FIGURE 2** that is in New South Wales.
8. Calculate the area of the land shown on **FIGURE 2** that is subject to inundation.

Apply your understanding

9. Use evidence from the map to list the changes that people have brought to the Tar-Ru Wetlands.
10. Providing evidence (such as elevation and proximity to waterways) discuss why the buildings on Moorna Station might have been built in their given location.
11. Imagine you are planning to hike from the airstrip (landing ground) around Pink Lake and back. The weather forecast suggests that heavy rain is expected from mid-afternoon.
    a. Outline the route you would take, including grid references, distances and appropriate positional language.
    b. Explain why you chose this route.

Challenge your understanding

12. If a dam was built two kilometres upstream of the most north-easterly point of the Murray River shown on **FIGURE 2**, predict what changes are likely to occur to:
    a. the wetlands
    b. peoples' access to homes, towns and settlements
    c. peoples' ability to access the parks and reserves of the area

To answer questions online and to receive **immediate feedback** and **sample responses** for every question, go to your learnON title at www.jacplus.com.au.

15.11 Thinking Big research project — Menindee Lakes murder — news report

online only

Scenario

The content in this subtopic is a summary of what you will find in the online resource.

During the summer of 2018–19, massive fish kills occurred in the Menindee Lakes in western New South Wales. More than a million native fish died; community outrage ensued.

A 'fish kill' event involves the death of a large number of fish or other aquatic animals (such as crabs or prawns) over a short period of time and often within a defined area.

Such events can occur due to a wide range of factors including:

- natural spawning and migration events
- diseases, including susceptibility to disease due to stress or poor water quality
- low dissolved oxygen, which can be caused by decay of algal blooms, decay of other organic matter, coral spawning, or poor mixing of a water body
- sudden change in water quality, such as salinity, pH, turbidity, dissolved solids or temperature
- contaminants such as hydrogen sulfide, carbon dioxide, ammonia, methane and others, including metals
- physical irritants, such as suspended sediment, algal cells and bacteria that interfere with fish gills
- algal toxins, which are produced by some species of algae under certain conditions.

The Menindee Lakes act as a storage facility for water in the Darling River, part of the extensive Murray–Darling Basin, so the amount of water in the lakes at any one time is actually controlled. There is considerable controversy and blame-laying over the management of the water in the Darling River, especially between upstream users (particularly those irrigating large cotton farms) and the downstream users of water, which include the city of Broken Hill and farmers and towns all the way to the mouth of the Murray–Darling River in South Australia.

Task

learnON

As a reporter for a city newspaper, you have been sent to the small town of Menindee to investigate the fish kills. What caused this horrifying event and what can be done to stop it happening again? It's your job to uncover the truth! You will research the Menindee Lakes fish kills and the surrounding controversy to write a front-page investigative report for your newspaper (print or online edition).

Go to learnON to access the resources you need to complete this research project.

on Resources

ProjectsPLUS Thinking Big research project — Menindee Lakes murder — news report (pro-0213)

15.12 Review

15.12.1 Key knowledge summary

15.2 Inland water

- Inland water covers a range of different landforms and environments.
- Water that is stored in rivers, lakes and groundwater provides a wide range of environmental services.
- Changes as a result of human activities can alter the environmental functions of inland water bodies.

15.3 Damming rivers

- Dams provide many benefits to societies from supplying water and electricity to preventing floods and providing irrigation water.
- At the same time, dams also create river fragmentation, displace communities and change river flows.
- Large-scale mega dams have always been associated with economic development and progress.
- Mega dams have brought significant environmental and social impacts.
- Globally, there are questions about the economic, social and environmental worth of mega dams.
- Indigenous and environmental groups have challenged the construction of a mega dam, the Belo Monte in Brazil.
- Partly as a result of the controversy, costs and corruption involved in the dam's construction, the Brazilian government will cease to build mega dams.

15.4 Alternatives to dams

- More attention is now being paid to small-scale, community-based water management schemes.
- Rainwater harvesting schemes and micro hydro-dams are two alternatives.
- Use of traditional small water harvesting dams (johads) in India are providing significant benefits — both environmental and social.

15.6 Using groundwater reserves

- Groundwater is an important section of the watertable used by more than 2 billion people across the world.
- There are many benefits to the use of groundwater, particularly for water and food security.
- It can take up to several thousands of years to replenish groundwater if overused.
- Water availability is unevenly distributed in China, with much more water available in the south than the north.
- In China's north, unsustainable use of water is lowering groundwater reserves.
- A large-scale transfer of water from the south to the north of China has been constructed.
- There are many social, economic and environmental impacts from such a scheme.

15.7 The impacts of drainage and diversion

- Wetlands are a very important biome.
- Wetlands are constantly under threat from a range of human activities.
- The overuse and diversion of water is causing over one-third of the world's major surface water supplies to dry up.
- Lake Urmia in Iran is an example of where over-extraction of water has led to the decline in the health and size of the lake.
- It is possible to restore Lake Urmia given enough funds and more sustainable farming practices.

15.8 Case study — Managing the Murray–Darling

- Environmental changes have developed because of the overuse of water resources in the Murray Darling Basin
- Several government plans have been put in place to provide environmental flows to improve the health of the river, but there are ongoing issues in balancing environmental, economic and social needs.
- Drought and water mismanagement contributed to a major fish kill in the Menindee Lakes.

15.12.2 Key terms

base flow water entering a stream from groundwater seepage, usually through the banks and bed of the stream

environmental flows the quantity, quality and timing of water flows required to sustain freshwater ecosystems

environmental impact assessment a tool used to identify the environmental, social and economic impacts, both positive and negative, of a project prior to decision-making and construction

ephemeral describes a stream or river that flows only occasionally, usually after heavy rain

eutrophication a process where water bodies receive excess nutrients that stimulate excessive plant growth

flood mitigation managing the effects of floods rather than trying to prevent them altogether

green energy sustainable or alternative energy (e.g. wind, solar and tidal)

groundwater water held underground within water-bearing rocks or aquifers

icon sites six sites located in the Murray–Darling Basin that are earmarked for environmental flows. They were chosen for their environmental, cultural and international significance.

impervious a rock layer that does not allow water to move through it due to a lack of cracks and fissures

infrastructure the facilities, services and installations needed for a society to function, such as transportation and communications systems, water and power lines

micro hydro-dams produce hydro-electric power on a scale serving a small community (less than 10 MW). They usually require minimal construction and have very little environmental impact.

perennial describes a stream or river that flows permanently

rainwater harvesting the accumulation and storage of rainwater for reuse before it soaks into underground aquifers

recharge the process by which groundwater is replenished by the slow movement of water down through soil and rock layers

reservoir large natural or artificial lake used to store water, created behind a barrier or dam wall

river fragmentation the interruption of a river's natural flow by dams, withdrawals or transfers

river regime the pattern of seasonal variation in the volume of a river

subsidence the gradual sinking of landforms to a lower level as a result of earth movements, mining operations or over-withdrawal of water

terminal lake a lake where the water does not drain into a river or sea. Water can leave only through evaporation, which can increase salt levels in arid regions. Also known as an endorheic lake.

weir wall or dam built across a river channel to raise the level of water behind. This can then be used for gravity-fed irrigation.

wetland an area covered by water permanently, seasonally or ephemerally. They include fresh, salt and brackish waters such as rivers, lakes, rice paddies and areas of marine water, the depth of which at low tide does not exceed 6 metres.

15.12.3 Reflection

Complete the following to reflect on your learning.

Revisit the inquiry question posed in the Overview:

Humans would find life very hard without healthy inland water sources. Are we being careful with how we use and change them?

1. Now that you have completed this topic, what is your view on the question? Discuss with a partner. Has your learning in this topic changed your view? If so, how?
2. Write a paragraph in response to the inquiry question, outlining your views.

Subtopic	Success criteria			
15.2	I can describe the different types of inland water.			
	I can explain the importance of inland water.			
	I can identify human activities that threaten inland water.			
15.3	I can explain why dams are used.			
	I can explain the positive and negative effects of dams.			
15.4	I can describe the alternatives to dams.			
15.5	I can create a fishbone diagram			
15.6	I can describe how groundwater is used.			
	I can describe the importance of groundwater.			
	I can explain how groundwater is being mismanaged in China.			
	I can explain strategies that have been used to fix it.			
15.7	I can explain the importance of wetlands.			
	I can describe the environmental changes that happen because of draining and diverting water.			
15.8	I can explain how people use the Murray–Darling Basin.			
	I can explain how the Murray–Darling Basin is being managed.			
	I can evaluate strategies to improve the health of the Murray–Darling Basin.			
15.9	I can calculate gradient, relief and area on a topographic map.			
15.10	I can explain the importance of managing wetlands.			

on Resources

eWorkbook Chapter 15 Student learning matrix (ewbk-8576)
Chapter 15 Reflection (ewbk-8577)
Chapter 15 Extended writing task (ewbk-8578)

Interactivity Chapter 15 Crossword (int-8724)

ONLINE RESOURCES

on Resources

Below is a full list of **rich resources** available online for this topic. These resources are designed to bring ideas to life, to promote deep and lasting learning and to support the different learning needs of each individual.

eWorkbook

- **15.1** Chapter 15 eWorkbook (ewbk-8107) ☐
- **15.2** Inland water (ewbk-10121) ☐
- **15.3** Dam it (ewbk-10125) ☐
- **15.4** Alternatives to dams (ewbk-10129) ☐
- **15.5** SkillBuilder — Creating a fishbone diagram (ewbk-10133) ☐
- **15.6** Using groundwater reserves (ewbk-10137) ☐
- **15.7** The impacts of drainage and diversion (ewbk-10141) ☐
- **15.8** Case study — Managing the Murray–Darling (ewbk-10153) ☐
- **15.9** SkillBuilder — Reading topographic maps at an advanced level (ewbk-10149) ☐
- **15.10** Investigating topographic maps — Wetlands along the Murray River (ewbk-10145) ☐
- **15.12** Chapter 15 Student learning matrix (ewbk-8576) ☐
  - Chapter 15 Reflection (ewbk-8577) ☐
  - Chapter 15 Extended writing task (ewbk-8578) ☐

Sample responses

- **15.1** Chapter 15 Sample responses (sar-0164) ☐

Digital documents

- **15.9** Topographic map of Wentworth, New South Wales (doc-11569) ☐
  - Topographic map of Berri, South Australia (doc-11570) ☐
- **15.10** Topographic map of the Tar-Ru Wetlands (doc-36369) ☐

Video eLessons

- **15.1** Draining away (eles-1709) ☐
  - inland water management — Photo essay (eles-5252) ☐
- **15.2** Inland water — Key concepts (eles-5253) ☐
- **15.3** Damming rivers — Key concepts (eles-5254) ☐
- **15.4** Alternatives to dams — Key concepts (eles-5255) ☐
- **15.5** SkillBuilder — Creating a fishbone diagram (eles-1748) ☐
- **15.6** Using groundwater reserves — Key concepts (eles-5256) ☐
- **15.7** The impacts of drainage and diversion — Key concepts (eles-5257) ☐
- **15.8** Case study — Managing the Murray–Darling — Key concepts (eles-5258) ☐
- **15.9** SkillBuilder — Reading topographic maps at an advanced level (eles-1749) ☐
- **15.10** Investigating topographic maps — Wetlands along the Murray River (eles-5259) ☐

Interactivities

- **15.2** Inland water (int-8678) ☐
- **15.3** Dam it (int-3292) ☐
  - The advantages and disadvantages of large-scale dams (int-5592) ☐
  - Distribution of downstream communities affected by large dams (int-5599) ☐
  - Degree of river fragmentation in the world's major drainage basins (int-7956) ☐
  - Location of the Belo Monte Dam and the related environmental changes (int-7958) ☐
- **15.4** Alternatives to dams (int-8679) ☐
- **15.5** SkillBuilder — Creating a fishbone diagram (int-3366) ☐
- **15.6** That sinking feeling (int-3293) ☐
- **15.7** Wetland wonderlands (int-3294) ☐
- **15.8** Case study — Managing the Murray–Darling (int-8680) ☐
- **15.9** SkillBuilder — Reading topographic maps at an advanced level (int-3367) ☐
- **15.10** Investigating topographic maps — Wetlands along the Murray River (int-8681) ☐
- **15.12** Chapter 15 Crossword (int-8724) ☐

ProjectsPLUS

- **15.11** Thinking Big Research project — Menindee Lakes murder — news report (pro-0214) ☐

Fieldwork

- **15.2** Features of your local catchment (fdw-0037) ☐
- **15.3** Impacts of weirs and walls (fdw-0038) ☐

Google Earth

- **15.10** Tar-Ru Wetlands (gogl-0142) ☐

myWorldAtlas

- **15.8** Murray–Darling Basin (mwa-4538) ☐

Teacher resources

There are many resources available exclusively for teachers online.

16 Managing change in coastal environments

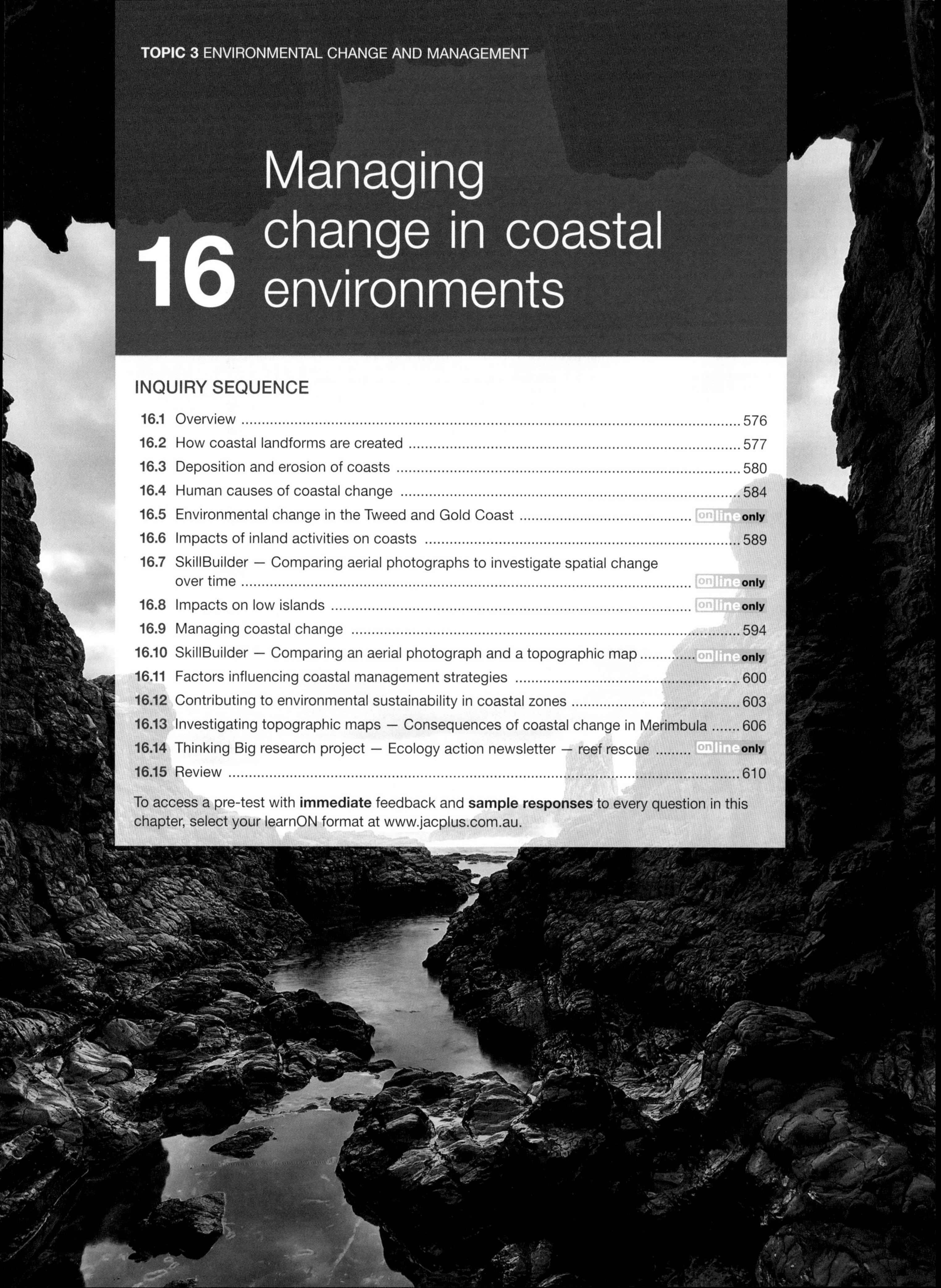

INQUIRY SEQUENCE

To access a pre-test with **immediate** feedback and **sample responses** to every question in this chapter, select your learnON format at www.jacplus.com.au.

16.1 Overview

Numerous **videos** and **interactivities** are embedded just where you need them, at the point of learning, in your learnON title at www.jacplus.com.au. They will help you to learn the content and concepts covered in this topic.

What are the causes and consequences of coastal change, and how can this change be managed?

16.1.1 Introduction

The coast is home to 80 per cent of the world's population, and it is a popular place to settle for reasons of climate, water resources, land for agriculture and industry, access to transportation systems, and recreation. Hence, it is essential to understand the changes that are occurring to coastal environments and how they will affect human settlements. The changes are both natural and human-induced. They are sometimes short term (for example as a result of storms and tsunamis) and sometimes long term (for example climate change leading to rising sea levels). To cope with these changes, careful planning and management is needed to ensure a sustainable future for human activity at the coast.

STARTER QUESTIONS

1. How do people use coastal places?
2. What changes have people brought to the coastal area in the image below (**FIGURE 1**)?
3. What could people do in the place shown in **FIGURE 1** to help reduce damage to the coast and danger for people?
4. Watch the **Managing change in coastal environments — Photo essay** video in your online Resources. What changes to coastal areas have you observed when visiting a beach?

FIGURE 1 Houses along Malibu Beach in California are regularly threatened by severe storms. Is housing the most suitable land use for this area?

on Resources

eWorkbook	Chapter 16 eWorkbook (ewbk-8108)
Video eLessons	Washed away (eles-1710) Managing change in coastal environments — Photo essay (eles-5260)

16.2 How coastal landforms are created

LEARNING INTENTION

By the end of this subtopic, you will be able to explain the biophysical processes essential to the functioning of coasts.

16.2.1 Types of coastal landforms?

Marine and terrestrial structures found at the coast include beaches, bays, dunes and cliffs (see **FIGURE 1**). Others such as fjords are unique to polar regions. Structures found under the sea can include the continental shelf, canyons and trenches.

int-5594

FIGURE 1 The range of coastal landforms

1 **Dune blowouts** — loose sand is blown from the dune because vegetation has been removed.
2 **Caves** — formed where weak rocks are eroded on each side of a headland as a result of wave refraction.
3 **Arch** — caves will erode on either side of a headland and join to form an arch.
4 **Cliff** — created when erosion undercuts a rock platform and the weakened rock collapses.
5 **Longshore drift** — moves sand and other material along a beach.
6 **Estuaries** — of a river that are tidal and occur at the mouth of the river where it meets the sea.
7 **Lagoon** — formed when a sandbar begins to develop, eventually closing an estuary.
8 **Beaches** — formed when material is brought to the shore by waves. Spits can develop when deposited sand accumulates perpendicular to the beach.
9 **Dunes** — formed when sand on a beach is stabilised by vegetation.
10 **Stack** — created by ongoing erosion of an arch, where one section of the arch collapses.
11 **Blowhole** — formed when the roof of a cave collapses as a result of the action of waves.
12 **Tombolo** — a spit joining two land areas.
13 **Headlands** — when coastal rocks are very hard and resist erosion from the waves.

16.2.2 Ocean processes

Waves, tides, currents, rips, storm surges and tsunamis form and transform coastal landscapes.

Winds generate waves and create swell. Destructive waves cause erosion of coastal landforms and are often associated with storm conditions (see **FIGURE 2**).

Constructive waves deposit sediments to form beaches and sand dunes and are associated with calm weather (see **FIGURE 3**).

Swash is the water that is washed onto the beach when a wave breaks. Backwash is the water that runs back down the beach. Destructive waves have more powerful backwash than swash. Constructive waves have more powerful swash than backwash.

Longshore drift occurs when the ongoing swash and backwash moves in a sideways direction and moves material in a zig-zag pattern along the coastline (see **FIGURE 4**).

Erosion and deposition are the main natural processes that create coastal change. These processes are examined in more detail in section 16.3.

longshore drift a current that moves sediment parallel to the shoreline, created by the backward and forward motion of waves

FIGURE 2 Destructive waves: the backwash is more powerful than the swash.

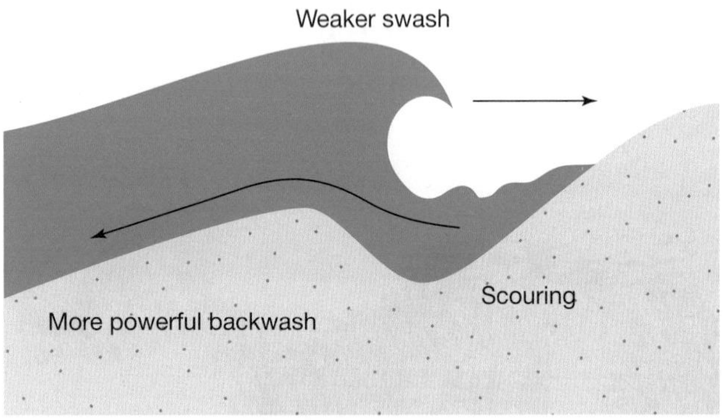

FIGURE 3 Constructive waves: the swash is more powerful than the backwash.

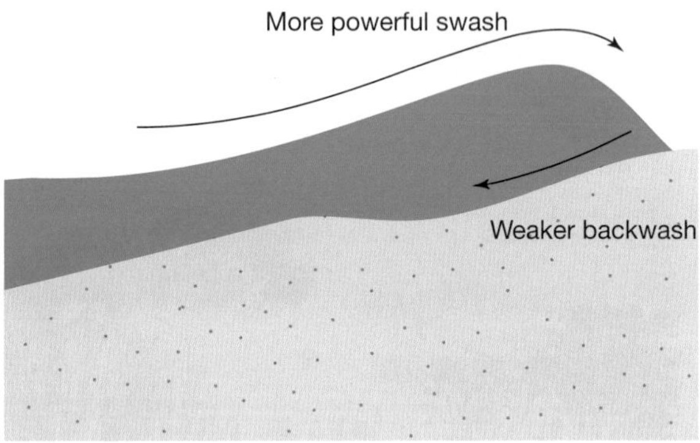

FIGURE 4 Longshore drift

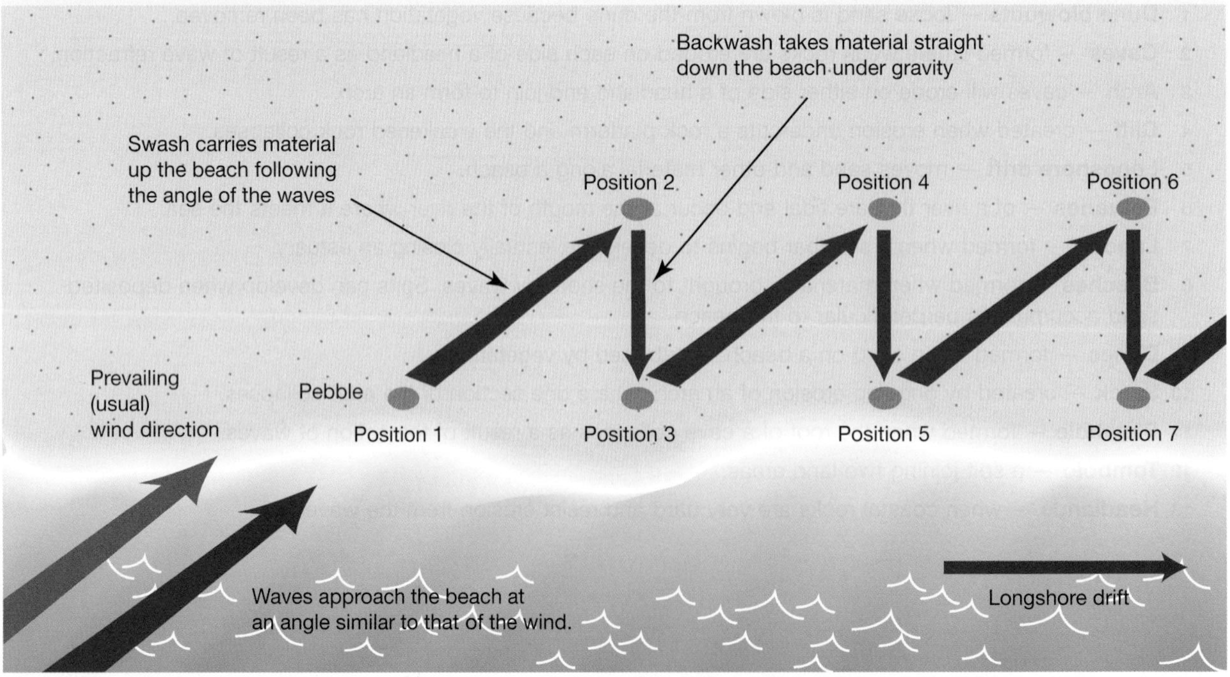

Resources

eWorkbook	How coastal landforms are created (ewbk-10161)
Video eLesson	How coastal landforms are created — Key concepts (eles-5261)
Interactivities	The range of coastal landforms (int-5594) How coastal landforms are created (int-8722)

16.2 ACTIVITIES

1. In groups, create a presentation that explains how a particular coastal landform is created. Include images of the landform, and use and define relevant geographical terms in your presentation.
2. Fraser Island is located off the coast of Queensland. It has been recognised as the largest sand island in the world and contains many specialised landforms, such as perched lakes, as well as unique coastal ecosystems. Once mined for sand and logged for timber, it has become a popular eco-tourism destination. An issue for today is how to sustainably manage the impacts of eco-tourism and protect the environment of this World Heritage listed island.
   a. Who are the traditional custodians of Fraser Island? What is their name for the island?
   b. Research the past and present uses of Fraser Island, before and after European colonisation.
   c. When did it become a World Heritage site?
   d. What are some of its unique features?
   e. What does the future hold for Fraser Island?

16.2 EXERCISE

Learning pathways

■ LEVEL 1	■ LEVEL 2	■ LEVEL 3
1, 3, 7, 8, 13	2, 4, 9, 10, 14	5, 6, 11, 12, 15

Check your understanding

1. Outline the two main coastal processes that form coastal landforms.
2. Describe the difference between constructive and destructive waves.
3. Define *swash* and *backwash*.
4. What is a dune blowout?
5. What is the difference between a marine landform and a terrestrial landform?
6. Where would you expect to find a fjord?

Apply your understanding

7. Explain how longshore drift moves sand along the coastline.
8. Explain why the sand transported by longshore drift moves in a zig-zag pattern up the coastline.
9. Explain how a landform might develop from a headland to an arch and then to a stack.
10. What role does the wind play in the development of waves?
11. Would you be safer swimming in constructive waves or destructive waves? Give reasons for your decision.
12. What impact does the prevailing wind direction have on the transformation of a beach?

Challenge your understanding

13. Predict what would happen on a beach if a council constructed a rock barrier at a right angle to the beach.
14. Suggest one strategy that might help to reduce dune blowouts.
15. Consider the beach shown in **FIGURE 1**. Predict how this beach would change if a large breakwater was built from the headland.

To answer questions online and to receive **immediate feedback** and **sample responses** for every question, go to your learnON title at www.jacplus.com.au.

16.3 Deposition and erosion of coasts

LEARNING INTENTION

By the end of this subtopic, you will be able to explain how the operation of biophysical processes maintains the functioning of coasts.

16.3.1 Deposition and coastal landforms

Coasts are a dynamic natural system. The forces of nature are constantly at work, either creating new land or wearing it away.

Beaches

Sediments transported down rivers and eroded from cliffs provide the material for beaches to develop. Constructive waves move sand and sediment onto the shore to create beaches.

Dunes

Wind moves dried out sediments onshore and inland. A fore dune will form close to the beach. Further from the shore a back dune will develop. Between the fore dune and back dune a depression will form called a swale. Dune vegetation helps to stabilise the sand and the landforms themselves. Vegetation becomes larger and more varied as freshwater and soil conditions improve further away from the beach (see **FIGURE 1**). This progression of plants is known as the **coastal dune vegetation succession**.

Other coastal landforms created by deposition include spits, bars and barriers.

Sand dunes erode when storms cause strong winds and waves to hit the coast with greater force and higher onto beaches than usual. The front face of the sand dune becomes very steep as sand is removed. Blowouts develop when sand is removed by the wind if there is little protection to block the wind. Usually, coastal vegetation such as grasses, shrubs and trees helps reduce the wind speed, and the root systems of these plants stabilise the dunes.

coastal dune vegetation succession the process of change in the plant types of a vegetation community over time — moving from pioneering plants in the high tide zone to fully developed inland area vegetation

FIGURE 1 Transect showing the beach and stable and well-vegetated dunes

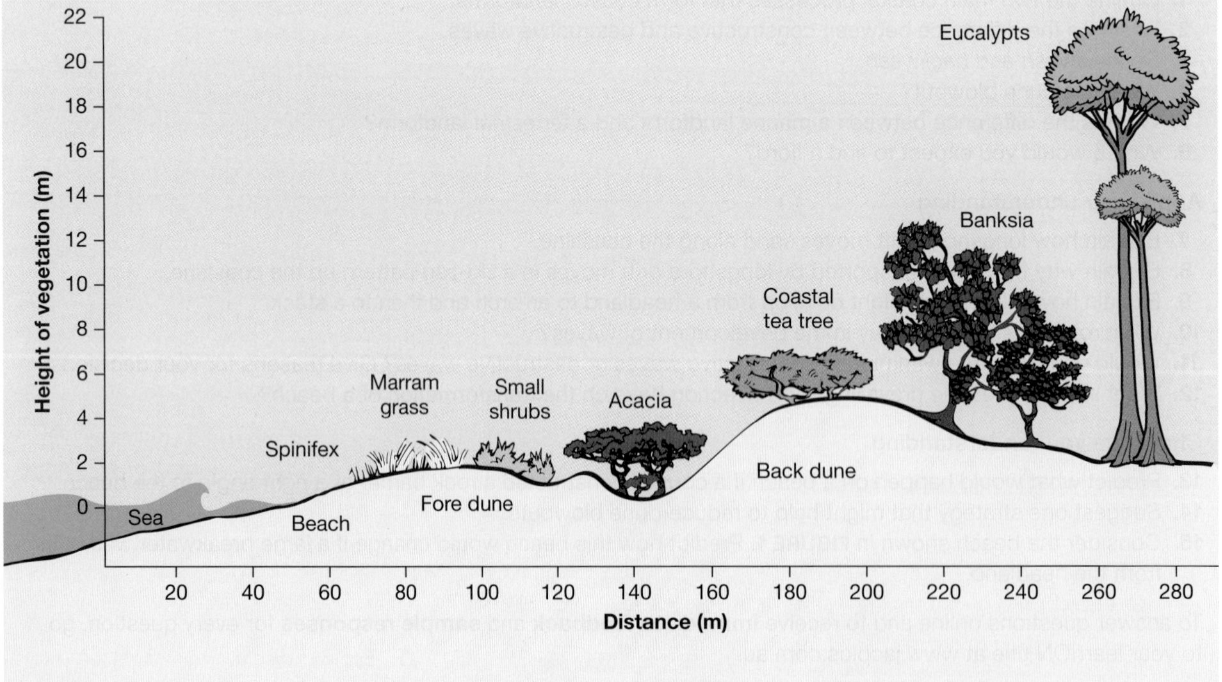

Source: Wiley Art

fdw-0039

FOCUS ON FIELDWORK

Measuring dunes

A measurement of the size and angles of coastal dunes can contribute to understanding how deposition and erosion have impacted on the dunes. To do this you need to identify the berm (the flat part of the beach just before the dune starts), fore dune and back dune. Use a tape measure to determine the length of each landform. A clinometer can be used to determine the steepness of the dunes. Use the clinometer to determine the angle of the slopes for each landform. You can make your own clinometer and practise using it without ever visiting a beach.

Learn more about making and using a clinometer with the **Measuring dunes** activity in your online fieldwork Resources.

16.3.2 Erosion and coastal landforms

The weight and pressure of sea water hitting coastal rocks can lead to weathering. When loose sediments or pebbles are carried by the water this can exacerbate weathering. Erosional landforms include headlands, bays, cliffs, platforms, caves, arches, blowholes and stacks.

Headlands and bays

Harder rocks or rocks with fewer fractures form headlands that tend to resist erosion. Bays are composed of softer rocks or rocks with more fractures that are more easily eroded. This leads to the retreat of the coastline (see **FIGURE 2**).

FIGURE 2 Wave refraction results in the concentration of waves on a headland.

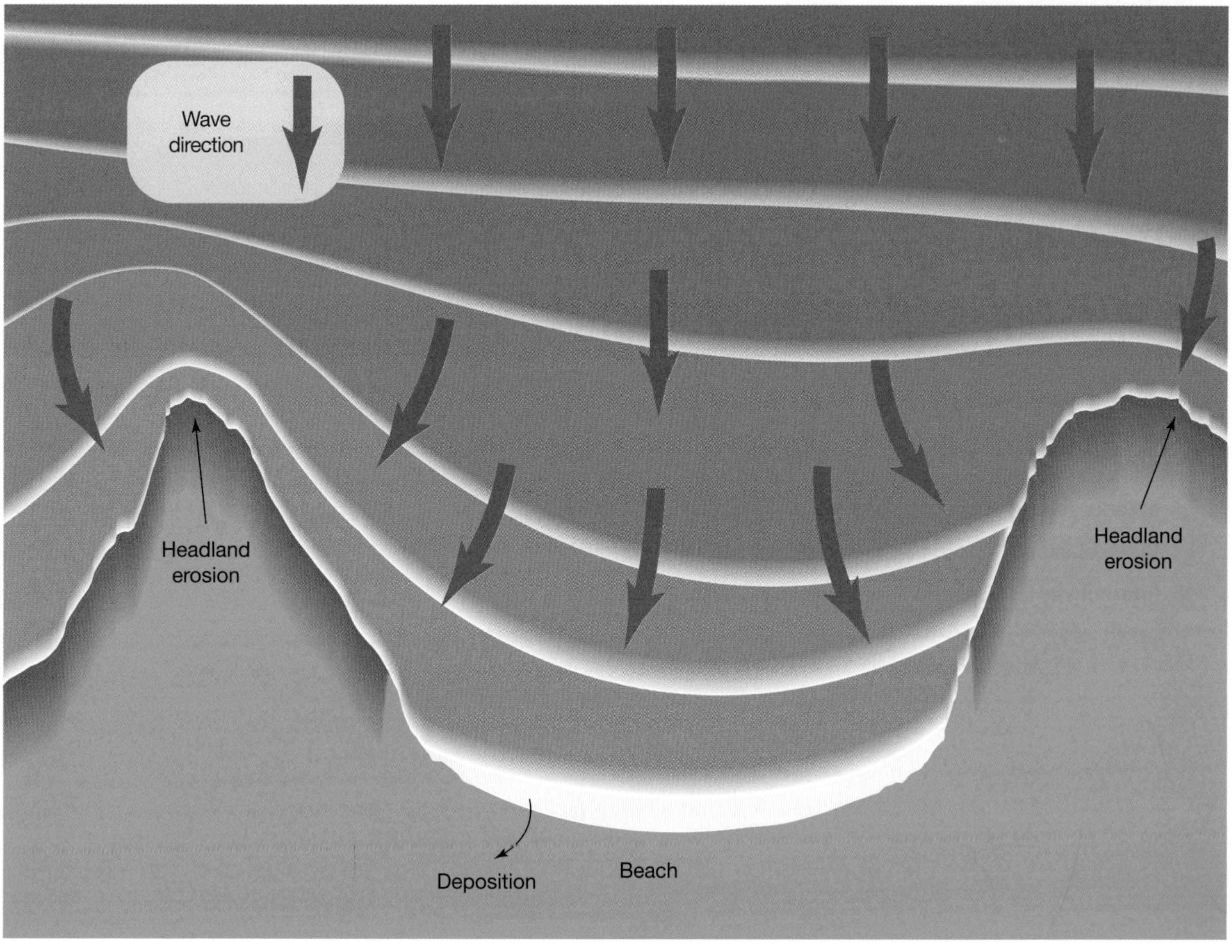

Cliffs and platforms

Platforms are flat, rocky, horizontal structures that have been worn down by the action of waves. Cliffs are raised, rocky structures. Waves wearing away the base of a cliff will create a notch, which will result in collapse and the coastline moving further inland (see **FIGURE 3**).

Caves, arches, blowholes and stacks

Caves are parts of the coast that have been more actively eroded from the surrounding rocky area, leaving a hollow in the cliff section. If part of the roof section of the cave collapses, in-rushing waves may be channelled up this chimney structure, forcing water and air out at the horizontal land surface above. Stacks are simply remnants of cliff areas that have resisted erosion and been left stranded out to sea.

FIGURE 3 Formation of cliffs and rocky platforms

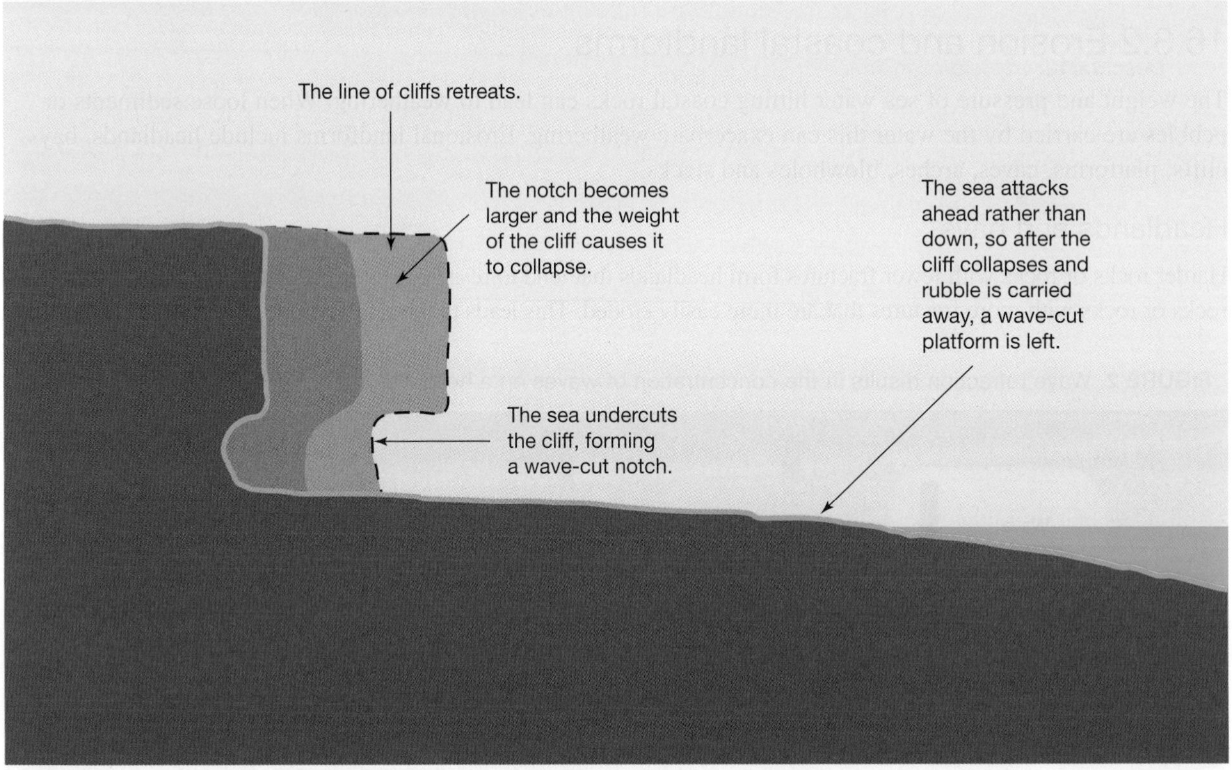
The line of cliffs retreats.
The notch becomes larger and the weight of the cliff causes it to collapse.
The sea attacks ahead rather than down, so after the cliff collapses and rubble is carried away, a wave-cut platform is left.
The sea undercuts the cliff, forming a wave-cut notch.

on Resources

eWorkbook	Deposition and erosion of coasts (ewbk-10165)
Video eLesson	Deposition and erosion of coasts — Key concepts (eles-5262)
Interactivity	Deposition and erosion of coasts (int-8682)

16.3 ACTIVITY

Using the internet, find examples of well-known coastal landforms that have been created by erosion and deposition. Present your findings to your class as a digital poster or presentation.

16.3 EXERCISE

Learning pathways

■ LEVEL 1	■ LEVEL 2	■ LEVEL 3
2, 4, 8, 10, 13	1, 3, 7, 9, 15	5, 6, 11, 12, 14

Check your understanding

1. Describe how erosion can shape coastal landforms.
2. Define the term *deposition*.
3. Describe how a headland is formed.
4. Outline how vegetation changes from the front of a sand dune to the back of the dune.
5. How is a wave-cut platform created?
6. What factors might contribute to a section of coastline becoming a headland? Identify and describe two.

Apply your understanding

7. Explain why the vegetation changes from the front of the dune to the back of the dune. Identify the natural processes that create this change.
8. Could a coastline be described as a *static* system? Give reasons for your answer.
9. Explain the organisation of dunes along a beach.
10. Why is the coastline of a bay more likely to retreat than a headland?
11. With the aid of a diagram, explain how cliffs and rocky platforms are formed.
12. Using examples from **FIGURE 1**, determine which form of coastal vegetation is most likely to be negatively affected by a severe storm.

Challenge your understanding

13. Predict which areas of **FIGURE 2** will be the most significantly changed by a severe storm.
14. Suggest one strategy that might prevent the wave-cut notches shown in **FIGURES 3** and **4** from becoming larger.
15. Create a three-cell cartoon to show the process of how a spit forms.

FIGURE 4 Tamarama, New South Wales

To answer questions online and to receive **immediate feedback** and **sample responses** for every question, go to your learnON title at www.jacplus.com.au.

16.4 Human causes of coastal change

LEARNING INTENTION

By the end of this subtopic, you will be able to explain some of the ways that humans cause coastal change.

16.4.1 Human impact on coastal processes

Human activities along the coast can interfere with natural coastal processes, resulting in significant changes to coastal environments. Human impacts on coastlines include the construction of ports, boat marinas and sea walls; changes in land use (for example from natural to agricultural, industrial or urban environments); and the disposal of waste from coastal and other settlements. Consequences of coastal changes can vary in severity from coastal footpaths crumbling and collapsing into the sea to whole unit blocks being undermined by pounding waves and collapsing, farmland being inundated with salty water, and whole communities being destroyed.

Some examples of human impacts on coasts are listed below:

• Land-based pollution • Overfishing • Marine accidents • Oil spills • Off-shore drilling • Trawl nets • Litter • Tourism developments • Shark nets • Storm water • Residential developments • Constructing fences	• Sewage outfalls • Anchor damage • Developing marinas and ports • Run-off • Removing mangroves • Removing natural vegetation on sand dunes • Toxic wastes • Industrial developments • Dredging • Groundwater discharge • Changing shorelines • Constructing roads

CASE STUDY 1

Cronulla sand dunes, Sydney

The Cronulla sand dunes at Kurnell and Wanda Beach, provide an example of a site where the land use has changed over time, impacting the sand dunes. The Cronulla sand dunes were originally covered in vegetation and were culturally significant to the Aboriginal Peoples of the area. Clearing began in the early 1800s for harvesting timber and continued in the late 1800s, when they began to be used for farming.

FIGURE 1 Sand dunes at Wanda Beach

The successive grazing of sheep and then cattle removed the stabilising grass cover on the dunes. In the 1930s they were used for sand mining, and a few decades later became the site of an oil refinery. They were also used in films such as *Mad Max: Beyond Thunderdome* and *40 000 Horsemen*.

Greenhills, an area that had man-made lakes as a result of sand mining, is also the site of a beachside housing development. Multiple construction companies have also tried to develop surrounding sites.

CASE STUDY 2

Tamarama, Sydney

Tamarama Beach is another example of a site where the land use has changed over time, impacting the sand dunes. Aboriginal cave and rock engravings and shell middens on the Tamarama Park site provide evidence of the Aboriginal heritage and connection to Country of Dharug people for thousands of years.

FIGURE 2 Bondi Beach pleasure park at Tamarama, c.1890

Source: Dictionary of Sydney

On the eastern part of Tamarama the Royal Aquarium, Pleasure Grounds or Bondi Aquarium opened in 1887. This included an aquarium building, a seal pond, a shark pool and a skating rink. In 1906 a new operation opened on the site, known as Wonderland City. This provided a range of recreational activities in addition to the existing aquarium and animal exhibits: fairground rides and attractions (such as a haunted house, a maze and a fun factory), an artificial lake, a movie house and theatre, and a miniature railway.

In 1907, Tamarama Beach was declared a public park. The Tamarama Surf Live Saving Club was established in 1906 and the club house was constructed in 1908. Wonderland City closed in 1911, and it was not until 1920 that Waverley Council was able to purchase and claim an area for a public park providing beach access and parkland. In the 1920s and 30s a sea wall and coastal drive were constructed. Picnic shelters and further facilities were provided in the 1960s and 70s. Bushcare volunteers continue to revegetate the lower southern slopes of Tamarama Gully.

CASE STUDY 3

Clovelly and Gordon's Bay, Sydney

Many areas along Australia's eastern coastline demonstrate the environmental change that results from human impact.

FIGURE 3 Clovelly beachfront

Clovelly Beach is a small beach located at the end of a narrow bay. Concrete platforms and walkways have been built on either side of the bay, with steps providing access to the water. The concrete foreshore at Clovelly was constructed in the 1920s to provide easy access to the bay for swimmers.

Some human actions have been taken to protect the Clovelly coast. The April 2015 Sydney storms resulted in significant erosion of Clovelly Headland near Shark Point, and as a result rock stabilisation works were required. Randwick City Council undertook this stabilisation work including excavation, installation of sandstone boulders and some landscaping.

Gordon's Bay is located between Clovelly and Coogee Beaches and provides an idea of what Clovelly may have resembled in the past. Gordon's Bay has a more natural landscape, with cliffs, rock platforms and some natural vegetation. Human impacts can still be seen around the bay with dense housing, walkways and a boatshed.

FIGURE 4 Remediation work at Clovelly Beach

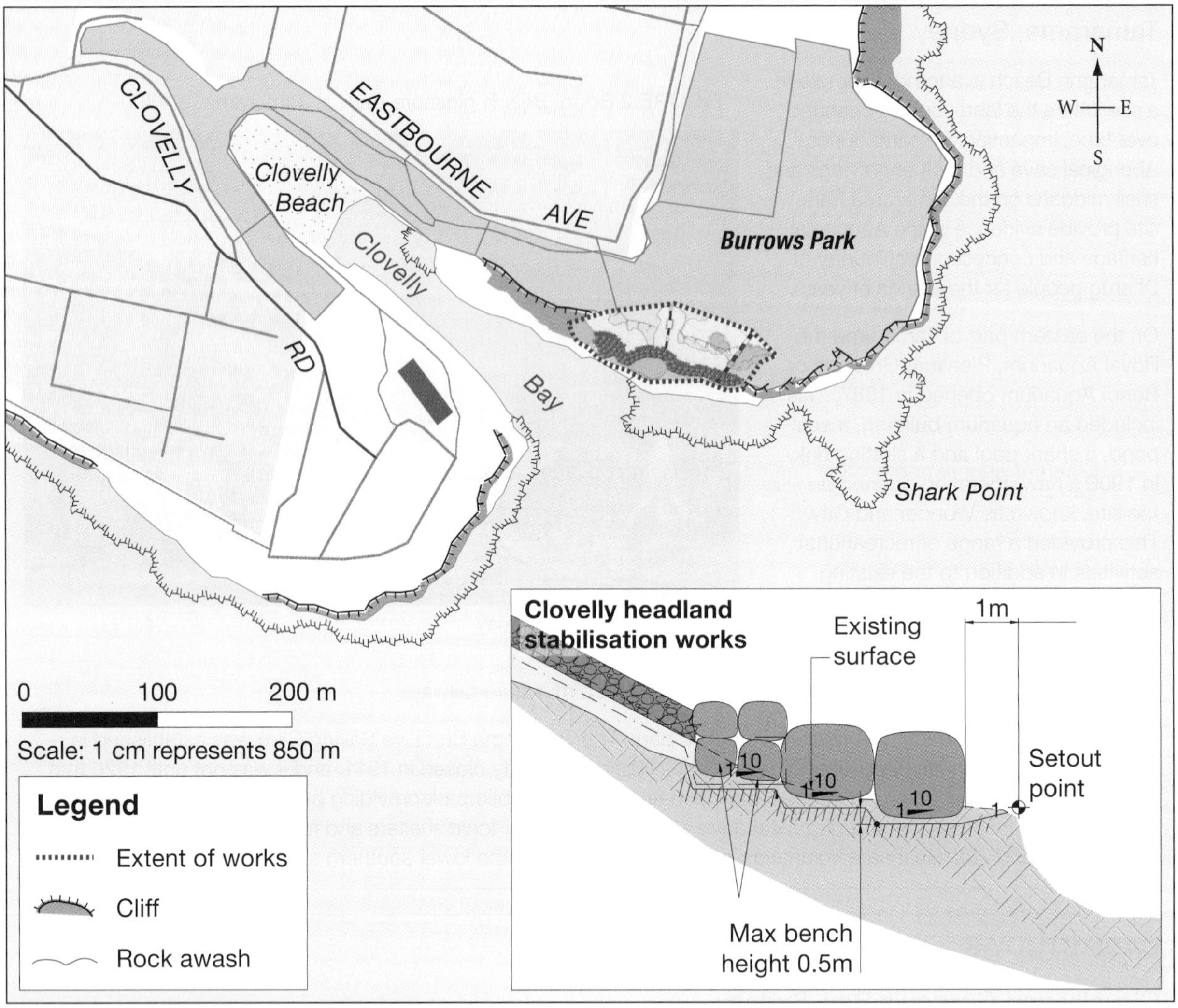

Source: Image courtesy Randwick City Council

FIGURE 5 Gordon's Bay

Resources

eWorkbook	Human causes of coastal change (ewbk-10169)
Video eLesson	Human causes of coastal change — Key concepts (eles-5263)
Interactivity	Human causes of coastal change (int-8683)

16.4 ACTIVITIES

1. As a class, discuss whether all forms of environmental change along coasts have negative impacts on natural coastal environments.
2. The Gold Coast is a popular tourist destination in Queensland. In small groups, complete the following tasks.
   a. Research past and ongoing developments along the Gold Coast coastal zone. Display these developments in the form of a photographic essay.
   b. Explore why these changes have taken place. Describe whether they have been mostly natural changes or human-induced changes.
   c. Different people and groups can have opposing opinions about change. List the various groups of people affected by coastal change in the Gold Coast. Describe how their opinions may differ.
3. Locate Clovelly Beach and Gordon's Bay using Google Maps. Examine the location as a satellite image.
   a. Describe the environmental change that has occurred at Clovelly Bay and Clovelly Beach.
   b. List the advantages and disadvantages of the concrete foreshore at Clovelly Beach.
   c. Describe whether you think Clovelly Beach or Gordon's Bay would be a more popular area for visitors.
   d. Predict how you think these areas will change over the next 20 years.
4. Investigate how human activities have caused coastal change at one location outside of Australia. Describe these human activities and analyse the consequences of these activities on the coastal zone.
5. Create a flow chart that shows the impacts of successive land uses that have impacted on Cronulla, Kurnell and Wanda Beach.

16.4 EXERCISE

Learning pathways

■ LEVEL 1	■ LEVEL 2	■ LEVEL 3
1, 3, 7, 8, 13	2, 4, 9, 10, 14	5, 6, 11, 12, 15

Check your understanding

1. Describe the impact of farming on vegetation on the dunes in Cronulla and Kurnell.
2. Describe the changes made to the coastline of Clovelly Bay.
3. Identify advantages of creating a cement promenade along a coastline.
4. Describe what Clovelly Beach may have looked like before 1788.
5. Identify the natural processes that most likely formed Clovelly Bay.
6. Outline why revegetating sand dunes is important.

Apply your understanding

7. Examine **FIGURE 2** closely. What evidence is shown to suggest this park is in a coastal location?
8. Examine **FIGURE 3** closely. Can you identify features of this coastal location that have not been affected by human actions? Explain how you arrived at your conclusion using examples from the image.
9. Which of the impacts listed in the table in section 16.4.1 would have the greatest impact on the native animals living in coastal dunes? Give reasons for your decision.
10. Based on **FIGURE 4**, explain what factors might have contributed to the choice of this specific remediation site.
11. Explain how the remediation work at Clovelly is expected to reduce the impact of coastal change.
12. Compare **FIGURES 3** and **5**. Analyse the extent of human impact, and assess which has had the most significant impact on the natural processes of the coastline. Justify your response.

Challenge your understanding

13. Suggest an alternative to creating a concrete foreshore that allows people easier access to water for swimming but has a less significant impact on the natural environment.
14. Present a case for the Cronulla sand dunes at Kurnell to be closed to public access in order to protect the Aboriginal cultural heritage of the area.
15. Predict whether the remediation works underway at Clovelly Beach will achieve their aims. Provide reasons for your assessment.

To answer questions online and to receive **immediate feedback** and **sample responses** for every question, go to your learnON title at www.jacplus.com.au.

16.5 Environment change in the Tweed and Gold Coast

online only

LEARNING INTENTION

By the end of this subtopic, you will be able to provide examples of the causes, extent and consequences of environmental change in the Tweed and Gold Coast.

The content in this subtopic is a summary of what you will find in the online resource.

Wind, wave and current action are responsible for moving sand on and off a beach and along a coastline, but human-induced environmental changes along the coast can interrupt these natural processes. This can create long-term problems.

To learn more about how the areas surrounding the Tweed and Gold Coast are being changed by human activity, go to your learnON resources at www.jacPLUS.com.au.

learnON

Contents

FIGURE 2 The effect of destructive storm waves on Duranbah Beach

on Resources

eWorkbook	Environmental change in the Tweed and Gold Coast (ewbk-10177)
Video eLesson	Environmental change in the Tweed and Gold Coast — Key concepts (eles-5264)
Interactivity	Pumping sand (int-3295)

16.6 Impacts of inland activities on coasts

LEARNING INTENTION

By the end of this subtopic, you will be able to describe the causes and extent of change to coasts in Bangladesh.

16.6.1 Bangladesh

The country of Bangladesh is a large **alluvial plain** crossed by three rivers: the Ganges, Brahmaputra and Meghna. Each river carries massive volumes of water from its source in the Himalayas, spreads out along the **deltaic plain**, and empties into the world's biggest delta, the Bay of Bengal (**FIGURE 1**). This makes Bangladesh's coastline one of the most flood-prone in the world.

Apart from flooding by rivers in the delta, sea level rises caused by global warming will lead to the expansion of ocean waters and additional inflows from melting Himalayan snow. Scientists predict a one-metre sea level rise by 2100 if global warming continues at the current rate. The United Nations Intergovernmental Panel on Climate Change (IPCC) predicts rising sea levels will overtake 17 per cent of Bangladesh by 2050, displacing at least 20 million people.

alluvial plain an area where rich sediments are deposited by flooding

deltaic plain flat area where a river(s) empties into a basin

FIGURE 1 Flooding in Bangladesh — some causes

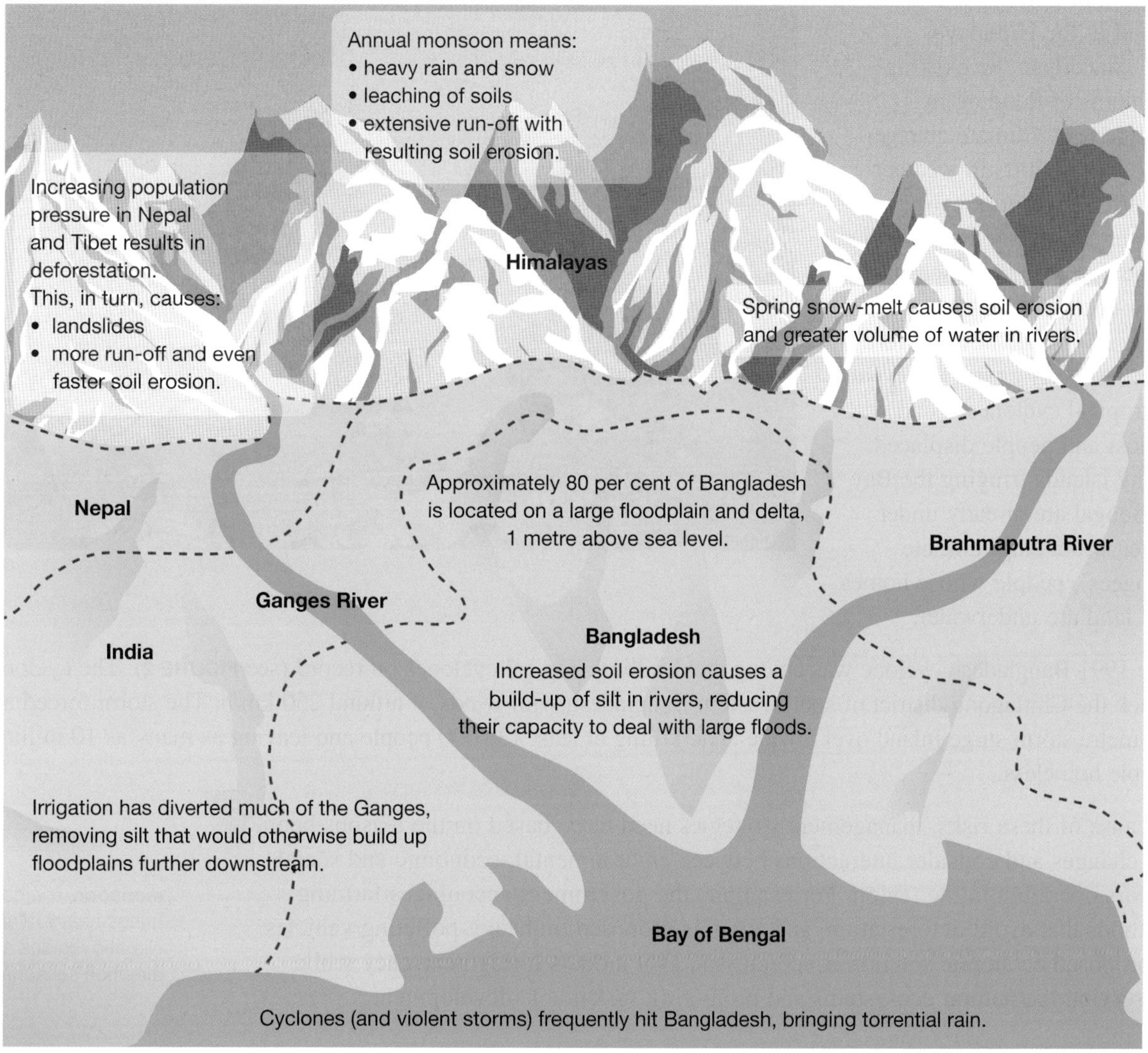

Annual monsoon means:
• heavy rain and snow
• leaching of soils
• extensive run-off with resulting soil erosion.
Increasing population pressure in Nepal and Tibet results in deforestation.
This, in turn, causes:
• landslides
• more run-off and even faster soil erosion.
Himalayas
Spring snow-melt causes soil erosion and greater volume of water in rivers.
Nepal
Approximately 80 per cent of Bangladesh is located on a large floodplain and delta, 1 metre above sea level.
Brahmaputra River
Ganges River
India
Bangladesh
Increased soil erosion causes a build-up of silt in rivers, reducing their capacity to deal with large floods.
Irrigation has diverted much of the Ganges, removing silt that would otherwise build up floodplains further downstream.
Bay of Bengal
Cyclones (and violent storms) frequently hit Bangladesh, bringing torrential rain.

16.6.2 The Sundarbans

The Sundarbans region, a World Heritage site, is just one area of Bangladesh at risk from increased flooding. The Sundarbans are the largest intact mangrove forests in the world. Mangroves protect against coastal erosion and land loss. They play an important role in flood minimisation because they trap sediment in their extensive root systems. Mangroves also defend against storm surges caused by tropical cyclones or king tides, both common in the Sundarbans.

The Sundarbans also provide a breeding ground for birds and fish, as well as being home to the endangered Royal Bengal tiger. By sheltering juvenile fish, the mangrove forest provides a source of protein for millions of people in South Asia. Recently, the Sundarbans have also attracted a growing human population as people move from the crowded capital city, Dhaka, or to avoid flooding and lack of work opportunities in other rural areas.

The increasing human population poses a severe threat to the Sundarbans. People in the region generally rely on wood as a source of energy, and mangroves are being cleared to make charcoal for cooking. Aquaculture industries also have a negative impact. Mangroves are cleared to accommodate huge ponds for fish breeding, which quickly become polluted by antibiotics, waste products and toxic algae. This damage to the Sundarbans contributes to the destruction of the mangroves, which provide the coastline with a natural defence against flooding.

16.6.3 The impact of flooding

The increase in temperature that has led to an increased melting of glaciers and snow inland in the Himalayas will exacerbate the existing problems of flooding in Bangladesh. Climate change also causes shifts in weather patterns. If the **monsoon** season (from June to October) coincided with an unseasonal snow-melt, flooding would occur on a scale never before seen, especially with the event of tropical cyclones. Land will be lost and people displaced. Many islands fringing the Bay of Bengal are already under water, producing 'climate refugees', people whose homes and land are underwater.

FIGURE 2 A village on one of Bangladesh's coastal islands devastated by a cyclone in 1991.

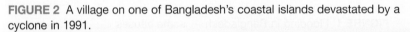

The 1991 Bangladesh cyclone was among the deadliest tropical cyclones on record (see **FIGURE 2**). The cyclone struck the Chittagong district of south-eastern Bangladesh with winds of around 250 km/h. The storm forced a six-metre storm surge inland over a wide area, killing at least 138 000 people and leaving as many as 10 million people homeless.

Because of these risks, management strategies need to be based on the reasons behind the changes and consider interactions between environmental, economic and social factors operating in the region. For example, the government encourages farming methods that avoid deforestation, and a ban is proposed on heavy-polluting vehicles. A proposed economic solution is ecotourism, as it attracts foreign currency while preserving the natural ecosystems and promoting sustainable development.

monsoon a wind system that brings heavy rainfall over large climatic regions and reverses direction seasonally

on Resources

eWorkbook	How inland activities affect coasts (ewbk-10173)
Video eLesson	How inland activities affect coasts — Key concepts (eles-5265)
Interactivity	Bangladesh awash (int-3297)

16.6 ACTIVITIES

1. Discuss in small groups to what extent you think economic goals and objectives are important with respect to environmental goals such as reducing greenhouse gas emissions and societal change. Listen respectfully to one another's views. Decide as a group what policy direction you would push if you were in a position of influence in government, and present this view jointly to the class.
2. Research a coastal case study in Australia or Asia of a location impacted by flooding and managed by multiple authorities.
   a. Define the geographical location of the case study and the physical processes that shape it.
   b. Describe how the area is impacted by change currently and how it may change in the future.
   c. Outline the different groups that play a role in managing change in the area.

16.6 EXERCISE

Learning pathways

■ LEVEL 1	■ LEVEL 2	■ LEVEL 3
2, 4, 7, 8, 13	1, 5, 9, 10, 14	3, 6, 11, 12, 15

Check your understanding

1. Define the terms *alluvial plain* and *deltaic plain*.
2. State two reasons mangroves are being cleared in the Sundarbans.
3. Define the term *climate refugees*.
4. Which three rivers that cross Bangladesh make the country so prone to flooding?
5. What are the two main reasons why people are moving to the Sundarban region?
6. Describe how cyclones contribute towards flooding in Bangladesh.

Apply your understanding

7. Refer to **FIGURE 1**. Explain how the geography of Bangladesh makes it so vulnerable to the threat posed by climate change.
8. Explain how mangroves minimise the impact of floods and coastal erosion.
9. Why is it important for the government of Bangladesh to consider social and economic factors when managing the Sundarbans?
10. What is ecotourism and how can it play a role in preserving Bangladesh's ecosystems?
11. Identify and explain two short-term and two long-term actions that neighbouring nations Tibet, India and Nepal could implement to lessen the impact of flooding in places such as Bangladesh.
12. Create a table with three columns, with the headings 'Food production', 'Transport' and 'Settlement', and list the consequences of flooding for each category. Which of the consequences you have identified do you think will have the most significant negative impact on the people of Bangladesh?

Challenge your understanding

13. Suggest one strategy that a farmer could put in place to reduce pollution from their farm.
14. Climate change is expected to cause glaciers to melt more rapidly. Predict how this will affect people living in Bangladesh.
15. Propose one strategy that the government of Bangladesh might implement to reduce the number of people impacted by extreme weather events in the region.

To answer questions online and to receive **immediate feedback** and **sample responses** for every question, go to your learnON title at www.jacplus.com.au.

16.7 SkillBuilder — Comparing aerial photographs to investigate spatial change over time

online only

LEARNING INTENTION

By the end of this subtopic, you will be able to identify examples of how a place has changed over time by examining aerial photographs.

The content in this subtopic is a summary of what you will find in the online resource.

16.7.1 Tell me

What is an aerial photo?

Aerial photos are images taken above the Earth from an aircraft or satellite. Aerial photos — either oblique or vertical — record how a place looks at a particular moment in time. Some aerial photos are also satellite compilations; that is, they have been created by a number of images transmitted from a satellite.

16.7.2 Show me

How to compare aerial photographs

eles-1750

Step 1

Identify patterns or features that are similar over time; that is, they appear in both of the aerial photographs being studied.

Step 2

Identify patterns and features that have changed over time; that is, they appear altered from one photograph to the next, when the photographs were taken at different times.

Step 3

Explain the processes at work that have changed the environment.

FIGURE 1 Lake Urmia (a) in 1988 and (b) in 2016

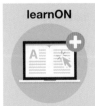

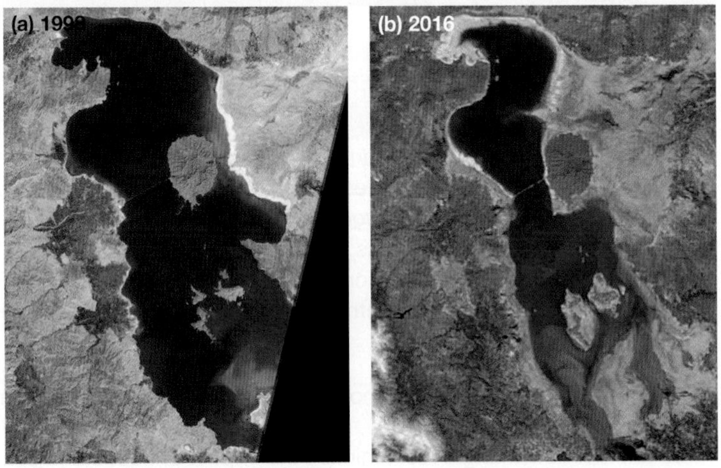

16.7.3 Let me do it

int-3368

learnON

Go to learnON to access the following additional resources to help you build this skill:

- a longer explanation of this skill and its application in Geography (Tell me)
- a video demonstrating the step-by-step process of this skill (Show me)
- an activity and interactivity for you to practise the skill (Let me do it)
- self-marking questions to help you understand and use the skill.

on Resources

eWorkbook	SkillBuilder — Comparing aerial photographs to investigate spatial change over time (ewbk-10181)
Video eLesson	SkillBuilder — Comparing aerial photographs to investigate spatial change over time (eles-1750)
Interactivity	SkillBuilder — Comparing aerial photographs to investigate spatial change over time (int-3368)

16.8 Impacts on low islands

online only

LEARNING INTENTION

By the end of this subtopic, you will be able to explain the long-term consequences of coastal change for low islands.

The content in this subtopic is a summary of what you will find in the online resource.

As a result of climate change, many low-lying islands will be flooded by the sea. Thermal expansion of oceans and melting ice leads to rising sea levels, threatening many coastal communities. Many island groups in the Pacific and Indian Oceans will be almost completely inundated by 2050.

FIGURE 4 Malé, the capital of the Maldives, occupies an entire island of its own. Why is there a need for a sea wall?

To learn more about how low-lying islands are being affected by rising sea levels and predictions for how they are likely to be affected in the future, go to your learnON resources at www.jacPLUS.com.au.

Contents

learnON

- 16.8.1 Low-lying islands
- 16.8.2 Rising sea levels in the Pacific
- 16.8.3 Rising sea levels in the Maldives

on Resources

eWorkbook	Impacts on low islands (ewbk-10185)
Video eLesson	Impacts on low islands — Key concepts (eles-5266)
Interactivity	Impacts on low islands (int-8684)
Google Earth	Maldives (gogl-0104)

16.9 Managing coastal change

LEARNING INTENTION

By the end of this subtopic, you will be able to describe management strategies to address coastal change.

16.9.1 Changing coastlines

Coastal environments have not always been managed sustainably. In the past, decision-makers had limited knowledge about the fragile nature of many coastal ecosystems, and they had limited environmental world views about the use of coastal areas. Their aim was to develop coastal areas for short-term economic gains. This was based on the belief that nature's resources were limitless. Building high-rise apartments and tourist resorts on sand dunes seemed a good idea — until they fell into the sea when storms eroded the shoreline. Over time, people have realised that coastal management requires an understanding of the processes that affect coastal environments.

To manage the coast sustainably we need to understand:

- the coastal environment and the effect of physical processes
- the effect of human activities within the coastal zone
- the different perspectives of coastal users
- how to achieve a balance between conservation and development
- how decisions are made about the ways in which coasts will be used
- how to evaluate the success of individuals, groups and the levels of governments in managing coastal issues.

16.9.2 Protecting the coast

The protection of the coast through management programs is a costly business that aims to overcome problems associated with land loss, waterlogging and incursions of **groundwater salinity**.

The Netherlands and Germany together spend 250 million euros on coastal works each year. The Netherlands, a country with two-thirds of its land below sea level, has proven that protecting the coastline is possible through a large investment of capital. The most common form of coastal protection in the Netherlands is **dykes** to hold back the sea. However, **floating settlements** that can rise and fall up to 5 metres as sea levels change (see **FIGURE 1**) also provide protection from rising sea levels and tides.

groundwater salinity presence of salty water that has replaced fresh water in the subsurface layers of soil

dykes an embankment constructed to prevent flooding by the sea or a river

floating settlements anchored buildings that float on water and are able to move up and down with the tides

FIGURE 1 Floating houses anchored to the embankments along the waterfront at Maasbommel, the Netherlands

TABLE 1 The ten countries with (a) the largest populations living in low-lying coastal regions and (b) the highest proportions of population living in low-lying coastal regions

(a) Top 10 countries classified by total population in low-lying coastal regions			(b) Top 10 countries classified by proportion of population in low-lying coastal regions		
Nation	Population in low-lying coastal regions ('000)	% of total population in low-lying coastal regions	Nation	Population in low-lying coastal regions ('000)	% of total population in low-lying coastal regions
1. China	127038	10	1. Maldives	291	100
2. India	63341	6	2. Bahamas	267	88
3. Bangladesh	53111	39	3. Bahrain	501	78
4. Indonesia	41807	20	4. Suriname	325	78
5. Vietnam	41439	53	5. Netherlands	9590	60
6. Japan	30827	24	6. Macau	264	59
7. Egypt	24411	36	7. Guyana	419	55
8. United States	23279	8	8. Vietnam	41439	53
9. Thailand	15689	25	9. Djibouti	250	40
10. Philippines	15122	20	10. Bangladesh	53111	39

16.9.3 Coastal management in Australia

If coastlines are to be protected, a wide range of strategies must be employed to combat changes to the coastline and, in particular, flooding of low-lying areas and increased erosion of beaches and bluffs. The techniques shown in **TABLE 2** are used in Australia.

The strategies are designed to protect the coastline by reducing the wave intensity or removing some human activities from the direct pathways of the waves. Each location requiring management will have different physical characteristics and will be used by the community in different ways. As a result there is no one solution to coastal management. In 2016, the wild storms associated with the collapse of homes along the Narrabeen-Collaroy coast was a case study in the natural forces that these strategies aim to address.

FIGURE 2 Wollongong breakwater

TABLE 2 Possible management solutions to reduce impacts of sea level rise and erosion

Description	Diagram	Advantages	Disadvantages
Beach nourishment			
Artificial placement of sand on a beach, which is spread along the beach by natural processes	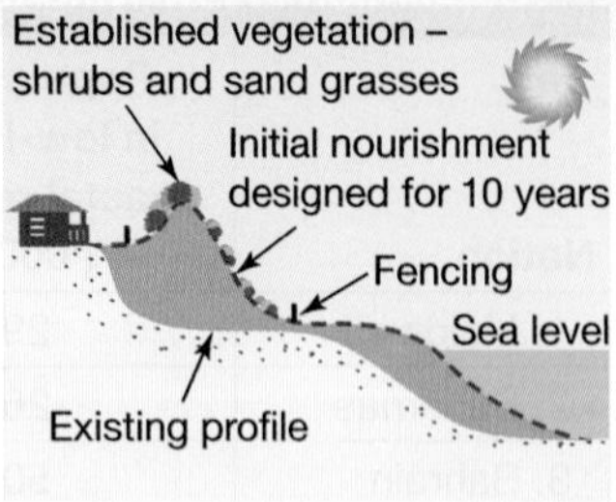	• Sand is used that best matches the natural beach material. • The environmental impact at the beach is low.	• The sand must come from another beach and may have a negative environmental impact in that location. • Must be carried out on a continuous basis and therefore requires ongoing funds.
Groyne			
Artificial structure designed to trap sand being moved by longshore drift (protects the beach) — can be timber, concrete, steel pilings and/or rock	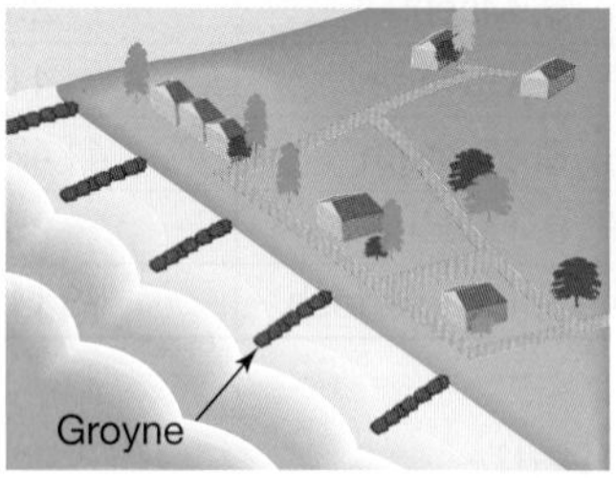	• Traps sand and maintains the beach.	• Does not stop sand movement that occurs directly offshore. • Visual eyesore
Sea wall			
Structure placed parallel to the shoreline to separate the land area from the water		• Prevents further erosion of the dune area and protects buildings	• The base of the sea wall will be undermined over time. • Visual eyesore • Will need a sand nourishment program • High initial cost and ongoing maintenance cost
Offshore breakwater			
Structure parallel to the shore and placed in a water depth of about 10 metres	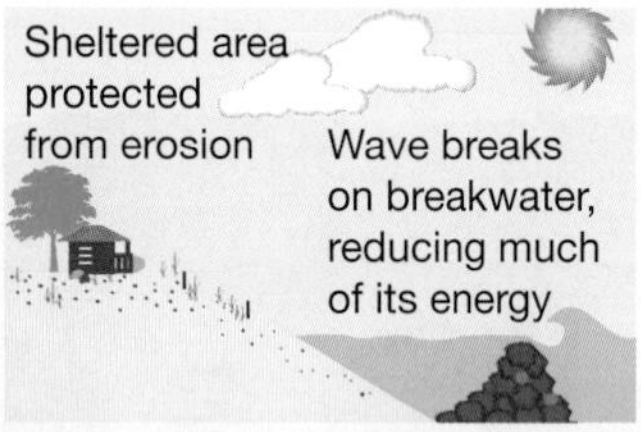	• Waves break in the deeper water, reducing their energy at the shore.	• Destroys surfing amenity of the coast • Requires large boulders in large quantities • Cost extremely high
Purchase property			
Buy the buildings and remove structures that are threatened by erosion.	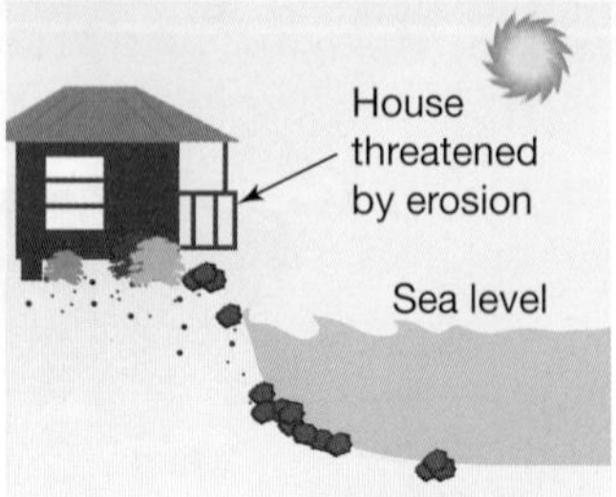	• Allows easier management of the dune area • Allows natural beach processes to continue • Increases public access to the beach	• Loss of revenue to council • Social problems with residents • Exposes back dune area, which will need protection • Cost extremely high • Does not solve sand loss

Resources

eWorkbook Managing coastal change (ewbk-10189)

Video eLesson Managing coastal change — Key concepts (eles-5267)

Interactivity Managing coastal change (int-8685)

16.9 ACTIVITIES

1. As a class, choose one of the case studies of coastal change that you have studied. Discuss the advantages and disadvantages of the coastal management strategies in this subtopic for addressing this coastal change.
2. Working in a group, design a criteria you could use to assess coastal management strategies. Use this criteria to judge the strategies you discussed as a class in question 1.

16.9 EXERCISE

Learning pathways

■ LEVEL 1	■ LEVEL 2	■ LEVEL 3
2, 6, 8, 10, 15	3, 4, 7, 9, 14	1, 5, 11, 12, 13

Check your understanding

1. Refer to **TABLE 1**. Identify which country is most susceptible to changing coastlines in terms of absolute population numbers.
2. What is the purpose of a 'floating settlement'?
3. Outline three factors that contribute to the way that coastal change is managed.
4. Which of the strategies used in Australia does not require the building of new structures?
5. List two negative consequences that might occur because of coastal erosion.
6. Using **TABLE 1**, identify which countries have a high number of people living in low-lying coastal regions and a high percentage of their population living in those regions.

Apply your understanding

7. Should we be trying to stop or reduce the impacts of coastal erosion on people's homes? Give reasons for your answer.
8. Explain why the Netherlands spends money on coastal protection.
9. Explain how a coastal defence system such as a dyke works.
10. If you lived in a home on the coast that was threatened by coastal erosion, which of the management strategies shown in **TABLE 2** would you least like to see used on your beach? Give reasons why.
11. What negative impacts might a 'beach nourishment' strategy have on the beach from which the sand is taken? Explain two possible impacts.
12. Evaluate the strengths and weaknesses of two of the management strategies shown in **TABLE 2**.
    a. Which strategy would have the least environmental impact?
    b. Which strategy would be the most costly to maintain?

Challenge your understanding

13. Predict the impact that a rise in sea level and erosion could have on future food security.
14. Predict possible impacts of sea level rise and coastal erosion on the tourist industries of Australia. What strategies of coastal protection, as mentioned in this chapter, could help solve the problems, and how might they work?
15. Predict how the Netherlands' 'floating houses' will withstand the impacts of climate change on sea levels.

To answer questions online and to receive **immediate feedback** and **sample responses** for every question, go to your learnON title at www.jacplus.com.au.

16.10 SkillBuilder — Comparing an aerial photograph and a topographic map

online only

LEARNING INTENTION

By the end of this subtopic, you will be able to analyse and compare an aerial photograph and a topographic map to gain information about the physical characteristics of an area.

The content in this subtopic is a summary of what you will find in the online resource.

16.10.1 Tell me

What comparisons can be made between aerial photographs and topographic maps?

Comparing an aerial photograph with a topographic map enables us to see what is happening in one place. Photographs and maps may be from the same date, but they may also be from different dates and will thus show different information.

16.10.2 Show me

eles-1751

How to compare an aerial photograph and a topographic map

FIGURE 3 Topographic map extract, Noosa, 2019

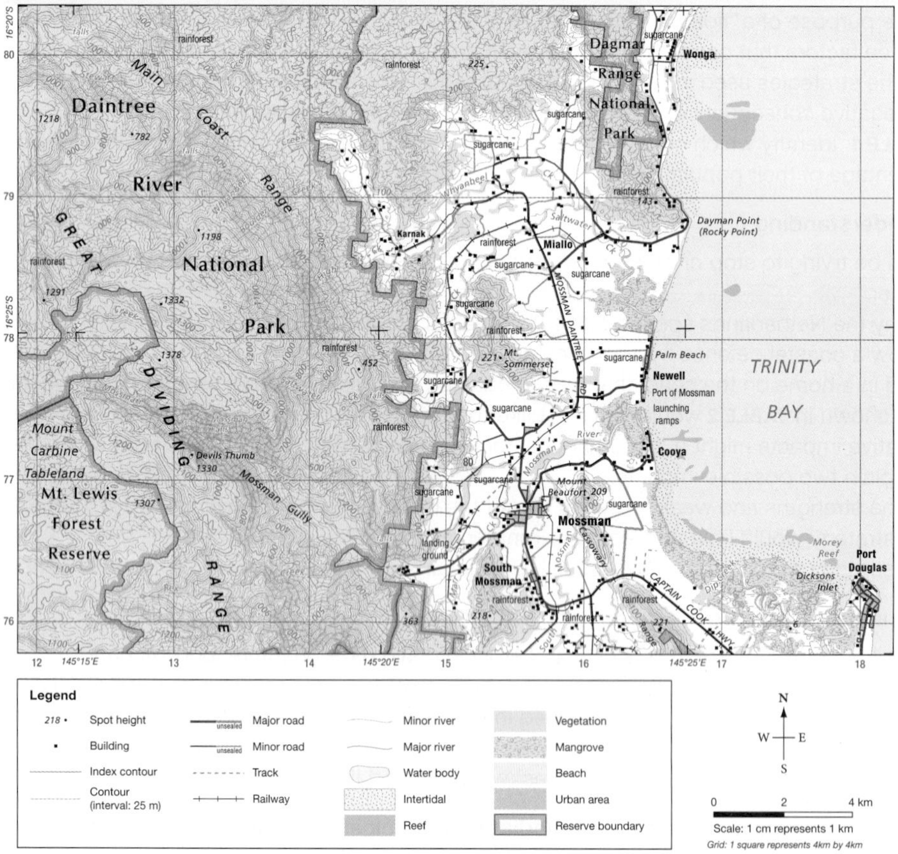

Source: Geoscience Australia; Data based on QSpatial, State of Queensland (Department of Natural Resources, Mines and Energy, Department of Environment and Science), http://qldspatial.information.qld.gov.au/catalogue/

Step 1

Check the titles of both the topographic map and the aerial photograph to ensure they are of the same place. If the titles or areas do not exactly match, work out which part of one relates to the other by identifying common features in both.

FIGURE 3 Aerial photograph of Noosa, 2012

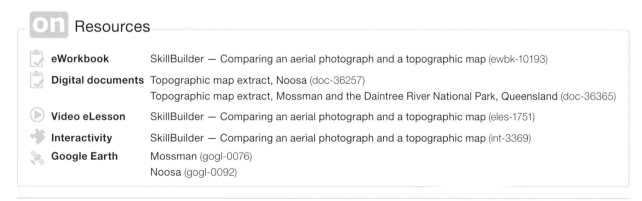

Step 2

Confirm the dates of both pieces of information, so that you are aware of any differences that exist between the photograph and the map as a result of being created at different times.

Step 3

Scan back and forth from the map to the photograph, looking for similarities and differences. Clarify any information that you are not sure about. Begin a paragraph by comparing the two sets of data, ensuring that you mention place names, dates and any available statistics. In some circumstances you could say 'it has changed significantly' or 'it has changed minimally'.

Step 4

Identify the changes that you see. What aspects did you find interesting and why? Continue your paragraph describing the differences that you see between the two sets of data. Conclude your paragraph with a summary sentence.

16.10.3 Let me do it

int-3369

learnON

Go to learnON to access the following additional resources to help you build this skill:
- a longer explanation of this skill and its application in Geography (Tell me)
- a video demonstrating the step-by-step process of this skill (Show me)
- an activity and interactivity for you to practise the skill (Let me do it)
- self-marking questions to help you understand and use the skill.

on Resources

eWorkbook	SkillBuilder — Comparing an aerial photograph and a topographic map (ewbk-10193)
Digital documents	Topographic map extract, Noosa (doc-36257) Topographic map extract, Mossman and the Daintree River National Park, Queensland (doc-36365)
Video eLesson	SkillBuilder — Comparing an aerial photograph and a topographic map (eles-1751)
Interactivity	SkillBuilder — Comparing an aerial photograph and a topographic map (int-3369)
Google Earth	Mossman (gogl-0076) Noosa (gogl-0092)

16.11 Factors influencing coastal management strategies

LEARNING INTENTION

By the end of this subtopic, you will be able to identify and discuss the factors that influence management responses to coastal change.

16.11.1 Worldviews

Environmental worldviews are the varying viewpoints of how the world works and where people fit into the world. Environmental worldviews shape the assumptions and values that guide an individual's actions towards coastal change and its management.

Five worldviews are shown in the table.

Egocentric	• People's needs are the most important factor to consider.
Anthropocentric	• Humans have a variety of needs and wants that often must be placed above the desire to protect environments. Environments are valued for the use they provide to humans.
Stewardship	• Recognises that although humans need to make use of environments for survival and development, they have a responsibility to care for the Earth to ensure that future generations will have access to environments of similar quality
Biocentric	• Recognises the significant role that the Earth and its environments play in sustaining life, including human life. It strives to minimise the impact of human activities on environments and species.
Ecocentric	• Places the preservation of environments above all other needs and wants.

An individual with an egocentric or anthropocentric worldview is more likely to support a new housing development along a coastline, because they will value the employment, the increase in property prices, and additional local businesses associated with the development. An individual with a biocentric or ecocentric worldview is more likely to be concerned about the destruction of natural vegetation, increased sewage, pollution and run-off, and the subsequent decline in water quality and aquatic organisms. An individual's worldview is likely to impact on their perspective on a range of issues related to coastal change.

16.11.2 Competing demands

Different individuals and groups use the coast in different ways, and as a result they have different needs or demands of the coast. Examples of the individuals or groups who use the coast include boat users, swimmers, divers, property owners, fishers, environmental groups, local business owners, property developers, resort owners and workers, holiday makers, surfers, conservationists and the different levels of government. Each of these groups wants the coast to be a specific way, so that they can carry out their activities. Swimmers, snorkellers, holiday makers and fishers are likely to demand clean, clear water, easy access to the beach and an abundance of sea life. Property developers and local business owners may demand access to a large population of customers, close proximity to the water and the development of transport infrastructure. Balancing the demands of different groups can be a challenge for decision makers.

Each level of government has some level of responsibility related to coastal change. The federal government oversees customs, immigration, quarantine, trade and defence. The state government oversees law and order, roads, railways, conservation, forestry and state resource management. The local government oversees land use zoning, building approvals, garbage collection, recreational facilities and the enforcement of laws regarding illegal dumping.

16.11.3 Technology

Technological developments continue to improve coastal management responses. Existing strategies have been improved, and some new developments are beginning to replace existing strategies. Computer modelling and simulation is an important field being shaped by technological developments. Developments are allowing greater accuracy in being able to predict changes to the coast by using computer simulation models and Geographic Information Systems (GIS) that take into account movement of sediment, winds and waves, and assess and map vulnerable coastal environments. Increasingly, engineering solutions are being explored that are adaptive and responsive to changing conditions (they respond differently under different conditions). Wave mitigation devices and the creation of artificial reefs are being used to slow down tidal surges. Although historically 'hard' engineering solutions have been implemented, developments in geotextiles are making 'soft' engineering solutions more popular. Marine mattresses made of geotextiles filled with a combination of stone and top soil and seeded with plants are becoming more widely used than gabion walls, as are coir (coconut fibre) logs and mats.

16.11.4 Climate change

Climate change is a global phenomenon impacting on coastal change. The greenhouse gases produced in one country spread through the atmosphere and affect other countries. Action by only a few countries to reduce greenhouse gases will, therefore, not resolve the problem — it requires international cooperation, especially by the largest polluters.

FIGURE 1 Projected global average temperature outcomes for three greenhouse gas emissions cases, 1990–2100

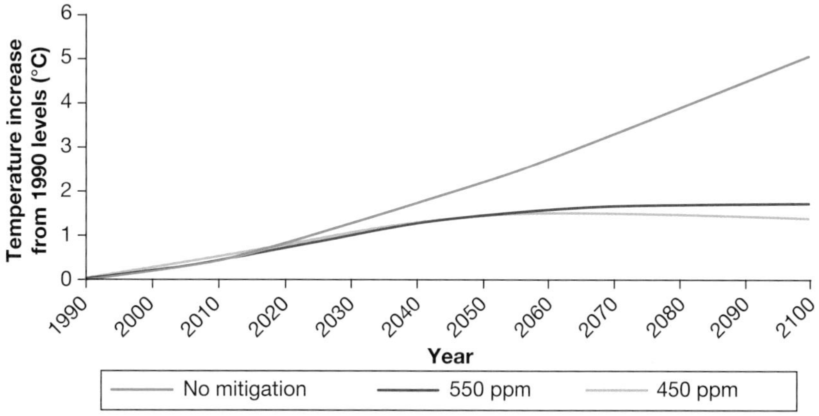

Source: The Garnaut Climate Change Review 2008, p. 88.

Since the 1990s, countries have met at United Nations Intergovernmental Panel of Climate Change (IPCC) conferences and agreed to take steps to reduce emissions of greenhouse gases. An early conference developed the **Kyoto Protocol**, an agreement that sets targets to limit greenhouse gas emissions, and 128 countries have agreed to this Protocol. To date, 192 of the world's 196 countries have signed the Kyoto Protocol; however, close to half have modified their commitment to reach targets for greenhouse emission reductions set for 2020.

To meet the greenhouse gas emissions targets defined by these agreements, countries must make changes that reduce their level of emissions. They can also meet the targets in two other ways:

1. A country can carry out projects in other countries that reduce greenhouse gas emissions and offset these reductions against their own target.
2. Companies can buy and sell the right to emit carbon gases. For example, a major polluter, such as a coal power station, is allowed to emit a certain amount of greenhouse gases. If it is energy efficient and emits less than its limit, it gains carbon credits. It has the right to sell these credits to another company that is having difficulty reducing its emissions. This is often called an **emissions trading scheme**. Companies can also gain credits by investing in projects that reduce greenhouse gases (such as renewable energy), improve energy efficiency or act as carbon sinks (such as tree planting and underground storage of CO_2).

Kyoto Protocol an agreement negotiated in 1999 between 160 countries designed to bring about reductions in greenhouse gas emissions

emissions trading scheme a market-based, government-controlled system used to control greenhouse gas as a cap on emissions. Firms are allocated a set permit or carbon credit, and they cannot exceed that cap. If they require extra credits, they must buy permits from other firms that have lower needs or a surplus.

Resources

eWorkbook Factors influencing coastal management strategies (ewbk-10197)

Video eLesson Factors influencing coastal management strategies — Key concepts (eles-5268)

Interactivity Factors influencing coastal management strategies (int-8686)

16.11 ACTIVITIES

1. Research the Garnaut Report 2011 and the State of the Climate 2014 Report.
   a. What has been done in Australia to mitigate the projected impacts that were outlined in these reports?
   b. How many reports have been produced by the government since 2014 on this issue? What were the findings of each report and what actions have been taken as a result?
   c. Research the Intergovernmental Panel on Climate Change (IPCC) sixth assessment report into climate change. The report is released in four sections from August 2021: physical science; mitigation; impacts, adaptation and vulnerability; and a synthesis report. Evaluate the policies, actions and plans of the Australian government based on the data and recommendations of this report.
2. Conduct a survey to determine the worldviews that your classmates hold.

16.11 EXERCISE

Learning pathways

■ LEVEL 1	■ LEVEL 2	■ LEVEL 3
2, 4, 7, 10, 13	1, 5, 8, 9, 14	3, 6, 11, 12, 15

Check your understanding

1. What is the Kyoto Protocol?
2. Refer to **FIGURE 1**. Determine how much temperatures will increase by 2070 with no mitigation. Which action will reduce temperature change the most by 2100?
3. Outline how an emissions trading scheme might work.
4. How might someone with an ecocentric worldview differ from someone with an anthropocentric worldview when it comes to considering an emissions trading scheme?
5. Which level of government in Australia is responsible for the sustainability of coastal areas?
6. Why is only a few countries reducing their greenhouse gases not the solution to climate change?

Apply your understanding

7. Explain how the two basic strategies developed by the Kyoto Protocol could sustainably reduce the amount of greenhouse gases in the atmosphere.
8. Explain why organisations such as the Conservation Council of Australia might have different views about climate change from companies that produce electricity.
9. What is expected to happen to average global temperatures if no action is taken to reduce greenhouse gas emissions? Refer to **FIGURE 1** in your response.
10. Identify five different groups or individuals who see the coast differently. Identify the needs or demands that each group or individual would have of the coast.
11. Provide one argument for and one argument against emissions trading schemes.
12. Many of the countries that signed the Kyoto Protocol have modified the goals that they had planned to reach by 2020. What does this suggest about their original goals?

Challenge your understanding

13. Present your environmental worldview, giving examples to explain why you think the way you do.
14. How successful do you think international forums are in helping to solve climate change?
15. Should countries be allowed to conduct sustainability projects in other countries to balance out their emissions debt? Suggest what some of the ethical difficulties with this arrangement might be.

To answer questions online and to receive **immediate feedback** and **sample responses** for every question, go to your learnON title at www.jacplus.com.au.

16.12 Contributing to environmental sustainability in coastal zones

LEARNING INTENTION

By the end of this subtopic, you will be able to provide examples of how community groups and individuals can contribute to environmental sustainability in coastal zones.

16.12.1 Community groups

In any community there will be a range of views and perspectives about how coasts should be managed.

An interest group is a group of people who **campaign** on a cause or interest they have in common. Interest groups can sometimes be major players in politics. Examples of groups involved in coastal management are resident action groups, Landcare groups, conservationists, local business groups.

Coastcare was established in 1995 and is an extension of the Landcare movement. Coastcare groups are made up of volunteers who put into action coastal management responses to protect sand dunes, protect endangered species and habitats, and revegetate coastal environments. An example is the Umina Community Dunecare group, which works along Umina Beach each week to weed and revegetate the dunes.

16.12.2 Individual actions

Individuals can influence decision-making processes related to coastal management. They can write to or meet with their local member of parliament or councillor, campaign or become involved in **lobbying**, or write a letter/email to a newspaper or media website. They can also engage in direct action (such as protests, demonstrations and sit ins).

Individual action for climate change

Greenhouse gases contribute to global warming, which contributes to rising sea levels, and this has a significant impact on the sustainability of coastal environments. Australian households produce about one-fifth of Australia's greenhouse gases through their use of transport, household energy, and the decay of household waste in landfill. This amounts to about 15 tonnes of CO_2 per household per year. (A tonne of CO_2 would fill one family home.) The Australian Conservation Foundation has suggested a ten-point that every Australian household can follow to reduce its level of greenhouse gas pollution.

FIGURE 1 The Australian Conservation Foundation Plan

GreenPower®

Source: GreenPower

1. **Switch to renewable power.**
   Choose renewable energy from your electricity retailer and support investment in sustainable, more environmentally friendly energies. Make sure it is accredited GreenPower (electricity produced using renewable resources) — see www.greenpower.gov.au to find out more about the program.
2. **Get rid of one car in your household.**
   A car produces seven tonnes of greenhouse pollution each year (based on travelling 15 000 kilometres per year). This does not include the energy and water used to build the car — 83 000 litres of water and eight tonnes of greenhouse pollution. So share a car with your family.

campaign an organised course of action for a specific purpose, for example, to create public interest in an issue

lobbying the activity individuals and organisation engage in to influence decision makers

3. **Take fewer air flights.**
A return domestic flight in Australia creates about 1.5 tonnes of greenhouse emissions (based on Melbourne to Sydney return). A return international flight creates about 9 tonnes (based on Melbourne to New York return). Holiday closer to home.
4. **Use less power to heat your water.**
A conventional electric household water heater produces about 3.2 tonnes of greenhouse pollution in a year. Using less hot water will reduce your pollution. Using the cold cycle on your washing machine will save 3 kg of greenhouse pollution. Switching off your water heater when you're away will also reduce your energy use.
5. **Eat less meat.**
Meat, particularly beef, has a very high environmental impact, using a lot of water and land to produce it, and creating significant greenhouse pollution. If you reduce your red meat intake by two 150-gram serves a week, you'll save 20 000 litres of water and 600 kg of greenhouse pollution a year.
6. **Heat and cool your home less.**
Insulate your walls and ceilings. This can cut heating and cooling costs by 10 per cent. Each degree change can save 10 per cent of your energy use. A 10 per cent reduction is 310 kg of greenhouse pollution saved.
7. **Replace your old showerhead with a water-efficient alternative.**
This will save about 44 000 litres of water a year and up to 1.5 tonnes of greenhouse pollution from hot water heating (on average).
8. **Turn off standby power.**
Turning appliances off at the wall could reduce your home's greenhouse emissions by up to 700 kg a year.
9. **Cycle, walk or take public transport rather than drive your car.**
Cycling 10 kilometres to work (or school) and back twice a week instead of driving saves about 500 kg of greenhouse pollution each year and saves you about $770. Besides, it's great for your health and fitness!
10. **Make your fridge more efficient.**
Ensure the coils of your fridge are clean and well ventilated — that will save around 150 kg of greenhouse pollution a year. Make sure the door seals properly — this saves another 50 kg. Keep fridges and freezers in a cool, well-ventilated spot to save up to another 100 kg a year. If you have a second fridge, turn it off when not in use.

FIGURE 2 Huskisson Beach, Jervis Bay, New South Wales

Resources

eWorkbook	Contributing to environmental sustainability in coastal zones (ewbk-10201)
Video eLesson	Contributing to environmental sustainability in coastal zones — Key concepts (eles-5269)
Interactivity	Contributing to environmental sustainability in coastal zones (int-8687)

16.12 ACTIVITIES

1. Create a poster to communicate the main points of the Australian Conservation Foundation's 10-point strategy to reduce greenhouse gases. Use the **GreenPower** weblink in your online Resources as a starting point.
2. Investigate the idea of an eco-friendly house. Start with the **Eco system homes** weblink in your online Resources. Design an eco-friendly house that minimises greenhouse gas emissions.
3. Undertake research on a local Coastcare group. Evaluate the effectiveness of this community group in addressing coastal change.

16.12 EXERCISE

Learning pathways

■ LEVEL 1	■ LEVEL 2	■ LEVEL 3
1, 2, 11, 12, 13	3, 4, 9, 10, 14	5, 6, 7, 8, 15

Check your understanding

1. How would selling one of your family's cars reduce your greenhouse gas emissions?
2. What does it mean to 'lobby' the government?
3. What proportion of Australia's greenhouse gas emissions is produced by households?
4. Describe how an interest group could influence government decisions about coastal management.
5. What role does Coastcare play in the environmental sustainability of coasts?
6. List the ways that individuals can influence coastal management.

Apply your understanding

7. Assess how effective individual and community action can be in addressing coastal change. Explain one aspect of coastal change that might require action on a national or international scale, as well as on a local scale.
8. Explain the interconnection between reducing greenhouse gas emissions and coastal sustainability.
9. The Australian Conservation Foundation ten-point plan addresses how households can reduce their greenhouse gas emissions. Discuss which of these suggestions could equally apply to businesses.
10. Which individual action from the ten-point plan might achieve the greatest reduction in greenhouse gas emissions for the least possible effort? Give reasons to justify your answer.
11. What is the role of community action and volunteer groups in caring for the coast?
12. What is 'green' power?

Challenge your understanding

13. Which of the suggestions provided in the ten-point plan could you implement the most easily where you live? Outline why it would be easy and how you would put the strategy in action.
14. Do you think the ten strategies for individual action are equally possible for everyone in Australia? Suggest what factors might influence whether the strategies are achievable, providing examples to support your view.
15. What do you think is the greatest barrier to Australia reducing its greenhouse gas emissions? Justify your response.

To answer questions online and to receive **immediate feedback** and **sample responses** for every question, go to your learnON title at www.jacplus.com.au.

16.13 Investigating topographic maps — Consequences of coastal change in Merimbula

LEARNING INTENTION

By the end of this subtopic, you will be able to use a map and aerial photograph to investigate coastal change.

16.13.1 Coastal change in Merimbula

Merimbula is a coastal resort town on the south-east coast of New South Wales. The 'Sapphire Coast' is a popular tourist destination because of its array of beautiful beaches, stunning scenery and mild, sunny weather. Similar to any other popular coastal location, the main pressures on this coastal system relate to the development of the town and tourist facilities. Careful management has enabled growth of the urban area while at the same time protecting many of the natural features of the coastline.

The natural landform features along this coastline include a series of headlands separated by bay head beaches. Merimbula Lake has formed from a slow and gradual build up of a sand barrier, leaving only a narrow channel for salt water to enter and fresh water to exit (see **FIGURE 1**). The shallow and sheltered waters of the lake provide an ideal environment for oyster farming and recreation.

FIGURE 1 Aerial view over Merimbula Lake

on Resources

eWorkbook	Investigating topographic maps — Consequences of coastal change in Merimbula (ewbk-10205)
Digital document	Topographic map of Merimbula (doc-36370)
Video eLesson	Investigating topographic maps — Consequences of coastal change in Merimbula — Key concepts (eles-5270)
Interactivity	Investigating topographic maps — Consequences of coastal change in Merimbula (int-8688)
Google Earth	Merimbula (gogl-0075)

FIGURE 2 Topographic map extract of Merimbula

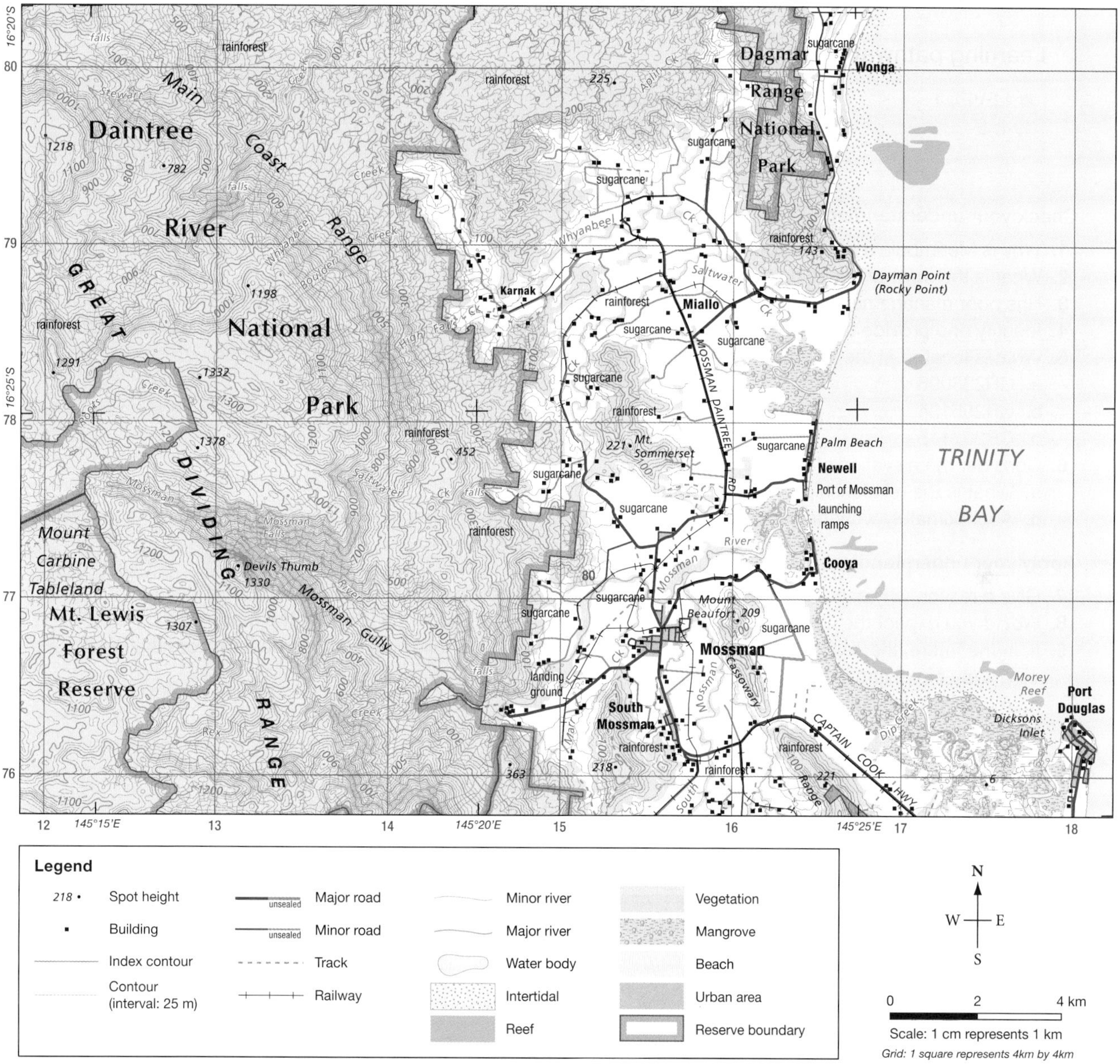

Source: Geoscience Australia; Data based on QSpatial, State of Queensland (Department of Natural Resources, Mines and Energy, Department of Environment and Science), http://qldspatial.information.qld.gov.au/catalogue/

16.13 EXERCISE

Learning pathways

■ LEVEL 1	■ LEVEL 2	■ LEVEL 3
1, 4, 5	2, 5, 8, 9	3, 6, 7, 10

Check your understanding

1. Why is Merimbula a popular tourist location?
2. What is the main pressure on the coastal system of Merimbula?
3. Has poor management resulted in the destruction of the Merimbula coastline?
4. Identify and provide the grid references for three natural landforms that feature along the Merimbula coastline.
5. What is located at the following grid references?
   a. GR233096
   b. GR255124
   c. GR267115
6. Consider the elevation of built-up areas.
   a. What is the relationship between elevation and built-up areas?
   b. Approximately what percentage of built-up areas would be on land higher than 20 metres above sea level?

Apply your understanding

7. Give a reason why sand has built up to form a beach at Middle Beach and not at Merimbula Point.
8. Would you expect the water in Back Lake to be fresh or salty? Use evidence from the map.
9. If, in the future, the sea level was to rise by 10 metres, would the following features be safe from the rising sea? Why or why not?
   a. A caravan park located on Short Point
   b. Merimbula Airport

Challenge your understanding

10. In what direction(s) is Merimbula likely to expand in the future? Justify your decision.

FIGURE 3 The shallow waters of Merimbula Lake

To answer questions online and to receive **immediate feedback** and **sample responses** for every question, go to your learnON title at www.jacplus.com.au.

16.14 Thinking Big research project — Ecology action newsletter — reef rescue

online only

Scenario

The content in this subtopic is a summary of what you will find in the online resource.

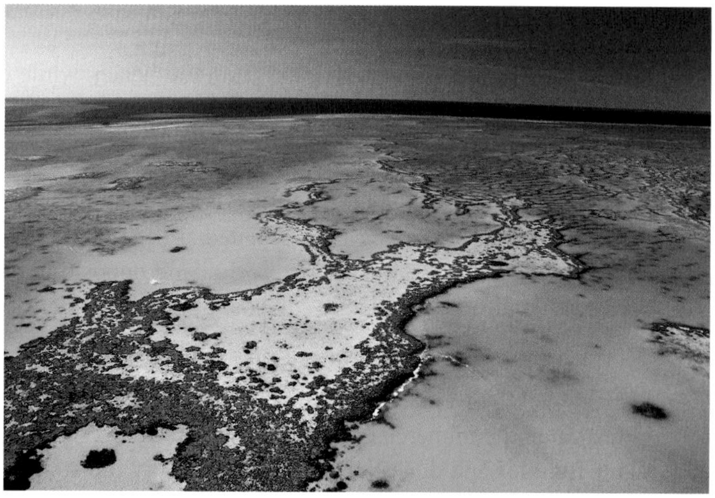

Located in the Coral Sea off Australia's north-east coast, the iconic Great Barrier Reef is the world's largest coral reef system. It supports a wide diversity of marine life. It is composed of over 2900 individual reefs and 900 islands, and stretches more than 2300 kilometres in length across an area of approximately 344 400 square kilometres. The reef was selected as a World Heritage Site in 1981 but, in addition to global warming, is currently threatened by increased shipping and its popularity as a tourism site.

You are a member of your school's ecology action group. Each term the group publishes a newsletter highlighting various environmental issues. This month your focus is on the Great Barrier Reef. How can we protect the reef from environmental threats and ensure its health for now and all time?

Task

You will research and create a newsletter focusing on the Great Barrier Reef and the challenges it faces related to human activity. Your newsletter can be in print or digital form and should include the following elements:

- an overview of the environmental characteristics of the reef, including a location map (with BOLTSS) and appropriate images annotated to provide relevant information
- interesting facts and figures
- details of threats to the reef from shipping and tourism
- details of social, environmental and economic approaches to tackling the identified threats
- an evaluation of the approaches identified, in terms of
  - economic viability (affordability)
  - social justice (fairness for all people)
  - environmental benefit (minimal negative environmental impact and with future sustainability)
- concluding recommendations for action at individual, local, national and international levels in response to the threats, based on your research and evaluation.

learnON

Go to learnON to access the resources you need to complete this research project.

on Resources

ProjectsPLUS Thinking Big research project — Ecology action newsletter — reef rescue (pro-0214)

16.15 Review

16.15.1 Key knowledge summary

16.2 How coastal landforms are created

- Waves, tides, currents, rips, storm surges and tsunamis form and transform coastal landscapes.
- Coastal landforms include caves, arches, cliffs, estuaries, lagoons, beaches, dunes, stacks, blowholes, tombolos and headlands.
- Longshore drift occurs when ongoing swash and backwash moves in a sideways direction and transport sediment parallel to the shoreline.
- Destructive waves can remove sand from the beach, while constructive waves are likely to deposit sand onto the beach.

16.3 Deposition and erosion of coasts

- Deposition is when sediment is dropped or deposited onto a site. An example is when sediment from cliffs is transported down a river and deposited at the mouth of a river.
- Dunes are shaped by the constant erosion and deposition of sand.
- Erosional landforms include headlands, bays, cliffs, platforms, caves, arches, blowholes and stacks.

16.4 Human causes of coastal change

- Human causes of coastal change include overfishing; litter; removal of natural vegetation; residential, industrial and commercial developments; dredging; stormwater; sewage outfalls; and off-shore drilling.
- The Cronulla sand dunes have been changed by human activities such as timber harvesting, farming, mining, industrial use and residential development.

16.5 Environmental change in the Tweed and Gold Coast

- In the past, sand would often bock the mouth of the Tweed River. To address this, a pair of training walls were constructed 400 metres out to sea.
- Sand destined for beaches further north was trapped by the training walls, and beaches north of the river mouth were stripped of sand by natural wave action and storm activity.
- To restore Gold Coast beaches, a sand bypass system was established to transport sand from south of the mouth of the Tweed River to the eroded beaches.

16.6 Impacts of inland activities on coasts

- The Sundarbans region in Bangladesh is a World Heritage site at risk of increased flooding. The Sundarbans are the largest intact mangrove forest in the world. The mangroves trap sediment in their root systems, slow water movement, and play an important role in flood mitigation and defending against storm surges.
- Mangrove clearing and the construction of aquaculture ponds are destroying Bangladesh's natural defence against flooding.
- Increased melting of glaciers and snow inland in the Himalayas and shifting weather patterns as a result of climate change will exacerbate flooding problems in Bangladesh.

16.8 Impacts on low islands

- Climate change will result in many low-lying islands being flooded by the sea. Many island groups in the Pacific and India Oceans will be almost completely inundated by 2050.
- Islands such as Kiribati, Tuvalu and the Marshall Islands in the south-west Pacific, which are only a few metres above sea level, are particularly vulnerable to rising sea levels. The Maldives in the Indian Oceans are also susceptible.
- Action on climate change may be able to slow the need for populations to abandon their islands. Other strategies may include construction of sea walls or raising the level of some key islands.

16.9 Managing coastal change

- Effective management of coasts requires understanding of physical processes, the effect of human activity on coasts, and an understanding of different perspectives of coastal users.
- A balance should be achieved between conservation and development of coasts.
- Management strategies include beach nourishment, groyne construction, sea walls, offshore breakwaters and acquisition of properties.

16.11 Factors influencing coastal management strategies

- Environmental worldviews are the varying viewpoints of how the world works and where people fit into the world. An individual's worldview is likely to impact on their perspective on a range of issues related to coastal change.
- Different individuals and groups use the coast in different ways, and as a result they have different needs or demands of the coast. Balancing the demands of different groups can be a challenge for decision makers.
- Technological developments continue to improve coastal management responses. Existing strategies have been improved, and new developments are beginning to replace existing strategies.

16.12 Contributing to environmental sustainability in coastal zones

- In any community, there will be a range of views and perspectives about how coasts should be managed.
- Coastcare groups are made up of volunteers who put into action coastal management responses to protect sand dunes, protect endangered species and habitats, and revegetate coastal environments
- Individuals can influence decision-making processes by engaging in political processes and taking action on climate change.

16.15.2 Key terms

alluvial plain an area where rich sediments are deposited by flooding

atoll a coral island that encircles a lagoon

campaign an organised course of action for a specific purpose, for example, to arouse public interest in an issue

coastal dune vegetation succession the process of change in the plant types of a vegetation community over time — moving from pioneering plants in the high tide zone to fully developed inland area vegetation

deltaic plain flat area where a river(s) empties into a basin

dykes an embankment constructed to prevent flooding by the sea or a river

emissions trading scheme a market-based, government-controlled system used to control greenhouse gas as a cap on emissions. Firms are allocated a set permit or carbon credit, and they cannot exceed that cap. If they require extra credits, they must buy permits from other firms that have lower needs or a surplus.

floating settlements anchored buildings that float on water and are able to move up and down with the tides

groundwater salinity presence of salty water that has replaced fresh water in the subsurface layers of soil

groynes a structure (e.g. a rock wall) that is built perpendicular to the shoreline to interrupt the flow of slow and the movement of sediment

Kyoto Protocol an agreement negotiated in 1999 between 160 countries designed to bring about reductions in greenhouse gas emissions

lobbying the activity individuals and organisation engage in to influence decision makers

longshore drift a current that moves sediment parallel to the shoreline, created by the backward and forward motion of waves

monsoon a wind system that brings heavy rainfall over large climatic regions and reverses direction seasonally

storm surges a temporary increase in sea level from storm activity

training walls walls or jetties that are constructed to direct the flow of a river or tide

tsunami a powerful ocean wave triggered by an earthquake or volcanic activity under the sea

16.15.3 Reflection

Complete the following to reflect on your learning.

Revisit the inquiry question posed in the Overview:

What are the causes and consequences of coastal change, and how can this change be managed?

1. Now that you have completed the topic, describe how coastal environments function.
2. Write an extended response: Compare and evaluate the effectiveness of coastal management responses in two countries.
3. Propose ways individuals and communities can contribute to more sustainable management of coasts.

Subtopic	Success criteria			
16.2	I can explain the biophysical processes essential to the functioning of coasts.			
16.3	I can explain how the operation of biophysical processes maintains the functioning of coasts.			
16.4	I can explain some of the ways that humans cause coastal change.			
16.5	I can analyse the short-term and long-term consequences of climate change on coasts.			
16.5	I can provide examples of the causes, extent and consequences of environmental change in the Tweed and Gold Coast.			
16.6	I can describe the causes and extent of change to coasts in Bangladesh.			
16.7	I can identify examples of how a place has changed over time by examining aerial photographs.			
16.8	I can describe the long-term consequences of coastal change.			
16.9	I can describe management strategies to address coastal change.			
16.10	I can compare an aerial photograph and a topographic map to identify coastal change.			
16.11	I can discuss the factors that influence management responses to coastal change.			
16.12	I can provide examples of how community groups and individuals can contribute to environmental sustainability in coastal zones.			

on Resources

eWorkbook Chapter 16 Student learning matrix (ewbk-8580)
Chapter 16 Reflection (ewbk-8581)
Chapter 16 Extended writing task (ewbk-8582)

Interactivity Chapter 16 Crossword (int-8689)

ONLINE RESOURCES

on Resources

Below is a full list of **rich resources** available online for this topic. These resources are designed to bring ideas to life, to promote deep and lasting learning and to support the different learning needs of each individual.

eWorkbook

- **16.1** Chapter 16 eWorkbook (ewbk-8108) ☐
- **16.2** How coastal landforms are created (ewbk-10161) ☐
- **16.3** Deposition and erosion of coasts (ewbk-10165) ☐
- **16.4** Human causes of coastal change (ewbk-10169) ☐
- **16.5** Environmental change in the Tweed and Gold Coast (ewbk-10177) ☐
- **16.6** Bangladesh awash (ewbk-10173) ☐
- **16.7** SkillBuilder — Comparing aerial photographs to investigate spatial change over time (ewbk-10181) ☐
- **16.8** Impacts on low islands (ewbk-10185) ☐
- **16.9** Managing coastal change (ewbk-10189) ☐
- **16.10** SkillBuilder — Comparing an aerial photograph and a topographic map (ewbk-10193) ☐
- **16.11** Factors influencing coastal management strategies (ewbk-10197) ☐
- **16.12** Contributing to environmental sustainability in coastal zones (ewbk-10201) ☐
- **16.13** Investigating topographic maps — Consequences of coastal change in Merimbula (ewbk-10205) ☐
- **16.15** Chapter 16 — Student learning matrix (ewbk-8580) ☐
  - Chapter 16 — Reflection (ewbk-8581) ☐
  - Chapter 16 — Extended writing task (ewbk-8582) ☐

Sample responses

- **16.1** Chapter 16 Sample responses (sar-0165) ☐

Digital documents

- **16.10** Topographic map extract, Noosa (doc-36257) ☐
  - Topographic map extract, Mossman and the Daintree River National Park, Queensland (doc-36365) ☐
- **16.13** Topographic map of Merimbula (doc-36370) ☐

Video eLessons

- **16.1** Washed away (eles-1710) ☐
  - Managing change in coastal environments — Photo essay (eles-5260) ☐
- **16.2** How coastal landforms are created — Key concepts (eles-5261) ☐
- **16.3** Deposition and erosion of coasts — Key concepts (eles-5262) ☐
- **16.4** Human causes of coastal change — Key concepts (eles-5263) ☐
- **16.5** Environmental change in the Tweed and Gold Coast — Key concepts (eles-5264) ☐
- **16.6** Impacts of inland activities affect coasts — Key concepts (eles-5265) ☐
- **16.7** SkillBuilder — Comparing aerial photographs to investigate spatial change over time (eles-1750) ☐
- **16.8** Impacts on low islands — Key concepts (eles-5266) ☐
- **16.9** Managing coastal change — Key concepts (eles-5267) ☐
- **16.10** SkillBuilder — Comparing an aerial photograph and a topographic map (eles-1751) ☐
- **16.11** Factors influencing coastal management strategies — Key concepts (eles-5268) ☐
- **16.12** Contributing to environmental sustainability in coastal zones — Key concepts (eles-5269) ☐
- **16.13** Investigating topographic maps — Consequences of coastal change in Merimbula — Key concepts (eles-5270) ☐

Interactivities

- **16.2** How coastal landforms are created (int-8722) ☐
  - The range of coastal landforms (int-5594) ☐
- **16.3** Deposition and erosion of coasts (int-8682) ☐
- **16.4** Human causes of coastal change (int-8683) ☐
- **16.5** Pumping sand (int-3295) ☐
- **16.6** Bangladesh awash (int-3297) ☐
- **16.7** SkillBuilder — Comparing aerial photographs to investigate spatial change over time (int-3368) ☐
- **16.8** Impacts on low islands (int-8684) ☐
- **16.9** Managing coastal change (int-8685) ☐
- **16.10** SkillBuilder — Comparing an aerial photograph and a topographic map (int-3369) ☐
- **16.11** Factors influencing coastal management strategies (int-8686) ☐
- **16.12** Contributing to environmental sustainability in coastal zones (int-8687) ☐
- **16.13** Investigating topographic maps — Consequences of coastal change in Merimbula (int-8688) ☐
- **16.15** Chapter 16 Crossword (int-8689) ☐

ProjectsPLUS

- **16.14** Thinking Big Research project — Ecology action newsletter — reef rescue (pro-0214) ☐

Weblinks

- **16.7** Hurricane Sandy (web-3355) ☐
- **16.8** Maldives (web-3356) ☐
- **16.12** Eco system homes (web-3922) ☐
  - GreenPower (web-6378) ☐

Fieldwork

- **16.3** Measuring dunes (fdw-0039) ☐

Google Earth

- **16.8** Maldives (gogl-0104) ☐
- **16.10** Mossman (gogl-0076) ☐
  - Noosa (gogl-0092) ☐
- **16.13** Merimbula (gogl-0075)

Teacher resources

There are many resources available exclusively for teachers online.

17 Marine environments and change management

INQUIRY SEQUENCE

online**only**

To access a pre-test with **immediate feedback** and **sample responses** to every question in this chapter, select your learnON format at www.jacplus.com.au.

17.1 Overview

online only

This chapter is included in your learnON title at www.jacplus.com.au.

Numerous **videos** and **interactivities** are embedded just where you need them, at the point of learning, in your learnON title at www.jacplus.com.au. They will help you to learn the content and concepts covered in this topic.

Exactly how much plastic ends up in oceans and waterways, and why should we care if it does?

Imagine you are on a beach. You are looking out to sea at the endless, constantly moving mass of water that stretches to the horizon. Why does it move, how does it move, what lies beneath?

Life on Earth would not be possible without our oceans. Humans are interconnected with the oceans, which provide or regulate our water, oxygen, weather, food, minerals and resources. Oceans also provide a surface for transport and trade and a habitat for 80 per cent of all life on Earth. Our oceans are under threat; as we use them to extract resources and dump waste, we destroy them. This image shows just a tiny fraction of the many thousands of tonnes of marine debris floating at sea. The health of our oceans is at risk. In this chapter, we look at this problem in more detail.

FIGURE 1 So much rubbish and marine debris can be found floating in the world's oceans.

STARTER QUESTIONS

1. What are your first thoughts when you view this photograph?
2. Suggest items that might be floating in this rubbish.
3. Where do you think this waste has come from, and how did it get here?
4. What waste does your family generate, and what happens to it?
5. Watch the **Marine environments and change management — Photo essay** video in your online Resources. What is your experience of the ocean?

on Resources

eWorkbook Chapter 17 eWorkbook (ewbk-8109)

Video eLessons Thrown overboard (eles-1711)
Marine environments and change management — Photo essay (eles-5271)

ONLINE RESOURCES

Resources

Below is a full list of **rich resources** available online for this topic. These resources are designed to bring ideas to life, to promote deep and lasting learning, and to support the different learning needs of each individual.

eWorkbook

- **17.1** Chapter 17 eWorkbook (ewbk-8109) ☐
- **17.2** Marine processes (ewbk-10213) ☐
- **17.3** Marine pollution and debris (ewbk-10217) ☐
- **17.4** Responses to marine debris (ewbk-10221) ☐
- **17.5** SkillBuilder — Using geographic information systems (GIS) (ewbk-10225) ☐
- **17.6** Marine pollution (ewbk-10229) ☐
- **17.7** SkillBuilder — Describing photographs (ewbk-10233) ☐
- **17.8** Investigating topographic maps — Coral bleaching on Lizard Island (ewbk-10237) ☐
- **17.10** Chapter 17 Student learning matrix (ewbk-8584) ☐
  Chapter 17 Reflection (ewbk-8585) ☐
  Chapter 17 Extended writing task (ewbk-8586) ☐

Sample responses

- **17.1** Chapter 17 Sample responses (sar-0166) ☐

Digital document

- **17.8** Topographic map of Lizard Island (doc-20456) ☐

Video eLessons

- **17.1** Thrown overboard (eles-1711) ☐
  Marine environments and change management — Photo essay (eles-5271) ☐
- **17.2** Marine processes — Key concepts (eles-5272) ☐
- **17.3** Marine pollution and debris — Key concepts (eles-5273) ☐
- **17.4** Responses to marine debris — Key concepts (eles-5274) ☐
- **17.5** SkillBuilder — Using geographic information systems (GIS) (eles-5348) ☐
- **17.6** Marine pollution — Key concepts (eles-5275) ☐
- **17.7** SkillBuilder — Describing photographs (eles-1660) ☐
- **17.8** Investigating topographic maps — Coral bleaching on Lizard Island — Key concepts (eles-5276) ☐

Interactivities

- **17.2** Motion in the ocean (int-3298) ☐
- **17.3** Garbage patch (int-3299) ☐
  The sources of marine pollution (int-5596) ☐
  Quantity of plastic released into the ocean for selected countries (int-7963) ☐
- **17.4** Plastic bag bans around the world (int-7964) ☐
- **17.5** SkillBuilder — Using geographic information systems (GIS) (int-3370) ☐
- **17.6** Oil slick (int-3300) ☐
  Marine pollution (int-8690) ☐
- **17.7** SkillBuilder — Describing photographs (int-3156) ☐
- **17.10** Chapter 17 Crossword (int-8692) ☐

ProjectsPLUS

- **17.9** Thinking Big research project — 'Plastic not-so-fantastic' media campaign (pro-0215) ☐

Weblinks

- **17.3** Plastiki Expedition (web-3493) ☐
- **17.5** ReefBase GIS (web-3497) ☐
- **17.7** Kibera slum (web-1573) ☐

Fieldwork

- **17.4** Plastics survey (fdw-0040) ☐

Google Earth

- **17.6** Gulf of Mexico (gogl-0078) ☐
- **17.8** Lizard Island (gogl-0106) ☐

Teacher resources

There are many resources available exclusively for teachers online.

18 Geographical inquiry — Developing an environmental management plan

18.1 Overview

LEARNING INTENTION

By the end of this subtopic, you will be able to identify and propose strategies to manage a specific environmental threat.

18.1.1 Scenario and task

Task: Prepare an environmental management plan that deals with a specific environmental threat.

There are many environmental changes that have an impact on different environments. Organisations or their specialist consultants often prepare environmental management plans (EMPs). EMPs recommend the steps to be undertaken to solve identified problems in managing the environment. They are also useful for predicting and minimising the effects of potential future changes. These strategies are designed to either remove or control the problem(s).

Your task

Each class team will research and prepare an EMP that deals with a specific environmental threat and then present it to the class. Decide on an environment and the threat it faces and then devise three key inquiry questions you would like to answer.

18.2 Inquiry process

18.2.1 Process

Open the ProjectsPLUS application for this project located in your online Resources. Watch the introductory video lesson and then click the 'Start project' button and set up your class group. Save your settings and the project will be launched.

- **Planning:** In pairs or groups, decide on a particular environmental issue and devise a series of three key inquiry questions that will become a focus of your study and a means of dividing the workload. Download the EMP planning template to help you think about and decide which environments your team will choose to research. Navigate to your Research Forum. Use the research topics to select the environments your team has chosen. The following steps will act as a guide for your report writing.

18.2.2 Collecting and recording data

Find out about the issue and why an EMP is needed. Identify potential environmental threats or changes that may occur. Describe the issue, the scale of potential changes and their significance. Prepare a map or series of maps to show the location of the issue. This may be sourced from a street directory, an atlas, Google maps or an online reference. Additional data can be researched and collected; for example, you may wish to survey people's opinions on the issue, use census data to determine the number of people affected in the region or find climatic data for the area. (Your teacher may guide you at this point.) Decide on the most suitable presentation method for your data, for example graphs, maps and annotated photographs. You may wish to refer to relevant SkillBuilders to help you present your data.

18.2.3 Processing and analysing your information and data

Review and discuss with your team members the information that you have collected. Has it come from reliable sources? What patterns, trends and interconnections can you identify from your data? Are there any anomalies? Can you identify the reasons for these anomalies?

Come up with two or three possible options that will address the issue(s) you have collected information about. It would be beneficial to include diagrams and/or photographs of strategies currently operating in different places that could be used or adapted to your site.

Evaluate which option would be most effective based on the following criteria:

- economic viability (affordable)
- social justice (fair to all people involved)
- environmental benefit (minimal environmental impact and with future sustainability).

Make concluding recommendations based on your research and evaluation of options. These should be in the form of a suggested course of action to follow in managing the environment and reducing any negative changes.

18.2.4 Communicating your findings

Present your report to the class and be prepared to answer questions from the audience. Use the EMP template to help you structure your report. Use graphics such as maps, graphs, images and charts in your EMP.

18.3 Review

18.3.1 Reflecting on your work

Review your participation in the production of your EMP by completing the reflection document in your Media Centre.

Print out your Research Report from ProjectsPLUS and hand it in with your EMP and reflection notes.

on Resources

ProjectsPLUS Geographical inquiry — Developing an environmental management plan (pro-0150)

Some of the environmental threats you might consider include the release of toxic substances into waterways, invasive species, fire, or industrial air pollution.

UNIT

4 Human wellbeing

19 Human wellbeing and development

INQUIRY SEQUENCE

To access a pre-test with **immediate** feedback and **sample responses** to every question in this chapter, select your learnON format at www.jacplus.com.au.

19.1 Overview

Numerous **videos** and **interactivities** are embedded just where you need them, at the point of learning, in your learnON title at www.jacplus.com.au. They will help you to learn the content and concepts covered in this topic.

Everyone wants a good life, but what does that mean for different people? Can wellbeing actually be measured and how can we improve it if it's not measuring up?

19.1.1 Introduction

We all want a better life for ourselves, our families and our children, no matter where we live. We care about the progress of our communities, our state or territory, and our country. But how can we measure this progress? What does progress really mean? What do we count when we measure progress? How do we know if we are succeeding in our efforts to create a better life for everyone?

STARTER QUESTIONS

1. How would you define wellbeing?
2. What does it mean to have a good life?
3. Do all Australians have a good life?
4. What could be improved in your life or community?
5. Watch the **Human development and wellbeing — Photo essay** video in your online Resources. What photographs would you take to represent wellbeing in your life?

FIGURE 1 Wellbeing can be measured in both qualitative (measurable) and quantitative (descriptive) ways. Do children have the opportunity to have fun and play with their friends?

on Resources

eWorkbook	Chapter 19 eWorkbook (ewbk-8111)
Video eLessons	The good life? (eles-1713) Human development and wellbeing — Photo essay (eles-5277)

19.2 Features of wellbeing

LEARNING INTENTION

By the end of this subtopic, you will be able to describe the tools, known as indicators, that allow us to gauge what makes a good life and measure wellbeing.

19.2.1 Defining wellbeing

In the past decade, a new global movement has emerged seeking to produce measures of progress that go beyond a country's income. Driven by citizens, policy makers and statisticians around the world, and endorsed by international organisations like the United Nations, the concept of **wellbeing** offers us a new perspective on what matters in our lives.

Wellbeing is experienced when people have what they need for life to be good. But how do we measure a good life? We can use **indicators** of wellbeing to help us. Indicators are important and useful tools for monitoring and evaluating progress, or lack of it. There are **quantitative** indicators and **qualitative** indicators.

Traditionally, development has been viewed as changing one's environment in order to enhance economic gain. Today, the concept of **development** is not only concerned with economic growth, but also includes other aspects such as providing for people's basic needs, equity and social justice, sustainability, freedom and safety. We have built on this traditional concept for measuring progress by considering wellbeing, which emphasises what is positive and desirable rather than what is lacking. The most successful development programs address all areas of wellbeing rather than simply focusing on economic, health or education statistics. There is a growing awareness that human beings and their happiness cannot simply be reduced to a number or percentage. We can measure development in a variety of ways, but the most common method remains to use economic indicators that measure economic progress using data such as **gross domestic product** (GDP).

wellbeing a good or satisfactory condition of existence; a state characterised by health, happiness, prosperity and welfare

indicator usually consists of a complex set of indices that measure a particular aspect of quality of life or describe living conditions; useful in analysing features that are not easily calculated or measured, such as freedom or security

quantitative indicator easily measured and can be stated numerically, such as annual income or how many doctors there are in a country

qualitative indicator usually consists of a complex set of indices that measure a particular aspect of quality of life or describe living conditions; useful in analysing features that are not easily calculated or measured, such as freedom or security

development according to the United Nations, defined as 'to lead long and healthy lives, to be knowledgeable, to have access to the resources needed for a decent standard of living and to be able to participate in the life of the community'

gross domestic product (GDP) a measurement of the annual value of all the goods and services bought and sold within a country's borders; usually discussed in terms of GDP per capita (total GDP divided by the population of the country)

FIGURE 1 Examples of quantitative and qualitative wellbeing data

Quantitative indicators	Qualitative indicators
• Gross national income (GNI)	• Happy Planet Index (HPI)
• Gross domestic product (GDP)	• Freedom of speech
• Literacry and numeracy rates	• Safety
• Human Development Index (HDI)	• Sustainability
• Life expectancy	

19.2.2 Using indicators

A wellbeing-centred approach to development takes into account a variety of quantitative and qualitative indicators. Quantitative indicators, based on numerical values, are commonly used because it is easy to compare changes over time and between places. Qualitative indicators, which measure quality of life, are harder to quantify and compare as the data collection can be subjective and time consuming. Indicators can be classified into a range of broad categories (see **FIGURE 2**).

FIGURE 2 Categories of wellbeing indicators

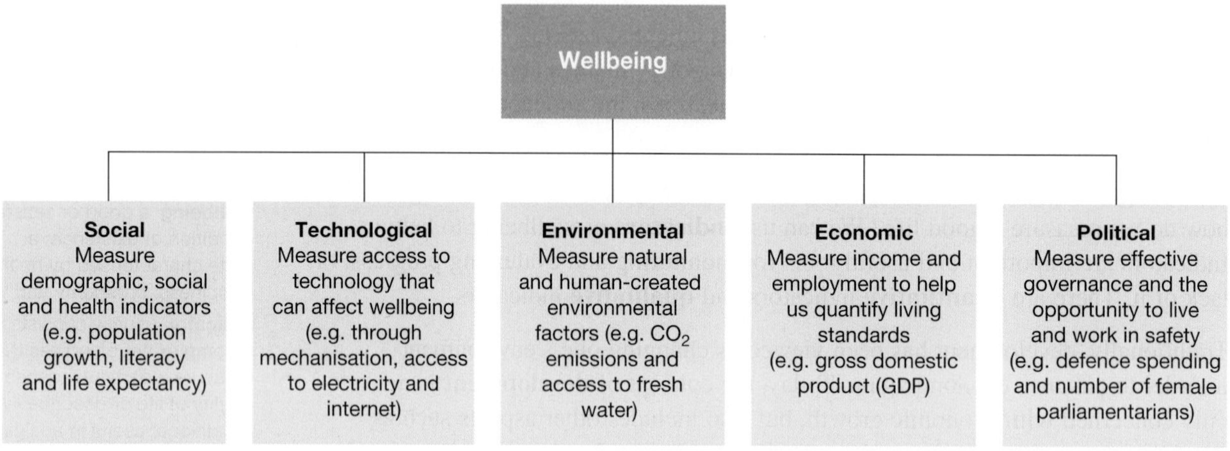

- Social indicators include demographic, social and health measures.
- Technological indicators in such fields as transport, industry, agriculture, mining and communications also contribute to wellbeing.
- Environmental indicators assess resources that provide us with the means for social and economic development, and gauge the health of the environment in which we live.
- Economic indicators measure aspects of the economy and allow us to analyse its performance.
- Political indicators look at how effective governments are in helping to improve people's **standard of living** by ensuring access to essential services.

Using patterns to describe wellbeing

Geographers use the spatial dimension, which helps us to identify patterns of where things are located over Earth's space and attempt to explain why these patterns exist. Identifying patterns across the globe may help to explain why the world is so unequal. Factors that affect equality across areas in a positive way may include the availability of natural resources or an educated workforce, whereas susceptibility to natural disasters or corruption may create more inequality.

Inequalities may exist between individuals, but also within and between countries, regions and continents (often referred to as 'spatial inequality'). Just as each person has their own unique strengths and weaknesses, places are either endowed with or lack various resources. They may have an excess of some resources that create wellbeing (such as a culture of strong extended family support) or a lack of other resources that create wellbeing in other ways (such as little access to fresh, clean water sources).

standard of living a level of material comfort in terms of goods and services available to someone or some group; continuum, for example a 'high' or 'excellent' standard of living compared to a 'low' or 'poor' standard of living

on Resources

eWorkbook	Features of wellbeing (ewbk-10253)
Video eLesson	Features of wellbeing — Key concepts (eles-5278)
Interactivity	Features of wellbeing (int-8693)

19.2 ACTIVITIES

1. Indicators can also convey information about a country's progress, rate of change or development. Could wellbeing indicators be clues to the factors affecting the development of a country? What else do they tell you?
2. Use the **Gauging interconnections** weblink in your online resources to discover some of the interconnections that exist between indicators. List two strong interconnections.
3. The concept of wellbeing is relative to who you are and the place where you live. Consider the following statements. Does the term *wellbeing* have any relevance to these people? Does wellbeing hold any relevance for people in the direst poverty?
   - **Person A:** 'We live in constant fear and starvation; there is a lack of government. Personal safety is crucial, so wellbeing is not there yet. Things are very difficult as people are living in despair.'
   - **Person B:** 'Before, we always talked of improving living standards, which mostly meant material needs. Now we talk of the importance of relationships among people and between people and the environment.'
   - **Person C:** 'The land looks after us. We have plenty to eat, but things are changing. There are no fish now, not like when my father was a boy.'

4. Select one of the indicator categories: social, economic or environmental. In pairs or small groups, brainstorm the various indicators that you think might be used to measure it. Create a short list of at least five before checking the World Statistics section of your atlas to see which indicators are commonly used.

19.2 EXERCISE

Learning pathways

■ LEVEL 1	■ LEVEL 2	■ LEVEL 3
1, 2, 5, 9, 13	3, 4, 7, 8, 14	6, 10, 11, 12, 15

Check your understanding

1. Define a quantitative indicator.
2. Define a qualitative indicator.
3. Classify the following as either quantitative or qualitative indicators:
   a. proportion of seats held by women in national parliaments
   b. unemployment
   c. quality of teaching at your school
   d. how safe you feel walking in the city at night
   e. how much you trust your neighbours

4. Using **FIGURE 2** as a guide, categorise the indicators below:
   a. obesity
   b. freedom of speech
   c. access to parks or bushland
   d. motor vehicle ownership
   e. internet speeds and access
5. Although indicators measure different aspects of quality of life, they are also interconnected. For example, if a country goes through an economic recession, other indicators will be affected. Describe one other example of how indicators are interconnected.
6. What does the term 'development' mean? Give examples to support your description.

Apply your understanding

7. Explain when you might use a quantitative indicator to assess wellbeing. Give an example.
8. Explain when you might use a qualitative indicator to assess wellbeing. Give an example.
9. Does a pet you know (dog, cat or other) have a good life? What indicators would you use to measure the wellbeing of your pet? Write a selection of 10 quantitative and qualitative indicators to help determine their wellbeing.
10. Discuss the following statement: 'Although indicators measure different aspects of quality of life, they are also interconnected.'
11. Indicators can suggest information about a country's progress, rate of change or development. Could these indicators be clues to the factors affecting the development of a country? If so, what else do they tell you?
12. What might be some of the flaws in judging wellbeing based on economic indicators?

Challenge your understanding

13. Are you better off or worse off? As a teenager in Australia, you might think you have it tough. But, when we look at the indicators, is that really the case? Decide whether you are better off or worse off for each indicator in **TABLE 1** by evaluating the data. What reasons could account for these differences?

TABLE 1 Australia compared with countries with a similar population for a selection of qualitative indicators

Indicator	Australia	Cote d'Ivoire	Venezuela	DPR Korea	Romania
Population (2019)	25 364 307	25 716 540	28 515 829	25 666 161	19 356 544
Life expectancy (years, 2018)	82.7	57	72.1	72	75.3
Population density (ppl/sq km, 2019)	3.2	78.8	32.7	212.2	84.6
Adolescent fertility rates (births/1000 women aged 15–19, 2018)	11	116	85	0	36
% seats held by women (national parliament 2020)	30	11	22	18	22
Gross National Income (per capita, US$, 2019)	55 190	2180	13 080	N/A	11 430
Mobile phone subscriptions (per 100 ppl, 2018)	111	134.9	71.8	15	116

Source: The World Bank

14. Which indicators are the most difficult to measure, and why do you think this is the case?
15. Suggest which of the indicators of wellbing might be the most reliable. Justify your response by explaining on what basis you would consider it reliable.

To answer questions online and to receive **immediate feedback** and **sample responses** for every question, go to your learnON title at www.jacplus.com.au.

19.3 SkillBuilder — Constructing and interpreting a scattergraph

online only

LEARNING INTENTION

By the end of this subtopic, you will be able to create a scattergraph from two or more data sets, and identify and describe correlation patterns in your scattergraph.

The content in this subtopic is a summary of what you will find in the online resource.

19.3.1 Tell me

What is a scattergraph?

A scattergraph is a graph that shows how two or more sets of data, plotted as dots, are interconnected. This interconnection can be expressed as a level of correlation. Scattergraphs are used to show us a visual image of the interconnection of factors. Sometimes it is difficult to see the relationship until the sets of data are presented visually. You will find that the graphs clearly show the interconnection of factors where clusters of dots form, while other dots stand out alone.

19.3.2 Show me

Creating a scattergraph

Step 1

To complete a scattergraph, you need two sets of information about which you want to test the interconnection.

Step 2

Decide which factor you will place on the horizontal axis and which factor you will place on the vertical axis. In **FIGURE 1**, road density is on the horizontal axis and food supply is on the vertical axis. (You can view the full data set used for this scattergram in your online Resources.)

Step 3

Look at the range of numbers in the data to be plotted and decide on a scale for each axis. Ensure that the maximum and minimum numbers will fit on the scale. Draw a graph outline and label the axes, including the units of measurement.

FIGURE 5 Drawing the line of best fit

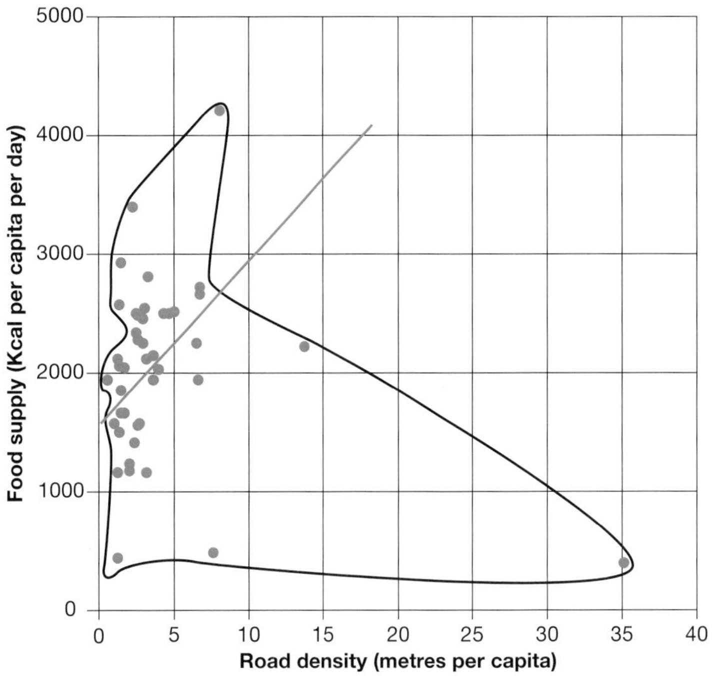

Step 4

Plot all the data on the graph. Mark with a small dot the point where both data values intersect.

Step 5

Draw a pencil outline around all the dots. This will show you the trend of the data and identify the anomalies. These anomalies occur where the outline bulges.

Step 6

Draw a line of best fit — that is, a line that has equal points either side of it. To do this, sit your ruler on its narrow edge on the graph and move it around until there is a roughly equal number of dots on both sides of the ruler. Draw along the ruler's edge to create the line of best fit, or trend line.

Step 7

Determine whether the shape shows:

- positive correlation — the line of best fit goes from bottom left to top right
- negative correlation — the line of best fit goes from top left to bottom right
- no correlation — the dots are randomly scattered rather than in a straight line
- a perfect correlation — all dots sit on the line of best fit rather than on either side of it.

The closer the points are to the line, the stronger the relationship. Note that the 'odd' points are considered anomalies.

Step 8

Give your graph a title.

Interpreting a scattergraph

To interpret a scattergraph is to write a few sentences explaining your findings. Use the following format:

- State the type of correlation.
- Describe what is happening on the graph regarding the two factors.
- Discuss any anomalies.
- Be specific about any particular places or countries you want to use as an example.
- Write a concluding statement.

19.3.3 Let me do it

learnON

Go to learnON to access the following additional resources to help you build this skill:

- a longer explanation of this skill and its application in Geography (Tell me)
- a video demonstrating the step-by-step process of this skill (Show me)
- an activity and interactivity for you to practise the skill (Let me do it)
- self-marking questions to help you understand and use the skill.

on Resources

eWorkbook	SkillBuilder — Constructing and interpreting a scattergraph (ewbk-10257)
Video eLesson	SkillBuilder — Constructing and interpreting a scattergraph (eles-1756)
Interactivity	SkillBuilder — Constructing and interpreting a scattergraph (int-3374)

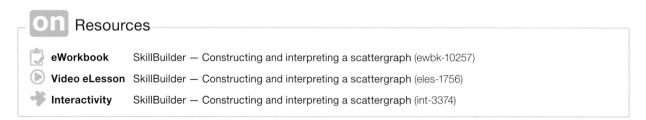

19.4 Examining quantitative indicators

LEARNING INTENTION

By the end of this subtopic, you will be able to identify and explain quantitative measures of development and wellbeing.

19.4.1 Defining quantitative indicators

A wellbeing approach to development takes into account a variety of quantitative and qualitative indicators. Quantitative indicators are the most common as they allow for the easiest possible way of comparing changes (hopefully improvements) between time periods and countries.

One quantitative indicator of wellbeing is the distribution of wealth, which varies significantly.

FIGURE 1a shows the distribution of the world's wealth in 1992, and **FIGURE 1b** shows the distribution in 2019. Today, the top 1 per cent of income earners have 50 per cent of the world's income; within that, the 80 richest people have as much wealth as the bottom half of the world combined — nearly 4 billion people.

FIGURE 1 'Champagne-glass' distribution of the world's wealth in (a) 1992 and (b) 2019

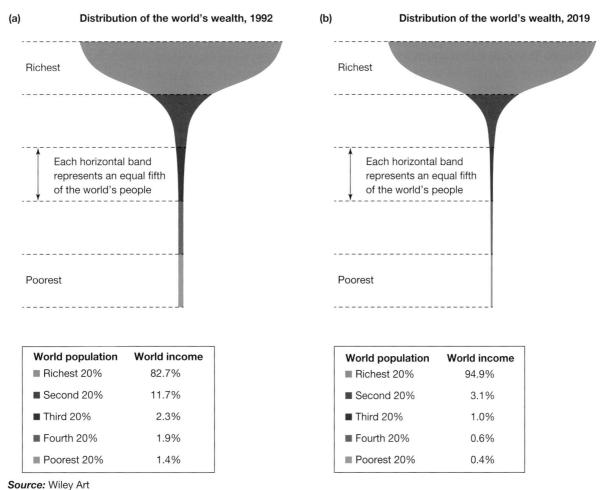

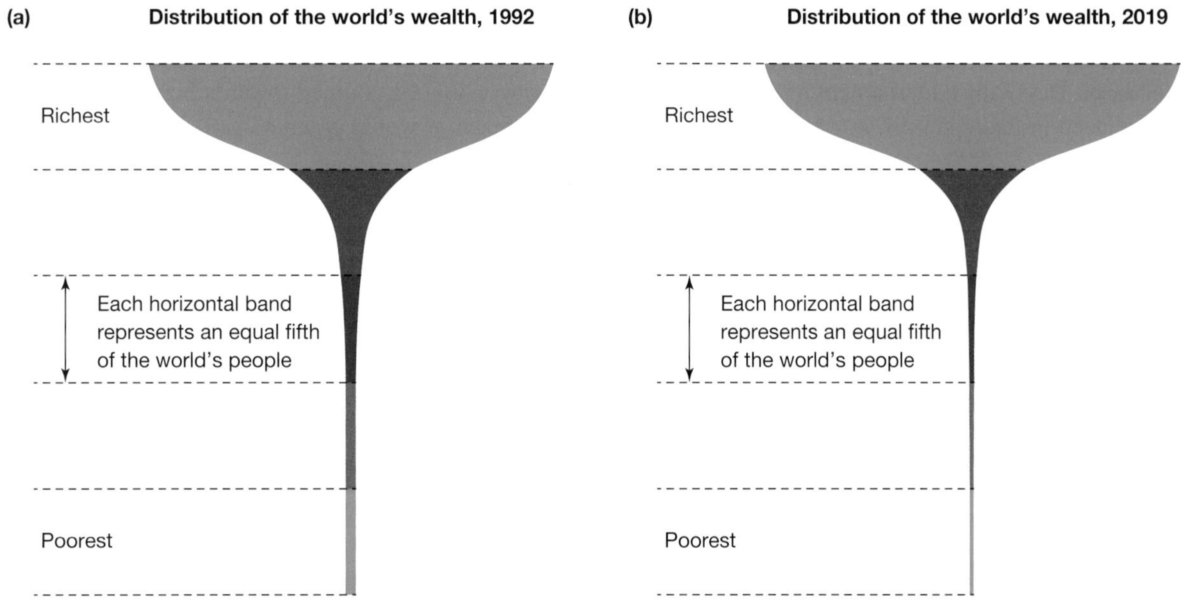

World population	World income
Richest 20%	82.7%
Second 20%	11.7%
Third 20%	2.3%
Fourth 20%	1.9%
Poorest 20%	1.4%

World population	World income
Richest 20%	94.9%
Second 20%	3.1%
Third 20%	1.0%
Fourth 20%	0.6%
Poorest 20%	0.4%

Source: Wiley Art

19.4.2 Economic indicators

The major economic indicators used globally are gross domestic product (GDP) and **gross national income** (GNI). Although these seem similar and are sometimes used interchangeably, they measure slightly different things. GDP measures the economic activity within a country (regardless of who generates it); GNI measures the economic activity of the residents and businesses of a country (regardless of where the activity takes place).

Defining GDP

GDP is a measurement of the value of all goods and services bought and sold within a country's borders. It includes goods and services that are produced within a country but that may actually be produced by an international business that has operations within that country. As a result, profits made by the business may not stay in the country where the good was made, but rather will return to the country where the business headquarters are located. For example, major transnational corporations (TNCs) such as Nike will sell footwear in Australia, which will contribute to Australia's GDP. However, the profits made from the sale will find their way back to the United States.

Problems with GDP measures

Over time, different societies have measured progress in different ways. A GDP-led development model focuses solely on boundless economic growth on a planet with limited resources — and this is not a balanced equation.

Benefits of GNI measures

GNI takes the value of GDP, then removes the income that is payable to residents or business that are from overseas and includes incomes and profits earned overseas by local residents or businesses. For example, if golfer Jason Day wins a tournament in Scotland, the prize money would be counted towards Scotland's GDP (it was produced in that country); however, as he is an Australian citizen, it would count as part of Australia's GNI.

Problems with both measures

One concern with GDP and GNI is that these measures make no distinction between transactions that add to wellbeing and those that detract from it. More generally, they do not recognise some of the greatest environmental, social and humanitarian challenges of the twenty-first century, such as pollution and stress levels.

19.4.3 Multiple component index

A single indicator gives us only a narrow picture of the development of a country. A country may have a very high GDP, but if we dig a little deeper and look at each individual's share in that country's income or their **life expectancy**, we may not find what we expected. Inequalities may be revealed. A combination of many indicators will create a more accurate picture of the level of wellbeing in a particular place. Much like using our five senses to try a new cuisine, a combination of indicators will give us better insight into a country's wellbeing. The **Human Development Index** (HDI) is one such index. It was developed in 1990 to measure wellbeing according to three indicators (see **FIGURE 2**).

FIGURE 2 The HDI measures quality of life according to three key factors.

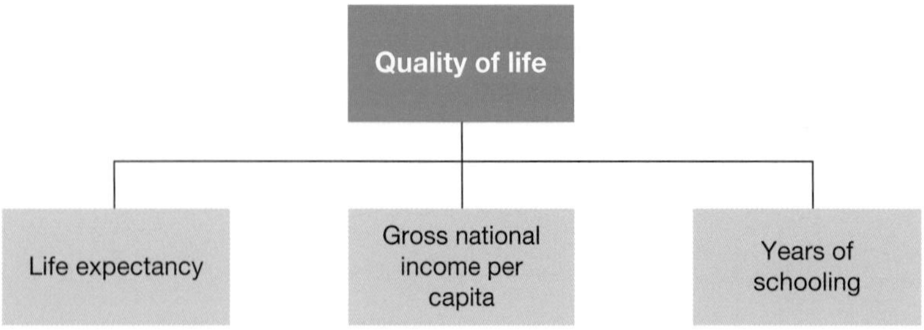

gross national income (GNI) a measurement of the value of goods and services produced by citizens and firms of a specific country no matter where they take place, normally discussed as GNI per capita

life expectancy the number of years a person is expected to live, based on the average living conditions within a country

Human Development Index (HDI) measures the standard of living and wellbeing in terms of life expectancy, education and income

FIGURE 3 HDI choropleth map

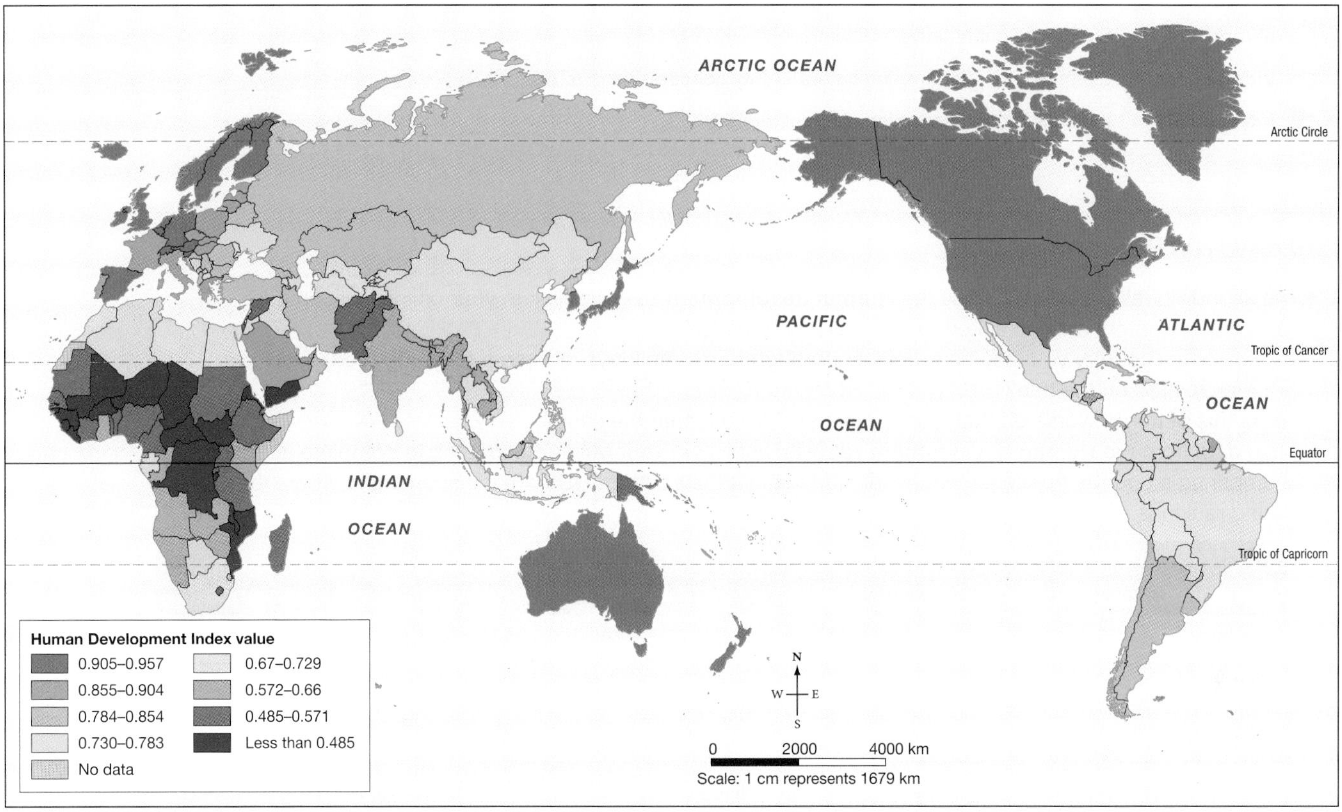

Source: Human Development Reports, United Nations Development Programme

CASE STUDY

Wellbeing and development indicators in Switzerland

Switzerland consistently rates well in international wellbeing rankings. The Federal Statistical Office (Section Environment, Sustainable Development) used the MONET indicator system to map progress towards the UN Sustainable Development Goals. The MONET 2030 system also categorises over 100 indicators according to their environmental, social and economic impacts (**TABLE 1**).

TABLE 1 Examples of some of the indicators in the MONET 2030 indicator system

Environmental	Social	Economic
• Nitrogen balance in agricultural land • Deaths from natural events • Greenhouse gas emissions • Energy consumption • Renewable energy use • Urban sprawl • Noise pollution • Waste generation • Invasive species numbers	• Poverty rate • Social security spending • Fruit and vegetable consumption • Deaths from natural events • Levels of alcohol consumption • Rates of smoking • Ability to afford healthcare • Personal satisfaction with life • Levels of domestic violence • Levels of education • Gender wage gap • Trust in government institutions	• Poverty rate • Social security spending • Export income • Ability to afford healthcare • Gender wage gap • Youth unemployment • Levels of working poor • Housing costs • Public debt • Investment in developing countries

Source: Federal Statistical Office

To learn more about Switzerland's MONET indicator system, use the **MONET 2030 system** weblink in your online resources.

Resources

eWorkbook Examining quantitative indicators (ewbk-10261)

Video eLesson Examining quantitative indicators — Key concepts (eles-5279)

Interactivity Examining quantitative indicators (int-8694)

19.4 ACTIVITIES

1. As an individual or as a class, use the **Human development tree** weblink in your online Resources.
   a. What indicators are you asked to give assessments for?
   b. Choose a country you may have links to (through family, holidays or even a favourite food). How is it represented on the tree?
   c. Is this better or worse off than your prediction? Why is that the case?
2. Use data from the **Human development reports** weblink in your online Resources to complete a table comparing a country from each continent. Include in it their current HDI ranking scores and where they are in world rankings.

Continent	Asia	Africa	Europe	North America	South America
Country					
HDI score					
HDI rank					

3. Consider the indicators included in Switzerland's MONET indicator system — use the **MONET 2030 system** weblink in your online Resources. Create a Venn diagram that categorises the indicators according to whether they are economic, environmental or social. Use colour coding to show which are qualitative and which are quantitative measures.

19.4 EXERCISE

Learning pathways

■ LEVEL 1	■ LEVEL 2	■ LEVEL 3
1, 3, 7, 8, 13	2, 4, 9, 11, 14	5, 6, 10, 12, 15

Check your understanding

1. What do the letters GNI stand for?
2. Define GNI.
3. Refer to **FIGURE 2**. In your own words, explain three indicators used to calculate the HDI.
4. Describe the pattern of wealth distribution shown in **FIGURE 1a** and **1b**.
5. Define the wellbeing approach and show how it is a multiple component index.
6. A multiple component index is significant because it gives us a measure of the ______________ of the people in a place across multiple important dimensions. It provides a more detailed understanding of the factors influencing wellbeing.
7. Add the correct terms to complete each description of each of the indicators used to calculate the HDI.
   a. ____________ is the number of years that a person is expected to live.
   b. ____________ is measured using years of schooling.
   c. ____________ is measured using gross national income per person.

Apply your understanding

8. Is the HDI the best indicator of a country's development? Give reasons for your answer.
9. Provide a detailed explanation of each of the indicators used to calculate the HDI.
10. Explain why a multiple component index is significant.
11. Why is global wealth so unevenly distributed (e.g. between Australia, the United States, Norway and parts of the developing world)?
12. Explain how life expectancy and wellbeing are interconnected.

Challenge your understanding

13. Use **FIGURE 1a** and **1b** to predict what will happen to the shape of this 'glass' in the future.
14. What are the factors that you believe are responsible for the uneven distribution of wealth? Suggest how this imbalance might be rectified.
15. In 2011 the HDI was revised to use GNI as its economic measurement rather than GDP. Suggest why you think this was done.

To answer questions online and to receive **immediate feedback** and **sample responses** for every question, go to your learnON title at www.jacplus.com.au.

19.5 Examining qualitative indicators

LEARNING INTENTION

By the end of this subtopic, you will be able to identify and explain qualitative measures of development and wellbeing.

19.5.1 Defining qualitative indicators

Qualitative indicators are used to determine particular aspects of quality of life or to describe living conditions. These indicators are often difficult to measure and compare, as the information used is time consuming to collect. This is because they are made up of surveys and interviews, rather than simple numerical values such as income or life expectancy.

19.5.2 Measuring happiness

The Happy Planet Index (HPI) maps the extent to which 151 countries across the globe produce long, happy and sustainable lives for the people that live in them (**FIGURE 1**). Each of the four component measures — life expectancy, **experienced wellbeing, inequality of outcomes** and **ecological footprint** — is given a traffic-light score based on thresholds for good (green), middling (amber) and bad (red) performance results (see **FIGURE 2**). These scores are combined into an expanded six-colour traffic light for the overall HPI score. To achieve bright green (the best of the six colours), a country would have to perform well on all three individual components.

experienced wellbeing an individual's subjective perception of personal wellbeing

inequality of outcomes inequalities between people in the one country

ecological footprint the amount of productive land needed on average by each person in a selected area for food, water, transport, housing and waste management

FIGURE 1 How the Happy Planet Index is calculated

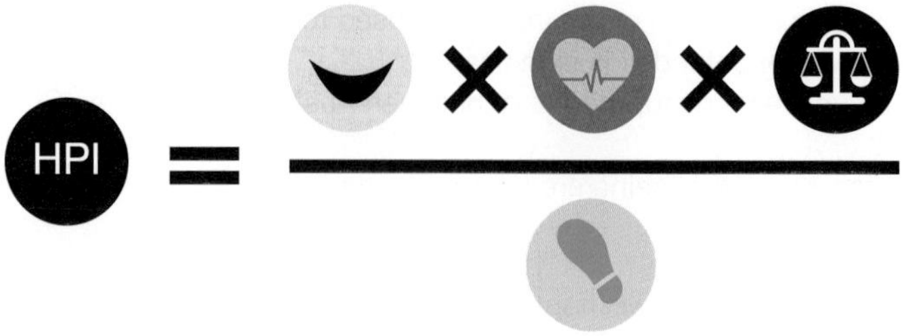

	**Wellbeing:** How satisfied the residents of each country say they feel with life overall, on a scale from zero to ten, based on date collected as part of the Gallup World Poll.
	**Life expectancy:** The average number of years a person is expected to live in each country based on data collected by the United Nations.
	**Inequality of outcomes:** The inequalities between people within a country, in terms of how long they live, and how happy they feel, based on the distribution in each country's life expectancy and wellbeing data. Inequality of outcomes is expressed as a percentage.
	**Ecological footprint:** The average impact that each resident of a country places on the environment, based on data prepared by the Global Footprint Network. Ecological footprint is expressed using a standardised unit: global hectares (gha) per person.

Source: New Economics Foundation

int-8467

FIGURE 2 Cartogram showing relative global populations and Happy Planet Index scores, 2016

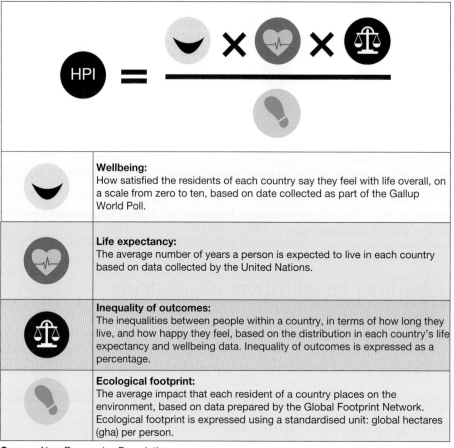

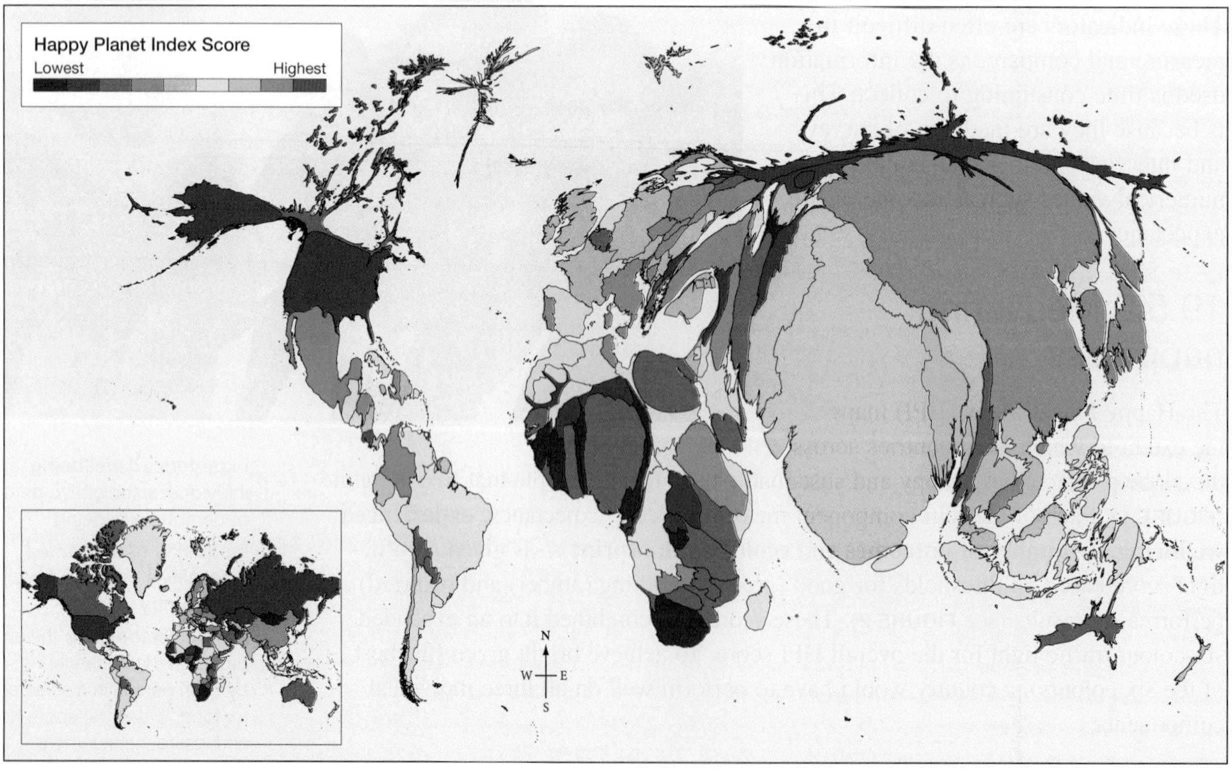

Source: http://happyplanetindex.org/

Gross National Happiness

In 2011, the Prime Minister of Bhutan (Central Asia) demonstrated his country's commitment to its wellbeing by developing the world's first measure of national happiness, and he encouraged world economies to do the same. UN Secretary-General Ban Ki-moon supported this innovation:

'Gross national product (GNP) … fails to take into account the social and environmental costs of so-called progress … Social, economic and environmental wellbeing are indivisible. Together they define gross global happiness.'

Australia's ranking on the Happy Planet Index in 2020 was 105th out of 140 ranked countries. Our Happy Planet Index (HPI) score was 21.2. Australia's results for specific indicators were:

- life expectancy: 82.1 years (7th out of 140 countries)
- wellbeing: 7.2 out of a possible score of ten (12th out of 140 countries)
- ecological footprint: 9.3 gha/p (139th out of 140 countries)
- inequality: 8 per cent (11th out of 140 countries).

Freedom of speech

Freedom, and in particular freedom of speech, is something we often take for granted in Australia. The access to this type of freedom is a significant qualitative measure and one which can be difficult to calculate. One way in which freedom is measured is through the World Press Freedom Index (WPFI), which measures the freedoms allowed to media outlets in a country. According to the WPFI, the five countries with the best scores in 2020 were Finland, Netherlands, Norway, Denmark and Sweden, while the five countries with the worst scores were China, Djibouti, Turkmenistan, North Korea and Eritrea.

on Resources

eWorkbook Examining qualitative indicators (ewbk-10265)

Video eLesson Examining qualitative indicators — Key concepts (eles-5280)

Interactivity Examining qualitative indicators (int-8695)

19.5 ACTIVITIES

1. Do Australia's HPI results surprise you? Use the **Happy Planet Index** weblink in your online Resources to learn more about the Happy Planet Index and explore the results. Find a country that you think has the kind of balance of results that might make you happy. Compare your choice with a partner. What similarities or differences did you find?
2. Many countries have already adopted national measures of wellbeing. Either individually or in pairs, research the history of one of the following indexes, identify the indicators used to measure it, and evaluate its success.
   - Gross National Happiness (Bhutan)
   - Key National Indicator System (USA)
   - Canadian Index of Wellbeing (Canada)
3. Should wellbeing or happiness be a core goal of a country's government? Debate this in a small group.
4. Locate Bhutan on a world map.
   a. Describe its location.
   b. Find Bhutan in **FIGURE 2**. How does it rate on the Happy Planet Index?
5. Without referring to **FIGURE 2**, name three places you would expect to rank high on the Happy Planet Index and three you would expect to rank low. Now, check your predictions with **FIGURE 2** or use the **Happy Planet Index** weblink in your online Resources. Were you correct? What assumptions or misconceptions might your guesses have been based on?

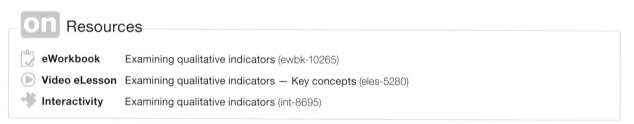

19.5 EXERCISE

Learning pathways

■ LEVEL 1	■ LEVEL 2	■ LEVEL 3
1, 4, 7, 8, 14	2, 3, 9, 10, 13	5, 6, 11, 12, 15

Check your understanding

1. What is an ecological footprint?
2. How is an ecological footprint used in relation to a composite or multicomponent indicator?
3. Using the Happy Planet Index and **FIGURE 2**, explain what wellbeing conditions you might find in the following countries.
   a. Australia
   b. The United States
   c. New Zealand
   d. China
4. What does the term 'experienced wellbeing' mean?
5. Outline why press freedom is considered to be an indicator of wellbeing.
6. Why are qualitative measures more difficult to measure than quantitative measures?

Apply your understanding

7. Is the HDI the best indicator of a country's development? Give reasons for your answer.
8. Comment on the distribution of the happiest and unhappiest countries across the world according to the data in **FIGURE 2**.
9. Explain the interconnection between life expectancy and inequality of outcomes.
10. Give an example of a Sustainable Development Goal that would be measured by both qualitative and quantitative indicators. Explain how this is the case.
11. Compare the locations of the best countries for WPFI to those of the worst. What trends can you see?
12. Discuss the following statement: 'The measurement of happiness has become less important in the twenty-first century.'

Challenge your understanding

13. Suggest why a range of indices is being developed in the twenty-first century to measure wellbeing.
14. Image you have been given the task to create a qualitative index with six indicators that measure the wellbeing of Australians. The purpose of your study is to compare levels of wellbeing in different Australian states before, during and after COVID-19 lockdowns. What indicators would you include and why?
15. What do you think would make a country unhappy? Comment on the distribution of the happiest and unhappiest countries across the world according to the data in **FIGURE 2**. Does there seem to be an interconnection between the size of a country's population and its levels of happiness?

To answer questions online and to receive **immediate feedback** and **sample responses** for every question, go to your learnON title at www.jacplus.com.au.

19.6 Measuring development

LEARNING INTENTION

By the end of this subtopic, you will be able to explain some of the different ways of measuring spatial distribution of wellbeing that have been used over time.

19.6.1 Describing development

Whichever method of classifying development or wellbeing we choose, it is important to understand the terms that have been used, the values that underpin it, and what perspective we take. With an overwhelming amount of data available to us, the world is often divided simplistically into extremes such as 'rich' or 'poor'.

The annotated classifications in **FIGURE 1** have been used in the past century, but they are very general and as such have been questioned by geographers for their accuracy (and sometimes offensiveness).

FIGURE 1 World map showing various definitions of development

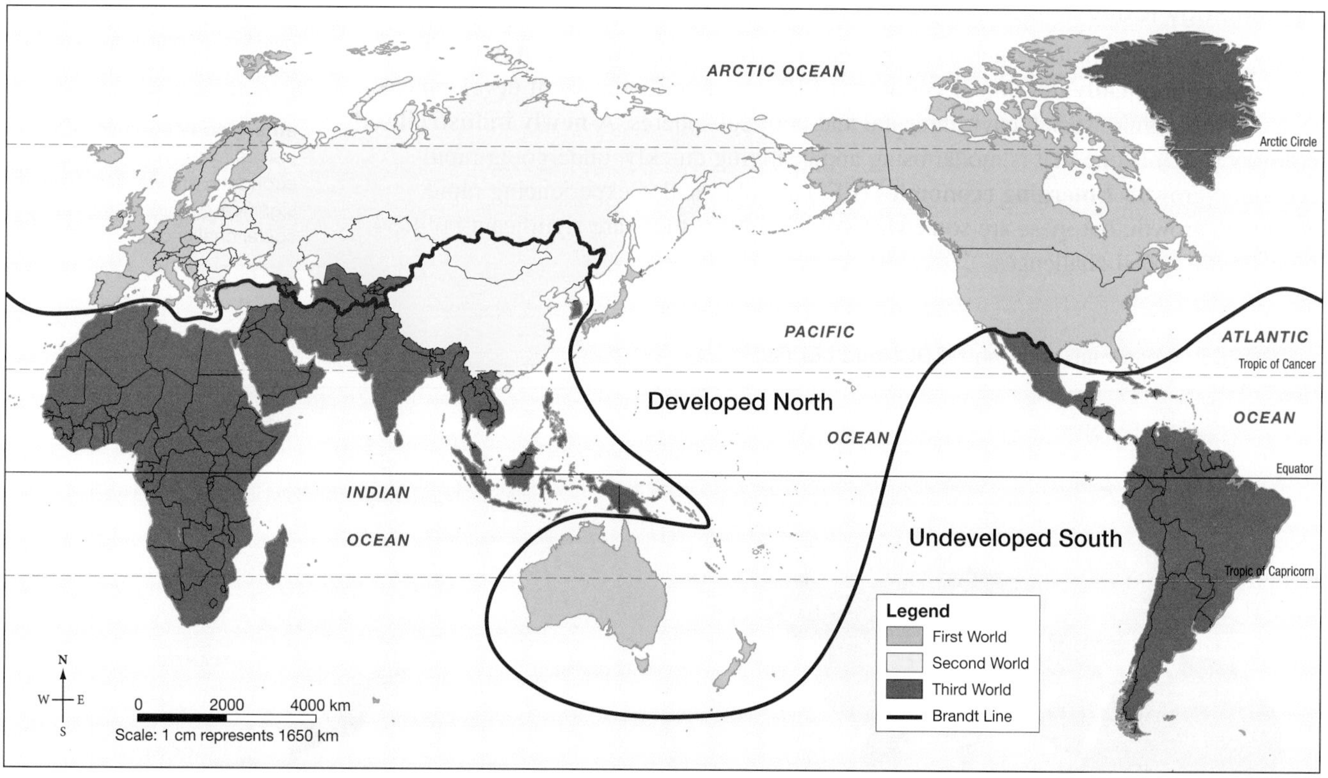

Source: United Nations Development Report.

Choosing your terminology

Developed or developing

One of the most common ways of talking about the level of development in various places is to label them as 'developed' or 'developing' (previously often referred to as 'undeveloped'). These terms assume that development is a linear process of growth, so each country can be placed on a continuum of development. Countries that are developing are still working towards achieving a higher level of living standard or economic growth, implying that the country could ultimately become 'developed'.

North or South

In 1980, the Chancellor of West Germany, Willy Brandt, chaired a study into the inequality of living conditions across the world. The imaginary Brandt Line divided the rich and poor countries, roughly following the line of the equator. The North included the United States, Canada, Europe, the USSR, Australia and Japan. The South represented the rest of Asia, Central and South America, and all of Africa. Once again, these terms have become obsolete as countries have developed differently and ignored these imaginary boundaries

First World or Third World

The terminology First, Second and Third Worlds was a product of the Cold War. The Western, industrialised nations and their former colonies (North America, western Europe, Japan and Australasia) were the First World. The Soviet Union and its allies in the Communist bloc (the former USSR, eastern Europe and China) were the Second World. The Third World referred to all of the other countries. However, over time this term became more commonly used to describe the category of poorer countries that generally had lower standards of living. The Second World ceased to exist when the Soviet Union collapsed in 1991.

Terminology today

Today, we use terminology such as **more economically developed country** (MEDC) and **less economically developed country** (LEDC) to describe levels of development — in the economic, social, environmental and political spheres. A **newly industrialised country** (NIC) is one that is modernising and changing quickly, undergoing rapid economic growth. **Emerging economies** (EEs) are places also experiencing rapid economic growth, but these are somewhat volatile in that there are significant political, monetary or social challenges.

more economically developed country (MEDC) higher levels of development — in the economic, social, environmental and political spheres

less economically developed country (LEDC) lower levels of development — in the economic, social, environmental and political spheres

newly industrialised country places modernising and developing quickly from a wide range of industries and rapid economic growth

emerging economies (EEs) places experiencing rapid economic growth, but that have significant political, monetary or social challenges

FIGURE 2 World map showing MEDCs and LEDCs

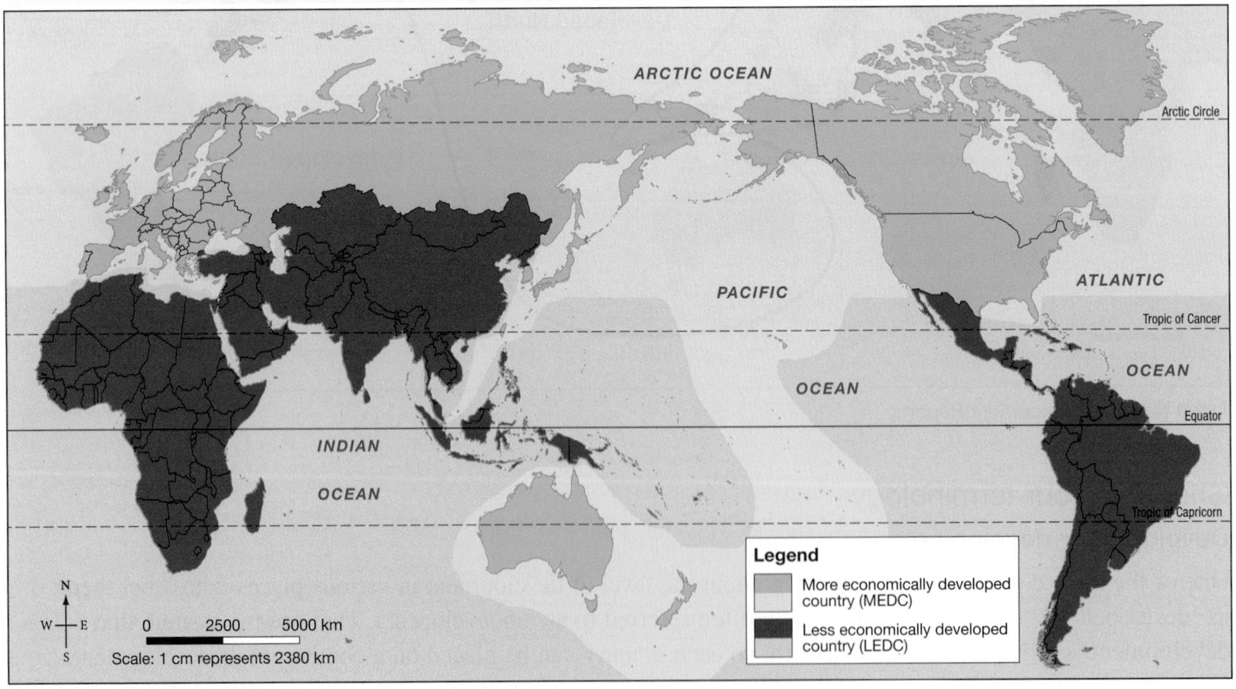

Source: CantGeoBlog

19.6.2 Defining poverty

There is a strong interconnection between development and poverty. The United Nations defines poverty as 'a denial of choices and opportunities, a violation of human dignity. It means not having enough to feed and clothe a family, not having a school or clinic to go to, not having the land on which to grow one's food or a job to earn one's living. It means susceptibility to violence, and it often implies living in marginal or fragile environments, without access to clean water or sanitation.' However, poverty is most often measured using solely economic indicators. More than 1 billion people live in poverty, as shown by **FIGURE 3**.

FIGURE 3 The proportion of the world's population living on less than US$1.90 per day, the World Bank's global poverty line indicator

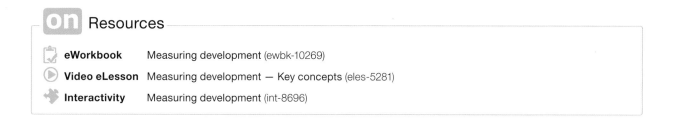

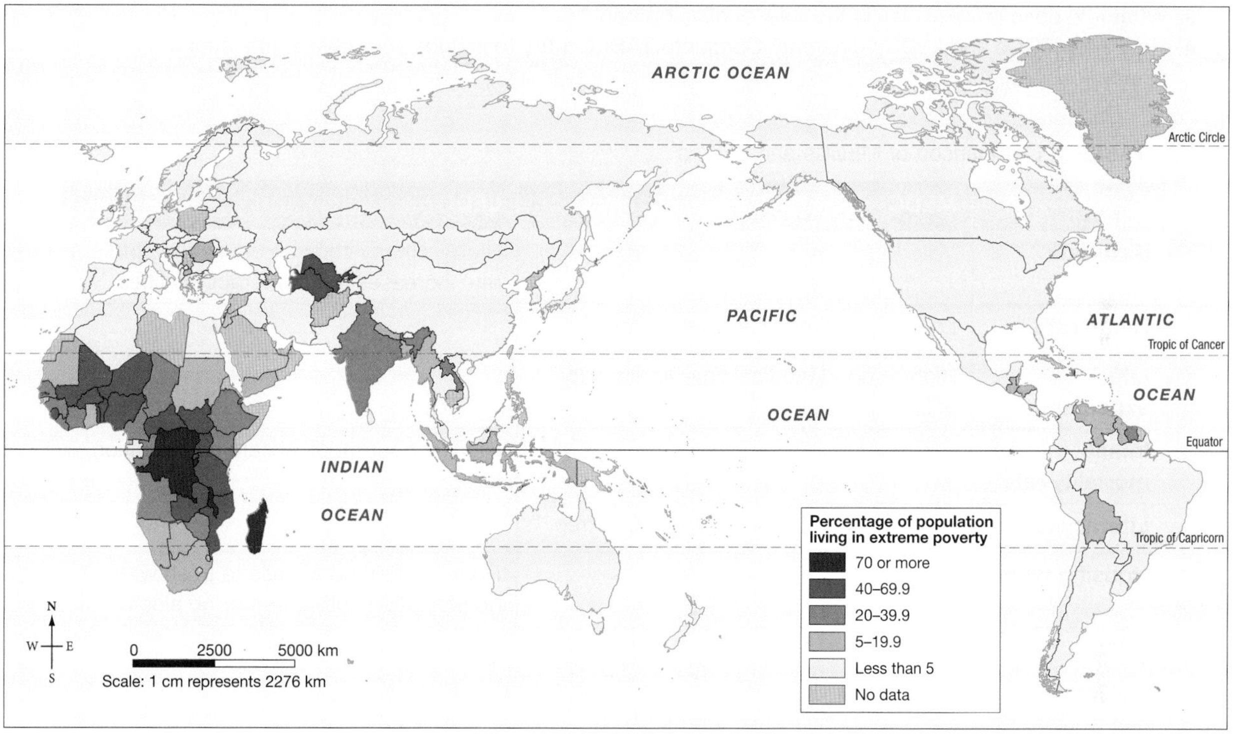

Source: World Bank – World Development Indicators (2021)

Resources

eWorkbook Measuring development (ewbk-10269)

Video eLesson Measuring development — Key concepts (eles-5281)

Interactivity Measuring development (int-8696)

19.6 ACTIVITIES

1. Indicators measure different aspects of quality of life, but they are also interconnected. For example, if a country goes through an economic recession, other indicators will be affected. Explain with examples. (A flow chart may be useful to step out your thinking.)
2. Investigate what connections exist, either in the presently or historically, that may link those countries of the 'First World'.

19.6 EXERCISE

Learning pathways

■ LEVEL 1	■ LEVEL 2	■ LEVEL 3
1, 7, 8, 9, 15	3, 4, 5, 10, 13	2, 6, 11, 12, 14

Check your understanding

1. Define the term *industrialised*.
2. Why is industrialisation important when considering levels of development?
3. What role does poverty play in the level of development?
4. How are MEDCs and LEDCs different? Complete **TABLE 1** (try to include your own explanations where possible).

TABLE 1 Comparison of MEDCs and LEDCs

	MEDC	LEDC
Birth rate		High — many children die so the birth rate increases to counteract fatalities
Death rate	Low — good medical care available	
Life expectancy	High — good medical care and quality of life	
Infant mortality rate		High — poor medical care and nutrition
Literacy rate	High — access to schooling, often free	
Housing type		Poor — often no access to fresh water, no sanitation, infrequent or no electricity

5. What is an emerging economy?
6. What was the Brandt line, and why is it no longer used?
7. Identify five MEDCs and five LEDCs using **FIGURE 2**.

Apply your understanding

8. Identify two examples of places that would have been classified as 'developed North' and two that would have been classified as 'undeveloped South'.
9. What do you think about Australia being labelled a part of the 'developed North'?
10. Describe the relationship between **FIGURE 1** and **FIGURE 2**.
11. Has the notion of North versus South changed? Why or why not?
12. Discuss the following question: Should poverty be assessed differently in each country according to its levels of development?

Challenge your understanding

13. The term *poverty* can be defined and measured in many ways, but is often referred to only in economic terms. Suggest why this might be the case.
14. Are terms to categorise parts of the world based on their levels of development or income necessary or just useful? Discuss why or why not.
15. Propose a set of indicators to determine the definition of 'poverty' in Australia.

To answer questions online and to receive **immediate feedback** and **sample responses** for every question, go to your learnON title at www.jacplus.com.au.

19.7 SkillBuilder — Interpreting a cartogram

online only

LEARNING INTENTION

By the end of this subtopic, you will be able to interpret the information shown in a cartogram and describe how the features are interconnected.

The content in this subtopic is a summary of what you will find in the online resource.

19.7.1 Tell me

What is a cartogram?

A cartogram is a diagrammatic map; that is, it looks like a map but is not a map as we usually know it. These maps use a single feature, such as population, to work out the shape and size of a country. Therefore, a country is shown in its relative location, but its shape and size may be distorted. Cartograms show patterns that are not identifiable on traditional maps.

FIGURE 1 Cartogram showing estimated world population, 2050

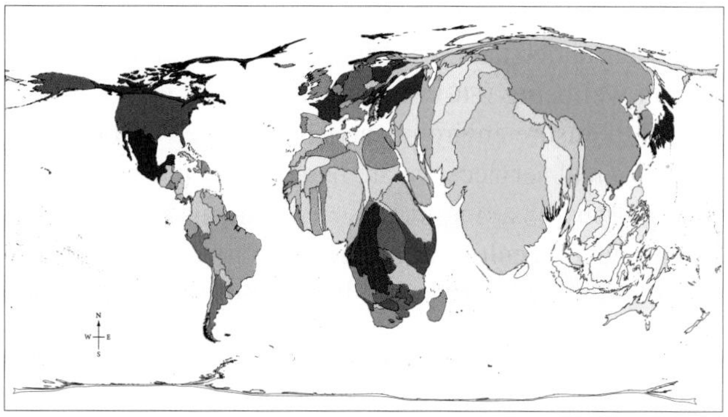

Source: Spatial Vision

19.7.2 Show me

Step 1

Read the title; check that the meaning is clear.
In **FIGURE 1**, the world's population in 2050 is mapped.

Step 2

Study the cartogram, looking for the largest and the smallest shapes on it. With your knowledge of the world map or by using an atlas, identify those countries and continents that are distorted in size and shape. For example, in **FIGURE 1**, Africa and Asia are expanded, indicating a large estimated growth in population, but Australia has almost disappeared, indicating a small expected growth in population.

Step 3

Interpreting the cartogram requires a description of the interconnection between the feature that has been mapped and the proportional size of a country. Look for countries that appear larger, countries that appear smaller, and countries and continents whose shapes have been distorted. Write a few sentences describing the feature mapped.

19.7.3 Let me do it

learnON

Go to learnON to access the following additional resources to help you build this skill:
- a longer explanation of this skill and its application in Geography (Tell me)
- a video demonstrating the step-by-step process of this skill (Show me)
- an activity and interactivity for you to practise the skill (Let me do it)
- self-marking questions to help you understand and use the skill.

on Resources

eWorkbook SkillBuilder — Interpreting a cartogram (ewbk-10273)
Video eLesson SkillBuilder — Interpreting a cartogram (eles-1757)
Interactivity SkillBuilder — Interpreting a cartogram (int-3375)

19.8 Contemporary trends in wellbeing — Sustainability

LEARNING INTENTION

By the end of this subtopic, you will be able to explain how sustainability and development are interconnected, and outline related trends and ideas related to their interconnection.

19.8.1 Stewardship

Understanding the link between ecological services, human actions and wellbeing is important as it can lead to more sustainable practices. By having a greater focus on the sustainability of our environment, we can potentially develop higher levels of human wellbeing. Undertaking the approach of **stewardship** in the management of environments benefits not only the environment, but also the wellbeing of those people who use the environment.

A greater understanding of the way in which all aspects of the environment link to human wellbeing (**FIGURE 1**) means that to increase the wellbeing of people, there is a need to ensure environments are managed sustainably. With a greater understanding of ecological services we are better placed to be able to manage them in a sustainable way, thereby ensuring increased wellbeing for citizens of the world. **FIGURE 2** shows the way in which **ecological services** are linked to one another and how managing them is important in maintaining the habitability and livelihood of people on the planet.

stewardship an ethic that embodies the responsible planning and management of resources

ecological services any beneficial natural process arising from healthy ecosystems, such as purification of water and air, pollination of plants and decomposition of waste

FIGURE 1 Interaction of environmental change and human wellbeing.

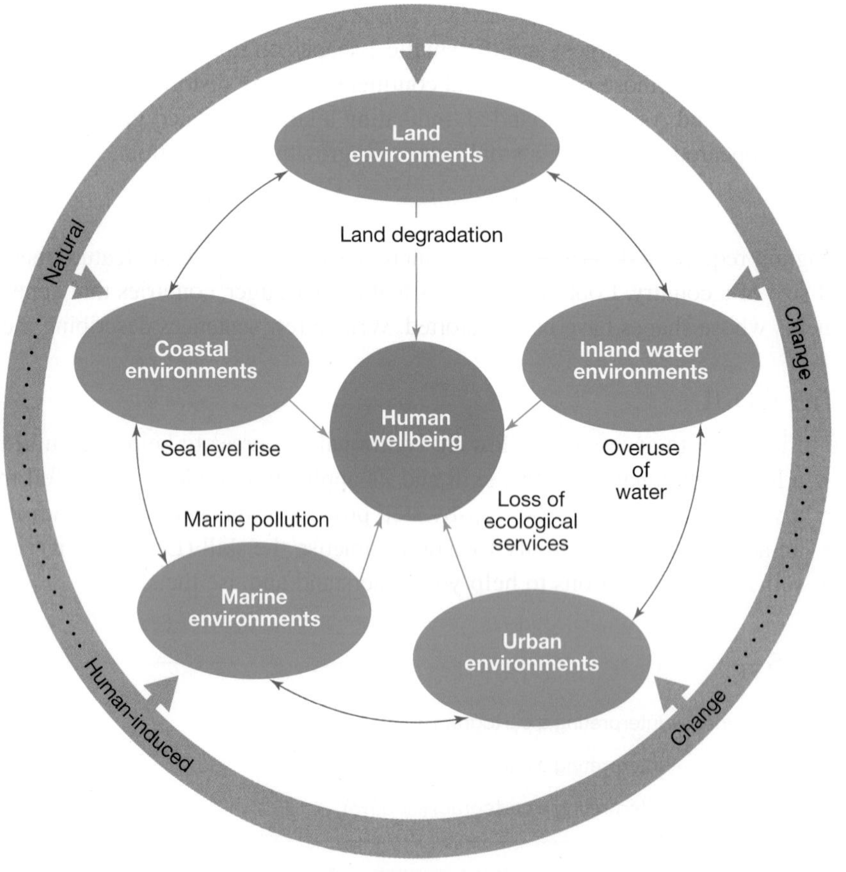

FIGURE 2 Ecological services model

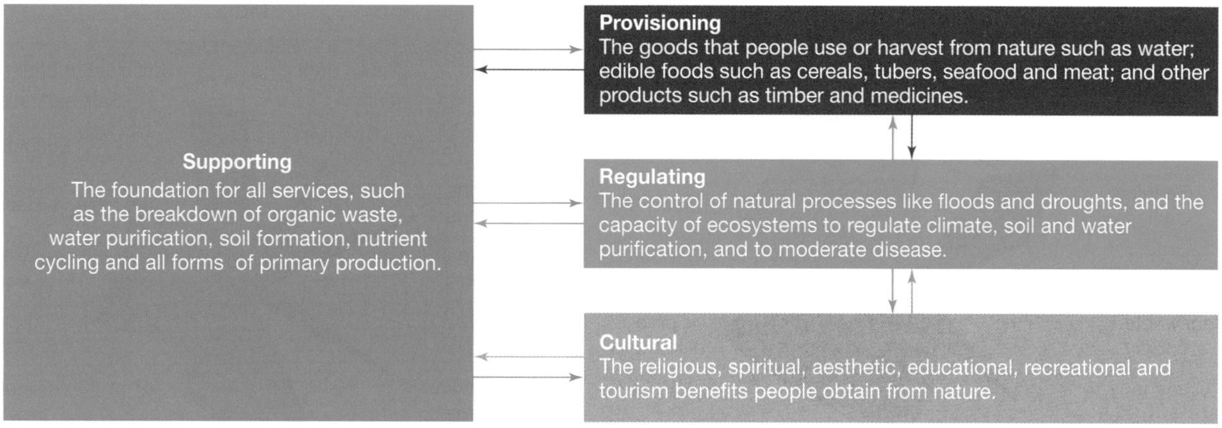
Supporting
The foundation for all services, such as the breakdown of organic waste, water purification, soil formation, nutrient cycling and all forms of primary production.
Provisioning
The goods that people use or harvest from nature such as water; edible foods such as cereals, tubers, seafood and meat; and other products such as timber and medicines.
Regulating
The control of natural processes like floods and droughts, and the capacity of ecosystems to regulate climate, soil and water purification, and to moderate disease.
Cultural
The religious, spiritual, aesthetic, educational, recreational and tourism benefits people obtain from nature.

The Sustainable Society Index (SSI, discussed in section 13.10.3) originally placed a value on 21 different factors that could be categorised as either social, political, economic or environmental (see **FIGURE 3**).

FIGURE 3 Sustainable Society Index (SSI) framework

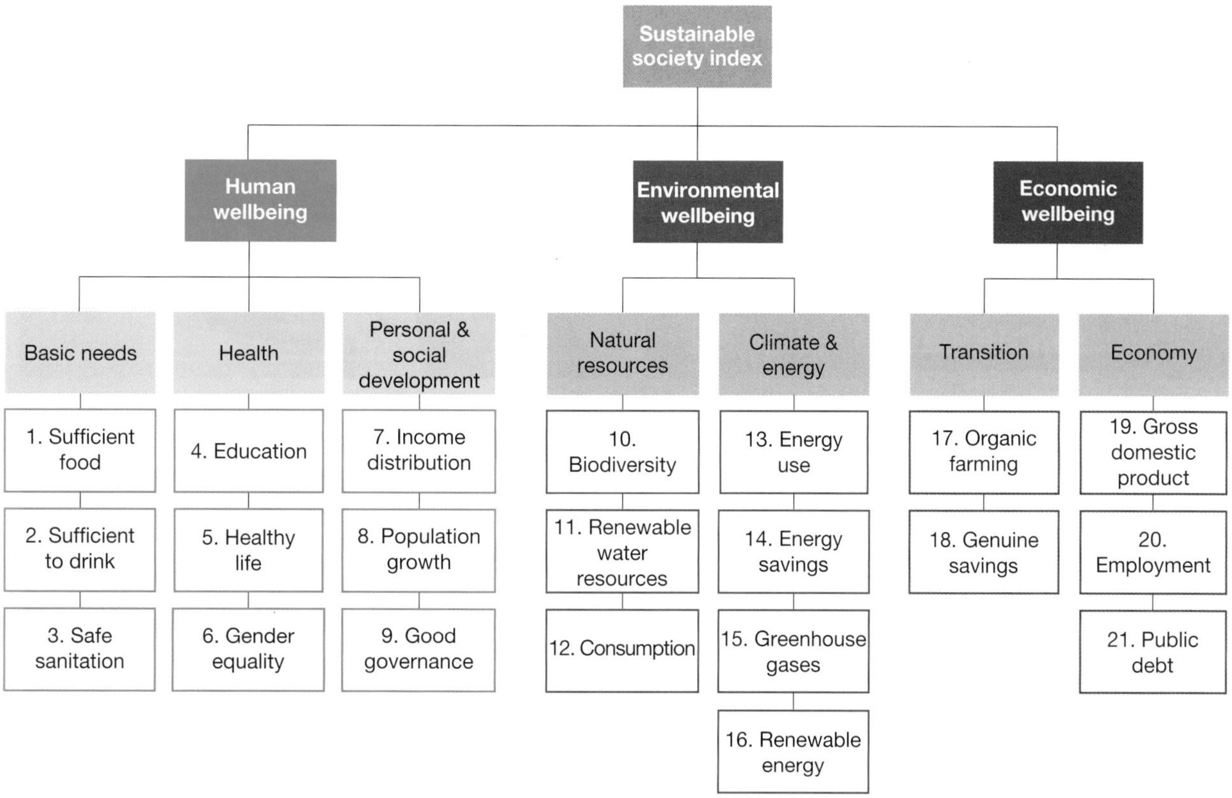
Sustainable society index
Human wellbeing
Environmental wellbeing
Economic wellbeing
Basic needs
Health
Personal & social development
Natural resources
Climate & energy
Transition
Economy
1. Sufficient food
2. Sufficient to drink
3. Safe sanitation
4. Education
5. Healthy life
6. Gender equality
7. Income distribution
8. Population growth
9. Good governance
10. Biodiversity
11. Renewable water resources
12. Consumption
13. Energy use
14. Energy savings
15. Greenhouse gases
16. Renewable energy
17. Organic farming
18. Genuine savings
19. Gross domestic product
20. Employment
21. Public debt

SSI can be used to calculate a 'score' for nations in each of the 21 different categories, with the score being a representation of how well that particular area is being managed. The aim is for nations to see their graph as large and round as possible. The spider web reveals the score for each indicator: the centre of the web represents a score of 0 or no sustainability, while the outer circle represents a score of 10, or full sustainability. **FIGURES 4** and **5** show the difference in SSI between two very different nations in Asia: Japan (developed) and India (developing). It can be seen that Japan has a more regular and much broader shape.

FIGURE 4 SSI for Japan

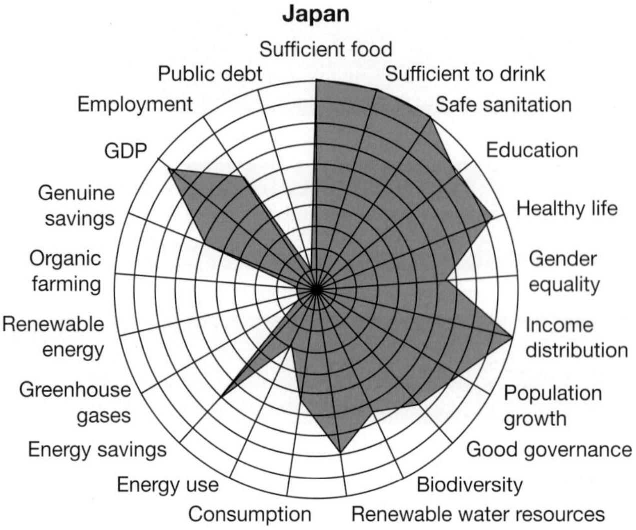

Source: Sustainable Society Foundation

FIGURE 5 SSI for India

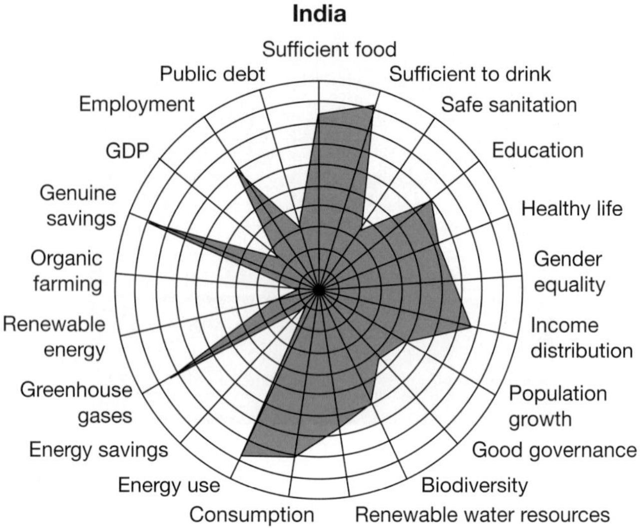

Source: Sustainable Society Foundation

FIGURE 6 Compare the working conditions in these two factories in (a) Japan and (b) India. How do you think the condition in each factory contributes to the economic wellbeing of its workers?

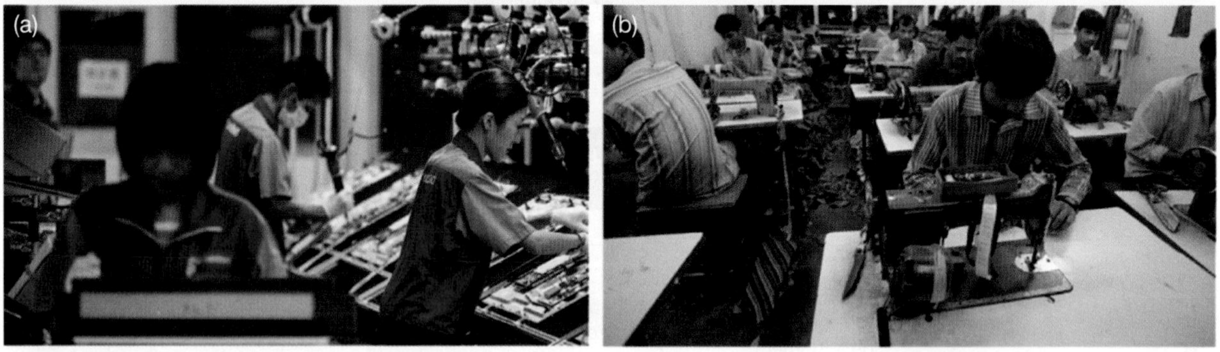

fdw-0041

FOCUS ON FIELDWORK

SSI spiderwebs

How does your class rate the SSI of your local area? Is your place a sustainable society? Or is there work that needs to be done?

Learn more about how to collect data and present your findings in a spider web using the **SSI spiderwebs** Fieldwork activity in your online resources.

on Resources

eWorkbook Contemporary trends in wellbeing — Sustainability (ewbk-10277)

Video eLesson Contemporary trends in wellbeing — Sustainability — Key concepts (eles-5282)

Interactivity Contemporary trends in wellbeing — Sustainability (int-8697)

19.8 ACTIVITIES

1. **Create your own index**
   The SSI has been updated for use by Dutch cities and now has 24 different indicators (see **FIGURE 7**). Investigate which other countries have developed their own versions of the SSI. What similarities or differences are there in the indicators they have added or removed?

FIGURE 7 The weighted average scores of the 24 indicators for all 393 cities in The Netherlands (a developed country) in 2014 and 2015

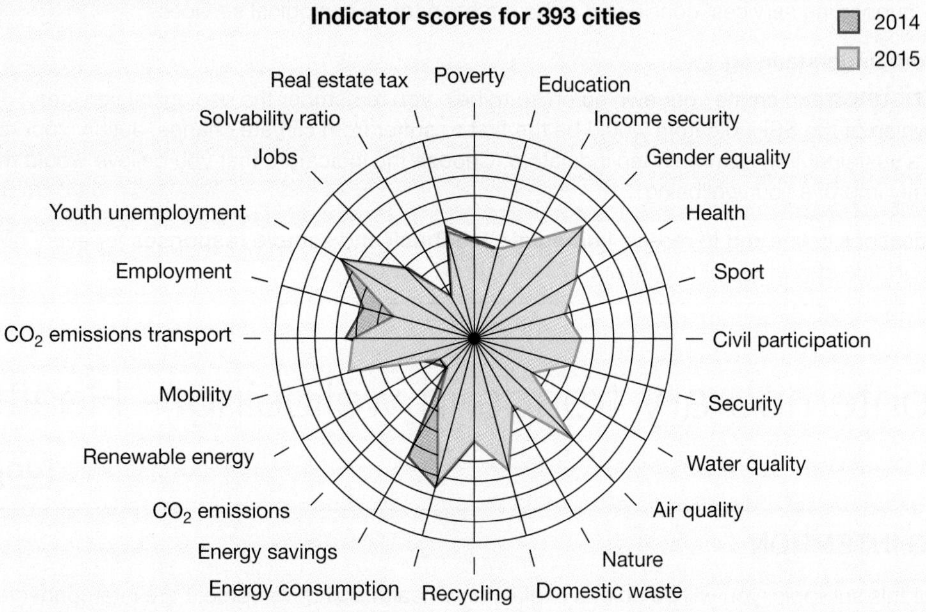

Source: Sustainable Society Foundation

2. Investigate an SSI for Australia. How does this compare with those in **FIGURES 4**, **5** and **7**?

19.8 EXERCISE

Learning pathways

■ LEVEL 1	■ LEVEL 2	■ LEVEL 3
1, 2, 9, 10, 13	3, 4, 8, 11, 15	5, 6, 7, 12, 14

Check your understanding

1. Define *stewardship.*
2. Refer to **FIGURE 1** and outline how each of the environments has a positive impact on human wellbeing.
3. Refer to **FIGURE 1** and outline how human wellbeing might be negatively affected if each of the environments is destroyed or polluted.
4. Outline what the SSI is.
5. Describe the ideal SSI graph shape for a country that is operating very sustainably.
6. Outline the differences between provisioning, regulating and cultural services.

Apply your understanding

7. How are the ecological services interconnected?
8. Using **FIGURES 4** and **5**, compare the indicators for Japan and India.
   a. Which components does Japan score highest in? Why is this the case?
   b. Which components does India score highest in? Why is this the case?
9. Select your five most important and three least important indicators from the list used in **FIGURE 4** and **5** or **FIGURE 7**. Justify why you have chosen these to be the most and least important.
10. Explain how working conditions can contribute to wellbeing indicators.
11. Explain how the health of the world's marine environments can affect human wellbeing.
12. Why are 'supporting services' considered to be the base for all ecological services?

Challenge your understanding

13. Refer to **FIGURE 2** and create your own acronym to help you remember the ecological services.
14. Predict which of the SSI indicators would be the first to suffer from climate change. Justify your response.
15. Develop a sustainability index with ten indicators. Choose the indicators that you believe would most significantly improve your wellbeing.

To answer questions online and to receive **immediate feedback** and **sample responses** for every question, go to your learnON title at www.jacplus.com.au.

19.9 Contemporary trends in wellbeing – Health

online only

LEARNING INTENTION

By the end of this subtopic, you will be able to explain how health and development are interconnected, and outline related trends and ideas related to their interconnection.

The content in this subtopic is a summary of what you will find in the online resource.

Traditionally, a major concern of wellbeing and development focused on health, specifically food and nutrition. Towards the later part of the twentieth century and into the early twenty-first century, we have been concerned with obesity as well as extreme hunger and malnutrition.

To learn more about the impacts of undernourishment and obesity, go to your learnON resources at www.jacPLUS.com.au.

Contents

learnON

- 19.9.1 Health

on Resources

eWorkbook	Contemporary trends in wellbeing — Health (ewbk-10281)	
Video eLesson	Contemporary trends in wellbeing — Health — Key concepts (eles-5283)	
Interactivity	Contemporary trends in wellbeing — Health (int-8698)	

19.10 Contemporary trends in wellbeing — Malaria and tuberculosis

online only

LEARNING INTENTION

By the end of this subtopic, you will be able to provide examples of how deaths and illness from preventable or treatable diseases and levels of development are interconnected, and outline related trends and ideas related to their interconnection.

The content in this subtopic is a summary of what you will find in the online resource.

One of the major disparities in health and wellbeing between nations of the developed and developing world has to do with the existence of disease. The United Nations Millennium Development Goal 6, which was set for 2000 and 2015, has since been replaced by Sustainable Development Goal 3 — to ensure healthy lives and promote wellbeing for all at all ages. Both of these goals had a specific aim of reducing infectious diseases including malaria and tuberculosis (TB).

FIGURE 4 Prevalence of malaria across the world (adapted from WHO)

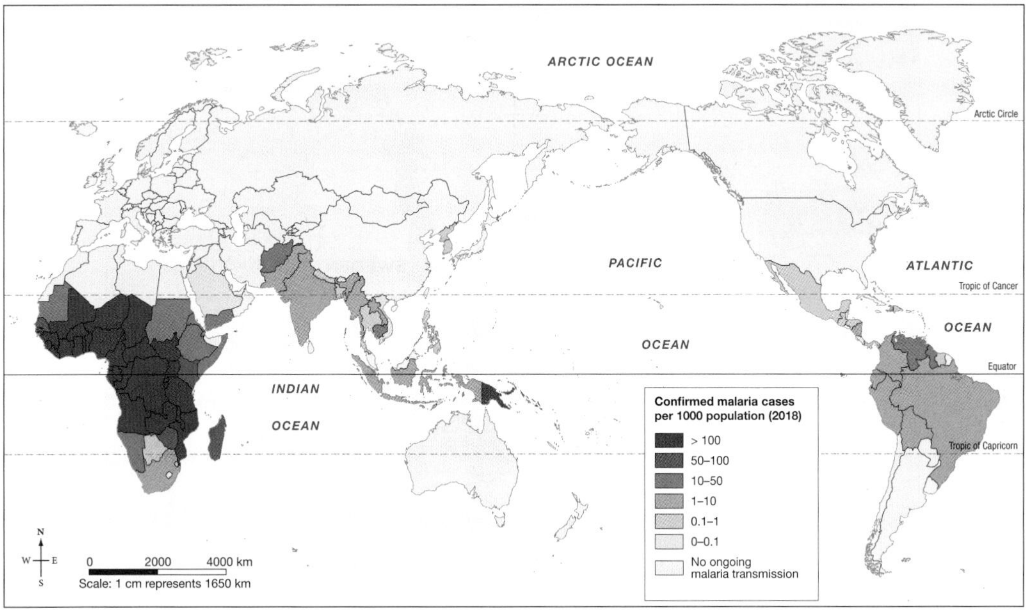

Source: World Bank (2018)

To learn more about malaria and TB spread and treatment, go to your learnON resources at www.jacPLUS.com.au.

Contents

learnON

- 19.10.1 Disparities in wellbeing
- 19.10.2 Malaria
- 19.10.3 Tuberculosis

on Resources

eWorkbook Contemporary trends in wellbeing — Malaria and tuberculosis (ewbk-10285)

Video eLesson Contemporary trends in wellbeing — Malaria and tuberculosis — Key concepts (eles-5284)

Interactivity Contemporary trends in wellbeing — Malaria and tuberculosis (int-8699)

19.11 Investigating topographic maps — Norway — the best place on Earth

LEARNING INTENTION

By the end of this subtopic, you will be able to provide examples of why Norway is well ranked on the Human Development Index and identify specific features on a topographic map of part of the country.

19.11.1 Norway's HDI ranking

Norway has consistently held the number one position in the Human Development Index (HDI) rankings for 16 of the past 18 years. This is largely because of its high levels of development in health, education and the economy. Norway makes up the western part of Scandinavia and shares borders with Sweden, Finland and Russia (see **FIGURE 1**).

FIGURE 1 Norway and Scandinavia

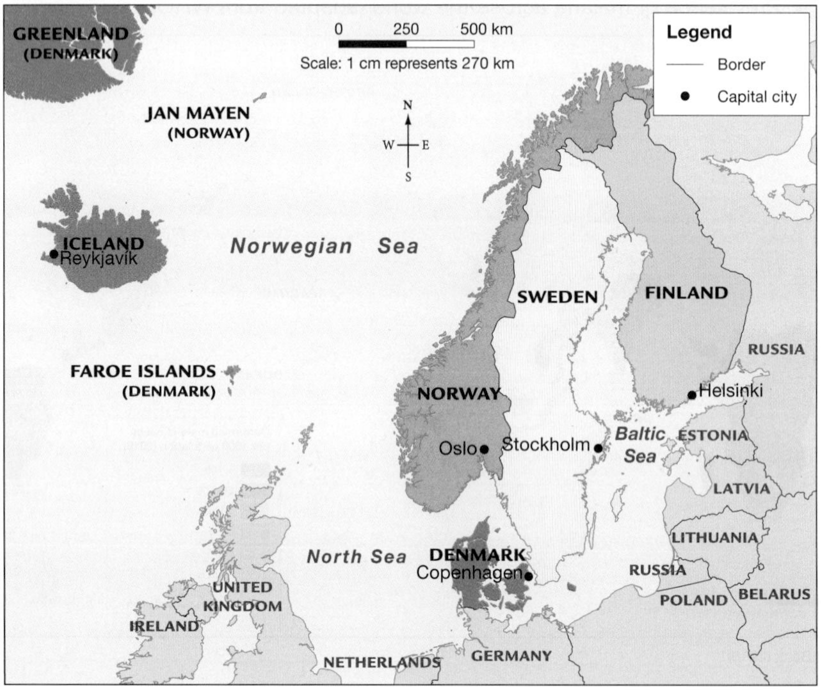

Source: Spatial Vision

Much of Norway's wealth is derived from its location on the North Sea and its proximity to oil. In 2019 Norway ranked fifteenth in the world in oil production, producing almost two million barrels per day. The value to the economy is around A$88 billion, which is 46 per cent of their exports.

Norway has around 40 accredited higher education institutions and several private ones. With the exception of some private university colleges, all higher education institutions are state-run and in general, tuition fees are not required.

Norway spends just under US$10 000 per person per year on health care, the highest in the world. Health care is free for children aged 16 or younger, and for pregnant and/or nursing women. Everyone else must pay a fee, which is currently on average US$325 a year. This entitles them to coverage of all immediate healthcare costs in the event of having to be admitted to a hospital's emergency department.

FIGURE 2 Topographic map of Ulvik

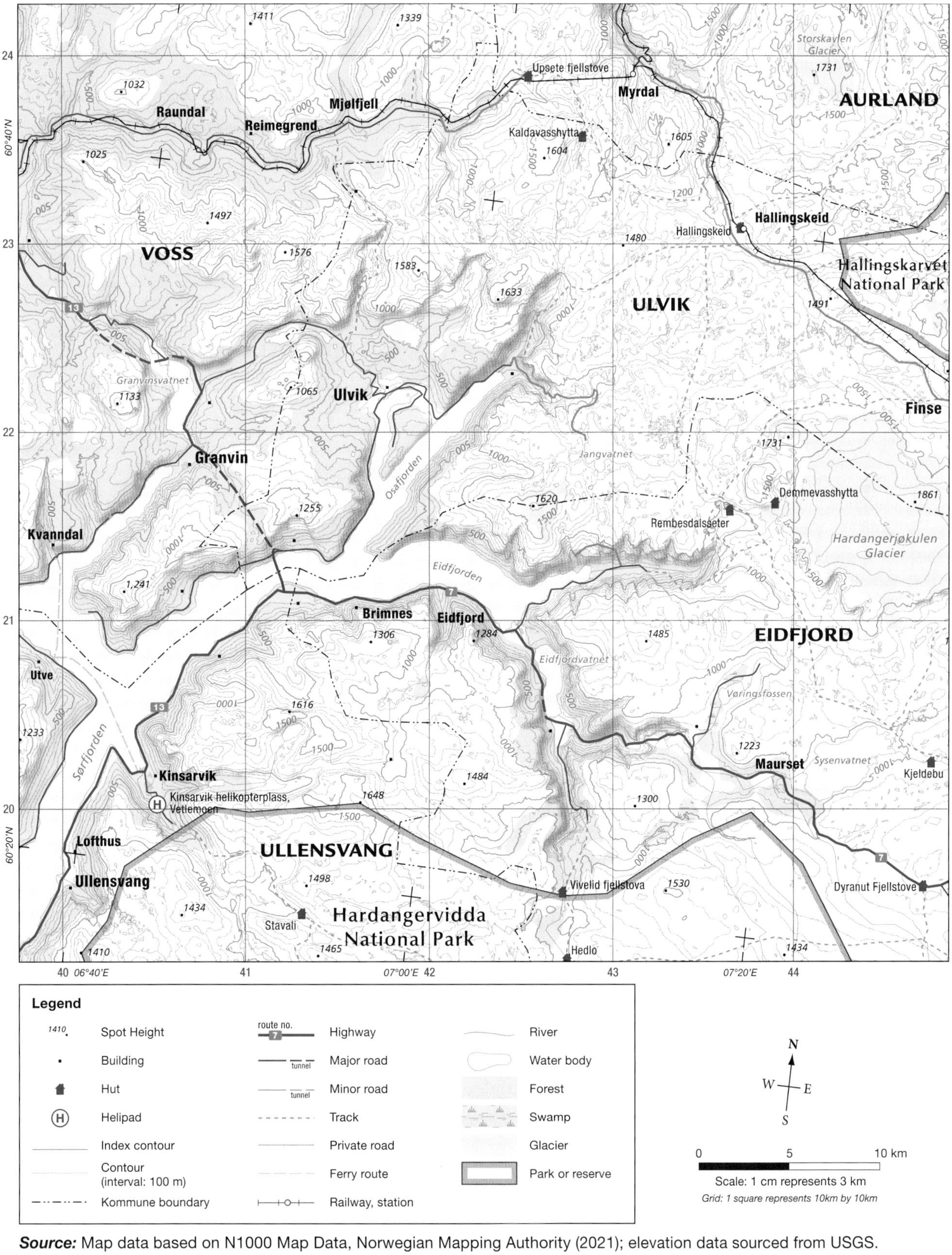

Source: Map data based on N1000 Map Data, Norwegian Mapping Authority (2021); elevation data sourced from USGS.

Resources

eWorkbook	Investigating topographic maps — The best place on Earth — Norway (ewbk-10289)
Digital document	Topographic map of Ulvik (doc-36371)
Video eLesson	Investigating topographic maps — The best place on Earth — Norway — Key concepts (eles-5285)
Interactivity	Investigating topographic maps — The best place on Earth — Norway (int-8700)
Google Earth	Ulvik, Norway (gogl-0145)

19.11 EXERCISE

Learning pathways

■ LEVEL 1	■ LEVEL 2	■ LEVEL 3
1, 2, 11, 12	3, 4, 5, 10, 13	6, 7, 8, 9, 14

Check your understanding

1. Refer to **FIGURE 1**. Identify the capital cities of three Scandinavian countries.
2. Describe the location of Norway.
3. Outline some of the keys features of the Norwegian government's income and expenditure.

Refer to **FIGURE 2** to answer questions 4 to 8.

4. Which quadrant has the highest elevation?
5. What is the aspect of the slope at Brimnes?
6. Estimate the area of the Hardanger Glacier (Hardangerjøkulen) on the eastern side of the map.
7. What is the gradient between the highest point of the Hardanger Glacier and the spot height 1306 metres just south of Brimnes?
8. What is the local relief between the highest point of the glacier and the spot height 1306 metres just south of Brimnes?

Apply your understanding

9. Explain how Norway's healthcare system functions.
10. To what extent is Norway's wealth is due to its proximity to oil?
11. Norway scores highly in a number of other categories as well as those on the HDI. Construct an argument to convince a person to emigrate to Norway.

Challenge your understanding

12. Suggest impacts that drilling for oil could have upon human wellbeing and the environment. How might these impacts affect the HDI of a country?
13. If you were moving to Norway, where would you prefer to live? Provide information about the location you chose from **FIGURES 1** and **2** to help explain your decision.
14. Norway is the best place on Earth. Propose a case for and a case against this statement.

To answer questions online and to receive **immediate feedback** and **sample responses** for every question, go to your learnON title at www.jacplus.com.au.

19.12 Thinking Big research project — UN report — Global wellbeing comparison

online only

19.12.1 Scenario

The content in this subtopic is a summary of what you will find in the online resource.

The world's population continues to grow, even though the rate of growth has slowed. The distribution of this population is not uniform across the Earth, nor does everyone enjoy the same standard or feelings of wellbeing.

Improvements in food production, nutrition, health care, education and hygiene have resulted in rapidly declining death rates, especially in our children. In the year 1870, the global average life expectancy for a person was 29.7 years, and this has steadily risen. Life expectancy in 1950 was 48 years, in 1973 it had risen to 60 years, in 2000 it had reached 66.4 years, and in 2020 it was 72.6 years. It is predicted that by the year 2050, life expectancy will be between 86 and 90 years, with around 3.7 million people living beyond the age of 100.

19.12.2 Task

learnON

The United Nations has asked you to report on changes and variations in human wellbeing found across one developed and one developing country. Life expectancy, child mortality and the prevalence of disease will all come under your microscope as you investigate and prepare your report.

Go to learnON to access the resources you need to complete this research project.

on Resources

ProjectsPLUS Thinking Big research project — UN report — Global wellbeing comparison (pro-0218)

19.13 Review

19.13.1 Key knowledge summary

19.2 Features of wellbeing

- Wellbeing may be defined as a good or satisfactory condition of existence; a state characterised by health, happiness, prosperity and welfare.
- We can use quantitative and qualitative Indicators to measure wellbeing.
- Indicators include social, technological, environmental, economic and political measures.

19.3 Examining quantitative indicators

- A wellbeing approach to development takes into account a variety of quantitative and qualitative indicators.
- Quantitative indicators are objective indices that are easily measured and can be stated numerically, such as annual income or the number of doctors in a country.
- Many quantitative indicators relate to the economy or wealth-related features of a country.

19.4 Examining qualitative indicators

- Quantitative indicators are subjective measures that cannot easily be calculated or measured, for example indices that measure a particular aspect of quality of life or that describe living conditions such as freedom or security
- The Human Development Index (HDI) is one such index. It measures wellbeing according to life expectancy, income and education.
- Other measures of wellbeing include the Happy Planet Index and the Gross National Happiness.

19.5 Measuring development

- Old descriptions of different levels of development throughout the world used terms such as 'developed North' and 'undeveloped South' or 'First World' and 'Third World'.
- Today we use terminology such as 'more economically developed country' (MEDC) and 'less economically developed country' (LEDC) to describe levels of development.
- There is a strong interconnection between development and poverty.
- Poverty is most often measured using solely economic indicators, but it may be taken to encompass many other aspects of life.

19.8 Contemporary trends in wellbeing — Sustainability

- Stewardship benefits the environment and aids wellbeing by ensuring we better manage the benefits we receive from nature.
- Sustainability is measured by the SSI, which maps social, political, economic and environmental factors.

19.9 Contemporary trends in wellbeing — Health

- Malnutrition or hunger affect about 1 in 8 people across the world.
- Food and water insecurity can also lead to other illnesses and diseases, such as cholera or typhoid.
- Rates of obesity are increasing due to rising calorie intake and decreased physical activity.

19.10 Contemporary trends in wellbeing — Malaria and tuberculosis

- Malaria is a blood disease spread by mosquitoes; risk factors include poverty (especially when it stops prevention strategies). Stopping mosquitoes from breeding or biting people are the most effective prevention strategies, but the most effective insecticide (DDT) is harmful to people, and many people at risk cannot afford prevention.
- Tuberculosis (TB) is caused by bacteria. It is a preventable and treatable disease, but over 10 million people a year become ill from it.

19.13.2 Key terms

Australian National Development Index (ANDI) an Australian index measuring elements of progress in 12 areas

development according to the United Nations, defined as 'to lead long and healthy lives, to be knowledgeable, to have access to the resources needed for a decent standard of living and to be able to participate in the life of the community'

ecological footprint the amount of productive land needed on average by each person in a selected area for food, water, transport, housing and waste management

ecological services any beneficial natural process arising from healthy ecosystems, such as purification of water and air, pollination of plants and decomposition of waste

emerging economies (EEs) places experiencing rapid economic growth, but that have significant political, monetary or social challenges

experienced wellbeing an individual's subjective perception of personal wellbeing

gross domestic product (GDP) a measurement of the annual value of all the goods and services bought and sold within a country's borders; usually discussed in terms of GDP per capita (total GDP divided by the population of the country)

gross national income (GNI) a measurement of the value of goods and services produced by citizens and firms of a specific country no matter where they take place, normally discussed as GNI per capita

Human Development Index (HDI) measures the standard of living and wellbeing in terms of life expectancy, education and income

indicator usually consists of a complex set of indices that measure a particular aspect of quality of life or describe living conditions; useful in analysing features that are not easily calculated or measured, such as freedom or security

inequality of outcomes inequalities between people in the one country

less economically developed country (LEDC) lower levels of development — in the economic, social, environmental and political spheres

life expectancy the number of years a person is expected to live, based on the average living conditions within a country

more economically developed country (MEDC) higher levels of development — in the economic, social, environmental and political spheres

newly industrialised country places modernising and developing quickly from a wide range of industries and rapid economic growth

qualitative indicator usually consists of a complex set of indices that measure a particular aspect of quality of life or describe living conditions; useful in analysing features that are not easily calculated or measured, such as freedom or security

quantitative indicator easily measured and can be stated numerically, such as annual income or how many doctors there are in a country

standard of living a level of material comfort in terms of goods and services available to someone or some group; continuum, for example a 'high' or 'excellent' standard of living compared to a 'low' or 'poor' standard of living

stewardship an ethic that embodies the responsible planning and management of resources

wellbeing a good or satisfactory condition of existence; a state characterised by health, happiness, prosperity and welfare

19.13.3 Reflection

Complete the following to reflect on your learning.

Revisit the inquiry question posed in the Overview:

Everyone wants a good life, but what does that mean for different people? Can wellbeing actually be measured, and how can we improve it if it's not measuring up?

1. Now that you have completed this topic, what is your view on the question? Discuss with a partner. Has your learning in this topic changed your view? If so, how?
2. Write a paragraph in response to the inquiry question, outlining your views.

Subtopic	Success criteria			
19.2	I can identify different features of wellbeing.			
19.3	I can construct and analyse scattergraphs.			
19.4	I can identify and outline quantitative indicators.			
19.5	I can identify and outline qualitative indicators.			
19.6	I can describe differences in measuring development.			
19.7	I can interpret and analyse cartograms.			
19.8	I can describe contemporary trends in wellbeing and sustainability.			
19.9	I can describe contemporary trends in wellbeing and health.			
19.10	I can describe contemporary trends in wellbeing related to malaria and tuberculosis.			
19.11	I can explain why Norway is considered the best place on Earth.			

on Resources

eWorkbook Chapter 19 Student learning matrix (ewbk-8592)
Chapter 19 Reflection (ewbk-8593)
Chapter 19 Extended writing task (ewbk-8594)

Interactivity Chapter 19 Crossword (int-8701)

ONLINE RESOURCES

on Resources

Below is a full list of **rich resources** available online for this topic. These resources are designed to bring ideas to life, to promote deep and lasting learning and to support the different learning needs of each individual.

eWorkbook

- **19.1** Chapter 19 eWorkbook (ewbk-8111) ☐
- **19.2** Features of wellbeing (ewbk-10253) ☐
- **19.3** SkillBuilder — Constructing and interpreting a scattergraph (ewbk-10257) ☐
- **19.4** Examining quantitative indicators (ewbk-10261) ☐
- **19.5** Examining qualitative indicators (ewbk-10265) ☐
- **19.6** Measuring development (ewbk-10269) ☐
- **19.7** SkillBuilder — Interpreting a cartogram (ewbk-10273) ☐
- **19.8** Contemporary trends in wellbeing — Sustainability (ewbk-10277) ☐
- **19.9** Contemporary trends in wellbeing — Health (ewbk-10281) ☐
- **19.10** Contemporary trends in wellbeing — Malaria and tuberculosis (ewbk-10285) ☐
- **19.11** Investigating topographic maps — Norway — the best place on Earth (ewbk-10289) ☐
- **19.13** Chapter 19 Student learning matrix (ewbk-8592) ☐
- Chapter 19 Reflection (ewbk-8593) ☐
- Chapter 19 Extended writing task (ewbk-8594) ☐

Sample responses

- **19.1** Chapter 19 Sample responses (sar-0167) ☐

Digital document

- **19.11** Topographic map of Ulvik (doc-36371) ☐

Video eLessons

- **19.1** The good life? (eles-1713) ☐
- Human wellbeing and development — Photo essay (eles-5277) ☐
- **19.2** Features of wellbeing — Key concepts (eles-5278) ☐
- **19.3** SkillBuilder — Constructing and interpreting a scattergraph (eles-1756) ☐
- **19.4** Examining quantitative indicators — Key concepts (eles-5279) ☐
- **19.5** Examining qualitative indicators — Key concepts (eles-5280) ☐
- **19.6** Measuring development — Key concepts (eles-5281) ☐
- **19.7** SkillBuilder — Interpreting a cartogram (eles-1757) ☐
- **19.8** Contemporary trends in wellbeing — Sustainability — Key concepts (eles-5282) ☐
- **19.9** Contemporary trends in wellbeing — Health — Key concepts (eles-5283) ☐
- **19.10** Contemporary trends in wellbeing — Malaria and tuberculosis — Key concepts (eles-5284) ☐
- **19.11** Investigating topographic maps — Norway — the best place on Earth — Key concepts (eles-5285) ☐

Interactivities

- **19.2** Features of wellbeing (int-8693) ☐
- **19.3** SkillBuilder — Constructing and interpreting a scattergraph (int-3374) ☐
- **19.4** Examining quantitative indicators (int-8694) ☐
- **19.5** Examining qualitative indicators (int-8695) ☐
- Cartogram showing relative global populations and Happy Planet Index scores, 2016 (int-8467) ☐
- **19.6** Measuring development (int-8696) ☐
- **19.7** SkillBuilder — Interpreting a cartogram (int-3375) ☐
- **19.8** Contemporary trends in wellbeing — Sustainability (int-8697) ☐
- **19.9** Contemporary trends in wellbeing — Health (int-8698) ☐
- **19.10** Contemporary trends in wellbeing — Malaria and tuberculosis (int-8699) ☐
- **19.11** Investigating topographic maps — Norway — the best place on Earth (int-8700) ☐
- **19.13** Chapter 19 Crossword (int-8701) ☐

ProjectsPLUS

- **19.12** Thinking Big Research project — UN report — Global wellbeing comparison (pro-0218) ☐

Weblinks

- **19.2** Gauging interconnections (web-3960) ☐
- **19.3** Human Development Report 2012 Africa (web-3961) ☐
- **19.4** Human development tree (web-1236) ☐
- Human development reports (web-2354) ☐
- MONET 2030 system (web-6412) ☐
- **19.5** Happy Planet Index (web-2963) ☐
- **19.10** Bill and Melinda Gates Foundation (web-6379) ☐
- WHO Global Malaria Program Tuberculosis (WHO) (web-6380) ☐
- Tuberculosis (WHO) (web-6381) ☐

Fieldwork

- **19.8** SSI spiderwebs (fdw-0041) ☐

Google Earth

- **19.11** Ulvik, Norway (gogl-0145) ☐

Teacher resources

There are many resources available exclusively for teachers online.

20 Spatial variations in human wellbeing

INQUIRY SEQUENCE

To access a pre-test with **immediate feedback** and **sample responses** to every question in this chapter, select your learnON format at www.jacplus.com.au.

20.1 Overview

Numerous **videos** and **interactivities** are embedded just where you need them, at the point of learning, in your learnON title at www.jacplus.com.au. They will help you to learn the content and concepts covered in this topic.

The world's population is constantly increasing. Can we fit so many people in the space we have without affecting our quality of life?

20.1.1 Introduction

Human wellbeing is different right across the globe. In some nations the reality of not having two cars may be seen as a problem, while in others it may be not having enough to eat. There are many reasons for differences in wellbeing between and within countries. In 2050, it is estimated that the world's population will be between 8 and 11 billion. Our wellbeing is affected not just by how many people we can fit in a particular place, but also the manner in which we live.

STARTER QUESTIONS

1. Why would the estimates for the world population for 2050 vary so widely?
2. How might the number of people in a given place be interconnected with their wellbeing?
3. How could the wellbeing of a particular place have an impact on the number of people living at that location?
4. Watch the **Spatial variations in human wellbeing — Photo essay** video in your online Resources. What places have you visited where the wellbeing levels of the community seemed much higher or lower than where you live?

FIGURE 1 Holi celebrations, Gopinath Temple in Vrindavan, Uttar Pradesh, India

on Resources

eWorkbook	Chapter 20 eWorkbook (ewbk-8112)
Video eLessons	A long life (eles-1714) Spatial variations in human wellbeing — Photo essay (eles-5308)

20.2 Variations in wellbeing between countries

LEARNING INTENTION

By the end of this subtopic, you will be able to explain and give examples of how human wellbeing can vary between countries by using key indicators.

20.2.1 Life expectancy

Spatial variations exist across the globe for a range of wellbeing indicators. One of the major indicators of wellbeing is **life expectancy**. On average, across the globe, people are now expected to live longer than any time in history. This has a significant impact on current and future population levels. This increased life expectancy is linked to a fall in child mortality rates across the world.

How long we can expect to live when we are born is referred to as our life expectancy and is calculated according to the conditions in a particular country in that year. A child born in Japan in 2019 can expect to live 84 years, whereas one born in the African country of Sierra Leone can expect to live only 54 years. **FIGURE 1** shows variation in life expectancy worldwide.

int-5603

FIGURE 1 Global life expectancy

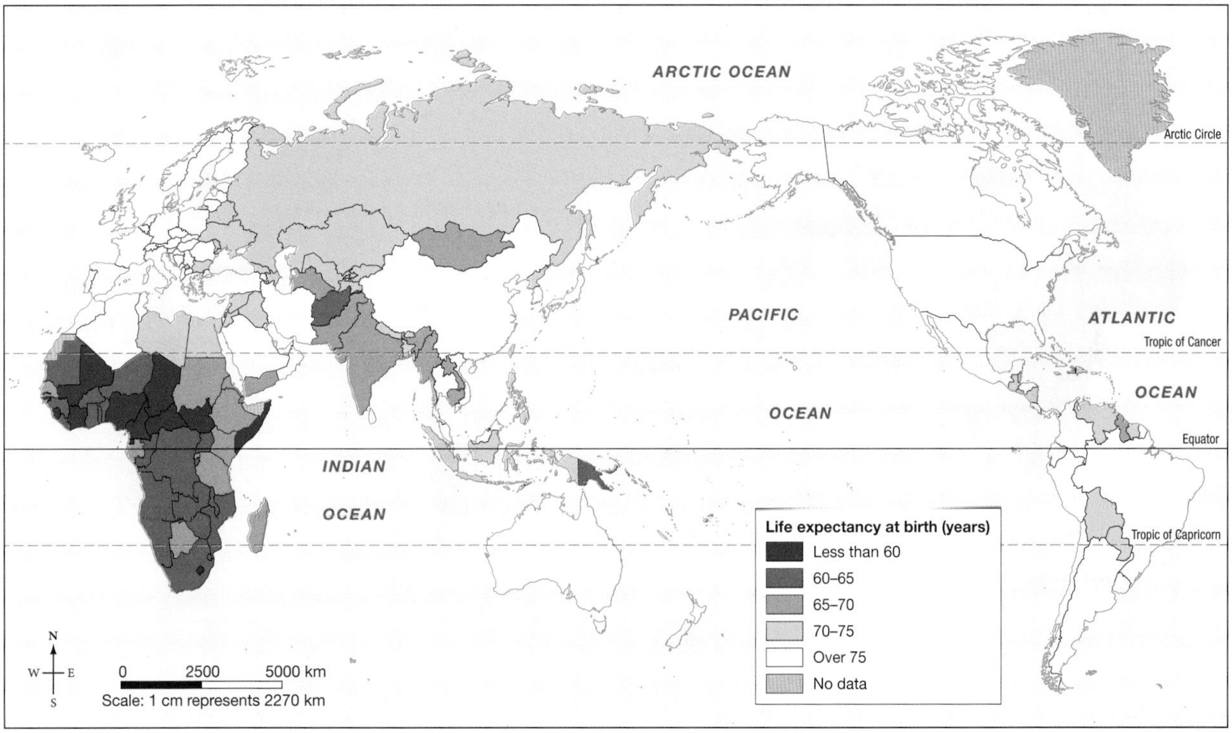

Source: World Health Organization (2019)

Life expectancy around the world started to increase in the mid 1700s due to improvements in farming techniques, working conditions, nutrition, medicine and hygiene. There is a clear interconnection between wealth and life expectancy: wealthier people in all countries can expect to live longer than poorer people. In general, women outlive men. A higher income enables people to have better access to education, food, clean water and health care. One region where life expectancy is decreasing rather than increasing is sub-Saharan Africa, where many countries have been affected by HIV and AIDS.

life expectancy the number of years a person can expect to live, based on the average living conditions within a country

20.2.2 Birth and death rates

Every minute there are an estimated 250 births and 105 deaths worldwide. This natural increase equates to an extra 145 people at a global scale every minute. However, the rate of population change varies considerably across the world, with some places experiencing a decline rather than an increase in numbers of people. Rates of population change have an impact on wellbeing, both now and in the future.

FIGURE 2 shows the global distribution of birth rates. The continent of Africa clearly stands out here with the highest figures. In 2010, Africa had a population of one billion; the United Nations projects that by 2050 the African population will be 2.5 billion — three times that of Europe. The majority of this growth will occur in sub-Saharan Africa, where the average **fertility rate** in 2018 was five children per woman. Countries such as Niger and Somalia have fertility rates as high as 7.2 and 6.2 respectively, while Tunisia has the lowest fertility rate in Africa with 2.2 births per woman. Europe has very low birth rates with a fertility rate of 1.6. Ireland and France are slightly higher than this average with fertility rates of 2.0, while the poorest country in Europe, Moldova, has the lowest fertility rate of 1.3. Taiwan recorded the lowest fertility rate in the world in 2018 with 1.2 births per woman. Australia's fertility rate is 1.8.

int-7976

FIGURE 2 Global distribution of birth rates

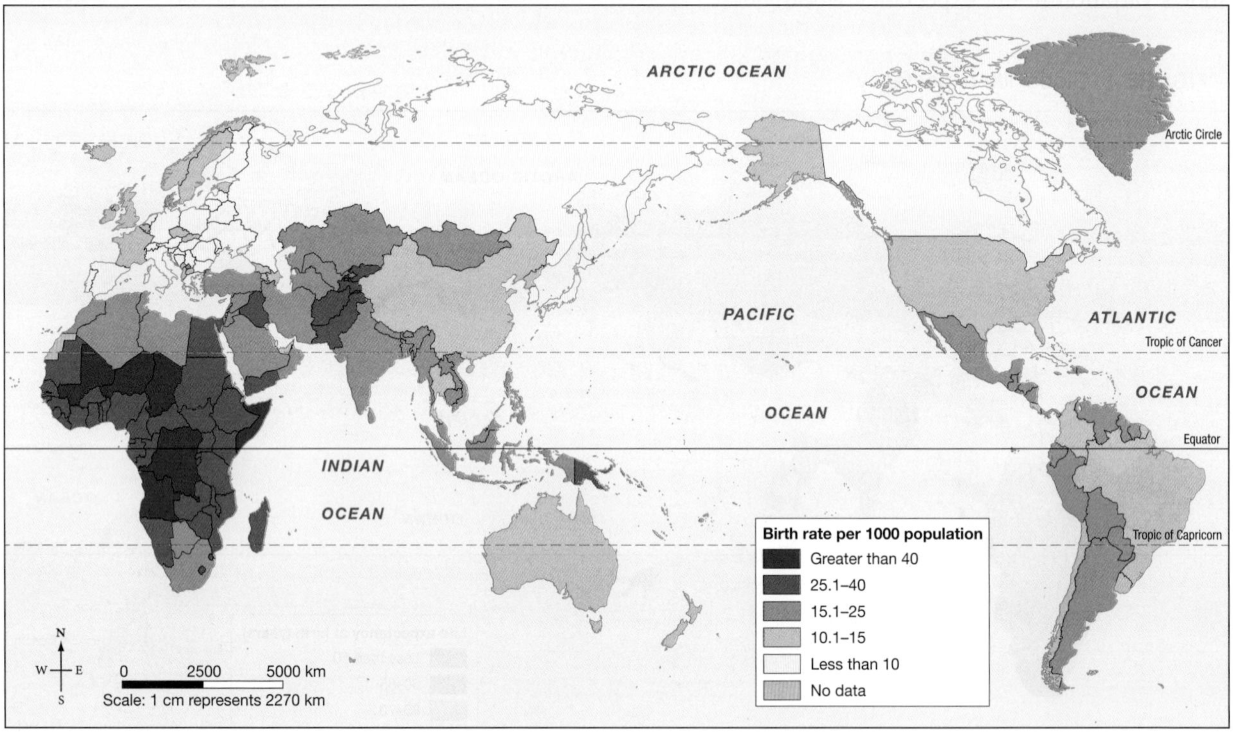

Source: © United Nations Publications; The World Bank (2019)

For death rates, as **FIGURE 3** illustrates, sub-Saharan Africa is again at the high end of the spectrum, with many places in that region experiencing death rates above 10 per 1000. However, high death rates are more dispersed, with many European countries, such as Bulgaria and Ukraine, included. Low death rates are widely distributed across the regions of the Americas, much of Asia and Oceania. These typical figures, however, might not reflect changes from significant local or regional events such as natural disasters, conflict or outbreaks of disease.

fertility rate the average number of children born per woman

int-7977

FIGURE 3 Global distribution of death rates

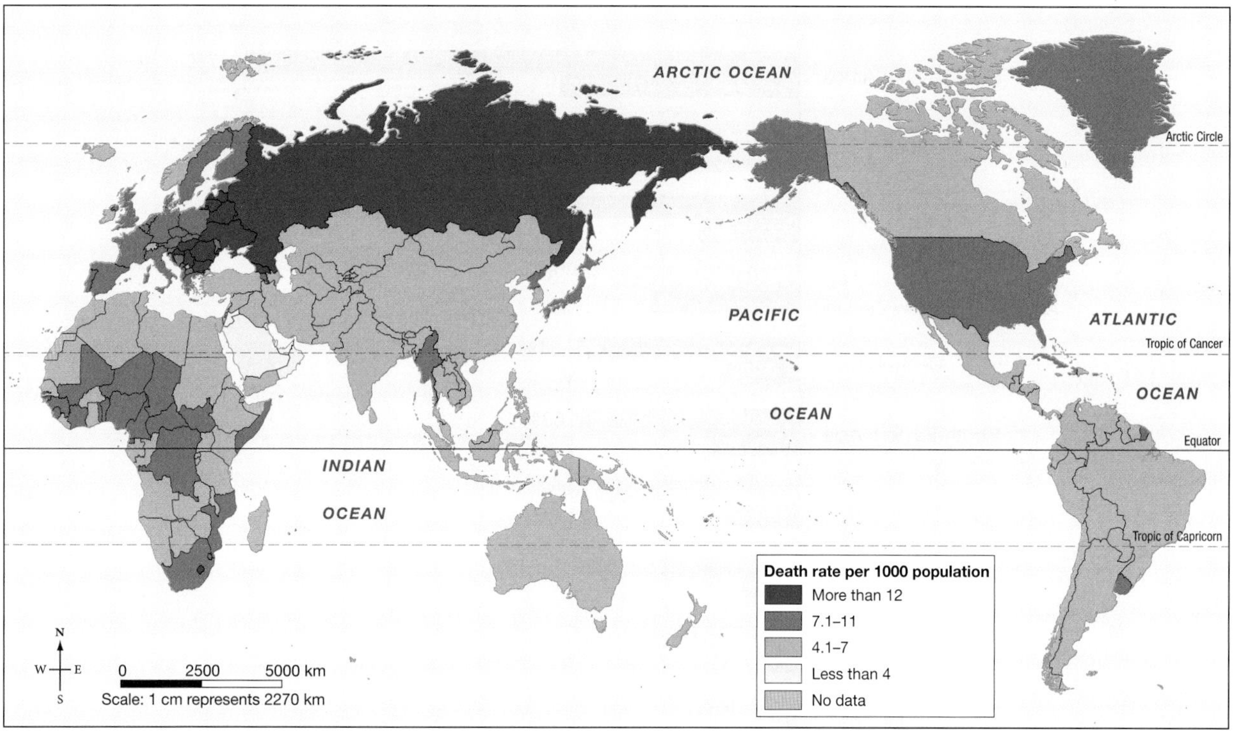

Source: © United Nations Publications; The World Bank (2019)

Whether a population increases or decreases is largely dependent on variations in births and deaths producing a **natural increase**. Where a fertility level is well above the **replacement rate** of 2.1 children, population growth will occur. Conversely, fewer births over a period of time will ultimately result in a declining population.

20.2.3 Child mortality

Life expectancy is closely interconnected to child mortality: countries with high death rates for children under five years of age have low life expectancy. Young children are particularly vulnerable to infectious diseases due to their lower levels of immunity. Major causes of death include pneumonia, diarrhoea, measles and malnutrition. In wealthier households, child deaths are lower as these children are likely to have better nutrition and be immunised, and parents are more likely to be educated and aware of how to prevent disease.

Under the United Nations' Millennium Development Goals program (MDG), which operated from 1990 to 2015, child mortality was reduced considerably. The number of deaths of children under the age of five declined from 12.7 million in 1990 to 6 million in 2015, the equivalent of nearly 12 000 fewer children dying each day. The greatest success was achieved in northern Africa and eastern Asia, with respective declines in under-five mortality of 68 per cent and 58 per cent.

natural increase the difference between the birth rate (births per thousand) and the death rate (deaths per thousand). This does not include changes due to migration

replacement rate the number of children each woman would need to have in order to ensure a stable population level — that is, to 'replace' the children's parents. This fertility rate is 2.1 children.

FIGURE 4 Under-five mortality rate (deaths per 1000 live births) and percentage change, 1990–2016

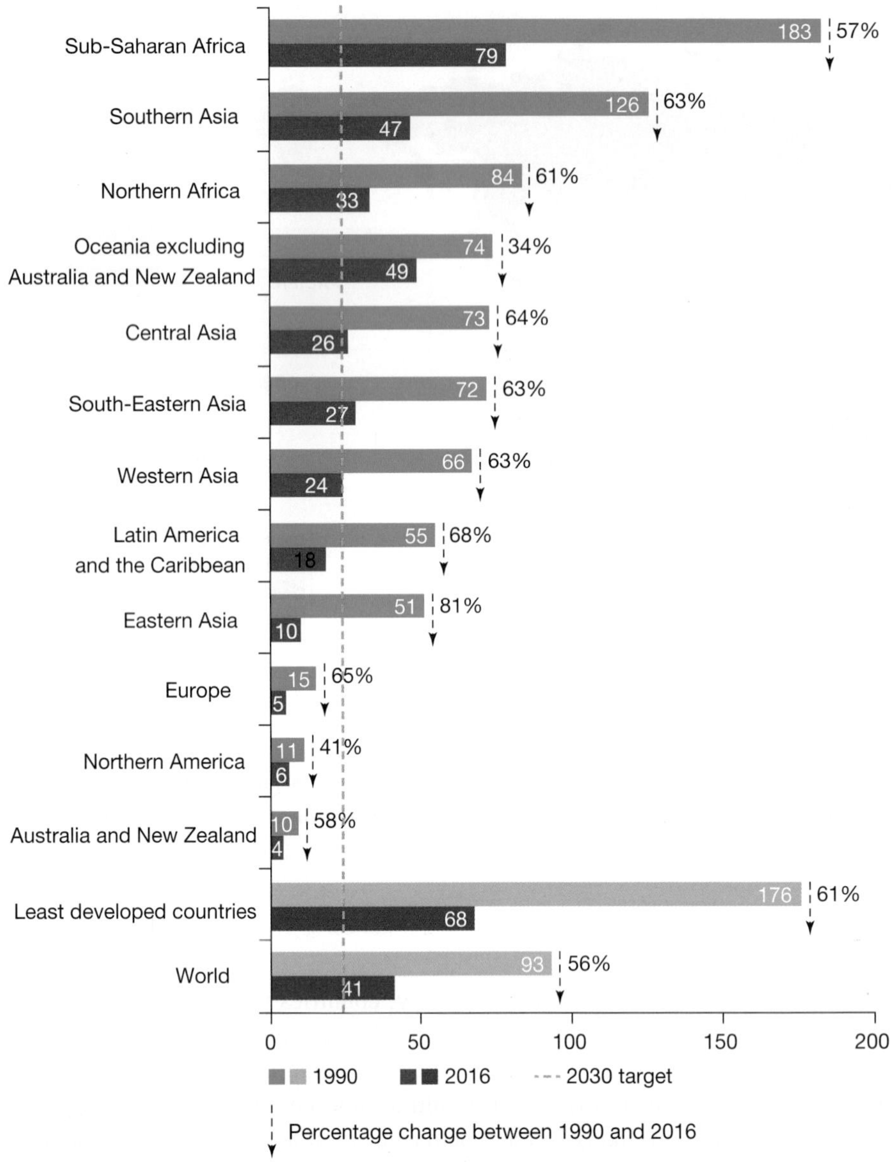

Source: United Nations Inter-agency Group for Child Mortality Estimation (UN IGME), 'Levels & Trends in Child Mortality: Report 2017, Estimates Developed by the UN Inter-agency Group for Child Mortality Estimation', United Nations Children's Fund, New York, 2017.
Note: Percentage change calculations are based on unrounded numbers.

The highest levels of under-five mortality continue to be found in sub-Saharan Africa, where one in eight children die before the age of five, 18 times the average in developed regions. However, substantial improvements are being made in this region, with four of these countries reducing child deaths by more than 50 per cent between 1990 and 2015. For example, increased measles vaccination coverage has been a relatively simple but effective way of reducing child deaths.

The Sustainable Development Goals (SDG), which followed the MDGs, will run from 2015 to 2030. The SDGs also have a focus on child mortality: SDG 3.2 aims to end preventable deaths of newborns and children under five years of age by 2030, with all countries aiming to reduce neonatal mortality to at least as low as 12 per 1000 live births and under-five mortality to at least as low as 25 per 1000 live births.

FIGURE 5 The future: the likely future of a child born today

I'll have at least a one in three chance of living to 100
$2.9m What it will cost me for a mid-price house in Melbourne.
I'll have a one in four chance of being obese at age 16.
There's a 70% chance that I will live in a city.
My chances of suffering dementia will be more than three times greater than my mother's.
I will never see a Tasmanian devil in the wild but there's a good chance I'll see a live mammoth.
When I learn to drive, my car will tell other cars about hazards on the road ahead. It will run on smart roads programmed to keep it a safe distance from others.
35.9m other people will be sharing Australia with me in 2050.
Cancer is unlikely to kill me thanks to vaccines and viral gene therapy.
Smoking will be SO last century.
After at least 20 jobs and four career changes, I'll still be working in my 60s.

Source: Wiley Art

Resources

eWorkbook Variations in wellbeing between countries (ewbk-10297)

Video eLesson Variations in wellbeing between countries — Key concepts (eles-5309)

Interactivity Changes in child mortality by region (int-8737)

20.2 ACTIVITIES

1. Use the **MDGs** and **SDGs** weblinks in your online Resources. Construct a table or Venn diagram showing the differences between the SDGs and MDGs. What is the same and what has been added?
2. Allocate a different country to each person (or pairs) in your class.
   a. Use the **Demographic indicators** weblink in your online Resources to examine interactive maps and charts from the OECD. You will find current data about life expectancy, infant mortality, fertility rates and other demographic indicators of wellbeing.
   b. Use the **Gauging interconnections** weblink in your online Resources to examine the correlation between life expectancy and other indicators such as infant mortality over place and time.
   c. Discuss your findings as a class. What did you find? How have the rates changed in different places over time? What factors might have influenced these changes?

20.2 EXERCISE

Learning pathways

■ LEVEL 1	■ LEVEL 2	■ LEVEL 3
1, 4, 7, 9, 15	2, 3, 10, 11, 13	5, 6, 8, 12, 14

Check your understanding

1. Define *life expectancy*.
2. Describe the distribution of life expectancy shown in **FIGURE 1**.
3. Refer to **FIGURE 4**. What was the percentage change in under-five mortality rates in the 'developing regions'?
4. Outline three factors that lead to a rise in life expectancy.
5. Outline the difference between fertility rate and replacement rate.
6. Quantify the differences between changes in under-five mortality rate between 1990 and 2016 in Australia and New Zealand and changes in:
   a. Oceania
   b. South-east Asia.

Apply your understanding

7. What age groups are likely to have the highest mortality rates? Why?
8. How does the overall figure in under-five mortality rate compare to different regions? Explain why Oceania might have had the lowest fall.
9. Will increased incomes always lead to increased life expectancy? Justify your answer.
10. Explain why reducing levels of child mortality is an indicator of increasing wellbeing.
11. Explain the significance of a national fertility level of 2.1 children per woman.
12. What might be the implications for levels of wellbeing in a country if its fertility rate is below the replacement rate for several generations?

Challenge your understanding

13. Predict how life expectancies in Sub-Saharan Africa may change if a cure for HIV/AIDS is discovered.
14. What implications does an increase in life expectancy have on the provision of health care?
15. Write your own version of 'The future' (**FIGURE 5**), imagining what your life will be like when you are 35 years old.

To answer questions online and to receive **immediate feedback** and **sample responses** for every question, go to your learnON title at www.jacplus.com.au.

20.3 Variations in wellbeing within countries

LEARNING INTENTION

By the end of this subtopic, you will be able to explain and give examples of how human wellbeing can vary between regions within a country by using key indicators.

20.3.1 Population change in India

You probably know that China has the highest population in the world. The population of China is 1.4 billion. With some 1.36 billion people in 2019, India is set to surpass China's population by 2025, when its population will reach an estimated 1.45 billion. With a predicted 1.6 billion by 2050, what happens to India's population will have major implications in terms of the wellbeing of the people in that country.

India's population was growing at a rate of 1.1 per cent in 2019. Improvements in water supply, a decrease in infectious diseases and an increase in education levels have resulted in a generally decreasing death rate since the 1950s, while the birth rate has not declined to the same extent. The COVID-19 pandemic presented an anomaly in these trends of declining death rates.

Infant mortality remains high; over two-thirds of the population are rural dwellers who may not have ready access to health and reproductive services. Children remain a vital part of the family's labour force both on farms (as shown in **FIGURE 1**) and for old age support, so it is essential for families to have more children to improve the chance of them surviving to adulthood. Thirty-one per cent of the population is under 15 years of age, creating huge momentum for future growth (see **FIGURE 2**).

FIGURE 1 Indian children assisting with rice planting

FIGURE 2 (a) Population pyramid for India, 2019 and (b) population pyramid for India, 2050

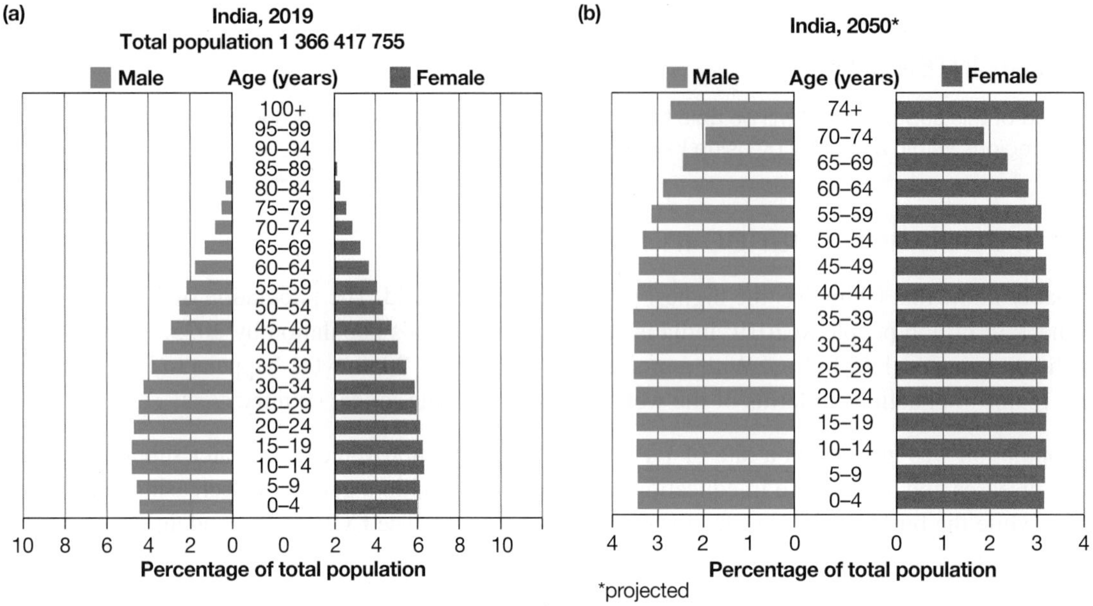

Source: © December 2019 by PopulationPyramid.net, made available under a Creative Commons license CC BY 3.0 IGO: http://creativecommons.org/licenses/by/3.0/igo/

int-8740

20.3.2 Regional variation in wellbeing and population

The number of children born per woman in India has declined substantially from 5 in the 1970s to 2.2 in 2020. There is considerable regional variation. Levels of literacy and poverty shown in **FIGURES 5** and **6** reflect a varying distribution of wellbeing in India. For information about the Indian government's moves to reduce poverty and improve wellbeing across the country in response to census results, use the **Poverty challenge** weblink in your online Resources.

FIGURE 3 The COVID-19 pandemic brought online learning and school closures for children

FIGURE 4 Proportion of children 0–6 years to total population by state, India, 2011

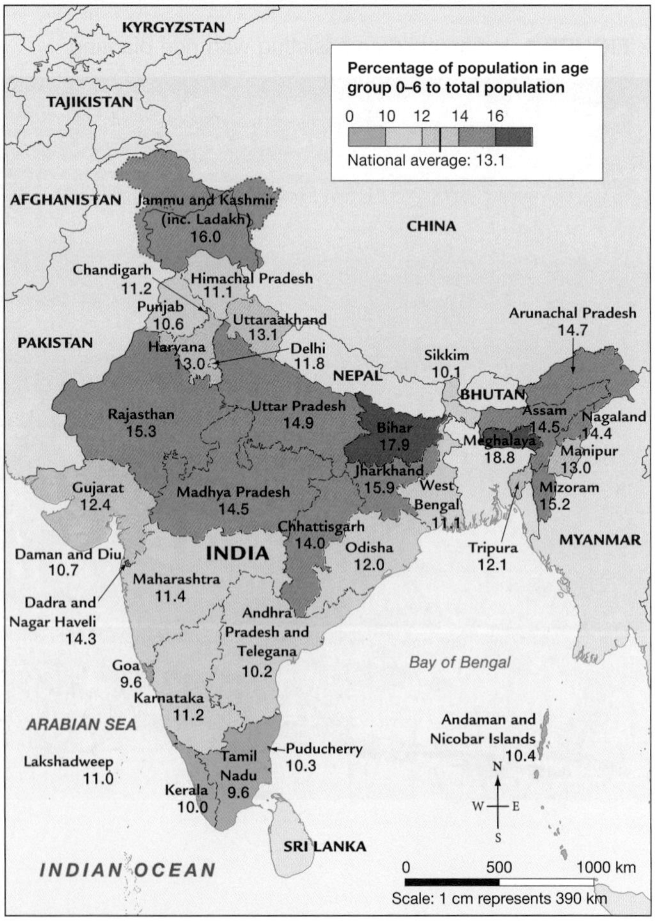

Source: Spatial Vision

Note: Most recent Indian census data by state, 2011

int-8738

int-8739

FIGURE 5 Literacy rates (percentage), India, 2011

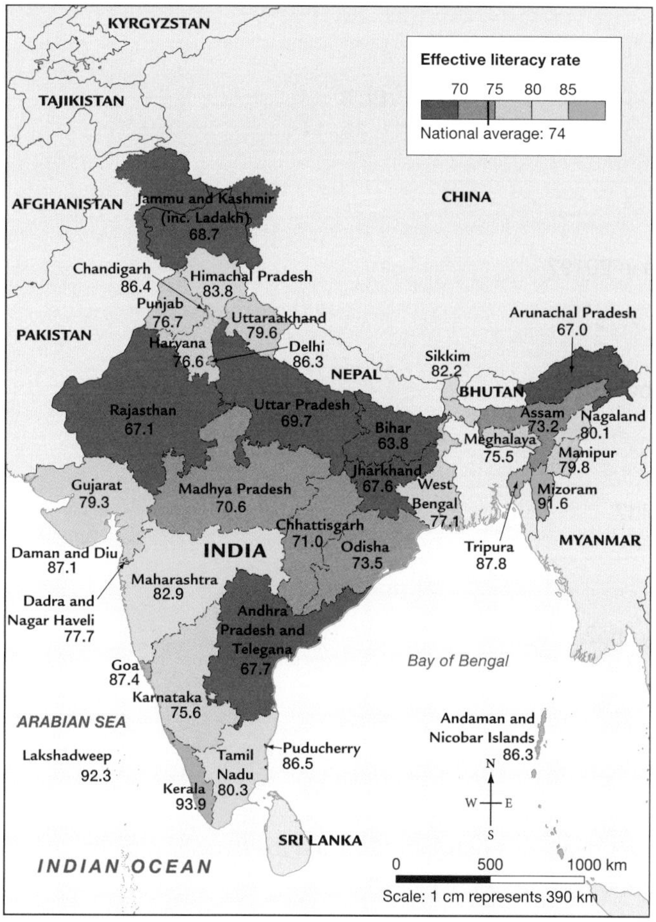

Source: Government of India, Ministry of Home Affairs, Office of Registrar General

Note: Most recent Indian census data by state, 2011

FIGURE 6 Poverty levels (percentage), India, 2011

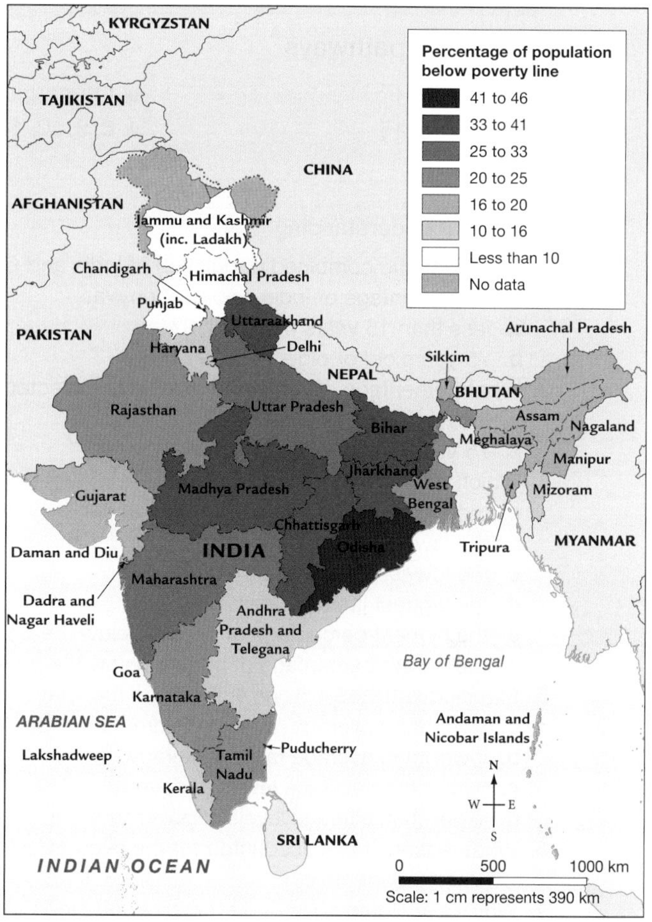

Source: Spatial Vision

Note: Most recent Indian census data by state, 2011

on Resources

eWorkbook Variations in wellbeing within countries (ewbk-10301)

Video eLesson Variations in wellbeing within countries — Key concepts (eles-5310)

20.3 ACTIVITIES

Consider the demographics of Indian regions.

1. Using **FIGURE 6**, pick two states that fall into different categories on the map. Use the **Census India** weblink in your online Resources to compare the 2011 demographic characteristics for those two states.
2. Share your findings with other members of your class who selected different states.
3. India conducts its census every ten years. Discuss as a class what you would expect the next results to show.

20.3 EXERCISE

Learning pathways

■ LEVEL 1	■ LEVEL 2	■ LEVEL 3
1, 2, 3, 8, 15	4, 6, 9, 10, 13	5, 7, 11, 12, 14

Check your understanding

1. What was the combined population of India and China in 2019?
2. What percentage of India's population was:
   a. less than 15 years old in 2019?
   b. 75 years old or older in 2019?
3. What percentage of India's population is expected to be:
   a. less than 15 years old in 2050?
   b. 75 years old or older in 2050?
4. Which region(s) of India has:
   a. the highest proportion of children under six years old?
   b. the lowest proportion of children under six years old?
   c. the lowest literacy rates?
   d. the highest literacy rates?
   e. the highest percentage of people below the poverty line?
   f. the lowest percentage of people below the poverty line?
5. Examine **FIGURES 4**, **5** and **6**. Describe the spatial pattern of:
   a. lower than average rates of children under six years old
   b. lower than average rates of literacy
   c. higher than average levels of poverty
   d. lower than average levels of poverty.
6. What reasons have accounted for the overall fall in India's death rate between 1950 and the beginning of the COVID-19 pandemic?
7. Why are children so important in rural locations?

Apply your understanding

8. What factors might account for the lower proportion of the population in the southern regions of India under the age of six?
9. Identify and explain any correlation that might exist between the following factors. Use data in your response.
   a. Literacy levels and poverty levels
   b. Percentages of children under six and literacy levels
   c. Percentages of children under six and poverty levels
10. With reference to the population pyramids shown in **FIGURE 2**, account for India's changing population growth.
11. Explain why India is set to overtake China in terms of total population.
12. What impact might significantly different levels of poverty and proportions of young people in neighbouring states have on internal migration (people moving from state to state)?

Challenge your understanding

13. Predict what a population pyramid (showing millions of people) for India might look like in 2070 if the current trends in population growth continue.
14. Suggest which state of India might have the highest levels of wellbeing. Use data in your answer.
15. How might the literacy and poverty rates change in India if children were no longer allowed to work? Choose two states and predict:
   a. what impact might this effect the data each state
   b. how wellbeing in each state might change because of this.

To answer questions online and to receive **immediate feedback** and **sample responses** for every question, go to your learnON title at www.jacplus.com.au.

20.4 SkillBuilder — Using Excel to construct population profiles

online only

LEARNING INTENTION

By the end of this subtopic, you will be able to create a population profile graph using Excel.

The content in this subtopic is a summary of what you will find in the online resource.

20.4.1 Tell me

Why do we use Excel to construct population profiles?

When we construct population profiles, there is a large amount of data and large numbers to handle. The use of an Excel spreadsheet simplifies the process. Excel allows actual population figures, which are generally large numbers, to be handled simply. Once the data is placed in the spreadsheet, the computer can create the graph.

20.4.2 Show me

How to construct a population profile

Step 1: Set up your spreadsheet

Open an Excel spreadsheet and create a layout with 5 columns.

- Column 1 (A) is for the age groups.
- Column 2 (B) is for the raw population figures for each age group of males.
- Column 3 (C) is for a percentage calculation.
- Column 4 (D) is for the raw population figure for each age group of females.
- Column 5 (E) is for a percentage calculation.

FIGURE 5 Calculating percentage using Excel

	A	B	C	D	E
1	Age group	Number of males	Males	Number of females	Females
2	0-4	10,319,427	-6.8%	9,881,935	6.3%
3	5-9	10,389,638		9,959,019	
4	10-14	10,579,862		10,097,332	
5	15-19	11,303,666		10,736,677	
6	20-24	11,014,176		10,571,823	

The number of rows required is one for the column titles and one for each age group. (This should come to 22 rows.)

Step 2: Enter the data

Enter the raw numbers of males and females in each age group into columns B and D.

Step 3: Calculate the total

Type 'Total' in cell A23. Click on cell B23 and click on the Greek letter Σ (AutoSum) found in the toolbar on your screen. This command will produce a display that asks you to check if these are the row numbers that you wish to total. If it is correct then press Enter and the total will appear. Repeat the process for the Column D.

Step 4: Calculate your percentages

Calculate the percentages, click on the cell C2. Type =.

Click on cell B2. Insert a division symbol (/) and click on the cell that shows the total number of males (C23) and press Enter. Convert the resulting to a percentage (click on the % symbol in the toolbar).

Convert your percentage for males to a negative number (type a minus sign in the formula bar).

Change your formula for the C2 cell so that the address SUM reads as =-B2/B23. Select all the cells in Row C for each age group (excluding the 'Total' row). Click on the Fill button in the Editing menu on the toolbar and select Down.

Repeat the process for female data.

Step 5: Create the bar chart

Select the data in the column with age groupings (for labels on the vertical axis), the percentage column for males (do not include the totals), and the percentage column for females (do not include the totals).

Select the data in column A. Hold down the Control key. Select the data in column C (including its heading, 'Males'). Keeping the control key down, select the data in column E (including its heading, 'Females'). Let go of the Control key and press F11. The graph should come out like **FIGURE 4**.

Step 6: Refine the chart type

Refine your graph to suit geographic conventions. In the toolbar, go to Design and select Change Chart Type. Select Horizontal bar graph.

Step 7: Line up the bars

Click and highlight just one bar. Right-click; select 'Format Data Point'. Select the slider under Series Overlap and move it right until the window reads 100%. Your male and female bars should now be aligned.

Step 8: Tidy up the data

Remove the negative signs in the formatting of the male column.

Click the horizontal (%) axis at the bottom of the graph so that the axis is highlighted. Right-click and a pop-up menu will appear. Click on Format Axis… and select Number. You will see a little Format Code window displaying 0.0%. Change this to #0.0%;#0.0%. Click Add and then Close.

FIGURE 12 Removing the negative signs

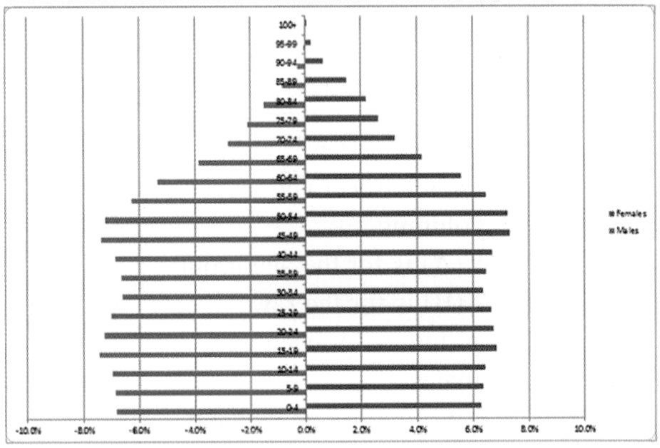

Step 9: Add the title

In the main menu bar, click on the Layout tab and click on Chart Title.

Step 10: Label your axes

Open Chart Tools, select Layout tab and click on the Axis Titles button. Select Primary Horizontal Axis Title and Title Below Axis. Type the word 'Percentage' into the box that appears below the horizontal axis. Click on the Axis Titles button and then select Primary Vertical Axis Title and Horizontal Title. Type the words 'Age group' into the box that appears beside the vertical axis. Manually move the vertical axis title to sit above the population profile. You can also manually adjust the position of the chart title.

20.4.3 Let me do it

learnON

Go to learnON to access the following additional resources to help you build this skill:

- a longer explanation of this skill and its application in Geography (Tell me)
- a video demonstrating the step-by-step process of this skill (Show me)
- an activity and interactivity for you to practise the skill (Let me do it)
- self-marking questions to help you understand and use the skill.

on Resources

eWorkbook	SkillBuilder — Using Excel to construct population profiles (ewbk-10305)
Video eLesson	SkillBuilder — Using Excel to construct population profiles (eles-1758)
Interactivity	SkillBuilder — Using Excel to construct population profiles (int-3376)

20.5 Internal reasons for variations in wellbeing

LEARNING INTENTION

By the end of this subtopic, you will be able to identify factors within a country that can influence levels of wellbeing and explain each factor's impact.

20.5.1 Understanding internal reasons for variations in wellbeing

Internal factors are those which take place within a country and are seen to have a greater degree of influence over its citizens and government. Factors such as the population, geographical location and resources, economy and political systems all contribute to a country's level of development and the wellbeing of its people.

20.5.2 Population growth

Most of the global population growth is taking place in developing countries, particularly the poorest nations (see **FIGURE 1**). By 2050, with an estimated nine billion people in the world, some eight billion people (86 per cent) will be in developing countries, with two billion of those in the least developed nations. Despite continued global population growth, global **fertility rates** are falling. Declines in fertility have coincided with improvements in living conditions, greater access to education (particularly for women), improved health care and access to contraception. It is anticipated that fertility rates in developing regions will continue to fall, particularly with increasing rural–urban migration. In the cities, a child is more likely to be an economic burden than an asset, and there is better access to health services and family planning programs.

FIGURE 1 World population growth per region

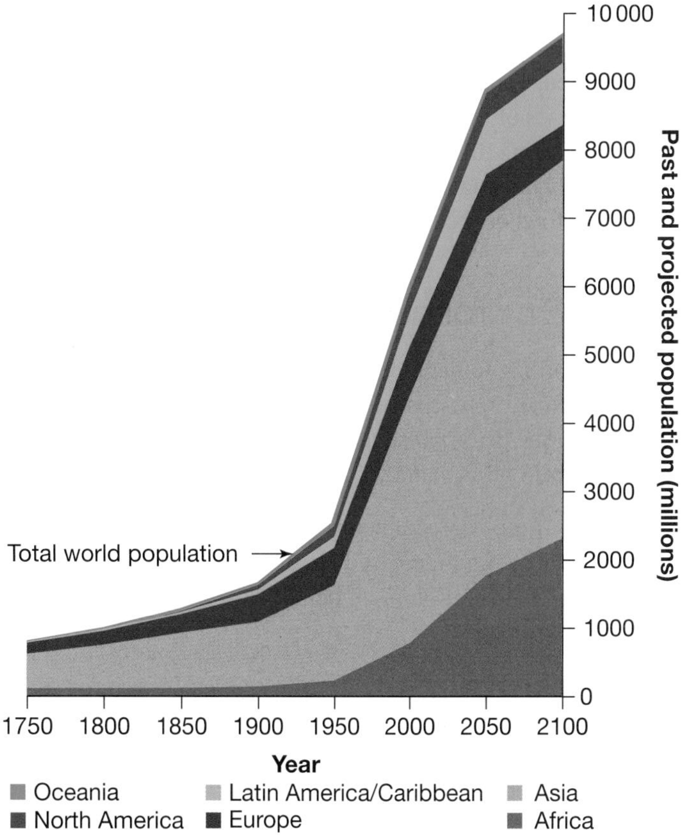

Source: Data based on United Nations, Department of Economic and Social Affairs, Population Division (2019). World Population Prospects 2019

20.5.3 Geographic location

Where a place is situated will contribute to variations in wellbeing. Landlocked countries are particularly affected. Geographic isolation from coastal regions leads to additional transportation costs and fewer economic opportunities to access higher market share of world markets. Botswana is an exception to this after investing the money made during a mining boom into infrastructure and inland services.

Another factor that leads to spatial variations is for those countries located in the tropics. These countries are more prone to infectious disease and natural disasters like cyclones, reducing wellbeing and quality of life.

fertility rate the average number of children born per woman

20.5.4 Political instability

War and civil unrest have a major impact on wellbeing and development of countries. Those nations that experience higher levels of political instability find it much harder to develop. The need to support police and military forces can often see funds diverted away from much needed development and infrastructure projects that would directly enhance wellbeing. Much of a nation's infrastructure can be destroyed through conflict, either indirectly, as was the case with the Syrian conflict in 2015 (**FIGURE 2**), or directly, such as when Indonesia withdrew from Timor-Leste in 1999.

FIGURE 2 Conflict in Syria

Struggles for governmental control between groups are not the only political causes of variations in internal rates of wellbeing in a country. There are many examples of minority groups being persecuted by a politically or economically dominant cultural or ethnic group. This often results in people being forced to leave their homes to find safety, food or shelter. This has impacts on both the regions they leave and the regions where they seek refuge.

20.5.5 Environmental degradation

Nations continue to degrade the environment as a result of a continued push for development. Researchers say that of the nine processes needed to sustain life on Earth, four have exceeded 'safe' levels. According to the United Nations, 130 000 square kilometres of forests are cut down or burned every year — equivalent to about twice the size of Tasmania.

As a result of this, the wellbeing of citizens is compromised with reductions in areas such as biodiversity, air and water quality, increased erosion and runoff, as well as increased occurrence of salinisation and desertification. This is an issue being addressed by all nations at different levels as its importance is significant.

20.5.6 Resources

Variations occur within many nations as their access to natural resources is uneven. A nation's biophysical environment will determine how well it can undertake agriculture and if it can produce excess amounts and then export the surplus. Some nations, such as Japan, cannot produce sufficient agricultural products to sustain their populations, so they are net importers of food products.

Natural resources such as gas and coal can be exploited in order to make significant economic gain, as can be seen in Qatar in the Middle East. Qatar's GDP (PPP) reached over US$133 000 in the mid 2010s when oil prices were high. (The average GDP (PPP), globally, is about US$18 381.) In 2020, the GDP (PPP) of Qatar was $91 897.

Then there are countries, such as those in Sub-Saharan Africa, which do not have known deposits of resources or do not have the infrastructure to mine available resources that are considered valuable on the global market. For example, the Central African Republic has some mineral deposits including some diamond, gold and uranium, but has one of the world's lowest GDP (PPP) of US$972. Political conflict and lack of infrastructure mean the Central African Republic has limited opportunities to benefit from the deposits.

FIGURE 3 The city of Doha, Qatar has expanded and developed rapidly from the income derived from oil

FIGURE 4 (a) Aerial view and (b) satellite image of Bangui, capital of the Central African Republic

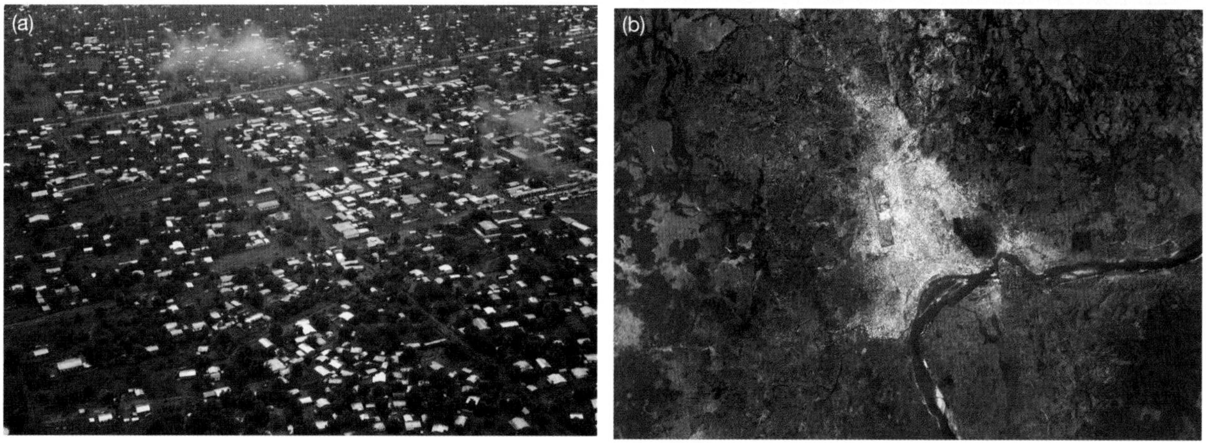

fdw-0047

FOCUS ON FIELDWORK

Advantage and disadvantage survey

There are many reasons why people in some parts of a community experience higher levels of wellbeing than others. This doesn't always depend on financial security. Sometimes strong community and family ties can provide a sense of comfort and security that boost wellbeing much more than a healthy bank balance. In this fieldwork activity, you will take a virtual walk through two neighbourhoods and assess what helps foster wellbeing in a community.

Learn how to respectfully compare communities using the **Advantage and disadvantage survey** fieldwork activity in your learnON Fieldwork Resources.

on Resources

eWorkbook Internal reasons for variations in wellbeing (ewbk-10309)

Video eLesson Internal reasons for variations in wellbeing — Key concepts (eles-5311)

Interactivity Internal reasons for variations in wellbeing (int-8741)

20.5 ACTIVITIES

1. Select a country located within in the tropics and use the **Gapminder** weblink in your online Resources to create a table showing its score/rank in the areas of health and education. How do these compare with Australia?
2. Use the **Displacement** weblink in your online Resources to view a video about displacement in the Central African Republic. What problems do these people face?
3. Choose one nation or region currently involved in a conflict.
   a. Research the causes of the political instability and its impacts.
   b. Share your findings as a class. Are there any patterns or common causes of why political instability occurs that emerge from your collective research?
   c. Discuss the ways that these conflicts have and continue to affect the wellbeing of people in that place and the people who have fled to find safety in other countries.
4. Find a collection of comments from world leaders on the importance of biodiversity. Place these quotes on a world map based on their locations.

20.5 EXERCISE

Learning pathways

■ LEVEL 1	■ LEVEL 2	■ LEVEL 3
1, 3, 10, 11, 15	2, 4, 6, 9, 14	5, 7, 8, 12, 13

Check your understanding

1. What are 'fertility rates'?
2. List three internal factors that influence wellbeing.
3. What proportion of people in the world are expected to be living in developing countries in 2050?
4. Describe how world population growth has changed over time.
5. What other social and economic factors have been affected by falling fertility rates?
6. Which continent will account for the greatest population in 2100?
7. Which continent has experienced the greatest rate of growth since 1900? Include the rate as a percentage in your answer.

Apply your understanding

8. What is meant by the phrase 'In cities a child is more likely to be an economic burden than an asset'?
9. How does the biophysical environment determine natural resources?
10. Explain the relationship between fertility rates and wellbeing.
11. How does environmental degradation influence wellbeing?
12. To what extent can the biophysical environment of a country influence wellbeing?

Challenge your understanding

13. Suggest whether there might be a correlation between political instability and economic disadvantage. Give reasons for your view.
14. Both very high and very low fertility rates have the capacity to negatively impact wellbeing. Propose where the ideal rate for wellbeing within a country lies, and give reasons for your view.
15. Respond to the following statement: 'Valuable deposits of natural resources result in higher wellbeing levels within a country.'

To answer questions online and to receive **immediate feedback** and **sample responses** for every question, go to your learnON title at www.jacplus.com.au.

20.6 External reasons for variations in wellbeing

LEARNING INTENTION

By the end of this subtopic, you will be able to identify factors that influence a country's levels of wellbeing that are caused by forces outside of the country's control, and explain each factor's impact.

20.6.1 Understanding external reasons for variations in wellbeing

External factors refer to those that take place outside the control of a country. Many of these are historical and have played a significant role in determining the level of development experienced by different countries in certain parts of the world. These factors include the decisions and actions of other governments or international/regional organisations, trends in international markets or politics, and the availability of resources. Some of these factors are outlined in the following sections.

FIGURE 1 Demonstrators against colonial rule, Mumbai (then Bombay), 1928

20.6.2 Colonisation

Colonisation was a direct contributor to wellbeing levels as European empires expanded throughout the world in the sixteenth, seventeenth, eighteenth and nineteenth centuries. This saw them forcibly take control over what are today some of the least-developed countries on Earth. Often this was done to benefit the colonial powers by acquiring a source of raw materials as well as a potential market for their products, and to obtain cheap or slave labour. For many peoples living in countries that were once colonies, the legacy of colonisation's economic and resource drain is compounded by ongoing effects of cultural, religious and/or political oppression.

20.6.3 Trade

Poorer and less developed nations traditionally had a far greater proportion of their economy based in primary production, often centred on raw materials. These agricultural commodities have a far lower value than high-cost manufactured goods that are required within the country. Countries can find themselves in debt, as the cost of the goods they import is more than what they receive for the goods they export. To fund this shortfall, they must borrow funds and go into debt. As time passes, this debt can increase and takes a greater proportion of the nation's income to pay it off. Global events, such as the COVID-19 pandemic, also have a significant impact on trade.

Subsistence farming is also increasingly under threat, not only from farmers looking to grow cash crops rather than produce food, but also transnational corporations and more developed nations attempting to secure land and agricultural production in these poorer areas.

TABLE 1 Breakdown of economies by sectors

		GDP ($US millions)	Agriculture	Industrial	Services
	World	**80 934 711**	**4.9%**	**31.2%**	**63.9%**
1	United States	21 433 228	0.9%	19.1%	80.0%
2	China	14 327 359	7.9%	40.5%	51.6%
3	Japan	5 078 679	1.1%	30.1%	68.7%
28	United Arab Emirates	421 077	0.9%	49.8%	49.2%
31	Thailand	543 798	8.2%	36.2%	55.6%
35	Venezuela	210 100	4.7%	40.4%	54.9%

Source: CIA Factbook 2019

20.6.4 TNCs

Transnational corporations (TNCs) are attracted to less developed nations because of their cheaper labour and often more relaxed or less regulated rules governing working conditions, such as minimum and maximum hours of work, rates of pay, penalty rates and other conditions. The irony is that in many of these developing nations the workers are producing goods that they would be unable to afford themselves, and the profits of TNCs often return to the developed nations in which they were founded.

Transnational corporations (TNCs) a large business that operates across multiple countries, but is run from a central office or offices in a single developed country

FIGURE 2 Australian Navy personnel unloading international disaster relief

FIGURE 3 Multilateral aid involvement helps these primary school students in Nepal.

20.6.5 Aid

Less developed countries are offered aid from other countries as a means of helping their citizens. However, there is the possibility of countries becoming aid dependent. Aid is usually delivered as either bilateral, multilateral or non-government organisation (NGO)/charity aid. Each of which has both advantages and disadvantages (**FIGURE 4**).

With any type of aid, there is always the threat that corruption will prevent food, money, medicine or other supplies from reaching the people who need it most. People's wellbeing is more likely to be improved by humanitarian aid if it empowers them in the longer term as well as providing the immediate help that is needed.

FIGURE 4 Advantages and disadvantages of aid

Bilateral aid	Multilateral aid	NGO/charity aid
+ Helps expand infrastructure: roads, railways, ports, power generation.	+ The organisations have clear aims around what they are trying to achieve (e.g. WHO combats disease and promotes health).	+ Usually targeted at long-term development within a country.
+ Aid which directly supports economic, social or environmental policies can result in successful programs.	+ Leading experts in their field work to help achieve multilateral aid program objectives.	+ Raises awareness of specific situations in a country or region.
– 'Tied aid' obliges the country receiving aid to spend it on goods and services from the donor country (may be expensive).	– Sometimes directed only towards specific areas or organisations, leaving many without benefit.	– The greatest source of need may not be prioritised (e.g. the 2006 tsunami devastation received many donations, but areas in Sub-Saharan Africa on a daily basis were just as much in need).
– Inappropriate technology may be given (e.g. tractors are of little use if there are no spare parts or fuel).	– May come with conditions to make big changes to structures, which can be difficult to manage once aid has 'finished'.	– Up to 30% of donations may be 'eaten up' by administration costs.

on Resources

eWorkbook External reasons for variations in wellbeing (ewbk-10313)

Video eLesson External reasons for variations in wellbeing — Key concepts (eles-5312)

Interactivity External reasons for variations in wellbeing (int-8742)

20.6 ACTIVITIES

1. Select a transnational corporation and create a table showing where their goods or services are manufactured (made) and consumed (sold).
2. Create a map showing which European nations had ties to various countries within Africa. Annotate the map with the time span of colonial control.
3. In groups, select a type of aid agency and create a marketing campaign to support your new organisation. You should create an infographic outlining your aim, name, logo and the key roles and responsibilities of your organisation.

20.6 EXERCISE

Learning pathways

■ LEVEL 1	■ LEVEL 2	■ LEVEL 3
1, 5, 7, 11, 15	3, 4, 8, 10, 13	2, 6, 9, 12, 14

Check your understanding

1. What are 'external factors' that affect a country? Write a definition in your own words.
2. How can trade improve wellbeing within a country?
3. Categorise the following according to whether they are bilateral, multilateral or NGO/charity aid.
   a. UNICEF providing help to young people living in refugee camps, allowing them to attend school
   b. A volunteer group going to a former war zone to help rebuild community facilities
   c. The Australian government sending food aid to a Pacific island after a cyclone
   d. Donating clothing to the Red Cross.
4. Identify which of the three economic sectors shown in **TABLE 1** earns the following countries the highest percentage of income.
   a. The United States
   b. China
   c. Japan
   d. United Arab Emirates
   e. Thailand
   f. Venezuela
5. What is a TNC?
6. Outline one reason why TNCs might choose to manufacture their goods in a developing nation.

Apply your understanding

7. Explain the key difference between external and internal factors on wellbeing.
8. How does international trade influence economic development within a country?
9. Refer to **TABLE 1**. Plot the data for the United States, United Arab Emirates and Thailand onto a ternary graph.
10. What role does debt play in a country's development? What do you think is meant by a 'cycle of debt'?
11. Explain the difference between the three types of aid.
12. Analyse the data in **TABLE 1**. What interconnections exist between the breakdown of economies by sector and the overall GDP of a country?

Challenge your understanding

13. Which type of aid do you believe is the most effective type of aid for Australia to provide to other countries? Give reasons for your opinion.
14. Since India achieved independence from Britain in 1947, many of the indicators of wellbeing have risen. Life expectancy was approximately 31 years in 1947 and steadily rose to around 70 by 2020. To what extent do you think this could be attributed to the end of colonial rule? What other factors might need to be considered when analysing this figure?
15. Suggest how a global pandemic may affect variation in wellbeing. Which factors connected with a pandemic may have more impact and which may have less impact? Give reasons to justify your answer.

To answer questions online and to receive **immediate feedback** and **sample responses** for every question, go to your learnON title at www.jacplus.com.au.

20.7 Variations in wellbeing within the middle class

online only

LEARNING INTENTION

By the end of this subtopic, you will be able to identify reasons why levels of wellbeing among the middle class vary around the world.

The content in this subtopic is a summary of what you will find in the online resource.

Generally speaking, standards of living across the world have improved over time, and wellbeing has improved as well. As a result of this we are now seeing the emergence of a larger middle class, with aspirations and purchasing power — and not only in the developed but also the developing world.

To learn more about why levels of wellbeing vary among people who are considered 'middle class' around the world, go to your learnON resources at www.jacPLUS.com.au.

FIGURE 4 What are the advantages and disadvantages of an expanding middle class?

Contents

learnON

- 20.7.1 Defining 'middle class'
- 20.7.2 Factors affecting wellbeing in the middle class

on Resources

eWorkbook	Variations in wellbeing within the middle class (ewbk-10317)
Video eLesson	Variations in wellbeing within the middle class — Key concepts (eles-5313)
Interactivity	Variations in wellbeing within the middle class (int-8743)

20.8 SkillBuilder — Developing structured and ethical approaches to research

online only

LEARNING INTENTION

By the end of this subtopic, you will be able to create a structured research plan and conduct ethically sound research.

The content in this subtopic is a summary of what you will find in the online resource.

20.8.1 Tell me

What is a structured and ethical approach to research?

A structured and ethical approach to research involves organising your work clearly and meeting research standards.

20.8.2 Show me

How to develop a structured and ethical approach to research

Step 1

Determine the features that you are going to explore in the community (your primary data) or to research (your secondary data). Set up an inquiry question that covers all of the areas you wish to research.

Step 2

Undertake fieldwork to gather primary data on the factors. Photographs and sketches may be needed to support ideas. As you gather data, always maintain privacy and respect the wishes of your sources, never coerce anyone to answer a question, and never damage the environment.

Step 3

Summarise and paraphrase or quote and attribute your secondary data appropriately.

Step 4

Consider the following questions as you write your report:
- Are you using reliable, unbiased data?
- Have your own opinions influenced your conclusions?
- Have you used a wide range of data?
- Are you communicating your findings clearly?
- Have you acknowledged any potential problems with your data?
- Have you provided a full reference list?

20.8.3 Let me do it

learnON

Go to learnON to access the following additional resources to help you build this skill:
- a longer explanation of this skill and its application in Geography (Tell me)
- a video demonstrating the step-by-step process of this skill (Show me)
- an activity and interactivity for you to practise the skill (Let me do it)
- self-marking questions to help you understand and use the skill.

on Resources

eWorkbook	SkillBuilder — Developing structured and ethical approaches to research (ewbk-10321)
Video eLesson	SkillBuilder — Developing structured and ethical approaches to research (eles-5314)
Interactivity	SkillBuilder — Developing structured and ethical approaches to research (int-8744)

20.9 Population management and wellbeing

LEARNING INTENTION

By the end of this subtopic, you will be able to demonstrate the interconnection between population management and levels of development, and explain how population management can affect levels of wellbeing.

20.9.1 Population

The size of a family can directly affect everyone's wellbeing. In Australia we may be concerned with the cost of raising children. In other countries, children may make a contribution to family income by completing simple jobs such as collecting firewood. At a national scale, the numbers of children also affect the wellbeing of the country as a whole.

TABLE 1 Selected demographic characteristics for Japan and Kenya

Demographic characteristic	Japan	Kenya
Population (2019)	125 507 472	52 573 967
GNI per capita (US$)	44 810	4 430
Population aged 0-14	13%	42%
Population aged 15-64	61%	55%
Population aged 65+	26%	3%
Annual population growth	–0.1%	2.6%
Life expectancy	83	62
Fertility rate	1.4	4.3
Infant mortality	2 per 1000	36 per 1000
Long-term unemployed (% of labour force)	3.7%	9.2%

Source: World Bank

On a wider level within a country, the fertility rate and proportions of very young and very old people has an impact on the state of the economy, employment and many other factors that significantly influence wellbeing. This impact can be seen by comparing the demographics of Kenya and Japan (**TABLE 1**).

20.9.2 Kenya: response to a youthful population

Although Kenya's fertility rate has fallen substantially through the second half of the twentieth century, from 8.1 in 1965 to 3.5 in 2020, it is important to note that it still has a relatively high rate of population growth. By 2030, it is estimated that the population of Kenya will be over 65 million, and its population structure has a high proportion of young people (see **TABLE 1** and **FIGURE 2**). This increase will put pressure on Kenya's resources in terms of providing food, services and employment. With a predominantly rural population, the amount of arable land per person is falling.

FIGURE 1 Children in class at Maji Mazuri Children's Centre and School, Nairobi, Kenya

FIGURE 2 Population pyramid for Kenya, (a) 2019 and (b) 2050

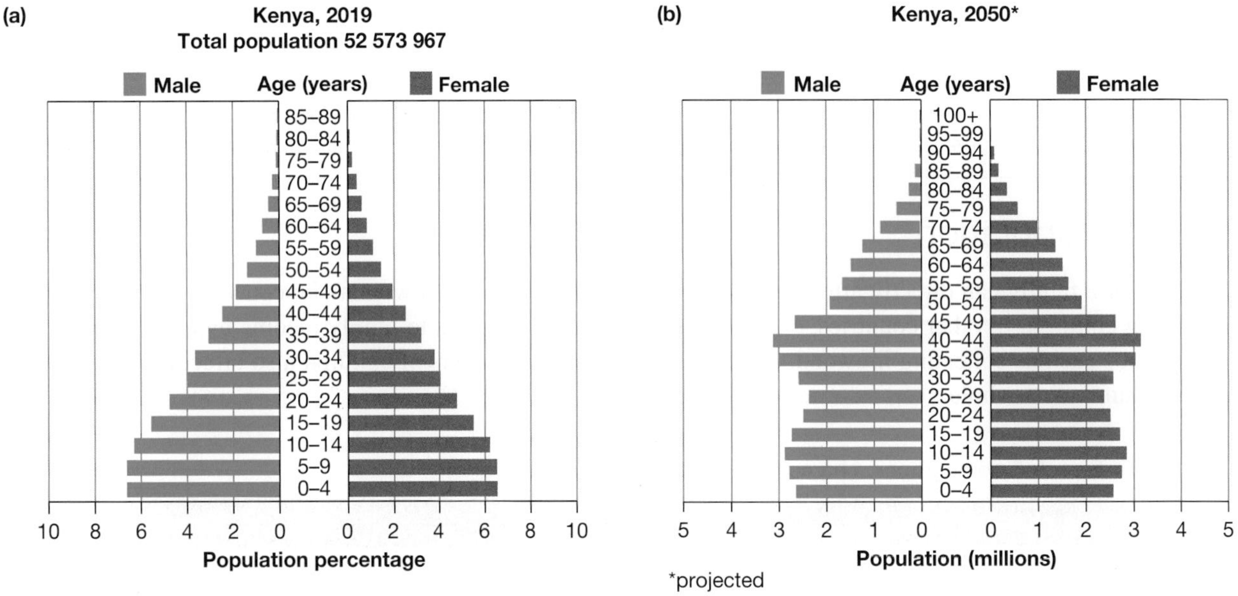

Source: © December 2019 by PopulationPyramid.net, made available under a Creative Commons license CC BY 3.0 IGO: http://creativecommons.org/licenses/by/3.0/igo/

Under Kenya's Vision 2030, a national framework for development, population management is an essential component of achieving wellbeing goals for health, poverty reduction, gender equality and environmental sustainability. The United Nations Population Fund (UNFPA) has been working with the Kenyan government since the 1970s to help improve wellbeing in the country. Between 2014 and 2018 UNFPA contributed $32.5 million to Kenya. This financed a range of maternal and newborn health services, services to prevent the contraction of HIV and sexually transmitted infections, advocacy for the education of girls, and elimination of gender-based violence.

Unfortunately, despite this work, there is still a huge unmet need for family planning in Kenya, particularly among the poorest women, where almost half report they have unplanned pregnancies.

20.9.3 Japan: response to an ageing population

Japan has one of the highest life expectancies in the world (84.7 in 2020). Combined with a very low fertility rate (1.37 in 2020), this has led to an ageing population. Over one-quarter of Japan's population is in the 65-plus age group (see **TABLE 1** and **FIGURE 4**). Fertility in Japan has been consistently below replacement level since the 1970s. A high standard of living, increased participation of women in the workforce, high costs of raising children and lack of childcare facilities have all contributed to this. Japan's total population is expected to decline from its current 126 million to an estimated 117 million in 2030.

FIGURE 3 Japan's population is ageing due to low fertility rates and high life expectancy

FIGURE 4 Population pyramid for Japan, (a) 2019 and (b) 2050

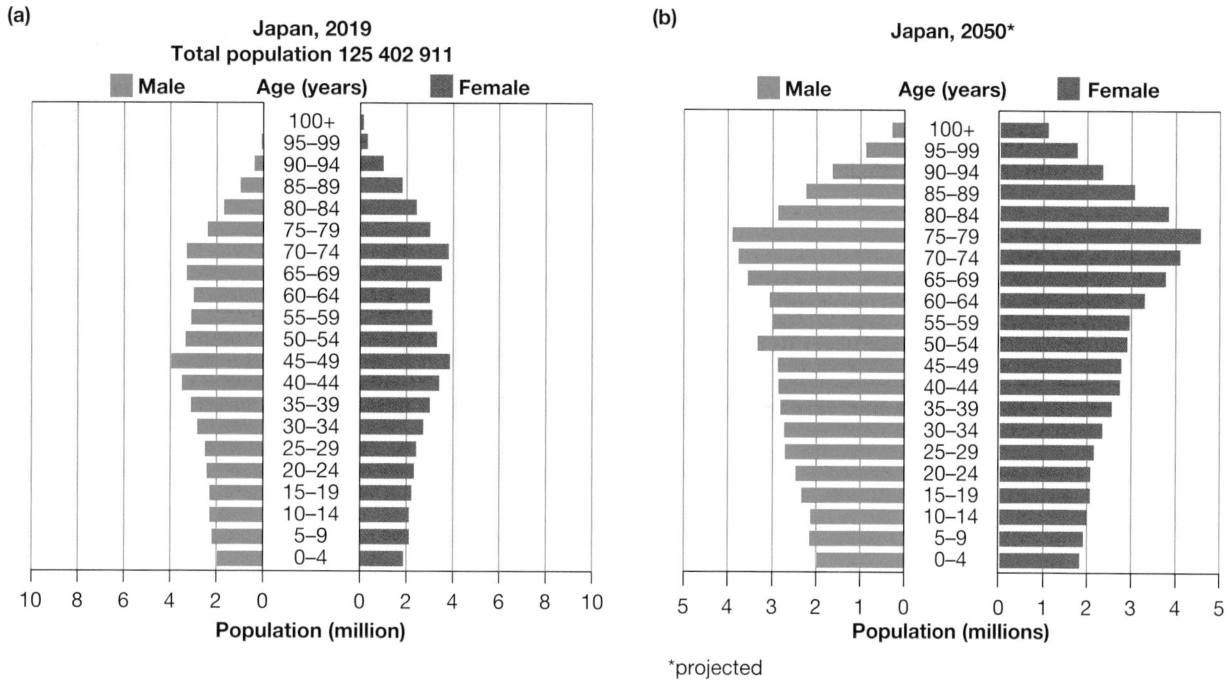

Source: © December 2019 by PopulationPyramid.net, made available under a Creative Commons license CC BY 3.0 IGO: http://creativecommons.org/licenses/by/3.0/igo/, United States Census Bureau

Accordingly, the workforce is expected to fall by 15 per cent over the next 20 years and halve in the next 50 years. In terms of impact on the Japanese economy, this means that by 2025 it is expected that there will be three working people for every two retirees. This means that the Japanese government also faces rising pension and healthcare costs. These economic concerns led to the Japanese government implementing a number of measures in 1994, such as subsidised child care and bonus payments for childbirth through a policy known as the Angel Plan (revised in 1999), as well as the Plus One policy aimed at encouraging parents to have one extra child.

These policies have been largely ineffective. Although the fertility rate rose slightly initially, it has remained well below replacement level — at approximately 1.4 in 2020. The Japanese government has historically been reluctant to use immigration to fill labour shortages; solutions may include increased female workforce participation, increased automation of the workforce and more working by older people.

on Resources

eWorkbook	Population management and wellbeing (ewbk-10325)
Video eLesson	Population management and wellbeing — Key concepts (eles-5315)
Interactivity	Analysing population pyramids — Japan and Kenya (int-8745)

20.9 ACTIVITIES

1. Use the **Aging population: Japan** weblink in your online Resources to determine what problems Japan faces with a large proportion of aged population.
2. Research a day in the life of a 15-year-old living in Kenya and compare this to that of a 15-year-old in Japan. Predict what their respective lives may look like by 2050, when they are in their mid-40s.

20.9 EXERCISE

Learning pathways

■ LEVEL 1	■ LEVEL 2	■ LEVEL 3
1, 2, 7, 11, 14	3, 5, 8, 9, 10, 15	4, 6, 12, 13, 16

Check your understanding

1. Identify three demographic characteristics for which Japan has a higher level than Kenya.
2. What areas will be under greater pressure as a result of Kenya's rapidly growing population? Identify two and describe how these will impact on wellbeing.
3. Calculate the change in percentage of aged population between 2019 and 2050 in both Kenya and Japan.
4. By what percentage is the Japanese population expected to decrease between 2019 and 2050?
5. Quantify the difference between the following:
   a. the life expectancy of a person in Kenya and in Japan in 2020
   b. the fertility rates in Kenya and in Japan in 2020.
6. What parts of Japanese society will be under greater pressure as a result of Japan's decreasing population? Identify two and describe how pressure on these will impact on wellbeing across the country.

Apply your understanding

7. How does an improvement in living conditions lead to a change in population structure?
8. Account and give reasons for the variation in shape of the population pyramids for Japan and Kenya in 2019 and 2050.
9. Explain the interconnection between educating girls and lower fertility rates.
10. What problems does the Japanese government face because of its ageing population?
11. Consider the challenges faced by Kenya and Japan because of their population structure.
    a. How do these issues affect the wellbeing of people in those countries?
    b. Which problem do you think is most significant for each country? Present an argument for which of the impacts of their population structures will have the greatest impact on wellbeing.
12. Consider the actions taken by Kenya to lower fertility rates. Which do you think will be the most successful?
13. Consider the actions taken by Japan to increase fertility rates. Which do you think will be the most successful?

Challenge your understanding

14. Traditional attitudes in Japan regarding male and female roles in a family, such as women caring for children and completing the housework, are cited as causes of the ageing population.
    a. Explain how these factors are connected.
    b. What strategies could a government implement to change long-standing cultural expectations?
15. Predict what impact a growing younger population would have on the natural environment of a country. How might this change affect wellbeing?
16. Do you think that the impacts of an ageing population or a growing young population are felt equally across all parts of a society? Or do some groups of people suffer greater negative impact while others benefit? Choose one factor that affects wellbeing to use as your basis for your answer.

To answer questions online and to receive **immediate feedback** and **sample responses** for every question, go to your learnON title at www.jacplus.com.au.

20.10 Gender, development and wellbeing

LEARNING INTENTION

By the end of this subtopic, you will be able to demonstrate the interconnection between levels of development and women's wellbeing.

20.10.1 Women's health and wellbeing

Approximately every 90 seconds, somewhere around the world, a woman dies from complications from pregnancy or childbirth. Most of these deaths are from preventable complications such as severe bleeding, infections and complications from unsafe abortions. The incidence of **maternal mortality** and related illness is interconnected to poverty and lack of accessible and affordable quality health care.

FIGURE 1 For every 100 000 births in India in 2019, 113 mothers died from complications of pregnancy or childbirth.

Eighty-five per cent of maternal deaths are in sub-Saharan Africa and southern Asia, with the former accounting for over half of all these deaths. The highest maternal mortality rates are recorded in South Sudan (1150 deaths per 100 000 live births) and Chad (1140 deaths per 100 000 live births).

At a national scale, two countries account for one-third of global maternal deaths: India at 19 per cent (56 000 annually) followed by Nigeria at 14 per cent (40 000 annually).

Sustainable Development Goal (SDG) 3 set targets for 2030:

- reducing the global maternal mortality ratio to less than 70 per 100 000 live births
- ensuring universal access to sexual and reproductive health-care services, including for family planning, information and education, and the integration of reproductive health into national strategies and programs.

Maternal mortality fell by 45 per cent between 1990 and 2015, with 10 countries meeting the MDG target of a three-quarter reduction. However, globally the target was not met, particularly in countries in southern Africa where AIDS had a major impact during that period. In 2017, the global rate was still higher than this target, at 211 deaths per 100 000 live births.

maternal mortality the death of a woman while pregnant or within 42 days of termination of pregnancy

20.10.2 Maternal mortality in India

In Australia, where maternal mortality rates are low (5 deaths per 100 000 women giving birth in 2019) being pregnant and giving birth are relatively safe in comparison to some parts of the world, in which giving birth could be one of the most life-threatening activities in which woman can engage.

India accounts for the single biggest percentage of global maternal deaths. On average, maternal mortality rates have declined by over 15 per cent in the five-year period from 2011 to 2015. It then fell again from 174 deaths in 2015 to 113 deaths per 100 000 births by 2019. However, there is substantial variation within India, as **FIGURE 2** indicates.

Maternal mortality is strongly interconnected with poverty in both rural and urban areas: places with poor provision of sanitation and a lack of affordable health services are associated with high levels of maternal mortality. In addition, women are likely to be less well nourished than males in a household, especially in communities where men are more commonly the primary income earner. According to the 2011 Indian **Census**, women also have much lower literacy levels — a 65 per cent literacy rate compared to 82 per cent for men — so they are less likely to be able to access information on health and contraception. In addition to these factors, women in rural areas may also have long distances to travel to seek medical attention.

FIGURE 2 Maternal mortality rates in India

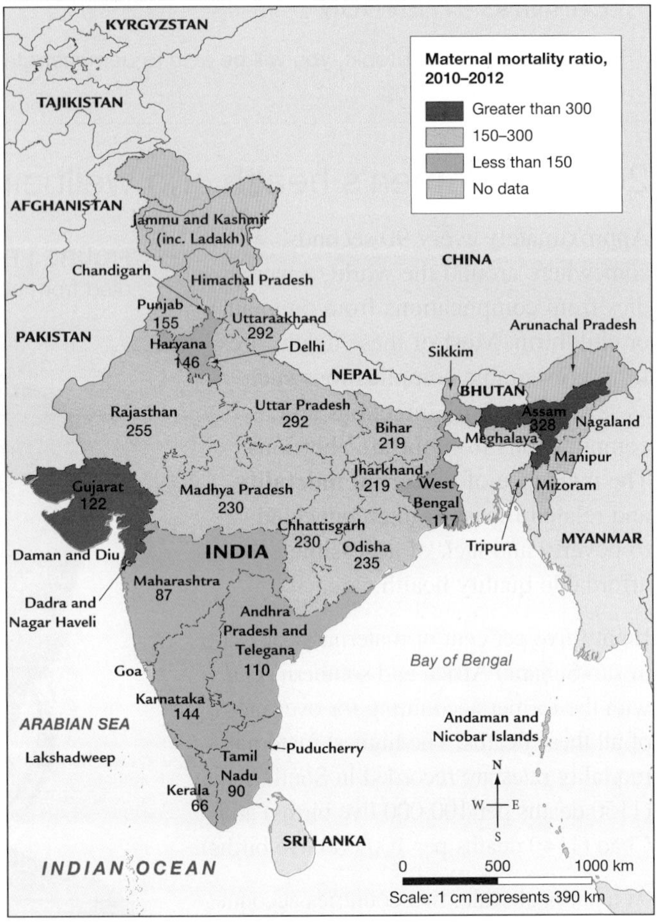

Source: Published and issued by Office of the Registrar General, India, Ministry of Home Affairs http://www.censusindia.gov.in/vital_statistics/SRS_Bulletins/MMR_Bulletin-2010-12.pdf

Note: Most recent Indian census data by state, 2011

FIGURE 3 Rural pregnancy clinic in India

census a survey or count usually conducted of a population looking at a number of specific characteristics such as age, income, occupation, residence, health, religion and many more.

The government of India launched the national Rural Health Mission in 2005, with a specific focus on maternal health. Efforts have been focused on those districts that account for 70 per cent of all infant and maternal deaths. Under this program, community workers have been trained to deliver babies, and 10 million women have been provided with a cash incentive to enable them to give birth in clinics rather than at home. Maternal mortality has fallen, but Human Rights Watch reports that many women are being charged for services as they are unaware of these entitlements.

20.10.3 Cultural factors

A related issue for pregnant women in India is the pressure to produce a son. Census data in 2011 revealed the number of female children (0–6 years) has decreased from 927 to 914 girls per 1000 boys in the past decade, despite some overall improvement in the **sex ratio** across all age groups (see **FIGURE 4**).

Traditionally in Indian society, male children were preferred over female children: sons are seen as the breadwinners who carry the family name, whereas daughters are often perceived as an economic burden who will eventually marry into another family. This is still the case in many communities and regions.

Although **female infanticide** is illegal, use of ultrasound for sex-determination tests has led to sex-selective abortions, with 500 000 more girls than boys aborted each year out of a total of 6.7 million abortions.

The pressure to produce a son means that many Indian women have multiple pregnancies, thereby increasing their risk of maternal mortality over their reproductive years. **FIGURES 2** and **4** indicate a strong interconnection between the places with high maternal mortality and those with a large imbalance in the sex ratio.

This perception of women also shapes attitudes to educating girls and other factors that lift wellbeing indicators for girls and women throughout their lives.

FIGURE 4 Variation in sex ratio within India

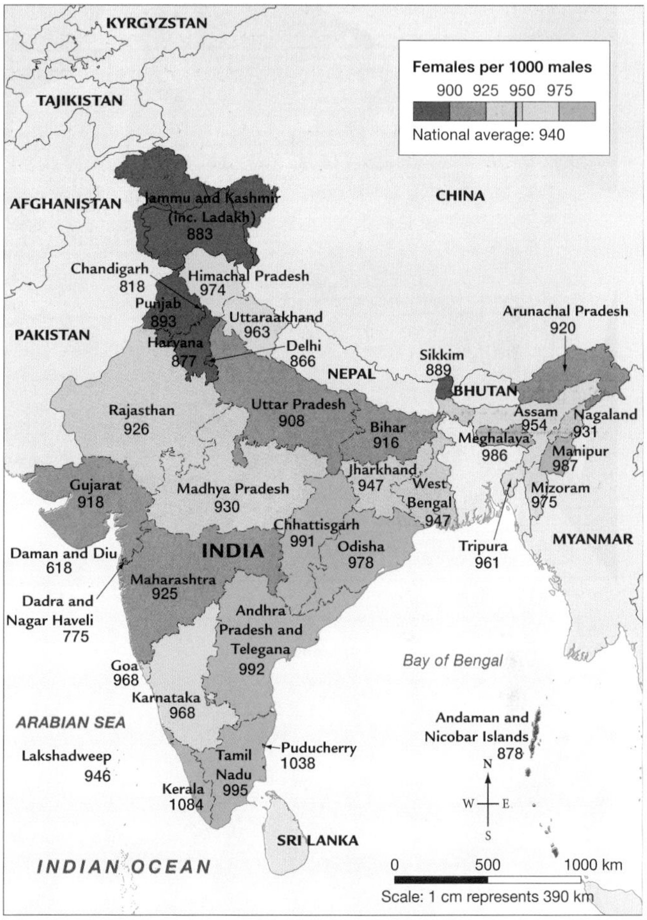

Source: Government of India, Ministry of Home Affairs, Office of Registrar General

Note: Most recent Indian census data by state, 2011

sex ratio the number of males per 1000 females

female infanticide the killing of female babies, either via abortion or after birth

FIGURE 5 Son preference has resulted in an imbalance in India's sex ratio.

Resources

eWorkbook	Gender, development and wellbeing (ewbk-10329)
Video eLesson	Gender, development and wellbeing — Key concepts (eles-5316)
Interactivity	Gender, development and wellbeing (int-8746)

20.10 ACTIVITIES

1. Use the **Women's health** weblink in your online Resources to learn more about the UN Global Strategy for women and children's health. Note any key information provided.
2. Using the blank **World map** in your online Resources:
   a. indicate the regions that account for 85 per cent of maternal deaths
   b. label the countries that have:
      i. the highest risk of maternal mortality
      ii. the greatest number of total deaths.

20.10 EXERCISE

Learning pathways

■ LEVEL 1	■ LEVEL 2	■ LEVEL 3
1, 2, 7, 11, 15	3, 4, 8, 9, 14	5, 6, 10, 12, 13

Check your understanding

1. Define *maternal mortality*.
2. List four key factors that contribute to India's maternal mortality rate being so high.
3. Outline two factors that help to improve maternal mortality rates.
4. What was the difference between Australia's and India's maternal mortality rates in 2019? Use data in your answer.
5. Describe the variations in maternal mortality in different parts of India as shown in **FIGURE 2**.
6. Identify two reasons why families in India might have a preference for male children.

Apply your understanding

7. Explain the interconnection between poverty and maternal mortality.
8. Explain how the data collected in a census can help governments formulate strategies to improve maternal mortality and women's wellbeing levels.
9. Why might cash payments be a good strategy to encourage women in rural areas with high maternal mortality rates to attend a birth clinic?
10. Areas with poor sanitation typically have higher maternal mortality rates. Do you think this is because the poor sanitation is a direct cause of women's deaths, or is it more likely to be an indicator of something else?
11. Explain why a larger proportion of the babies born in India are boys.
12. Of the factors that affect maternal mortality rates, which do you think would be the most significant contributor to the rates in India?

Challenge your understanding

13. Predict the shape of India's population pyramid if the trends in India's sex ratio continue. What are some long-term implications of this trend?
14. Suggest measures that could be introduced by the Indian government to help Indian parents see the value of female babies as equal to that of males.
15. The Indian census is held every 10 years; in Australia we hold it every 5 years. Why do you think countries have different approaches? Which is better for mapping trends in factors that affect wellbeing and why?

To answer questions online and to receive **immediate feedback** and **sample responses** for every question, go to your learnON title at www.jacplus.com.au.

20.11 Investigating topographic maps — Spatial variations in wellbeing in Tokyo, Japan

LEARNING INTENTION

By the end of this subtopic, you will be able to discuss ways that densely populated urban areas might influence wellbeing levels.

20.11.1 Japan

Japan, as a developed nation with an ageing population, faces a very specific set of issues pertaining to development and wellbeing in both rural and urban areas. The greater Tokyo–Yokohama conurbation is the largest urban area in the world with a population of just under 38.2 million people in 2020.

FIGURE 1 Images of Tokyo (a) Aerial view of the famous Shibuya Crossing, one of the busiest crossings in the world (b) Shinjuku's Kabuki-cho district, known as Tokyo's 'nightlife' area (c) Meiji-Jingu Shinto Shrine and surrounding parkland in central Tokyo near Shinjuku business district (d) Residential streets, Daizawa, Setagaya-ku, Tokyo (e) Shimokitazawa district, Tokyo, known for its independent fashion and design stores, and cafes (f) Shinagawa station, Tokyo, during morning rush hour; just under 400 000 people pass through the station on an average working day.

FIGURE 2 Topographic map extract of Tokyo, 2021

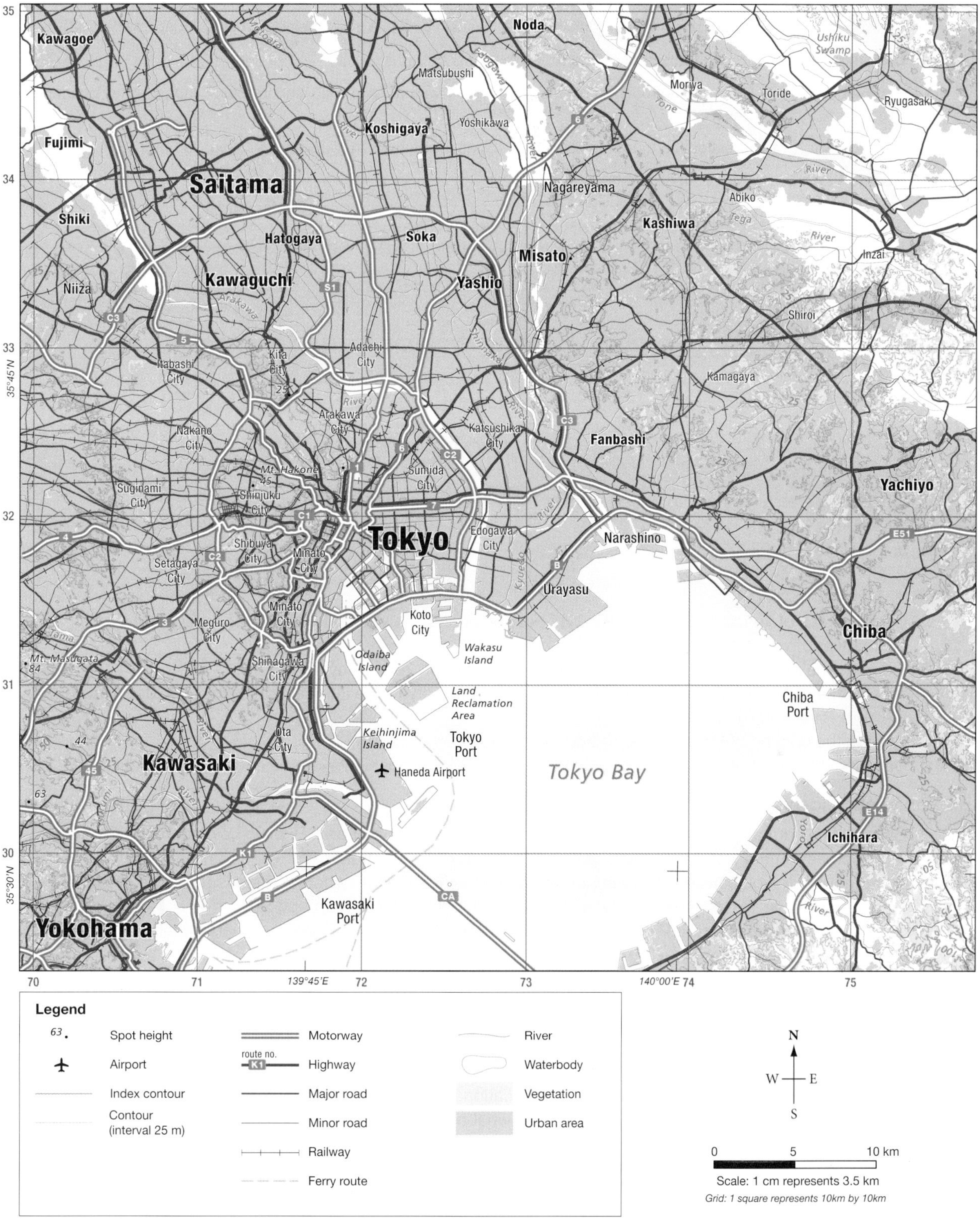

Source: Map data © OpenStreetMap contributors, https://openstreetmap.org. Data is available under the Open Database Licence, https://opendatacommons.org/licenses/odbl/; urban area data sourced from Atlas of Human Expansion (2016); elevation data sourced from USGS.

Resources

eWorkbook Investigating topographic maps — Spatial variations in wellbeing in Tokyo, Japan (ewbk-10333)

Digital document Topographic map of Tokyo, Japan (doc-36373)

Video eLesson Investigating topographic maps — Spatial variations in wellbeing in Tokyo, Japan — Key concepts (eles-5317)

Interactivity Investigating topographic maps — Spatial variations in wellbeing in Tokyo, Japan (int-8747)

Google Earth Tokyo, Japan (gogl-0063)

20.11 EXERCISE

Learning pathways

■ LEVEL 1	■ LEVEL 2	■ LEVEL 3
1, 4, 7, 12	2, 3, 5, 8	6, 9, 10, 11

Check your understanding

1. Identify the area references for these parts of Tokyo mentioned in **FIGURE 1**.
   a. Shibuya City
   b. Shinjuku City
   c. Setagaya City
   d. Shinagawa City
2. Identify areas of vegetation in **FIGURE 2**. What observations can you make about the extent and spatial distribution of vegetation in Tokyo?
3. Estimate the length of the natural (not built) coastline around Tokyo Bay.
4. What is the distance by road from Mt Hakone in Shinjuku City to the intersection of the E14 Motorway and the Yoro River in Ichikawa?
5. What is the bearing of Haneda Airport from Noda?
6. Describe the topography of the greater Tokyo area, providing specific examples from **FIGURE 2**.

Apply your understanding

7. Choose one natural and one human feature shown in **FIGURES 1** and/or **2**. Explain how each of these features might impact on the wellbeing of people living in the surrounding areas.
8. What factors do you consider to be important for wellbeing? Identify whether there is evidence of these factors existing in Tokyo, based on the information provided in **FIGURES 1** and **2**.
9. Would wellbeing be greater for you in Tokyo than where you live? Justify your answer by using examples from **FIGURES 1** and **2**.
10. Identify from **FIGURE 2** which areas of the city might be reclaimed land. Explain the reasons why you think that specific area is reclaimed based on evidence from **FIGURE 2**.

Challenge your understanding

11. Suggest which area shown in **FIGURE 2** would be the most likely to be reclaimed for more residential land. Justify your response. How would living in this location impact on the wellbeing of the residents?
12. Based on your observations of **FIGURES 1** and **2**, what challenges would living in Tokyo have posed for its residents' wellbeing during the COVID-19 epidemic?

To answer questions online and to receive **immediate feedback** and **sample responses** for every question, go to your learnON title at www.jacplus.com.au.

20.12 Thinking Big research project — The displaced Rohingya children

online only

The content in this subtopic is a summary of what you will find in the online resource.

Scenario

Between August 2017 and September 2019, nearly 915 000 people from Myanmar, mostly from the western state of Rakhine, crossed the border into Bangladesh. These are the displaced Rohingya Muslims, fleeing the conflict between their ethnic group and the Myanmar armed forces. Most of these people arrived in Bangladesh with few or no possessions and are reliant on humanitarian aid to survive. More than 380 000 children are caught in this situation.

Kutupalong–Balukhali Expansion Site, in Cox's Bazaar District, has become the largest refugee site, with a population of over 730 000 displaced Rohingya persons.

Task

learnON

Research the situation in the Kutupalong–Balukhali refugee camp in Bangladesh. Create an annotated photographic essay to outline the conditions and challenges to daily wellbeing that these children face. Present your photo essay to the class as an oral presentation.

Go to learnON to access the resources you need to complete this research project.

on Resources

ProjectsPLUS Thinking Big research project — The displaced Rohingya children (pro-0220)

20.13 Review

20.13.1 Key knowledge summary

20.2 Variations in wellbeing between countries

- Life expectancy is the length of time we can expect to live from birth. The average global figure has been steadily rising.
- Rates of population change (births and deaths) have short-term and long-term impacts on wellbeing.

20.3 Variations in wellbeing within countries

- Wellbeing in India is affected by its large population (e.g. decreasing death rates with slower declines in birth rates).
- Literacy, poverty and birth rates vary between urban and rural regions.

20.5 Internal reasons for variations in wellbeing

- Internal factors take place within a country.
- Population growth: access to better education, healthcare and family planning leads to improved wellbeing.
- Geographical location: coastlines improve transport and trade; infectious diseases thrive in tropical climates.
- Political instability: reduces access to food/healthcare/personal safety; can divert funds from essential services and destroy infrastructure; some persecuted people may need to flee for their safety.
- Environmental degradation: reduces access to food and arable land; increases erosion, pollution, salinity.
- Resources: determine the amount of food available; provide economic benefits from export.

20.6 External reasons for variations in wellbeing

- External factors are outside the control of a country.
- Colonisation: drained resources (natural and human); created oppression and disadvantage.
- Trade: covering a shortfall in production of essentials (such as food) leads to national debt; some products are more lucrative than others, leading to inequalities.
- Transnational corporations: use cheap labour in developing countries; most profits go to developed countries.
- Aid: can help to redress imbalance, but is vulnerable to corruption and may be ineffectual in the long term without building local expertise and training.

20.7 Variations in wellbeing within the middle class

- Middle class is often defined as anyone outside the bottom or top quintiles of income in a society. This segment of society is growing globally but shrinking in the developed world, creating greater strain on the environment.
- Factors affecting wellbeing of the middle class include economic indicators related to growth/decline such as income, consumer culture, employment rates and government services.

20.9 Population management and wellbeing

- The sizes of families and populations impact wellbeing and have a significant impact on age distribution, which affects key wellbeing indicators
- Kenya has a large, young population due to high birth rates and low life expectancy, which places stress on the country's environmental resources, arable land, health and education services, and economy.
- Japan has an ageing population because of low fertility rates and high life expectancy, which leads to increased healthcare costs, lower income tax revenue and fewer workers.

20.10 Gender, development and wellbeing

- Factors affecting levels of wellbeing for women include complications of pregnancy and childbirth.
- Maternal mortality rates are connected to poverty, healthcare and education.
- In some cultures, male children are more prized than female children, creating gender imbalance and a greater likelihood of women having multiple pregnancies, which increases their risk of complications.

20.13.2 Key terms

census a survey or count usually conducted of a population looking at a number of specific characteristics such as age, income, occupation, residence, health, religion and many more.

female infanticide the killing of female babies, either via abortion or after birth

fertility rate the average number of children born per woman

life expectancy the number of years a person can expect to live, based on the average living conditions within a country

maternal mortality the death of a woman while pregnant or within 42 days of termination of pregnancy

natural increase the difference between the birth rate (births per thousand) and the death rate (deaths per thousand). This does not include changes due to migration

quintile any of five equal groupings used to measure and compare values

replacement rate the number of children each woman would need to have in order to ensure a stable population level — that is, to 'replace' the children's parents. This fertility rate is 2.1 children.

sex ratio the number of males per 1000 females

Transnational corporations (TNCs) a large business that operates across multiple countries, but is run from a central office or offices in a single developed country

20.13.3 Reflection

Complete the following to reflect on your learning:

Revisit the inquiry question posed in the Overview:

The world's population is constantly increasing. Can we fit so many people in the space we have without affecting our quality of life?

1. Now that you have completed this topic, what is your view on the inquiry question above? Discuss with a partner. Has your learning in this topic changed your view? If so, how?
2. Write a paragraph in response to the inquiry question, outlining your views.

Subtopic	Success criteria			
20.2	I can describe spatial variations between countries.			
20.3	I can describe spatial variations within countries.			
20.4	I can use Excel to construct population profiles.			
20.5	I can outline internal reasons for spatial variations.			
20.6	I can outline external reasons for spatial variations.			
20.7	I can describe consequences of spatial variations for the middle class.			
20.8	I can develop structured and ethical approaches to research.			
20.9	I can explain population's effect on development and wellbeing.			
20.10	I can compare wellbeing between Japan and Kenya.			
20.11	I can explain gender's effect on development and wellbeing.			

on Resources

eWorkbook Chapter 20 Student learning matrix (ewbk-8596)
Chapter 20 Reflection (ewbk-8597)
Chapter 20 Extended writing task (ewbk-8598)

Interactivity Chapter 20 Crossword (int-8748)

ONLINE RESOURCES

on Resources

Below is a full list of **rich resources** available online for this topic. These resources are designed to bring ideas to life, to promote deep and lasting learning and to support the different learning needs of each individual.

eWorkbook

- **20.1** Chapter 20 eWorkbook (ewbk-8112) ☐
- **20.2** Variations in wellbeing between countries (ewbk-10297) ☐
- **20.3** Variations in wellbeing within countries (ewbk-10301) ☐
- **20.4** SkillBuilder — Using Excel to construct population profiles (ewbk-10305) ☐
- **20.5** Internal reasons for variations in wellbeing (ewbk-10309) ☐
- **20.6** External reasons for variations in wellbeing (ewbk-10313) ☐
- **20.7** Variations in wellbeing within the middle class (ewbk-10317) ☐
- **20.8** SkillBuilder — Developing structured and ethical approaches to research (ewbk-10321) ☐
- **20.9** Population management and wellbeing (ewbk-10325) ☐
- **20.10** Gender, development and wellbeing (ewbk-10329) ☐
- **20.11** Investigating topographic maps — Spatial variations in wellbeing in Tokyo, Japan (ewbk-10333) ☐
- **20.13** Chapter 20 Student learning matrix (ewbk-8596) ☐
  Chapter 20 Reflection (ewbk-8597) ☐
  Chapter 20 Extended writing task (ewbk-8598) ☐

Sample responses

- **20.1** Chapter 20 Sample responses (sar-0168) ☐

Digital documents

- **20.10** Blank World Map, Countries (doc-36711) ☐
- **20.11** Topographic map of Tokyo, Japan (doc-36373) ☐

Video eLessons

- **20.1** A long life (eles-1714) ☐
  Spatial variations in human wellbeing — Photo essay (eles-5308) ☐
- **20.2** Variations in wellbeing between countries — Key concepts (eles-5309) ☐
- **20.3** Variations in wellbeing within countries — Key concepts (eles-5310) ☐
- **20.4** SkillBuilder — Using Excel to construct population profiles (eles-1758) ☐
- **20.5** Internal reasons for variations in wellbeing — Key concepts (eles-5311) ☐
- **20.6** External reasons for variations in wellbeing — Key concepts (eles-5312) ☐
- **20.7** Variations in wellbeing within the middle class — Key concepts (eles-5313) ☐
- **20.8** SkillBuilder — Developing structured and ethical approaches to research (eles-5314) ☐
- **20.9** Population management and wellbeing — Key concepts (eles-5315) ☐
- **20.10** Gender, development and wellbeing — Key concepts (eles-5316) ☐
- **20.11** Investigating topographic maps — Spatial variations in wellbeing in Tokyo, Japan — Key concepts (eles-5317) ☐

Interactivities

- **20.2** Global life expectancy (int-5603) ☐
  Global distribution of birth rates (int-7976) ☐
  Global distribution of death rates (int-7977) ☐
- **20.3** Proportion of children 0–5 years to total population, India, 2011 (int-8740) ☐
  Literacy rates (percentage) in India, 2011 (int-8738) ☐
  Poverty levels in India (int-8739) ☐
- **20.4** SkillBuilder — Using Excel to construct population profiles (int-3376) ☐
- **20.5** Internal reasons for variations in wellbeing (int-8741) ☐
- **20.6** External reasons for variations in wellbeing (int-8742) ☐
- **20.7** Variations in wellbeing within the middle class (int-8743) ☐
- **20.8** SkillBuilder — Developing structured and ethical approaches to research (int-8744) ☐
- **20.9** Analysing population pyramids — Japan and Kenya (int-8745) ☐
- **20.10** Gender, development and wellbeing (int-8746) ☐
- **20.11** Investigating topographic maps — Spatial variations in wellbeing in Tokyo, Japan (int-8747) ☐
- **20.13** Chapter 20 Crossword (int-8748) ☐

ProjectsPLUS

- **20.12** Thinking Big research project — The displaced Rohingya children (pro-0220) ☐

Weblinks

- **20.2** MDGs (web-0175) ☐
  Demographic indicators (web-1242) ☐
  Gauging interconnections (web-2351) ☐
  SDGs (web-5971) ☐
- **20.3** Poverty challenge (web-2359) ☐
  Census India (web-2361) ☐
- **20.5** Displacement (web-1244) ☐
  Gapminder (web-2962) ☐
- **20.7** OECD (web-4329) ☐
- **20.9** Aging population Japan (web-0170) ☐
- **20.10** Women's health (web-3341) ☐

Fieldwork

- **20.5** Advantage and disadvantage survey (fdw-0047) ☐

Google Earth

- **20.11** Tokyo, Japan (gogl-0063) ☐

Teacher resources

There are many resources available exclusively for teachers online.

21 Human wellbeing in Australia

INQUIRY SEQUENCE

To access a pre-test with **immediate feedback** and **sample responses** to every question in this chapter, select your learnON format at www.jacplus.com.au.

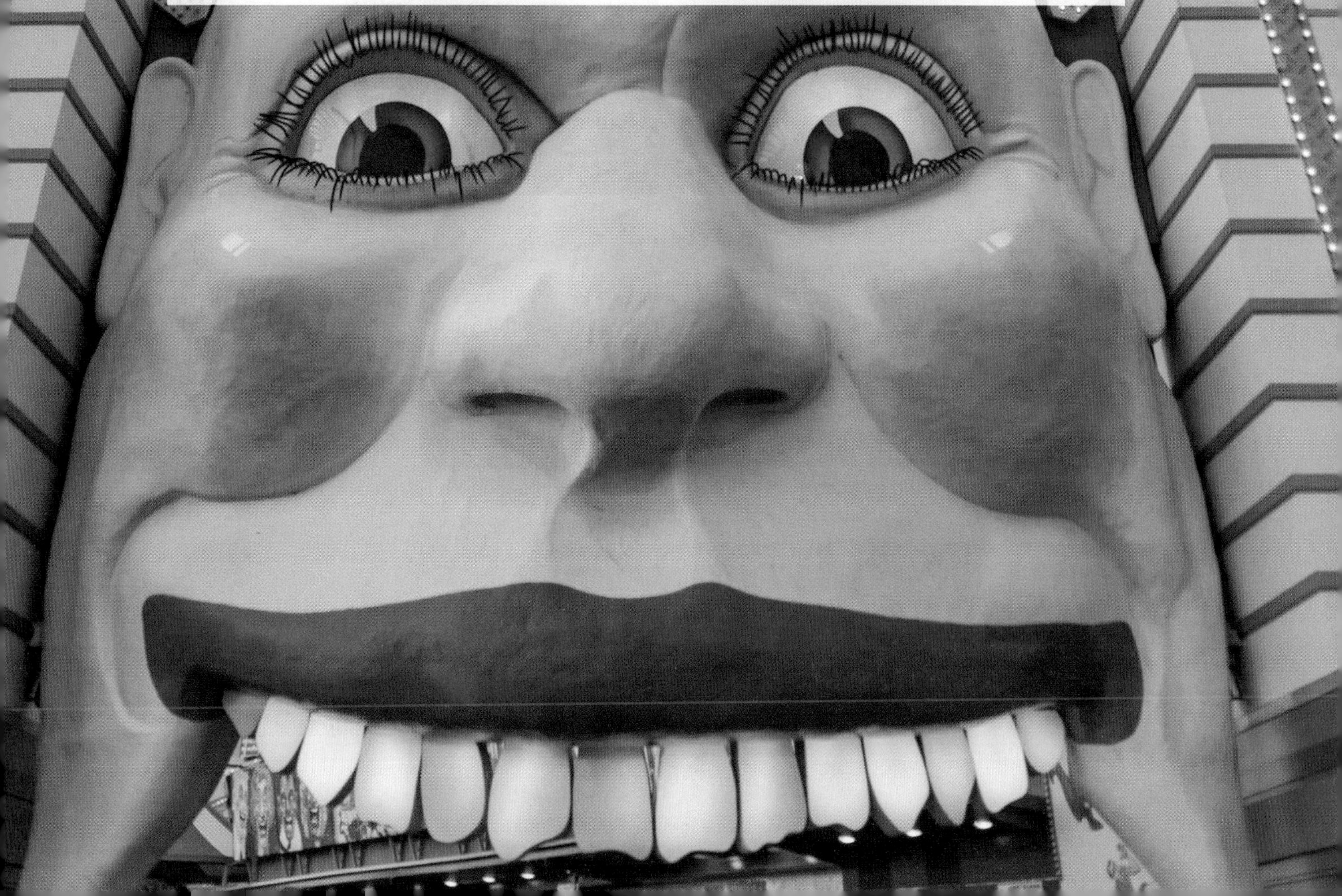

21.1 Overview

Numerous **videos** and **interactivities** are embedded just where you need them, at the point of learning, in your learnON title at www.jacplus.com.au. They will help you to learn the content and concepts covered in this topic.

How is wellbeing impacted by place and space, and what accounts for the variations in wellbeing within Australia?

21.1.1 Introduction

We have all travelled to different places during our lives. These places may be within our own suburb, within our town or city, in another state of Australia, or in another country. Although we tend to be more conscious of differences between our own country and other countries, **variations** also occur at local and regional scales. Variation may occur between urban and rural environments, or even within the one city or town, and levels of wellbeing can vary significantly because of these variations. Think about how the various spaces and places near where you live reflect or affect levels of wellbeing. What are the possible reasons for these differences?

variation (in terms of wellbeing) differences experienced between groups of people due to factors such as their location, demographic or socioeconomic status

FIGURE 1 Contrast between two places in NSW: (a) Bourke and (b) Sydney

STARTER QUESTIONS

1. What are the characteristics of the particular rural or urban environment in which you live (your suburb, town or place)?
2. In what ways do these characteristics vary from those of the neighbouring environments, whether it is a farm, town or suburb?
3. What interconnection is there between these characteristics and the wellbeing of the people in these places?
4. Why do similarities or variations in wellbeing occur at a local or regional scale?
5. Watch the **Human wellbeing in Australia — Photo essay** video in your online Resources. How can wellbeing vary between groups of people living in the same place?

on Resources

eWorkbook	Chapter 21 eWorkbook (ewbk-8113)
Video eLessons	The other side of life (eles-1715) Human wellbeing in Australia — Photo essay (eles-5318)

21.2 Characteristics of Australia's population

LEARNING INTENTION

By the end of this subtopic, you will be able to describe the spatial patterns and characteristics of Australia's population and explain why Australia's population is continuing to grow.

21.2.1 The spatial distribution of Australia's population

Within Australia, variations in wellbeing can occur for many different reasons. In order to understand the variations in wellbeing, you must first understand the spatial patterns of Australia's population. Before British contact in 1788, Aboriginal Peoples lived within specific geographical areas known as Nations and rarely moved from those areas. This spatial distribution still exists today; however, many individuals have relocated to more urbanised areas for a variety of reasons.

Since 1788, Australia's population has increased dramatically. According to the 2016 Census, Australia's population in that year was 23 401 892. Statistically speaking, a typical Australian in that year would be female, born in Australia, aged 38 years, and living in a household consisting of a couple and children (although only 44.7% of families in Australia are couples with children). Of course, Australia's demographic characteristics are much more diverse than this. To what extent do you fit the 'typical' profile?

Most of Australia's population is concentrated in coastal regions in the south-east and east and, to a lesser extent, in the south-west. The population within these regions is concentrated in urban centres, particularly the capital cities (see **FIGURE 1**).

FIGURE 1 Australia's population distribution

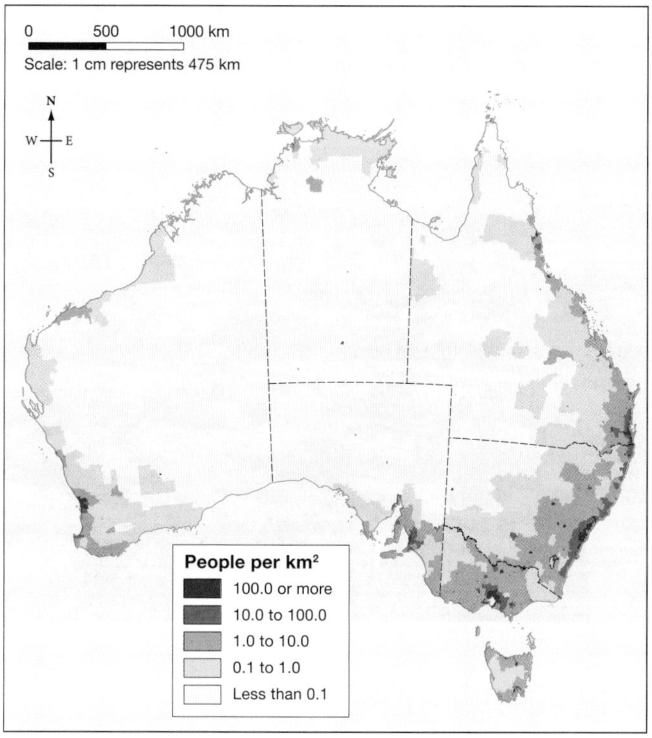

Source: Spatial Vision, 2021

This disparity between the high population densities in coastal areas and low population densities inland affect the wellbeing of residents in many ways: positive and negative. Consider and compare the facilities and resources likely to be found in the two places shown in **FIGURES 2** and **3**.

FIGURE 2 Aerial view over St Clair in Sydney's western suburbs

FIGURE 3 Aerial view over Tibooburra in northwestern NSW

21.2.2 General population trends

Australia's population has increased considerably over time and is continuing to grow. Between 2011 and 2016, the population increased by 1 894 182 people, or 8.8 per cent (see **FIGURE 4**). Our typical population growth, outside pandemic border closures, is due to immigration rather than **natural increase**. The level of migration is set annually by the Federal Government and is currently around 160 000 people per year.

natural increase the changes in population as a result of the difference between birth rates and death rates

FIGURE 4 Australia's changing population growth

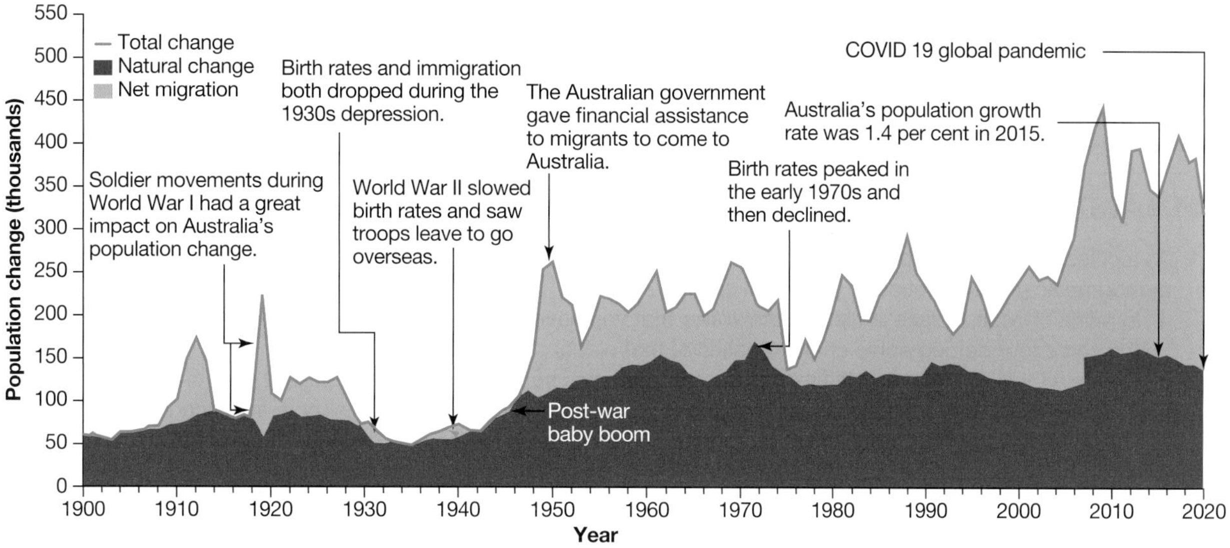

Source: © Commonwealth of Australia. Australian Bureau of Statistics (2020) 'Table: Components of annual population change(a)', National, state and territory population, accessed 20 April 2020.

Our rate of fertility has declined steadily since the 1970s and is now well below replacement rate. Despite attempts to increase the number of children via a Federal Government baby bonus of approximately $5000 per baby, which was in place between 2003 and 2013, our fertility rate was 1.65 in 2019.

The decline in fertility and increased life expectancy has resulted in an **ageing population** (see **FIGURE 5**). The proportion of the Australian population aged 65 years and over increased from 11.8 per cent to 15.7 per cent between 1994 and 2019.

ageing population the average age of a population rising as a result of increases in life expectancy. This can result in social issues such as a need for increased government spending on health care for the elderly population.

FIGURE 5 Australia's changing population structure

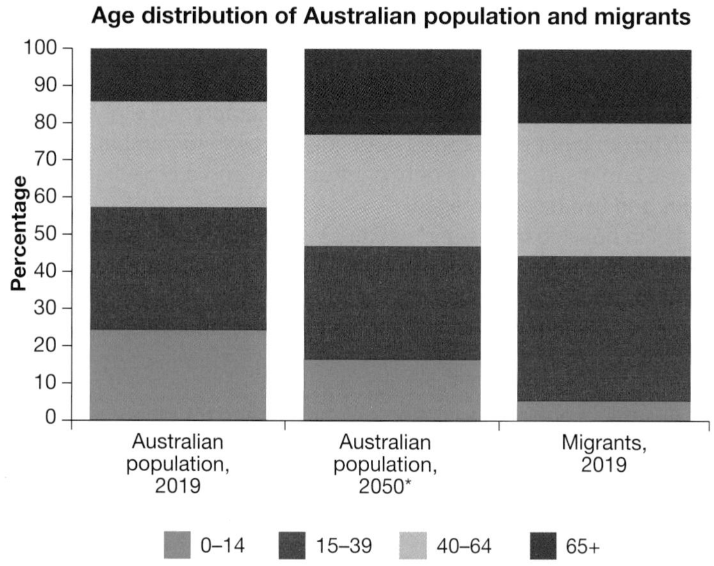

*projected

Source: Unpublished ABS data, Treasury projections and Australian Bureau of Statistics (2020), Migration Australia, 2018-19 financial year, accessed 20 April 2021

on Resources

eWorkbook The characteristics of Australia's population (ewbk-9802)

Video eLesson The characteristics of Australia's population — Key concepts (eles-5319)

Interactivity The characteristics of Australia's population (int-8749)

21.2 ACTIVITIES

1. As a class, use the **Australian Bureau of Statistics** weblink in your Resources to access statistics on four demographic characteristics of your local area.
   a. In pairs, choose a place in New South Wales that you have visited or would like to visit. Find the data for the same four demographic characteristics of that place.
   b. Create a table to show the differences in characteristics between your local area and your chosen area.
2. Investigate social and/or health issues that impact rural and remote communities more significantly than urban populations of Australia. Consider the factors that created these issues and brainstorm potential solutions. Discuss the issues, the causes and your proposals as a class.

21.2 EXERCISE

Learning pathways

■ LEVEL 1	■ LEVEL 2	■ LEVEL 3
2, 3, 7, 9, 15	1, 5, 8, 11, 14	4, 6, 10, 12, 13

Check your understanding

1. Outline the general spatial patterns of Australia's population.
2. List three factors that have accounted for Australia's changing population growth over time.
3. Summarise how Australia's population is expected to change in the future.
4. Define the term *ageing population* and give one reason why this is occurring in Australia.

Apply your understanding

5. Compare the population distribution of the east and west coast of Australia. Use statistics from **FIGURE 1** to support your answer.
6. Account for the variation in Australia's population distribution.
7. Explain how a declining fertility rate changes the population characteristics of a country.
8. How has the Australian government encouraged population growth to combat the declining fertility rate?
9. What are the advantages and disadvantages of a 'big Australia' and a projected population of 35 million? Explain two advantages and two disadvantages.
10. Account for the changes to housing trends and needs as Australia's population has grown.
11. Give three reasons why many Australians choose to live in either suburban areas or more densely populated urban areas as demand for housing increases.
12. Identify and account for one reason why Australia's most densely populated areas are on or near the coast.

Challenge your understanding

13. Predict the impact of Australia's ageing population on our demand for different facilities and services.
14. Predict what a map of Australia's population distribution will look like in 50 years.
15. What other methods could the Australian government use in order to encourage population growth in Australia? Suggest a strategy that the government could implement that would encourage people to have bigger families.

To answer questions online and to receive **immediate feedback** and **sample responses** for every question, go to your learnON title at www.jacplus.com.au.

21.3 Variations in rural and urban wellbeing

LEARNING INTENTION

By the end of this subtopic, you will be able to explain the reasons why human wellbeing varies between major cities and regional and remote areas of Australia.

21.3.1 The impact of where we live

Wellbeing varies considerably from one place to another within the one country and also within urban and rural environments. Sometimes these variations are quite distinct, but at other times they may be quite minor. Particular **indicators** used to show variations in human wellbeing may show a more realistic picture of the quality of life experienced in these places. For example, an indicator of wellbeing may be rates of hospitalisation for people aged 18–25 years old.

FIGURE 1 Percentage of Australia's population by remoteness classification

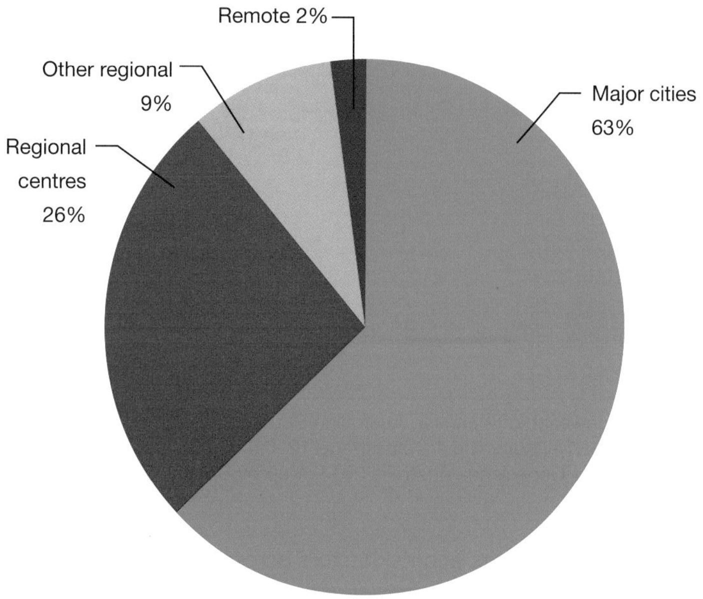

Source: Australian Bureau of Statistics

21.3.2 Health and lifespan

According to the Australian Institute of Health and Welfare (AIHW), people living in rural places tend to have shorter lives and higher levels of some illnesses than those in major cities. The level of health of Australians is much lower in **regional and remote areas** than in major cities. The population percentages for these categories are shown in **FIGURE 1**. People from regional and remote areas tend to be more likely than their counterparts in major cities to smoke and drink alcohol in harmful or hazardous quantities. This is reflected in higher mortality rates than for those living in major cities. Use the **100 people** weblink in your online Resources to find out more about the health of the Australian population.

indicators measurable characteristics of a population

regional and remote areas areas classified by their distance and accessibility from major population centres

FIGURE 2 Australia's population by remoteness classification

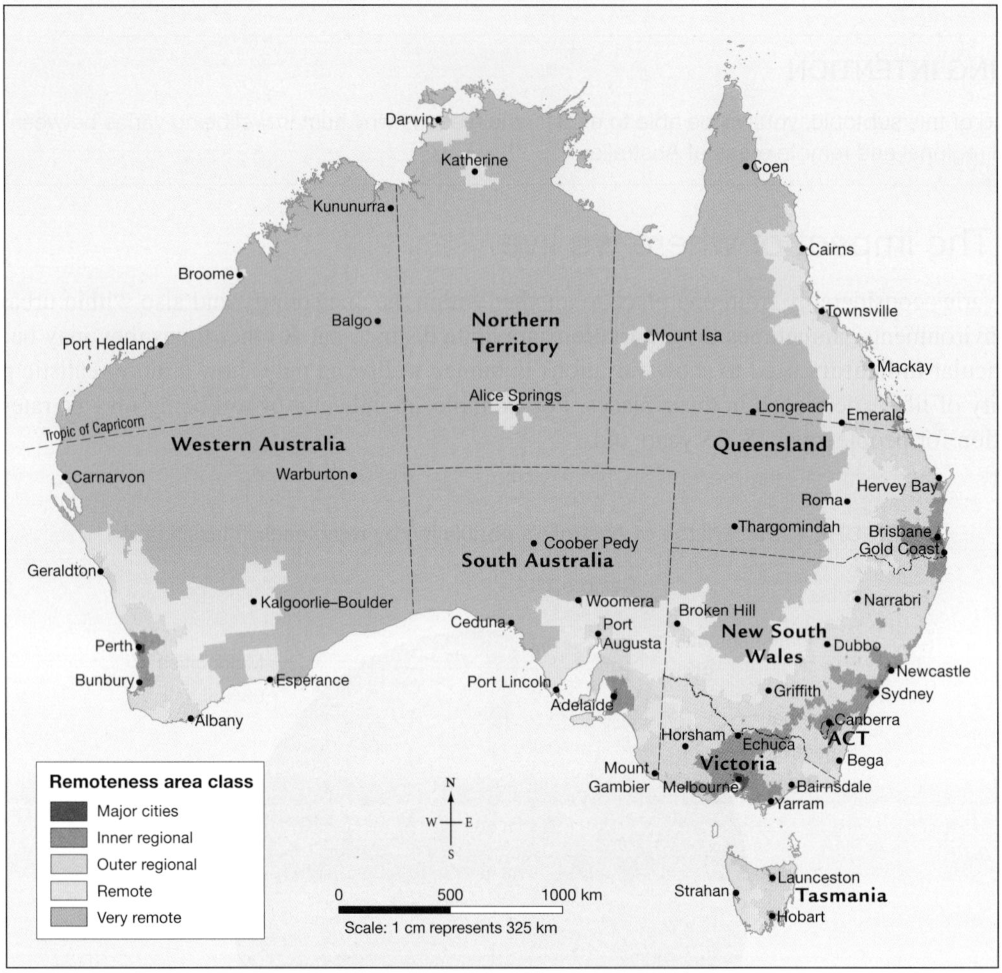

Source: Copyright Commonwealth of Australia, Australian Bureau of Statistics
http://www.abs.gov.au/AUSSTATS/abs@.nsf/DetailsPage/1270.0.55.005July%202011?OpenDocument

TABLE 1 Rates of different health behaviours and risk factors across different living areas

Health risk factor	Major cities	Inner regional	Outer regional/remote
Daily smoker	13%	18%	22%
Overweight or obese	61%	67%	68%
No/low levels of exercise	64%	69%	72%
Exceed lifetime alcohol risk guideline	16%	18%	24%
High blood pressure	22%	24%	22%

Notes:

1. % represents prevalence of risk factor in each region (excluding very remote areas of Australia).
2. Proportions were age standardised to the 2001 Australian Standard Population.

Source: Based on Australian Institute of Health and Welfare material.

These higher death rates may relate to differences in access to basic health services, increased risk factors and the regional/remote environment. More physically dangerous occupations in rural areas lead to higher accident rates. Factors associated with driving such as long distances, greater speed and animals on roads contribute to increased road accident rates in country areas.

21.3.2 Accessing health and disability support services

In general, people living in rural Australia do not always have the same opportunities for good health as those living in major cities. Residents of more inaccessible areas of Australia are generally disadvantaged in their access to health facilities with skilled personnel. People with disability living outside major cities are significantly less likely to access disability support services. Where health services are provided, regional and remote residents also face higher out-of-pocket expenses. These challenges highlight the difference between **equality** and **equity** of access.

FIGURES 3 and **4** show that access to services is at least partially affected by the number of available health workers per person. Medical personnel in rural areas have a higher average age and work longer hours than their city counterparts. Recruitment difficulties in rural areas also affect the **sustainability** of such services and consistency of medical care provided. Research by the Australian Institute of Health Workers showed that, in 2018, the number of working health professionals was highest in major cities (1927 per 100 000 people) compared with all other remoteness areas.

FIGURE 3 Distribution of medical practitioners, Australia, 2018

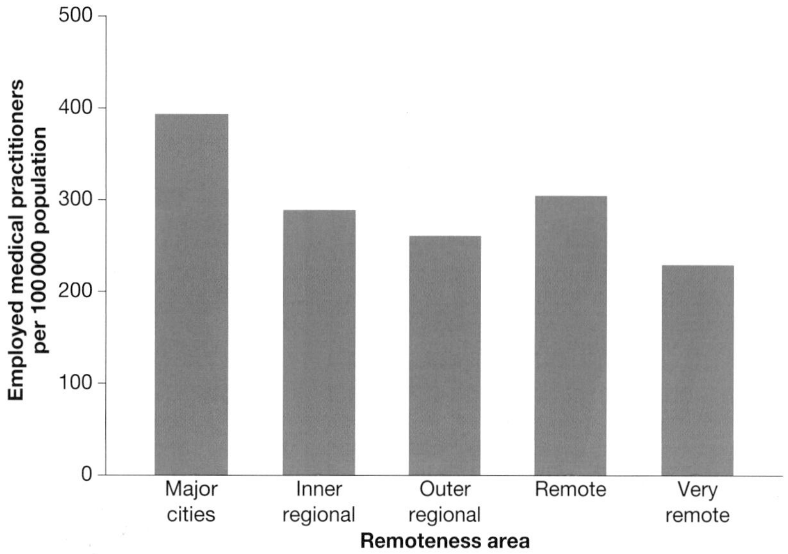

Source: Australian Institute of Health Workers, https://www.aihw.gov.au/reports/australias-health/health-workforce

FIGURE 4 Distribution of nurses and midwives, Australia, 2019

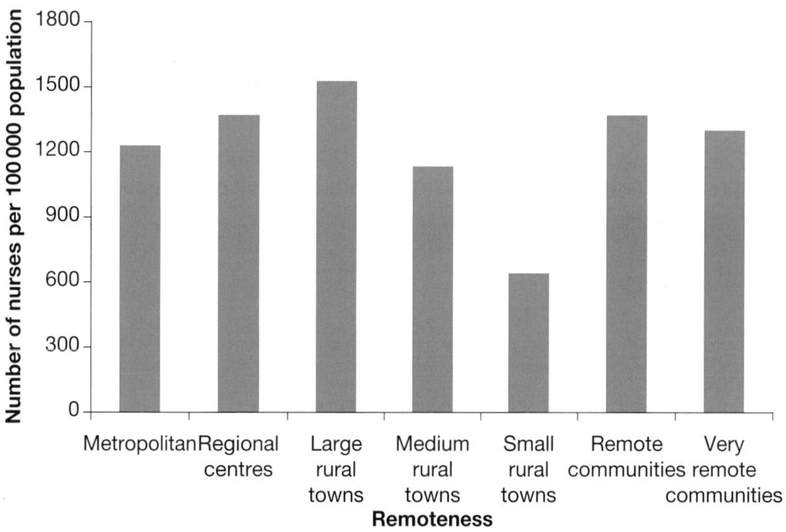

Source: Health Workforce Data, Department of Health, https://hwd.health.gov.au/publications

Higher costs and more limited availability of products such as fresh fruit and vegetables also impact on health and wellbeing. A government survey found that absence of competition in remote areas led to mark-ups of up to 500 per cent on some foods, particularly fresh fruit and vegetables, which took up to two weeks to reach their destination. A typical packet of pasta cost approximately five times more than in metropolitan stores. Fewer educational and employment opportunities are other challenges faced by those in regional and remote places.

equality providing all people with the same opportunities and treatment

equity providing people with different opportunities and treatment based on level of need

sustainability use of a service, product, item or material without depleting its value or accessibility for future generations

On the positive side, in terms of wellbeing, rural Australians tend to have higher levels of **social cohesiveness**, as reflected in higher rates of participation in volunteer work and feelings of safety in their community. The Country Women's Association is one such volunteer organisation (see **FIGURE 5**).

social cohesiveness the positive feeling of connectedness to a group of people or community

FIGURE 5 Members of the Canberra Evening Branch of the Country Women's Association held a charity bake-off for the Women's Refuge of the ACT.

FIGURE 6 (a) and (b) The Royal Flying Doctor Service provides vital healthcare services for people living, travelling and working in remote areas of Australia from bases across Australia.

Resources

eWorkbook	Variations in rural and urban wellbeing (ewbk-9806)
Video eLesson	Variations in rural and urban wellbeing — Key concepts (eles-5320)
Interactivity	Australia's population by remoteness classification (int-8750)

21.3 ACTIVITIES

1. As a class, create a mind map of the benefits of living in regional areas compared to the benefits of urban areas.
2. Separate the class into two groups and debate the topic 'The benefits to wellbeing of living in regional areas outweigh the disadvantages.'
3. Visit the **Equity versus equality** weblink in your online Resources.
   a. As a class, discuss the difference between the terms *equality* and *equity*.
   b. Using this information, discuss the reasons why regional and remote areas of Australia should have greater access to healthcare services than urban areas.
4. Using the **RFDS** weblink in your online Resources, assess the positive impact the introduction of the flying doctors service had on access to health services for people living in remote areas of Australia. Determine what the impact would be today if the services was no longer available to Australians living or travelling in remote areas.

21.3 EXERCISE

Learning pathways

■ LEVEL 1	■ LEVEL 2	■ LEVEL 3
1, 4, 9, 13	2, 3, 8, 11, 12	5, 6, 7, 10, 14, 15

Check your understanding

1. Define the terms *city*, *regional* and *remote*.
2. Using **FIGURE 2**, list the regions of Australia in order of most to least populated.
3. Using **FIGURE 3**, list the regions of Australia in order of most to least access to medical practitioners.
4. Draw a table to show the advantages and disadvantages of rural versus urban areas in terms of wellbeing in Australia.
5. What factor contributes positively to human wellbeing in rural Australia?
6. What is the difference between equality and equity?

Apply your understanding

7. Compare **FIGURES 3** and **4** showing the distribution of doctors and nurses within Australia. Suggest reasons to account for this variation.
8. Explain why regional and remote areas of Australia require increased access to medical services as a matter of equity.
9. Why might rates of nurses be higher in regional and remote areas compared to medical practitioners (doctors)?
10. Explain the reasons why levels of wellbeing vary between urban and rural Australia. Use data in your answer.

Challenge your understanding

11. Predict what the potential long-term outcomes of the distribution of medical practitioners and nurses might be for wellbeing.
12. Suggest an alternative measure of wellbeing not mentioned in this section that could highlight the variation in wellbeing between rural and urban areas.
13. Recommend a strategy for improving the accessibility of health services for people living in rural Australia.
14. Propose other ways wellbeing can be improved in remote and regional areas of Australia.
15. Suggest what the positive and negative impacts of the mining boom may have been on levels of wellbeing in remote communities in Australia.

To answer questions online and to receive **immediate feedback** and **sample responses** for every question, go to your learnON title at www.jacplus.com.au.

21.4 Human wellbeing in Sydney

LEARNING INTENTION

By the end of this subtopic, you will be able to identify and explain some of the reasons for variations in wellbeing in urban areas and to describe spatial patterns of wellbeing within Sydney.

21.4.1 Variation in wellbeing within Sydney

Although, according to many measures, people in urban places generally have a higher standard of wellbeing than those living in rural areas, levels of wellbeing are not uniform across towns and cities. If you live in a town or city yourself, you would be aware that not all parts of that location have the same access to facilities or the same types of housing. Variations in wellbeing occur on a local scale as well as at national and global scales.

21.4.2 Factors affecting urban wellbeing

Housing affordability

FIGURE 1 shows **housing affordability** as a measure of wellbeing in Sydney. The cost of housing is a major expenditure for people, so its affordability directly impacts on people's living standards. Based on this information, it would appear that most of Sydney's population are living in areas where housing affordability is low, particularly those in the eastern and northern suburbs. In Australia, those spending more than 30 per cent of their income on housing while earning in the bottom 40 per cent of the income range are considered to be in housing affordability stress.

housing affordability a person's ability to pay for their housing

FIGURE 1 Changing housing affordability in Sydney, 2015–2020

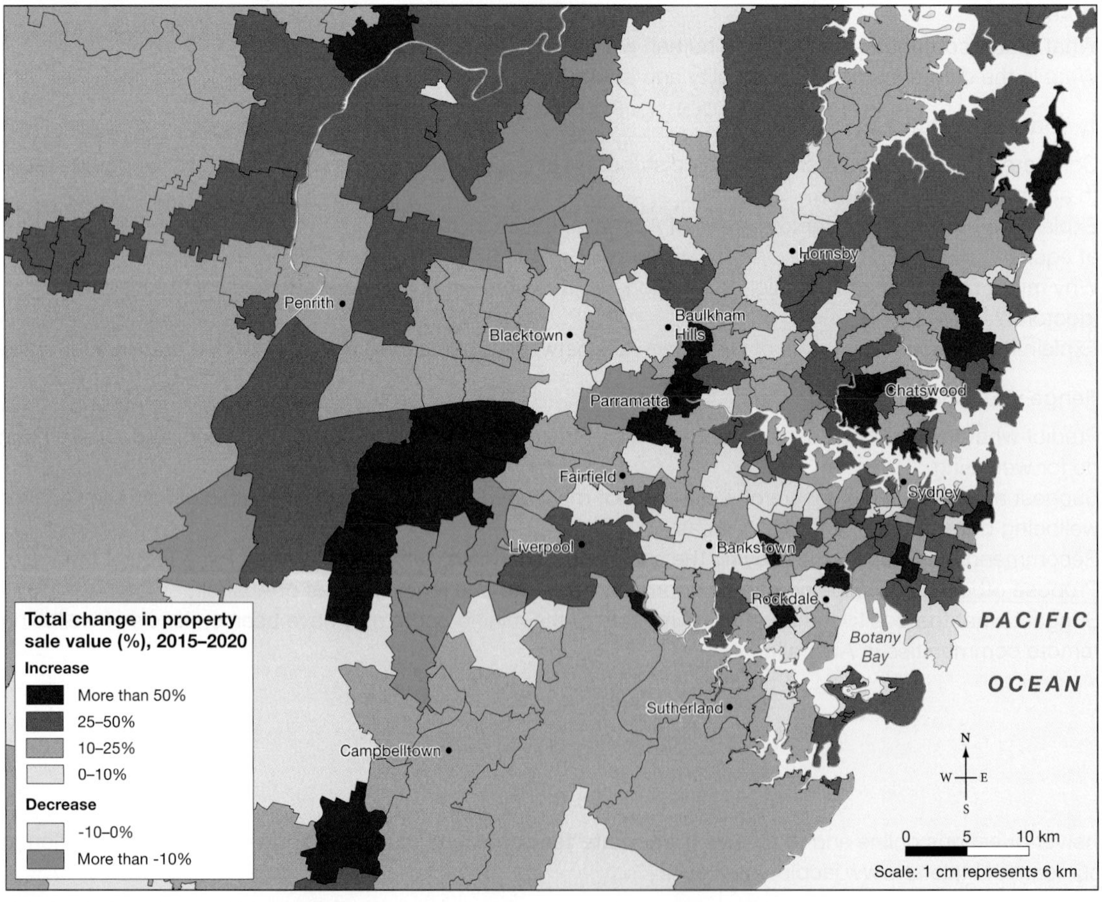

Source: Valuer General NSW (2020)

Housing affordability also has secondary impacts on wellbeing. For example, if a young couple decide to move to a regional area of New South Wales to access more affordable housing, they may also move away from their support network of family and friends. This may reduce their levels of wellbeing.

Income

FIGURE 2 shows Sydney's average wage growth from 1998 to 2020. The percentage increase varies greatly across Greater Metropolitan Sydney and since 2012 has declined consistently. Housing affordability is a key driver of the growing income divide in Sydney. Rising house prices are pushing low-income workers out of inner-city harbourside areas.

There are significant spatial variations within Sydney, as indicated in **FIGURE 1** which reflect the varying level of human wellbeing among Sydney-siders. Some suburbs are more affordable than others. Residents of some suburbs earn higher taxable incomes. The relationship between housing affordability and **taxable income** affects the way people can improve their own wellbeing. The New South Wales government allocates funding to allow people to gain access to adequate housing, healthcare and education without being disadvantaged by their income.

FIGURE 2 Sydney's average wage growth, 1998–2020

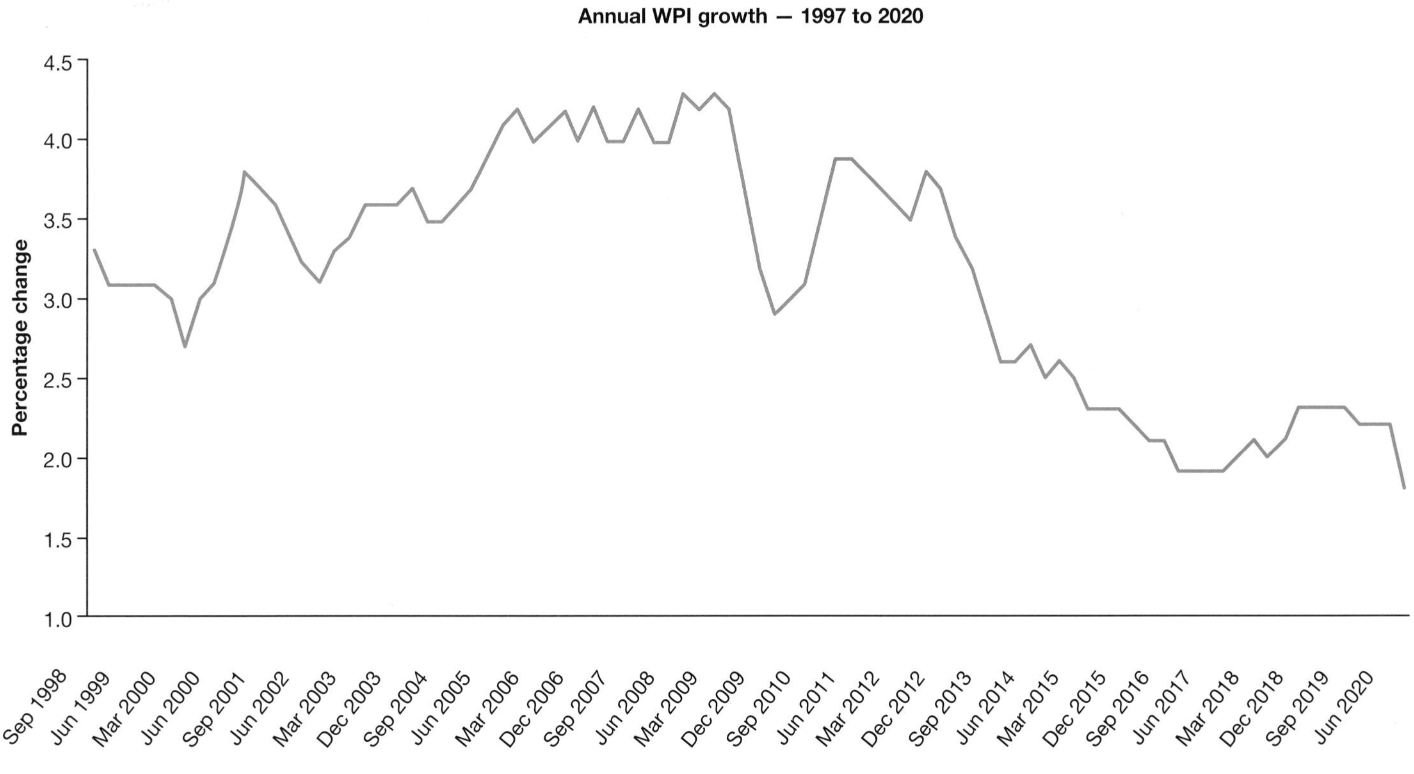

Source: Australian Bureau of Statistics (2020) 'Table – Annual WPI growth – 1997 to 2020', Annual wage growth 1.8% in June quarter 2020, accessed 20 April 2021

FIGURE 3 shows the average salaries within municipalities of Sydney. There are very large differences, which would affect a person's ability to purchase a house and to have access to education opportunities and healthcare.

taxable income the amount of money a person earns (salary), as a result of their employment, and is used as an indicator of personal wealth

FIGURE 3 Sydney's average taxable income brackets by postcode

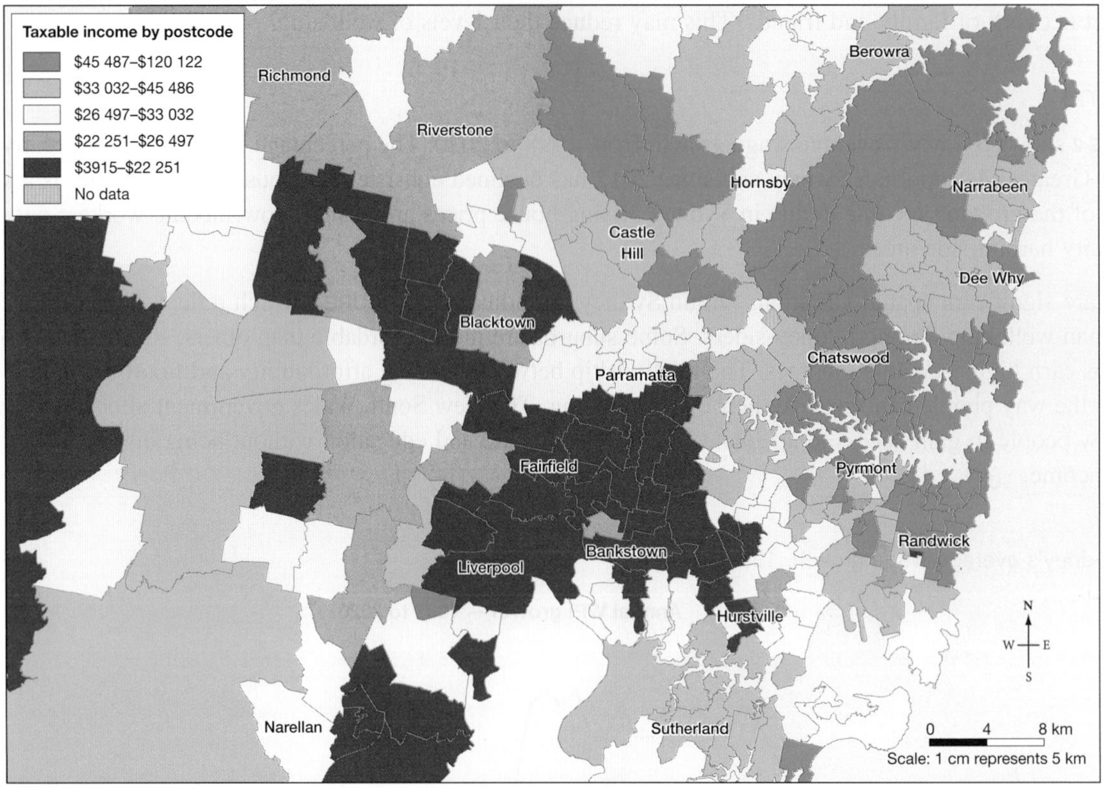

Source: Australian Tax Office, Australian Bureau of Statistics, 2021

on Resources

eWorkbook	Human wellbeing in Sydney (ewbk-9810)
Video eLesson	Human wellbeing in Sydney — Key concepts (eles-5321)
Interactivity	Human wellbeing in Sydney (int-8751)
myWorldAtlas	Deepen your understanding of this topic with related case studies and questions. • Wellbeing in Western Sydney

21.4 ACTIVITIES

1. As a class, use the **Australian Bureau of Statistics** weblink in your online Resources to research the average weekly earnings in Australia. Create a table to show the percentage difference in weekly earnings between men and women in each state of Australia.
2. Brainstorm the challenges that specific groups of people might face depending on which area of Sydney they live in. For example, a person living in Western Sydney but working in the CBD will need to commute to and from work each day, whereas a person living close to the CBD may have higher living costs.
   a. In groups, combine the challenges of living in Sydney for each of your groups and sort them in order of most to least impact on human wellbeing.
   b. As a class, collate your lists of challenges and brainstorm solutions. For example, a single parent living in Western Sydney and working in the CBD might receive access to a travel concession card from the government to assist them in commuting to the CBD for work.

21.4 EXERCISE

Learning pathways

■ LEVEL 1	■ LEVEL 2	■ LEVEL 3
1, 2, 11, 12, 14	3, 7, 8, 9, 15	4, 5, 6, 10, 13

Check your understanding

1. Define the term *housing affordability*.
2. Give three examples of how levels of income can affect wellbeing.
3. Describe the change in housing affordability shown in **FIGURE 1**.
4. Describe the distribution of taxable income across Sydney.
5. What challenges might low-income workers in Sydney face as the cost of housing continues to increase? Outline three, using data in your answer.
6. Outline one way in which wage growth and housing affordability are interconnected.
7. Using **FIGURE 2**, summarise the trend in wage growth in Sydney between 2009 and 2020.

Apply your understanding

8. Explain why the information shown in **FIGURES 1**, **2** and **3** is considered to be a measure of wellbeing.
9. How do income and housing affordability contribute to variations in wellbeing in Sydney?
10. Why might housing have become less affordable between 2012 and 2017?
11. Use **FIGURES 1**, **2** and **3** to discuss the interconnection between housing affordability and income.
12. Identify one change that someone living and working in the inner city might make to improve their access to affordable housing. Analyse the impacts of the decision, including at least two potential secondary impacts.

Challenge your understanding

13. Suggest an alternative measure of wellbeing not mentioned in this section. Highlight how this measure would show the variation in wellbeing within an urban area.
14. Suggest a possible solution to the continuing issue of unaffordable housing in Sydney.
15. Does wellbeing hold relevance for all people equally? Write a paragraph considering the term *human wellbeing* from at least two perspectives.

FIGURE 4 Camp Cove, Watsons Bay, NSW

To answer questions online and to receive **immediate feedback** and **sample responses** for every question, go to your learnON title at www.jacplus.com.au.

21.5 SkillBuilder — Using multiple data formats

online only

LEARNING INTENTION

By the end of this subtopic, you will be able to present multiple sets of data in the most appropriate presentation format and show the interconnection between sets of data in a geographical report.

The content in this subtopic is a summary of what you will find in the online resource.

21.5.1 Tell me

What are multiple data formats?

Multiple data formats are varied forms of data presentation, used when a range of data needs to be shown. All the information must be read before the data can be interpreted. Multiple data formats are useful in major reports when a range of ideas needs to be pulled together and presented as a united document.

FIGURE 1 Multiple data sources in different formats showing information about Fiji (exports and population)

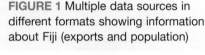

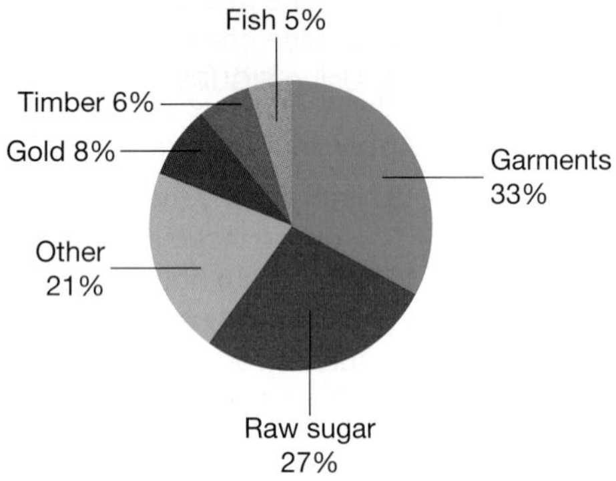

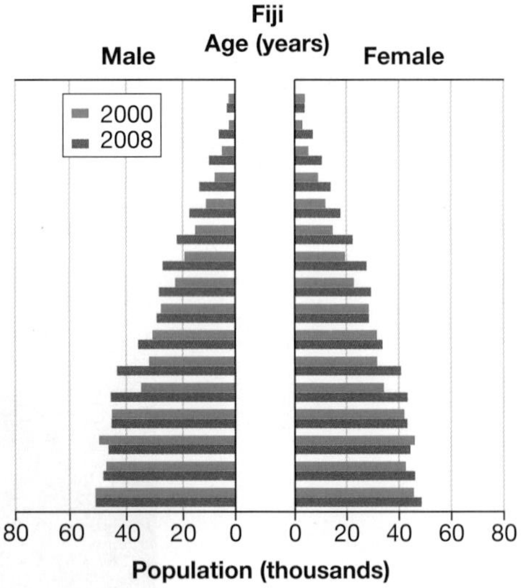

21.5.2 Show me

Step 1

Study the data and the type of format. Read titles, labels, units of measurement, dates and legends.

Step 2

Consider an approach that allows you to link data in a logical flow of ideas about the topic. Introduce your topic with a short sentence stating the intention of your paragraph.

Step 3

Find the figures related to the comparison you are making. To show the interconnection or relationships between the data, choose your connecting words and phrases carefully.

- To show cause and effect: therefore, because, thus, consequently
- To compare: similarly, likewise, in the same way
- To contrast: in contrast, however, conversely, yet still

Step 4

End with a concluding statement that summarises the overall trends or point.

21.5.3 Let me do it

learnON

Go to learnON to access the following additional resources to help you build this skill:

- a longer explanation of this skill and its application in Geography (Tell me)
- a video demonstrating the step-by-step process of this skill (Show me)
- an activity and interactivity for you to practise the skill (Let me do it)
- self-marking questions to help you understand and use the skill.

on Resources

eWorkbook SkillBuilder — Using multiple data formats (ewbk-9814)

Video eLesson SkillBuilder — Using multiple data formats (eles-5322)

Interactivity SkillBuilder — Using multiple data formats (int-8752)

21.6 Wellbeing of Aboriginal Peoples and Torres Strait Islander Peoples

LEARNING INTENTION

By the end of this subtopic, you will be able to demonstrate variations in wellbeing experienced by Aboriginal Peoples and Torres Strait Islander Peoples using statistical data, and to explain the social, economic and health factors contributing to these variations.

21.6.1 Inequality in wellbeing in Australia

In a society such as Australia's, which prides itself on its legal and political principles of equality, we would expect that everyone is able to experience a similar standard of living. It would be unfair for one sector of a community to experience significant disadvantage when the other sectors of the community enjoy the privileges of a 'good life'. However, inequality exists in many parts of Australian society.

Aboriginal Peoples and Torres Strait Islander Peoples generally experience lower levels of wellbeing than the non–Aboriginal and Torres Strait Islander Australian population, including lower levels of health, education, employment and economic security. These disadvantages result in an overall lower **socioeconomic** status, which limits opportunities for both Aboriginal and Torres Strait Islander Peoples to access the same quality of services as non-Aboriginal and Torres Strait Islander Australians.

socioeconomic relating to or involving a combination of social and economic factors

Many historical factors have resulted in Aboriginal Peoples and Torres Strait Islander Peoples being denied access to socially valued resources, and this lack of access has resulted in contemporary issues including diminished wellbeing and equality. However, with reclamation, revitalisation and maintenance of languages and cultural practices today, both groups are at the forefront of social and emotional wellbeing initiatives that improve their quality of life and build social justice and equality.

The age distribution among Aboriginal Peoples and Torres Strait Islander Peoples does not follow the same pattern as the general trend for the Australian population. As **FIGURE 1** shows, there is a higher proportion of younger people in Aboriginal and Torres Strait Islander communities, and fewer people in each age category over 30.

FIGURE 1 Population pyramid for Aboriginal and Torres Strait Islander Peoples and non–Aboriginal and Torres Strait Islander people

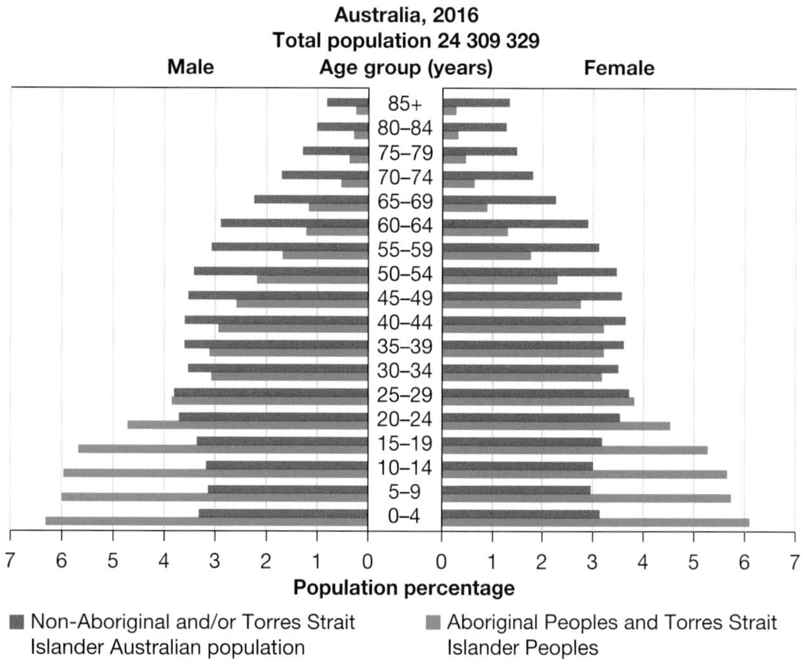

Source: Australian Bureau of Statistics

21.6.2 Measuring the wellbeing of First Nations Australians

The National Aboriginal and Torres Strait Islander Health Survey conducted by the Federal Government aimed to measure the emotional and social health of Aboriginal and Torres Strait Islander adults. In this, more than half the respondents reported being happy (71 per cent), calm and peaceful (56 per cent), and/or full of life (55 per cent) all or most of the time. Just under half (47 per cent) said they had a lot of energy all or most of the time. Respondents in remote areas were more likely to report having had these positive feelings all or most of the time than people living in non-remote areas. This could be related to the intrinsic spiritual and emotional links with Country that Aboriginal Peoples and Torres Strait Islander Peoples have and the important role being on Country holds. However, the health, economic and education outcomes of First Nations Australians still remain relatively low in comparison with the wider population.

With this, and much of the other data that the Australian Bureau of Statistics collects, the ABS includes notes to suggest the limitations of their collection processes and results. Some of the factors to consider when interpreting data about any specific group or culture include possible variations in understandings of terms and ideas (for example, what one group of people considers a disability might not be considered so by a different culture), language differences (for example using words that have different connotations in a person's first language) and cultural perceptions of the topic being discussed (for example it may be culturally inappropriate to discuss a topic with people of another culture or gender).

FIGURE 2 Indicators of Aboriginal Peoples' and Torres Strait Islander Peoples' wellbeing

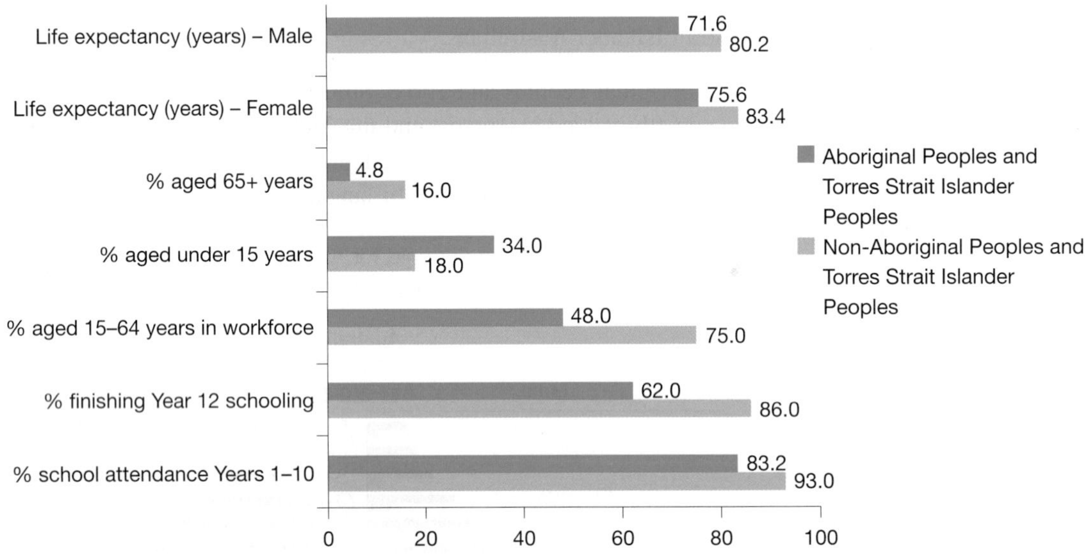

Source: Australian Bureau of Statistics, 2016

21.6.3 Differences in wellbeing

The inequalities experienced by Aboriginal Peoples and Torres Strait Islander Peoples can be attributed to three main causes:

- the dispossession of land
- the displacement of people
- discrimination.

Throughout history, Aboriginal Peoples and Torres Strait Islander Peoples have been denied or have been unable to access the same services and opportunities as other Australians, creating ongoing negative consequences — social, economic and health-related.

Until the 27 May 1967 referendum, Aboriginal Peoples and Torres Strait Islander Peoples were not counted in the national census and were therefore excluded from accessing services, as they were not counted as citizens. There were also numerous policies and decisions that contributed to social inequalities in all areas relating to socioeconomic status that are still evident today. Disadvantage in one area often leads to disadvantage or reduced opportunity in other areas. For example, lower levels of educational attainment reduce opportunities to gain high quality employment and adequate or good quality housing. Additionally, inability to access health services will affect a young person's ability to attend and participate at school, which in turn will affect their options for further study or employment. This often leads to an entrenched cycle of poverty that is the result of exclusion and discrimination.

The National Aboriginal and Torres Strait Islander Social Survey (NATSISS) is conducted by the Federal Government and the ABS every 6 years. It aims to measure the health and wellbeing of Aboriginal and Torres Strait Islander communities. Some of the data gathered from the survey is highlighted in **FIGURE 3**. The data reveals some of the key wellbeing issues facing Aboriginal and Torres Strait Islander Peoples, but also highlights the cultural and community connections and the importance of these connections in relation to overall sense of life satisfaction and wellbeing. Aboriginal Peoples and Torres Strait Islander Peoples generally have improved social and emotional wellbeing from being on Country and from close interactions with family and community as well as being able to maintain their cultural identity in westernised ways of living and doing.

The ABS also conducts the National Aboriginal and Torres Strait Islander Health Survey (NATSIHS). This shows that Aboriginal and Torres Strait Islander Peoples (as a combined statistical group) experience disadvantage not only in social terms, but also in areas of health. In 2018–19, the survey showed that in some measures the situation is getting worse. For example, 46% of Aboriginal and Torres Strait Islander Peoples had at least one chronic condition that posed a significant health problem in 2018–19. This figure was an increase of 40% from 2012–13 (**FIGURE 4**).

FIGURE 3 Some findings from the National Aboriginal and Torres Strait Islander Social Survey (NATSISS) 2014–15

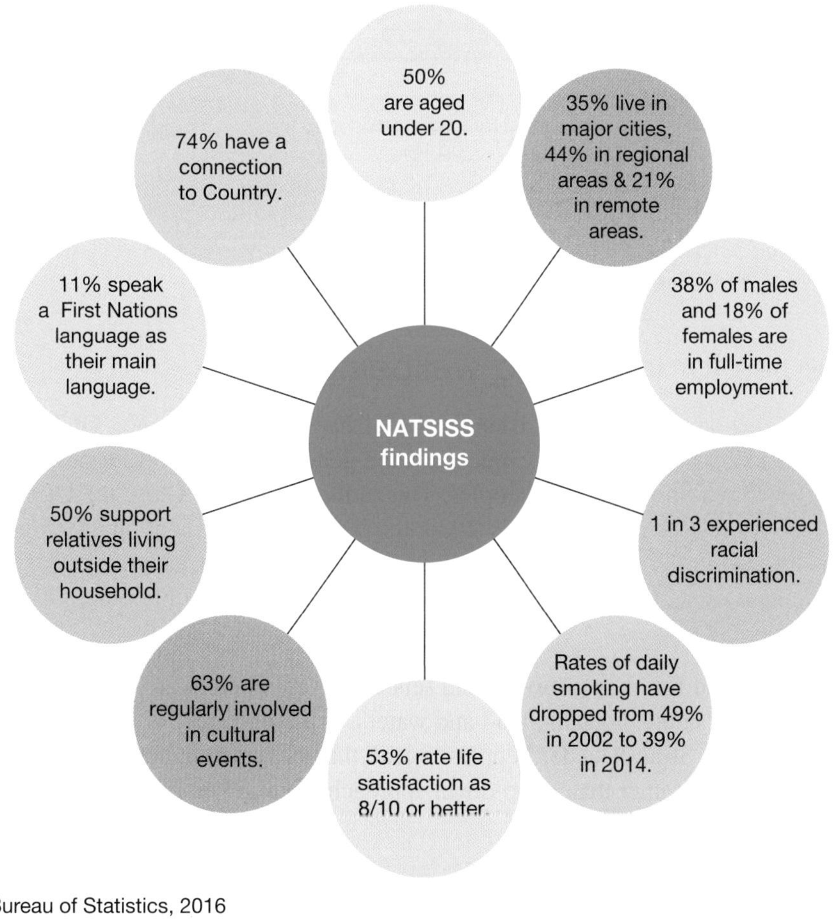

Source: Australian Bureau of Statistics, 2016

FIGURE 4 Some findings from the National Aboriginal and Torres Strait Islander Health Survey (NATSIHS), comparison of 2018–19 and 2012–13 data

Source: Australian Bureau of Statistics, 2019

21.6.4 Strategies for improving wellbeing

In 2005, the then Aboriginal and Torres Strait Islander Social Justice Commissioner, Dr Tom Calma, highlighted the need to address the lower life expectancy and generally poorer health outcomes experienced by Aboriginal Peoples and Torres Strait Islander Peoples. This led to the initial 'Close the Gap' campaign being launched in April 2007. More recently, on 27 July 2020, the National Agreement on Closing the Gap came into effect. This agreement between the Coalition of Aboriginal and Torres Strait Islander Peak Organisations and all Australian Governments seeks to create equity and improve wellbeing and life outcomes for Aboriginal and Torres Strait Islander Peoples.

The agreement is based around four key priorities and sets 16 targets in eight areas (education, employment, health and wellbeing, justice, safety, housing, land and waters, and languages). It also highlighted the need for Aboriginal Peoples and Torres Strait Islander Peoples to lead the conversation about the best way forward on issues directly relating to them, rather than a top-down approach from governments.

FIGURE 5 National Agreement on Closing the Gap — priority reform and targets

4 PRIORITY REFORMS

At the heart of the National Agreement are four Priority Reforms to change the way governments work with Aboriginal and Torres Strait Islander people supported by specific targets.

1. **Formal partnerships and shared decision making** — Building and strengthening structures to empower Aboriginal and Torres Strait Islander people to share decision-making with governments.
2. **Building the community-controlled sector** — Building formal Aboriginal and Torres Strait Islander community-controlled sectors to deliver services to support Closing the Gap.
3. **Transforming government organisations** — Systemic and structural transformation of mainstream government organisations to improve accountability and better respond to the needs of Aboriginal and Torres Strait Islander people.
4. **Shared Access to Data and Information at a Regional Level** — Enable shared access to location specific data and information to support Aboriginal and Torres Strait Islander communities and organisations achieve the first three Priority Reforms.

16 TARGETS

The Agreement establishes 16 socio-economic targets to measure progress in the outcomes experienced by Aboriginal and Torres Strait Islander people.

1. People enjoy long and healthy lives
2. Children are born healthy and strong
3. Early childhood education is high quality and culturally appropriate
4. Children thrive in their early years
5. Students achieve their full learning potential
6. Students reach further education pathways
7. Youth are engaged in education or employment
8. Strong economic participation and development
9. People can secure appropriate and affordable housing
10. Adults are not overrepresented in incarceration
11. Young people are not overrepresented in detention
12. Children are not overrepresented in out-of-home care
13. Families and households are safe
14. Social and emotional wellbeing
15. People maintain distinctive relationships with land and waters
16. Cultures and languages are strong

Four additional targets – on family violence, access to information, community infrastructure and inland waters – will be developed over the next year to further strengthen the National Agreement.

Source: Closing the Gap, https://www.closingthegap.gov.au

Plans to meet these targets include funding for community-controlled organisations that improve justice outcomes, promote social and emotional wellbeing, and strengthen Aboriginal and Torres Strait Islander languages and cultures. Part of the agreement is acknowledging that it is a working or 'living' document that can and likely will change over time. Read more about the process for developing the agreement and the current details of the plan using the **Closing the Gap** weblink in your online Resources.

on Resources

eWorkbook Wellbeing of Aboriginal Peoples and Torres Strait Islander Peoples (ewbk-9818)

Video eLesson Wellbeing of Aboriginal Peoples and Torres Strait Islander Peoples — Key concepts (eles-5323)

Interactivity Wellbeing of Aboriginal Peoples and Torres Strait Islander Peoples (int-8753)

21.6 ACTIVITIES

1. As a class, create a mind map of all the wellbeing issues facing Aboriginal and Torres Strait Islander Australians.
   a. In small groups, choose nine issues and rank them, using the diamond ranking template, in order of what you think would have the most to least impact on someone's wellbeing.
   b. Use the **National Aboriginal and Torres Strait Islander Social Survey** (NATSISS) weblink in your online Resources to watch a video about the survey. While you watch the video, write down statistics that relate to the issues in your diamond.
   c. Discuss your findings as a class.

2. Use the **Closing the Gap** weblink in your online Resources to create an infographic poster outlining the new targets relating to each of the key focus areas.
3. Research one of the following organisations that have experienced success in combating some of the health, social and educational disadvantages experienced by Aboriginal and Torres Strait Islander Peoples. Determine their aims and summarise how they improve wellbeing.
   - Aboriginal Women Against Violence (New South Wales)
   - MPower — Family Income Management Plan (Queensland)
   - Indigenous Enabling Program at Monash University (Victoria)

21.6 EXERCISE

Learning pathways

■ LEVEL 1	■ LEVEL 2	■ LEVEL 3
1, 2, 3, 8, 13	4, 5, 7, 10, 15	6, 9, 11, 12, 14

Check your understanding

1. List some of the reasons why disadvantages exist for Aboriginal Peoples and Torres Strait Islander Peoples.
2. Refer to **FIGURE 1**.
   a. What is the average life expectancy for Aboriginal Peoples and Torres Strait Islander Peoples?
   b. What is the average life expectancy for non-First Nations Australians?
   c. What is the difference (in years) between these average life expectancies?
3. Using **FIGURE 4**, list the five key wellbeing issues facing Aboriginal Peoples and Torres Strait Islander Peoples.
4. How does connection to Country affect wellbeing of Aboriginal Peoples and Torres Strait Islander Peoples?
5. What is the purpose of the Closing the Gap Agreement?
6. Refer to **FIGURE 5**.
   a. Identify three of the targets of the Closing the Gap Agreement that relate to wellbeing.
   b. Outline why achieving these targets will help to improve wellbeing of Aboriginal and Torres Strait Islander Peoples.

Apply your understanding

7. Explain what is meant by the sentence 'Aboriginal and Torres Strait Islander Peoples are culturally and linguistically diverse.' How might the degree to which policy makers understand this diversity have affected their decisions in the past?
8. How might the NATSISS and NATSIHS surveys help to improve the wellbeing of Aboriginal Peoples and Torres Strait Islander Peoples?
9. How might cultural views influence people's perception of their own wellbeing or the wellbeing of others?
10. What do you consider the most significant wellbeing issue facing Aboriginal Peoples and Torres Strait Islander Peoples today? Use data from this subtopic to support your view.
11. Why might the Closing the Gap agreement have been established as a 'living' document?
12. Provide a one-paragraph justification for the need to the Closing the Gap initiative.

Challenge your understanding

13. Did any of the statistics about Aboriginal Peoples and Torres Strait Islander Peoples surprise you? Explain your reaction to them and how they may have either changed or reinforced your own opinions or beliefs.
14. Social justice means fair and equitable access to a community's resources. Do you think Aboriginal Peoples and Torres Strait Islander Peoples experience social justice in Australia? Explain your answer.
15. Another factor contributing to disadvantage may be the remoteness of some Aboriginal Peoples' and Torres Strait Islander Peoples' communities. What innovative solutions can you come up with to try to solve these accessibility problems?

To answer questions online and to receive **immediate feedback** and **sample responses** for every question, go to your learnON title at www.jacplus.com.au.

21.7 SkillBuilder — Understanding policies and strategies

online only

LEARNING INTENTION

By the end of this subtopic, you will be able to identify the relevance and importance of policies and strategies and show the role they play in communication within an organisation.

The content in this subtopic is a summary of what you will find in the online resource.

21.7.1 Tell me

What are policies and strategies?

Policies are principles and guidelines that allow organisations to shape their behaviour and decisions, and to clarify future directions. **Strategies** are actions to ensure that the key components of a plan are implemented. Policies and strategies are particularly useful in large organisations, where information needs to be spread to all employees.

21.7.2 Show me

Step 1

Seek out the general statement of an organisation's aims. These should be big picture guiding rules, aims or principles: the policies.

Step 2

A policy is not achieved without a set of strategies to make it become reality. Look for a series of specific actions that will be taken.

FIGURE 1 An example of a school policy document

Policy ← big picture guiding rules, aims or principles

The Metcalfe Boys' High School must maximise student numbers in order to remain viable and to offer a broad range of subjects.

Strategies ← actions that will be taken to achieve the big aims

Long term

- The Metcalfe Boys' High School is to become a co-educational school. It will do this over a six-year period, beginning with Year 7.
- It must achieve a gender balance in the classes within six years.
- Awards for girls should be developed.
- Associations for past students should be created.

Short term

- Design and create a uniform for the girls.
- Create facilities such as toilets and gymnasium change rooms.
- Consider school camp facilities.
- Consider the need to join other sporting organisations such as softball and girls' competitions.

Step 3

Each strategy is likely to have a set of programs within it that help make the strategy successful. Identify these and assess how well they will achieve the main aims.

21.7.3 Let me do it

learnON

Go to learnON to access the following additional resources to help you build this skill:

- a longer explanation of this skill and its application in Geography (Tell me)
- a video demonstrating the step-by-step process of this skill (Show me)
- an activity and interactivity for you to practise the skill (Let me do it)
- self-marking questions to help you understand and use the skill.

on Resources

eWorkbook SkillBuilder — Understanding policies and strategies (ewbk-9822)

Video eLesson SkillBuilder — Understanding policies and strategies (eles-5324)

Interactivity SkillBuilder — Understanding policies and strategies (int-8754)

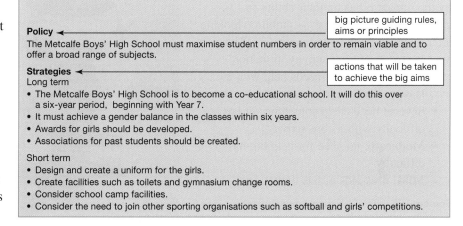

21.8 Wellbeing for people living with disabilities

LEARNING INTENTION

By the end of this subtopic, you will be able to objectively demonstrate the variations in wellbeing experienced by people with disabilities, provide examples of the types of disadvantage and discrimination experienced by people with disabilities, and discuss strategies to limit these variations.

21.8.1 Defining the term disability

A **disability** is any condition that restricts a person's mental, sensory or mobility functions. It may be caused by accident, trauma, genetics or disease. A disability may be temporary or permanent, total or partial, lifelong or acquired, visible or invisible. Disabilities restrict or prevent a person from completing everyday activities, such as mobility, self-care or communication.

FIGURE 1 Dylan Alcott, Tennis Grand Slam champion, Paralympic Gold medallist and founder of the Dylan Alcott Foundation, which provides pathways to study and sport for young people living with disabilities, and Get Skilled Access, which works to build inclusion in the workplace

The ABS categorises the extent of a disability by the impact it has on their ability to complete 'core activities' — mobility, looking after themselves and communicating. There are four levels of core activity limitation.

- **Profound:** always needs help with at least one core activity
- **Severe:** needs help sometimes or has difficulty with a core activity
- **Moderate:** no need for help but has difficulty
- **Mild:** uses aids or has limitations

Source: ABS, https://www.abs.gov.au/articles/disability-and-labour-force

Causes of disability

A disability can result from any of the following:

- accident — including loss of limbs and paralysis
- trauma — such as brain damage
- genetic conditions — such as cystic fibrosis and Down syndrome
- disease and disorders — including Alzheimer's, arthritis, diabetes and cancer.

21.8.2 The impact of disabilities on wellbeing

Around 6.8 million Australians aged 18 years and over report having a disability or long-term health condition. As a direct result of their disability, people with disabilities are more likely to experience poverty, live in poor quality or insecure housing, and have low levels of education reducing their level of wellbeing compared to other Australians. Some daily activities can also become more difficult; for example, 1.2 million people with disabilities report difficulty using public transport.

disability any condition that restricts a person's mental, sensory or mobility functions

FIGURE 2 As a consequence of inadequate support, the wellbeing of Australians living with a disability can suffer.

Disability Rights
4 in 10 Australians aged 18 yrs and over report having a disability or long-term health condition
Mental health problems and mental illness are among the greatest causes of disability, diminished quality of life and reduced productivity
AUSTRALIA RANKS LOWEST AMONG OECD COUNTRIES FOR THE RELATIVE INCOME OF PEOPLE WITH DISABILITIES
27th
1 2 3
ICELAND MEXICO SWEDEN
WORKFORCE PARTICIPATION OF PEOPLE WITH DISABILITIES AND WITHOUT DISABILITIES
54% with disabilities
83% without disabilities
1.2 million people with disabilities report difficulties using public transport
6% Live in non-private dwellings
20% Live independently in private dwellings
74% Live with others in private dwellings
82% Children with disabilities
77% Children without disabilities
2009 RATES OF PARTICIPATION IN SCHOOL
1 in 4 people who report sexual assault are people with disabilities
9 in 10 WOMEN WITH INTELLECTUAL DISABILITIES HAVE BEEN SEXUALLY ABUSED
94% OF PEOPLE WITH DISABILITIES HAVE THE SUPPORT THEY NEED TO LIVE IN PRIVATE RESIDENCES
2014 Face the Facts www.humanrights.gov.au/face-facts
Australian Human Rights Commission

Mental health problems and illness are among the greatest cause of disability. They reduce a person's productivity and therefore their ability to maintain a job in the long term. Young Australians with mental health disorders are at least six times more likely to be in prison than counterparts without mental health disorders. **FIGURE 2** illustrates many more of the issues that affect people with disabilities.

The Australian government enacted the Disability Discrimination Act in 1992 to protect Australians with disabilities from unfair treatment and allow them to have equal opportunities and access to education, employment and healthcare.

More recently the government has implemented the National Disability Insurance Scheme (NDIS) to assist people with disabilities and their families. The NDIS is a holistic approach to care, aimed at empowering people living with a permanent and significant disability. The NDIS provides access to support and services, as well as intervention to help prevent worsening or future difficulties.

The NDIS is part of the Australian Government's National Disability Strategy, and is a national strategy supported by the Council of Australian Governments (CoAG), which comprises the state and territory premiers or chief ministers and a representative of the Australian Local Government Association.

fdw-0048

FOCUS ON FIELDWORK

Accessing community resources

Explore a place in your community such as a train station, ferry wharf, bus terminal, school, shopping centre or sporting field to determine how accessible it is to people with disabilities.

Learn how to survey your local area's accessibility using the **Accessing community resources** fieldwork activity in your online Resources.

on Resources

eWorkbook	Wellbeing for people with disabilities (ewbk-9826)	
Video eLesson	Wellbeing for people with disabilities — Key concepts (eles-5325)	
Interactivity	Wellbeing for people with disabilities (int-8755)	

21.8 ACTIVITIES

1. In small groups, discuss the wellbeing issues faced by people with disabilities and then complete the following activities.
   a. Develop a survey of five to ten questions that could be used to gather primary data on the levels of wellbeing in people with disabilities either at your school or within your local community. Consider types and levels of disability, remembering that the experiences of two people with different disabilities will often be very different.
   b. Compare your survey with other groups in the class.
   c. Discuss the types of questions that would gather the best primary research data and why.
2. You have been tasked by the principal of your school to develop a report on the wellbeing of students with disabilities at your school. You need to develop a Research Action Plan (RAP) to show how you are going to report objectively (and appropriately) on this topic. Remember to be tactful in your approach. Use the following subheadings to construct your plan.
   a. Develop a clear and concise aim.
   b. Write a one-sentence hypothesis (expected result).
   c. Develop five primary research questions — you may like to use your survey questions from the previous task.
   d. Identify the types of secondary research required.
   e. List how you will collect the primary and secondary data.
   f. Identify the most appropriate method of presenting the data, for example charts, graphs or maps.
   g. Select a presentation method and justify its effectiveness.
3. Use the **ABS — Disability and the Labour Force** weblink in your online Resources to find reliable evidence and statistics to support your answers to the following questions.
   a. Research what the most common types of employment are for people with disabilities.
   b. Explain the reasons why the public sector is a larger employer of people with disabilities than the private sector.
   c. What levels of education do people with disability in the labour force have?
   d. How many people with disabilities experienced discrimination in the workplace? Explain why this might be the case.
   e. Using the evidence you have now gathered, make an educated judgement on the level of access people with disabilities have to employment and how this will impact their wellbeing.
   f. Research organisations and strategies that help to build inclusion in the workplace for people with disabilities.

21.8 EXERCISE

Learning pathways

■ LEVEL 1	■ LEVEL 2	■ LEVEL 3
1, 2, 4, 7, 13	3, 5, 8, 12, 14	6, 9, 10, 11, 15

Check your understanding

1. Study **FIGURE 2** and answer the following.
   a. What percentage of Australians aged 18 and over report having a disability or long-term health condition?
   b. What is the workforce participation of people with disabilities compared to those without disabilities?
   c. How many people with disabilities live in a private dwelling?
   d. What is the workforce participation rate of people with disabilities?
2. What are the four categories of disabilities?
3. People with disabilities are more likely to experience lower levels of wellbeing. List three reasons why.
4. What is the purpose of the Disability Discrimination Act (1992)?
5. Categorise the following levels of disability according to the ABS descriptions of profound, severe, moderate or mild.
   a. A person who wears hearing aids to be able to communicate in noisy situations
   b. A person with limited mobility who needs assistance showering or bathing
   c. A person with cystic fibrosis who sometimes experiences difficulty breathing and completing everyday activities
   d. A person experiencing chronic pain who takes longer than most people to board a bus
6. Describe the aims of the NDIS and give one example of how it could improve the wellbeing of people with disabilities and their families.

Apply your understanding

7. Why it is difficult for some people to use public transport? Give an example to support your explanation.
8. Explain why it is difficult for people with a mental health disorder to maintain a steady job.
9. According to the Australian Bureau of Statistics, 53.4% of people with a disability are in the labour force, compared with 84.1% of people without disability. Discuss this statement by identifying the impact reduced employment rates has on a person's wellbeing.
10. Explain why there is often an interconnection between disability and poverty.
11. Why might a higher percentage of children with a disability be attending school than those without a disability?
12. Why might providing customised seating and other support on public transport help to build wellbeing of people with disabilities? Explain one immediate reason and one secondary reason.

Challenge your understanding

13. Propose a solution that your community and/or government can adopt to better support people living with disabilities. Pick one specific disability as your focus.
14. Predict two significant ways that the wellbeing levels of people living with disabilities would change if the NDIS was no longer available.
15. Do you think that publicity of the Paralympics and the success of sports and media personalities such as Dylan Alcott play a role in building the sense of wellbeing of people living with a disability? Give reasons for your answer.

To answer questions online and to receive **immediate feedback** and **sample responses** for every question, go to your learnON title at www.jacplus.com.au.

21.9 Investigating topographic maps — Wellbeing in the Northern Territory

LEARNING INTENTION

By the end of this subtopic, you will be able to explain challenges associated with living in both urban and regional areas of the Northern Territory through the analysis of topographic maps.

21.9.1 Darwin and the Northern Territory

With some of the most diverse landscapes in Australia, the Northern Territory has an abundance to offer locals and tourists. Covering an area of 1 350 000 square kilometres, the landscape is vast with only a small population of approximately 245 000. Most of the population lives in the more urbanised centres of Darwin, Katherine and Alice Springs. There are many rural towns with populations under 1000 people, such as Pine Creek (473), Adelaide River (237) and Noonamah (528).

Wellbeing can be difficult to manage in rural areas due to their remote location. People in rural areas have lower life expectancies. Higher death rates may relate to the difference in access to services and other risk factors associated with living in remote areas. There are also lower rates of hospital surgical procedures and GP consultations, and higher rates of hospital admissions.

However, rural Australians generally have higher levels of social cohesiveness than their urban counterparts. People living in rural areas have higher rates of participation in volunteer work and feelings of safety in their community, which contributes to an individual and community wellbeing.

FIGURE 1 Many people living in the Northern Territory have to travel vast distances for services.

FIGURE 2 Topographic map of Darwin and the surrounding area

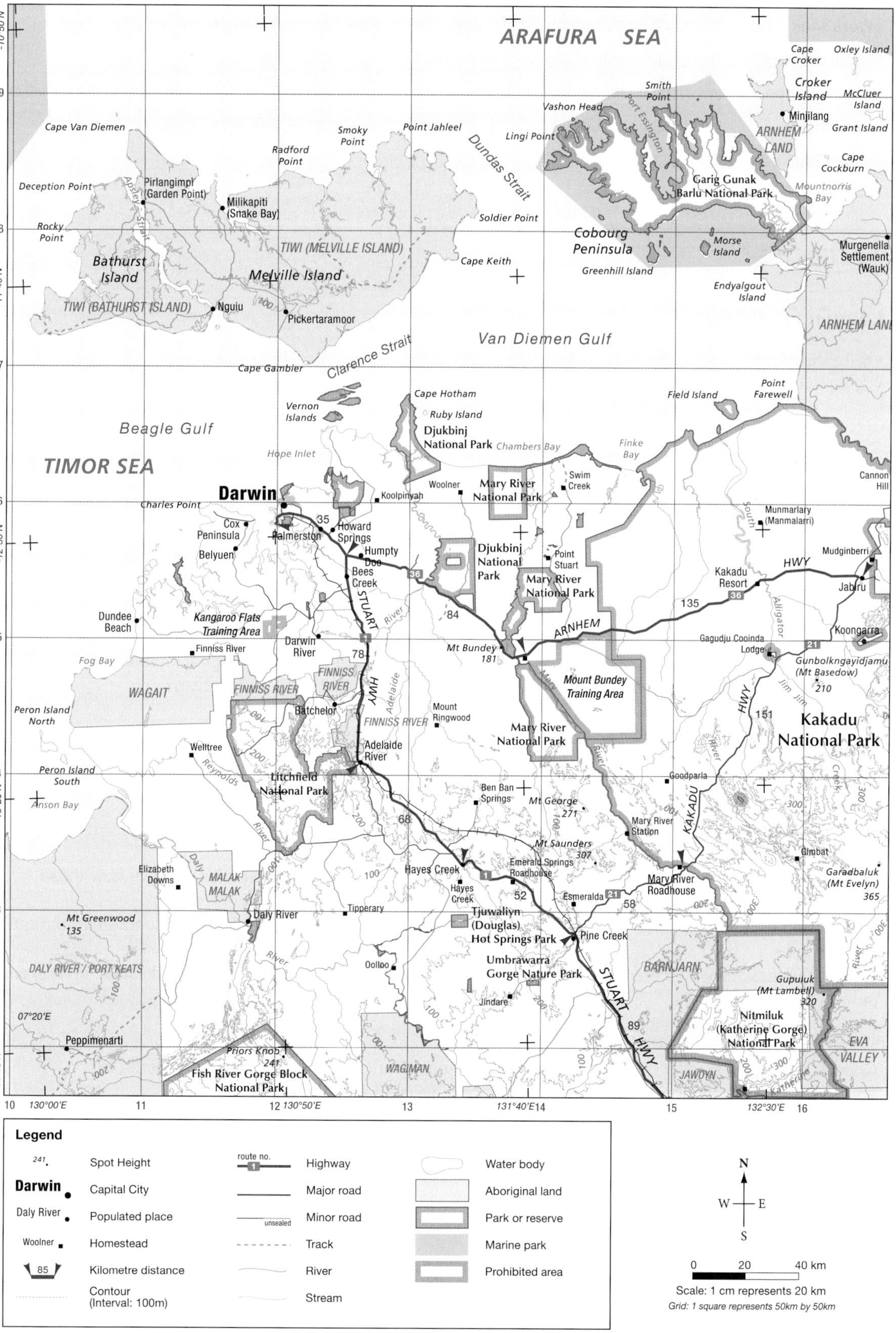

Source: Map data based on Geoscience Australia; © OpenStreetMap contributors, https://openstreetmap.org. Data is available under the Open Database Licence, https://opendatacommons.org/licenses/odbl/.

Resources

eWorkbook	Investigating topographic maps — Wellbeing in the Northern Territory (ewbk-9830)
Digital document	Topographic map of Darwin and the surrounding area (doc-36374)
Video eLesson	Investigating topographic maps — Wellbeing in the Northern Territory — Key concepts (eles-5326)
Interactivity	Investigating topographic maps — Wellbeing in the Northern Territory (int-8756)
Google Earth	Darwin (gogl-0109)

21.9 EXERCISE

Learning pathways

■ LEVEL 1	■ LEVEL 2	■ LEVEL 3
1, 3, 4, 9	2, 5, 8, 10	6, 7, 11, 12

Check your understanding

Use **FIGURE 1** to answer the following questions.

1. Give an approximate latitude and longitude for Darwin.
2. Calculate the approximate area of Litchfield National Park.
3. Give the direction of Adelaide River from Darwin.
4. Calculate the bearing of Pine Creek from Litchfield.
5. Calculate the distance between Darwin and Pine Creek along the main highway.
6. Calculate the population density of the Northern Territory.
7. You are staying at the Emerald Springs Roadhouse and you need to get to a hospital located in Darwin.
   a. What is the distance by road between the Emerald Springs Roadhouse and Darwin?
   b. How long would it take you if you travelled at 60 kilometres per hour?
   c. How long would it take you if you travelled at 80 kilometres per hour?

Apply your understanding

8. Provide reasons why people in remote areas:
   a. have lower life expectancies than those in urban areas
   b. do more volunteer work than those who live in urban areas.
9. List ten services and facilities that would improve wellbeing in remote communities. Explain how these services and facilities improve wellbeing.
10. Identify and explain one factor that leads to higher levels of wellbeing for people living in remote and rural communities.
11. Explain the reasons why there are higher rates of health issues, such as coronary heart disease, in rural and remote areas of Australia.

Challenge your understanding

12. Choose one of the smaller communities in the north-east Northern Territory. Predict what social, environmental and economic challenges might affect wellbeing in that area.

To answer questions online and to receive **immediate feedback** and **sample responses** for every question, go to your learnON title at www.jacplus.com.au.

21.10 Thinking Big research project — SDG progress infographic

online only

The content in this subtopic is a summary of what you will find in the online resource.

Scenario

Introduced in 2015, the United Nations (UN) Sustainable Development Goals (SDGs) are reported on annually by the UN at mid year. The UN SDG Report highlights the areas of progress toward the goals and areas in which ongoing action needs to be taken to ensure no person or country is left behind.

In the 2018 report, for example, conflict and climate change were found to be major contributing factors leading to more people facing hunger and displacement from their homes, and reducing progress toward access to clean water and improved sanitation.

Since the SDGs apply to *all* countries — developed and developing — there are also reports that outline how each country is responding to the SDGs. Australia's progress can also be seen in these reports.

Task

Help the UN make its annual SDG Report more understandable! Working in small groups, you will create an engaging infographic detailing the global progress towards achieving one of the SDGs, as reported in the latest UN SDG Report, and look at Australia's response to that goal.

learnON

Go to learnON to access the resources you need to complete this research project.

on Resources

ProjectsPLUS Thinking Big research project — SDG progress infographic (pro 0217)

21.11 Review

21.11.1 Key knowledge summary

21.2 Characteristics of Australia's population

- Australia's population in 2016 was 23401892. It is unevenly distributed and has varying levels of density.
- Most of Australia's population is concentrated in cities in the coastal regions in the south-east and east and, to a lesser extent, in the south-west.
- Australia's population is growing. At the time of the 2016 census, Australia had a population growth rate of 1.7%. However, Australia's fertility has declined steadily since the 1970s and is now well below replacement rate.
- Our current population growth is due to immigration rather than natural increase.

21.3 Variations in rural and urban wellbeing

- Significant variations in human wellbeing occur between urban and rural places within Australia.
- People living in rural places tend to have shorter lives and higher levels of some illnesses than those in major cities.
- People in rural and remote regions of Australia have less access to health services and health professionals.
- Rural areas have more limited availability of products such as fresh fruit and vegetables, which also impacts on health and wellbeing.
- However, rural Australians tend to have higher levels of social cohesiveness, as reflected in higher rates of participation in volunteer work and feelings of safety in their community.

21.4 Human wellbeing in Sydney

- Wellbeing can vary in urban areas due to issues such as cost of living, crime and access to green spaces.
- The cost of housing in Sydney is a major expenditure, so its affordability directly impacts on people's living standards.
- Rising house prices are pushing low-income workers out of inner-city harbourside areas.

21.6 Wellbeing of Aboriginal Peoples and Torres Strait Islander Peoples

- Aboriginal Peoples and Torres Strait Islander Peoples consistently experience lower levels of health, education, employment and economic independence than those enjoyed by most Australians.
- Their health, economic and education outcomes still remain low relative to the general trends across the Australian population.
- 46% of Aboriginal Australians had at least one chronic condition that posed a significant health problem in 2018–19, up from 40% in 2012–13.
- However, more than half of the adult population who are of Aboriginal and/or Torres Strait Islander heritage reported being happy (71 per cent), calm and peaceful (56 per cent), and/or full of life (55 per cent) all or most of the time.

21.8 Wellbeing for people living with disabilities

- Around 6.8 million Australians over 18 report having a disability or long-term health condition.
- People with disabilities are more likely to experience poverty, live in poor quality or insecure housing, and have low levels of education reducing their level of wellbeing compared to other Australians
- Young Australians with mental health disorders are at least six times more likely to be in prison than counterparts without mental health disorders.
- Recently the government has implemented the National Disability Insurance Scheme (NDIS) to assist people with disabilities and their families.

21.9 Wellbeing in the Northern Territory

- The Northern Territory covers an area of 1 350 000 square kilometres; however, it has only a small population of approximately 245 000.
- Most of the population lives in the more urbanised centres of Darwin, Katherine and Alice Springs.
- There are many rural towns with populations under 1000 people, such as Pine Creek (473), Adelaide River (237) and Noonamah (528).
- Wellbeing can be difficult to manage in rural areas due to their remote location.

21.11.2 Key terms

ageing population the average age of a population rising as a result of increases in life expectancy. This can result in social issues such as a need for increased government spending on health care for the elderly population.
disability any condition that restricts a person's mental, sensory or mobility functions
equality providing all people with the same opportunities and treatment
equity providing people with different opportunities and treatment based on level of need
First Nations Australians a generic, collective term that refers all Aboriginal Peoples and Torres Strait Islander Peoples
housing affordability a person's ability to pay for their housing
indicators measurable characteristics of a population
natural increase the changes in population as a result of the difference between birth rates and death rates
regional and remote areas areas classified by their distance and accessibility from major population centres
social cohesiveness the positive feeling of connectedness to a group of people or community
socioeconomic relating to or involving a combination of social and economic factors
sustainability use of a service, product, item or material without depleting its value or accessibility for future generations
taxable income the amount of money a person earns (salary), as a result of their employment, and is used as an indicator of personal wealth
variation (in terms of wellbeing) differences experienced between groups of people due to factors such as their location, demographic or socioeconomic status

21.11.3 Reflection

Complete the following to reflect on your learning:

Revisit the inquiry question posed in the Overview:

How is wellbeing impacted by place and space, and what accounts for the variations in wellbeing within Australia?

1. Now that you have completed this topic, what is your view on the inquiry question? Discuss with a partner. Has your learning in this topic changed your view? If so, how?
2. Write a paragraph in response to the inquiry question, outlining your views.
3. When you have finished, discuss the questions below with your partner.

Given the variations in wellbeing that you are aware of within Australia, what possible solutions might improve levels of wellbeing for the following:

- remote and regional areas of Australia
- Aboriginal Australians
- people with disabilities?

Subtopic	Success criteria			
21.2	I can describe the spatial patterns and characteristics of Australia's population.			
	I can explain why Australia's population is continuing to grow.			
21.3	I can explain the reasons why human wellbeing varies between major cities and regional and remote areas of Australia.			
21.4	I can identify and explain why variations in wellbeing occur within urban areas.			
	I can use statistics and research to show the spatial patterns of wellbeing within Sydney.			
21.5	I can analyse and compare data on one topic when it is presented in multiple formats.			
21.6	I can use statistical data to show the variations in wellbeing experienced by Aboriginal Peoples and Torres Strait Islander Peoples.			
	I can explain the social, economic and health reasons contributing to variations in wellbeing within Aboriginal and Torres Strait Islander communities.			
21.7	I can read policies and strategies and identify the core aim, the strategies that will help to achieve that aim, and the elements of the plan to put the strategies into practice.			
21.8	I can provide data to show how disadvantage and discrimination are experienced by people living with disabilities.			
	I can discuss strategies being implemented to limit these variations.			
21.9	I can explain the challenges associated with living in both urban and regional areas of Northern Territory through the analysis of a topographic map.			

on Resources

eWorkbook Chapter 21 Student learning matrix (ewbk-8600)
Chapter 21 Reflection (ewbk-8601)
Chapter 21 Extended writing task (ewbk-8602)

Interactivity Chapter 21 Crossword (int-8757)

ONLINE RESOURCES

on Resources

Below is a full list of **rich resources** available online for this topic. These resources are designed to bring ideas to life, to promote deep and lasting learning and to support the different learning needs of each individual.

eWorkbook

- **21.1** Chapter 21 eWorkbook (ewbk-8113) ☐
- **21.2** Characteristics of Australia's population (ewbk-9802) ☐
- **21.3** Variations in rural and urban wellbeing (ewbk-9806) ☐
- **21.4** Human wellbeing in Sydney (ewbk-9810) ☐
- **21.5** SkillBuilder — Using multiple data formats (ewbk-9814) ☐
- **21.6** Wellbeing of Aboriginal Peoples and Torres Strait Islander Peoples (ewbk-9818) ☐
- **21.7** SkillBuilder — Understanding policies and strategies (ewbk-9822) ☐
- **21.8** Wellbeing for people with disabilities (ewbk-9826) ☐
- **21.9** Investigating topographic maps — Wellbeing in the Northern Territory (ewbk-9830) ☐
- **21.11** Chapter 21 Student learning matrix (ewbk-8600) ☐
  Chapter 21 Reflection (ewbk-8601) ☐
  Chapter 21 Extended writing task (ewbk-8602) ☐

Sample responses

- **21.1** Human wellbeing in Australia — Sample responses (sar-0169) ☐

Digital document

- **21.9** Topographic map of Darwin and the surrounding area (doc-36374) ☐

Video eLessons

- **21.1** The other side of life (eles-1715) ☐
  Human wellbeing in Australia — Photo essay (eles-5318) ☐
- **21.2** The characteristics of Australia's population — Key concepts (eles-5319) ☐
- **21.3** Variations in rural and urban wellbeing — Key concepts (eles-5320) ☐
- **21.4** Human wellbeing in Sydney — Key concepts (eles-5321) ☐
- **21.5** SkillBuilder — Using multiple data formats (eles-5322) ☐
- **21.6** Wellbeing of Aboriginal Peoples and Torres Strait Islander Peoples — Key concepts (eles-5323) ☐
- **21.7** SkillBuilder — Understanding policies and strategies (eles-5324) ☐
- **21.8** Wellbeing for people with disabilities — Key concepts (eles-5325) ☐
- **21.9** Investigating topographic maps — Wellbeing in the Northern Territory — Key concepts (eles-5326) ☐

Interactivities

- **21.2** The characteristics of Australia's population (int-8749) ☐
- **21.3** Australia's population by remoteness classification (int-8750) ☐
- **21.4** Human wellbeing in Sydney (int-8751) ☐
- **21.5** SkillBuilder — Using multiple data formats (int-8752) ☐
- **21.6** Wellbeing of Aboriginal Peoples and Torres Strait Islander Peoples (int-8753) ☐
- **21.7** SkillBuilder — Understanding policies and strategies (int-8754) ☐
- **21.8** Wellbeing for people with disabilities (int-8755) ☐
- **21.9** Investigating topographic maps — Wellbeing in the Northern Territory (int-8756) ☐
- **21.11** Chapter 21 Crossword (int-8757) ☐

ProjectsPLUS

- **21.10** Thinking Big research project — SDG progress infographic (pro-0217) ☐

Weblinks

- **21.2** Australian Bureau of Statistics (web-4015) ☐
- **21.3** 100 people (web-1249) ☐
  Equity versus equality (web-6398) ☐
  Royal Flying Doctors Association (RFDA) (web-6399) ☐
- **21.4** Australian Bureau of Statistics (web-4015) ☐
- **21.6** Closing the Gap (web-6414) ☐
  National Aboriginal and Torres Strait Islander Social Survey (web-6415) ☐
- **21.7** Aboriginal Education Policy (web-6411) ☐
- **21.8** Human Rights Commission Australia (web-6400) ☐
  ABS — Disability and the labour force (web-6401) ☐

Fieldwork

- **21.8** Accessing community resources (fdw-0048) ☐

Google Earth

- **21.9** Darwin (gogl-0109) ☐

myWorldAtlas

- **21.4** Wellbeing in Western Sydney (mwa-7352) ☐

Teacher resources

There are many resources available exclusively for teachers online.

22 Improving human wellbeing

INQUIRY SEQUENCE

To access a pre-test with **immediate feedback** and **sample responses** to every question in this chapter, select your learnON format at www.jacplus.com.au.

22.1 Overview

Numerous **videos** and **interactivities** are embedded just where you need them, at the point of learning, in your learnON title at www.jacplus.com.au. They will help you to learn the content and concepts covered in this topic.

Who should be responsible for improving human wellbeing standards, and what is the most effective strategy to reduce the spatial inequalities in human wellbeing?

22.1.1 Introduction to improving human wellbeing

Human wellbeing and quality of life can be improved with targeted programs by governments, non-government organisations and individuals. Societies pressure governments for change. Improvements are sought in living conditions, better access to services such as education and health care, and opportunities for jobs and economic advancements. You too can have an impact on the quality of life of people in your community.

FIGURE 1 Families collecting water rations in South Sudan

STARTER QUESTIONS

1. Think of the best things in your life — those things that really make you content. List them in order of importance to your own wellbeing.
2. In the media there is often reference to people whose wellbeing has been affected. In the past month, recall those places and people that you have heard or seen mentioned.
3. What can governments do to improve wellbeing for a community?
4. Watch the **Improving human wellbeing — Photo essay** video in your online Resources. Can you think of any non-government organisations whose purpose is to improve the wellbeing of other people?
5. What can you as an individual do to help others?

on Resources

eWorkbook	Chapter 22 eWorkbook (ewbk-8114)
Video eLessons	Improving human wellbeing — Photo essay (eles-5327) Improving wellbeing (eles-5351)

22.2 The role of governments in improving human wellbeing

LEARNING INTENTION

By the end of this subtopic, you will be able to describe the terms *humanitarianism* and *aid*, and explain how these are used by governments to improve human wellbeing globally.

22.2.1 Humanitarianism and aid

Governments play a national and international role in improving the wellbeing of people. They have the ability to fund projects and facilities to improve health, happiness, prosperity and welfare not only in their own country but across the world. This most often comes in the form of **aid** or **humanitarianism.** Humanitarianism is the most significant motivation for the giving of aid, but it may be motivated by other functions as well:

- as a sign of friendship between two countries
- to strengthen a military ally
- to reward a government for actions approved by the donor
- to extend the donor's cultural influence
- to gain some kind of business or commercial access to a country.

The different types of aid

Bilateral aid is aid given by governments to donor countries. Multilateral aid is provided through international institutions such as UNICEF. Non-government organisation (NGO) or charity aid is provided by organisations such as the Red Cross. Aid takes many forms: money, food, medicine, equipment, expertise, scholarships, training, clothing or military assistance (to name just a few). Large-scale aid (top-down aid) is usually given to the government of a developing country so that it can spend it on the projects that it needs. Small-scale aid projects (bottom-up aid) target the people most in need of the aid and help them directly, without any government interference. Aid from NGOs tends to be bottom-up aid.

FIGURE 1 The Friendship Bridge across the Mekong River, which connects Thailand with Laos, was built with Australian aid.

Aid can increase the dependency of recipient countries on donor countries. Sometimes aid is not a gift but a loan, and poor countries may struggle to repay the money. Aid may also be used to put political or economic pressure on a country, which may leave its people feeling like they owe their donors a favour. There is also the threat that corruption among politicians and officials will prevent aid from reaching the people who need it most. If aid does not provide for and empower citizens, then wellbeing will not improve.

aid (also referred to as international aid) the use of one country's resources (such as military, food or materials) or money (normally in the form of investment) to help another country that is in need

humanitarianism is the concern for the welfare of human beings

22.2.2 Australia's international aid

The Australian government will provide an estimated $4 billion between 2020 and 2021 to many regions in the world (see **FIGURE 3**). Australian aid helps developing countries reduce poverty and achieve **sustainable development**. It focuses on:

- the Asia–Pacific region
- helping over 58 million people annually through various programs
- distributing funds through bilateral and multilateral aid programs as well as emergency relief, such as to the crisis in Syria.

Aid is not always about giving large sums of cash to other governments. It is about investing in various industries to increase the number of jobs available, giving access to education and health services, and setting up sustainable industries to improve all aspects of wellbeing (see **FIGURE 4**).

FIGURE 2 Students study in a school room built with Australian aid in Tawi-Tawi, Bangsamoro Autonomous Region in Muslim Mindanao, The Philippines

sustainable development development that meets the needs of the current generation without the depletion of resources

FIGURE 3 The countries to which the Australian government gives: the size of the dot represents the quantity of aid given.

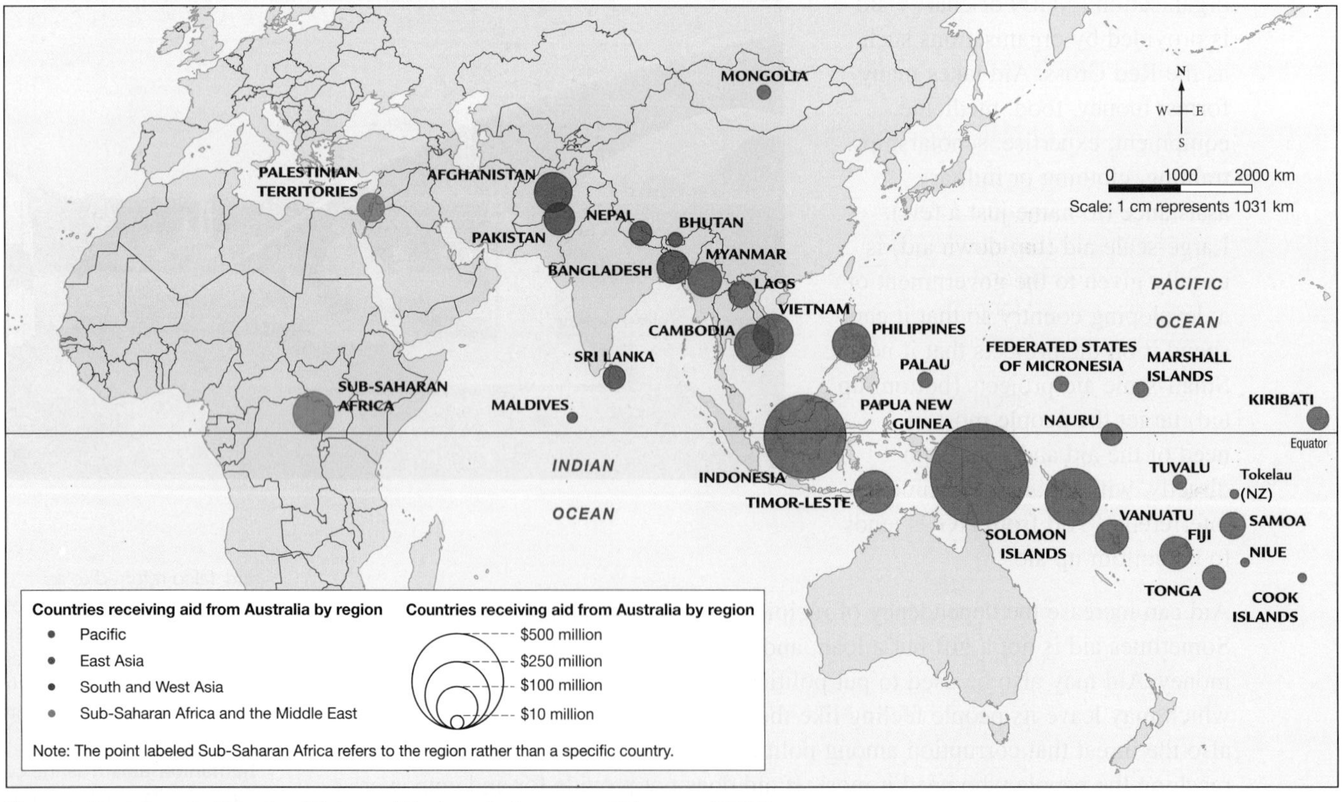

Source: Department of Foreign Affairs and Trade website – www.dfat.gov.au (2021)

FIGURE 4 How Australia provides aid to other countries

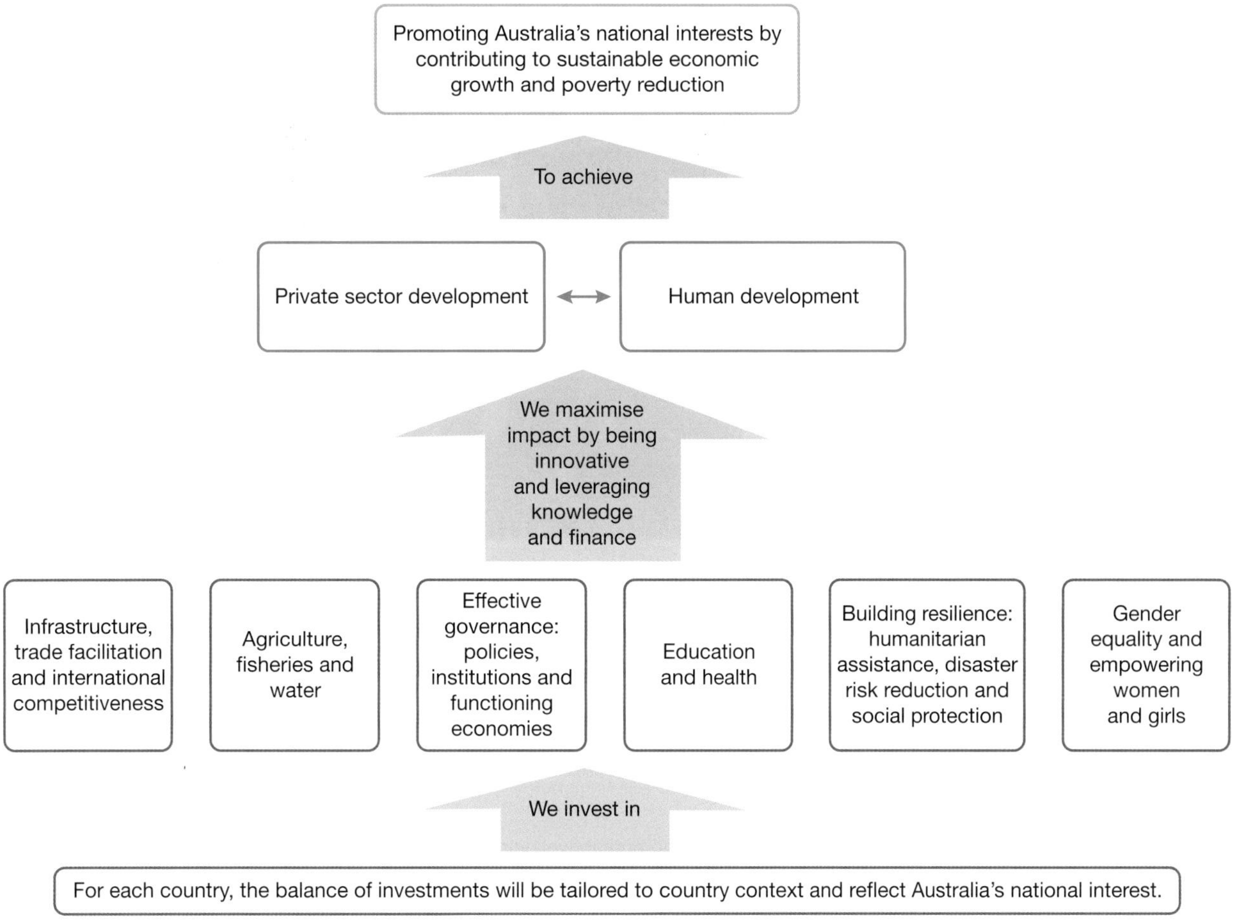

CASE STUDY

Australian aid to Indonesia

Indonesia is a close neighbour and an important partner in our bilateral, regional and global interests. A prosperous, stable and growing Indonesia is good for regional stability, security, trade and cooperation. Australia works with Indonesia to develop effective economic institutions and infrastructure, human development, effective governance and poverty reduction. Australia is pledging $375.7 million to the various sectors presented in **FIGURE 5**).

FIGURE 5 Australian aid to Indonesia by investment priorities, 2015–2016

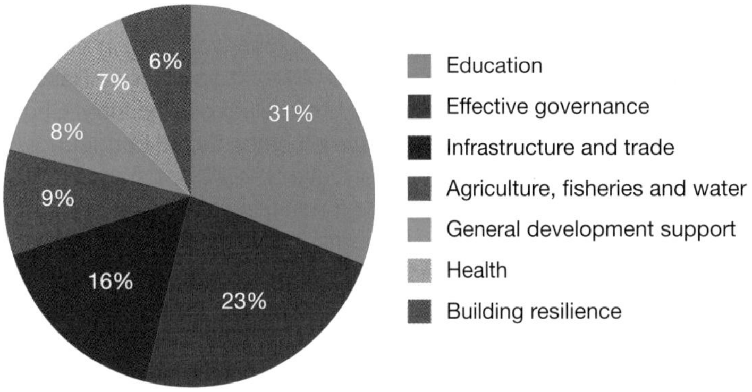

Source: Department of Foreign Affairs and Trade website — www.dfat.gov.au

Australia's aid program with Indonesia has resulted in improvements in wellbeing including the following:

- 866 Australian Awards for Indonesian students to study in Australia
- connecting 464 034 people in rural areas to water and sanitation services in 2014–2015
- supporting the formation of 950 local women's groups across Indonesia, which will help influence policy in areas such as access to jobs and reducing violence against women
- funding skilled birth attendants to reduce maternal mortality by 40 per cent between 2009 and 2014.

In the 2021–22 financial year, the Australian Government allocated $255.7 million for aid specifically to Indonesia. The plans for this funding were adjusted as the extent and severity of the impact of COVID-19 in Indonesia became apparent.

FIGURE 6 The Indonesian government employed food delivery drivers to distribute food aid to vulnerable people during the pandemic, Bogor, Indonesia, 2020.

Working closely with the Indonesian government and health authorities, Australia provided aid to support three key goals in pandemic management and recovery:

- health security: providing help with virus testing and health services to mitigate the spread and impact of the virus, including providing ventilators, emergency assistance and improved sanitation services
- stability: providing aid to assist food security, continuing education and social safety as a means of protecting levels of political and social stability, especially for isolated, marginalised or vulnerable communities
- economic recovery: providing support to businesses affected by the pandemic, especially businesses run by people from vulnerable communities.

on Resources

eWorkbook	The role of governments in improving human wellbeing (ewbk-10347)
Video eLesson	The role of governments in improving human wellbeing — Key concepts (eles-5328)
Interactivity	Helping others (int-3305)

22.2 ACTIVITIES

1. Aid: Is it good or bad? As a class, create a four-column table highlighting the positives and negatives for both the donor and recipient nations of aid, as well as the advantages and disadvantages for both the donor and recipient of aid.
2. How does Australian aid contribute to human wellbeing in other countries? Use the link to the most recent **Australia Aid Budget Summary** from DFAT in your online Resources to answer the following questions. (Open the Australian Aid Budget Summary and select one country. Choose different countries to other groups in your class to get an idea of how Australia allocates its aid budget.)
   a. How much is Australia's Official Development Assistance (ODA) in your selected country?
   b. List the three main priorities the funding is supporting in your country.
   c. Provide two examples of programs, activities or initiatives that the Australian Government is supporting and what they are hoping to achieve.
   d. As a class, discuss your findings and decide on the main priorities of Australia's aid. Develop a one-sentence mission statement for the Australian Department of Foreign Affairs and Trade.
3. In pairs or small groups, design a country-specific program that you think would help alleviate chronic poverty in one of sub-Saharan Africa's most poverty-stricken countries.
4. In two groups, debate the contention 'Australia should do more to support global human wellbeing'.
5. Investigate the impact a cyclone might have on an LEDC compared to a MEDC. Do some research online on the impact of Cyclone Yasi on Northern Queensland in comparison to Vanuatu. Write a one-paragraph justification on why Vanuatu would require different types of aid in order to recover from the impacts of Cyclone Yasi.
6. Investigate how the Australian government has supported countries in the Pacific region during the COVID-19 pandemic. Discuss the extent and types of aid provided.

22.2 EXERCISE

Learning pathways

■ LEVEL 1	■ LEVEL 2	■ LEVEL 3
1, 2, 7, 10, 14	4, 5, 9, 11, 15	3, 6, 8, 12, 13

Check your understanding

1. List the two main types of aid and the different forms aid can come in.
2. List the countries Australia gives aid to.
3. Provide reasons why Australia would give aid to other countries.
4. What is Australia's total aid contribution for 2020–2021?
5. Refer to **FIGURE 3** and the case study above to answer the following questions.
   a. How much aid did Australia give to Indonesia in 2015–2016?
   b. In what sectors was the aid invested?
   c. How much money was given to develop the education sector?

Apply your understanding

6. Using **FIGURE 4**, suggest reasons why aid is given in infrastructure and development and not cash.
7. Explain how aid improves the wellbeing of people.
8. Assess the positive and negative impacts of aid on recipient countries.
9. Using **FIGURE 3**, examine the map of Australia's aid and write a short paragraph describing the spatial patterns of Australia's aid.
10. Provide some reasons why Australia's aid might be focused on neighbouring countries.
11. What are the benefits to Australia in providing aid to our neighbouring countries?
12. Use **FIGURE 4** to answer the following questions.
    a. Choose one area that Australia's aid invests in. Justify the benefit to Australia in investing in this area in another country.
    b. Distinguish the difference between private sector development and human development.
    c. What is the benefit to Australia in reducing poverty in other countries?

Challenge your understanding

13. Recommend how Australia could change or improve their foreign aid allocations as a result of a significant conflict in a neighbouring country.
14. Do you think the Australian government does enough to assist countries in need? Justify your answer.
15. Might it be inappropriate for Australia to provide aid in some circumstances? Write one paragraph explaining why.

FIGURE 7 An Australian Army private teaches first aid to members of the Papua New Guinea Defence Force.

To answer questions online and to receive **immediate feedback** and **sample responses** for every question, go to your learnON title at www.jacplus.com.au.

22.3 The role of NGOs in improving human wellbeing

LEARNING INTENTION

By the end of this subtopic, you will be able to explain how NGOs improve human wellbeing in LEDCs and justify why a bottom-up approach to managing human wellbeing issues is effective.

22.3.1 NGOs

Non-government organisations (NGOs), also known as charities, play a critical role in improving wellbeing for many different groups around the world. NGOs are not-for-profit organisations. Most NGOs aim to improve the standard of living and quality of life for people, animals and the environment. Charity aid is voluntary; governmental, private and individual donations are collected by organisations such as the Red Cross, Greenpeace, Sea Shepherd and the World Wildlife Fund (WWF) to meet their aims.

22.3.2 Aid from NGOs

Aid takes many forms, including money, food, medicine, equipment, expertise, scholarships, training, clothing and military assistance. Aid received from NGOs is generally for small-scale aid projects (bottom-up aid) that target the people most in need of assistance. These small-scale projects help people directly without any government interference. UNICEF, UNHCR, Red Cross, Oxfam and OzHarvest are just some examples of NGOs that work domestically and internationally to improve people's wellbeing.

non-government organisation (NGO) a citizen-based association that operates independently of government, usually to deliver resources or serve some social or political purpose

CASE STUDY

OzHarvest and Oxfam

OzHarvest

OzHarvest is an Australian NGO that redistributes excess and unwanted food from supermarkets, bakeries, cafés and restaurants to people in need at refuge centres, homeless shelters, youth groups and other organisations that service people with addictions, people with mental health issues, people with disabilities and the elderly. OzHarvest focuses on rescuing food, educating people, engaging the community and promoting innovation to combat food waste. With assistance from a large group of volunteers, OzHarvest successfully rescues and redelivers approximately 56 tonnes of food each week. Food is an important component in a person's wellbeing. It contributes to good health and welfare of an individual. According to Foodbank, nearly one million Australian children go without breakfast or dinner.

FIGURE 1 OzHarvest redistributes excess and unwanted food to people in need.

Oxfam

Oxfam is an international NGO that focuses on fighting poverty through the provision of education, food, healthcare and infrastructure development. It has assisted approximately 25 million people in more than 85 countries around the world during 2014–2015. It also has partnerships with more than 138 fair trade and ethical producers, which contributes to the wellbeing of those producers.

Oxfam relies on donations to support its wellbeing projects and the majority of the money donated goes directly to the program (see **FIGURE 2**). A $300 donation allows Oxfam to provide a water harvesting system to supply clean water for 200 families in drought-stricken southern Africa. Clean water is essential in preventing diseases such as cholera. Water-poor communities are usually economically poor as well, because of the waterborne illnesses. Education levels also suffer because sick children are unable to attend school.

FIGURE 2 How Oxfam spends the money that is donated to its wellbeing projects:

(a) Education resources

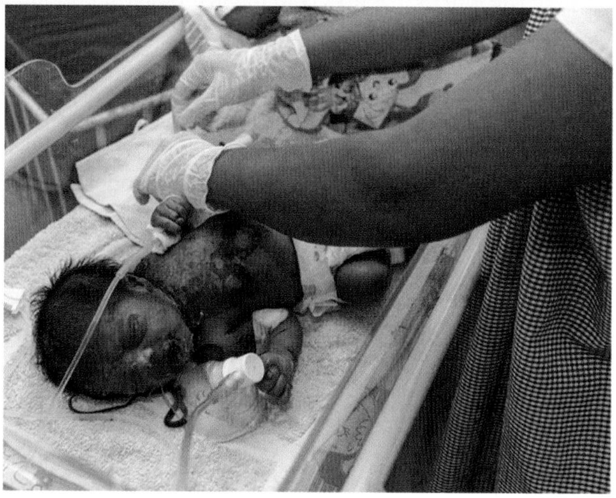

(b) Healthcare and health facilities

(c) Food programs and nutrition

(d) Access to clean water sources to ensure water security

on Resources

eWorkbook	The role of NGOs in improving human wellbeing (ewbk-10351)
Video eLesson	The role of NGOs in improving human wellbeing — Key concepts (eles-5329)

22.3 ACTIVITIES

1. Research an international NGO and show how it is working towards peace in areas of very low peacefulness. Focus on a country not studied in this chapter.
2. In groups, do some research online to find a list of NGOs that help in your local area.
   a. What do they do for people?
   b. Develop your own NGO aimed at making a positive change on a human wellbeing issue in your local area. The aim of this task is to make a positive change on human wellbeing within your area. Structure your task using the subheadings below.
      i. The issue
      ii. Examples, statistics and research to explain the issue
      iii. Strategies/programs/initiatives you have developed to resolve the issue
   c. Develop your idea further by focusing on its sustainability. Think about any negative economic, social and environmental impacts of your NGO and find ways to minimise them. For example, how can your idea be funded sustainably, or what will you do with any waste created?

22.3 EXERCISE

Learning pathways

■ LEVEL 1	■ LEVEL 2	■ LEVEL 3
1, 2, 7, 8, 13	3, 4, 9, 10, 14	5, 6, 11, 12, 15

Check your understanding

1. Describe the role of NGOs in your own words. (Include what NGO stands for in your answer.)
2. Identify at least three different types of aid that NGOs can give.
3. Outline the benefits of small-scale aid.
4. What role do NGOs play in restoring wellbeing to a country?
5. Define the terms *bottom-up* and *top-down* in terms of approaches to managing human wellbeing issues.
6. Describe why NGOs are often described as using a bottom-up approach to human wellbeing issues.

Apply your understanding

7. Provide reasons why education levels drop in areas where children are exposed to health issues such as cholera.
8. With the use of statistics and examples, write a paragraph to explain how OzHarvest and Oxfam improve human wellbeing.
9. Write a paragraph outlining what NGOs do and how and why they do it. Use examples to illustrate your explanation.
10. Explain why a bottom-up approach to managing human wellbeing issues is effective.
11. Choose an image from **FIGURE 2** and explain how that image shows that Oxfam has had a positive impact on human wellbeing.
12. Develop a flow chart to show the potential flow on effect from the positive impact of one example of aid. For example, improvements in successful birthing rates may lead to a reduction in population growth in LEDCs, resulting in less strain on primary resources such as food.

Challenge your understanding

13. Choose an important issue that affects wellbeing of people in Australia. Justify why a bottom-up focused program would be the most effective approach to improving your issue.
14. Suggest why it might be important for NGOs and governments to work collaboratively on a human wellbeing issue.
15. Predict whether you think communities and individuals in need of aid will rely on NGOs more or less in the next 50 years. Give reasons for your view.

To answer questions online and to receive **immediate feedback** and **sample responses** for every question, go to your learnON title at www.jacplus.com.au.

22.4 The role of individuals in improving human wellbeing

LEARNING INTENTION

By the end of this subtopic, you will be able to discuss the role individuals play in improving standards of human wellbeing.

22.4.1 Factors that improve the wellbeing of individuals

Individuals have a role to play in their own state of wellbeing, but they also contribute to the wellbeing of others. All aspects of life influence a person's health, happiness, prosperity and welfare.

Happiness and contentment enhance a person's wellbeing. The Victorian Government published a list of factors that influence a positive state of wellbeing for individuals. The list includes:

- a close network of friends
- an enjoyable and fulfilling career
- regular exercise
- nutritional diet
- sufficient sleep
- fun hobbies and leisure pursuits
- a healthy self-esteem
- realistic and achievable goals
- a sense of belonging
- the ability to adapt to change
- living in a fair and democratic society
- enough money
- optimistic outlook
- spiritual or religious beliefs.

FIGURE 1 Factors that influence wellbeing are interconnected. For example, participating in regular exercise allows you to meet people to form a close network of friends.

22.4.2 Individual action to improve human wellbeing

Helping another person not only makes you feel good; it also contributes to the other person's wellbeing. There are many ways individuals can get involved in larger programs through governments and NGOs to assist other people and improve their wellbeing.

Volunteering

Volunteering your time in a project that interests you can be one of the most rewarding and eye-opening experiences. Many non-government organisations offer volunteer positions domestically and internationally. Getting involved in projects allows you to engage and connect with your local community or even the international community. For example, Doctors Without Borders (also known as Médecins Sans Frontières) are an international NGO who deliver emergency medical aid to people affected by conflict, epidemics, disasters or exclusion from healthcare.

Regardless of the type of volunteering work and the time you have available, you are making a difference to a person or family. If you are interested in becoming involved, a good place to start is Volunteering Australia (www.volunteeringaustralia.org).

volunteering giving up time to commit to an idea or organisation without being paid

FIGURE 2 (a) Localised volunteering includes activities such as serving food in homeless shelters. (b) Participating in local and overseas programs can involve teaching children how to read and write or helping with afterschool programs. (c) Juan Mann made it his mission to reach out and hug strangers to brighten up their day. He was often located in Pitt Street Shopping Mall, Sydney. Community service volunteers also play an important role in keeping Australians safe, such as volunteer emergency services (d) and surf life savers (e).

fdw-0049

FOCUS ON FIELDWORK

Wellbeing at school

Your task is to produce a fieldwork report that you could present to your school that outlines student wellbeing. You need to focus on what your school already does for student wellbeing and areas that could be improved. You also need to make suggestions on how individuals and groups can get involved to improve wellbeing at your school. Tip: A good way to measure wellbeing is to create a student satisfaction survey based on what makes them happy.

FIGURE 3 Article extract, from *The Conversation* on measuring wellbeing

Beyond GDP: are there better ways to measure well-being?

Kubiszewski, I. (2014, December 2). 'Beyond GDP: are there bettern ways to measure well-being?' The Conversation

'At present, we are stealing the future, selling it in the present, and calling it GDP.' — Paul Hawken

Imagine if a corporation used Gross Domestic Product (GDP) accounting to do its books: it would be adding all its income and expenses together to get a final number. Nobody would think that's a very good indication of how well that business was doing. Herman Daly, a former senior economist at the World Bank, said that, 'the current national accounting system treats the earth as a business in liquidation.' He also noted that we are now in a period of 'uneconomic growth'; where GDP is growing but societal welfare is not.

The good news is that there are several alternatives to GDP being actively developed, discussed, and used. One of these is the Genuine Progress Indicator (GPI).

GPI starts with personal consumption expenditures — a major component of GDP — and adjusts it using 25 components. These adjustments include incorporating the negative effects of income inequality on welfare; adding positive elements not considered in GDP, such as the benefits of household work, volunteer work, and higher education; and subtracting environmental costs and social costs like the costs of crime, unemployment, and pollution. In doing so, it paints a more accurate picture of how far we've come over the last three decades.

Regional level

In the United States, the states of Maryland and Vermont officially report their GPI annually. In 2010, Maryland was the first state to officially adopt the GPI as an alternative to GDP. The state's goal was 'to measure whether or not economic progress results in sustainable prosperity'.

Since its beginning, Maryland governor Martin O'Malley has actively implemented policies to encourage the increase of GPI. The media has also taken up the challenge of shedding light on the true picture of societal welfare, with media coverage now regularly reporting on changes in GPI.

In 2012, the state of Vermont passed legislation mandating the calculation of GPI. Since then, GPI has been estimated for other states including Colorado, Hawaii, Massachusetts, Michigan, Ohio, Oregon, and Utah, while ten others have expressed interest in developing their own studies. This movement towards GPI is part of an international trend, and GPI has been calculated in around 20 countries worldwide. The international research community has begun to develop what is being called 'GPI 2.0'. GPI 2.0 seeks to improve standardization and robustness of the current GPI.

National level

GPI is not the only new measure of societal welfare being adopted around the world. The Kingdom of Bhutan began using Gross National Happiness (GNH) as an alternative to GDP in 1972 after fourth King, Jigme Singye Wangchuck, stated that 'Gross National Happiness is more important than Gross National Product.'

GNH is estimated using a survey that takes approximately seven hours per person. In 2013, it was taken by more than 10% of the Bhutanese population. The Bhutanese government also established the GNH Commission to assess all new policies for their impact on the 'happiness' or well-being of the population.

The Australian Bureau of Statistics also began moving in this direction with the Measures of Australia's Progress (MAP) initiative. MAP was established to address the question, 'Is life in Australia getting better?' MAP provides Australians with 26 indicators related to society, economy, environment, and governance. Unlike the GPI, it does not aggregate the indicators into one overall measure, but allows viewers to make their own assessment regarding the well-being of the Australian population based on the individual indicators. However, funding was discontinued for MAP in early 2014.

Global effort

Currently, no global consensus exists regarding alternatives to GDP. However, there is growing agreement that the continued use of GDP as a proxy for overall well-being is not appropriate. A range of national indicators exist and are being used around the world.

Alternative national indicators of welfare and well-being		
Indicator	**Explanation**	**Coverage**
Index of Sustainable Economic Welfare (ISEW) & Genuine Progress Indicator (GPI) **Type: GDP modification** **Unit: dollar**	Personal consumption expenditures weighted by income distribution, with volunteer and household work added and environmental and social costs subtracted.	• 17 countries, several states and regions • 1950-various years
Genuine savings **Type: Income accounts modification** **Unit: dollar**	Level of saving after depreciation of produced capital, investments in human capital, depletion of minerals, energy, and forests, and damages from local and global air pollutants are accounted for.	• 140 countries • 1970–2008
Inclusive Wealth Index **Type: Capital accounts modification** **Unit: dollar**	Asset wealth including built, human, and natural resources.	• 20 countries • 1990–2008
Australian Unity Well-Being Index **Type: Survey based index** **Unit: Index**	Annual survey of various aspects of well-being and quality of life	• Australia • 2001-present
Gallup-Healthways Well-Being Index **Type: Survey based index** **Unit: Index**	Annual survey in taking into account five elements purpose (employment, etc), social, financial, community and physical (health).	• 50 states of the USA, expanded to 135 countries in 2013. • 2008-present
Gross National Happiness **Type: Survey based index** **Unit: Index**	Detailed in-person survey around nine domains psychological well-being, standard of living, governance, health, education, community vitality, cultural diversity, time use, and ecological diversity.	• Bhutan • 2010
Human Development Index **Type: Composite index** **Unit: Index**	Index of GDP per person, spending on health and education, and life expectancy.	• 177 countries • 1980-present
Happy Planet Index **Type: Composite index** **Unit: Index**	A calculation based on subjective well being multiplied by life expectancy divided by ecological footprint.	• 153 countries • 3 years
OECD Better Life Index **Type: Composite index** **Unit: Index**	Includes housing, income, jobs, community education, environment, civic engagement, health, life satisfaction, safety, and work-life balance.	• 36 OECD countries • 1 year

Source: The Conversation (Ida Kubiszewski)

The Millennium Development Goals, which are ending in 2015, look primarily at health, poverty, and education. To replace them, a UN-led initiative has developed a set of 17 Sustainable Development Goals (SDGs). These new goals have a broader agenda that includes the environment, inequality, and sustainable and equitable economic growth, amongst other aspects.

Although metrics are being developed for each of the 17 goals and their sub-goals, as of yet, no overall indicator has been developed to assess the overall success of the SDGs.

Robert F. Kennedy once said that a country's GDP measures 'everything except that which makes life worthwhile'. The only way to dethrone GDP from its current role, is to start measuring all those things that do 'make life worthwhile'.

on Resources

eWorkbook — The role of individuals in improving human wellbeing (ewbk-10355)

Video eLesson — The role of individuals in improving human wellbeing — Key concepts (eles-5330)

Interactivity — The role of individuals in improving human wellbeing (int-8759)

22.4 ACTIVITIES

1. As a class, discuss the terms *wellbeing* and *development*. Brainstorm key words for each of the terms using a Venn diagram.
2. Discuss the differences in human wellbeing and development between MEDCs and LEDCs.
3. Think of how wellbeing as a concept can be measured and quantified. What indicators could be used to measure wellbeing? Discuss what the issues might be with measuring wellbeing using quantitative indicators?
4. Refer to the article from The Conversation on measuring wellbeing **(FIGURE 3)** to complete the following.
   a. Choose **one** alternative indicator of wellbeing from the article.
   b. Explain what it measures.
   c. Identify the top 10 and bottom 10 countries in the world using this index.
   d. Compare your findings with the person sitting next to you.
   e. How does the ranking of countries based on wellbeing change when different indicators are used?
   f. How might cultural or social factors affect the way people respond to surveys or questions about their own wellbeing? Discuss how you could avoid bias (of any kind such as racial, gender, age, ability) when measuring qualitative indicators of wellbeing.

FIGURE 4 How might your social or cultural views impact on your understanding of wellbeing

22.4 EXERCISE

Learning pathways

■ LEVEL 1	■ LEVEL 2	■ LEVEL 3
2, 4, 7, 11	1, 3, 6, 10	5, 8, 9, 12

Check your understanding

1. Define *wellbeing*.
2. Create a list of what makes you happy. Would a list of what helps your wellbeing be different to this list? Outline why or why not.
3. What factors influence a person's wellbeing?
4. What is a volunteer? What do they do?
5. Give three reason why people volunteer.

Apply your understanding

6. Explain how volunteering can improve human wellbeing.
7. What could your school do to improve:
   a. student wellbeing
   b. the wellbeing of other community groups?
8. Think of a human wellbeing issue in your local area and justify why an individual approach to improving this issue would be the most effective.
9. Read the extract below and write justifications for the following.
   'Human wellbeing is the recognition that everyone around the world, regardless of geography, age, culture, religion or political environment, aspires to live well. Wellbeing is not necessarily bound by income, rather, it is an individual's thoughts and feelings about how well they are doing in life, contentment with material possessions and having relationships that enable them to achieve their goals.'
   Source: © Commonwealth of Australia, 2013. *Geographies of Human Wellbeing*. Geography Teachers' Association of Victoria Inc. (Global Education Project Victoria).
   a. Why human wellbeing is subjective in nature
   b. Why human wellbeing should be a compulsory subject in school
   c. Why a deeper understand of human wellbeing issues will have a positive impact to the world

Challenge your understanding

10. What makes human wellbeing a geographical issue?
11. What can you do to contribute to and improve:
    a. your own wellbeing
    b. someone else's wellbeing
    c. community wellbeing?
12. Examine the alternative indicators of welfare and wellbeing presented in this subtopic. Suggest which would be the most appropriate way to measure the wellbeing of people in your community, and justify your decision with examples.

To answer questions online and to receive **immediate feedback** and **sample responses** for every question, go to your learnON title at www.jacplus.com.au.

22.5 Improving wellbeing for Aboriginal Peoples and Torres Strait Islander Peoples

LEARNING INTENTION

By the end of this subtopic, you will be able to explain inequalities in wellbeing within the Aboriginal and Torres Strait Islanders populations, describe programs aimed to improve the wellbeing of these communities, and propose action to further reduce the inequality.

22.5.1 Recognising the divide

Significant disadvantages are experienced by Aboriginal Peoples and Torres Strait Islander Peoples across all socioeconomic indicators in Australia. These disadvantages are the result of historical influences that have contributed to social inequality, which in many ways is still present in contemporary society. Aboriginal Peoples and Torres Strait Islander Peoples have experienced exclusion from socially valued resources and services as a result of historical policies and laws. This exclusion is a key contributing factor to these issues still being prevalent in society today. Australian governments and other agencies such as Oxfam are continuing to push initiatives aimed at improving some of the problems that many Aboriginal communities face.

Ultimately, all Australians benefit from a united effort to address disadvantage, and benefit when disadvantage is overcome. However, Aboriginal Peoples and Torres Strait Islander Peoples will specifically be better placed to fulfil their own cultural, social and economic aspirations if this disadvantage is addressed.

22.5.2 Closing the Gap

The Closing the Gap partnership is a national agreement between federal and state governments and the Coalition of Aboriginal and Torres Strait Islander Peak Organisations. The core principle of the agreement is outlined in **FIGURE 1**.

FIGURE 1 The principle and focus of Closing the Gap, as explained by the Australian government

What we know

Closing the Gap acknowledges the ongoing strength and resilience of Aboriginal and Torres Strait Islander people in sustaining the world's oldest living cultures.

Closing the Gap is underpinned by the belief that when Aboriginal and Torres Strait Islander people have a genuine say in the design and delivery of policies, programs and services that affect them, better life outcomes are achieved. It also recognises that structural change in the way governments work with Aboriginal and Torres Strait Islander people is needed to close the gap.

What we are doing differently

All Australian governments are working with Aboriginal and Torres Strait Islander people, their communities, organisations and businesses to implement the new National Agreement on Closing the Gap at the national, state and territory, and local levels.

This is an unprecedented shift in the way governments have previously worked to close the gap. It acknowledges that to close the gap, Aboriginal and Torres Strait Islander people must determine, drive and own the desired outcomes, alongside all governments.

This new way of working requires governments to build on the strong foundations Aboriginal and Torres Strait Islander people have, through their deep connection to family, community and culture.

Source: https://www.closingthegap.gov.au/

The National Agreement is divided into eight outcome areas:

1. Education
2. Employment
3. Health and wellbeing
4. Justice
5. Safety
6. Housing
7. Land and waters
8. Languages.

FIGURE 2 The opening of the food store in Ngukurr, Northern Territory

Under the Closing the Gap program, these outcomes have specific targets, many of which relate directly or indirectly to wellbeing, including:

- Target 1: closing the life expectancy gap between Aboriginal and non-Aboriginal people by 2031
- Target 2: ensuring 91 per cent of Aboriginal and Torres Strait Islander babies are born at a healthy birthweight by 2031
- Target 4: ensuring 55 per cent of children are meeting development standards for their age
- Target 13: reducing family violence by at least 50 per cent by 2031
- Target 14: reducing rates of suicide and self-harm
- Target 16: supporting Aboriginal Cultures and Languages to flourish.

Aspects of these targets also involve significant public health initiatives. Child mortality rate targets are currently on track to be met due to improvements in **antenatal care**, sanitation and public health; better **neonatal intensive care**; and the development of immunisation programs.

antenatal care the branch of medicine that deals with the care of women during pregnancy, childbirth and recovery after childbirth

neonatal intensive care the specialised nursing practice of caring for newborn infants

Meeting the life expectancy target will be challenging, particularly as overall life expectancy for the population as a whole is increasing. To meet the life expectancy target, average Indigenous life expectancy gains of 0.6 and 0.8 years per year are needed—that is almost 21 years by 2031 to close the gap.

22.5.3 Programs to improve wellbeing

The following initiatives provide examples as to how both government and non-government agencies are working to improve the health of the Aboriginal Peoples of Australia.

- **The National Partnership Agreement on Closing the Gap in Indigenous Health Outcomes.** For example, the Many Rivers Aboriginal Medical Service Alliance in northern New South Wales brings together 10 Aboriginal-controlled health organisations that share resources and programs servicing 35 000 people.
- **The Australian Government licensing scheme for community stores in the Northern Territory.** This scheme requires store managers to offer a range of healthy food and drinks and to make these attractive to customers. Before this, people in remote Indigenous communities often had little choice. Goods and food were of poor quality, and basic consumer protection was lacking. In December 2011, 90 Northern Territory stores, such as that pictured in FIGURE 2, were licensed.
- **Oxfam's Indigenous Health and Wellbeing Program.** Oxfam works with Aboriginal and Torres Strait Islander organisations to hold governments to account over the Closing the Gap program. It also supports the Fitzroy Stars Football Club, which competes in Melbourne's Northern Football League. This club brings together 300 Aboriginal men with the aim of nurturing a culture that promotes a healthy lifestyle, fitness, nutrition and self-esteem. It also aims to build bridges between Aboriginal and non-Aboriginal communities.

CASE STUDY

Lombadina community program

Lombadina community is home to the Bardi people. It is located on the north-western coast of Western Australia (see **FIGURE 3**). Lombadina and the neighbouring Djarindjin community are home to approximately 200 people. The Lombadina community is working towards self-sufficiency through ventures that include tourism operations, a general store, an artefact and craft shop, a bakery and a garage. The tourist ventures centre on sharing knowledge of Bardi lifestyle and culture.

In addition to serviced accommodation, many tours are offered, including cultural tours, fishing charters, kayaking and bushwalking. Lombadina has received a number of tourism awards. The considerable success of these businesses has contributed substantially to the wellbeing of this community. Use the **Lombadina** weblink in your online Resources to experience a kayaking tour led by a Lombadina community member.

Lombadina is also involved in the EON Thriving Communities Project. EON is a non-government organisation operating by invitation in Aboriginal communities in Western Australia. It aims to close the gap in terms of health; for example, with the provision of practical knowledge about growing and preparing healthy food in schools and communities. The project has community ownership and is designed to be sustainable, thus improving wellbeing in the long term.

FIGURE 3 Location of Lombadina, Western Australia

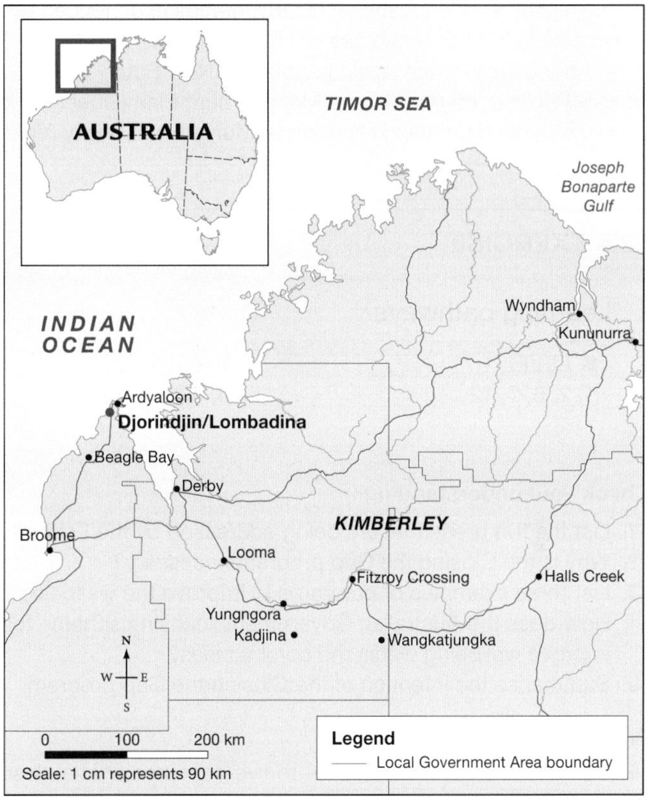

Source: Spatial Vision GAT-45; © Commonwealth of Australia Geoscience Australia 2013. © Commonwealth of Australia Australian Bureau of Statistics 2013.

on Resources

eWorkbook Improving wellbeing for Aboriginal Peoples and Torres Strait Islander Peoples (ewbk-10359)

Video eLesson Improving wellbeing for Aboriginal Peoples and Torres Strait Islander Peoples — Key concepts (eles-5331)

Interactivity Improving wellbeing for Aboriginal Peoples and Torres Strait Islander Peoples (int-8760)

22.5 ACTIVITIES

1. National Close the Gap Day is held in March each year to improve community awareness of the issue of Indigenous disadvantage and to publicise Federal Government action. Research when the next National Close the Gap day is and what activities are taking place in your state and/or local area for Close the Gap Day.
2. Using the **Lombadina Census QuickStats** weblink in your online Resources, find some statistics to assess the wellbeing of the Lombadina community relative to Western Australia and Australia. Justify why you chose the statistics as indicators of wellbeing.
3. The history of aid and intervention in communities from governments and other organisations is a complex issue that is a matter of deep pain and frustration for many Aboriginal Peoples in Australia. As a class, discuss what it means to be a good ally to Aboriginal Peoples in Australia today. Consider the best ways to empower other people rather than imposing your own ideas of what they might need or want.

4. Using an online infographic generator such as Canva or Piktochart, create a poster promoting National Close the Gap day to improve awareness of the issue at your school. Include statistics and research on the issue of disadvantage within the indigenous Australian community and give some examples of some initiatives being introduced to reduce this inequality.
5. Investigate one of the following organisations that have experienced success in combating some of the health, social or educational disadvantages experienced by First Nations Australians. Why have they been successful? What outcomes will be changed?
   - Aboriginal Women Against Violence (New South Wales)
   - MPower — Family Income Management Plan (Queensland)
   - Indigenous Enabling Program at Monash University (Victoria)

22.5 EXERCISE

Learning pathways

■ LEVEL 1	■ LEVEL 2	■ LEVEL 3
1, 2, 6, 7, 12	3, 4, 8, 9, 13	5, 10, 11, 14

Check your understanding

1. List the five areas that are being addressed by the Federal Government's Closing the Gap program.
2. Why is the Closing the Gap program necessary?
3. List three examples of programs to improve the wellbeing of Aboriginal peoples in Australia.
4. How does the Australian Government licensing scheme for community stores in the Northern Territory improve wellbeing within the communities?
5. Summarise the intention of the Closing the Gap program.

Apply your understanding

6. Using **FIGURE 3**, write a three- to five-sentence spatial description for the location of Lombadina.
7. Explain how tourism initiatives such as those run by the Lombadina community improve the wellbeing of people beyond that community.
8. Justify the reasons why all Australians benefit from a united effort to address Aboriginal and Torres Strait Islander disadvantage.
9. Write a one-paragraph discussion on how the Lombadina community is working towards self-sufficiency in improving wellbeing.
10. Explain how the tourism industry has improved wellbeing within the Lombadina community.
11. Explain the purpose of the Thriving Communities Program.

Challenge your understanding

12. Propose a sustainable, bottom-up solution to improve the wellbeing of Aboriginal Australians from each of the following.
    a. Individuals
    b. A non-government organisation
    c. A government organisation
13. Write a one-paragraph justification of how one bottom-up solution to help close the gap is sustainable economically, socially, and environmentally.
14. Why might some programs such as the Thriving Communities Program be more successful than others? Suggest and explain three reasons why a project to improve wellbeing might be successful.

To answer questions online and to receive **immediate feedback** and **sample responses** for every question, go to your learnON title at www.jacplus.com.au.

22.6 SkillBuilder — Debating like a geographer

online only

LEARNING INTENTION

By the end of this subtopic, you will be able to analyse and discuss geographical issues from an objective perspective using accurate and reliable evidence to support your view.

The content in this subtopic is a summary of what you will find in the online resource.

22.6.1 Tell me

What does debating like a geographer mean?

Debating like a geographer involves being able to give the points for and against any issue that has a geographical basis, and supporting the ideas with arguments and evidence of a geographical nature. Debating like a geographer is useful for showing the different points of view on a wide range of global, national and local issues that affect our lives.

22.6.2 Show me

Step 1

Determine the topic to be debated and the rules of your debate such as timings.

Step 2

Create two teams of three debaters. One team must argue for the topic (the affirmative) and one team must argue against the topic (the negative).

Step 3

Appoint a chairperson, judges and a timekeeper. The chairperson introduces the speakers and keeps order during the debate. The judges use a set of criteria to score each of the speakers' points. The timekeeper ensures that each speaker has equal time to speak.

Step 4

Research and prepare arguments based on geographical information: maps, statistics, graphs and data.

Step 5

Prepare the classroom for a formal debate. The adjudicated outcome is given when the judges have considered three key aspects of the debate: geographical matter, method and manner.

22.6.3 Let me do it

learnON

Go to learnON to access the following additional resources to help you build this skill:

- a longer explanation of this skill and its application in Geography (Tell me)
- a video demonstrating the step-by-step process of this skill (Show me)
- an activity and interactivity for you to practise the skill (Let me do it)
- self-marking questions to help you understand and use the skill.

on Resources

eWorkbook	SkillBuilder — Debating like a geographer (ewbk-10363)
Video eLesson	SkillBuilder — Debating like a geographer (eles-1762)
Interactivity	SkillBuilder — Debating like a geographer (int-3380)

22.7 Improving wellbeing in Brazil

LEARNING INTENTION

By the end of this subtopic, you will be able to describe the spatial patterns of wellbeing within Brazil and evaluate the effectiveness of government initiatives to reduce the variations in human wellbeing for people living in Rio de Janeiro.

22.7.1 Variations in wellbeing within a city

How would you like to live with spectacular views over one of the world's most beautiful coastlines? The only problem is that you could be living in a slum without running water and your only access in or out is via hundreds of stairs and laneways. This is life in a typical **favela** in Rio de Janeiro.

In 2011, Brazil overtook the United Kingdom as the world's seventh-largest economy in terms of **GDP**. Despite its development over the past 30 years, the benefits of economic growth have not trickled down to the poor, resulting in large differences in wellbeing across the nation.

A relatively small group of Brazilians live extremely well, while 11 per cent of the approximate population of 202 million inhabitants remain in poverty. Of these, almost half continue to experience **extreme poverty** (see **FIGURE 1**). It is important to note that

favela an area of informal housing usually located on the edge of many Brazilian cities. Residents occupy the land illegally and build their own housing. Dwellers often live without basic infrastructure such as running water, sewerage or garbage collection

GDP (gross domestic product) the total value of all goods and services produced in a country in a given year, equal to total consumer, investment and government spending, plus the value of exports minus the value of imports

FIGURE 1 Distribution of wealth per state of Brazil

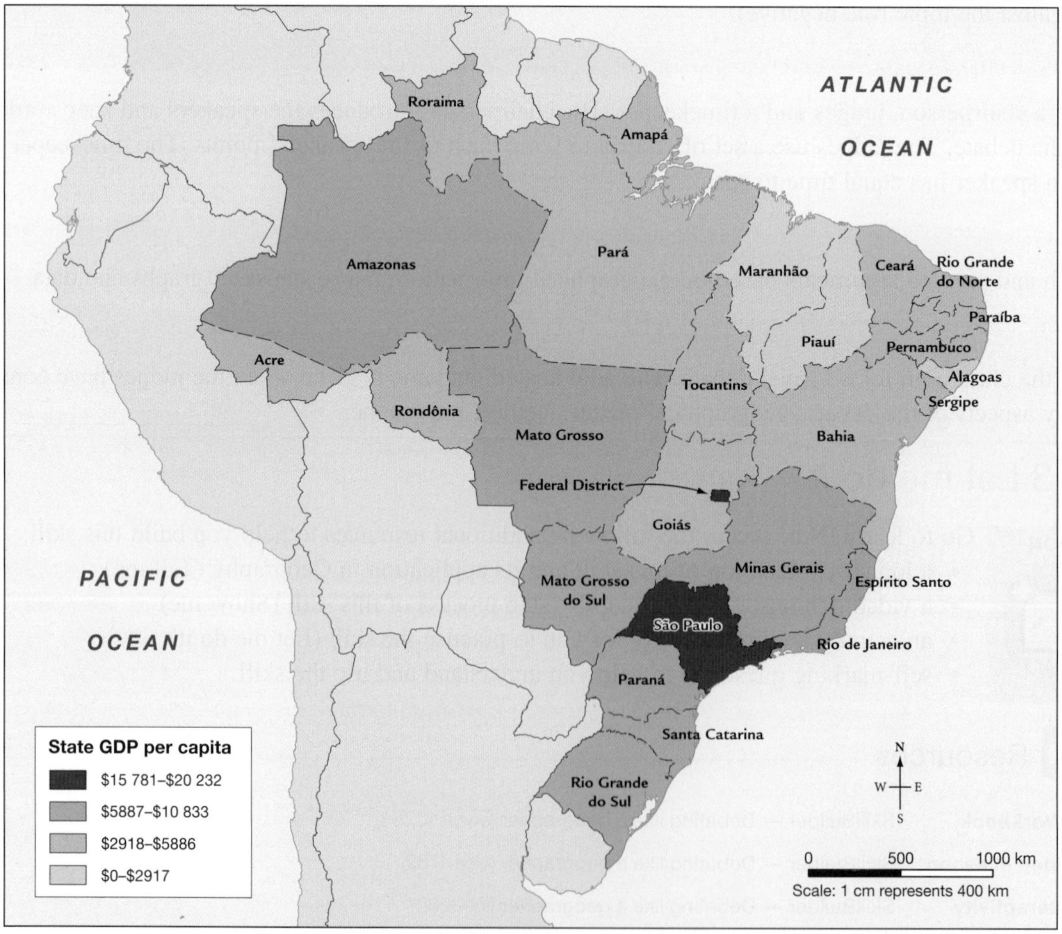

Source: Instituto Brasileiro de Geografia e Estatística. Made with Natural Earth. Map by Spatial Vision.

Brazil conducts a national census only every ten years, unlike Australia, which conducts a census every five years. As a result, it can be difficult to find accurate and up-to-date population data, particularly in the less economically developed areas of Brazil such as the favelas.

extreme poverty under the United Nations' Millennium Development Goals, this is defined as living on less than AU$1.30 per day

There is considerable spatial variation in wellbeing between regions in Brazil. The majority of industrial development in Brazil has occurred in the south and south-east regions, generating more wealth there. This contrasts markedly with the agriculturally based north-east region, which has higher rates of poverty and infant mortality and lower rates of nutrition.

22.7.2 The impact of development on wellbeing in Rio de Janeiro

Rio de Janeiro is a well-known tourist destination in Brazil, famous for its beautiful beaches, spectacular scenery and carnivals. However, for many local people, this is not the reality. Even within one of the wealthiest cities in the wealthiest region in Brazil, there is considerable variation in wellbeing and living conditions.

The city has experienced rapid growth, starting back in the eighteenth century when freed slaves who had originally worked on plantations came into the city in search of employment. This rural–urban migration still exists today, with thousands flocking to the city in search of opportunity and a new life. New settlers faced the dual problems of low wages and high housing costs, thus forcing them to construct illegal shanties on wasteland or vacant land. Over time, these have developed into full-blown suburbs, or favelas. Typically, these slums are located on steep slopes on the edges of the city, although, as the city has expanded, it has wrapped itself around the favelas.

Ironically, the poorest citizens live on unstable slopes with spectacular million-dollar views (see **FIGURE 2**), while the wealthier citizens tend to live on the more stable flatter land closer to the city centre.

According to Brazil's last census, 22 per cent of Rio de Janeiro's population of over 6.35 million people live in some 763 favelas. Rocinha (shown in **FIGURE 3**) is considered Rio's largest favela, with its population estimated in the range of 150 000 to 300 000 people. It is located in the south zone of the city, in close proximity to the famous beaches of Rio's Ipanema and Copacabana districts.

FIGURE 2 A favela located on a steep slope in Rio de Janeiro

22.7.3 The wellbeing of people living in a favela

Living conditions in the favelas are extremely difficult, as they have developed without any type of planning or government regulations and the housing is generally substandard. As a result, the issues that have arisen are affecting the development of the city as well as the wellbeing of its citizens. Issues affecting wellbeing include:

- lack of infrastructure such as sanitation and piped water; for example, almost one-third of favela households lack sanitation, leading to higher rates of disease. Garbage has to be put in sectioned-off dumping sites.
- vulnerability to weather extremes; for example, heavy rainfall creates landslides and floods on steep slopes. Timber shacks are more vulnerable to collapse than houses built of concrete bricks.
- lack of access: there is often only one main road, so movement around the favelas is via narrow lanes and steep staircases (see **FIGURE 3**)
- long commuting times: the average time to travel into the city centre of Rio is 1 1/2 hours by bus. The cost of public transport also takes a sizeable proportion of the average worker's salary. This in turn limits both educational and employment opportunities.
- lower household income: the average household income for people living in the favelas is approximately half that of those people living in the inner suburbs
- high crime rates such as homicide, particularly linked to the influence of drug trafficking and criminal gangs who have established themselves within the relative safety of the favelas. In 2013, an estimated 37 per cent of favelas were controlled by drug traffickers.
- a sense of inferiority and insecurity felt by residents: most people do not have legal title to their land or dwellings and can be moved by the government at any time.

FIGURE 3 Street view of Rocinha favela, Rio de Janeiro

22.7.4 Improving wellbeing in favelas

In order to reduce crime and the control of the favelas by drug traffickers, and to improve safety, in 2008 the government introduced Pacifying Police Units (UPP), installing 37 UPPs by 2014. These have been very successful in the places where they have been implemented.

To help improve access for favela residents, the government has installed cable cars to transport people up and down the steep hillsides quickly and effectively, with local residents entitled to one free round trip per day. It is also hoped that the cable cars will allow for expansion of tourism. However, one favela community criticised the government's priorities, maintaining that locals were not properly consulted and that basic services such as sewerage and education should have come first.

As Brazil hosted the 2016 Olympic Games and the 2014 World Cup, the city expanded infrastructure and built new facilities. **FIGURE 4** highlights what was a major issue: many of the planned Olympic Zones were located on favela sites. Many residents were very unhappy at the prospect of being relocated to make way for new sporting venues. They claimed that the financial compensation offered was insufficient for a new home and that communities that had existed for generations were being destroyed. Over 3000 families were forcibly relocated and another 11 000 threatened with eviction. Use the **Favelas** weblink in the Resources tab to view a video clip about this issue.

Preventing the continued growth of favelas by providing adequate low-income housing is the most cost-effective means of improving wellbeing. The cost of upgrading a favela with basic infrastructure is estimated to be two to three times as much as the cost of providing new high-rise housing estates. However, with 65 per cent of Rio's population growth coming from rural–urban migration, it is difficult for authorities to keep up with demand for housing and space.

Nationally, the government aimed to eliminate extreme poverty by 2014 with its Brazil Without Misery Plan. It involved the expansion of cash transfer payments to low-income families in exchange for them keeping their children in school and following a health and vaccination program. Improved infrastructure, vocational training and **micro-credit** were also part of this plan. Although the plan reduced the numbers of those living in extreme poverty from 10 per cent in 2004 to 4 per cent by 2012, extreme poverty remains. The government's strategies are continuing, so the wellbeing of those living in favelas should also improve.

micro-credit the provision of small loans to borrowers who usually would not be eligible to obtain loans due to having few assets and/or irregular employment

FIGURE 4 Location of Rio's favelas and Olympic venues

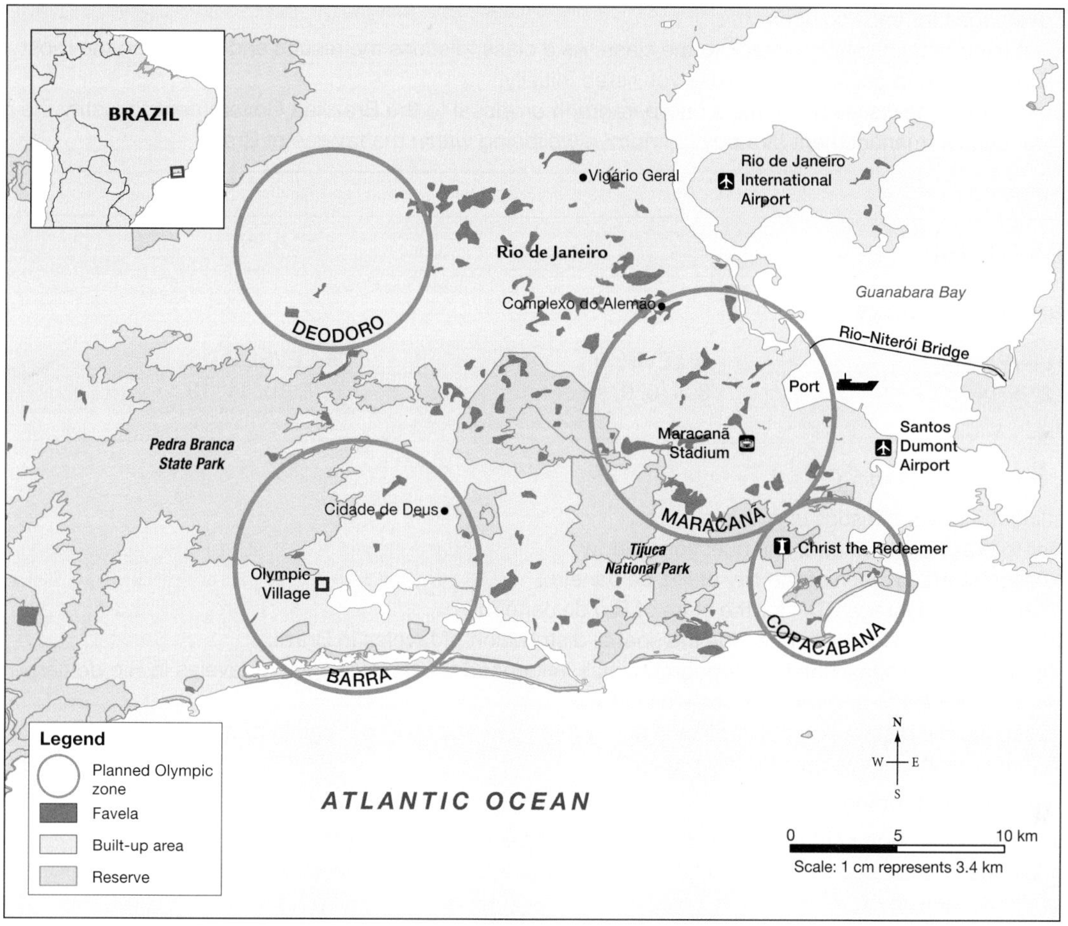

Source: UNEP-WCMC 2012. Made with Natural Earth.

on Resources

eWorkbook	Improving wellbeing in Brazil (ewbk-10367)
Video eLesson	Improving wellbeing in Brazil — Key concepts (eles-5332)
Interactivity	Improving wellbeing in Brazil (int-8761)

22.7 ACTIVITIES

The following activities are designed for you to think creatively, innovatively, and sustainably about how best to reduce the variations in human wellbeing within the favelas. The important part of this activity is that you are open to all ideas, regardless of your own opinion on the effectiveness of them. Great ideas come from people who are willing to think outside the box — now is your chance to do this! Before you begin, use the **Favelas** weblink in your online Resources to prepare.

1. As a class, list as many of the issues you can think of that contribute to poor wellbeing within the favelas of Brazil, for example high crime rates.

2. Choose your top ten issues and put them in order of importance. For example, access to clean, reliable drinking water is more important than education facilities, although they are both critical to achieving acceptable human wellbeing levels.
3. In groups, choose one of the issues in your list and come up with two solutions. You may think of your own solution, or research online to find an organisation that is already solving the issues. Your two solutions should be:
   a. government-led, large-scale, top-down approach to improving human wellbeing
   b. a non-government-led, small-scale, bottom-up approach to improving human wellbeing.
4. Swap your solutions with another group in your class and analyse them. List the advantages and disadvantages for each solution.
5. Present each other's solutions back to the class. As a class, discuss the results and decide on the most effective solution to each of your issues you listed initially.
6. Choose one of the issues and write a one-paragraph proposal to the Brazilian Government to justify the need for your chosen management strategy to improve wellbeing within the favelas of Brazil.

22.7 EXERCISE

Learning pathways

■ LEVEL 1	■ LEVEL 2	■ LEVEL 3
1, 2, 8, 14	3, 4, 6, 9, 12	5, 7, 10, 11, 13

Check your understanding

1. Outline the characteristics of a favela.
2. Refer to **FIGURE 1** to answer the questions below.
   a. Describe the location of the city of Rio de Janeiro.
   b. What is the average GDP for the state of Rio de Janeiro?
3. With reference to **FIGURE 4**, describe the spatial distribution of favelas in Brazil.
4. Refer to **FIGURE 2**. Describe how topography has influenced the development of favelas in Rio de Janeiro.
5. Refer to **FIGURE 1** to answer the questions below.
   a. Describe the distribution of Brazilian states with an average GDP per capita of more than $12 000.
   b. What reason can you give for this pattern?

Apply your understanding

6. How has the development of Rio de Janeiro affected people's wellbeing?
7. Explain the interconnection between the development of favelas and movement of people from rural areas into Rio de Janeiro.
8. Study **FIGURE 3**.
   a. What difficulties would exist for people of different ages living in this street?
   b. Why would the government find it cheaper to build new high-rise housing rather than upgrading existing favelas?
9. What changes might people face if they are moved from living in a favela to living in a high-rise housing estate?
10. Why would the estimated population of Rocinha (150 000 to 300 000) vary to such an extent?
11. Discuss government action to improve wellbeing in favelas on two different scales.

Challenge your understanding

12. How might hosting the Olympic Games and World Cup have affected the wellbeing of the residents of Rio?
13. It is 2012 and the Brazilian Government are beginning their planning for the Olympic Games in 2016. Propose an investment plan for the Brazilian Government that would effectively improve the wellbeing of Brazilian people living in the favelas as well as prepare the city to host the Olympic Games. Think of some resources or infrastructure that would benefit both target markets.
14. Visiting favelas is increasingly popular among tourists.
    a. Suggest what the positive and negative impacts of tours of favelas might be.
    b. Is such tourism exploiting or helping locals?

To answer questions online and to receive **immediate feedback** and **sample responses** for every question, go to your learnON title at www.jacplus.com.au.

22.8 SkillBuilder — Writing a geographical essay

online only

LEARNING INTENTION

By the end of this subtopic, you will be able to structure a written response that is logical, coherent and concise, and is objective through the use of reliable geographical data and research.

The content in this subtopic is a summary of what you will find in the online resource.

22.8.1 Tell me

What is a geographical essay?

A geographical essay is an extended response structured like any essay, but it focuses on geographical facts and data, particularly data that can be mapped. A geographical essay may indicate change over time, refer to the scale of activities, or look to the future in discussing sustainability.

22.8.2 Show me

Step 1

Brainstorm all the ideas you can think of about the topic. Group the ideas into three or four themes.

Step 2

Plan your essay using the following structure:

- **Introduction:** Three reasons, or themes, are listed with a clear contention.
- **Theme 1:** Reason 1
- **Theme 2:** Reasons 2
- **Theme 3:** Reason 3
- **Conclusion:** Strong statement of your contention

Step 3

Find some geographic facts and figures to support your ideas. Using case studies and giving examples of particular places add value to your writing. Quoting organisations gives authority to your work. Keep your work organised according to the key ideas so you can find information when writing.

Step 4

Introduction: Begin with a powerful fact that captures the reader's interest. In the next sentence, outline the aspects that are going to be discussed in the following paragraphs. Make sure that you list these in the order in which you wish to present the paragraphs.

Step 5

Paragraphs: Each paragraph that you write needs to have a distinct and powerful opening sentence that summarises the facts you are going to present in the following sentences. The factual sentences need to be presented in an organised manner. The last sentence should link clearly to the next paragraph.

Step 6

Conclusion: This should be only one or two sentences. It must contain no new data. It needs to leave the reader in no doubt about what your opinion on the topic is.

Step 7

Provide a list of the references you have used. Your school will have a preferred system for bibliographies and reference lists.

22.8.3 Let me do it

learnON

Go to learnON to access the following additional resources to help you build this skill:
- a longer explanation of this skill and its application in Geography (Tell me)
- a video demonstrating the step-by-step process of this skill (Show me)
- an activity and interactivity for you to practise the skill (Let me do it)
- self-marking questions to help you understand and use the skill.

on Resources

eWorkbook	SkillBuilder — Writing a geographical essay (ewbk-10371)
Video eLesson	SkillBuilder — Writing a geographical essay (eles-1763)
Interactivity	SkillBuilder — Writing a geographical essay (int-3381)

22.9 Investigating topographic maps — Improving wellbeing in Cumborah

LEARNING INTENTION

By the end of this subtopic, you will be able to outline key wellbeing challenges associated with living in remote and regional areas in New South Wales through the analysis of topographic maps.

22.9.1 Cumborah, New South Wales

Cumborah is located in northern New South Wales, approximately 50 kilometres north-west of Walgett and 50 kilometres south-west of Lightning Ridge. The town is situated in Walgett Shire, on the Country of the Kamilaroi/Gamilaraay people.

At the time of the 2016 Australian Census, there were 249 people living in Cumborah. The town recorded a median age of 55, higher than the New South Wales state and Australian medians of 38. The gender distribution was 58 per cent male and 42 per cent female, a contrast with the New South Wales and Australian distributions of 49.3 per cent male and 50.7 per cent female.

For such a small population, it is interesting to consider whether the measures of wellbeing — such as income, employment, secure housing and access to services such as healthcare — paint an accurate picture of the wellbeing of people in the community.

The median incomes for households are well below state and national medians: Cumborah ($491), New South Wales ($1486) and Australia ($1436); however, 64.8 per cent of residents own their own homes outright, compared with only 32.2 per cent of people across New South Wales and 31 per cent nationwide. Other signs of a strong community and wellbeing are evident in the data comparing Cumborah with the rest of Australia. For example, in the two weeks before the census, 14 per cent of Cumborah residents provided unpaid help to someone with a disability, compared with 11.6 per cent across New South Wales. Similarly, in the 12 months leading up to the census, 28.9 per cent of Cumborah residents did volunteer work through a local group or organisation, compared with only 18.1 per cent of New South Wales residents as a whole.

Explore the ABS census data for Cumborah using the **Cumborah Census Quickstats** weblink in your online Resources.

FIGURE 1 Topographic map extract of Cumborah, New South Wales

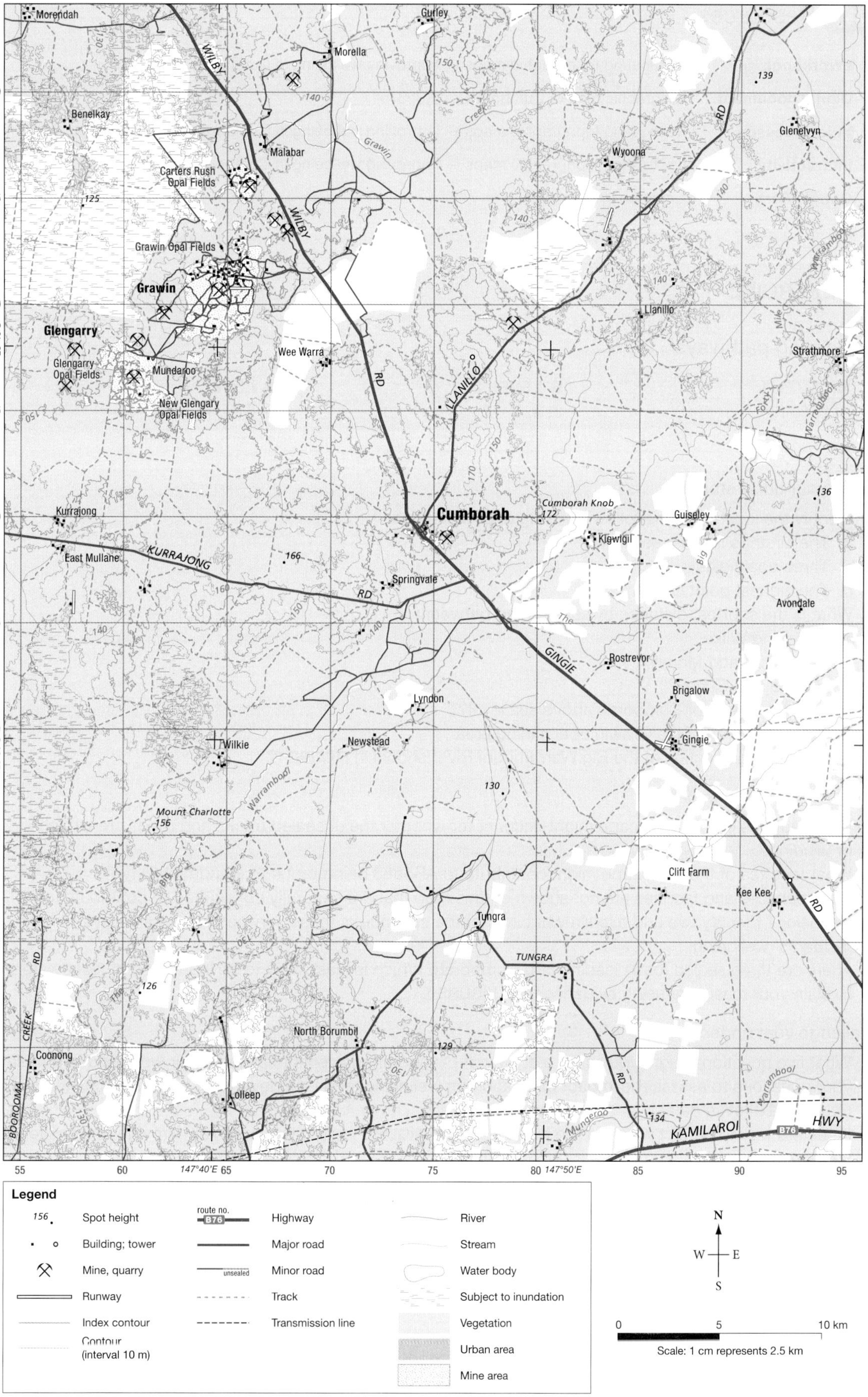

Source: Data based on Spatial Services 2019.

on Resources

eWorkbook	Investigating topographic maps — Improving wellbeing in Cumborah (ewbk-10375)
Digital document	Topographic map of Cumborah (doc-36375)
Video eLesson	Investigating topographic maps — Improving wellbeing in Cumborah — Key concepts (eles-5333)
Interactivity	Investigating topographic maps — Improving wellbeing in Cumborah (int-8762)
Google Earth	Cumborah (gogl-0146)

22.9 EXERCISE

Learning pathways

■ LEVEL 1	■ LEVEL 2	■ LEVEL 3
1, 2, 6, 10	3, 4, 7, 12	5, 8, 9, 11

Check your understanding

1. Identify the location of the following using grid references.
   a. The airstrips in the area shown
   b. Three opal fields
   c. The highest point shown
2. What is the shortest distance by sealed road between:
   a. Gigie and Wilke
   b. Comborah and Morella
   c. Morella and Gurley?
3. What is the height of Cumborah above sea level?
4. Calculate the area of the Grawin opal mining area.
5. How long is the section of the Big Warrambool River shown in **FIGURE 1**?

Apply your understanding

6. What features of this landscape might account for whether the unsealed tracks and roads are straight or winding?
7. What features of the landscape might contribute to AR5588 being prone to inundation?
8. What services and facilities are not shown on this map that are generally considered to contribute to wellbeing? Identify two and explain what impact the absence of these services and facilities might have on residents' wellbeing.
9. There are three sealed roads leading into Cumborah. Which is least likely to be inundated by floodwaters? Explain your reasoning using evidence from **FIGURE 1**.

Challenge your understanding

10. What five questions might you ask in a survey of Cumborah residents to assess their wellbeing? Formulate your questions to assess the impact of location on wellbeing, and include at least two questions that would return quantitative data.
11. Is there a way to predict which of the rivers shown in **FIGURE 1** might be perennial? Provide reasons to support your answer.
12. Predict what the landscape might be like in AR5888. Provide reasons for your answer based on the features shown in **FIGURE 1**.

To answer questions online and to receive **immediate feedback** and **sample responses** for every question, go to your learnON title at www.jacplus.com.au.

22.10 Thinking Big research project — Improving wellbeing in a low-HDI-ranked country

online only

The content in this subtopic is a summary of what you will find in the online resource.

Scenario

Geographers use many different demographic indicators to rank, analyse and discuss the development status of countries. The Human Development Index (HDI) is a meta-indicator that provides a ranking based on a number of different development indicators. The HDI is a composite ranking that includes measures of life expectancy, education and gross national income (per capita) within a country, thus providing insight into the wellbeing of the country's citizens.

The highest possible HDI score is 1.0. HDI rankings are categorised into four main groups: very high human development (scores from 0.8 to 1.0), high human development (scores of at least 0.7 but less than 0.8), medium human development (scores of less than 0.7 but at or above 0.55) and low human development (scores below 0.55). The majority of countries that are categorised as having low human development are found in Africa; others are located in the Middle East, Central Asia and the Pacific Islands.

You may have only been in the job for a week, but you already have your first major assignment. Your boss, Australia's representative to the United Nations, has asked for a report on countries with the lowest HDI rankings. As part of the team working on this project, you will choose one country to investigate, report on, and make suggestions as to what the Australian government can do to help alleviate the situation.

Task

learnON

Choose one country from those with the lowest HDI rankings. Prepare a report that includes:

- a summary of the country's current development status — under the Sustainable Development Goals framework, what areas have been targeted and what progress has been achieved?
- a discussion of what you have identified as the country's three most pressing problems
- suggestions as to what the Australian government could do to help alleviate these problems and assist this country in raising its HDI ranking.

Go to learnON to access the resources you need to complete this research project.

on Resources

ProjectsPLUS Thinking Big research project — Improving wellbeing in a low-HDI-ranked country (pro-0219)

22.11 Review

22.11.1 Key knowledge summary

22.2 The role of governments in improving human wellbeing

- Governments improve wellbeing by funding projects and facilities to improve health, happiness, prosperity and welfare in their own country as well as other countries.
- Aid is not always about giving large sums of cash to other governments. It can also be about investing in various industries to increase the number of jobs available, giving access to education and health services, and setting up sustainable industries to improve all aspects of wellbeing
- Aid can increase the dependency of recipient countries on donor countries.
- The Australian government provided an estimated $4 billion between 2020 and 2021.

22.3 The role of NGOs in improving human wellbeing

- Non-government organisations aim to improve the standard of living and quality of life for people, animals and the environment.
- NGO aid is mostly voluntary. Individual donations are collected by organisations such as the Red Cross, Greenpeace, Sea Shepherd and the World Wildlife Fund (WWF) to meet their aims.
- NGO aid takes many forms, including money, food, medicine, equipment, expertise, scholarships, training, clothing and military assistance
- It is generally for small-scale aid projects (bottom-up aid) that target the people most in need of assistance, helping people directly without any government interference.

22.4 The role of individuals in improving human wellbeing

- Individuals play an important role in contributing to the wellbeing of others, which in turn improves their own personal wellbeing.
- Individuals can get involved in programs through governments and NGOs to volunteer to assist other people and improve their wellbeing
- Volunteering allows an individual to engage and connect with their local community or even the international community.

22.5 Improving wellbeing for Aboriginal Peoples and Torres Strait Islander Peoples

- Australian governments and NGOs such as Oxfam are continuing to promote initiatives aimed at improving human wellbeing within Aboriginal communities.
- All Australians benefit from improving the wellbeing of Aboriginal Peoples and Torres Strait Islander Peoples as when disadvantage is overcome, the need for government expenditure is decreased.
- Aboriginal communities are working hard to improve their wellbeing
- The Lombadina community is self-sufficient in improving their wellbeing through ventures that include tourism operations, a general store, an artefact and craft shop, a bakery and a garage.
- Lombadina is also involved in the EON Thriving Communities Project. EON is a non-government organisation operating by invitation in Aboriginal communities in Western Australia.

22.7 Improving wellbeing in Brazil

- There is considerable spatial variation in wellbeing between regions in Brazil.
- Eleven per cent of the approximate population of 202 million remain in poverty. Of these, almost half continue to experience extreme poverty, living in favelas.
- The city of Rio de Janeiro has experienced rapid population growth as a result of rural–urban migration. New settlers face dual problems of low wages and high housing costs, thus forcing them to construct illegal shanties on wasteland or vacant land.
- Approximately 6.35 million people live in some 763 favelas. Living conditions in the favelas are extremely difficult as they have developed without any type of planning or government regulations.
- The Brazilian government aimed to eliminate extreme poverty by 2014 with its Brazil Without Misery Plan.

22.11.2 Key terms

aid (also referred to as international aid) the use of one country's resources (such as military, food or materials) or money (normally in the form of investment) to help another country that is in need

antenatal care the branch of medicine that deals with the care of women during pregnancy, childbirth and recovery after childbirth

extreme poverty under the United Nations' Millennium Development Goals, this is defined as living on less than AU$1.30 per day

favela an area of informal housing usually located on the edge of many Brazilian cities. Residents occupy the land illegally and build their own housing. Dwellers often live without basic infrastructure such as running water, sewerage or garbage collection

GDP (gross domestic product) the total value of all goods and services produced in a country in a given year, equal to total consumer, investment and government spending, plus the value of exports minus the value of imports

humanitarianism the concern for the welfare of human beings

micro-credit the provision of small loans to borrowers who usually would not be eligible to obtain loans due to having few assets and/or irregular employment

neonatal intensive care the specialised nursing practice of caring for newborn infants

non-government organisation (NGO) a citizen-based association that operates independently of government, usually to deliver resources or serve some social or political purpose

sustainable development development that meets the needs of the current generation without the depletion of resources

volunteering giving up time to commit to an idea or organisation without being paid

22.11.3 Reflection

Complete the following to reflect on your learning.

Revisit the inquiry question posed in the Overview:

Who should be responsible for improving human wellbeing standards and what is the most effective strategy to reduce the spatial inequalities in human wellbeing?

1. Now that you have completed this topic, what is your view on the question? Discuss with a partner. Has your learning in this topic changed your view? If so, how?
2. Write a paragraph in response to the inquiry question, outlining your views on the following:
   a. What are the reasons that variations in wellbeing exist?
   b. Who should be responsible for improving human wellbeing standards?
   c. What is the most effective strategy to reduce the spatial inequalities in human wellbeing?

FIGURE 1 Our wellbeing is determined by many factors. (a) Durbar Square, Kathmandu, Nepal

(continued)

Subtopic	Success criteria			
22.2	I can describe the terms *humanitarianism* and *aid*.			
	I can explain how these are used by governments to improve human wellbeing.			
22.3	I can explain how NGOs improve human wellbeing in LEDCs.			
	I can justify why a bottom-up approach to managing human wellbeing issues is effective.			
22.4	I can discuss the role individuals play in improving standards of human wellbeing.			
22.5	I can explain inequalities that have damaged the wellbeing of Aboriginal Peoples and Torres Strait Islander Peoples.			
	I can describe programs aimed to improve the wellbeing of Aboriginal Peoples and Torres Strait Islander Peoples.			
	I can propose action to reduce the inequality experienced by Aboriginal Peoples and Torres Strait Islander Peoples.			
22.7	I can describe the spatial patterns of wellbeing within Brazil.			
	I can evaluate the effectiveness of government initiatives to reduce the variations in human wellbeing for people living in Rio de Janeiro.			
22.9	I can describe the challenges associated with living regional areas through the analysis of topographic maps.			

FIGURE 2 Changarawe, Kilosa district, Tanzania

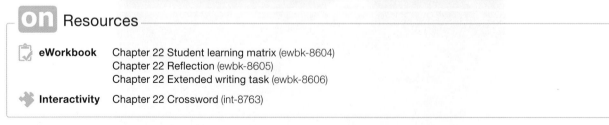

on Resources

eWorkbook Chapter 22 Student learning matrix (ewbk-8604)
Chapter 22 Reflection (ewbk-8605)
Chapter 22 Extended writing task (ewbk-8606)

Interactivity Chapter 22 Crossword (int-8763)

ONLINE RESOURCES

on Resources

Below is a full list of **rich resources** available online for this topic. These resources are designed to bring ideas to life, to promote deep and lasting learning and to support the different learning needs of each individual.

eWorkbook

- **22.1** Chapter 22 eWorkbook (ewbk-8113) ☐
- **22.2** The role of governments in improving human wellbeing (ewbk-9802) ☐
- **22.3** The role of NGOs in improving human wellbeing (ewbk-9806) ☐
- **22.4** The role of individuals in improving human wellbeing (ewbk-9810) ☐
- **22.5** Improving wellbeing for Aboriginal Peoples and Torres Strait Islander Peoples (ewbk-9814) ☐
- **22.6** SkillBuilder — Debating like a geographer (ewbk-9822) ☐
- **22.7** Improving wellbeing in Brazil (ewbk-9826) ☐
- **22.8** SkillBuilder — Writing a geographical essay (ewbk-10371) ☐
- **22.9** Investigating topographic maps — Improving wellbeing in Cumborah (ewbk-9830) ☐
- **22.11** Chapter 22 Student learning matrix (ewbk-8604) ☐
  Chapter 22 Reflection (ewbk-8605) ☐
  Chapter 22 Extended writing task (ewbk-8606) ☐

Sample responses

- **22.1** Chapter 22 Sample responses (sar-0170) ☐

Digital document

- **22.9** Topographic map of Cumborah (doc-36375) ☐

Video eLessons

- **22.1** Improving human wellbeing — Photo essay (eles-5327) ☐
  Improving wellbeing (eles-5351) ☐
- **22.2** The role of governments in improving human wellbeing — Key concepts (eles-5328) ☐
- **22.3** The role of NGOs in improving human wellbeing — Key concepts (eles-5329) ☐
- **22.4** The role of individuals in improving human wellbeing — Key concepts (eles-5330) ☐
- **22.5** Improving wellbeing for Aboriginal Peoples and Torres Strait Islander Peoples — Key concepts (eles-5331) ☐
- **22.6** SkillBuilder — Debating like a geographer (eles-1762) ☐
- **22.7** Improving wellbeing in Brazil — Key concepts (eles-5332) ☐
- **22.8** SkillBuilder — Writing a geographical essay (eles-1763) ☐
- **22.9** Investigating topographic maps — Improving wellbeing in Cumborah — Key concepts (eles-5333) ☐

Interactivities

- **22.2** Helping others (int-3305) ☐
- **22.4** The role of individuals in improving human wellbeing (int-8759) ☐
- **22.5** Improving wellbeing for Aboriginal Peoples and Torres Strait Islander Peoples (int-8760) ☐
- **22.6** SkillBuilder — Debating like a geographer (int-3380) ☐
- **22.7** Improving wellbeing in Brazil (int-8761) ☐
- **22.8** SkillBuilder — Writing a geographical essay (int-3381) ☐
- **22.9** Investigating topographic maps — Improving wellbeing in Cumborah (int-8762) ☐
- **22.11** Chapter 22 Crossword (int-8763) ☐

ProjectsPLUS

- **22.10** Thinking Big research project — Improving wellbeing in a low-HDI-ranked country (pro-0219) ☐

Weblinks

- **22.2** Department of Foreign Affairs and Trade (DFAT) (web-4285) ☐
  Australian Aid Budget Summary (web-6402) ☐
- **22.4** Volunteering Australia (web-6403) ☐
- **22.5** Closing the Gap (web-3259) ☐
  Lombadina (web-3341) ☐
  Lombadina Census QuickStats (web-6404) ☐
  EON Foundation (web-6405) ☐
- **22.7** Favelas (web-3402) ☐
- **22.9** Cumborah Census QuickStats (web-6416) ☐

Fieldwork

- **22.4** Wellbeing at school (fdw-0049) ☐

Google Earth

- **22.9** Cumborah (gogl-0146) ☐

Teacher resources

There are many resources available exclusively for teachers online.

23 Fieldwork inquiry — Comparing wellbeing in the local area

23.1 Overview

LEARNING INTENTION

By completing this subtopic, you will be able to plan fieldwork to investigate wellbeing in the local area, use relevant fieldwork techniques, and process and communicate the findings of fieldwork relevant to wellbeing.

23.1.1 Scenario and task

Task: Produce a fieldwork report you could present to your local council that outlines variations within your local area, reasons for the differences and strategies to improve the situation in the future.

You may have noticed that there are distinct variations across space in any city, suburb or regional community in terms of human wellbeing. Your council has asked for locals to inform them about differences in wellbeing they notice within their local areas, and what could or should be done about these in the future. Investigation of the topic will require you to undertake some fieldwork in order to make firsthand observations in the field and collect, process and analyse data.

Your task

The aim of the fieldwork is for you to explore some area variations by comparing two places at the local scale. The key inquiry questions the council wants to know answers to are the following:

- How does wellbeing vary between area X and area Y in the local area?
- What factors might explain the variations in wellbeing?
- How can wellbeing be improved in the local area?

23.2 Inquiry process

23.2.1 Process

Open the ProjectsPLUS application for this project located in your online Resources. Watch the introductory video lesson and then click the 'Start project' button and set up your class group. Save your settings and the project will be launched.

- **Planning:** As a class, discuss the types of indicators you would use as a basis for comparing wellbeing in your local area, for example surveys. Decide on teams and allocate tasks or different streets to each team member. Download the fieldwork planning document from the Media Centre to help you plan your fieldwork. Navigate to your Research Forum. Use the research topics to provide a framework for your fieldwork.

- You will need to determine the features of the houses and streets that you wish to gain data about. How will you record this data on the day (per house block or per street block)? Think carefully and plan your data record sheet so it is easy to use and also easy to summarise. Download the sample street analysis template from the Media Centre to help you plan and to record your housing data.
- If you are planning to survey people, you will need to plan and prepare survey questions. Download the community sample survey template from your Media Centre to help you plan and to record your data. You may wish to refer to the relevant SkillBuilder to help you with this.

23.2.2 Collecting and recording data

Before going on your field trip, prepare a simple map to show the location of your fieldwork site(s) relative to key features such as your school or city centre. You will need a separate location map if your second site is not in the same area.

Prepare a more detailed map of your fieldwork site(s). Use a street directory, Google Earth or a local council map as a guide. Include streets, street names, schools, preschools, shops and shopping centres, parks, public transport and other community facilities. Complete your map with BOLTSS.

During the field trip you may be required to survey houses whereby you record key features, take photographs and survey local residents in public places such as parks and shopping centres. (*Hint:* Keep a record of the location of photographs taken.)

Download the fieldwork report document from your Media Centre to help you prepare your report.

23.2.3 Processing and analysing your information and data

An important skill is the ability to analyse the information you have collected on your field trip and any other supplementary data in order to write the findings of your inquiry into a fieldwork report. A key part of your report is to determine any patterns or trends revealed in the data. At the same time, try to identify any anomalies (variations) from the patterns or trends. Download the analysis document from your Media Centre to help you further analyse the data you have collected.

23.2.4 Communicating your findings

Formally write your observations as a fieldwork report using these suggested sub-headings:

- Background and key inquiry question (include location descriptions and map(s))
- Conducting the fieldwork (planning and collection of data)
- Findings (results of data analysis)
- Future (How might wellbeing in the local community be improved upon? What could local councils and other community-based organisations do to improve living conditions? You might like to put forward a proposal to local council outlining your suggestions.)

You may wish to add your own headings.

23.3 Review

23.3.1 Reflecting on your work

Think about how you approached this fieldwork project and how you, personally, were able to organise yourself and contribute to the working of the team. Access and complete the reflection document in your Media Centre. Be honest in assessing your strengths and areas where you think you could do better next time.

Print out your Research Report from ProjectsPLUS and hand it in with your fieldwork report and reflection notes.

on Resources

ProjectsPLUS Fieldwork inquiry — Comparing wellbeing in the local area (pro-0151)

GLOSSARY

absolute location the latitude and longitude of a place
active consumerism a movement that is opposed to the endless purchase of material possessions and the pursuit of economic goals at the expense of society or the environment
affordability the quality of being affordable—priced so that people can buy an item without inconvenience
ageing population the average age of a population rising as a result of increases in life expectancy. This can result in social issues such as a need for increased government spending on health care for the elderly population.
agglomeration the extended area of a city, including its suburbs and adjoining populated areas
agribusiness business set up to produce, process and distribute agricultural products, or to support the agricultural sector
agroforestry the use of trees and shrubs on farms for profit or conservation; the management of trees for forest products
aid also be referred to as international aid; the use of one country's resources such as military, food or materials) or money (normally in the form of investment) to help another country that is in need
algal bloom rapid growth of algae caused by high levels of nutrients (particularly phosphates and nitrates) in water
alluvial plain an area where rich sediments are deposited by flooding
alpha world city a city generally considered to be an important node in the global economic system
anabranch section of a river or stream that diverts from the main channel
antenatal care the branch of medicine that deals with the care of women during pregnancy, childbirth and recovery after childbirth
anthropocentric describes the belief that humans needs and wants are more important than protecting the environment
anthropogenic resulting from human activity
aquaculture the farming of aquatic plants and aquatic animals such as fish, crustaceans and molluscs
aquaponics a sustainable food production system in which waste produced by fish or other aquatic animals supplies the nutrients for plants, which in turn purify the water
aquifer a body of permeable rock below the Earth's surface that contains water, known as groundwater
arable describes land that is suitable for growing crops
archipelago a chain or group of islands
atoll a coral island that encircles a lagoon
Australian National Development Index (ANDI) an Australian index measuring elements of progress in 12 areas
bankruptcy a legal process that declares that a person cannot pay their debts and allows them to make a fresh start
base flow water entering a stream from groundwater seepage, usually through the banks and bed of the stream
bi-articulated bus an extension of an articulated bus, with three passenger sections instead of two
biocapacity the capacity of a biome or ecosystem to generate a renewable and ongoing supply of resources and to process or absorb its wastes
biocentric describes the belief that humans should minimise their impact on the Earth to ensure it is able to sustain life for all environments and species
biodiversity the variety of plant and animal life within an area
biofuel fuel that has been produced from renewable resources, such as plants and vegetable oils
biophysical environment all elements or features of the natural or physical and the human or urban environment, including the interaction of these elements; made up of the Earth's four spheres — the atmosphere, biosphere, lithosphere and hydrosphere
bioremediation the use of biological agents, such as bacteria, to remove or neutralise pollutants
booms floating devices to trap and contain oil
brackish (water) water that contains more salt than fresh water but not as much as sea water

campaign an organised course of action for a specific purpose, for example to arouse public interest in an issue
carbon credit a tradable certificate representing the right of a company to emit one metric tonne of carbon dioxide into the atmosphere
carbon sequestration capturing and storing carbon dioxide from the air
carrying capacity the ability of the land to support livestock
cash crop a crop grown to be sold so that a profit can be made, as opposed to a subsistence crop, which is for the farmer's own consumption
census a survey or count usually conducted of a population looking at a number of specific characteristics such as age, income, occupation, residence, health, religion and many more
chlamydia a sexually transmitted disease infecting koalas
chlorophyll green pigment in plants that uses energy from the Sun to transform carbon dioxide into glucose
clearfelling the removal of all trees in an area
climate the long-term precipitation and temperature patterns of an area
climate change any change in climate over time, whether due to natural processes or human activities
coastal dune vegetation succession the process of change in the plant types of a vegetation community over time — moving from pioneering plants in the high tide zone to fully developed inland area vegetation
commercial an activity that is concerned with buying and or selling of goods or services
community a group of people who share something in common
condensation when water in the atmosphere cools and changes from a gaseous state into a liquid state. This occurs when the water vapour clusters around a solid particle (such as dust)
consumers organisms that eat other organisms
conurbations an urban area formed when two or more towns or cities (e.g. Tokyo and Yokohama) spread into and merge with each other
coral polyp a tube-shaped marine animal that lives in a colony and produces a stony skeleton. Polyps are the living part of a coral reef.
Coriolis force (or effect) force that results from the Earth's rotation. Moving bodies, such as wind and ocean currents, are deflected to the left in the southern hemisphere and to the right in the northern hemisphere.
council verges the land between the footpath and the road
Country the land and its features, which is bound to the concept of belonging to a place that is fundamental to Aboriginal Peoples' sense of identity and Culture; the place where an Aboriginal and Torres Strait Islander Australian comes from and where their ancestors lived; it includes the living environment and the landscape
crop rotation a procedure that involves changing or alternating crops, so that no bed or plot sees the same crop in successive seasons
cull selective reduction of a species by killing a number of animals
decomposers organisms that break down dead organisms
deforestation clearing forests to make way for housing or agricultural development
degradation reduced quality of land and water resources caused by overuse
deltaic plain flat area where a river(s) empties into a basin
desertification the transformation of land once suitable for agriculture into desert by processes such as climate change or human practices such as deforestation and overgrazing
developing nation a country whose economy is not well developed or diversified, although it may be showing growth in key areas such as agriculture, industries, tourism or telecommunications
development according to the United Nations, defined as 'to lead long and healthy lives, to be knowledgeable, to have access to the resources needed for a decent standard of living and to be able to participate in the life of the community'
development corridor area set aside for urban growth or development
disability any condition that restricts a person's mental, sensory or mobility functions
diurnal temperature the variation in high and low temperature on a given day
drainage area (or basin) an area drained by a river and its tributaries
dryland ecosystems characterised by a lack of water. They include cultivated lands, scrublands, shrublands, grasslands, savannas and semi-deserts. The lack of water constrains the production of crops, wood and other ecosystem services.
dyke an embankment or barrier constructed to prevent flooding by the sea or a river

ecocentric describes the belief that protecting the environment should come before all other needs and wants

ecological footprint the amount of productive land needed on average by each person for food, water, transport, housing and waste management; a measure of human demand on the Earth's natural systems in general and ecosystems in particular

ecological services any beneficial natural process arising from healthy ecosystems, such as purification of water and air, pollination of plants and decomposition of waste

ecology the environment as it relates to living organisms

economic downturn a recession or downturn in economic activity that includes increased unemployment and decreased consumer spending

edible fit to be eaten as food; eatable

emerging economies (EEs) places experiencing rapid economic growth, but that have significant political, monetary or social challenges

emigrant a person who migrates from a place

emissions trading scheme a market-based, government-controlled system used to control greenhouse gas as a cap on emissions. Firms are allocated a set permit or carbon credit, and they cannot exceed that cap. If they require extra credits, they must buy permits from other firms that have lower needs or a surplus.

endemic describes species that occur naturally in only one region

enhanced greenhouse effect the observable trend of rising world atmospheric temperatures over the past century, particularly during the last couple of decades

environmental ethics an individual's beliefs about what is right or wrong behaviour in relation to the Earth and its environments and communities

environmental flows the quantity, quality and timing of water flows required to sustain freshwater ecosystems

environmental impact assessment a tool used to identify the environmental, social and economic impacts, both positive and negative, of a project prior to decision-making and construction

environmental refugees people who are forced to flee their home region due to environmental changes (such as drought, desertification, sea-level rise or monsoons) that affect their wellbeing or livelihood

environmental worldview varying viewpoints of how the world works and where people fit into the world. The worldview will form the assumptions and values that guide an individual's actions towards the environment.

ephemeral describes a stream or river that flows only occasionally, usually after heavy rain

equality providing all people with the same opportunities and treatment

equity providing people with different opportunities and treatment based on level of need

erosion the wearing down of rocks and soils on the Earth's surface by the action of water, ice, wind, waves, glaciers and other processes

eutrophication a process where water bodies receive excess nutrients that stimulate excessive plant growth

evaporation when water contained in water bodies is heated by the Sun and the liquid changes into a gaseous state and rises into the atmosphere

evapotranspiration the process by which water is transferred to the atmosphere from surfaces such as the soil and plants

experienced wellbeing an individual's subjective perception of personal wellbeing

extensive farm a farm that extends over a large area and requires only small inputs of labour, capital, fertiliser and pesticides

extinction when all individuals of a species have died

extreme poverty under the United Nations' Millennium Development Goals, this is defined as living on less than AU$1.30 per day

factory farming the raising of livestock in confinement, in large numbers, for profit

fair trade trading exchanges that are sustainable and fair for both the seller and the buyer

family household two or more persons, one of whom is at least 15 years of age, who are related by blood, marriage (registered or de facto), adoption, step-relationship or fostering

famine a drastic, widespread food shortage

FAO Food and Agricultural Organization of the United Nations

fault an area on the Earth's surface that has a fracture

favela an area of informal housing usually located on the edge of many Brazilian cities. Residents occupy the land illegally and build their own housing. Dwellers often live without basic infrastructure such as running water, sewerage or garbage collection.

feedlot farming intensive farming practice in which high concentrations of animals are kept in confined areas to facilitate rapid growth and weight gain

female infanticide the killing of female babies, either via abortion or after birth

fertility rate the average number of children born per woman

First Nations Australians a generic, collective term that refers to all Aboriginal Peoples and Torres Strait Islander Peoples

floating settlements anchored buildings that float on water and are able to move up and down with the tides

flood mitigation managing the effects of floods rather than trying to prevent them altogether

fly-in, fly-out (FIFO) system in which workers fly to work in places such as remote mines and after a week or more fly back to their home elsewhere

fodder food such as hay or straw for cattle and other livestock

food aid food, money, goods and/or services given for the specific purpose of helping those in need

food loss takes place at the production, post-harvest, processing, and distribution stages of food production

food miles the distance food is transported from the time it is produced until it reaches the consumer

food waste takes place at the retail and consumption stages food production and consumption

foreclosure a legal process that allows a lender to sell a borrower's asset to recover their money if the borrower has stopped making repayments

GDP (gross domestic product) the total value of all goods and services produced in a country in a given year, equal to total consumer, investment and government spending, plus the value of exports minus the value of imports

genetically modified describes seeds, crops or foods whose DNA has been altered by genetic engineering techniques

geographic processes the physical forces that form and transform our world

geographical factors reasons for spatial patterns, including patterns noticeable in the landscape, topography, climate and population

global warming increased ability of the Earth's atmosphere to trap heat

glucose sugar used by plants as energy source

green energy sustainable or alternative energy (e.g. wind, solar and tidal)

Green Revolution a significant increase in agricultural productivity resulting from the introduction of high-yield varieties of grains, the use of pesticides and improved management practices

green wedges land in urbanised areas that is protected from development and kept as farming, parkland or green space

greenfield developments when undeveloped land is released by the government for housing developments

greenhouse gases any of the gases that absorb solar radiation and are responsible for the greenhouse effect. These include water vapour, carbon dioxide, methane, nitrous oxide and various fluorinated gases.

gross domestic product (GDP) a measurement of the annual value of all the goods and services bought and sold within a country's borders; usually discussed in terms of GDP per capita (total GDP divided by the population of the country)

gross national income (GNI) a measurement of the value of goods and services produced by citizens and firms of a specific country no matter where they take place, normally discussed as GNI per capita

groundwater water held underground within water-bearing rocks or aquifers

groundwater salinity presence of salty water that has replaced fresh water in the subsurface layers of soil

groyne a structure (e.g. a rock wall) that is built perpendicular to the shoreline to interrupt the flow of water and the movement of sediment

gyre swirling circular ocean current (similar to water swirling around a plug hole)

heat island effect structures in urban environments absorb heat from the Sun and raise the temperature of the city environment compared to rural surrounds

high-density housing residential developments with more than 50 dwellings per hectare

horticulture the practice of growing fruit and vegetables

housing affordability a person's ability to pay for their housing.

Human Development Index (HDI) measures the standard of living and wellbeing in terms of life expectancy, education and income

human–environment systems thinking using thinking skills such as analysis and evaluation to understand the interaction of the human and biophysical or natural parts of the Earth's environment
humanitarian aid assistance provided in response to a human crisis caused by natural or man-made disasters, to save lives and alleviate suffering
humanitarianism humanitarianism is the concern for the welfare of human beings
humus an organic substance in the soil needed for plant growth that is formed by the decomposition of leaves and other plant and animal material
hunger the sensation felt when a person does not have enough to eat to meet their body's energy needs
hybrid a plant or animal bred from two or more different species, breeds or varieties, usually to attain the best features from each type
hydroponic describes a method of growing plants using mineral nutrients, in water, without soil
icon sites six sites located in the Murray–Darling Basin that are earmarked for environmental flows. They were chosen for their environmental, cultural and international significance.
immigrant a person who migrates to a place
impervious describes a rock layer that does not allow water to move through it due to a lack of cracks and fissures
improved pasture pasture that has been specially selected and sown, which is usually more productive than the local native pasture
incentive something that motivates or encourages a person to do something
income diversity income that comes from many sources
indicators a set of measurable characteristics that informs us of a condition or progress; useful in analysing features that are not easily calculated or measured, such as freedom or security
industrial effluents waste from manufacturing and other industrial processes
industrial materials primary industry sources such as forestry (wood) or mining (iron ore) that can be used in the manufacturing of other goods such as furniture or steel
Industrial Revolution the period from the mid-1700s into the 1800s that saw major technological changes in agriculture, manufacturing, mining and transportation, with far-reaching social and economic impacts
inequality of outcomes inequalities between people in the one country
infiltration the process in which absorbed into the ground, flows downward and collects above an impermeable layer or rock
infrastructure the facilities, services and installations needed for a society to function, such as transportation and communications systems, water and power lines
innovation new and original improvement to something, such as a piece of technology or a variety of plant or seed
insolation the level of solar energy that reaches the Earth's surface
intensive farm farm that requires a lot of inputs, such as labour, capital, fertiliser and pesticides
internal migration the movement of people temporarily or permanently within parts of a country
invasive plant species commonly referred to as weeds; any plant species that dominates an area outside its normal region and requires action to control its spread
invasive species any plant or animal species that colonises areas outside its normal range and becomes a pest
investment an item that is purchased or has money dedicated to it with the hope that it will generate income or be worth more in the future
irrigation the supply of water by artificial means to agricultural areas
jatropha any plant of the genus *Jatropha*, but particularly *Jatropha curcas*, which is used as a biofuel
JMA Jakarta metropolitan area
kenaf a plant in the hibiscus family that has long fibres; useful for making paper, rope and coarse cloth
Kyoto Protocol an agreement negotiated in 1999 between 160 countries designed to bring about reductions in greenhouse gas emissions
land reclamation expansion onto land created by humans
land degradation a decline in the quality of land, which makes it less able to support agriculture or native vegetation
leaching the process in which water runs through soil, dissolving minerals and carrying them into the subsoil
leeward describes the area behind a mountain range, away from the moist prevailing winds
less economically developed country (LEDC) country with lower levels of development — in the economic, social, environmental and political spheres

life expectancy the number of years a person can expect to live, based on the average living conditions within a country

lithosphere the Earth's crust, including landforms, rocks and soil

lobbying the activity individuals and organisation engage in to influence decision makers

lock a gateway across a channel of water that can be used to raise or lower water on either side; allows boats to travel between waterways at different levels

logging the cutting down, processing and removal of trees from an area

longshore drift a current that moves sediment parallel to the shoreline, created by the backward and forward motion of waves

low-density housing residential developments with around 12–15 dwellings per hectare; usually located in outer suburbs

mallee vegetation areas characterised by small, multi-trunked eucalypts found in the semi-arid areas of southern Australia

malnourished describes a person who is not getting the right amount of vitamins, minerals and other nutrients to maintain healthy tissues and organ function

malnourishment a condition that results from not getting the right amount of vitamins, minerals and other nutrients needed to maintain healthy tissues and organ function

maternal mortality the death of a woman while pregnant or within 42 days of termination of pregnancy

Mediterranean (climate) characterised by hot, dry summers and cool, wet winters

medium-density housing low rise (up to 3 or 4 storeys) residential developments with about 25–50 dwellings per hectare; definitions of 'medium density' can vary between council and government areas

megacity a settlement with 10 million or more inhabitants

metropolitan region an urban area that consists of the inner urban zone and the surrounding built-up area and outer commuter zones of a city

microclimate the climate of a small area

micro-credit the provision of small loans to borrowers who usually would not be eligible to obtain loans due to having few assets and/or irregular employment

micro hydro-dams produce hydro-electric power on a scale serving a small community (less than 10 MW). They usually require minimal construction and have very little environmental impact.

migrant a person who moves from living in one location to another

mitigate to reduce negative impacts or severity

monoculture the cultivation of a single crop on a farm or in a region or country

monsoon a wind system that brings heavy rainfall over large climatic regions and reverses direction seasonally

more economically developed country (MEDC) country with higher levels of development — in the economic, social, environmental and political spheres

mulch organic matter such as grass clippings

national park an area set aside for the purpose of conservation

natural increase change in population as a result of the difference between birth rates and death rates

neonatal intensive care the specialised nursing practice of caring for newborn infants

newly industrialised country places modernising and developing quickly from a wide range of industries and rapid economic growth

nocturnal active during the night

nomadic describes a group of people who have no fixed home and move from place to place according to the seasons, in search of food, water and grazing land

non-government organisation (NGO) a citizen-based association that operates independently of government, usually to deliver resources or serve some social or political purpose

off-shoring the practice of relocating a business's processes from one country to another, to take advantage of lower costs

old-growth forests natural forests that have developed over a long period of time, generally at least 120 years, and have had minimal unnatural disturbance such as logging or clearing

ores raw material that minerals can be extracted from

organic products grown or created with ingredients that were grown without any contact with artificial fertilisers or chemicals

organic matter decomposing remains of plant or animal matter
paddy field a growing area that is flooded to farm semi-aquatic plants, such as rice
pastoral run an area or tract of land for grazing livestock
per capita per person
perennial describes a stream or river that flows permanently
permafrost permanently frozen ground
photochemical smog smog that is produced by a reaction between chemicals in the air and ultraviolet light from the Sun
plantation an area in which trees or other large crops have been planted for commercial purposes
pneumatophores exposed root systems of mangroves, which enable them to take in air when the tide is in
population density the number of people living within one square kilometre of land; it identifies the intensity of land use or how crowded a place is
population distribution the pattern of where people live
potable drinkable; safe to drink
prairie native grasslands of North America
precipitation water droplets or ice crystals become too heavy to be suspended in the air and fall to Earth as rain, snow, sleet or hail
prevailing winds winds that blow from the direction that is typical at that time of year and place
producer a plant that can photosynthesise
pull factor favourable quality or attribute that attracts people to a particular location
pulp the fibrous material extracted from wood or other plant material to be used for making paper
push factor unfavourable quality or attribute of a person's current location that drives them to move elsewhere
qualitative indicator usually consists of a complex set of indices that measure a particular aspect of quality of life or describe living conditions; useful in analysing features that are not easily calculated or measured, such as freedom or security
quality of life your personal satisfaction (or dissatisfaction) with the conditions under which you live
quantitative indicator an indicator that is easily measured and can be stated numerically, such as annual income or how many doctors there are in a country
quintile any of five equal groupings used to measure and compare values
rain shadow the dry area on the leeward side of a mountain range
rainwater harvesting the accumulation and storage of rainwater for reuse before it soaks into underground aquifers
Ramsar site a wetland of international importance, as defined by the Ramsar Convention—an intergovernmental treaty on the protection and sustainable use of wetlands
rebate a partial refund on something that has been bought or paid for
recharge the process by which groundwater is replenished by the slow movement of water down through soil and rock layers
regional and remote areas areas classified by their distance and accessibility from major population centres
relative location the direction and distance from one place to another
replacement rate the number of children each woman would need to have in order to ensure a stable population level — that is, to 'replace' the children's parents. This fertility rate is 2.1 children.
reservoir large natural or artificial lake used to store water, created behind a barrier or dam wall
retrofitting adding a component or accessory to something that did not have it when it was originally built or manufactured
Return and Earn vending machines 'reverse' vending machines where people can return recyclable bottles and cartons for a refund (part of the NSW government scheme launched in 2017)
ringbark removing the bark from a tree in a ring that goes all the way around the trunk. The tree usually dies because the nutrient-carrying layer is destroyed in the process.
river fragmentation the interruption of a river's natural flow by dams, withdrawals or transfers
river regime the pattern of seasonal variation in the volume of a river
road intersection place where two or more roadways cross
Royal Commission a public judicial inquiry into an important issue, with powers to make recommendations to government

run-off water that is unable to be absorbed into the ground, flows over its surface and collects in nearby waterways or reaches stormwater drains

rural–urban fringe the transition zone where rural (country) and urban (city) areas meet

Sahel a semi-arid region in sub-Saharan Africa. It is a transition zone between the Sahara Desert to the north and the wetter tropical regions to the south. It stretches across the continent, west from Senegal to Ethiopia in the east, crossing 11 borders.

salinity the presence of salt on the surface of the land, in soil or rocks, or dissolved in rivers and groundwater

salt scald the visible presence of salt crystals on the surface of the land, giving it a crust-like appearance

sanitation services provided to remove waste such as sewage and rubbish

sea change movement of people from major cities to live near the coast to achieve a change of lifestyle

seasonal crops crops that are harvested in a certain season of the year, rather than all year round

sex ratio the number of males per 1000 females

shifting agriculture system in which small parcels of land are used to produce food for a period and abandoned when they become less productive so they can recover naturally, while the farmers move to another plot of land

sluice a constructed channel for water

slum a run-down area of a city characterised by poor housing and poverty

social cohesiveness the positive feeling of connectedness to a group of people or community

socioeconomic relating to or involving a combination of social and economic factors

space geographical concept concerned with location, distribution (spread) and how we change or design places

standard of living a level of material comfort in terms of goods and services available to someone or some group; continuum, for example a 'high' or 'excellent' standard of living compared to a 'low' or 'poor' standard of living

staple an important food product or item that people eat or use regularly

stewardship the belief that humans have a responsibility to care for the Earth to protect its future

storm surges a temporary increase in sea level from storm activity

subsidence the gradual sinking of landforms to a lower level as a result of earth movements, mining operations or over-withdrawal of water

subsistence describes farming that provides food only for the needs of the farmer's family, leaving little or none to sell

sustainability use of a service, product, item or material without depleting its value or accessibility for future generations

sustainable describes the use by people of the Earth's environmental resources at a rate such that the capacity for renewal is ensured

sustainable development development that meets the needs of the current generation without the depletion of resources

taxable income the amount of money a person earns (salary) as a result of their employment; is used as an indicator of personal wealth

terminal lake a lake where the water does not drain into a river or sea. Water can leave only through evaporation, which can increase salt levels in arid regions. Also known as an endorheic lake.

thermohaline circulation refers to the flow of ocean water caused by changes in water density due to temperature and salinity

topsoil the top layers of soil that contain the nutrients necessary for healthy plant growth

training walls walls or jetties that are constructed to direct the flow of a river or tide

transnational corporations (TNCs) a large business that operates across multiple countries, but is run from a central office or offices in a single developed country

transpiration when water contained in plants is heated and changes from a liquid into a gaseous state and rises into the atmosphere

tree change movement of people from major cities to live near forests to achieve a change of lifestyle

tsunami a powerful ocean wave triggered by an earthquake or volcanic activity under the sea

turbid describes water that contains sediment and is cloudy rather than clear

undernourished describes a person who is not getting enough calories in their diet; that is, not enough to eat

undulating describes an area with gentle hills

urban consolidation to develop and upgrade existing urban space
urban environment the humanmade or built structures and spaces in which people live, work and recreate on a day-to-day basis
urban expansion the increasing size of urban areas
urban farming growing plants and raising animals within and around cities
urban infilling the division of larger house sites into multiple sites for new homes
urban renewal repurposing land and improving services to better meet the needs of more people
urban sprawl the spreading of urban areas into surrounding rural areas to accommodate an expanding population
urbanisation the process of economic and social change in which an increasing proportion of the population of a country or region lives in urban areas
utilities services provided to a population, such as water, natural gas, electricity and communication facilities
variation (in terms of wellbeing) differences experienced between groups of people due to factors such as their location, demographic or socioeconomic status
volunteering a volunteer is a person who offers to give up time to commit to an idea or organisation without being paid
water rights refers to the right to use water from a water source such as a river, stream, pond or groundwater source
water scarcity situation that occurs when water supplies drop below 1000 m^3 per person per year
water security the reliable availability of acceptable quality water to sustain a population
water stress situation that occurs when water demand exceeds the amount available or when poor quality restricts its use
waterlogging saturation of the soil with groundwater such that it hinders plant growth
watertable the upper level of the groundwater, below which all pores in the soils and rock layers are saturated with water
weathering the breaking down of rock through the action of wind and water and the effects of climate, mainly by water freezing and cooling as a result of temperature change
weed any plant species that dominates an area outside its normal region and requires action to control its spread
weir wall or dam built across a river channel to raise the level of water behind. This can then be used for gravity-fed irrigation.
wellbeing a good or satisfactory condition of existence; a state characterised by health, happiness, prosperity and welfare
wetland an area covered by water permanently, seasonally or ephemerally. They include fresh, salt and brackish waters such as rivers, lakes, rice paddies and areas of marine water, the depth of which at low tide does not exceed 6 metres.
windward describes the side of a mountain that faces the prevailing winds
yield amount of agriculture produced or provided
yield gap the gap between a certain crop's average yield and its maximum potential yield

INDEX

I

T

U